D0460974

Toxic Air Pollution Handbook

Toxic Air Pollution Handbook

Edited by David R. Patrick

VNR VAN NOSTRAND REINHOLD
_____ New York

Copyright © 1994 by Van Nostrand Reinhold

Library of Congress Catalog Card Number 93-21545
ISBN 0-442-00903-8

I(T)P   Van Nostrand Reinhold is an International Thomson Publishing company.
         ITP logo is a trademark under license.

Printed in the United States of America

Van Nostrand Reinhold   ITP Germany
115 Fifth Avenue   Königswinterer Str. 418
New York, NY 10003   53227 Bonn
                                         Germany

International Thomson Publishing   International Thomson Publishing Asia
Berkshire House,168-173   38 Kim Tian Rd., #0105
High Holborn, London WC1V 7AA   Kim Tian Plaza
England   Singapore 0316

Thomas Nelson Australia   International Thomson Publishing Japan
102 Dodds Street   Kyowa Building, 3F
South Melbourne 3205   2-2-1 Hirakawacho
Victoria, Australia   Chiyada-Ku, Tokyo 102
                                          Japan

Nelson Canada
1120 Birchmount Road
Scarborough, Ontario
M1K 5G4, Canada

16 15 14 13 12 11 10 9 8 7 6 5 4 3 2 1

Library of Congress Cataloging in Publication Data
Toxic air pollution handbook / edited by David R. Patrick.
        p.   cm.
    Includes bibliographical references and index.
    ISBN 0-442-00903-8
    1. Air—Pollution. 2. Air—Pollution—Law and legislation.
  3. Air quality management. I. Patrick, David R.
TD883.T65   1994
363.73'92—dc20   93-21545
                                      CIP

To the late Bernie Steigerwald, without whose vision and dedication I might never have become interested in solving the problems of toxic air pollutants, and

To Barbara, my wife and best friend, without whose constant love, support, and assistance this book would not have been possible.

"All substances are poisons:
there is none which is not a poison.
The right dose differentates
a poison and a remedy."

Paracelsus (1493–1541)

Contents

APPENDICES

Acknowledgment

The editor would like to express his appreciation to the following individuals who helped guide the development of the various sections of the handbook: Dr. Annette Shipp, Health Assessment; Dr. Paul C. Chrostowski, Exposure Assessment; Kathy D. Bailey, Esquire, Regulatory Strategies; Dr. Joseph Laznow, Control Methods; and Dr. C. Shepherd Burton, Other Relevant Issues. The many authors and contributors also are recognized with great appreciation for the long hours volunteered and the keen insights into their fields of endeavor. A special thanks goes to Mr. Robert Esposito of Van Nostrand Reinhold for his continued faith and patience.

Disclaimer

Portions of this handbook were prepared or reviewed by employees of the U.S. Environmental Protection Agency (EPA). Their participation does not represent the endorsement of this handbook by the EPA, nor do their personal views represent the policies of the EPA.

Preface

In the final days of 1970, two actions were taken that irrevocably changed the way America would deal with its environment. First, the Environmental Protection Agency (EPA) was created. More than anything else, this signalled a commitment by the political leadership of the nation to improve the environment. Second, the 1970 Amendments to the Clean Air Act were enacted. This was the first comprehensive federal environmental regulatory program and it would serve as the blueprint for future environmental legislation.

Shortly after the formation of EPA, I joined EPA's Office of Air Quality Planning and Standards (OAQPS) in the Hazardous Pollutants Branch. Our job was to carry out the hazardous air pollutant requirements of section 112 of the new Clean Air Act. This was to be a difficult task. The program enacted by Congress was not well defined, the arena of federal environmental regulation was new, and there was considerable uncertainty on the part of EPA's leaders in how to go about regulating hazardous air pollutants.

In these early years, OAQPS was led by Dr. Bernard J. Steigerwald, known to all as Bernie. Bernie was the unquestioned intellect behind EPA's air pollution regulatory program in its first decade and possessed keen insights into the responsibilities, authorities, and limitations of this new federal agency and its air pollution regulatory mandate. More than anyone else in those early years, he also understood the potential scope and complexities of nationwide air pollution regulation. In particular, he was the first to recognize the size and complexity of the potential hazardous air pollutant problem and he directed that the first studies be conducted to identify important sources of organic chemical air pollutants.

Another individual who significantly influenced the direction and tone of EPA's early air pollution regulatory program was Walter Barber. Walt was Director of OAQPS in the late 1970s and early 1980s and was both technically and politically astute, a fact supported by his later elevation to Acting Administrator of EPA. Walt recognized the growing need for the agency to deal with hazardous air pollutants and he directed that work be initiated to understand better the requirements and processes for dealing with hazardous air pollutants and implementing section 112 of the Act.

Although EPA made considerable progress in understanding hazardous air pollutants in the early 1980s, there is little argument that the agency's implementation of the hazardous air pollutant requirements of section 112 was unsuccessful, at least in terms of the numbers of pollutants listed and regulations promulgated. In its defense, many things hindered success. One of the most important was that section 112 did not provide detailed guidance on how to regulate these pollutants; for example, it was mute on whether and how to consider the economic and other effects of regulation. It is axiomatic that government agencies rarely are successful in carrying out legislative mandates that have substantial political and economic impact when there is less than complete legislative guidance. In this case, EPA was required to develop and carry out a regulatory program to control pollutants that adversely affect public health but whose control could significantly affect the nation's economy.

Another factor contributing to EPA's limited success in dealing with hazardous air pollutants was the inconsistency between the short deadlines imposed by the Act for the regulation of hazardous air pollutants and the rapidly growing scope and complexity of the federal regulatory process. During the late 1970s and early 1980s, a complex and lengthy regulatory process resulted from bipartisan actions to ensure that federal regulations were technically and legally correct, were based on broad technical consensus, and resulted in minimum social and economic impacts. Direction was provided in the Administrative Procedures Act, the Freedom of Information Act, and several Executive Orders. The increasingly powerful Office of Management and Budget in the 1980s also expanded Executive Office oversight to new levels. Although all of these changes were appropriate in concept, the federal regulatory process that resulted became excruciatingly detailed, time consuming, and resource intensive.

Timely regulation of hazardous air pollutants also was hampered by the enormity of the potential hazardous air pollutant problem. In the late 1970s, EPA data bases indicated that there were at least sixty-five thousand industrial chemicals that had been in commercial production since 1945 and many more chemicals were being developed each year. Scientists also recognized that other toxic chemicals could be formed in atmospheric reactions involving even seemingly innocuous emissions to the air. Emitting these potential hazardous air pollutants were tens of thousands of industrial and commercial sources and literally hundreds of thousands of smaller sources, including such ubiquitous sources as automobiles, dry cleaners, and residential wood stoves. Deciding which hazardous air pollutants to regulate and on what schedule was an almost insurmountable problem to the agency.

If that were not enough, the regulators came to realize what legislators did not understand, or presumed would be resolved, namely that there are substantial uncertainties in the ability of scientists to quantify adverse effects associated with exposure to hazardous air pollutants. In fact, it is likely that some of these uncertainties cannot be resolved in the foreseeable future.

On the positive side, hazardous air pollutant program successes resulted from EPA's efforts to understand the scope of the hazardous air pollutants problem and use that knowledge in constructive ways. For example, the National Air Toxics Information Clearinghouse (NATICH) was developed in the mid-1980s for use by state and local agencies. NATICH was funded by EPA and developed by EPA staff, under my direction, with guidance from members of STAPPA (the State and Territorial Air Pollution Program Administrators) and ALAPCO (the Association of Local Air Pollution Control Officials). NATICH was developed to provide on-line and hard-copy information services to state and local agencies dealing with local toxic air pollutant problems; it since has been made available to any user through the National Technical Information Service. Priority ranking schemes for hazardous air pollutants and their sources also were developed in the 1980s; these later helped define the hazardous air pollutants and sources to be regulated under the

1990 Clean Air Act Amendments. Lastly, guidelines were developed and published by EPA describing appropriate methods for conducting exposure and risk assessments for hazardous substances.

Throughout the 1970s and 1980s, many people and organizations worked to facilitate a broader understanding of the problems associated with, and potential solutions to, hazardous air pollutants. A focal point for analysis and debate was the Air Pollution Control Association (APCA), which in 1988 changed its name to the Air & Waste Management Association (A&WMA). Under their sponsorship, a Critical Review of Air Toxics was published in 1983, written by Dr. Bernard Goldstein, a past EPA Assistant Administrator for Research and Development. The first APCA Air Toxics Committee also was formed in 1985 under my chairmanship. In 1986, the APCA Board of Directors identified air toxics as an important theme for the association and, in that same year, a second Critical Review of Air Toxics was published in 1986, written by Joseph Cannon, a past Assistant Administrator for EPA's Office of Air and Radiation; Dr. Goldstein again contributed. The Air Toxics Committee also reorganized that year into a broader Intercommittee Task Force on Air Toxics, again under my chairmanship.

Over the next four years, A&WMA sponsored a number of workshops and seminars on air toxics issues. The purpose of these meetings was to provide regulators, environmental specialists, researchers, plant managers and operators, equipment suppliers, and others with timely technical information on air toxics issues of interest. During the last half of the 1980s, toxic air pollutants remained an important technical issue in the A&WMA. Then, with passage of the Clean Air Act Amendments in 1990, A&WMA's Inter-committee Task Force on Air Toxics was restructured into the Air Toxics Division. Association members clearly view air toxics as a critical air pollution issue and A&WMA as a focal point for improving the understanding and control of air toxics.

Throughout the short twenty-year history of air toxics regulatory concerns, both regulators and those being regulated have attempted to understand and control air toxics using general air pollution and technical literature and resources. To this point, no single resource was available to provide the necessary information; in fact, understanding and acting on toxic air pollutant concerns span such a wide range of scientific and technical disciplines that many different resources were needed. Some of the disciplines required to understand and act on toxic air pollutant issues include engineering, chemistry, toxicology, epidemiology, biology, meteorology, ecology, demography, law, economics, and policy analysis. In the past, professionals in these disciplines had to assess air toxics problems and develop solutions frequently using incomplete and often inadequate documentation. This more than any other reason led me to conceive, and Van Nostrand Reinhold editors to agree to publish, this handbook.

In order to carry out the mandates of the Clean Air Act Amendments of 1990 concerning the assessment and control of toxic air pollution, EPA and state and local agencies must organize their technical and administrative resources and develop appropriate regulations. Simultaneously, owners and operators of sources of toxic air pollutants must plan their control strategies; consultants and attorneys will market their skills at solving problems and negotiating solutions; and researchers will delve more deeply into the technologies and the many uncertainties. Each group requires up-to-date and complete information that describes appropriate methodologies and resources and is designed specifically to deal with toxic air pollutants and their unique concerns. It is the sincere hope of those who prepared this handbook that it will help all of these people better understand the issues, act more responsibly on the needs, and in those actions find solutions that are both technically effective and economically and socially appropriate.

This handbook focuses on the methodologies and information used in the assessment, regulation, and control of toxic air pollutants; it does not deal in depth with individual sources of toxic air pollutants. This is done for two reasons. First, because there are no other available reference books that contain the full breadth of information necessary to understand and deal with air toxics issues, it is important initially to fill that gap. Second, at the time of this writing EPA had just begun publishing the guidance, schedules, and emission standards that are required by the 1990 Clean Air Act Amendments. As such, the scope and stringency of the specific actions that are to be taken by EPA necessarily were uncertain. A second handbook that focuses more specifically on sources of air toxics and their controls is envisioned for later in this decade.

Contributor List

Part I – Introduction

David R. Patrick, P.E., Vice President, ICF Kaiser Engineers, Fairfax, VA.
The Honorable Henry A. Waxman, Representative of the 24th District of California, Chairman of the Subcommittee on Health and Environment, U.S. House of Representatives, Washington, D.C.

Part II – Health Assessment

Bruce C. Allen, Staff Scientist, K.S. Crump Division, Clement International Corporation, Ruston, LA.
Annie M. Jarabek, Staff Scientist, Environmental Criteria and Assessment Office, U.S. Environmental Protection Agency, Research Triangle Park, NC.
Sharon A. Segal, Project Manager, Environmental Health Division, Clement International Corporation, Fairfax, VA.
Annette M. Shipp, Director, K.S. Crump Division, Clement International Corporation, Ruston, LA.

Part III – Exposure Assessment

Gerald E. Anderson, Principal Atmospheric Scientist, Systems Applications International, San Rafael, CA.
Paul C. Chrostowski, Principal, The Weinberg Consulting Group, Washington, D.C.
Judi L. Durda, Consultant, The Weinberg Consulting Group, Washington, D.C.
Sarah A. Foster, Consultant, The Weinberg Consulting Group, Washington, D.C.
James F. Lape, Consultant, The Weinberg Consulting Group, Washington, D.C.
Mary P. Ligocki, Senior Scientist, Systems Applications International, San Rafael, CA.
David R. Patrick, P.E., Vice President, ICF Kaiser Engineers, Fairfax, VA.
Arlene S. Rosenbaum, Senior Scientist, Systems Applications International, San Rafael, CA.

Robin R. Segall, Section Chief, Emission Measurement Branch, Technical Support Division, Office of Air Quality Planning and Standards, Office of Air and Radiation, EPA, Research Triangle Park, NC.

Peter R. Westlin, Section Chief, Emssion Measurement Branch, Technical Support Division, Office of Air Quality Planning and Standards, Office of Air and Radiation, EPA, Research Triangle Park, NC.

Part IV – Regulatory Strategies

Kathy D. Bailey, Esquire, Senior Attorney, American Automobile Manufacturers Association, Washington, D.C.

Jack R. Farmer, Principal Scientist, Research Triangle Institute, Research Triangle Park, NC.

David R. Patrick, P.E., Vice President, ICF Kaiser Engineers, Fairfax, VA.

Part V – Control Methods

James Cummings-Saxton, Principal, Industrial Economics Incorporated, Cambridge, MA.

Seshasayi Dharmavaram, Senior Consultant Engineer, E.I. duPont de Nemours Company, Newark, DE.

Robert K. Henderson, Quality Specialist, E.I. duPont de Nemours Company, Newark, DE.

Paul R. Jann, Senior Consultant, E.I. duPont de Nemours Company, Newark, DE.

Thomas A. Kittleman, Senior Consultant, E.I. duPont de Nemours Company, Newark, DE.

Joseph Laznow, Vice President, SRS Technologies, Arlington, VA.

Lewis H. Mitchell, Staff Engineer, E.I. duPont de Nemours Company, Newark, DE.

Avi Patkar, Staff Engineer, Radian Corporation, Cincinnati, OH.

Jerry M. Schroy, Fellow, Monsanto Corporation, St. Louis, MO.

Gerald N. Vander Werff, Environmental Specialist, E.I. duPont de Nemours Company, Newark, DE.

Part VI – Other Relevant Issues

Sharon M. Friedman, Iacocca Professor and Chairperson, Department of Journalism and Communication, Lehigh University, Bethlehem, PA.

Si Duk Lee, Chairman, A&WMA International Affairs Committee, Research Triangle Park, NC.

Toni Schneider, National Institute of Public Health and Environmental Protection, The Netherlands.

David R. Patrick, P.E., Vice President, ICF Kaiser Engineers, Fairfax, VA.

Michael P. Walsh, Consultant, Arlington, VA.

I

Introduction

1

Background

David R. Patrick

INTRODUCTION

The 1990 Clean Air Act Amendments were signed into law on November 15, 1990. This culminated many years of deliberation and work and represented the first major revision to air pollution legislation since the passage of the 1970 Clean Air Act Amendments. Importantly, a new program was enacted in section 112 to deal with air pollutants of a hazardous or toxic nature. The program aims at substantially reducing emissions from stationary sources of 189 'hazardous air pollutants'[1] listed in the Act. Ultimately, section 112 requires emissions to be reduced to levels that provide an 'ample margin of safety to protect public health.' At the time of writing of this handbook in 1992, regulatory actions carrying out this program were just beginning to be undertaken by the U.S. Environmental Protection Agency (EPA).

The hazardous air pollutant program descri-bed in the 1990 Amendments adds to and significantly expands the requirements under section 112 of the 1970 Amendments that served as the basis for EPA's activities on these pollutants for twenty years. As such, much is known about the pollutants that are to be regulated, specifically about their health and environmental effects, the manner in which people and the environment are exposed to them, how they are formed, and how they can be controlled. The purpose of this handbook is to build on what was learned, draw on the current state of the art, and describe the procedures and information needed to evaluate, regulate, and control these pollutants. Resources available for use by those engaged in these activities also are identified. The handbook is divided into six sections, each representing a major area of scientific or policy interest.

A BRIEF HISTORY OF AIR POLLUTION

Air pollution always has been a part of human life. Early humans contended with natural air pollution from forest fires, volcanic eruptions, and decay of organic matter. Humans later

---

[1] The term 'hazardous air pollutant' applies to those pollutants specifically regulated under section 112 of the Clean Air Act; the term 'air toxics' is broader, applying to any toxic substance in the air.

created air pollution by using fire for cooking and heating, generating wastes, producing tools and materials, and manufacturing and using other materials necessary for existence. Air pollution was a part of that early life and was accepted with little thought or attempt to mitigate it.

The gathering of humans into larger communities exacerbated the instances and frequent unpleasant nature of air pollution. Sources of air pollution increased in number and became more concentrated. Fledgling enterprises arose, often with their own peculiar air pollutants, and there was greater potential for the production and spread of airborne pathogens. Increasing use of fire for heating and food preparation and as part of early manufacturing led to significant increases in air pollution.

Although combustion of fuels is essential to life and progress, it typically is a significant source of air pollution. Highly efficient combustion of clean organic fuels can result in the conversion of most combustible organic matter into carbon dioxide and water, that generally are innocuous at the point of emission. Less efficient combustion, however, can result in the formation and release to the air of undesirable and often dangerous waste products including carbon monoxide, polycyclic aromatic hydrocarbons, formaldehyde, and inhalable particulate matter. Impurities in fuels also can be released as air pollutants. Important impurities include sulfur and various trace elements.

Historical records describe early concerns with the undesirable effects of air pollution and society's attempts to reduce those effects. One of the first efforts to restrict the use of 'dirty' coal occurred in England in 1272 when the use of 'sea coal' was banned by King Edward I (Stern 1984). Richard III and Henry V also limited the use of coal in England in the late fourteenth and early fifteenth centuries, respectively. Coal perhaps more than any other material led to the Industrial Revolution, but with it also came significant air pollution.

The first documented serious human health effects resulting at least in part from air pollution were coal related. In 1775, cancer of the scrotum was observed in chimney sweeps in England and investigators ultimately concluded that it resulted from contact with coal tar residues deposited from air emissions (CEQ 1979). Wark and Warner (1976) describe the serious effects of air pollution incidents in the Meuse Valley, Belgium (1930), in Manchester and Salford, England (1931), in Donora, Pennsylvania (1948), in London (1952 and 1956), and in other areas.

Significant increases in air pollution from sources unrelated to coal combustion also began to result from the industrial growth in the nineteenth and early twentieth centuries. Krier and Ursin (1977) discuss how many industrial areas began experiencing serious air pollution episodes during this period. This air pollution, however, usually was accepted as a necessary consequence of the local industry and those whose livelihoods depended on the industry ignored it or said 'it looks [or smells] like money.' Nevertheless, the health and aesthetic concerns led many cities and states in the U.S. to begin enacting smoke abatement programs (Stern 1984). For example, it is reported in IUAPPA (1991) that Chicago and Cincinnati passed 'smoke emission control' regulations in the 1880s, and similar regulations were passed in Pittsburgh and New York in the 1890s. In fact, growing concerns with and regulation of air pollution led to the formation of the Smoke Prevention Association of America. This organization later became the Air Pollution Control Association and today is known as the Air & Waste Management Association.

In the early 1940s, an entirely new air pollution phenomenon arose when a brownish, irritating haze formed in the air above the sun-drenched Los Angeles basin. Termed 'smog,' a combination of the words smoke and fog that in this case was wholly inappropriate, this brown haze soon was related to its primary source – emissions from the urban automobile fleet. This growing urban nature of air pollution and America's postwar affluence resulted in an increased sensitivity by the public to the problems of air pollution and growing appeals to governmental officials to clean up the air. These efforts began at State and local levels.

EARLY EFFORTS TO REGULATE AIR POLLUTION

As urban air pollution worsened in the mid-twentieth century, it became apparent that governmental action of some type was necessary. Scientists studying the air pollution identified sources and possible solutions, and recognized very early that air pollution is a complex problem with complex and potentially expensive solutions. Legislators and policymakers debated the issues and searched for politically acceptable solutions. Owners and operators of the sources of the air pollutants often took issue with the scientists because control would be both expensive and potentially upsetting to the emitting processes. The public reacted to the problems and the possibility of governmental regulation of air pollution in many ways. At the extremes, some viewed government regulation as the only certain way to improve health and aesthetic values; others saw it as an unnecessary limitation on life-style and individual freedom.

Clearly though, solutions were needed. As America's industrial machine grew through and beyond the war years and as Americans became more affluent and educated, air quality was worsening. People began to realize that visibility was decreasing, that adverse health effects related to air pollution were increasing, and that property and lifestyles were being degraded by air pollution. People also began to realize that the air, like the water and land, was not limitless in its ability to cleanse itself of contamination.

Reacting to the growing problems in Southern California, the state government passed the Air Pollution Control Act of 1947. This Act authorized the formation of local air pollution control districts, drawn largely on county lines. It also contained two important statutory prohibitions. The first prohibited the discharge of emissions more opaque than prescribed standards; this prohibition continues to serve as the foundation of many air pollution regulatory programs. The second prohibited discharges that would constitute a 'nuisance;' this prohibition was particularly significant because it provided the initial basis for regulation of pollutants that cause adverse health effects.

At about the same time, several heavily industrialized areas in the U.S. began programs to reduce air pollution. A notable example was Pittsburgh, the center of the steel industry in the country. Air pollution from that industry largely was uncontrolled before the 1950s and Pittsburgh frequently experienced air pollution episodes so severe that city street lights were required at midday. Growing concerns in that community with the effects of air pollution on public health and the potential negative effects on regional growth and economics because of its unpleasant nature led to an aggressive, and largely successful, program to reduce air pollution (Goldman 1967).

Legislative attention also began to surface in the U.S. Congress in the early 1950s. Initially, committees were formed, hearings were held, and there were calls for further research (Krier and Ursin 1977). These early efforts to regulate air pollution were based largely on attempts to reduce visible emissions and noticeable eye and respiratory irritation, the obvious adverse effects of air pollution. At that time, although concerns were being raised by some scientists, there was little clear evidence that lasting adverse human health effects resulted from the air pollution levels typically seen in urban and industrial areas.

In 1959, important new legislation was passed in California. This provided for ambient air quality standards to be established to 'reflect the relationship between the intensity and composition of air pollution and the health, illness, including irritation to the senses, and death of human beings, as well as damage to vegetation and interference with visibility.'[2] The necessary levels for these standards and the processes by which they were to be derived were not well defined. However, it generally was accepted that the processes should not include cost as a factor. It also generally was accepted that the legislation was aimed at

---

[2] California Health and Safety Code, section 426.1.

protecting the most sensitive members of the population (Krier and Ursin 1977).

In the U.S. Congress, a consensus was growing that the federal government should support research into the causes, effects, and control of air pollution. However, there was much less agreement on the appropriateness of federal regulation of releases to the air. Reflecting this, in 1955 Congress passed the Air Pollution Control Act. This Act focused largely on requiring the U.S. Public Health Service to conduct research on the effects of air pollution and air pollution control technologies and to provide states with technical assistance and training in air pollution.

The Air Pollution Control Act was amended by Congress in 1960. In these amendments, the Surgeon General was required to conduct thorough studies of the effects of motor vehicle exhaust on humans and to report on the results. The Surgeon General did submit a report to Congress in June 1962 describing the available evidence concerning the adverse effects on humans, animals, and materials from exposure to motor vehicular exhaust. However, the report reached no definitive conclusions and the Surgeon General requested additional time to obtain information to make more 'equitable and appropriate judgments' concerning potential limitations of discharges from motor vehicles.

At that time, the administration of President Kennedy was contemplating a broader federal role in dealing with air pollution, in opposition to the previously accepted views that pollution was better handled at the state and local level. The Second National Conference on Air Pollution, held in December 1962, served as a springboard for those urging greater federal responsibilities. Advocates were aided unfortunately by a serious air pollution episode in London at that time, said to have claimed 700 lives. Shortly thereafter, President Kennedy became convinced that both federal enforcement and federal abatement authority were necessary, although his administration offered no immediate legislation. However, Senator Ribicoff of Connecticut, who earlier had served as Secretary of Health, Education and Welfare under President Kennedy, decided to focus on

air pollution as a political issue and introduced legislation that included federal air pollution abatement authority. After lengthy debate, and soon after the death of President Kennedy, a compromise bill was signed into law by President Johnson.

The compromise Clean Air Act of 1963 continued the earlier focus on research and technical assistance. It also provided for the development of air quality criteria by the Department of Health, Education and Welfare, and it created federal investigative and abatement programs much like those already in effect for water pollution. However, it specified that the air quality criteria were only advisory and a lengthy and complex set of requirements were specified for abatement.

Perhaps the most important outcome of this legislation was the formation of the Senate Special Subcommittee on Air and Water Pollution. Senator Edmund Muskie was named its chairman and he became the dominant congressional player in the regulation of air pollution for more than a decade. He is credited, along with leaders in California, with focusing early congressional attention on automotive emissions. In an unprecedented action soon after his appointment as Subcommittee chairman, Senator Muskie held a series of public hearings on air pollution across the nation. The first hearing, appropriately held in Los Angeles in January 1964, concentrated attention on the automobile. At that hearing, Governor Brown of California argued persuasively that the automobile industry was 'in interstate commerce and the Federal Government clearly had jurisdiction.' (Sundquist 1968) Senator Muskie heard much the same story at the other hearings and subsequently concluded that vehicular air pollution was appropriate to be dealt with at the federal level. Not surprisingly, the automotive industry disagreed. They argued that air pollution control should be based on the needs of each state.

Nonetheless, Senator Muskie quickly introduced legislation to establish emission standards for gasoline-powered vehicles. President Johnson remained unconvinced but stated publicly that he would initiate discussions with the automobile industry who continued to argue that air pollution was not severe enough to require federal intervention. When no discussions ensued after several months, editorials in the *New York Times*, the *Washington Post*, the *Los Angeles Times* and the *Wall Street Journal* all hinted at an association between the President and the automobile industry on this issue (Krier and Ursin 1977).

In an odd turnabout, the automotive industry suddenly reversed its position and announced that it would equip new automobiles with some exhaust controls, provided sufficient lead time was allowed to design and engineer the changes. Some observers theorized that the change of heart by the automotive industry resulted from the passage, or pending passage, of tough, new automotive emissions control bills in several states. For whatever reasons, developments in dealing with motor vehicle pollution control from that point were rapid.

A motor vehicle compromise was reached by Congress in 1965. It provided for federal discretion in administering the law rather than explicit standards and deadlines. In addition, the effective date of the standards was set for 1968. The resulting Motor Vehicle Air Pollution Control Act of 1965, which amended the Clean Air Act of 1963, required federal standards to mitigate the confusion that would arise from different standards in different states, and it specifically recognized technological and economic feasibility in setting automotive emission standards. The federal government subsequently promulgated emissions standards for new motor vehicles.

In 1967, national air pollution emission standards for stationary sources also were being debated in Congress. Early in that year, President Johnson proposed national emission standards for industries that 'contribute heavily' to air pollution. The Senate Committee, however, opted for a less controversial approach and the Clean Air Act Amendments of 1967 established a more modest federal control program. Importantly, the 1967 Amendments established for the first time the concept of the Air Quality Control Region, which was defined as an interstate or intrastate region specified for coordinated control. It also required the development by the federal government of air quality criteria and recommended air pollution control techniques, and it required states to establish air quality standards within a fixed time schedule.

By 1970, however, little regulatory headway had been made by the federal government in carrying out the 1967 Amendments. This lack of action, and the growing political realization that air pollution control was not only publicly acceptable but largely expected, hastened significant new actions by the administration and Congress. Not the least of these was the establishment of the new Environmental Protection Agency and the passage of the landmark Clean Air Act Amendments of 1970. Federal regulation of air pollution had begun.

THE 1970 CLEAN AIR ACT AMENDMENTS

The 1970 Clean Air Act Amendments defined two principal types of air pollutants for regulation: criteria air pollutants and hazardous air pollutants. The term criteria air pollutant had been in use since the 1963 Amendments. These pollutants were defined in section 108 as those that 'cause or contribute to air pollution that may reasonably be anticipated to endanger public health or welfare ... the presence of which in the ambient air results from numerous or diverse mobile or stationary sources.' EPA was required to establish national ambient air quality standards for pollutants for which air quality criteria had been issued. The criteria pollutant control program was defined in detail and based on the premise that the federal government should deal with issues that were national in scope, or where air pollution crossed political boundaries, and that state and local governments should deal with more localized issues.

Under section 109, EPA first identified pollutants meeting the definition and prescribed national primary air quality standards 'the attainment and maintenance of which ... allowing an adequate margin of safety, are requisite to protect the public health.' National secondary air quality standards also were prescribed 'the attainment and maintenance of which … is requisite to protect the public welfare from any known or anticipated effects associated with the presence of the air pollutant.' Welfare effects include injury to agricultural crops and livestock, damage to and the deterioration of property, and hazards to air and ground transportation.

The national ambient air quality standards (usually referred to as NAAQS) were to be attained and then maintained, even in the face of growth in the population and the economy, through regulation of stationary and mobile sources of the pollutants or their precursors. Stationary source regulation was to be accomplished largely under section 110 by state and local agencies (existing sources), and under section 111 by EPA (new sources). Mobile source regulation was accomplished largely by EPA under Title II of the 1970 Amendments. Together, these programs resulted in substantial improvements in air quality in the 1970s.

The 1970 Amendments also required, under section 112, regulation of hazardous air pollutants. The regulatory approach specified, however, was very different from that for criteria air pollutants. Section 112 defined a hazardous air pollutant as one 'to which no ambient air standard is applicable and that … causes, or contributes to, air pollution which may reasonably be anticipated to result in an increase in mortality or an increase in serious irreversible,

or incapacitating reversible, illness.' Congress did not expand on this definition but did identify mercury, beryllium, and asbestos as initial pollutants of concern.

The process for regulation of hazardous air pollutants was described in section 112 in general terms only. EPA was to list substances that meet the definition as hazardous air pollutants and publish national emission standards for these pollutants providing 'an ample margin of safety to protect the public health from such hazardous air pollutant[s].' The use of the term 'ample margin of safety' for hazardous air pollutants, as opposed to 'adequate margin of safety' for criteria air pollutants, obviously implied more stringent control, although the distinction was not clarified. As worded, section 112 relied largely on EPA to develop and carry out the hazardous air pollutant control program although it provided for the delegation of enforcement authority to state and local agencies.

Although the criteria air pollutant control program in the 1970s generally was considered successful, the program for the hazardous air pollutants was considered by most to be far less successful. This lack of success in regulating hazardous air pollutants in part resulted from the lack of detail in the Act concerning the regulatory process. Another reason was that this was the first instance where a federal regulatory initiative was being undertaken for 'toxic' pollutants, and there was little regulatory precedent on which EPA could draw.

REGULATION OF HAZARDOUS AIR POLLUTANTS FROM 1970 TO 1990

Soon after the passage of the Clean Air Act of 1970 and the formation of EPA, implementation of section 112 began with the listing of asbestos, beryllium, and mercury as hazardous air pollutants. Table 1-1 provides the citation and publication dates for all hazardous air pollutants listed under the 1970 Amendments. Standards for the first three hazardous air pollutants were promulgated in 1973 (EPA 1973). There was no further action until 1975 when, in the face of litiga-

tion, EPA listed vinyl chloride as a hazardous air pollutant and proposed standards (EPA 1975). The vinyl chloride standards were promulgated in 1976 (EPA 1976c). Shortly thereafter, EPA was sued by the Environmental Defense Fund (EDF), who argued that the regulations were not sufficiently protective. EDF's view was that the regulations were based in part on the technical feasibility of control and that this was not allowed under section 112.

Table 1-1   Published listing decisions on hazardous air pollutants

Hazardous air pollutant	Citation and date of listing
Asbestos	36 FR 23239, Dec. 7, 1973
Beryllium	36 FR 23239, Dec. 7, 1973
Mercury	36 FR 23239, Dec. 7, 1973
Vinyl chloride	40 FR 59477, Dec. 24, 1975
Benzene	42 FR 29332, June 8, 1977
Radionuclides	44 FR 76738, Dec. 27, 1979
Inorganic arsenic	45 FR 37886, June 5, 1980
Coke oven emissions	49 FR 36560, Sep. 18, 1984

At that time, public concerns with environmentally caused cancer were growing rapidly and public interest groups were beginning to argue that any involuntary cancer risk was unacceptable and that our society should eliminate the emissions of carcinogens. EPA and EDF subsequently reached a settlement on the vinyl chloride suit with EPA agreeing to propose new, more stringent standards and to set zero emissions of vinyl chloride as an eventual objective. EPA proposed the revised standards for vinyl chloride the next year (EPA 1977c). In those standards, EPA halved the earlier emission limit and set a target of zero emissions. However, EPA also reiterated its earlier view that the inability to identify a health threshold did not require a zero emission standard; instead, EPA stated that it should be permitted to regulate to a level of best control technology. Interestingly, no further action was taken on the vinyl chloride standards for eight years.

EPA's next published hazardous air pollutant action following vinyl chloride was the listing of benzene as a hazardous air pollutant in 1977, again in the face of litigation. Then, in 1979 EPA proposed an air cancer policy (EPA 1979d) that was intended to establish, among other things, EPA's approach for dealing with pollutants, such as vinyl chloride, for which there is no identified health effects threshold. This policy served as a useful focus for public

deliberation and came close to being promulgated. However, it had not been when President Reagan took office in 1981. No further action was taken on the air cancer policy.

In 1977, Congress also revised the Clean Air Act Amendments and added minor amendments to the hazardous air pollutant provisions. One change expanded the definition of emission standard to include design, equipment, work practice, and operational standards. This was done to allow the agency to regulate sources where emissions could not be measured directly; such was the case with asbestos. The 1977 Amendments also required in section 122 that EPA make listing decisions for four 'unregulated' pollutants: radioactive materials, arsenic, cadmium, and polycyclic organic matter (POM). In response to this requirement, EPA listed radionuclides and inorganic arsenic as hazardous air pollutants. However, EPA declined to list POM as a hazardous air pollutant (EPA 1984e), although in doing so the Agency acknowledged that coke oven emissions, a well-studied subset of POM, should be regulated; listing of coke oven emissions as a hazardous air pollutant occurred shortly thereafter. Finally, EPA announced in 1985 its 'intent' to list cadmium as a hazardous air pollutant (EPA 1985g).

During this time, 'intent-to-list decisions' for several other potentially toxic air pollutants also were published. This process was viewed by the agency as a way in which to move toward regulation while not triggering the regulatory deadlines specified in section 112.[3] These pollutants, the decisions, and the citations are shown in Table 1-2. Although EPA announced its 'intent to list' several pollutants as hazardous air pollutants, no formal listings occurred before the passage of the 1990 Amendments.

EPA did propose standards for radionuclides in 1984 (EPA 1984f), later withdrew the standards (EPA 1984g), and then promulgated them in 1989 (EPA 1989l). Standards also were proposed for some benzene sources in 1980,

---

[3] Section 112 required, among other things, that regulations be proposed within 180 days of listing a substance as a hazardous air pollutant and then promulgated within another 180 days.

Table 1-2   Published listing decisions on potential hazardous air pollutants

Substance	EPA decision	Citation and date
Acrylonitrile	Referral[a]	50 FR 24319, June 10, 1985
1,3-Butadiene	Intent-to-list	50 FR 41466, Oct. 10, 1985
Cadmium	Intent-to-list	50 FR 42000, Oct. 16, 1985
Carbon tetrachloride	Intent-to-list	50 FR 32621, Aug. 13, 1985
Chlorofluorocarbon 113	No listing	50 FR 24313, June 10, 1985
Chlorinated benzenes	No listing	50 FR 32628, Aug. 13, 1985
Chloroform	Intent-to-list	50 FR 39626, Sep. 27, 1985
Chloroprene	No listing	50 FR 39632, Sep. 27, 1985
Chromium	Intent-to-list	50 FR 24317, June 10, 1985
Copper	No listing	52 FR 5496, Feb. 23, 1987
Ethylene dichloride	Intent-to-list	50 FR 41994, Oct. 16, 1985
Epichlorohydrin	No listing	50 FR 24575, June 11, 1985
Ethylene oxide	Intent-to-list	50 FR 40286, Oct. 2, 1985
Hexachlorocyclopentadiene	No listing	50 FR 40154, Oct. 1, 1985
Manganese	No listing	50 FR 32627, Aug. 13, 1985
Methyl chloroform	Intent-to-list	50 FR 24314, June 10, 1985
Methylene chloride	Intent-to-list	50 FR 42037, Oct. 17, 1985
Nickel	Intent-to-list	51 FR 34135, Sep. 25, 1986
Perchloroethylene	Intent-to-list	50 FR 52880, Dec. 26, 1985
Phenol	No listing	51 FR 32854, June 23, 1986
Polycyclic organic matter	No listing	49 FR 31680, Aug. 8, 1984
Toluene	No listing	49 FR 22195, May 25, 1984
Trichloroethylene	Intent-to-list	50 FR 52422, Dec. 23, 1985
Vinylidene chloride	No listing	50 FR 32632, Aug, 13, 1985
Zinc and zinc oxide	No listing	52 FR 32597, Aug. 28, 1987

[a] Referral to states for regulatory decisions.

partially withdrawn in 1984 (EPA 1984h), and promulgated in 1989 (EPA 1989k). EPA also proposed standards for coke oven emissions in 1987 (EPA 1987i); however, coke oven emission standards never were promulgated.

One of the biggest difficulties facing EPA at this time, and a major reason for their lack of success in regulating more hazardous air pollutants, was the potentially enormous size of the toxic air pollutant problem. In the late 1970s, it was estimated that 65,000 chemicals had been in commerce and that many more were being introduced each year (EPA 1984a). Scientists also recognized that other toxic substances could be formed in atmospheric reactions and that there were literally hundreds of thousands of potential sources of emissions of these chemicals. Regulators were faced with a daunting task of deciding the air toxics to regulate and how to do so within reasonable time and resource limitations. Thus, in the early 1980s,

EPA focused its study of hazardous air pollutants on identifying those with the greatest potential adverse health effects, their sources, and their geographic concentrations. Studies of various industries known to be major emitters of these pollutants also were conducted. However, there was little substantive movement toward publishing regulations to deal with the problems that were identified.

What was expected to be a major stimulus to air toxics regulation came in August of 1983 when the General Accounting Office (GAO), at the request of Congressman John Dingell, Chairman of the House Energy and Commerce Committee, conducted a study of EPA's hazardous air pollutant program and released a critical report (GAO 1983). A Congressional hearing was held in November 1983. In his testimony (Ruckelshaus 1983b), EPA Administrator Ruckelshaus acknowledged the problems identified by the GAO and publicly committed the

Agency to aggressive new actions on air toxics. Administrator Ruckelshaus also argued strongly for a more flexible test for regulating hazardous air pollutants, one that included factors such as the costs of the controls and the benefits of the risks reduced.

One result of Administrator Ruckelshaus' commitment before Congress was the release in 1985 of EPA's 'Air Toxics Strategy.' This document was described as the blueprint for implementation of the 'aggressive' new air toxics regulatory program. In fact, however, little occurred in terms of substantive regulation or policy decision after release of the strategy. There were no further listings of hazardous air pollutants and the only regulations promulgated during that time were for inorganic arsenic (EPA 1986u).

In 1985, EPA also withdrew the 1977 vinyl chloride proposal, citing unreasonable costs, the unavailability of technology to reduce emissions significantly and consistently below the original emission limits, and the minor risks posed by vinyl chloride emissions. The Natural Resources Defense Council (NRDC) quickly petitioned the courts to review the decision. In 1986, D.C. Circuit Judge Robert Bork wrote a decision for a three-judge panel that stated that EPA could consider cost and technological feasibility in setting emission standards for hazardous air pollutants. However, the full U.S. Court of Appeals, on rehearing, significantly modified that interpretation and the final opinion,[4] and significantly curtailed but did not eliminate the extent to which EPA could consider cost and technical feasibility.

The full court rejected NRDC's contention that section 112 requires zero emissions for nonthreshold pollutants but prohibited the consideration of 'practical factors' (e.g., cost and feasibility) in setting standards. The court found no evidence that Congress intended the 'massive economic and social dislocations' that a zero emissions interpretation could cause. The court remanded the 1985 decision to EPA but

declined to tighten the standard. However, the court ruled that EPA's approach in regulating vinyl chloride did not meet the requirements of section 112 because it failed to determine that a certain level of emissions is 'safe' or that the chosen level would provide an 'ample margin of safety.' To remedy this, the court set forth a two-step process for EPA to follow in making these determinations.

In the first step, EPA had to establish a 'safe' or 'acceptable' risk level, which under no circumstances could consider cost and technological feasibility. The court emphasized that this did not mean the identification of a 'risk-free' level or that it must be free of uncertainty. Instead, the decision must be based on expert judgment and a definition of safe as meaning 'acceptable in the world in which we live.' In the second step, EPA had to set standards at the level – that could be equal to or lower, but not higher, than the safe or acceptable level – to protect the public with an 'ample margin of safety.'

In late 1989, after taking public comment and in conjunction with benzene regulations, EPA published its response to the Court's ruling (EPA 1989k). EPA's stated approach for protecting public health with an ample margin of safety under section 112 was to 'provide maximum feasible protection against risks to health from hazardous air pollutants by (1) protecting the greatest number of persons possible to an individual lifetime risk level no higher than approximately 1×10^{-6} [one in one million][5] and (2) limiting to no higher than approximately 1×10^{-4} [one in ten thousand] the estimated risk that a person living near a plant would have if he or she were exposed to the maximum pollutant concentrations for 70 years.' This latter risk is known as the maximum individual risk (MIR).

EPA applied this new approach in promulgating final regulations for benzene emissions from two source categories. Further actions by the agency were expected; however, none were taken. The enactment of the Clean Air Act

---

[4] Natural Resources Defense Council, Inc. v. EPA, 824 F.2d 1146[1987]

[5] The scientific notation 1×10^{-6} means one divided by one million; one million is one followed by six zeroes.

Amendments on November 15, 1990, effectively superseded all that had been done, or not done, previously. The specific requirements in the 1990 Amendments with respect to hazardous or toxic air pollutants are described in detail in Chapter 2.

TOXIC AIR POLLUTANT REGULATORY RESPONSIBILITIES

Federal Regulation

Federal implementation of the Clean Air Act is the responsibility of EPA's Office of Air and Radiation. That office, headquartered in Washington, D.C., administers the air pollution control programs required by law, provides overall air pollution policy guidance, provides guidance to state and local air pollution control agencies, and coordinates with other offices in EPA. Overall

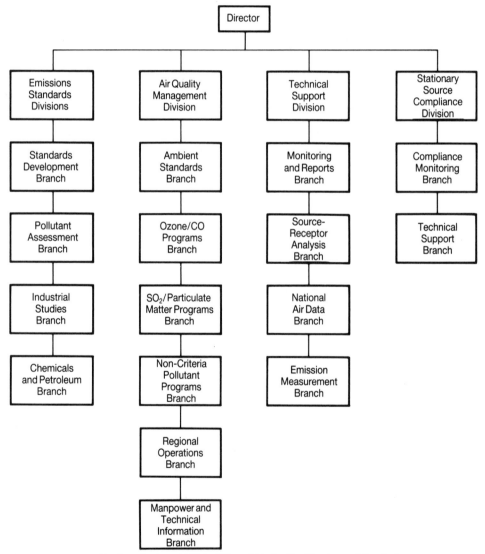

Figure 1-1.   The 1991 Organization of EPA's Office of Air Quality Planning and Standards.

Table 1-3   EPA air toxics decision-makers

Position	Responsibility	Phone
EPA Administrator	Directs EPA	(202) 260–4700
Deputy Administrator	Second in command	(202) 260–4700
Assistant Administrator for Air and Radiation	Overall direction of air program	(202) 260–7400
Deputy Assistant Administrator for Air and Radiation	Second in command of air program	(202) 260–7400
National Air Toxics Coordinator	Interfaces with other programs	(202) 260–7400
Director, Air Quality Planning and Standards	Manages stationary source air pollution	(919) 541–5616
Director, Emission Standards Division	Directs development of source standards	(919) 541–5571
Chief, Pollutant Assessment Branch	Directs air toxics support activities	(919) 541–5645

leadership is provided by an Assistant Administrator and a Deputy Assistant Administrator. A National Air Toxics Coordinator reports to the Assistant Administrator. As the name implies, the National Air Toxics Coordinator interfaces with Congress, industry and other outside interests, and coordinates on a day-to-day basis with EPA's other media programs.

Unlike most other EPA media program offices, the office specifically charged with stationary source air pollution regulations is not located in Washington. Before the formation of EPA, federal air pollution matters were the responsibility of the National Air Pollution Control Administration (NAPCA), a part of the Public Health Service and located in Cincinnati. In the late 1960s, however, apparently influenced by Luther Hodges, Secretary of Commerce under President Johnson and past Governor of North Carolina, NAPCA was moved to North Carolina to help anchor the newly state-chartered Research Triangle Park between Raleigh, Durham, and Chapel Hill. EPA's stationary source air pollution office remains there to this day. Although movement of the office to Washington may have been discussed, there apparently never has been a compelling need to do so.

EPA's Office of Air Quality Planning and Standards (OAQPS) provides the technical and administrative oversight for the development and implementation of regulations under the Clean Air Act for stationary sources of air pollution. Figure 1-1 shows the OAQPS organization, circa 1991. A list of key decision-makers in EPA concerned with toxic air pollutants is shown in Table 1-3. Some information sources relating to air toxics that are available from OAQPS are shown in Table 1-4.

EPA's Regional Offices also play a key role in interfacing with state and local agencies and in carrying out federal regulatory programs. As shown in Table 1-5, each EPA Regional Office has assigned an Air Toxics Coordinator. EPA's Office of Research and Development (ORD) also provides substantial support for the agency's air toxics program, principally in conducting research, preparing health assessment documentation for hazardous air pollutants, and developing guidance for exposure and risk assessment. Table 1-6 provides a listing of key ORD offices working on toxic air pollution. EPA's 1991 Research Priority Areas relating to toxic air pollutants are listed below.[6]

EPA Air Toxics Research Priority Areas

Health effects	Risk assessments
Ambient monitoring	Complex modeling
Source and ambient methods	Technology assessment
Integrated air cancer project	Technology transfer centers
Health assessment documents	Risk reference concentrations
Exposure modeling and monitoring	Transformation and fate
Area source emission inventory methods	

---

[6] From a presentation at the Clean Air Scientific Advisory Committee (CASAC), April 29, 1991, by the EPA Office of Research and Development Air and Radiation Research Committee.

Table 1-4   Important EPA air toxics information sources

Source[a]	Description	Phone
AirRISC[b]	Support and hotline	(919) 541–5344
NATICH[c]	On-line services	(703) 487–4807
	Reports	(703) 487–4650
	EPA assistance (hotline)	(919) 541–0850
CTC[d]	Support and hotline	(919) 541–5432
Bulletin board	Emission estimates/factors	(919) 541–5373
Modeling	Guidance	(919) 541–5381
Air/Superfund	Coordination	(919) 541–5589
Training	Classroom and self-help	(919) 541–2354
BACT/LAER Clearinghouse	Support	(919) 541–2736

[a] Generally free to governmental agencies; fees may be charged to non-government users.
[b] Air Risk Information Support Center.
[c] National Air Toxics Information Clearinghouse.
[d] Control Technology Center.

Table 1-5   EPA regional office air toxics contacts

Region	Contact	Phone
1	Air Management Division	(617) 565–3800
2	Air and Waste Management Division	(212) 264–2301
3	Air Management Division	(215) 597–9390
4	Air, Pesticides and Toxics Management Division	(404) 347–3043
5	Air Management Division	(312) 353–2213
6	Air, Pesticides and Toxics Division	(214) 655–7200
7	Air and Toxic Management Division	(913) 551–7020
8	Air and Toxics Division	(303) 293–1438
9	Air Division	(415) 744–1219
10	Air and Toxics Division	(206) 442–4152

Source: *Journal of the Air & Waste Management Association*, April 1991.

Table 1-6   EPA Office of Research and Development air toxics responsibilities

Office[a]	Responsibility	Phone
OHEA	Health/environmental assessment	(202) 260–7317
ECAO	Air toxics health documents	(919) 541–4173
HHAG	Human cancer assessment	(202) 260–5898
EAG	Exposure assessment	(202) 260–8909
HERL	Health research	(919) 541–2281
AEERL	Technology research	(919) 541–2821
AREAL	Atmosphere/exposure research	(919) 541–2106
OTTRS	Technology transfer	(202) 260–7669
OMMSQA	Modeling/monitoring research	(202) 260–5776
CERI	Information distribution center	(513) 569–7562

[a] Refer to list of acronyms for full name.

14

Table 1-7   State regulatory agency air program contacts

Agency	Contact	Phone
Alabama	Air Division, Dept. of Environmental Management	(205) 271–7861
Alaska	Department of Environmental Conservation	(907) 465–5100
Arizona	Department of Environmental Quality	(602) 207–2308
Arkansas	Department of Pollution Control and Ecology	(501) 562–7444
California	California Air Resources Board	(916) 322–2990
Colorado	Air Pollution Control Division	(303) 331–8500
Connecticut	Bureau of Air Management	(203) 566–2506
Delaware	Division of Air and Waste Management	(302) 736–4791
District of Columbia	Air Quality Control and Monitoring Branch	(202) 404–1180
Florida	Department of Environmental Regulation	(904) 488–0114
Georgia	Department of Natural Resources	(404) 363–7100
Hawaii	Clean Air Branch	(808) 543–4200
Idaho	Department of Health and Welfare	(208) 334–0502
Illinois	Division of Air Pollution Control	(217) 782–7326
Indiana	Department of Environmental Management	(317) 232–3210
Iowa	Department of Natural Resources	(515) 281–5145
Kansas	Bureau of Air and Waste Management	(913) 296–1593
Kentucky	Division of Air Quality	(502) 564–3382
Louisiana	Office of Air Quality and Nuclear Energy	(504) 765–0219
Maine	Bureau of Air Quality Control	(207) 289–2437
Maryland	Air Management Division	(410) 631–3255
Massachusetts	Division of Air Quality Control	(617) 292–5593
Michigan	Air Quality Division	(517) 373–7023
Minnesota	Air Quality Division	(612) 296–7331
Mississippi	Office of Pollution Control	(601) 961–5171
Missouri	Air Pollution Control Program	(314) 751–4817
Montana	Air Quality Bureau	(406) 444–3454
Nebraska	Department of Environmental Control	(402) 471–2189
Nevada	Bureau of Air Quality	(702) 687–5065
New Hampshire	Air Resources Division	(603) 271–1370
New Jersey	Air Pollution Control Program	(609) 292–6704
New Mexico	Air Quality Bureau	(505) 827–2850
New York	Division of Air Resources	(518) 457–7230
North Carolina	Division of Environmental Management	(919) 733–5317
North Dakota	Department of Health	(701) 221–5188
Ohio	Division of Air Pollution Control	(614) 644–2270
Oklahoma	Air Quality Service	(405) 271–5220
Oregon	Air Quality Division	(503) 229–5397
Pennsylvania	Bureau of Air Quality Control	(717) 787–9702
Puerto Rico	Environmental Quality Board	(809) 767–8071
Rhode Island	Division of Air and Hazardous Materials	(401) 277–2808
South Carolina	Bureau of Air Quality Control	(803) 734–4750
South Dakota	Division of Environmental Regulation	(605) 773–3351
Tennessee	Division of Air Pollution Control	(615) 741–3931
Texas	Texas Air Control Board	(512) 451–5711
Utah	Bureau of Air Quality	(801) 536–4000
Vermont	Air Pollution Control Division	(802) 244–8731
Virgin Islands	Department of Planning and Natural Resources	(809) 773–0565
Virginia	Department of Air Pollution Control	(804) 786–2378
Washington	Department of Ecology	(206) 459–6256
West Virginia	Air Pollution Control Commission	(304) 348–4022
Wisconsin	Department of Natural Resources	(608) 266–7718
Wyoming	Air Quality Division	(307) 777–7391
Canada	Department of the Environment, Conservation and Protection	(819) 997–1575

Source: *Journal of the Air & Waste Management Association*, April 1993.

State and Local Regulation

Because EPA largely was unsuccessful in carrying out a hazardous air pollutant program under the 1970 Clean Air Act Amendments, many state and local air pollution agencies initiated air toxics regulatory programs of their own. These were stimulated both by public concerns in their jurisdictions and by laws dealing with these problems. Again because little guidance was published by EPA and because state and local agencies varied markedly in their technical and resource capacity to deal with the issues, the state and local programs took a variety of forms. Because it was easier to negotiate changes as sources were constructed or modified, most air toxics programs became part of ongoing new source permitting programs; few state and local programs focused on existing sources (ENSR 1988).

By the time of passage of the 1990 Clean Air Act Amendments, most state and a few larger local agencies had air toxics programs in place. These programs frequently regulated air toxics through the use of ambient concentration limits, although many also attempted to use risk assessment. A 1989 survey (STAPPA/ALAPCO 1989) by the State and Territorial Air Pollution Program Administrators (STAPPA) and the Association of Local Air Pollution Control Officials (ALAPCO) reported that ten states had comprehensive air toxics regulatory programs in place, another twenty-two had comprehensive air toxics policies in place, three had proposed comprehensive regulations, and eleven had informal new source review programs in place for sources of air toxics. In addition, local agencies in eleven states had either comprehensive air toxics policies in place or formal regulatory programs.

Under the 1970 Clean Air Act Amendments, EPA delegated to state and local agencies the authority to enforce federal hazardous air pollutant standards. Title VII (Provisions Relating to Enforcement) of the 1990 Clean Air Act Amendments continues to provide for that delegation and Title V (Permits) provides for a comprehensive new operating permit program to oversee compliance with all requirements. It is anticipated that most, if not all, states and major local agencies will seek and continue to exercise that delegation authority since Title V also provides for the collection of emission fees by state and local agencies to 'cover all reasonable (direct and indirect) costs required to develop and administer the permit program requirements' of the Act. The level of the fee is suggested in the 1990 Amendments to be a minimum of $25 per ton of emission per year; some states (e.g., New Jersey) have indicated an intent to seek higher fees for toxic air pollutants.

State and local air toxics contacts are shown in Table 1-7.

THE DEFINITION OF TOXIC AIR POLLUTANTS

Before toxic air pollutants can be assessed and regulated, there should be consensus on their definition. This definition can be considered both scientifically and legally. Goldstein[7] (1983) provided a scientific definition. He stated 'there are three major criteria for a compound to be included under the rubric of toxic air pollutant:

1. It is measurable in the air,
2. It is for the most part produced by the activities of man, and
3. It is not a primary air quality pollutant as currently defined by EPA.'

Interestingly, this definition does not hinge on evidence of toxicity. Thus, this definition

---

[7]Chairman of the Department of Environmental and Community Medicine and Dentistry at Rutgers Medical School. Dr. Goldstein served as Assistant Administrator of EPA's Office of Research and Development in the early 1980s.

could encompass virtually any measurable substance produced by man that is not regulated by EPA in another way. Indeed, Goldstein admitted that many compounds meeting these three criteria would prove to be nontoxic. There is a logic in Goldstein's scientific definition because most substances can adversely affect humans if exposure is sufficiently long or concentrated. However, a regulatory definition of 'toxic air pollutant' would seem to require at least some presumption of toxicity, and Congress in the 1970 Clean Air Act Amendments clearly specified that hazardous air pollutants were substances that caused 'serious' adverse human health effects.

A legal definition would seem best derived from the environmental legislation. However, federal environmental legislation to date provides no consistent definition of a toxic chemical or pollutant. In early laws, 'toxic' appears to have been used mainly for substances that could cause adverse effects in humans and 'hazardous' used more broadly for substances that caused environmental damage. In the 1970 Clean Air Act Amendments, Congress specifically defined 'hazardous air pollutant' to be air pollutants that cause more serious human health effects, including mortality. Also in 1970, the Occupational Safety and Health Act (Public Law 91–596) was enacted and included a process for establishing health standards dealing with 'toxic materials or harmful physical agents.' Later, the 1976 Resource Conservation and Recovery Act described both hazardous and toxic wastes; the Toxic Substances Control Act of 1976 specifically dealt with 'toxic substances' that 'present an unreasonable risk of injury to health or the environment'; the Clean Water Act of 1977 referred to 'toxic' pollutants; and the Comprehensive Environmental Response, Compensation, and Liability Act of 1980 (Superfund) regulated 'hazardous' substances and wastes, and specifically included 'toxic' and 'hazardous' pollutants from other Acts in the definition.

The 1977 Clean Air Act Amendments provided little further insight into the definition of 'hazardous air pollutant.' Congress did not materially change section 112 but did add

section 122, as noted earlier, requiring EPA to review all available relevant information and determine whether or not emissions of radioactive pollutants, cadmium, arsenic, and polycyclic organic matter into the ambient air 'will cause, or contribute to, air pollution which may reasonably be anticipated to endanger public health.' Although this did not define 'hazardous' more precisely, it did provide specific additional examples of substances of concern to Congress.

It was not until the Emergency Planning and Community Right-to-Know Act of 1986 (EPCRA) that the term 'toxic' was specifically applied to air pollutants. In section 313 of that law, releases (including releases to the air) of 'toxic chemicals' are required to be reported to EPA and applicable state and local organizations. A group of toxic chemicals initially were specified by Congress, and a process was provided for the addition or deletion of chemicals from the list. In section 313(d)(2), additions are to be based upon the showing that there is sufficient evidence to establish any one of the following:

(A) The chemical is known to cause or can reasonably be anticipated to cause significant adverse acute human health effects at concentration levels that are reasonably likely to exist beyond facility site boundaries as a result of continuous, or frequently recurring, releases.

(B) The chemical is known to cause or can reasonably be anticipated to cause in humans

   (i) cancer or teratogenic effects, or
   (ii) serious or irreversible
        (I) reproductive dysfunctions,
        (II) neurological disorders,
        (III) heritable genetic mutations, or
        (IV) other chronic health effects.

(C) The chemical is known to cause or can reasonably be anticipated to cause, because of

   (i) its toxicity,
   (ii) its toxicity and persistence in the environment, or
   (iii) its toxicity and tendency to bioaccumulate in the environment,

a significant adverse effect on the environment of sufficient seriousness ... to warrant reporting under this section.

Deletions from the list are to be based only on a determination by the Administrator of EPA that 'there is not sufficient evidence to establish any of the criteria described above.'

The 1990 Clean Air Act Amendments retain the term 'hazardous air pollutant,' but now define the term similarly to section 313 of EPCRA. In the 1990 Amendments, Congress provides a list of 189 pollutants (see Table 3-1) specifically defined as hazardous and establishes a process for revising the list. It is in the revision process that hazardous air pollutants are defined precisely. Namely, section 112(b)(2) specifies that revisions to the list are to be made by rule:

... adding pollutants which present, or may present, through inhalation or other routes of exposure, a threat of adverse human health effects (including, but not limited to, substances which are known to be, or may reasonably be anticipated to be, carcinogenic, mutagenic, teratogenic, neurotoxic, which cause reproductive dysfunction, or which are acutely or chronically toxic) or adverse environmental effects whether through ambient concentrations, bioaccumulation, deposition, or otherwise.

In summary, as used in this handbook, a toxic air pollutant is one that meets both scientific and legal criteria. Specifically, a toxic air pollutant is assumed to meet three criteria:

1. Toxic air pollutants have the potential to cause serious adverse effects in the general population or the environment as a result of ambient exposures that may be expected to occur.
2. Toxic air pollutants are released from sources as a routine course of operation or business.
3. Toxic air pollutants do not include air pollutants regulated under sections 108–110 of the Clean Air Act or pollutants regulated in the work place through the Occupational Safety and Health Act.

Finally, although 'hazardous air pollutants' is the term used in the Clean Air Act to refer to pollutants to be regulated under section 112, these pollutants are commonly referred to as toxic air pollutants. In this handbook, the term toxic air pollutants generally is preferred. For convenience, toxic air pollutants often are referred to simply as 'air toxics.'

THE SCOPE OF THE TOXIC AIR POLLUTANT PROBLEM

A comprehensive study of the scope of the air toxics problem has never been conducted. One reason was the lack of consensus on what constitutes a toxic air pollutant. In response to the 1970 Clean Air Act Amendments, EPA did initiate studies to define substances of potential concern under section 112. The first of these studies (Fuller et al. 1976) identified 637 organic chemicals and compiled and summarized for these chemicals relevant information including chemical, physical and toxicological properties, estimated production, and estimated emissions.

Studies continued at EPA into the 1980s to identify potential hazardous air pollutants and their sources. However, little effort was expended at defining the nationwide scope of the air

toxics problem until the publication of the report that came to be known as 'The Six-Month Study' (Haemisegger et al. 1985). That study focused on cancer because it is the best studied serious adverse health effect. Unfortunately, that study and others of a similar nature all suffer from the same significant disadvantages. First, there are substantial uncertainties in both cancer risk estimation procedures and human exposure estimation procedures. These uncertainties largely are dealt with in these studies by making highly conservative (i.e., health protective) assumptions, that generally lead to significant overestimates in the risks. On the other hand, available data are limited to a small number of well-studied potential carcinogens; this obviously leads to underestimates in

Table 1-8   Reported air releases by industry of TRI chemicals

Industry	SIC code	Facilities reporting	Air releases (lbs)
Food	20	1,452	13,716,106
Tobacco	21	19	13,820,462
Textiles	22	401	35,617,602
Apparel	23	29	1,019,201
Lumber	24	616	27,813,245
Furniture	25	397	56,894,704
Paper	26	587	202,210,446
Printing	27	313	50,423,807
Chemicals	28	3,838	754,922,471
Petroleum	29	364	54,989,933
Plastics	30	1,293	158,832,600
Leather	31	132	14,255,347
Stone/clay	32	559	23,283,963
Primary metals	33	1,380	232,958,571
Fabr. metals	34	2,579	117,524,318
Machinery	35	870	51,754,691
Electrical	36	1,578	115,198,789
Transportation	37	1,054	201,297,144
Meas./photo.	38	344	45,076,520
Miscellaneous	39	372	26,171,054

Source: EPA 1990a.

the total population cancer risk. Because of these and other limitations, it is difficult to be confident in the conclusions drawn in any study aimed at defining the scope of the air toxic problem.

Another reason EPA did not publish a list of toxic air pollutants, and from that define the scope of the problem, was that there was no legislative requirement to do so. In 1986, however, both in response to the tragic accident in Bhopal, India, and EPA's longstanding inability to define the air toxics problem, Congress enacted the Emergency Planning and Community Right-to-Know Act. In section 313, Congress defined an initial list of 'toxic chemicals' and instructed EPA to establish a program whereby manufacturing facilities that release these and other toxic chemicals to the environment are required to report on a yearly basis

their releases to air, water and land. Reports were required to be submitted beginning in 1988. This formed the basis for the now well-known Toxic Release Inventory (TRI) Program.

EPA's TRI data base provides substantial information from which to define initially the scope of the nation's air toxics problem. Importantly, releases of 170 chemicals from the list of 189 hazardous air pollutants in the 1990 Clean Air Act Amendments are reported to TRI.[8] However, TRI reports are required only from the manufacturing sector of U.S. business and there are many nonmanufacturing sources of toxic air pollutants. A TRI summary report is published annually by EPA. At the time this handbook was prepared in 1992, the 1988 Toxic Release Inventory National Report (EPA 1990a) was the most recently available published

---

[8] The TRI Report referenced here was published before the enactment of the Clean Air Act Amendments of 1990 and referred to 191 chemicals listed in draft Amendments. When finally enacted, the list contained 190 entries – ammonia had been removed. Interestingly, hydrogen sulfide (H_2S) also was to have been removed from the list, but was inadvertently left in the bill signed by the President. However, H_2S finally was removed from the section 112(b) list in late 1991 in a correction passed by Congress and signed by the President.

Table 1-9   Reported air releases by state of TRI chemicals

State	Facilities reporting	Air releases (pounds)
Alabama	382	54,224,680
Alaska	6	21,996,215
American Samoa	2	29,500
Arizona	166	14,780,075
Arkansas	304	46,802,874
California	1,655	81,594,258
Colorado	170	11,129,196
Connecticut	403	23,950,336
Delaware	56	4,873,438
District of Columbia	1	250
Florida	467	53,107,471
Georgia	598	82,538,859
Hawaii	24	874,145
Idaho	48	3,982,578
Illinois	1,229	104,592,707
Indiana	756	110,075,627
Iowa	331	43,135,115
Kansas	186	24,631,652
Kentucky	298	43,739,655
Louisiana	286	133,070,512
Maine	98	16,553,882
Maryland	195	17,383,926
Massachusetts	587	26,494,040
Michigan	790	95,641,067
Minnesota	330	49,388,024
Mississippi	248	53,968,917
Missouri	495	48,634,108
Montana	24	2,384,167
Nebraska	134	17,499,077
Nevada	34	724,620
New Hampshire	137	11,654,930
New Jersey	824	36,522,739
New Mexico	34	1,883,293
New York	816	92,806,469
North Carolina	805	88,335,109
North Dakota	28	1,236,260
Ohio	1,360	136,453,929
Oklahoma	182	31,100,849
Oregon	194	19,887,098
Pennsylvania	1,030	80,987,641
Puerto Rico	171	15,562,284
Rhode Island	188	5,774,024
South Carolina	375	62,613,127
South Dakota	41	2,478,960
Tennessee	517	133,697,458
Texas	1,089	169,936,759
Utah	111	119,410,265
Vermont	55	1,519,293
Virgin Islands	1	1,490,792
Virginia	405	119,593,757
Washington	305	27,604,148
West Virginia	101	31,885,186
Wisconsin	664	44,412,636
Wyoming	26	2,923,126
TOTAL	19,762	2,427,570,103

Source: EPA 1990a.

20

summary, although 1989 and 1990 data were available through the National Library of Medicine's TOXNET system. The precise results of the TRI summary are not important to this handbook; however, the general magnitude of toxics releases to the air nationwide are instructive and are available from the TRI. This provides useful insight into the scope of the air toxics problem.

Key statistics from the TRI are summarized in Tables 1-8 and 1-9. TRI provides the results of reports detailing total (i.e., air, water, and land) releases of over 350 chemicals in 1988 from almost 20 thousand U.S. manufacturing facilities. Total releases of these chemicals to the air were reported in 1988 to be about 2.4 billion pounds. The chemical industry led all manufacturing categories with releases of over 750 million pounds. On the basis of geography, the leading states for total emissions were Texas, Ohio, Tennessee, and Louisiana. Preliminary results of the 1990 TRI, reported in early 1992, indicated that total air releases dropped to 2.2 billion pounds. Some of this reduction undoubtedly was real, although a portion likely was due to better estimation methods.

As noted above, 170 of the 189 hazardous air pollutants listed in the Clean Air Act Amendments are reported in the TRI database. These 170 account for over 80 percent of the total TRI emissions to the air in 1988. In fact, seven of these chemicals were released in quantities exceeding 100 million pounds each in 1988 and accounted for over 70 percent of the total air releases of the

Table 1-10   Reported air releases of the top 25 TRI chemicals also listed in the 1990 Clean Air Act Amendments[a]

Chemical	Point source emissions (pounds)	Fugitive emissions (pounds)	Total air emissions (pounds)
Toluene	173,833,803	89,614,660	263,448,463
Methanol	171,468,217	43,899,334	215,367,551
1,1,1-Trichloroethane	79,689,328	81,985,784	161,675,112
Xylene (mixed)	109,615,976	30,648,844	140,264,820
Chlorine	129,064,368	3,502,536	132,566,904
Methyl ethyl ketone	91,820,664	34,300,344	126,121,008
Dichloromethane	71,587,386	43,224,440	114,811,826
Carbon disulfide	79,186,318	3,094,001	82,280,319
Hydrochloric acid	51,212,395	5,003,543	56,215,938
Trichloroethylene	25,561,821	20,834,198	46,396,019
Glycol ethers	35,935,313	9,719,071	45,654,384
Styrene	22,317,645	11,298,375	33,616,020
Tetrachloroethylene	15,968,309	15,048,908	31,017,217
MIBK	17,873,943	12,121,805	29,995,748
Benzene	10,058,775	18,229,914	28,288,689
Chloroform	15,839,174	6,871,200	22,710,374
Carbonyl sulfide	19,117,592	7,643	19,125,235
Ethylene glycol	9,327,479	3,901,322	13,228,801
Formaldehyde	8,539,725	3,801,543	12,332,268
Phenol	5,818,364	4,336,737	10,155,101
Chloromethane	6,778,071	2,060,596	8,838,667
Acetaldehyde	4,132,695	2,540,410	6,673,114
Ethylbenzene	3,819,831	2,674,266	6,494,097
1,3-Butadiene	2,626,680	3,771,933	6,398,613
p-Xylene	4,107,772	2,198,210	6,305,982
TOTAL	1,165,301,644	454,689,617	1,619,991,261

[a] Ranked in order of total TRI air emissions.

Source: EPA 1990a.

Table 1-11   Summary of estimated cancer risks associated with air toxics emissions from various source categories

Source category	Annual cancer cases[a]	Percentage of total[b]	Typical pollutants
Motor vehicles	769–1461	54–58	PIC[c], butadiene
Formaldehyde[d]	106–154	6.1–7.4	Formaldehyde
Electroplating	120	4.7–8.4	Chromium[e]
Waste facilities	49–140	3.4–5.5	Dioxin
Woodsmoke	89	3.5–6.2	PIC
Asbestos Demol.	81	3.2–5.6	Asbestos
Unspecified point	27–92	1.9–3.6	Arsenic, formaldehyde
Cooling towers	0.01–111	0.0–4.4	Chromium[e]
Gasoline marketing	24–75	1.7–3.0	Gasoline vapors benzene
Solvent use	22–36	1.4–1.5	Perchloroethylene Methylene chloride
Unspecified area	21	0.8–1.5	Carbon tetrachloride
Vinyl chloride	19	0.7–1.4	Vinyl chloride
Iron and steel	17–18	0.7–1.2	Coke oven emissions, benzene, PIC
Sewage Sludge Inc.	13	0.5–0.9	Cadmium, vinyl chloride
Municipal Waste Inc.	2–22	0.9–1.0	Dioxin
Petroleum refinery	8–14	0.6–0.6	Gas vapor, formaldehyde
1,3-Butadiene prod.	10	0.4–0.7	1,3-Butadiene
Styrene-butadiene	10	0.4–0.7	1,3-Butadiene
Coal and oil comb.	8–10	0.4–0.6	Arsenic
POTWs	6	0.2–0.4	Vinyl chloride
Smelters	3–4	0.1–0.2	Formaldehyde
Ethylene oxide ster.	3–4	0.1–0.2	Ethylene oxide
Pesticide prod./use	3–4	0.1–0.2	Benzene
Dry cleaning	3	0.1–0.2	Perchloroethylene
Pulp and paper mfg.	2.1	0.08–0.1	Chloroform
Drinking water	1.7	0.08–0.1	Chloroform
Ethylene dibromide prod.	1.5	0.06–1.0	Ethylene dibromide
Polybutadiene prod.	1.2	0.05–0.08	1,3-Butadiene
Ethylene oxide prod.	1.2	0.05–0.08	Ethylene oxide
Ethylene dichloride prod.	0.8	0.05–0.08	Ethylene dichloride
Waste oil burning	0.6	0.02–0.04	Arsenic
Asbestos mfg.	0.5	0.02–0.04	Asbestos
Asbestos renov.	0.4	0.02–0.03	Asbestos
Glass mfg.	0.4	0.02–0.03	Arsenic
Hazardous Waste Inc.	0.3	0.01–0.02	Hexavalent chromium
Paint stripping	0.22	0.01–0.02	Hexavalent chromium
Pharmaceutical mfg.	0.2–0.4	0.01–0.02	Chloroform
Benzene fugitives	0.2	0.01–0.01	Benzene
Other	6–13	0.4–0.5	Hexavalent chromium, Radon, acrylonitrile

[a] Estimated cancer cases.
[b] Percent of total maximum and minimum.
[c] Products of incomplete combustion.
[d] Secondary formation of formaldehyde in the atmosphere.
[e] Hexavalent chromium.
Source: EPA 1990b.

170 chemicals. Appendix D in the TRI Report lists the chemicals on the Clean Air Act list, and provides their point source and fugitive release quantities. Table 1-10 summarizes the air emissions information for the top twenty-five chemicals on the Clean Air Act list.

More recently, EPA (1990b) has attempted to place the air toxic problem into a risk perspective, again utilizing estimated cancer risks as the primary indicator. In that report, EPA concluded that area sources contribute approximately 75 percent of the total number of annual

Table 1-12   Summary of cancer risk estimates for motor vehicles

Motor vehicle pollutant	U.S. cancer incidence/year		
	1986	1995	2005
Diesel particulate	178–860	106–662	104–518
Formaldehyde	46–86	24–43	27–30
Benzene	100–155	60–107	67–114
Gasoline vapors	17–68	24–95	30–119
1,3-Butadiene	236–269	139–172	144–171
Acetaldehyde	2	1	1
Gasoline particulate	1–176	1–156	1–146
Asbestos	5–33	0	0
Cadmium	<1	<1	<1
Ethylene dibromide	1	<1	<1
Total	586–1650	355–1236	374–1099

Source: EPA 1990b.

cancer cases (including those from secondary formaldehyde) with point sources contributing approximately 25 percent of the total. Of the area sources, the major contributor is mobile sources, contributing 78 percent of the total annual cancer incidence attributed to area sources. Table 1-11 summarizes the information from that report. Table 1-12 summarizes the risk estimates for mobile sources. It shows that diesel particulates and various organic compounds are major contributors. Mobile source emissions of air toxics are discussed in more detail in Chapter 25.

References

Council on Environmental Quality (CEQ). 1979. Environmental Quality in 1979. 10th Annual Report of the Council on Environmental Quality. U.S. Government Printing Office. Washington, D.C.

ENSR Consulting and Engineering. 1988. Air quality handbook: A guide to permitting and compliance under the Clean Air Act and air toxics programs. Tenth Edition.

Fuller, B., J. Hushon, M. Kornreich, R. Ouellette, L. Thomas, and P. Walker. 1976. Mitre Corporation. Scoring of organic air pollutants: chemistry, production and toxicity of selected synthetic organic compounds. Report # MTR-7248. Prepared for U.S. Environmental Protection Agency.

Office of Air Quality Planning and Standards. Research Triangle Park, NC.

General Accounting Office (GAO). 1983. Delays in EPA's regulation of hazardous air pollutants. Washington, D.C.

Goldman, Marshall I. 1967. *Ecology and Economics – Controlling Pollutants in the 70s.* Englewood Cliffs, NJ: Prentice-Hall.

Haemisegger, E., A. Jones, B. Steigerwald, and V. Thomson. 1985. The air toxics problem in the United States: An analysis of cancer risks for selected pollutants. U.S. Environmental Protection Agency. Washington, D.C.

International Union of Air Pollution Prevention Associations (IUAPPA). 1991. *Clean Air Around the World.* 2nd Edition. Loveday Murley, ed. International Union of Air Pollution Prevention Associations. Brighton, England.

Krier, J.E., and E. Ursin. 1977. *Pollution and Policy.* Berkeley and Los Angeles, CA: University of California Press.

Ruckelshaus, W.D. 1983b. Statement before the Subcommittee on Oversight and Investigations, Committee on Energy and Commerce, U.S. House of Representatives. November 7, 1983.

State and Territorial Air Pollution Program Administrators and the Association of Local Air Pollution Control Officials (STAPPA/ALAPCO). 1989. *Toxic Air Pollutants: State and Local Regulatory Strategies – 1989.* Washington, D.C.

Stern, A.C. ed., 1984. *Fundamentals of Air Pollution.* 2nd Edition. New York: Academic Press.

Sundquist, J. 1968. *Politics and Policy: The Eisenhower, Kennedy, and Johnson Years*. The Brookings Institution. Washington, D.C.

U.S. Environmental Protection Agency (EPA). 1973. National emission standards for hazardous air pollutants. 38 FR 8826. April 6, 1973.

U.S. Environmental Protection Agency (EPA). 1975. Proposed national emission standards for hazardous air pollutants. 40 FR 59532. December 24, 1975.

U.S. Environmental Protection Agency (EPA). 1976c. National emission standards for hazardous air pollutants. 41 FR 46560, October 21, 1976.

U.S. Environmental Protection Agency (EPA). 1977c. Proposed national emission standards for hazardous air pollutants. 42 FR 28154. June 2, 1977.

U.S. Environmental Protection Agency (EPA). 1979d. Proposed air cancer policy. 44 FR 58642. October 10, 1979.

U.S. Environmental Protection Agency (EPA). 1984a. Risk assessment and management: Framework for decision making. EPA 600/9-86-002. Office of Research and Development. Research Triangle Park, NC.

U.S. Environmental Protection Agency (EPA). 1984e. Decision on listing of polycyclic organic matter. 49 FR 31680. August 8, 1984.

U.S. Environmental Protection Agency (EPA). 1984f. Proposed national emission standards for hazardous air pollutants. 49 FR 15076. April 6, 1984.

U.S. Environmental Protection Agency (EPA). 1984g. Withdrawal of proposed national emission standards for hazardous air pollutants. 49 FR 43906. October 31, 1984.

U.S. Environmental Protection Agency (EPA). 1984h. Partial withdrawal of proposed national emission standards for hazardous air pollutants. 49 FR 8386. March 6, 1984.

U.S. Environmental Protection Agency (EPA). 1985g. Intent-to-list decisions. 50 FR 42000. October 10, 1985.

U.S. Environmental Protection Agency (EPA). 1986u. National emission standards for hazardous air pollutants. 51 FR 28025. August 4, 1986.

U.S. Environmental Protection Agency (EPA). 1987i. Proposed national emission standards for hazardous air pollutants. 52 FR 13586. April 23, 1987.

U.S. Environmental Protection Agency (EPA). 1989k. National emission standards for hazardous air pollutants. 54 FR 38073. September 14, 1989.

U.S. Environmental Protection Agency (EPA). 1989l. National emission standards for hazardous air pollutants. 54 FR 51694. December 15, 1989.

U.S. Environmental Protection Agency (EPA). 1990a. Toxics in the community: national and local perspectives. EPA 560/4-90-017. Office of Pesticides and Toxic Substances. Washington, D.C.

U.S. Environmental Protection Agency (EPA). 1990b. Cancer risk from outdoor exposure to air toxics. Volume 1, Final Report. EPA 450/1-90-004a. Office of Air Quality Planning and Standards. Research Triangle Park, NC.

Wark, K. and C.F. Warner. 1976. *Air Pollution: Its Origin and Control*. New York: Harper & Row.

2

Title III of the 1990 Clean Air Act Amendments

Henry A. Waxman

INTRODUCTION

The release of hazardous air pollutants is a problem of surprising magnitude. Data submitted as part of the Toxic Release Inventory under Title III of the 1986 Emergency Planning and Community Right-to-Know Act indicate that more than 2 billion pounds of hazardous air pollution are released into the nation's air supply each year. In addition, EPA estimates indicate that industrial facilities sometimes are associated with potentially high cancer risks.[1]

EPA has estimated that hazardous air pollutants cause some 1,600 to 3,000 cancer cases a year. There have been no quantitative assessments of the risks created by noncancerous toxic emissions, although it is known that toxic chemicals can cause birth defects, neurological injury, and genetic mutations. In 1987, EPA qualitatively ranked the noncancer risks created by over thirty environmental problems within the agency's jurisdiction. Toxic air emissions ranked as the second greatest threat to human health, exceeded only by the health risks attributable to ozone and other 'criteria' air pollutants.

Unfortunately, in the twenty years following passage of the original 1970 Clean Air Act, the hazardous air pollution control program in section 112 was not implemented effectively. Although many believed that hundreds of compounds released into the air met the Act's definition of 'hazardous,' EPA regulated only seven substances under section 112 in two decades. The release of numerous substances formally classified as carcinogens by the EPA itself, including chloroform, formaldehyde, carbon tetrachloride, and PCBs, remained unregulated.

---

[1] Source: EPA ATERIS data base. Many facilities are associated with lifetime cancer risks of greater than 1 in 100 to the most exposed individuals. One facility has been associated with a lifetime cancer risk to the most exposed individual of greater than 1 in 10. In a separate study of risks associated with coke oven emissions, EPA identified an additional twenty facilities associated with a lifetime cancer risk of greater than 1 in 1000 to the most exposed individual, including six facilities associated with a greater than 1 in 100 cancer risk.

REGULATION OF HAZARDOUS AIR POLLUTANTS UNDER THE 1990 AMENDMENTS

Title III of the 1990 Clean Air Act Amendments fundamentally restructures section 112 of the Clean Air Act to establish an aggressive new program for the regulation of hazardous air pollution. Specific programs are established for the control of major source and area source emissions, the control of nonbiodegradable compounds, the protection of the Great Lakes and coastal waters, the regulation of emissions from incinerators of all types, and the control of chemical accidents. Table 2-1 lists the key provisions of the 1990 Amendments relating to hazardous and toxic air pollutants.

The 1990 Amendments establish a statutory list of 189 substances that are formally designated hazardous air pollutants, thereby

Table 2-1   Key provisions of section 112 of the 1990 Clean Air Act Amendments

Section	Purpose
112(a)	Definitions (includes major and area source definitions)
112(b)	List of pollutants and provisions for revision to the list
112(c)	Requirements for publication of list of source categories
112(d)	Requirements for emission standards, including definition of MACT
112(d)(8)	Coke oven provisions
112(e)	Schedule for standards and review
112(f)	Standard to protect health and the environment
112(f)(2)	Emission standards
112(g)	Modifications and offset provisions
112(g)(2)	Construction and reconstruction
112(h)	Work practice standards
112(i)	Schedule for compliance
112(i)(4)	Presidential exemption
112(i)(5)	Early reduction
112(i)(8)	Additional coke oven provisions
112(j)	Equivalent emission limitation by permit
112(k)	Area source program, including National Strategy
112(l)	State programs, including delegation of authority
112(m)	Atmospheric deposition to Great Lakes and coastal waters, including Chesapeake Bay and Lake Champlain
112(n)	Other requirements
112(n)(1)	Electric utility steam generating plant study
112(n)(2)	Coke oven study
112(n)(3)	Publicly owned treatment works study
112(n)(4)	Oil well, gas well, and pipeline study
112(n)(5)	Hydrogen sulfide study
112(n)(6)	Hydrofluoric acid study
112(o)	National Academy of Sciences study of risk assessment
112(p)	Formation of Mickey Leland Urban Air Toxics Research Center
112(q)	Savings provision
112(r)	Prevention of accidental releases
112(r)(3)	Requirement to publish list of substances
112(r)(6)	Formation of Chemical Safety and Health Investigation Board
112(r)(7)	Provisions for accident prevention
303	Formation of Risk Assessment and Management Commission
304	Requirements for OSHA to prevent accidental releases in the workplace
305	Requirements for solid waste combustion (new section 129)
306	Requirements for ash management and disposal

short-circuiting the listing process that proved to be a major obstacle to regulatory action in past years. EPA is to produce a list of all categories of major sources and area sources of each listed pollutant, promulgate standards requiring installation of the maximum achievable control technology (MACT) at all new and existing major sources in accordance with a statutory schedule, and establish standards to protect the public health with an ample margin of safety from any residual risks remaining after application of MACT technology.

Definition of Major Source

Many of the most important requirements applicable to sources of hazardous air pollutants hinge on the designation of the facility as a 'major source' of air toxics. Examples include mandatory applicability of MACT standards, deadlines for regulation, and mandatory applicability of permit requirements.

Section 112(a) indicates that, at a minimum, the term 'major source' includes 'any stationary source or group of stationary sources located within a contiguous area and under common control that emits or has the potential to emit considering controls, in the aggregate, 10 tons per year or more of any hazardous air pollutant or 25 tons per year or more of any combination of hazardous air pollutants.' For purposes of defining a major source, all emissions of listed pollutants, including fugitive emissions, are counted from all points within a plant boundary. EPA is authorized to establish lower thresholds for designation as a major source, based on characteristics such as the potency of the air pollutant or its potential for bioaccumulation.

MACT Standards

The cornerstone of the new hazardous air pollution control program is the requirement mandating use of the MACT to reduce emissions. Under new section 112(d), major sources are to be subject to standards requiring the maximum degree of emission reduction that is deemed achievable. These standards are to be established by source category for all categories that release hazardous air pollutants. In cases where a source emits more than one hazardous air pollutant, EPA regulations are to require the maximum degree of emission reduction for each pollutant.

For new sources, section 112(d) standards are specifically required to be no less stringent than the level of emission control 'achieved in practice by the best controlled similar source.' The best controlled similar source is the source in the category or subcategory with the lowest emission rate. Existing source MACT standards may be less stringent than those applicable to new sources, but are required to be no less stringent than the average emission limit achieved by the best performing twelve percent of the similar sources, or the best performing five sources in a category or subcategory with fewer than thirty sources.

In some instances, there may be no source in the category or subcategory with a level of control as stringent as MACT. In this situation, section 112(d)(3) provides that the Administrator may establish MACT standards without regard to the performance of similar sources. Under no circumstances, however, may the Administrator set a MACT standard that is less stringent than the controls achieved in practice by the best controlled similar source (in the case of new sources) or the average achieved by the best controlled 12 percent of similar sources (in the case of existing sources).

Because the stringency of MACT standards under 112(d) is tied to the performance of other sources in the same category or subcategory, the categorization of sources is extremely important. Some may advocate establishment of a long list of narrow categories where, on the basis of limited differences, more stringently controlled sources can be separated from heavily polluting facilities. This approach could lead to far less stringent standards for more heavily polluting facilities, and tougher standards for facilities that are already better controlled. Those sources that already are clean would be penalized under such a reading, and requirements for the uncontrolled sources where tight restrictions are most needed would be relaxed. This was not Congress' intent, as evidenced by section 112(c)(1), which specifically directs that categories and subcategories established in the hazardous air pollutant program are to be consistent with the list of source categories established pursuant to the regulation of new sources under section 111 (New Source Performance Standards) and Part C (Prevention of Significant Deterioration) of the Clean Air Act.

An important provision in section 112(d)(3) excludes certain well-controlled sources from the calculation of the minimum stringency of existing source MACT standards. In determining the best performing sources under 112(d)(3), the Administrator is not required to consider sources that have within eighteen months of proposal of a standard (or within twenty-four months of final promulgation) first achieved the lowest achievable emission rate (LAER). All other sources within the category or subcategory are to be considered. This exclusion is intended to provide EPA with discretion to conclude that at times of rapid growth, where large numbers of new facilities are present, existing sources will not necessarily be required to meet a MACT standard reflecting LAER technologies required on new sources. Of course, since section 112(d)(3) establishes only a minimum stringency for the existing source MACT standards, EPA retains authority to establish more stringent standards, including those that do take LAER technologies into consideration. In fact, if new source LAER technologies are amenable for use in reducing existing source emissions, it is expected that EPA would take them into consideration.

In any case, over time the minimum stringency of MACT standards will increase as the eighteen-month LAER exclusion periods lapse, and LAER sources are brought into the calculation. The continual tightening of existing source standards will be assured under section 112(d)(6), which provides that all MACT standards are to be reviewed every eight years, and revised as necessary to take into account 'developments in practices, processes, and control technologies.'

Residual Risks

Section 112(f) provides for the regulation of source categories of hazardous air pollutants to address residual risks remaining after application of MACT under section 112(d). Within six years of enactment of the 1990 Amendments,[2] the Administrator is to report to Congress on residual risks, and make legislative recommendations. Based on this study, and other information available at that time, Congress may choose to amend section 112 to establish new standards governing the control of residual risks.

If Congress does not enact legislation establishing new residual risk provisions in response to the EPA study, the Administrator is required, within eight years after promulgation of MACT standards for a category or subcategory, to determine whether to promulgate standards for such category or subcategory in accordance with section 112 of the Clean Air Act as it existed before enactment of this legislation. Such standards are to be promulgated, if required, in order to prevent an 'adverse environmental effect' or to provide 'an ample margin of safety to protect the public health in accordance with [section 112] (as in effect before the date of enactment of the Clean Air Act Amendments of 1990).'

Concerning carcinogens, section 112(f)(2)(A) specifically defines the crucial phrase 'ample margin of safety to protect the public health in accordance with [section 112] (as in effect before the date of enactment of the Clean Air Act Amendments of 1990).' It provides that in the case of known, probable, or possible human carcinogens, if a section 112(d) standard does not reduce the lifetime cancer risk to the individual most exposed to emissions from that facility to less than one in one million, the 'ample margin of safety' standard is not met and the Administrator must promulgate residual risk emission standards under section 112(f).

---

[2]The 1990 Amendments were enacted on November 15, 1990.

Emission Limitations Established by the Permitting Authority

Section 112(j) establishes a program under which state permitting authorities are to impose MACT standards on their own if EPA does not provide them in a timely manner. The standards are to be established on a case-by-case basis for facilities in a source category, if EPA fails to issue applicable section 112(d) standards for that category within eighteen months of the deadline established under section 112(e). This requirement does not take effect until a state or federal permit program is in effect under Title V, and until at least forty-two months after the date of enactment.

The provisions for equivalent emissions limitations are intended to provide a fallback to assure that hazardous air pollutants will be effectively regulated, even if the agency does not issue standards. This provision is a reflection of congressional concern about potential EPA failure to take the mandated steps and stems from the lack of regulatory action under section 112 of the 1970 Amendments.

Modifications

Section 112(g) establishes a new program under which the modification of any major source of hazardous air pollution is to be subject to stringent new requirements. Once a permit program is in effect in any state, modifications of major air toxics sources are prohibited unless the source, as modified, meets an existing source MACT emission limitation. As in section 112(d), existing source MACT may be less stringent than new source MACT, but is not required to be. Section 112(g) also provides that the construction or reconstruction of major air toxics sources is prohibited unless the Administrator or the state determines that the source complies with the new source MACT.

Where EPA has not established emission limitations applicable to the modification, construction, or reconstruction, the state is to make the MACT determination on a case-by-case basis. An EPA emission limitation that otherwise applies to an industrial category may be determined not to be applicable if the modification or construction reflects the use of new processes or technologies capable of achieving greater emission reductions than those under consideration in the establishment of standards. In that instance, the state would be required to establish more stringent MACT standards on its own.

An offset program also is established in section 112(g)(1), under which a physical change that results in an emission increase may be considered not to be a modification if the increase is offset by a greater decrease in the release of the same hazardous air pollutant from the source, or another pollutant previously determined by EPA to be more hazardous. The offsetting reductions must occur within the same process unit as the emission increase. The offset program applies only to modifications, and not to the construction or reconstruction of new sources.

Area Sources

Air toxics sources too small to be considered major sources are termed 'area sources.' Although individually small, area sources are large in number, and can be associated with significant adverse health effects. Section 112(c)(3) puts the agency on a schedule for the regulation of area sources. Within five years of enactment, EPA is directed to list sufficient categories and subcategories of area sources to account for sources representing 90 percent of the aggregate area source emissions of the thirty most important hazardous air pollutants released from area sources. It is anticipated that the listed sources will reflect the conclusions of the National Urban Air Toxics Strategy required in section 112(k). EPA is required to develop this plan within five years after enactment and is required to identify a strategy for achieving a 75 percent reduction in the cancer incidence associated with emissions from area sources. Regulations are required to be promulgated within ten years of enactment for all listed categories. The regulations are to impose emission limitations reflecting use of MACT for each of the thirty pollutants, or in some cases reflecting use of generally available control technology (GACT). Unlike MACT, GACT is not specifically defined in the Amendments.

Accidental Releases of Hazardous Air Pollutants

Accidental releases occur with surprising frequency. EPA reported that 11,048 accidental releases of toxic chemicals occurred in the U.S. between 1980 and 1987. These releases killed 309 people and caused 11,341 injuries. They also caused the evacuation of nearly 500,000 people. Of these releases, 4,375 – an average of nearly two a day – involved hazardous pollutants. Although these releases were just 40 percent of the total, they represented 63 percent of the accidental releases causing death or injury and 75 percent of the releases requiring evacuations.

Although some accidental releases pose little threat to human health or the environment, others have the potential to be truly catastrophic. The most disastrous ever was the accidental release on December 3, 1984, of methyl isocyanate (MIC) from a chemical plant in Bhopal, India. In this accident, a storage tank ruptured and released 30 tons of MIC. The release killed over 3,000 people and injured more than 200,000.

According to EPA records, there have been seventeen accidental releases of toxic chemicals in the U.S. since 1980 that had potential toxic effects greater than the Bhopal release. In each case, the 'quantity/toxicity ratio' of the release, a measure of the release's potential for catastrophic injury, exceeded the Bhopal ratio. Fortunately, a number of fortunate factors prevented the U.S. releases from causing Bhopal-like injuries, including favorable weather, the remoteness of the releases, and conditions that kept some of the releases from becoming airborne. Nevertheless, five deaths did occur, a number that EPA called 'surprisingly lower than might be expected.'

Accidental releases are particularly dangerous when they involve substances that are heavier than air and remain on the ground when released. Examples of such substances include chlorine and hydrogen chloride – chemicals used at numerous industrial and commercial facilities around the country.

Law On Accidental Releases Before the 1990 Amendments

Before passage of the 1990 Amendments, existing law contained few provisions regulating the prevention, detection, or response to accidental releases. The Emergency Planning and Community Right-to-Know Act of 1986 (enacted as Title III of the Superfund Amendments and Reauthorization Act of 1986) established local emergency planning commissions and directed the local commissions to develop plans for responding to chemical accidents, including those that involve releases to the air. That legislation also required industrial facilities to notify the local commissions both when the facilities manufacture, use, or store toxic substances above threshold amounts and when accidental releases occur. However, it did not contain requirements for the prevention, detection, or response to accidental releases. In these areas, the actions of industry were essentially unregulated before the 1990 Amendments.

The Accident Prevention Program in the 1990 Amendments

Section 112(r) of the 1990 Amendments establishes a new program for the prevention, detection, and response to accidental releases. The Administrator is to promulgate within two years of enactment a list of not less than 100 substances that 'may reasonably be expected to cause serious adverse effects to human health or the environment in the event of an accidental release.' At the time a substance is listed under section 112(r), the Administrator is to establish a threshold quantity reflecting the minimum amount that, if accidentally released, would reasonably be anticipated to pose a serious adverse effect.

Sources having listed substances on site in greater than threshold quantities are required to comply with accident prevention regulations under section 112(r)(7). Each such source must prepare and carry out a risk management plan to detect and prevent or minimize accidental releases. The plan is to include a hazard assessment that evaluates possible worst case accidental releases, a history of any previous accidental releases over the prior five years, an accident prevention program, and a response program outlining actions to be taken in the event of a release.

Section 112(r) also includes provisions for the establishment of a new Chemical Safety and Hazard Investigation Board to investigate accidental chemical releases, and make recommendations for how future releases can be avoided. Patterned after the National Transportation Safety Board, the Chemical Safety and Hazard Investigation Board is to be an independent entity authorized to conduct investigations, issue periodic reports to Congress and federal agencies, establish reporting requirements, conduct research, hold hearings, and make recommendations. When the board submits a recommendation to EPA, the Administrator must respond within 180 days and indicate whether rule-making will be undertaken or not. A decision not to carry out a recommendation must be accompanied by an explanation.

Protection of the Great Lakes, the Chesapeake Bay, and Coastal Waters

Section 112(m) directs the Administrator to investigate the sources of atmospheric deposition of hazardous air pollutants and their transformation products into the Great Lakes, the Chesapeake Bay, and the nation's coastal waters, and to evaluate the adverse effects to human health and the environment caused by such deposition. This assessment is to include consideration of the tendency of such pollutants to bioaccumulate, and consideration of effects associated with indirect exposure pathways. The Administrator is to report to Congress on the results within three years of enactment.

In the required report to Congress, EPA is to determine whether other provisions of section 112 are adequate to prevent serious adverse effects to human health, and serious or widespread environmental effects, including effects from indirect exposure pathways, associated with atmospheric deposition of hazardous air pollutants on the Great Lakes, the Chesapeake Bay, and coastal waters. Within five years of enactment of the 1990 Amendments, the Administrator is to promulgate such further emission standards or control measures as may be necessary, based on the EPA report, to prevent such effects to human health or the environment, including effects due to bioaccumulation and indirect exposure pathways.

Radionuclides

The 1990 Amendments include several provisions that specifically address the regulation of radionuclide emissions. Probably the most important is Section 112(d)(9); it provides that the Administrator is not required to regulate radionuclide emissions from categories or subcategories of facilities licensed by the Nuclear Regulatory Commission (NRC), if the Administrator concludes by rule that NRC regulation of radionuclide emissions from that source category provides an ample margin of safety to protect the public health. This determination must be made independently for each source category, and must be made by rule. Regardless of whether this determination is made, the Administration retains authority to regulate radionuclides emissions from all sources.

Coke Ovens

Coke ovens are important sources of hazardous air pollution emissions that remain unregulated, despite the fact that coke oven emissions were listed in 1984 as a hazardous air pollutant under section 112. Under the 1990 Amendments, they are subject to a specifically tailored regime designed to assure that the most effective, available pollution control steps are utilized as quickly as possible, while also minimizing the possibility that coke ovens must close down because of inability to meet residual risk standards under section 112(f).

Section 112(d)(8) provides specifically that in establishing MACT standards for new coke ovens, the Administrator is to consider the extremely low emission Jewell design Thompson nonrecovery coke oven batteries. For existing coke oven batteries the use of sodium silicate luting compounds to prevent door leaks must be considered, and the Administrator is directed to establish work practice regulations requiring use of such luting compounds and other door and jam cleaning processes. The work practice regulations must be complied with within three years of enactment. No other source category is subject to section 112(d) control requirements that take effect so promptly.

In addition, section 112(i)(8) defers the application of residual risk requirements in section 112(f) until the year 2020 for coke ovens meeting specified additional requirements. To qualify for the extension, coke ovens must meet an 8 percent leaking door standard and other requirements in section 112(d)(8)(C) within three years of enactment. Facilities receiving an extension also must meet the requirements of section 112(i)(8)(B) and 112(i)(8)(C). Section 112(1)(8)(B) requires that by January 1, 1998, the coke oven must meet the requirements of LAER as defined in section 171 of the Act.

The LAER standard in section 171 applies to new pollution sources in nonattainment areas. It is the Act's most demanding technology-based pollution control standard, mandating that any technology that has been successfully utilized must be put in place. It has never before been applied to existing sources, and might well require that sources reconfigure to take advantage of the emission reductions achievable through the Jewell design facilities. Detailed minimum requirements for the 1998 coke oven LAER standard, including a 3 percent leaking door standard, are provided in section 112(i)(8)(B). EPA is to update the LAER standard by January 1, 2007, and coke ovens securing deferral of the section 112(f) standard also are required to comply with the updated LAER standard by the year 2010.

Incinerator Emissions

Incinerators are a source of hazardous air pollutant emissions, especially in urban areas. As landfill space has become limited, and waste disposal requirements have tightened, incineration has become an increasingly popular form of waste disposal. The wide array of materials subject to incineration, from rubber tires to newsprint to metals, gives rise to a comparably wide array of hazardous emissions, including heavy metals and organic chemicals. Without aggressive pollution controls, these emissions can present significant health risks.

The 1990 Amendments establish a broad new program in section 129 to assure that emissions from the full range of new and existing incinerators are controlled. Incinerator emissions are to be regulated both under Clean Air Act section 111, where EPA has already initiated a regulatory effort for new and existing facilities, and under section 129. The Amendments specifically provide that EPA regulations under section 111, part of an EPA consent decree, are to continue on schedule and subsequently be modified to conform to section 129 in accordance with a statutory schedule. The schedule calls for regulations for large municipal waste incinerators within one year of enactment; hospital, medical, and infectious waste incinerator regulations within two years; and commercial and industrial waste incinerator regulations within four years.

Emission limitations established for new and existing incinerators are to reflect the 'maximum degree of reduction in emissions,' a term defined to parallel the section 112(d) MACT standard. Existing units have five years to comply with the incinerator regulations, while regulations for new units are effective six months after promulgation. Residual risk requirements of section 112(f) apply eight years after promulgation of incinerator regulations, just as if such regulations were MACT standards under section 112(d).

3

Toxic Air Pollutants and Their Sources

David R. Patrick

INTRODUCTION

The 1990 Clean Air Act Amendments provide in section 112(b)(1) an initial list of 189 hazardous air pollutants and require in section 112(c)(1) that EPA publish a list of major and area source categories and subcategories of those substances to be regulated. The list of pollutants resulted after lengthy consideration both within the two houses of Congress and between Congress and EPA. Importantly, when enacted the list represented a consensus of the air pollutants of concern to humans for their hazardous or toxic nature and it expanded the hazardous air pollutants that were regulated under the 1970 Amendments by a factor of almost twenty-four. Congress left to EPA the job of identifying potential source categories of the listed hazardous air pollutants.

The purpose of this chapter is to discuss the derivation of the list of 189 hazardous air pollutants and describe the process provided in the 1990 Amendments to revise that list. This chapter also summarizes EPA's plans, as known in mid-1992, concerning regulation of major and area source categories and subcategories of these air pollutants. The 1990 Amendments also require studies or regulations for several special source categories of toxic air pollutants, including electric utilities, publicly owned treatment works, coke oven batteries, and municipal incinerators. These source categories are discussed more fully in Chapter 24.

THE LIST OF HAZARDOUS AIR POLLUTANTS

Derivation of the List of Hazardous Air Pollutants

As discussed in Chapter 1, one of the biggest difficulties that faced EPA in the early years was the potentially enormous size of the toxic air pollutant problem and the agency's inability to settle on the 'list' of hazardous air pollutants. Thus, in the 1980s, EPA focused its efforts on identifying those hazardous air pollutants with potential adverse health effects, substantial sources, and a high likelihood of release (e.g., the candidate is volatile). Studies attempting to identify and priority rank candidate hazardous air pollutants were conducted within EPA and for EPA by Argonne National Laboratories (Smith and Fingleton 1982) and the Radian Corporation (1987). As the work continued,

Table 3-1   List of hazardous air pollutants

Acetaldehyde	Dichlorvos
Acetamide	Diethanolamine
Acetonitrile	N,N-Diethylaniline (N,N-dimethylaniline)
Acetophenone	Diethyl sulfate
2-Acetylaminofluorene	3,3-Dimethoxybenzidene
Acrolein	Dimethyl aminoazobenzene
Acrylamide	3,3'-Dimethyl benzidene
Acrylic acid	Dimethyl carbamoyl chloride
Acrylonitrile	Dimethyl formamide
Allyl chloride	1,1-Dimethyl hydrazine
4-Aminobiphenyl	Dimethyl phthalate
Aniline	Dimethyl sulfate
o-Anisidine	4,6-Dinitro-o-cresol and salts
Asbestos	2,4-Dinitrophenol
Benzene (including from gasoline)	2,4-Dinitrotoluene
Benzidine	1,4-Dioxane (1,4-Diethyleneoxide)
Benzotrichloride	1,2-Diphenylhydrazine
Benzyl chloride	Epichlorohydrin (1-chloro-2,3-epoxypropane)
Biphenyl	1,2-Epoxybutane
Bis(2-ethlhexyl)phthalate (DEHP)	Ethyl acrylate
Bis(chloromethyl)ether	Ethyl benzene
Bromoform	Ethyl carbamate (urethane)
1,3-Butadiene	Ethyl chloride (chlorethane)
Calcium cyanamide	Ethylene dibromaide (dibromoethane)
Caprolactam	Ethylene dichloride (1,2-dichloroethane)
Captan	Ethylene glycol
Carbaryl	Ethylene imine (aziridine)
Carbon disulfide	Ethylene oxide
Carbon tetrachloride	Ethylene thiourea
Carbonyl sulfide	Ethylidine dichloride (1,1-dichloroethane)
Catechol	Formaldehyde
Chloramben	Heptachlor
Chlordane	Hexachlorobenzene
Chlorine	Hexachlorobutadiene
Chloroacetic acid	Hexachlorocyclopentadiene
2-Chloroacetophenone	Hexachloroethane
Chlorobenzene	Hexamethylene-1,6-diisocyanate
Chlorobenzilate	Hexamethylphosphoramide
Chloroform	Hexane
Chloromethyl methyl ether	Hydrazine
Chloroprene	Hydrochloric acid
Cresols/cresylic acid (isomers and mixture)	Hydrogen fluoride (hydrofluoric acid)
o-Cresol	Hydroquinone
m-Cresol	Isophorone
p-Cresol	Lindane (all isomers)
Cumene	Maleic anhydride
2,4-D, salts and esters	Methanol
DDE	Methoxychlor
Diazomethane	Methyl bromide (bromomethane)
Dibenzofurans	Methyl chloride (chloromethane)
1,2-Dibromo-3-chloropropane	Methyl chloroform (1,1,1-trichloroethane)
Dibutylphthlate	Methyl ethyl ketone (2-butanone)
1,4-Dichlorobenzene(p)	Methyl hydrazine
3,3-Dichlorobenzidene	Methyl iodide (iodomethane)
Dichloroethyl ether (bis(2-chloroethyl)ether)	Methyl isobutyl ketone (hexone)
1,3-Dichloropropene	Methyl isocyanate

Table 3-1   Continued

Methyl methacrylate	Toluene
Methyl-*t*-butyl ether	2,4-Toluene diamine
4,4-Methylene bis(2-chloroaniline)	2,4-Toluene diisocyanate
Methylene chloride (dichloromethane)	*o*-Toluidine
Methylene diphenyl diisocyanate (MDI)	Toxaphene (chlorinated camphene)
4,4'-Methylenedianiline	1,2,4-Trichlorobenzene
Naphthalene	1,1,2-Trichloroethane
Nitrobenzene	Trichloroethylene
4-Nitrobiphenyl	2,4,5-Trichlorophenol
4-Nitrophenol	2,4,6-Trichlorophenol
2-Nitropropane	Triethylamine
N-Nitroso-*N*-methylurea	Trifluralin
N-Nitrosodimethylamine	2,2,4-Trimethylpentane
N-Nitrosomorpholine	Vinyl acetate
Parathion	Vinyl bromide
Pentachloronitrobenzene (quintobenzene)	Vinyl chloride
Pentachlorophenol	Vinylidene chloride (1,1-dichloroethylene)
Phenol	Xylenes (isomers and mixtures)
p-Phenylenediamine	*o*-Xylene
Phosgene	*m*-Xylene
Phosphine	*p*-Xylene
Phosphorus	
Phthalic anhydride	Antimony compounds
Polychlorinated biphenyls (aroclors)	Arsenic compounds
1,3-Propane sultone	Beryllium compounds
β-Propiolactone	Cadmium compounds
Propionaldehyde	Chromium compounds
Propoxur (Baygon)	Cobalt compounds
Propylene dichloride (1,2-dichloropropane)	Coke oven emissions
Propylene oxide	Cyanide compounds
1,2-Propylenimine (2-methylaziridine)	Glycol ethers
Quinoline	Lead compounds
Quinone	Manganese compounds
Styrene	Mercury
Styrene oxide	Fine mineral fibers
2,3,7,8-Tetrachlorodibenzo-*p*-dioxin	Nickel compounds
1,1,2,2-Tetrachloroethane	Polycyclic organic compounds
Tetrachloroethylene (perchloroethylene)	Radionuclides
Titanium tetrachloride	Selenium compounds

Source: Section 112(b)(1).

EPA began to focus more on source categories than individual pollutants. The principal reason was that many sources emit more than one hazardous air pollutant and it would be inefficient and costly to regulate them one at a time. Still, EPA never was able to reach a consensus on either the hazardous air pollutants or the sources to be regulated. This inability to reach consensus as much as anything led Congress to publish in the 1990 Amendments a specific list of pollutants to be regulated.

The hazardous air pollutants listed in the 1990 Amendments are shown in Table 3-1. The list was derived from several sources.[1] It originated with a list of 224 chemicals proposed by Senator Mitchell in 1988. Those chemicals

---

[1] Personal contact with EPA staff.

came from three lists: (1) over three hundred substances manufactured, processed, or used in excess of thresholds established under section 313 of the Emergency Planning and Community Right-to-Know Act of 1986 (EPCRA, enacted as Title III of the Superfund Amendments and Reauthorization Act of 1986); (2) over one hundred substances reported to EPA as exceeding 'reportable quantities' as specified under section 104 of the Comprehensive Emergency Response and Compensation Liability Act (CERCLA, or Superfund); and (3) substances regulated by one or more states as listed in EPA's National Air Toxics Information Clearinghouse (NATICH).

Using the Mitchell list and other data sources, EPA staff recommended addition of eight and deletion of fifty-one substances. The criteria used by EPA included toxicity, air pollution potential (e.g., whether a chemical is manufactured in the U.S.), delistings from section 313, consolidation with other listed substances to form a single group, previous listing under the Clean Air Act (e.g., coke oven emissions and radionuclides were added), and whether the substances would be regulated under other provisions of the Act (e.g., stratospheric ozone depleters were removed). The resulting list contained 181 substances.

EPA refined this list further based on additional toxicity information obtained from EPA's Office of Research and Development, emissions and health data collected during the development of EPA's hazardous air pollutant standards and new source performance standards, U.S. production data on pesticides obtained from EPA's Office of Pesticides Programs, data from EPA's Office of Toxic Substances on emissions and delisting efforts for section 313 and section 302 of EPCRA (which deals with extremely hazardous substances), data on recently verified carcinogens, and data on the number of states regulating a substance through acceptable ambient concentrations (i.e., substances were added if four or more states had established acceptable ambient concentrations).

The listing of one substance was changed during this time. 'Mineral fibers' was on the original list but was changed to 'fine mineral fibers.' The original listing included fibers with an average diameter of 3 microns or less; the revised listing includes only fibers with an average diameter of 1 micron or less. This change resulted from EPA's evaluation of test data indicating that only fine mineral fibers were emitted in sufficient quantity to warrant listing.

The final list comprised 191 substances that were included in the bills passed by both the House and Senate in mid-1991. While in conference committee, however, hydrogen sulfide and ammonia were removed from the list; thus, the final list in the 1990 Amendments contained 189 substances. One interesting error occurred in the final bill signed by President Bush. Apparently through a clerical error, hydrogen sulfide reappeared on the list of pollutants. As such, the Amendments enacted on November 15, 1990, actually listed 190 hazardous air pollutants. The error later was discovered and in late 1991 an amendment was passed by Congress and signed by the President removing hydrogen sulfide from the list.

Table 3-2 provides the list of 189 hazardous air pollutants with general information on the sources of the pollutants and the nature of the known or suspected adverse health effects associated with exposure.

Key Pollutant Listing Issues

The development of the final list of 189 hazardous air pollutants naturally involved some controversy. One of the more contentious issues was the listing of seventeen groups of substances, including all compounds of eleven elements (antimony, arsenic, beryllium, cadmium, chromium, cobalt, lead, manganese, mercury, nickel, and selenium). Many argued that this was too inclusive, because not all compounds of an element are always toxic. Although EPA generally acknowledged that toxicity varies among compounds of an element, the lack of adequate toxicity data in most instances and the language defining hazardous air pollutant in section 112(b)(2) the 1990

Table 3-2   **Hazardous air pollutant characteristics**

Pollutant	Structure	Sources	Other
Acetaldehyde[b]	Organic	Chemical ind.	Secondary[a]
Acetamide	Organic	Chemical ind.	
Acetonitrile	Organic	Chemical ind.	
Acetophenone	Organic	Chemical ind.	
2-Acetylaminofluorene	Organic	Chemical ind.	
Acrolein	Organic	Chemical ind.	Secondary[a]
Acrylamide[b]	Organic	Chemical ind.	
Acrylic acid	Organic	Chemical ind.	
Acrylonitrile[b]	Organic	Chemical ind.	Plastics
Allyl chloride	Organic	Chemical ind.	
4-Aminobiphenyl	Organic	Chemical ind.	
Aniline	Organic	Chemical ind.	
o-Anisidine	Organic	Chemical ind.	
Asbestos[b]	Mineral	Insulation	
Benzene[b]	Organic	Chemical ind.	Gasoline
Benzidine[b]	Organic	Chemical ind.	
Benzotrichloride	Organic	Chemical ind.	
Benzyl chloride	Organic	Chemical ind.	
Biphenyl	Organic	Chemical ind.	
Bis(2-ethylhexyl)phthalate (DEHP)	Organic	Chemical ind.	Plasticizer
Bis(chloromethyl)ether[b]	Organic	Chemical ind.	
Bromoform[b]	Organic	Chemical ind.	
1,3-Butadiene[b]	Organic	Chemical ind.	Plastic
Calcium cyanamide	Organic	Chemical ind.	
Caprolactam	Organic	Chemical ind.	
Captan	Organic	Pesticide	
Carbaryl	Organic	Pesticide	
Carbon disulfide	Organic	Chemical ind.	
Carbon tetrachloride[b]	Organic	Chemical ind.	
Carbonyl sulfide	Organic	Chemical ind.	
Catechol	Organic	Chemical ind.	
Chloramben	Organic	Pesticide	
Chlordane[b]	Organic	Pesticide	
Chlorine	Element	Chemical ind.	Disinfectant
Chloroacetic acid	Organic	Chemical ind.	
2-Chloroacetophenone	Organic	Chemical ind.	
Chlorobenzene	Organic	Chemical ind.	
Chlorobenzilate	Organic	Chemical ind.	
Chloroform[b]	Organic	Chemical ind.	
Chloromethyl methyl ether[b]	Organic	Chemical ind.	
Chloroprene	Organic	Chemical ind.	Polymers
Cresols/cresylic acid (isomers and mixture)	Organic	Chemical ind.	Coke ovens
o-Cresol	Organic	Chemical ind.	Coke ovens
m-Cresol	Organic	Chemical ind.	Coke ovens
p-Cresol	Organic	Chemical ind.	Coke ovens
Cumene	Organic	Chemical ind.	
2,4-D, salts and esters	Organic	Herbicide	
DDE	Organic	Pesticide	
Diazomethane	Organic	Chemical ind.	
Dibenzofurans	Organic	Combustion prod.	
1,2-Dibromo-3-chloropropane	Organic	Pesticide	
Dibutylphthlate	Organic	Chemical ind.	
1,4-Dichlorobenzene	Organic	Chemical ind.	
3,3-Dichlorobenzidene	Organic	Chemical ind.	

Table 3-2 Continued

Pollutant	Structure	Sources	Other
Dichloroethyl ether[b]	Organic	Chemical ind.	
1,3-Dichloropropene	Organic	Pesticide	
Dichlorvos	Organic	Pesticide	
Diethanolamine	Organic	Chemical ind.	
N,N-Diethylaniline	Organic	Chemical ind.	
Diethyl sulfate	Organic	Chemical ind.	
3,3-Dimethoxybenzidene	Organic	Chemical ind.	
Dimethyl aminoazobenzene	Organic	Chemical ind.	
3,3'-Dimethyl benzidene	Organic	Chemical ind.	
Dimethyl carbamoyl chloride	Organic	Chemical ind.	
Dimethyl formamide	Organic	Chemical ind.	
1,1-Dimethyl hydrazine	Organic	Chemical ind.	Rocket fuel
Dimethyl phthalate	Organic	Chemical ind.	Plasticizer
Dimethyl sulfate[b]	Organic	Chemical ind.	
4,6-Dinitro-o-cresol and salts	Organic	Pesticide	
2,4-Dinitrophenol	Organic	Chemical ind.	
2,4-Dinitrotoluene	Organic	Chemical ind.	
1,4-Dioxane	Organic	Chemical ind.	
1,2-Diphenylhydrazine[b]	Organic	Chemical ind.	
Epichlorohydrin[b]	Organic	Chemical ind.	
1,2-Epoxybutane	Organic	Chemical ind.	Gas additive
Ethyl acrylate	Organic	Chemical ind.	
Ethyl benzene	Organic	Chemical ind.	
Ethyl carbamate	Organic	Chemical ind.	
Ethyl chloride	Organic	Chemical ind.	
Ethylene dibromide[b]	Organic	Chemical ind.	
Ethylene dichloride[b]	Organic	Chemical ind.	
Ethylene glycol	Organic	Chemical ind.	Antifreeze
Ethylene imine	Organic	Chemical ind.	
Ethylene oxide	Organic	Chemical ind.	Sterilizers
Ethylene thiourea	Organic	Chemical ind.	
Ethylidine dichloride	Organic	Chemical ind.	
Formaldehyde[b]	Organic	Chemical ind.	Secondary
Heptachlor[b]	Organic	Pesticide	
Hexachlorobenzene[b]	Organic	Pesticide	
Hexachlorobutadiene	Organic	Chemical ind.	
Hexachlorocyclopentadiene	Organic	Pesticide	
Hexachloroethane	Organic	Chemical ind.	
Hexamethylene-1,6-diisocyanate	Organic	Chemical ind.	
Hexamethylphosphoramide	Organic	Chemical ind.	
Hexane	Organic	Chemical ind.	Petroleum
Hydrazine[b]	Organic	Chemical ind.	Fuel
Hydrochloric acid	Inorganic	Chemical ind.	
Hydrogen fluoride	Inorganic	Chemical ind.	
Hydroquinone	Organic	Chemical ind.	
Isophorone	Organic	Chemical ind.	Solvent
Lindane	Organic	Pesticide	
Maleic anhydride	Organic	Chemical ind.	
Methanol	Organic	Chemical ind.	Fuel
Methoxychlor	Organic	Pesticide	
Methyl bromide	Organic	Pesticide	
Methyl chloride	Organic	Chemical ind.	
Methyl chloroform	Organic	Chemical ind.	Degreasing
Methyl ethyl ketone	Organic	Chemical ind.	Solvent
Methyl hydrazine	Organic	Chemical ind.	

Table 3-2   Continued

Pollutant	Structure	Sources	Other
Methyl iodide	Organic	Chemical ind.	
Methyl isobutyl ketone	Organic	Chemical ind.	Solvent
Methyl isocyanate	Organic	Chemical ind.	
Methyl methacrylate	Organic	Chemical ind.	
Methyl-*t*-butyl ether	Organic	Petroleum	Gas additive
4,4-Methylene bis (2-chloroaniline)	Organic	Chemical ind.	
Methylene chloride[b]	Organic	Chemical ind.	Solvent
Methylene diphenyl diisocyanate (MDI)	Organic	Chemical ind.	
4,4′-Methylenedianiline	Organic	Chemical ind.	
Naphthalene	Organic	Chemical ind.	Coke ovens
Nitrobenzene	Organic	Chemical ind.	
4-Nitrobiphenyl	Organic	Chemical ind.	
4-Nitrophenol	Organic	Chemical ind.	
2-Nitropropane	Organic	Chemical ind.	
N-Nitroso-N-methylurea	Organic	Chemical ind.	
N-Nitrosodimethylamine[b]	Organic	Chemical ind.	
N-Nitrosomorpholine	Organic	Chemical ind.	
Parathion	Organic	Pesticide	
Pentachloronitrobenzene	Organic	Chemical ind.	
Pentachlorophenol	Organic	Pesticide	
Phenol	Organic	Chemical ind.	
p-Phenylenediamine	Organic	Chemical ind.	
Phosgene	Organic	Chemical ind.	
Phosphine	Organic	Chemical ind.	
Phosphorus	Inorganic	Chemical ind.	
Phthalic anhydride	Organic	Chemical ind.	
Polychlorinated biphenyls (aroclors)	Organic	Dielectric	
1,3-Propane sultone	Organic	Chemical ind.	
β-Propiolactone	Organic	Chemical ind.	
Propionaldehyde	Organic	Chemical ind.	
Propoxur (Baygon)	Organic	Pesticide	
Propylene dichloride	Organic	Chemical ind.	
Propylene oxide[b]	Organic	Chemical ind.	
1,2-Propylenimine	Organic	Chemical ind.	
Quinoline	Organic	Chemical ind.	
Quinone	Organic	Chemical ind.	
Styrene	Organic	Chemical ind.	
Styrene oxide	Organic	Chemical ind.	
2,3,7,8-Tetrachlorodibenzo-*p*-dioxin	Organic	Combustion Prod.	
1,1,2,2-Tetrachloroethane	Organic	Chemical ind.	
Tetrachloroethylene	Organic	Chemical ind.	Dry cleaning
Titanium tetrachloride	Inorganic	Chemical ind.	
Toluene	Organic	Petroleum	Solvent
2,4-Toluene diamine	Organic	Chemical ind.	
2,4-Toluene diisocyanate	Organic	Chemical ind.	
o-Toluidine	Organic	Chemical ind.	
Toxaphene[b] (chlorinated camphene)	Organic	Pesticide	
1,2,4-Trichlorobenzene	Organic	Chemical ind.	
1,1,2-Trichloroethane	Organic	Chemical ind.	
Trichloroethylene	Organic	Chemical ind.	Degreasing
2,4,5-Trichlorophenol	Organic	Herbicide	
2,4,6-Trichlorophenol[b]	Organic	Herbicide	
Triethylamine	Organic	Chemical ind.	
Trifluralin	Organic	Chemical ind.	
2,2,4-Trimethylpentane	Organic	Chemical ind.	Petroleum

Table 3-2   Continued

Pollutant	Structure	Sources	Other
Vinyl acetate	Organic	Chemical ind.	
Vinyl bromide	Organic	Chemical ind.	
Vinyl chloride	Organic	Chemical ind.	Polymers
Vinylidene chloride	Organic	Chemical ind.	Polymers
Xylenes (isomers and mixtures)	Organic	Petroleum	Solvent
o-Xylene	Organic	Petroleum	Solvent
m-Xylene	Organic	Petroleum	Solvent
p-Xylene	Organic	Petroleum	Solvent
Antimony compounds	Inorganic	Metals	Chemicals
Arsenic compounds[b]	Inorg./org.	Smelting	Chemicals, pesticides
Beryllium compounds[b]	Inorganic	Metals	Ceramics
Cadmium compounds[b]	Inorganic	Chemicals	Smelting
Chromium compounds[c]	Inorganic	Metals	Plating
Cobalt compounds	Inorganic	Metals	
Coke oven emissions[b]	Inorg./Org.	Coke ovens	Steel
Cyanide compounds	Organic	Chemical ind.	
Glycol ethers	Organic	Chemical ind.	
Lead compounds[b]	Inorganic	Chemical ind.	Metals
Manganese compounds	Inorganic	Metals	Chemicals
Mercury	Inorg./org.	Chlor-alkali	Biogenic
Fine mineral fibers	Inorganic	Insulation	
Nickel compounds[c]	Inorganic	Metals	
Polycyclic organic compounds	Organic	Combustion	
Radionuclides[b]	Inorg./org.	Nuclear ind.	
Selenium compounds	Inorganic	Chemical ind.	Metals

[a]Photochemical reaction product.
[b] Considered Class A or B inhalation carcinogen by EPA (Source: EPA/IRIS).
[c] Some compounds considered Class A or B inhalation carcinogens by EPA (Source: EPA/IRIS).

Amendments[2] led EPA to conclude that all compounds of elements, where one or more are known to be toxic or carcinogenic, should be listed. EPA also reasoned that the petition process for removing substances from the list (see below) would stimulate the generation of the necessary data to remove nontoxic members from the list.

Another concern with the group listings was that while the Amendments and most observers consider the list of 189 hazardous air pollutants as the focus of this regulatory program, in fact thousands of individual pollutants are included. For example, at least fifty cadmium compounds

are listed in chemical handbooks, a large number of glycol ethers are known, and polycyclic organic matter includes literally thousands of multi-ring organic compounds composed of carbon, hydrogen, and other elements.

Another issue is the lack of health criteria for many of the listed substances. For example, as of mid-1992 EPA had developed cancer criteria (i.e., inhalation or oral cancer potency values and Class A, B, or C weight of evidence[3]) or chronic noncancer intake criteria (i.e., inhalation or oral risk reference doses[4]) for 152 of the 189 listed pollutants, although eighteen of the

---

[2] '... substances which are known to be, or may reasonably be anticipated to be, carcinogenic, mutagenic, teratogenic, neurotoxic, which cause reproductive dysfunction, or which are acutely or chronically toxic.'

[3] In general terms, Class A refers to known human carcinogens, Class B refers to probable human carcinogens, and Class C refers to possible human carcinogens. Chapter 6 discusses EPA's methods for treating carcinogens.

[4] Noncancer health criteria are described in more detail in Chapter 7.

substances without EPA criteria do have occupational exposure limits established.

EPA has numerous ongoing programs to develop cancer potency values and weights of evidence for carcinogens, as well as chronic reference dose criteria for noncarcinogens. These studies provide results that span a range of quality. In general, EPA's best health criteria are available in the Integrated Risk Information System (IRIS). Results in IRIS generally have been developed by EPA using formal procedures and are peer-reviewed, at least among EPA health scientists. Interestingly, for the 189 listed substances, in mid-1992 only thirty-six substances listed in IRIS have cancer potency values (by inhalation) and are Class A or B[5]; an additional twenty-six substances are listed in IRIS and have inhalation reference doses. In other words, only about thirty-three percent (52 of 189) of the listed substances had verified inhalation health criteria as of mid-1992.

Petitions for Revision to the Pollutant List

Congress clearly understood that the hazardous air pollutant list published in the 1990 Clean Air Act Amendments was derived from sources that varied greatly in quality and that unequivocal evidence of harm to human health or the environment from exposure was not always available. The list also could not be considered all inclusive. For these reasons, the 1990 Amendments require EPA to review and revise the list periodically and provide opportunity for others to petition EPA to add or delete substances from the list.

Hazardous air pollutants are defined in the 1990 Amendments through this process of revision. In section 112(b)(2), EPA is required to review and revise the list of pollutants, where appropriate, by adding pollutants that '. . . present, or may present, through inhalation or other routes of exposure, a threat of adverse human health effects (including, but not limited

to, substances which are known to be, or may reasonably be anticipated to be, carcinogenic, mutagenic, teratogenic, neurotoxic, which cause reproductive dysfunction, or which are acutely or chronically toxic) or adverse environmental effects whether through ambient concentrations, bioaccumulation, deposition, or otherwise . . .'

Beginning six months after enactment, section 112(b)(3)(A) also provides that any person may petition EPA to modify the list by adding or deleting substances, or removing certain unique substances. EPA is required to publish a written approval or denial of petitions within eighteen months of receipt. Petitions are required to provide a showing that there is adequate data on the health or environmental defects of the pollutant or other evidence adequate to support the petition. EPA is not allowed to deny a petition solely on the basis of inadequate resources or time for review. The following other specific criteria are included in section 112(b)(3).

- EPA shall add a substance to the list upon showing by the petitioner or EPA's determination that the substance is an air pollutant and meets the definition of a hazardous air pollutant.
- EPA shall delete a substance upon showing by a petitioner or EPA's determination that there is adequate data to determine that it does not meet the definition . . .
- EPA is authorized to delete unique chemical substances that contain a listed hazardous air pollutant not having a Chemical Abstract Service (CAS) registry number[6] (except coke oven emissions, fine mineral fibers, and polycyclic organic matter) upon showing by the petitioner or EPA's determination that the substance meets the . . . deletion criteria. EPA must approve or deny such deletion petition before publishing emissions standards applicable to any source category or subcategory of the listed pollutant.

---

[5]Five more substances on the list of 189 and listed in IRIS have Class C evidence. Some argue that chemicals should not be regulated on the basis of Class C evidence alone.

[6]The substances without CAS numbers are the seventeen groups of substances and compounds at the end of the list.

At the time of writing this handbook in 1992, EPA had not yet published formal guidance on the petition process or the type and nature of the data required to support approval or denial of a petition.

AIR TOXICS SOURCE CATEGORIES TO BE REGULATED

Source Categories to be Regulated

The 1990 Amendments specifically define two types of sources of hazardous air pollutants: major sources and area sources. A major source is one that emits 10 tons per year or more of a listed pollutant or 25 tons per year or more of a combination of listed pollutants. A wide variety of industrial, commercial, and other facilities and operations qualify as major sources. An area source is defined as one that does not meet the definition of major source. Practically speaking, area sources are small, numerous, and widespread facilities and operations such as dry cleaning establishments, metal degreasing operations, painting and coating shops, commercial sterilizers, and electroplaters.

EPA was required in section 112(c)(1) to publish by November 15, 1991, a list of categories and subcategories of major and area sources of the listed air pollutants. In mid-1991, EPA published for comment a draft list of toxic air pollutant source categories (EPA 1991l). That list contained about 750 categories and did not specifically distinguish between major and area sources. Eight months after the date mandated in the Act, EPA published the initial list of hazardous air pollutant source categories (EPA 1992f). The list is shown in Table 3-3. The published schedule for promulgation of MACT standards for these source categories also is shown in Table 3-3; the mandated and actual schedules are discussed below.

The initial list contains 166 major source categories and eight area source categories. The most significant change from the draft list was the consolidation of about 400 separate synthetic organic chemical manufacturing processes in the draft list into one source category.

Derivation of the List of Source Categories

As noted above, EPA's early hazardous air pollutant regulatory program under the 1970 Clean Air Act concentrated on identifying pollutants to be regulated. In the early 1980s, however, EPA realized that many source categories emitted more than one potential hazardous air pollutant. Because the 1970 Amendments also required emission standards on sources of these pollutants, EPA began focusing more on source categories and the potential for exposure to multiple hazardous air pollutants from sources. One of the first efforts, by Fuller et al. (1976), identified and scored organic chemicals that were toxic and were produced in sufficient volume to result in likely public exposures. Later, Argonne National Laboratories developed for EPA the Hazardous Air Pollutant Prioritization System (HAPPS) (Smith and Fingleton 1982), a ranking system that took into consideration production volume in addition to toxicity. HAPPS later was revised and adapted for personal computer use by Radian (1987); it was called the Modified Hazardous Air Pollutant Prioritization System (MHAPPS). Over 900 candidate chemicals were contained in the MHAPPS data base.

Perhaps the most important contribution from this work was the development of the Source Category Ranking System (SCRS) (Mohin et al. 1989, Radian 1988). The SCRS was developed to 'facilitate the transition to the multiple pollutant assessment approach.' The SCRS was designed to rank source categories relative to one another by building on MHAPPS but grouping pollutants by source category. The output from the SCRS is a list of source categories ranked by potential risk to public

Table 3-3   NESHAP source categories and schedule

Source category	Scheduled promulgation date
Fuel combustion	
Industrial boilers	November 15, 2000
Institutional/commercial boilers	November 15, 2000
Engine test facilities	November 15, 2000
Process heaters	November 15, 2000
Stationary turbines	November 15, 1997
Stationary internal combustion engines	November 15, 1997
Nonferrous metals	
Primary aluminum production	November 15, 1997
Secondary aluminum production	November 15, 1997
Primary copper smelting	November 15, 1997
Primary lead smelting	November 15, 1997
Lead acid battery manufacturing	November 15, 2000
Secondary lead smelting	November 15, 1994
Primary magnesium refining	November 15, 2000
Ferrous metals	
Ferroalloys production	November 15, 1997
Iron and steel manufacturing	November 15, 1997
Steel foundries	November 15, 1997
Iron foundries	November 15, 1997
Coke by-product plants	November 15, 2000
Coke ovens	December 31, 1992
Steel pickling – HCl process	November 15, 1997
Electric arc furnaces – non-stainless steel	November 15, 1997
Electric arc furnaces – stainless steel	November 15, 1997
Petroleum and gas production	
Oil and natural gas production	November 15, 1997
Petroleum refineries (catalytic crackers/catalytic reformers/sulfur plants)	November 15, 1997
Petroleum refineries (other sources)	November 15, 1994
Mineral products	
Portland cement manufacturing	November 15, 1997
Clay products manufacturing	November 15, 2000
Asphalt/coal tar application	November 15, 2000
Asphalt concrete manufacturing	November 15, 2000
Asphalt processing	November 15, 2000
Asphalt roofing manufacturing	November 15, 2000
Mineral wool production	November 15, 1997
Taconite iron ore processing	November 15, 2000
Chromium refractories production	November 15, 1997
Alumina processing	November 15, 2000
Lime manufacturing	November 15, 2000
Wool fiberglass manufacturing	November 15, 1997
Liquids distribution	
Gasoline distribution (stage 1)	November 15, 1994
Organic liquids distribution (nongasoline)	November 15, 2000

Table 3-3   Continued

Source category	Scheduled promulgation date
Surface coating processes	
Aerospace industries	November 15, 1994
Autos and light duty trucks	November 15, 1997
Flat wood paneling	November 15, 2000
Large appliances	November 15, 2000
Magnetic tapes	November 15, 1994
Manufacture of paints, coatings and adhesives	November 15, 2000
Metal cans	November 15, 2000
Metal coils	November 15, 2000
Metal furniture	November 15, 2000
Miscellaneous metal parts and products	November 15, 2000
Paper and other webs	November 15, 1997
Plastic parts and products	November 15, 2000
Printing, coating and dyeing of fabrics	November 15, 2000
Printing/publishing	November 15, 1994
Shipbuilding and ship repair	November 15, 1994
Wood furniture	November 15, 1994
Waste treatment and disposal	
Hazardous waste incineration	November 15, 2000
Sewage sludge incineration	November 15, 1997
Municipal landfills	November 15, 1997
Site remediation	November 15, 2000
Solid waste treatment, storage and disposal facilities (TSDF)	November 15, 1994
Publicly owned treatment works (POTW)	November 15, 1995
Fibers production	
Acrylic fibers/modacrylic fibers	November 15, 1997
Rayon production	November 15, 1997
Spandex production	November 15, 2000
Food and agriculture	
Baker's yeast manufacture	November 15, 2000
Cellulose food casing manufacture	November 15, 2000
Vegetable oil production	November 15, 2000
Agricultural chemical production	
2,4-D salts and esters production	November 15, 2000
4,6-Dinitro-*o*-cresol production	November 15, 2000
4-Chloro-2-methylphenoxyacetic acid production	November 15, 2000
Captafol production	November 15, 2000
Captan production	November 15, 2000
Chloroneb production	November 15, 1997
Chlorothalonil production	November 15, 2000
Dacthal™ production	November 15, 2000
Sodium pentachlorophenate production	November 15, 2000
Tordon™ acid production	November 15, 2000
Pharmaceutical production	
Pharmaceutical production	November 15, 1997

Table 3-3 Continued

Source category	Scheduled promulgation date
Polymers and resins production	
Acetal resins production	November 15, 1997
Acrylonitrile–butadiene–styrene production	November 15, 1994
Alkyd resins production	November 15, 2000
Amino resins production	November 15, 1997
Boat manufacturing	November 15, 2000
Butadiene–furfural cotrimer	November 15, 2000
Butyl rubber production	November 15, 1994
Carboxymethylcellulose production	November 15, 1997
Cellophane production	November 15, 1997
Cellulose ethers production	November 15, 2000
Epichlorohydrin elastomers production	November 15, 1994
Epoxy resins production	November 15, 1994
Ethylene–propylene elastomers	November 15, 1994
Flexible polyurethane foam production	November 15, 1997
Hypalon™ production	November 15, 1994
Maleic anhydride copolymers production	November 15, 2000
Methyl methacrylate–acrylonitrile–butadiene–styrene production	November 15, 1994
Methyl methacrylate–butadiene–styrene terpolymers	November 15, 1994
Methylcellulose production	November 15, 2000
Neoprene production	November 15, 1994
Nitrile butadiene rubber production	November 15, 1994
Nonnylon polyamide production	November 15, 1994
Nylon 6 production	November 15, 1997
Phenolic resins production	November 15, 1997
Polybutadiene rubber production	November 15, 1994
Polycarbonates production	November 15, 1997
Polyester resins production	November 15, 1997
Polyethylene terephthalate production	November 15, 1994
Polymerized vinylidene chloride	November 15, 2000
Polymethyl methacrylate resins production	November 15, 1997
Polystyrene production	November 15, 1994
Polysulfide rubber production	November 15, 1994
Polyvinyl acetate emulsions production	November 15, 1997
Polyvinyl alcohol production	November 15, 1997
Polyvinyl butyral production	November 15, 1997
Polyvinyl chloride and copolymers	November 15, 2000
Reinforced plastic composites production	November 15, 1997
Styrene–acrylonitrile production	November 15, 1994
Styrene–butadiene rubber and latex production	November 15, 1994
Inorganic chemical production	
Ammonium sulfate – caprolactam by-product plants	November 15, 2000
Antimony oxides manufacturing	November 15, 2000
Chlorine production	November 15, 1997
Chromium chemicals manufacturing	November 15, 1997
Cyanuric chloride production	November 15, 1997
Fume silica production	November 15, 2000
Hydrochloric acid production	November 15, 1997
Hydrogen cyanide production	November 15, 1997
Hydrogen fluoride production	November 15, 1997
Phosphate fertilizer production	November 15, 1997
Phosphoric acid manufacturing	November 15, 1997
Quaternary ammonium compounds production	November 15, 2000
Sodium cyanide production	November 15, 1997
Uranium hexafluoride production	November 15, 2000

Table 3-3   Continued

Source category	Scheduled promulgation date
Organic chemical manufacturing	
Synthetic organic chemical manufacturing	November 15, 1992
Miscellaneous sources	
Aerosol can-filling facilities	November 15, 1997
Benzyltrimethylammonium chloride production	November 15, 1997
Butadiene dimers production	November 15, 1997
Carbonyl sulfide production	November 15, 2000
Chelating agents production	November 15, 1997
Chlorinated paraffins production	November 15, 2000
Chromic acid anodizing	November 15, 1994
Commercial dry cleaning – transfer machines (perchloroethylene)	November 15, 1992
Commercial sterilization	November 15, 1994
Decorative chromium electroplating	November 15, 1994
Dodecanedioic acid production	November 15, 2000
Dry cleaning (petroleum)	November 15, 2000
Ethylidene norbornene production	November 15, 2000
Explosives production	November 15, 2000
Halogenated solvent cleaners	November 15, 1994
Hard chromium electroplating	November 15, 1994
Hydrazine production	November 15, 1997
Industrial dry cleaning – transfer machines (perchloroethylene)	November 15, 1992
Miscellaneous sources	
Industrial dry cleaning – dry to dry machines (perchloroethylene)	November 15, 1992
Industrial process cooling towers	November 15, 1994
OBPA/1,3-diisocyanate production	November 15, 2000
Paint stripper users	November 15, 2000
Photographic chemicals production	November 15, 1997
Phthalate plasticizers production	November 15, 2000
Plywood/particle board manufacturing	November 15, 2000
Polyether polyols production	November 15, 1997
Pulp and paper production	November 15, 1997
Rocket engine test firing	November 15, 2000
Rubber chemicals production	November 15, 1997
Semiconductor manufacturing	November 15, 1997
Symmetrical tetrachloropyridine production	November 15, 2000
Tire production	November 15, 2000
Wood treatment	November 15, 1997
Area source categories	
Asbestos processing	November 15, 1994
Chromic acid anodizing	November 15, 1994
Commercial dry cleaning – transfer machines (perchloroethylene)	November 15, 1992
Commercial dry cleaning – dry to dry machines (perchloroethylene)	November 15, 1992
Commercial sterilization facilities	November 15, 1994
Decorative chromium electroplating	November 15, 1994
Halogenated solvent cleaners	November 15, 1994
Hard chromium electroplating	November 15, 1994

health. (Note: The risk ranking is relative only among source categories; actual or estimated public health risks are not derived.) Importantly, the SCRS attempted to include area sources in addition to both stack and fugitive releases from point sources. It also considered exposure more explicitly by using simple dispersion algorithms to estimate typical community ambient air concentrations and coupling that information to national population statistics. In the end, the SCRS data base included about 500 candidate chemicals and ranked 360 source categories.

The enactment of the 1990 Clean Air Act Amendments curtailed further efforts at identifying and ranking source categories. EPA's prior efforts were aimed at combining pollutants and sources in a logical and technically supportable way. The list of 189 hazardous air pollutants in the 1990 Amendments eliminated the need, at least at this time, to combine the two. Instead, EPA was to identify major and area source categories of the listed pollutants.

EPA did so by using several available data bases. The first was EPA's National Emissions Data System (NEDS) that contains information on sources emitting more than 100 tons per year of criteria air pollutants, including volatile organic compounds (VOC), and particulate matter (PM). The sources in NEDS are assigned a unique identifier, called a source classification code (SCC), and speciation profiles have been assigned to each SCC. The speciation profiles are estimates of the total VOC and PM emissions for a category. In many cases, the chemical species are listed hazardous air pollutants. The quality of the NEDS data is specified by a ranking factor (A, B, C, or D) with A providing the highest confidence.

The second source of information was literature describing synthetic organic chemical manufacturing source categories. These source categories are important because a large majority of listed hazardous air pollutants are organic chemicals. Categories were included if they either manufactured or used a listed chemical in a process.

The third source of information was published production and consumption data for organic chemicals; it was used to identify end-user processes that could emit listed hazardous air pollutants. Five general categories were considered: foam blowing processes, process solvent use, polymerization processes, pesticide production, and pharmaceutical production.

The fourth information source was EPA's Toxic Release Inventory (TRI) (discussed in Chapter 1). The TRI results from the requirements of section 313 of the Emergency Planning and Community Right-to-Know Act of 1986, and contains data on sources in SIC codes 20 through 39, with ten or more full-time employees, and manufacturing, processing or otherwise using any of about 350 listed toxic chemicals.

Finally, the list of source categories was augmented with data from the past and ongoing standards development activities within EPA. Over twenty years of developing New Source Performance Standards and National Emission Standards for Hazardous Air Pollutants provided substantial additional data.

Source Listing Issues

Section 112(c)(1) requires EPA to list '... all categories and subcategories of major and area sources ...' The terms category and subcategory are not defined specifically in the 1990 Amendments or in EPA's preliminary list of source categories. However, in the July 1992 initial list of source categories EPA attempted to clarify the distinction. A clear understanding of these terms is essential because EPA also is required in section 112(e) to publish a schedule for all of the hazardous air pollutant standards to be promulgated; a schedule implies priority ranking of categories and subcategories.

EPA defines a category of sources as a group of sources having some common features suggesting that they should be regulated in the same way and on the same schedule. Based on this, a large plant or facility, such as a refinery or chemical manufacturing plant, would comprise multiple source categories. The term subcategory is more difficult to define. In fact, after reviewing comments submitted concerning the draft list published in 1991, EPA decide to exclusively (with one exception[7]) use the term 'category' in the source category listing, and to disaggregate source categories into subcategories, where appropriate, during development of emissions standards. Such disaggregation could be required based on variations in size, operations, raw materials, emissions, controllability, etc. This disaggregation provides EPA with scheduling flexibility and it should facilitate the petition process.

Section 112(e)(2) specifies that EPA set priorities for promulgating emissions standards by considering the known and anticipated adverse effects on humans and the environment, the quantity and location of emissions from each category or subcategory of sources, and the efficiency of grouping categories or subcategories according to the pollutants emitted or processes or technologies used. Clearly, a large amount of information is required to identify, categorize, and priority rank sources. EPA has admitted to having far less information than is needed to define properly categories and subcategories. All reasonably available data bases were used, but the magnitude of the total effort (i.e., considerably more than 189 pollutants and literally tens of thousands of potential sources) was daunting. As noted earlier, key tools were the Agency's National Emissions Data System (NEDS), studies of the synthetic organic chemical manufacturing industry (see EPA 1980c), published production and consumption data on organic chemicals, EPA's Toxic Release Inventory, and other standards development activities.

The distinction between categories and subcategories is important to industries that are made up of many facilities, often with operations and processes that are uniquely designed and controlled, and that uniquely pollute. Section 112(d)(1) allows EPA to distinguish among 'classes, types, and sizes of sources within a category or subcategory ...' Furthermore, the application of maximum achievable control technology implies that a process or operating unit be specified that is unique enough to be controlled using a clearly definable technology. Finally, because individual regulations must be supported with technical and economic information sufficient to ensure that they withstand technical and legal challenge, it is important that regulations cover legally and technically definable process or operating units.

Revisions to the List of Source Categories

Congress did not offer as much guidance on revising the source category list as it did for the pollutant list. EPA is required in section 112(c)(1) to revise the list of source categories no less often than every eight years in response to public comment or new information. The only major provision is that the list be consistent with the list of source categories established under section 111 (New Source Performance Standards) or Part C (Prevention of Significant Deterioriation).

Schedules for Regulation

Section 112(e)(1) requires EPA to publish emission standards for a specific quantity of the listed source categories according to a schedule provided in the Act. The mandated schedule is as follows:

Quantity of source categories	Publication date
At least 40 source categories	November 15, 1992
25 percent of the listed categories	November 15, 1994
25 percent of the listed categories	November 15, 1997
All remaining listed categories	November 15, 2000

In late-1992, when this chapter was written, EPA announced its schedule plans (EPA

---

[7] Separate subcategories for perchloroethylene dry cleaning facilities were listed, pursuant to previously published emissions standards for that category (56 FR 64382).

1992g). This schedule is shown in Table 3-3. Importantly, EPA has acknowledged that it will not meet most of the mandated schedules in the 1990 Amendments. There was no word at that time on how state and local agencies and environmental groups would react. Given past history, litigation is likely.

In the preamble to the published schedule, EPA described its rationale and method for establishing the schedule. The three criteria used were: (1) known or anticipated adverse effects on public health and the environment, (2) quantity and location of emissions or reasonably anticipated emissions from the category, and (3) efficiency of grouping the categories according to pollutants, processes, or technologies. The Source Category Ranking System was described along with the exposure scoring and health effects scoring systems used in establishing the priorities for regulation. EPA specifically requested comment on the schedule, the ranking criteria used to develop the schedule, and whether the statute affords EPA the flexibility to adjust the regulatory schedule after publication. This latter issue likely will be debated at length.

The cornerstone of the hazardous air pollutant regulatory program is intended to be the first major emissions standards likely to be published, the Hazardous Organic National Emission Standards for Hazardous Air Pollutants (NESHAP), or HON. EPA was to propose this regulation in late 1992. It consists of a number of different regulations that apply to various specifically or generically defined organic chemical industry sources and operations, including regulations for several specific

processes and operations, fugitive emissions, and secondary emissions.

In the 1992 final source category list, EPA listed eight area source categories for regulation. Other possible area sources and pollutants that they typically emit are shown in Table 3-4.

References

Fuller, B., J. Hushon, M. Kornreich, R. Ouellette, L. Thomas, and P. Walker. 1976. Mitre Corporation. Scoring of organic air pollutants: chemistry, production and toxicity of selected synthetic organic compounds. Report # MTR-7248. Prepared for U.S. Environmental Protection Agency. Office of Air Quality Planning and Standards. Research Triangle Park, NC.

Mohin, T.J., R. Rosensteel, and W. Maxwell. 1989. Multiple pollutant air toxics assessments – Development of a source category ranking system. Paper No. 89-45.2. Presented at the 82nd Annual Meeting of the Air & Waste Management Association. Anaheim, CA.

Radian Corporation. 1987. The modified hazardous air pollutant prioritization System (MHAPPS). Final Report. Contract No. 68-02-4330, Work Assignment 12. U.S. Environmental Protection Agency. Research Triangle Park, NC.

Radian Corporation. 1988. Source category ranking system (SCRS). Final Report. Contract No. 68-02-4330, Work Assignment 51. U.S. Environmental Protection Agency. Research Triangle Park, NC.

Smith, A.E. and D.J. Fingleton. 1982. Hazardous air pollutant prioritization system (HAPPS). EPA 450/5-82-008. U.S. Environmental Protection Agency, Office of Air Quality Planning and Standards. Research Triangle Park, NC.

U.S. Environmental Protection Agency (EPA). 1980c. Organic Chemical Manufacturing. Published in six volumes, EPA 450/3-80-023 through EPA 450/3-80-028. Office of Air Quality Planning and Standards. Research Triangle Park, NC.

U.S. Environmental Protection Agency (EPA). 1991l. Draft list of hazardous air pollutant source categories. 56 FR 28548. June 21, 1991.

U.S. Environmental Protection Agency (EPA). 1992f. Initial list of hazardous air pollutant source categories. 57 FR 31576. July 16, 1992.

U.S. Environmental Protection Agency (EPA). 1992g. Schedule for hazardous air pollutant standards. 57 FR 44147. September 24, 1992.

Table 3-4   Other potential area sources and chemicals

Potential area sources	Chemicals used and emitted
Painting and coating	Various organic solvents
Service stations and gasoline marketing	Gasoline, Benzene, Toluene, xylene
Metal cleaners	Petroleum solvents
Hospital sterilizers	Ethylene oxide
Paint removal	Methylene dichloride
Combustion sources	Formaldehyde

4

Assessment and Control of Toxic Air Pollutants

David R. Patrick

INTRODUCTION

Section 112 of the 1990 Clean Air Act Amendments establishes a new nationwide program whose purpose is to reduce significant routine emissions of toxic air pollutants in order to protect the public health. The new program first requires the application of 'maximum achievable control technology' (MACT) to significant sources of listed hazardous air pollutants. In a second phase some years later, the risks remaining after application of MACT are to be assessed and additional emissions reductions required if necessary to protect the public health with an 'ample margin of safety.'

This process is significantly different from that used in the past by EPA and state and local governments. In early EPA efforts, regulations were developed that accounted for cost and technical feasibility; later, risks to the population were estimated and regulations were developed to reduce those risks to levels considered 'acceptable.'[1] The degree of control required to achieve the acceptable risk could range widely depending on the initial emissions and how 'acceptable' was defined. The requirement in

section 112 of the 1990 Amendments that MACT be applied first and that risks be dealt with later in part resulted from Congress' concern that neither EPA nor the scientific community had clearly defined procedures for estimating risks and, as a result, many perceived sources of air toxics might not be regulated. As Congress apparently viewed it, MACT regulations could be established and applied more easily than those based on risk and, thus, could achieve more rapid reductions in toxic air pollutants. This process also would allow time to define better risk decision-making processes.

Regardless of the sequence of the process, however, several types of information must be obtained before regulating toxic air pollutants. First, to be called 'toxic' a substance must be shown to cause an adverse health or environmental effect. The potential adverse effects associated with exposure to the substance of concern then must be assessed. A toxic air pollutant by definition must present a hazard to human health or the environment. The hazard

---

[1] Some prefer the term 'not unacceptable' to avoid the connotation that some risk is acceptable.

assessment step involves: (1) the qualitative assessment of data from human, animal, or other studies to identify the types of adverse effects that can result and to determine whether or not exposure to the substance is causally related to a particular adverse effect, and (2) the quantitative evaluation of the potential of developing these adverse effects as a result of a specific exposure to the substance.

Second, the degree to which people or the environment are or can be exposed to the toxic air pollutant must be assessed to determine whether likely exposures are significant. This also is necessary to establish the extent to which exposure and, thus, emissions must be reduced. The exposure assessment step can include: (1) quantification of emissions, (2) estimation of how emissions disperse into the community from the source, (3) prediction of atmospheric transformations, (4) estimation of human uptake, and (5) analysis of population location and activity.

Third, regulatory agencies must identify appropriate regulatory strategies. An appropriate regulatory strategy generally is one that can reduce emissions to levels consistent with the hazard and exposure information and the mandates of the regulating agency, and can be carried out in a timely and cost-effective manner.

Fourth, control methods must be identified for reducing the emissions to appropriate levels. These must be consistent with the chosen regulatory strategy and also should be able to be carried out in a timely and cost-effective manner.

This handbook is intended primarily as a resource and reference book for use by those conducting these four major activities. It identifies and describes the key scientific, technical, legal, regulatory, and policy issues associated with carrying out air toxics control programs. The four major activities described above are addressed in separate sections of the handbook: Part II – Health Assessment, Part III – Exposure Assessment, Part IV – Regulatory Strategies, and Part V – Control Methods. In addition, Part I provides background information and Part VI provides information on other issues relevant to understanding and dealing with toxic air pollutants.

Although the handbook is designed to serve principally as a resource and reference book for those who must comply with toxic air pollutant regulations, it also is intended to be of use to those who develop the regulatory programs and to those who provide research and consulting assistance. Whatever the need of the end-user, the handbook was written under the assumption that the user is scientifically or technically trained in an area of specialization relevant to this endeavor. Typical scientific and technical disciplines that are involved on a routine basis with the assessment and control of toxic air pollutants are listed in Table 4-1.

It is not the intent of this handbook to serve as the definitive treatment on any broad scientific or technical subject; that will be left to publications specific to those subjects. It is the intent of this handbook to identify and describe issues of concern using language and terms that are familiar to the typical industrial environmental control specialist and governmental regulator, and to provide information needed by the control specialist or regulator to accomplish

Table 4-1   Typical disciplines and expertise required to assess and control toxic air pollutants

Activity	Disciplines
Pollutant/source identification	Engineering, chemistry
Health assessment	Toxicology, biology, epidemiology, biostatistics, chemistry
Exposure assessment	Engineering, chemistry, atmospheric sciences, pollutant sampling, analytical chemistry, environmental engineering, statistics, biology
Regulatory strategies	Engineering, law, economics, policy analysis
Control technologies	Engineering, chemistry
Other	Public communication

his or her job in an effective and timely manner. Specific references and resources are identified for those, such as researchers, who may desire to delve more deeply into individual subject areas. For example, Chapter 10 describes air dispersion modeling in general terms and dis-

cusses its use in the assessment and regulation of toxic air pollutants. However, the detailed understanding and application of the science of air dispersion modeling is left to the listed references and other publications specializing in that field.

HOW THIS HANDBOOK IS ORGANIZED

Assessment and control of toxic air pollutants, where appropriate, generally follow a logical stepwise sequence. Ideally, a regulatory official begins by identifying substances that can exist as air pollutants and that can cause adverse health or environmental effects sufficient to warrant reductions in emissions. When it is determined that an adverse effect is possible over a range of concentrations that can exist in the ambient air, the regulatory official or the emitter of the substance must determine whether the substance can be released to the ambient air and result in people being exposed at concentrations of concern. This requires the identification of sources of air emissions and the quantification and characterization of the emissions. Once it is determined that emissions can result in population exposures of possible concern, regulators publish regulations specifying technologies or emissions limits acceptable for reducing emissions and exposures to levels that achieve established public health or environmental needs. To comply with the regulations, the sources of the emissions must then identify and evaluate a range of control options and select and install processes or equipment such that the atmospheric concentrations of concern will not occur as a routine matter.

The process mandated in the 1990 Amendments differs from this sequence only in that certain health and exposure assessment steps have been predetermined by Congress. For example, 189 substances have been established, by definition, as having sufficient health and exposure information to warrant control. However, the Act reverts to the stepwise sequence for the later risk-based regulations and it provides a process whereby other toxic air pollutants may be identified, listed, and regulated.

This stepwise process was used to organize this handbook. The format is illustrated in Figure 4-1. Following the identification of candidate toxic air pollutants and sources, two separate assessment tracks are taken. On the first track, discussed in Part II, the specific adverse health effects associated with potential toxic air pollutants are assessed. This involves

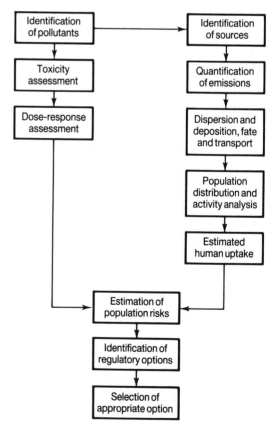

Figure 4-1.   Schematic diagram of the toxic air pollutant evaluation and regulation process.

both the qualitative determination that an effect is possible and the quantitative determination of the severity of the effect at various exposure levels. Health assessment typically is divided into two principal groupings of human health effects – carcinogenic effects and noncarcinogenic effects.

On the second track, discussed in Part III, potential sources of the pollutants are identified and assessed to determine whether emissions of the pollutants can result in significant population exposures. This can involve measuring the emissions from the sources or measuring the atmospheric concentrations resulting from the emissions, or it can involve estimation through mathematical air dispersion modeling. Important variables in exposure assessment are potential atmospheric transformations, human uptake, and population analyses.

The adverse effects and source paths then converge and, if the exposures and potential health effects warrant, regulatory strategies are identified and carried out. Depending on the legislative and legal requirements, regulations can take a variety of forms. Several common types of regulatory formats are described in Part IV.

After selection of the appropriate regulatory format, control methods are selected that comply with the regulation. Current control methods and processes are described in Part V.

Finally, Part VI provides relevant information concerning other regulatory programs related to air toxics. Special sources of toxic air pollutants are described, including several specifically mentioned in the 1990 Amendments (e.g., electric utility steam generators, publicly owned treatment works, coke ovens, and municipal incinerators); mobile source emissions of toxic air pollutants also are described. Other programs that reduce toxic air pollutant emissions are described, including: the criteria air pollution program under Title I, the accidental release program under Title III, indoor air pollution activities, and the toxics reporting program under the Emergency Response and Community Right-to-Know Act. Finally, recent international activities on air toxics are summarized and air toxics public communication needs are described.

II

Health Assessment

5

Evaluation of Toxicological Data

Staff of Clement International Corporation

INTRODUCTION

A complete health assessment consists of two major components. The first is the qualitative evaluation of data from epidemiological or toxicological studies to identify the types of adverse health effects associated with exposure to a chemical and to determine whether or not there is a causal relationship. The second component of the health assessment is the quantitative evaluation of the magnitude of these adverse health effects resulting from specific exposures to the chemical. The approaches to qualitatively evaluating and interpreting toxicological data are described in this chapter; the quantitative aspects of health assessment are described in Chapter 6 (carcinogens) and Chapter 7 (noncarcinogens).

The determination of a chemical's potential toxicity in humans following exposure is based ideally on observations in humans; however, such information frequently is limited and experimental data from animals often must be relied on. A critical assessment of the available toxicological data and an indication of its quality, quantity, and consistency is an integral part of the qualitative evaluation of the toxicity of a chemical. Data evaluation involves assessment of the scientific correctness of the available material, comparison against acceptable standards governing study methodologies and protocols, and validation of the conclusions that the evidence permits. Data interpretation, on the other hand, involves the assessment of the data as to its significance and possible relevancy to humans and the environment, based on known or estimated human exposure and the implications of identified uncertainties. Thus, the ultimate goal of interpreting and evaluating toxicological data is to determine their value for use in risk assessment. The level of confidence that can be placed in the data depends on the quality of the data used to derive them.

EPIDEMIOLOGICAL STUDIES

Description

Epidemiology is the study of the occurrence or nonoccurrence of disease in a population. There are two major types of epidemiological studies – the descriptive and the analytic. Descriptive studies include: (1) correlational studies in which disease rates in populations are compared to temporal or spatial distribution of suspected risk factors, and (2) case reports of individuals with interesting associations of disease entity and some suspected risk factor, such as a demographic, life-style, or environmental or occupational exposure. These types of investigations are used to generate and refine hypotheses, but rarely can be used to infer causal relationships.

The second type of epidemiology study is the analytic study. These may be: (1) case-control studies, in which individuals with the disease of interest are identified and the rate of occurrence of the risk factor of interest is compared to the rate in some appropriately selected control (nondiseased) group, or (2) cohort studies, in which groups of individuals are identified through some exposure of interest, and the rates of either morbidity or mortality in the exposed group are compared to the rates in some appropriately chosen reference population. General population morbidity and mortality rates frequently are used on the assumption that if the exposure of interest were not present the study cohort would experience the same risk of disease as the rest of the population. Certain caveats apply here, as comparisons must be made taking into account factors that are known to impact disease etiology, such as age, race, sex, birth cohort, and in some instances ethnic background.

Methods of Application

Epidemiological studies are observational in nature and, thus, far more subject to bias than controlled laboratory studies. The two major tasks in evaluating an epidemiologic study for risk assessment are to establish the quality of the study design and to establish the quality of the exposure measures.

Critical evaluation of epidemiological data requires careful weighing of potential biases and concise descriptions of the limitations and the merits of individual studies and reports. The key to successful use of epidemiological and other human data is the integration of these data with other types of data using a weight of

evidence approach. Standard practice in epidemiology has been to examine all of the animal, human, and in vitro data to establish causality and to determine the nature of the likely health effects in humans.

Characterizing a chemical's human toxicity under different exposure scenarios requires different uses of the various types of data available on humans. Case reports, especially if compiled by a single research group into a series, provide qualitative information on the target organ or nature of the effects likely to occur in humans and, for noncarcinogens, often provide data that bracket the range of exposures associated with acute and subchronic effects. Case-control studies are more likely to offer data demonstrating the association of exposure with an effect rather than offering quantitative information on exposure. Alternatively, case-control studies that are nested in a cohort are very likely to have useful dose-response data. Cohort studies often prove to have sample sizes that are too small to give support to a lack of association but may still include effect levels that can be used to identify exposure levels associated with responses. All types of population studies serve to identify risk factors and thereby identify high-risk or sensitive populations, even if the results are in terms of mortality, an end point that is too severe to use in risk assessment.

A confounding factor is a risk factor for the endpoint that, if not controlled, has the potential to confound or cast doubt on the purported association found in analyzing the data. Importantly, a characteristic that is not believed to affect the endpoint is not a potential confounder; in other words, if a factor such as age, race, or sex is not related to risk for disease it is not essential that it is controlled in the design or analysis. Often a screening analysis of the data will establish this if the information is not known from other studies. To reduce or remove a potentially confounding factor, one must control for it in the analysis. Many multivariate analytical approaches, such as logistic regression, enable the investigator to measure the effect of exposure to an agent after controlling for the effect of other potential risk factors.

A well-designed study, analyzed with statistical procedures appropriate for the design and data type in the study, may nevertheless provide exposure measures that are only categorical levels, such as high, medium, and low. Positive, statistically significant results of these studies offer strong support for the association between the exposure and effect, provided the exposure is verified and is not mixed with other chemicals. If exposure is only categorically measured, studies without positive results are of limited use. When results of these studies with well-quantified exposure measures are not positive, the interpretation hinges on the power, as reflected in the confidence interval of the effect measure (e.g., relative risk), standardized mortality ratios, or standardized incidence ratio. A study may, for example, provide evidence that the effect at this level is not greater than a relative risk of 1.5 or 2, or that the reduction in forced expiratory volume is no greater than a certain percentage, the smallest effect that a study of this size could detect. This information is particularly useful for those end points that cannot be evaluated completely in animal studies, such as an increased prevalence of coughing. This information should be considered in the analysis of critical effects and compared with human equivalent concentrations that may have been derived from other end points.

Complete characterization of health effects in an exposed population also requires identification of sensitive populations. People with high susceptibility are those who respond to the toxic and carcinogenic effects of chemicals more quickly or at lower exposure levels than the general population because of differences in developmental (i.e., age-specific), genetic, nutritional, behavioral, or life-style factors or to preexisting disease states. For example, fetuses and neonates are especially susceptible to pesticides and PCBs because their detoxifying enzyme systems are immature. Smoking, alcohol, and drug use not only increase an individual's exposure to chemicals but also increase susceptibility to the toxic effects of other chemicals. These groups may include a larger portion of the general population than is generally recognized.

ANIMAL STUDIES

In the review and evaluation of animal data, the important aspects to consider include experimental design and execution, clinical and pathological examinations, animal care, and test evaluation and analyses. The experimental design determines the nature and extent of the results and sets permanent limits on the interpretation and analysis of the data. Considerations in the evaluation of the experimental design include adequate information on the identity of the chemical in its original form and in a vehicle, the choice of animal model, the route of administration of the toxic agent and the end points chosen for study. The following should be included in the descriptions of toxicity studies.

■ A description of the test animals (sex, species, strain, age, body weight)
■ The dose(s) (expressed in terms of mg/kg per day, or mg/m^3 for inhalation studies) and the number of animals per dose level
■ The routes of compound administration
■ The duration and frequency of compound administration
■ Experimental methods, results (including statistical significance), conclusions, and limitations that diminish the study conclusions
■ Identification of no-observed-adverse-effect levels (NOAELs) and lowest-observed-adverse-effect levels (LOAELs) for non-carcinogenic effects

Characterizations of a chemical's potential toxicity to humans, based on the results of studies performed using animals, requires an assessment of the validity of such an extrapolation. Questions such as the following should be asked.

- Are target organs susceptible to chemical damage in humans likely to be those identified in animal studies?
- Are the types of neoplastic and non-neoplastic lesions found in animals likely to occur in humans exposed to the same chemical?
- Are experimental study results primarily an indication that adverse effects of some type may occur in humans?

When based on animal data, a comprehensive assessment of the health hazard from exposure to a potentially toxic chemical should include evaluation of data from carcinogenicity, teratogenicity, reproductive studies, and other toxicological results with reference to the reliability of the experimental design and toxicological interpretation of the results. In addition, consideration should be given to studies designed to evaluate the metabolism and toxicokinetics, to results of short-term in vitro tests, such as mutagenicity data, and to data from other studies that may elucidate mechanisms of action of the chemical. Finally, the chemical's physicochemical properties must be considered because they will influence the deposition and retention of the chemical within the body, its movement to other tissues, and its ultimate toxic effect(s).

The general approach to the evaluation of data is to apply expert scientific judgment in determining whether a reported result is adequate to support risk assessment. If a test is judged to be inadequate, the rationale is documented for this conclusion. If a test is judged to be adequate but to have important limitations (e.g., statistical insensitivity because sample sizes were too small), these limitations should be pointed out as well. The general approach to evaluating and interpreting studies in various disciplinary areas uses many of the same techniques. There are special concerns and problems that are unique to each type of study; therefore, the major types of studies are discussed separately.

Acute Toxicity Studies

Data from an acute toxicity study provide information on the health hazards that are likely to result from short-term exposures. The evaluation of an acute toxicity study involves examination of the relationship between the exposure to the test compound and the incidence and severity of the observed effects including behavioral and clinical abnormalities, gross lesions, body weight changes, and lethality, as well as the reversibility of any observed abnormalities. Because responses caused by a compound vary greatly among different species, toxicity tests ideally should be conducted in the species that will elicit the compound-related toxic response similar to what may occur in humans. Although it may not be known with certainty which animal species is the 'ideal' for any given compound, in general one species may be preferred for a variety of reasons. For example, the rat is the preferred species for acute inhalation studies.

Acute toxicity, even within a particular species, can vary with age, sex, genetic makeup, and differences in toxicokinetic or hormone influences. Therefore, acute studies should be evaluated for the appropriate choices of these parameters. In general, young animals in the appropriate weight range and an equal number of animals of both sexes should be used. Females should not have produced offspring and not be pregnant. At least five animals per dose group and at least three dose groups and a vehicle concurrent control group should be used. The dose groups should be appropriately spaced and sufficient to result in a dose-response curve and to permit an acceptable estimation of the median lethal concentration. The test substance should be administered over a four-hour period and animals should be observed daily and observation continued for at least fourteen days. The animals should be exposed in an inhalation chamber that sustains a dynamic air flow of twelve to fifteen air changes per hour, ensures an adequate oxygen content of 19 percent, and provides an evenly distributed exposure atmosphere. The test atmosphere must be adequately characterized. The rate of air flow should be continuously monitored and recorded every thirty minutes and measurement of exposure concentrations should be made in the breathing zone. The study report should contain a description of the exposure apparatus, including design, type, dimensions, source of air, system for generating particulates and aerosols, method of conditioning the air, treatment of exhaust air, and the method of housing the animals in the exposure chamber. For particulate matter and aerosols, the median aerodynamic diameter with the standard deviation from the mean should be reported.

Data should be presented for all toxic responses by sex and dosage level. The time at which toxicity appeared or disappeared, as well as the duration of a toxic effect and the time to death, should be recorded. In addition, gross pathology and histopathological information and clinical chemistry findings should be reported. Cage-side observations, such as skin, somatomotor activities and behavior, and pharmacotoxic signs, such as tremor, convulsions, and diarrhea, provide valuable clues to the target organ or system and may give some indication as to the potential mechanism of activity. For example, dyspnea or abnormal breathing may indicate involvement of the central nervous system, respiratory center, paralysis of rib cage muscles, or cholinergic inhibition.

A frequently determined endpoint in acute inhalation toxicity testing is the median lethal concentration (LC_{50}), which represents the concentration of a compound that is lethal to 50 percent of a population of treated animals under a specified set of experimental conditions. Although LC_{50} values have been used to classify and compare toxicity among chemicals, toxicity cannot be fully evaluated using the LC_{50} alone. A chemical can induce nonlethal adverse effects in a variety of biological functions. For animal studies, LC_{50} data for acute exposures are presented along with causes of death, if available.

Subchronic Toxicity Studies

Subchronic toxicity tests are designed to assess nonlethal adverse effects. The endpoints of concern in these tests can be defined by biochemical, hematological, and clinical measurements, as well as gross and histopathological observations. A variety of organ systems are evaluated in these tests including the nervous system, cardiovascular, respiratory digestive, urinary, musculoskeletal, lymphohematopoietic, endocrine, genital systems, and special sense organs. A subchronic study should not exceed ten percent of the animal's life span. In general, the subchronic study is carried out for three months in rodents and one year in dogs or monkeys. Subchronic studies should be conducted in at least two species, preferably young animals, with an equal number of males and females at each of three or four dose levels (at least ten per sex per dose group for rodents), along with a vehicle control. However, tests may have alternate durations and dose levels, depending on the intent of the study (e.g., range finding studies).

As part of the evaluation of the dose-response relationship, a NOAEL (if possible) or a LOAEL should be determined. The NOAEL is defined as the highest experimental dose of a chemical at which there is no statistically or biologically significant increase in frequency or severity of a toxicological effect between an exposed group and its appropriate control. Adverse effects are defined as any effect that results in functional impairment or pathological lesions that may affect the performance of the whole organism or that reduce an organism's ability to respond to an additional challenge. In general, NOAELs for several of these endpoints will differ. Everything else being equal, the critical endpoint is the one with the lowest NOAEL.

In some instances, the NOAEL for the critical toxic effect is simply referred to as the NOEL. This latter term, however, is ambiguous because there may be observable effects that are not considered to be of biological or toxicological significance and, thus, are not 'adverse.' This is often a matter of professional judgment. Furthermore, the magnitude of the NOAEL is dependent on the population size under study and dose selection. Studies using a small sample size generally are less sensitive to low-dose effects than studies with a larger number of subjects. In addition, if the interval between doses is large, then the NOAEL determined from that particular study may be lower than what would possibly be seen in a study that used intermediate doses.

In instances when a NOAEL cannot be demonstrated, a LOAEL is used to evaluate the critical toxic endpoints. A LOAEL is the lowest experimental dose of a chemical at which there is a statistically or biologically significant increase in frequency or severity of a toxicological effect. Again, an adverse effect is any effect that results in functional impairment or pathological lesions that may affect the performance of the whole organism or that reduce an organism's ability to respond to an additional challenge. The LOAEL generally is less useful than the NOAEL because the 'lowest' test concentration may be significantly higher than the no-effect level.

Unlike acute studies, the major endpoints in subchronic studies are not mortality but nonlethal adverse effects, that can be defined by biochemical, hematological, or clinical measurements, including changes in body and organ weight, food or water consumption, and histopathological examinations. The dose levels selected should result in no toxicity at the low dose, no or slight toxicity at the intermediate dose levels, and toxicity in the high dose group that should be a frank toxic effect without excessive mortality that prevents meaningful interpretation of the data. For subchronic studies, the species selected should maximize the biological comparability between the experimental animal and humans with regard to metabolism, toxicokinetics, and target organ toxicity (N.B. This information usually is not available).

The relevance of the route of exposure to the route by which humans are exposed should be considered. Animals should be observed daily and signs of toxicity should be recorded as they are observed, including the time of onset, the degree, and the duration. Cage-side (e.g., skin and fur, eyes, somatomotor activity, and behavior) and pharmacotoxic (e.g., tremor, convulsions, and diarrhea) observations should be made; both give valuable insights into the possible target organs and mechanism of action of chemicals. Clinical examinations should include hematological and clinical chemistry

determinations and should be carried out at least three times during the test period on at least five animals per group. The clinical chemistry tests that are appropriate evaluate electrolyte balance, carbohydrate metabolism, and liver and kidney function. For example, the major tests that have proved useful for evaluation of hepatic injury in experimental animals include the following serum enzyme tests: serum glutamic–pyruvic transaminase (SGPT); hepatic excretory tests, such as biliary excretion; alteration in chemical constituents of the liver, such as lipid content or alteration in liver metabolizing enzymes; and histological analysis of liver injury. The results of these tests can give insights into the mechanisms of action of the chemical. For example, carbon tetrachloride may be acting via a free radical mechanism that may disrupt lipid peroxidation resulting in alterations of normal structure and function. The biochemical parameters examined should be based on the class of chemicals and the expected toxicity. For example, cholinesterase activity should be considered if the test substance is an organophosphate or carbamate, both expected to be inhibitors of this enzyme. Alterations in biochemical parameters may suggest target organs and mechanisms of action. For example, statistically significant increases in SGPT levels may suggest hepatotoxicity and increases in lactate dehydrogenase (LDH) may be related to myocardial damage. In addition, gross necropsy and histopathological examinations should be comprehensive and follow recommended protocols. The statistical interpretation of the data should be considered so that it is possible to distinguish normal biological variations from treatment-specific changes.

The adequacy of subchronic toxicity tests may be judged on a number of criteria, for example:

- Rodent studies should test 20/sex/dose and nonrodent studies should test 4/sex/dose. Controls of the same number/sex also should be tested.
- Animals should be observed daily for signs of toxicity. Weekly body weight and food

consumption records should be kept. Hematology, clinical chemistries, and urinalysis should be performed at appropriate intervals.

■ Gross pathology should be performed on all animals dying during the course of the study and on all animals at terminal sacrifice. Histopathological examination of tissues from control and high dose animals should be performed.

■ Levels of the test material to which the animals are exposed should be verified. Purity of the test material must be specified.

Chronic Toxicity

Chronic toxicity studies are intended to assess the effects of chemicals after prolonged or repeated exposure over a substantial portion of an animal's lifetime. The increased duration of these studies allows effects to be observed that would not be detected in a subchronic study, such as effects that have a long latency or that are cumulative in nature. The criteria for judging chronic toxicity studies are essentially the same as for subchronic toxicity studies, with some exceptions. If interim sacrifices are planned, the number of animals should be increased accordingly. Furthermore, if mortality in the high dose group compromises the statistical analysis of histopathological findings in the high dose group, complete histopathological examination of the next lowest dose group should be performed.

Carcinogenicity Studies

The carcinogenicity bioassay is a type of chronic test and should include all of the components discussed above. The critical evaluation of carcinogenicity bioassays, however, is multi-faceted. Animal bioassays are evaluated to determine whether the maximum tolerated dose (MTD) was identified and considered, whether exposure affected survival or body weight, whether the duration of exposure and observation were sufficient to elicit or detect an effect, whether appropriate controls

and both sexes were used and exposure groups were of adequate size, and whether the spontaneous tumor incidences preclude detection of significant effects.

Evaluation of experimental data from animal bioassays is critical in elucidating a chemical's potential for human toxicity. An important part of the critical evaluation of bioassays is the analysis of the pathological observations. Pathology data generally available for review consist of published reports in the open scientific literature on toxicity and oncogenicity experimental studies in rodents and other species. For each chemical, the neoplasms and nonneoplastic lesions that were identified by the investigators as being related to chemical exposure should be analyzed for biological and statistical significance. In the case of carcinogens, attempts should be made to determine whether or not each report has specified criteria for particular tumor diagnoses and whether these criteria correspond to those currently in use in toxicologic pathology. This type of review is critical for all studies but particularly for older studies where inappropriate diagnostic criteria and terminology could result in under- or overdiagnosis of tumors related to chemical exposure.

Pathology incidence data should be comprehensively examined and used to determine whether all chemical-related endpoints have been identified in the report. For example, certain neoplasms are rare spontaneous findings. The presence of such tumors may be considered biologically significant, although not statistically significant, and may indicate a chemical effect. Conversely, certain common spontaneous neoplasms may have marginally statistically significant increases in treated animals related to normal biological variation rather than a chemical effect. The potential significance of such findings should be discussed with reference to relevant historical control data.

Nonneoplastic lesions may be of two types – proliferative and nonproliferative. The evaluation should determine whether proliferative lesions are related to chemical exposure (i.e., which target organs are involved and whether the lesions show a dose-related increase in incidence). If associated with chemical exposure, the review determines whether proliferative lesions represent a preneoplastic response, are potentially controversial (e.g., tumor versus hyperplasia), or are unrelated to any chemically induced neoplasms. For nonproliferative lesions, attempts should be made to determine whether these lesions were a result of chemical damage to tissues and cells or were related to other factors. As an example of the latter, in an inhalation study rats infected with *Mycoplasma pulmonis* can show a dose-related increase in pneumonia; however, pneumonia is related to effects of the chemical on the upper respiratory tract, facilitating growth of the infecting organism that then invades the lungs, rather than an effect of the chemical per se on pulmonary tissue.

Route of exposure and relationship to non-neoplastic and neoplastic lesions – such as forestomach ulceration, epithelial hyperplasia, and squamous cell carcinoma in gavage studies – should be evaluated. In gavage studies particularly, mortality in treated animals should be correlated, when possible, with gross and microscopic lesions in order to separate possible gavage errors from any effects of the chemical.

If nonneoplastic lesions are found to be chemical-related, lesion severity is analyzed by evaluating lesion description and grading and by correlating lesion incidences with body weight changes, survival data, and subchronic study data used to establish doses for the chronic study. Attempts are made to determine whether the MTD was attained or exceeded in the chronic study and how this relates to tumor endpoints, if any.

If statistical analysis of the data is necessary, the procedures adopted by the National Toxicology Program generally should be utilized. The weight of evidence approach developed by EPA (1986c) normally should be used as guidance in performing such assessments.

The key consideration in extrapolating the results of animal carcinogenicity studies using high exposure levels to human carcinogenicity, that may result from low environmental exposure levels, is an understanding of the mechanism of carcinogenesis of the substance in question. Many carcinogens are genotoxic and may act by causing mutations at critical sites on the genome, leading to tumor initiation. Alternatively, some carcinogens are tumor promoting agents and elicit their effects by altering the control of cellular proliferation and differentiation without causing mutations. The nature of a carcinogen's dose-response relationship is highly dependent on its mechanism of action. A careful evaluation of the target organs, tumor incidences, and tumor and related pathology observed, as well as any supplemental information (e.g., pharmacokinetic data), must be performed in order to provide a valid basis for the qualitative and quantitative assessment of human carcinogenicity on the basis of animal bioassays.

There may be several key carcinogenicity studies and the study or studies that form the basis of a quantitative risk assessment (if one exists) or should form the basis (if a risk assessment has not been performed) should be discussed in detail. Other supporting studies also should be summarized. The weight of evidence (i.e., whether data in humans and animals are adequate to consider the chemical a human carcinogen, a probable human carcinogen, a possible human carcinogen or a noncarcinogen) should be discussed. The summary should consider other routes of administration, data on the mechanism of carcinogenicity, results from in vitro studies, and the issues related to the weight of evidence.

Developmental Toxicity

Developmental toxicity is any adverse effect on the developing organism that may result from exposure before conception (either parent), during prenatal development, or postnatally to the time of sexual maturation. The major manifestations of developmental toxicity include death of the developing organism (either prenatal or early postnatal), structural abnormalities, altered growth (usually in the form of growth retardation), and functional deficits that may include a variety of problems such as reduced pulmonary function, reduced immunological competence, and neurobehavioral deficits, such as learning disorders or mental retardation.

For laboratory animals, standard developmental studies provide data on a number of potential developmental effects. Because each of the major developmental stages (i.e., the preimplantation, embryonic (organogenesis), fetal, and early postnatal periods) possess a characteristic vulnerability, complete evaluation of developmental toxicity of a chemical should be derived from groups of tests that administer the compound during these particular stages. For example, exposure to the chemical during the preimplantation period may result in embryolethality but rarely teratogenicity. Exposure during the embryonic stage may lead to structural birth defects, and the type of malformation produced may change if the same agent is applied at different times during the organogenesis period, given the narrow time span of vulnerability for individual organ systems. With the basic protocol, young pregnant rats, mice, rabbits, or hamsters are administered the test compound for at least part of the pregnancy, especially during the period of organogenesis. At least three exposure groups and a concurrent control should be used; a vehicle control should be used when necessary. The exposure levels may have been selected in a previous pilot study. Ideally, the lowest dose should not produce any maternal or developmental toxicity, while the highest dose should be large enough to induce overt maternal toxicity (unless limited by the physical or chemical properties of the substance), such as a slight weight loss. Shortly before delivery, the animals are sacrificed and the uterine contents examined for embryonic and fetal deaths and live fetuses. The degree of resorption should be described and the number of corpora lutea should be determined (except in mice). Gross examination for skeletal and soft tissue anomalies should be performed.

Ideally, these studies should provide evidence of a dose-response for the effects observed and should be able to identify a NOAEL and a LOAEL for developmental toxicity. The adequacy of this type of developmental toxicity test can be judged on a number of criteria such as:

- At least two species should be tested – 20 pregnant rats, mice, or hamsters, or 12 pregnant rabbits per dose.
- The high dose should cause some signs of maternal toxicity with no more than 10 percent mortality. The low dose should be without observed toxic effects on the mother.
- Animals should be observed daily. Food consumption and body weight data should be recorded weekly.
- Females aborting should be sacrificed and examined completely at necropsy.
- At the time of sacrifice, females should be subject to gross pathology; uteri should be removed and fetal viability, number of resorptions, fetal weight, and sex should be determined.
- One-third to one-half of fetuses from rats, mice, or hamsters should be examined for skeletal abnormalities, and the remainder should be examined for soft tissue anomalies; rabbit fetuses should be examined for both skeletal and visceral anomalies.

Developmental toxicity includes toxic effects on the fetus from exposure of the parent to chemicals before conception; therefore, maternal animals may be treated before conception. Also, in some studies effects of exposure of young animals to chemicals during the period before weaning is of concern. Furthermore, in other studies developmental effects on function may not be detectable if fetuses are sacrificed before delivery. Thus, in some studies females are allowed to deliver and neonates are observed for periods up to adolescence.

The manifestations of developmental toxicity are assumed to represent a continuum of effects. Thus, any of the manifestations of developmental toxicity are considered to be indicative of an agent's potential for disrupting development and constituting a developmental hazard. An adverse effect in animal studies is assumed to indicate a potential risk for humans following exposure during development. The types of developmental effects seen in animal studies are not assumed to be the same as those produced in the human. Because of species-specific differences in developmental patterns or mechanisms of action, the expression of developmental toxicity may vary across species. In fact, humans have been found to be the most sensitive species to developmental toxicants. Therefore, the most sensitive animal species is used to estimate the risk to humans, unless data are available to suggest a more appropriate species.

The level of concern is greatest when developmental effects occur at doses below those producing maternal toxicity, but developmental effects in the presence of maternal toxicity are not assumed to be secondary to the maternal effects and should not be discounted. Current information is inadequate to assume that developmental effects at maternally toxic doses result only from maternal toxicity; rather, when the LOAEL is the same for the adult and developing organism, it may indicate that both are sensitive to that dose. Moreover, the maternal effects may be reversible whereas effects on the offspring may be permanent.

Reproductive Toxicity Studies

Reproductive toxicity studies are concerned with the effects of chemicals on gonadal function, conception, parturition, and the growth and development of the offspring. The most common type of reproductive toxicity test is the two-generation reproduction test. The two-generation reproduction test is designed to investigate the adverse effects of a test compound on multiple aspects of reproductive function. In addition to effects on parenteral reproductive processes, the study also may provide information on neonatal mortality, toxicity, and preliminary data on developmental effects. In this type of study, the test compound is administered to the parental generation before mating and throughout gestation and lactation. The size of litters of the first generation should be standardized. The F_1 generation then is administered the test compound from weaning through mating and gestation and lactation of the F_2 generation. The rat is the preferred species, and at least three dose levels and a control group (twenty per sex per dose group) should be used. The range of doses selected should be such that

the lowest test dose does not produce observable toxicity. Animals should be examined daily and all cage-side and pharmacotoxic observations recorded, especially changes in pertinent behavioral aspects. Following parturition, the number of pups, stillbirths, live births, sex, and presence of gross anomalies are determined. Gross necropsy and histopathological examination of the reproductive organs and other organs and tissues should be reported.

Multigeneration studies sometimes provide information on developmental effects as well. Effects seen in fetuses at exposure levels that result in toxicity to the dams should be discussed as should effects observed in fetuses from dams with no overt toxicity.

Other studies, such as subchronic and chronic toxicity tests, that examine gonad pathology, also provide information on the reproductive effects of test materials. Sperm evaluations also play a role in assessing reproductive toxicity. Ideally, these studies should be able to identify a NOAEL and a LOAEL for reproductive effects. The adequacy of two-generation reproduction studies can be judged on a number of criteria such as:

- At least 20 males and a sufficient number of females to yield at least 20 pregnant females/dose should be used.
- The high dose should cause some toxicity but not high mortality in parental animals. The low dose should be without observed adverse effects.
- Animals should be observed daily, with special emphasis on recording abnormal mating and parturition.
- At necropsy, reproductive organs (vagina, uterus, ovaries, testes, epididymides, seminal vesicles, prostate, pituitary, and other target organs) should be examined in all P_1 and F_1 animals.
- Histopathology should be performed on all high dose and control P_1 and F_1 animals that were selected for mating. Organs demonstrating pathology, either at gross necropsy or in histopathological examination of P_1 and F_1 animals, should be microscopically examined for all dose groups.

Neurotoxicity Studies

Neurotoxicity is an adverse effect on the neurological system (i.e., brain and nervous system). A number of tests of neurotoxicity can be performed for chemicals which produce signs of neurotoxicity in animals. These include studies for which EPA has devised protocols and many other tests of specific nervous system-related endpoints. The tests for which EPA has guidelines include the functional observational battery, motor activity test, neuropathology, delayed NTE neurotoxicity assay, operant behavior, acute delayed neurotoxicity, subchronic delayed neurotoxicity, and peripheral nerve function.

The functional observational battery describes a checklist of symptoms used to describe possible neurotoxic effects. In these studies animals are administered test materials and are observed for changes in behavior, physiological symptoms of nervous system excitation or depression, and sensory function as it relates to visual, auditory, and pain reflexes. Eight animals per dose and controls are required for this assay.

The motor activity test examines the response of animals on an activity monitor to administration of test material. Activity may be measured after acute or subchronic exposure and both control and positive controls are necessary for this test.

The neuropathology referred to in the EPA guidelines refers not to a particular test but to the extended pathology protocol that may be used in other studies (i.e., subchronic toxicity) when neurotoxic effects are suspected. This protocol calls for more thorough examination of nervous system tissue than normally is undertaken in routine histopathology. Unless neurological symptoms suggest otherwise, the following samples are examined histologically: forebrain, center of cerebrum, midbrain, cerebellum and pons, medulla, spinal cord at C3–C6 and L1–L4, Gasserian ganglia, dorsal root ganglia, dorsal and ventral root fibers, proximal sciatic nerve, and tibial nerve.

The operant behavior test examines the ability of chemicals to affect schedule rein-forced behavior. In these tests animals are trained to respond to a schedule of reinforcement and the effects of chemicals on the response rate are measured. For each dose, six to twelve animals are used.

The acute delayed neurotoxicity test examines the ability of a single dose of chemical to produce behavioral changes, ataxia or paralysis in hens. After twenty-one days of observation, animals are sacrificed and sections of nervous system tissue (medulla, spinal cord, and peripheral nerves) are examined microscopically. If no response is observed in the first twenty-one days, dosage is repeated and animals are observed for an additional twenty-one days. At least six hens per dose should survive the observation period. Positive controls are used in this test. In the subchronic delayed neurotoxicity test, hens are administered test material for at least ninety days and are observed for signs of ataxia. At termination of the assay the nervous system tissue sections specified above are taken and subjected to microscopic examination. For these tests, ten hens per dose are necessary as is a positive control group.

Peripheral nerve function is assayed by measuring nerve conduction velocity of both sensory and motor nerves. For these tests, hindlimb or tail nerves are chosen and electrodes are placed with respect to landmarks that are constant across animals. Motor nerve conduction velocity is determined from the latency from stimulation to the observation of an action potential in muscle for stimulation at two sites along the nerve. The distance between the sites of stimulation is divided by the differences in latencies to give the conduction velocity. Sensory nerve conduction velocity is determined similarly, with the exception that latencies are determined from the observation of a somatosensory evoked potential recorded from the scalp. These studies require twenty animals/sex/dose and concurrent controls.

Genotoxicity Studies

Genotoxicity studies provide evidence of the effects of chemicals on chromosomal material. These studies may demonstrate evidence of gene mutations, abnormalities of chromosomal structure, or changes in DNA replication or repair. In addition, there are other alterations, such as effects on the spindle apparatus or changes in the cell cycle, that often are considered in conjunction with assays of damage to the genome. These changes, often called epigenetic effects, frequently indirectly affect chromosomal integrity or DNA replication and, thus, may produce the same sequence of events as a more direct genotoxic agent.

The use of genotoxicity data in human health assessment must rely heavily on epidemiological and clinical case report data. The correlation of genotoxicity with a recognized health outcome is not always easy. The occurrence of increased chromosomal aberrations in populations exposed to ionizing radiation, for example, has been linked with an increased occurrence of cancer. In fact, increased morbidity or mortality from cancer is one of the primary effects that is suspected following the identification of a human population exposed to a known genotoxic agent. This is because there is a moderately high correlation between the mutagenicity and carcinogenicity of many agents. For chemical exposures, the use of genetic toxicity assays to verify potential carcinogenicity is based on the somatic mutation theory of cancer and work that indicates that most chemical carcinogens are electrophilic in nature.

Ten years ago, the correlation between carcinogenicity results in rodent assays and results of short-term mutagenicity tests was about 90 percent. Now it is nearer 60 percent. The reasons include different priorities for choosing chemicals for testing. Previously, suspicion of carcinogenicity was most important and usually was based on rodent data or human occupational epidemiology data. Increased emphasis is now placed on production volumes and numbers of people occupationally or environmentally exposed. It seems likely that this shift in emphasis has resulted in chemicals being identified as rodent carcinogens that are not electrophilic and, thus, not positive on the standard genotoxicity assays. It is clear that not all carcinogens operate by a genotoxic mechanism,

so that negative mutagenicity data do not prove negative carcinogenicity. Nonetheless, positive genotoxicity data provide supportive evidence for carcinogenic potential for many substances and are considered in the overall weight of evidence for this important health endpoint.

Unlike the relationship between chromosome aberrations and increased cancer incidence, sister chromatid exchanges are more of an enigma in terms of human health effects. These chromosome effects have been shown to be associated with heavy cigarette smoking, with some chemical exposures, and in some industrial occupational settings. However, the mechanisms by which these exchanges are formed are not clear at this time. Specific detrimental health outcomes in populations with high levels of sister chromatid exchanges also have not been demonstrated to be causally related. Interesting data correlating the occurrence of increased sister chromatid exchanges with polycyclic aromatic hydrocarbon adducts in DNA in coke oven workers offers the possibility of elucidation of a mechanism for the process as well as a chance to link the chromosomal effect with specific health outcomes in workers known to be at increased risk of lung cancer and other respiratory ailments.

The recognition that there are chromosomal sites showing increased probability of breakage, either spontaneously or following exposure to genotoxic agents, has led to studies that attempt to associate specific health outcomes in individuals with particular distributions of these fragile sites. Data that show an association of increased chromosomal fragility at a specific site to some environmental exposure, in conjunction with data that show an association for that chromosomal marker with some deleterious health effect, would provide evidence for a health outcome associated with a genotoxic effect. For example, exposure to bis-chloromethyl ether (BCME) is known to produce a statistically significant increase in small cell carcinoma of the lung. This cancer also is associated with increased breakage at the fragile site on the fourteenth band of the short arm of chromosome 3. There are studies under way looking for evidence of increased breakage at 3p14 in human cells in vitro following exposure to BCME. Other environmental toxicants may be shown to affect specific chromosomal sites. As gene mapping progresses, the health effects that might be expected to occur from disruptions at particular sites will be identified. Evidence for increased deleterious health outcomes then could be sought through epidemiological methods in high-risk populations.

There are several known genetic diseases that affect DNA repair and replication. Much work has been done to elucidate the health outcomes in individuals affected with xeroderma pigmentosa, ataxia telangiectasia, and Fanconi anemia. It appears that among the clinical pictures associated with these diseases, an increased risk of cancers also is found. Genotoxicants from occupational or other environmental exposures that affect DNA repair and replication can be expected to affect the susceptibility of individuals to the effects of cancer initiators or promoters.

Genotoxic agents also can be expected to affect more than susceptibility to cancer. Reproductive and developmental effects also may result from alterations in the genetic material. Developmental effects may result from changes in the somatic genome of the developing fetus. These changes may result from mutations that affect either structural development or timing and sequence of development. Mutations that occur in germ cells either may affect the ability to reproduce or, in the case of mutations in viable germ cells, may result in heritable changes that may be deleterious to the offspring. Positive genotoxicity data provide supportive data for the occurrence of reproductive effects of an agent. Studies of reproductive or developmental effects in animals associated with exposure to genotoxicants also must be evaluated in light of the possible extrapolation to humans. Comparison of the observed effects in animals to epidemiological or case report data in humans putatively or definitely exposed to some toxic substance can be assessed for the evidence of the effect in humans.

For the human health assessment of genotoxic agents, the first step is an analysis of the evidence bearing on a chemical's ability to induce some genotoxic effect in vitro with either human or animal cells. Some highlights of genotoxicity testing are:

- Positive controls should be run concurrently with assays to test the detection sensitivity. Negative controls also should be included.
- Historical rates of mutation for each test system should be compared with the background mutation rates.
- In microbial assays, the highest dose should elicit some signs of toxicity.
- Replicate experiments should be performed.
- For bone marrow cytogenetic studies, the observations should be timed to cover two cell cycles.

Once there is confirmation of the mutagenic potential of the agent in question, evidence should be obtained that associates this potential with the development of adverse health effects. Toxicity data can be obtained from animal studies that indicate that the substance also causes toxic effects in animals known to be reasonable surrogates for humans. Studies that are conducted with routes and durations of exposure that are particularly relevant to the human situation are the most desirable. Lacking appropriate studies having been conducted, some extrapolation across routes or duration may be permitted. The mechanism relating the genotoxic effect (e.g., altered DNA repair enzymes or chromosome aberrations) should be elucidated and, if possible, described in relation to the systemic health effects noted in the exposed groups.

Structure–Activity Relationship Analysis

Inherent in the hazard evaluation of toxic air pollutants is the fact that there very likely will be little or no information on the toxicity of many of these substances. If assessments are required nonetheless, structure–activity relationship (SAR) and quantitative structure–activity relationship (QSAR) analyses may serve to predict the potential hazard posed by these toxic air pollutants. SAR and QSAR are used to predict the biological activity of an untested compound based on an examination of chemically analogous substances with well-documented health or environmental effects data. SAR and QSAR have been important tools in guiding the development of new drugs and pesticides for many years. EPA routinely uses SAR in assessing new chemicals before their manufacture or importation under the Pre-manufacture Notice (PMN) Program of the Toxic Substances Control Act.

In performing SAR analysis, a stepwise approach generally is employed, although it may not be possible to conform strictly to this hierarchy of analysis for some chemicals. First, SAR analysis usually begins with a considera-tion of the whole molecule. The structure of the substance is compared to a class (or classes) of well-defined, tested compounds with similar molecular structure and substituents. This com-parison should entail consideration of the lipo-philic (fat solubility), steric (spatial arrangement of the molecule) and electronic (bond strength) properties of the molecule. These factors are important because they deter-mine whether the molecule, following expo-sure, crosses biological membranes to reach various target sites. The electronic and steric factors also influence whether a molecule will interact with a substrate (i.e., enzymes, DNA, RNA, and other macromolecules).

The second step involves an analysis of how the functional groups or substituents of the compound being examined influence the reac-tivity of the molecule. For example, it is well known pharmacologically that compounds con-taining fluorine have increased stability in the body because of the difficulty of breaking the carbon–fluorine bond. This prolongs the time needed for the body to break down the bond. Thus, if the parent compound is biologically active, the presence of fluorine in the molecule will prolong its activity in the body. Alterna-tively, if the parent compound is inactive and must be metabolized to an active form, the presence of fluorine in the molecule will delay this biotransformation and the inactive fluori-nated parent compound often will be excreted unchanged.

The final step in employing SAR analysis usually involves a consideration of how the parent compound will be metabolized by the body. This is an important aspect of the SAR process because, in some cases, the metabolic transformation of the parent compound converts it from an inactive substance into a more active chemical. Here, one must consider how the enzymes present in the body interact with different functional groups. Steric hindrances, such as the presence of chlorine instead of hydrogen, can make a substance slow to react with enzyme systems because of changes in the size and shape of the molecule. This prolongs the presence of the parent compound and makes the biotransformation of the parent compound into an active metabolite less likely.

Although SAR is a useful tool for predicting biological and toxicological activity, it has substantial limitations. For example, it is diffi-cult to incorporate both positive and negative toxicity testing results of the analog into the SAR. The selection of analogs represents a compromise between optimal structural sim-ilarity to the compound of interest and selection of analogs for which there is a good likelihood of finding biological test data. Although analo-gies can be made to compounds with conflicting toxicity testing results (i.e., both positive and negative outcomes), it is difficult to determine how to incorporate the negative results in the SAR. Thus, the negative data often are ignored in the SAR, which could lead to an over-statement of the unknown compound's toxicity. The discounting of negative data is a problem common to any predictive procedure, including risk assessment. Risk assessment guidelines currently make no provisions for including negative data in the evaluation of a compound's potential toxicity. A consistent approach for evaluating and considering negative data must be devised so that the entire database for a compound can be used when employing pre-dictive procedures such as SAR analysis.

A second limitation of classical SAR analysis concerns the ability to predict the biological activity of a compound based solely on an analysis of isolated portions of the compound (i.e., its substituents). In fact, the action of the compound depends not only on its chemical properties but initially on its physical properties and the implications of these properties for target cell exposure. From a pharmaceutical point of view, for example, it is apparent that the lipophilic, electronic, and steric properties of a molecule influence its crossing of biological membranes and its disposition within the body. These physical parameters also influence the likelihood of its contact with target receptors such as enzymes and DNA.

A third limitation associated with SAR analysis is that its ability to predict different endpoints of toxicity varies. For example, SAR analyses have been employed successfully in predicting carcinogenic activity, hepatotoxicity, determining the potency of enzyme–substrate interactions, and in predicting mutagenic activity and carcinogenic activity based on information gleaned from the Ames test (an in vitro test using bacteria) for certain chemical classes. However, it has been less successful in predicting developmental or reproductive effects and the mutagenicity and carcinogenic potential of test chemicals using genotoxic tests other than the Ames test.

ROUTE OF EXPOSURE

Available toxicity studies on toxic air pollutants or related analogs may have employed a route of exposure that is not relevant to the expected route of human exposure to the air pollutant (i.e., inhalation exposure). In such instances, the feasibility of performing a route-to-route extrapolation should be explored. A primary consideration in attempting a route-to-route extrapolation is the possibility of portal of entry effects. For example, if the only available data are by the oral route of exposure, and the expected human route of exposure is inhalation, one must consider whether the possibility exists for adverse respiratory effects to occur when the substance is inhaled. If no effects on the gastrointestinal tract have been reported in oral studies, it may be feasible to presume that portal of entry effects do not occur. However, there generally is no firm basis to rule out respiratory system effects. Secondly, to perform a route-to-route extrapolation, one must have quantitative data on the pharmacokinetics (particularly absorption efficiency) of the substance for both routes of exposure. If the toxic air pollutant has a low water solubility, it is not possible to speculate on the extent of its absorption by the lungs. Therefore, a lack of data on pharmacokinetics and possible portal-of-entry effects would preclude conducting a route-to-route extrapolation. If it is determined that route-to-route extrapolation is not possible, then further testing may be necessary before determining the potential hazard of the toxic air pollutant.

DETERMINATION OF THE WEIGHT OF EVIDENCE

There are three major factors to be considered in characterizing the overall weight of evidence for toxicity: (1) the quality of the evidence from human studies, (2) the quality of evidence from animal studies, that are combined into a characterization of the overall weight of evidence for human toxicity, and (3) other supportive information, that is assessed to determine whether the overall weight of evidence should be modified. The process for establishing the weight of evidence for the noncarcinogenic effects of a chemical may not be as well characterized into a group of classification system as is that for carcinogens; however, the approach to developing a weight of evidence for noncarcinogens should encompass similar considerations.

Human Data

All human data available for use in risk assessment must be evaluated on its own merit and in the context of how it can augment the weight of evidence derived from the available toxicological and physiological information. The major categories of data include case reports, geographical correlation studies, cross-sectional studies, clinical studies, and cohort studies. There is a much larger body of human data in the literature for inhalation exposures than for ingestion because of occupational exposures where the exposure route primarily is inhalation. These data can be valuable in augmenting the information from animal studies for pulmonary endpoints, that tend to provide information on pathology, whereas human data tend to report subjective symptoms and pulmonary function.

Animal Data

In cases where there is inadequate epidemiological information, and this is the case for most toxic air pollutants, it is necessary to use animal data to infer that a human hazard might exist. Animal bioassays have come to play a central role in hazard identification. For example, the fact that a chemical causes liver toxicity in laboratory animals suggests that it also might have the potential for causing this effect in humans, even though relevant epidemiologic data do not exist or the data that do exist do not firmly establish such a relationship.

When based on animal data, the discussion of the weight of evidence for a chemical posing a potential hazard to human health should include a comprehensive evaluation of experimental data in several species and variety of endpoints. The end points include systemic toxicity following acute, intermediate, or chronic exposure by the several routes of exposure, developmental and reproductive effects, mutagenicity, and carcinogenicity. These data are used to elucidate potential mechanisms of biological activity and characterize the chemical's spectrum of potential human toxicity by identifying target organs and the dose ranges associated with adverse effects in animals. This evaluation should be made with particular reference to the reliability of the experimental design and toxicological interpretation of the results. In addition, consideration should be given to studies designed to evaluate the metabolism and toxicokinetics and to data from other studies that may elucidate mechanisms of action of the chemical.

Once the data have been critically reviewed, all the results from the various studies should be examined collectively to determine if a causal relationship exists between chemical exposure and the observed effects and also to evaluate the quality, consistency, and biological plausibility of the observed effects. In general, the strength of the overall evidence is enhanced when any the following occur.

- Similar effects are seen across sex, strain and species, or similar effects are observed by different investigators.
- There is a well-defined dose-response relationship.
- A plausible relationship exists between the observed effect and known information about the metabolism of the compound and the mechanism of action.
- Similar effects are observed in structurally similar compounds.
- There is some evidence that the chemical also causes the particular effects.

However, evaluation of this evidence when the route of exposure is inhalation may not be as straightforward as evaluating the evidence with the oral route of exposure.

When comparing results across species, in addition to species-specific sensitivity to the toxicity of a chemical or to known species-specific differences in metabolism to the putative toxicant and biodistribution to the target tissue, species differences in the anatomy and physiology of the respiratory system should be considered. When the route of exposure is the inhalation route, it is important to have a thorough understanding of the complexity of the respiratory system and its diversity across species. Differences in the anatomy and physiology of the respiratory system play an important role in the evaluation of any dose-response relationship. Transport and uptake mechanisms of inhaled toxicants and disposition and retention patterns of particulate toxicants are influenced by the anatomy and physiology of the respiratory system. Therefore, if a clear dose-response relationship is not apparent and cannot be fully explained by presumed metabolic and biodistribution processes of the absorbed chemical, then differences in respiratory function may need to be considered. For example, depending on the size, a particle may not deposit into the small, deep areas of the lung but may be moved by the mucociliary apparatus and eventually ingested, thereby introducing an oral exposure component to the inhalation exposure. In addition, for some particle toxicants the effect may be related more to the amount deposited in a particular area and the clearance rate than to the exposure concentration.

One of the key elements in the evaluation of toxicological data for inhalation exposure and the determination of weight of evidence is the evaluation of conflicting results. It is important to discern differences that are the result of study limitations and experimental design from those that are related to species-specific differences in respiratory function. Therefore, similarities and differences in experimental data should include consideration not only of possible metabolic and pharmacokinetic differences (some of which are route-specific) but also the differences in dynamics of the respiratory system.

Despite the apparent validity of the acceptability by most scientists of the use of animal data to determine the weight of evidence for chemical toxicity, there are occasions in which observations in animals may not be relevant to humans. However, unless there are data on human toxicity that refute a specific finding of toxicity in experimental animals or unless there are other biological reasons to consider certain types of animal data irrelevant to humans (for example data showing metabolism differences), it generally is assumed that toxicity to humans can be inferred from observations in experimental animals.

HEALTH INFORMATION SOURCES

Major elements of the health assessment process are a literature search and acquisition of relevant studies, identification of potential health hazards, including a comprehensive and critical review of experimental data in animals and epidemiological data from human studies, a determination of the weight of evidence that includes an assessment of the quality, consistency and reliability of the data. This section describes some of the key information sources necessary to conduct health assessments.

Information Sources

Two critical sources of information are technical libraries and other technical organizations.

Table 5-1   Important medical and scientific libraries[a]

Medical or scientific library	Location
Library of Congress	Washington, D.C.
Environmental Protection Agency	Washington, D.C.; Research Triangle Park, NC; Cincinnati, OH
Food and Drug Administration	Rockville, MD
National Center for Health Statistics	Hyattsville, MD
Centers for Disease Control	Atlanta, GA
National Institutes of Health	Bethesda, MD
National Academy of Sciences	Washington, D.C.
Harvard University Medical School	Cambridge, MA
Duke University Medical School	Durham, NC
Johns Hopkins Medical School	Baltimore, MD
New York Academy of Medicine	New York, NY
New York University Medical Center	Tuxedo, NY
University of California at Davis	Davis, CA
University of California at Los Angeles	Los Angeles, CA
British Library Lending Library	W. Yorkshire, England

[a] This listing contains examples only and is not intended to be complete.

Technical libraries at universities and at federal government departments and agencies provide some of the most complete medical and scientific library collections in the world. Membership in the Interlibrary Loan Network provides further access to hundreds of public, private, and academic libraries throughout the nation. Table 5-1 lists a number of important medical and scientific libraries available through selected reference acquisition vendors and the Interlibrary Loan Network. Tables 5-2 and 5-3 provide available toxicological, environmental, and regulatory data bases. Some federal and private organizations that frequently are valuable sources of toxicological information are listed below.

- Environmental Protection Agency
- Food and Drug Administration
- National Institutes of Health
- National Academy of Sciences
- Occupational Safety and Health Administration
- Pan American Health Organization

Table 5-2 Major data base systems

Data base system	Source
MEDLARS	National Library of Medicine
TOXNET	National Library of Medicine
DIALOG	Dialog Information Services, Inc.
CIS	Chemical Information Systems, Inc.
STN/CAS Online	STN International
NUMERICA	Mead Data General
LEXIS/NEXIS	Mead Data General
NATICH	Environmental Protection Agency
TSCATS	Toxic Substances Control Act Test Submissions (EPA)
ORBIT	System Development Corp.
National Pesticide Info. Service	Purdue University

Table 5-3 Bibliographic on-line data bases

Data base	Source[a]				Information type[b]											
	A	B	C	D	1	2	3	4	5	6	7	8	9	10	11	12
Agricola			x	x			x		x	x	x	x	x	x		
Analytical Abst.			x	x		x	x			x		x				x
BIOSIS Reviews			x							x	x	x	x	x	x	
CANCERLIT	x		x							x	x	x		x		
Chemical Exposure			x		x	x				x	x	x	x	x	x	
COMPENDEX			x				x	x	x	x	x					
Dissertation Abst.			x		x	x	x	x	x	x	x	x	x	x	x	
EMBASE			x							x	x	x		x		
ENVIRONLINE			x	x						x	x	x	x	x	x	
MEDLINE	x		x							x	x	x		x		
NTIS			x	x			x	x	x	x	x	x	x	x	x	
Pharm. News			x							x	x	x				
Pollution Abst.			x							x	x			x		
SCISEARCH			x		x	x	x		x	x	x	x	x	x	x	
TOXLINE	x									x	x	x	x	x	x	
TOXLIT	x									x	x	x	x	x	x	

[a] A = MEDLARS; B = CIS; C = DIALOG; D = ORBIT.
[b] 1 = Chemical structure; 2 = Nomenclature; 3 = Chemical/physical properties; 4 = Production data; 5 = Use data; 6 = Environmental fate and transport; 7 = Exposure data; 8 = Pharmacokinetics/metabolism data; 9 = Health effects data; 10 = Environmental effects data; 11 = Risk assessments; 12 = Analysis methods.

- World Health Organization
- Consumer Product Safety Commission
- National Institute for Environmental Health Sciences
- Chemical Industry Institute of Toxicology
- Agency for Toxic Substances and Disease Registry

Literature Search Strategies

A typical literature search strategy is presented graphically in Figure 5-1. It involves the following elements:

- Computerized data bases are used to identify bibliographic and nonbibliographic data, including ongoing research.
- Manual searching is used to supplement the computerized search. The emphasis of the manual search should be to identify important older literature and very recent literature not generally covered by computerized data bases.
- Personal and professional contacts within various federal and state agencies, academic institutions, research organizations, trade associations, and consulting firms are used to identify special reports, unpublished studies, and current ongoing research.

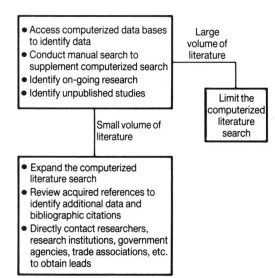

Figure 5-1.   Typical literature search strategy.

In addition, the literature search strategy can be modified according to the volume of literature that is available. The individual components of the literature search strategy, modifications to the strategy and the procedures for documentation and approval of the strategy are described below.

Computerized Literature Searches

Information specialists trained in computerized literature retrieval systems typically design search strategies for particular substances using appropriate search terms suggested by scientists. The specific search terms used may vary according to the chemical, subject, and data base. The development of an integrated literature search strategy involves selection of appropriate data bases, use of specific search terms such as chemical registry number, chemical names, synonyms, text words, medical subject headings, and use of specialized vocabulary adapted for each data base. Information specialists ensure that the search terms are compatible with the vocabularies of each of the automated systems to be searched. The computerized search strategy always should be saved in the computer memory to facilitate updating searches.

Important public (government) and private data base systems are shown in Table 5-3. These data base systems in turn provide access to literally hundreds of detailed data bases and files. The key databases include MEDLINE and its backfiles, TOXLINE, TOXLIT, TOXLIT 65, CANCERLIT, HSDB, RTECS, CCRIS, IRIS, EMIC, ETIC, NTIS, NIOSHTIC, EMBASE, GIABS, SUSPECT, and CHEMICAL EXPOSURE. Table 5-4 provides a sample list of nonbibliographic on-line data bases that typically may be used to acquire information. The general types of data available within each data base also are identified.

A manual search often is an essential component of a total literature search to ensure all relevant literature is identified and retrieved. Because of the high labor costs typically associated with manual searches, however, the extent of manual searching needs to be flexible.

Table 5-4   Nonbibliographic on-line data bases

Data base	Source[a]				Information type[b]										
	A	B	C	D	1	2	3	4	5	6	7	8	9	10	11
CCRIS	x	x												x	
CHEMDEX			x		x	x									
CHEMFATE		x					x			x					
CHEMLINE	x				x	x									
CHEMNAME			x			x	x								
CHEMSEARCH			x		x	x									
CHEMCIS			x		x	x									
CHEMZERO			x		x	x									
CTCP		x					x	x		x		x			
DERMAL		x									x	x	x		
GENOTOX		x											x		
HSDB	x					x	x	x	x	x	x	x	x	x	x
MERCK		x			x	x	x		x			x	x		
OHM/TADS		x				x	x	x	x	x	x	x	x	x	x
RTECS	x	x				x					x		x	x	
SANSS		x			x	x									

[a] A = MEDLARS; B = CIS; C = DIALOG; D = ORBIT.
[b] 1 = Chemical structure; 2 = Nomenclature; 3 = Chemical/physical properties; 4 = Production data; 5 = Use data; 6 = Environmental fate and transport; 7 = Exposure data; 8 = Pharmacokinetics/metabolism data; 9 = Health effects data; 10 = Environmental effects data; 11 = Risk assessments.

In some instances, the volume of literature on a particular chemical is substantial. In these cases, it may be more cost-effective to emphasize the computer search and to limit the manual search. In other instances, the volume of literature on a particular chemical may be limited. In these cases, it may be necessary to expand the manual search. Expanded manual searches also may be necessary if there is inadequate coverage in computerized data bases or if there are indications that relevant data are available in earlier journals.

Identification of On-going Research

In order to develop a complete hazard or risk assessment on a chemical, ongoing research should be identified. The sources of information used to identify ongoing research typically include the following.

- Federal Research in Progress data base (DIALOG) and Current Research Information System data base (DIALOG)
- CHEMTRAC data base (National Toxicology Program)

- International Registry of Potentially Toxic Chemical's computerized registry of chemicals currently being tested for toxic effects
- Organization for Economic Cooperation and Development chemical clearinghouse EXICHEM database
- National Toxicology Program yearly publication, 'Review of Current DHHS, DOE, and EPA Research Related to Toxicology'
- Direct contact with researchers, research institutions (e.g., CIIT), and trade associations
- World Health Organization annual publications, including 'Information Bulletin on the Survey of Chemicals Being Tested for Carcinogenicity' and 'Directory of On-going Research in Cancer Epidemiology'

Manual Literature Searches

A limited manual search often is necessary to supplement searches of computerized bibliographic data bases. Computerized data bases tend to be less comprehensive on coverage before 1970 and they tend to be approximately six months behind in current literature. Consequently, the focus of a manual search should be important older literature and very current literature.

Important pre-1970 literature can be identified manually by reviewing indexes of library holdings; Science Citation Index, Chem Abstract Index, or other science indices; primary references; secondary references; books, review articles, and published and unpublished literature reviews; and authoritative secondary sources (e.g., IARC monographs, EPA reports, and NIOSH documents). Very current literature can be identified by reviewing recent issues of selected journals, Current Contents (Agriculture, Biology, Environmental Sciences, Life Sciences) and CA Selects (Carcinogens, Mutagens, and Teratogens; Chemical Hazards, Health and Safety; Environmental Pollution). Both Current Contents and CA Selects are excellent and cost-effective means to close the literature gap between the date an article is published and the date the article is entered into a computerized literature retrieval system.

Search for Unpublished Material

In addition to a search of the published literature, a thorough evaluation of the health effects of a substance also should include identification and retrieval of any relevant unpublished materials. Mechanisms to do this include reviewing documents on chemicals prepared under the direction of:

- EPA – TSCA Interagency Testing Committee (ITC)
- EPA – Office of Toxic Substances (OTS)
- EPA – Office of Pesticide Programs (OPP)
- EPA – Office of Air and Radiation (OAR)
- EPA – Office of Research and Development (ORD)
- National Toxicology Program (NTP)
- National Institute for Occupational Safety and Health (NIOSH)

A second means for identifying unpublished data is to access and review EPA records. These records can include but are not necessarily limited to EPA Reports data base, the Toxic Substances Control Act Test Submissions (TSCATS) data base, and the Pesticide Document Management System (PDMS) data base.

Literature Searches Involving Large Volumes of Literature

For certain substances and situations, the volume of literature could be substantial. In those instances where the number of references numbers into the thousands, the literature search strategy must be flexible in order to limit the search to the most important and relevant articles. Steps typically include:

- Limiting the number of computerized data bases searched (e.g., emphasis placed on searching the National Library of Medicine or Dialog data bases).
- Limiting the number of search terms used (e.g., emphasis placed on using only the CAS number and key chemical terms).
- Limiting the period of coverage (e.g., emphasis placed on literature from 1970 to the present).
- Using search terms and controlled vocabulary identified by a vendor (e.g., the National Library of Medicine publishes a list of Medical Subject Headings, MESH, which can be used to 'keyword' search its MEDLARS system).
- Specifying 'keywords' or 'text words' that must appear in the title, subject, or abstract files of the bibliographic record
- Emphasizing the computer search over the manual search.

Literature Searches Involving a Small Volume of Literature

In some cases, the volume of reference literature may be limited. In the instances where the number of references is small and some are published in obscure journals, the literature search strategy must again be flexible to identify relevant information. This could necessitate expanding the literature search. A computerized data base search can be broadened to access additional data. This may include:

- Expanding the number of computerized data bases searched.
- Expanding the number of search terms used (e.g., use trade names and retired CAS numbers).
- Expanding the period of coverage.
- Conducting 'author searches' (e.g., using a computerized data base such as TOXLINE an author search can identify published papers by a particular investigator; this technique is particularly useful when the investigator is an expert on one of the chemicals of concern).
- Conducting 'backward tree searches' (e.g., using the Science Citation Index a background tree search will identify the background articles cited by a particular author in a published paper).
- Conducting 'forward tree searches' (e.g., using the Science Citation Index a forward tree search will identify various articles that cite a key article in their bibliographies).
- Searching structural analogs and related chemicals with similar chemical and biological properties.

Reference

U.S. Environmental Protection Agency (EPA). 1986c. Cancer assessment guidelines. 51 FR 33992, September 24.

6

Quantitative Methods for Cancer Risk Assessment

Annette M. Shipp and Bruce C. Allen

INTRODUCTION

Many of the chemicals used or produced by modern society may present hazards to human health. Although individuals or governmental regulatory agencies may try to minimize contact with potentially harmful chemicals, it is not always possible to eliminate all sources of chemical exposure. The issue then is not the hazard posed by a chemical, in other words the intrinsic ability of the chemical to cause harm to human health, but instead the risk associated with exposure to that chemical. Risk implies a probability statement, namely a probability that an adverse effect will occur under actual or hypothetical conditions. Risk assessment is the qualitative or quantitative determination of that probability.

Methods have been developed to estimate quantitatively the risk of adverse health effects occurring in humans exposed to hazardous chemicals in environmental or occupational settings. For example, the U.S. Environmental Protection Agency (EPA) developed guidelines for cancer risk assessment (EPA 1986c) based on the model proposed by the National Academy of Sciences (NAS 1983). In general, assessment of risk involves the estimation of the probability that an individual will develop a specified adverse effect as a result of exposure

to the specified chemical. Risks can be expressed relative to the baseline or background risks, as absolute or added risks per unit time, or as lifetime risks (Gilbert et al. 1989). Estimation of the lifetime risk of developing cancer is the most commonly used risk measure (EPA 1986c).

The procedure currently used by the EPA and other regulatory agencies to assess lifetime cancer risk assumes that exposure to a potential cancer-causing agent (i.e., carcinogen) is continuous and constant over the lifetime of the individual (EPA 1986c; NAS 1983; OSTP 1985). Therefore, estimates of lifetime human cancer risk are based on two interrelated factors – exposure and potency. The potency of a chemical is defined as the plausible upper-bound estimate of the probability of a response per unit intake of a chemical over a lifetime. The potency of the chemical is established through a qualitative evaluation of epidemiological or experimental data, as discussed in Chapter 5, and a quantitative estimate of the amount of chemical that may be expected to produce harm. This chapter considers the quantitative evaluation of the potential of developing cancer as a result of exposure to a chemical, in other words the relationship between the dose of an agent administered or received and the incidence of cancer in exposed populations (NAS 1983).

Estimates of the potency of a chemical usually are based on the results of animal bioassays and the dose-response information then is extrapolated to estimate the shape of the dose-response curve at the lower levels of exposure to which humans may be exposed. Although little is known about the true shape of the dose-response curve at low doses, two important assumptions are made, both of which are based on the understanding of the cancer process, and discussed in more detail below. First, it is assumed that the dose-response relationship is linear, or at least does not exceed linearity, in the low-dose region. Second, it is assumed that any dose, no matter how small, carries with it a probability that cancer may occur.

The purpose of this chapter is to describe the methods and techniques used to assess quantitatively the carcinogenicity data and to develop cancer dose-response relationships. This provides estimates of a chemical's potency and, as a result, the estimated risks associated with exposure to the chemical in the environment.

OVERVIEW OF CHEMICAL CARCINOGENESIS

Cancer is a broad term for a group of diseases distinguished by the uncontrolled proliferation of abnormal cells. Cancer is thought to be a multistage process in which cells progress through a series of stages with cells in the final stage recognizable as tumor cells (Armitage and Doll 1954; Farber 1982; OSTP 1985).

A distinguishing feature of carcinogenesis is that it appears to develop through molecular biologic processes in which a small number of rare events are required in a highly localized area to begin the process. This process is thought to consist of at least three distinct steps referred to as initiation, promotion, and progression. Although these steps can be defined operationally, the distinctions between the steps are not clear-cut and their exact mechanisms are not understood with certainty. There are numerous factors that may control and modify the carcinogenic process, including species, strain, sex, DNA repair, and hormonal, immunological, and mutational status.

Initiation

The first step in the multistage process has been termed initiation; this is thought to involve a heritable change in the genetic material (genome) of the cell. This change may occur by a direct, genotoxic mechanism that results in a mutation to cellular DNA or in other chromosomal damage. Numerous chemicals either directly or after metabolic activation form electrophilic compounds that interact with DNA causing the formation of DNA-adducts that can result in miscoding or other mutations at the time of DNA replication (Miller and Miller 1976). Chemicals also can cause other genotoxic effects (e.g., translocations, sister chromatid exchanges, and deletions) that may lead to transformation of cells (Hart and Turturro 1988).

It is postulated that the carcinogenic process also can be initiated by an indirect, epigenetic mechanism that produces alterations in normal gene regulation or expression. Epigenetic alterations may occur through hormone-like effects or by the action of other cellular macromolecules, such as enzymes, that control gene replication or expression. This initiation or transformation may occur spontaneously by some unknown mechanism, as is the case with endogenous cancers (i.e., those occurring without an identifiable outside cause) or age-related cancers, or it may be caused by the action of exogenous chemicals (i.e., chemicals introduced into the body). Initiation is assumed to be irreversible; in other words, initiated cells will not spontaneously revert after cessation of exposure to the chemical.

Promotion

It is assumed that alterations in genetic material alone are not enough to result in cancer. The probability of a genetic alteration becoming fixed as a heritable change is dependent on the relative rates of repair and cell division (Lutz 1990). The next stage in the carcinogenic process, then, is the production of new, altered cells through cell proliferation. This is called promotion. DNA replication and subsequent cell division may result in the formation of a transformed cell clone. Any process or agent that accelerates the rate of cell division, thus enhancing the formation of transformed cells and tumor production, has been termed a promoter. The mechanism by which promotion occurs is not understood with certainty but, as with the initiation step, promotion could occur by either genetic or epigenetic mechanisms. Promotion can occur at more than one stage in the carcinogenic process and certain stages in certain cell lines may be more responsive to promotional actions. Promotional effects also may have an impact on the rates of cell proliferation or on rates of cell death, both of which influence the risk of cancer development. The term promoter has been used to characterize chemicals that generally do not act on DNA and, thus, have weak or no carcinogenic activity by themselves, but which enhance the carcinogenic response when applied after exposure to an initiator has occurred (EPA 1987f). This definition may be both too limited (i.e., in the sense that it has been applied only to initiated cells) and too general (i.e., in the sense that a wide variety of mechanisms may lead to a promotional response).

Progression

The final stage in the multistage process – progression – is largely undefined. In this stage, transformed cells form recognizable tumors and may become malignant with the capacity to metastasize (OSTP 1985). The progression through these stages may occur spontaneously as in the case of endogenously-caused tumors or age-related tumors, or a cell may be transformed through chemical insult and then progress through a series of stages to cancer production.

QUANTITATIVE METHODS FOR CANCER RISK ASSESSMENT

Interpretation of experimental data and the determination that a chemical is or is not a carcinogen is a complex process that requires careful evaluation of experimental data gathered by scientists from several different disciplines, including statistics, toxicology, epidemiology, and pathology. OSTP (1985) stated that evaluation of the carcinogenicity of a chemical should consist of an evaluation of a variety of biological information including carcinogenicity bioassays, mutagenicity studies, and other toxicological studies. With carcinogenicity data in particular, evaluation of the statistical significance must be conducted in combination with the biological significance, and the relevance of the experimental findings to human health must be considered.

In a small number of instances, epidemiological data are suitable for quantitative estimates of risk and permit a dose-response relationship to be developed directly from human data. The strengths and weaknesses of various types of epidemiological studies should be considered, as well as their applicability for establishing causal relationships and their suitability for inclusion in quantitative risk assessment.

In the majority of cases, however, the available epidemiological studies are inadequate and assessment of human cancer risk is based on animal bioassays. Carcinogenicity bioassays usually are designed as screening procedures with the primary focus being hazard identification, rather than risk assessment. In such studies, a limited number of animals may be exposed to a maximum tolerated dose that is several factors of ten higher than that encountered by humans. That being the case, two extrapolations are necessary to convert the animal data to appropriate human risk estimates – the first from animals to humans and the second from high experimental doses to the lower doses encountered by humans. These are discussed below.

Animal to Human Extrapolation

Animal to human extrapolation includes the general problem of human and animal equivalence. In other words, the appropriate route, species, tumor type, and dose units (i.e., those that provide an adequate model of human carcinogenicity) are not always known with certainty. When several bioassays of a chemical are available, it is necessary to select for analysis those experiments that are most appropriate for making quantitative estimates.

Toxicological and statistical considerations that apply in that selection are shown in Table 6-1. Features of the experiments, such as experimental design and execution, clinical and pathological examination, animal care, and test chemical moiety, are basic factors that determine their toxicological relevance. Pattern and number of dose groups, sample size, and length of dosing or observation are aspects relating to the statistical evaluation of a bioassay.

Ideally the process of selecting data from among the various available bioassays, in which different species, strains, sexes, or routes of administration may have been tested, should maximize the biological correlations between

Table 6-1   Selection of data to use in dose-response modeling

Data	Selection factors
Length of experiment	Use data from any experiment but correct for short observation periods.[a]
	Use data from experiments which last no less than 90 percent of the standard experiment length of the test animal.
Length of dosing	Use data from any experiment, regardless of exposure duration and correct for length of dosing.[a]
	Use data from experiments that expose animals to the test chemical no less than 80 percent of the standard experiment length.
Route of exposure	Use data from experiments for which route of exposure is most similar to that encountered by humans.
	Use data from any experiment, regardless of route of exposure.
	Use data from experiments that exposed animals by gavage, inhalation, or any oral route.[a]
Units of dose assumed to give human–animal equivalence	mg/kg body weight/day
	ppm in diet, ppm in air
	mg/kg body weight/lifetime
	mg/m^2 surface area/day[a]
Animals to use in analysis	Use all animals examined for the particular tumor type.[a]
	Use animals surviving just prior to discovery of the first tumor of the type chosen.
Malignancy status to consider	Consider malignant tumors only.
	Consider both benign and malignant tumors.[a]
Tumor type to use	Use combination of tumor types with significant dose response.
	Use total tumor-bearing animals.
	Use response that occurs in humans.
	Use any individual response.[a]
Combining data from males and females	Use data from each sex within a study separately.[a]
	Average the results of different sexes within a study.
Combining data from different studies	Consider every study within a species separately.[a]
	Average the results of different studies within a species.
Combining data from different species	Average results from all available species.
	Average results from mice and rats.
	Use data from a single, preselected species.
	Use all species separately.[a]

[a] Approach typically used by EPA.

animal species and humans. Specific guidelines for evaluating studies for use in risk assessment were published by EPA (1986c). Studies with suitable dose-response data and that meet statistical and toxicological criteria generally are included in a quantitative risk assessment.

Once particular experiments have been selected for analysis, it is necessary to select the specific tumor responses that are used to estimate a dose-response relationship. Tumor responses that may be considered include: tumors located at sites related to the metabolism, storage, or elimination of the chemical; tumor types identified as associated with exposure to the chemical on the basis of other (particularly epidemiological) studies; and tumors that show a statistically significant dose-related trend or significant increased incidence in treated animals when compared to control. Available information on comparative metabolism, pharmacokinetics, and mechanisms of action should be considered when making the choice of data to use. For example, the kidney tumors that occur in male rats following exposure to certain types of hydrocarbons are considered by some scientists to be secondary to nephropathy (kidney damage) related to accumulation of $\alpha_2\mu$ globulin by a mechanism that may be unique to the male rat (Kanerva et al. 1987). EPA (1987f) recognizes that the renal syndrome and tumor endpoint may have limited significance in relationship to human exposure to specific hydrocarbons and generally does not use this endpoint in quantitative risk assessment. Another example is that a variety of drugs, chemicals, and physiological perturbations can result in the development of thyroid follicular cell tumors in rodents, particularly rats. Tumors are preceded by thyroid toxicity consisting of diffuse or nodular hyperplasia (goiter). The mechanism of tumor development appears to be a secondary effect of longstanding hypersecretion of thyroid stimulating hormone by the pituitary (Capen and Martin 1989). With this mechanism, a threshold level can be recognized and risks to humans would be considered minimal when normal thyroid-pituitary function exists (McClain 1989; Paynter et al. 1988).

Without quantitative information describing differences in metabolism, pharmacokinetics, or pharmacodynamics between animals and humans, quantitative estimates of human cancer risk are based on those tumor responses that show a statistically significant increased incidence in specific organs or tissues. In general, statistical significance for quantal data (i.e., the number of tumor-bearing animals in the test group compared to the control) is determined by either the Fisher exact test (Bickel and Doksum 1977) or the Cochran-Armitage trend test (Armitage 1955). However, there are a number of statistical issues in the analysis and interpretation of animal carcinogenicity studies that should be considered. For example, survival differences among groups should be taken into account in the analysis of tumor incidence data. It may be important to try to distinguish between tumors that contribute to the death of the animal and incidental tumors discovered in an animal dying of an unrelated cause. In the case where individual animal pathology reports or time-to-tumor data are available, statistical significance could be evaluated using an incidental tumor analysis and life table analysis as currently performed by the National Toxicology Program (Haseman 1984).

Although the statistical significance of an observed tumor increase is perhaps the most important evidence used in the evaluation process, a number of biological factors also must be taken into account. Therefore, no rigid statistical decision rule should be employed in the interpretation of carcinogenicity data. Even if a study has been carefully designed and appropriate statistical methodology employed, interpretation of results is a complex process. Carcinogenic responses should be evaluated carefully as to their biological relevance with respect to human carcinogenic risks. Special consideration should be given to the evaluation of rare tumors or to tumors at sites with a high spontaneous background.

In addition to the selection of the appropriate study and tumor response within a study, the appropriate expression of dose must be determined for inclusion in establishing the dose-response relationship. As noted earlier, the procedure currently used by EPA and other regulatory agencies to assess lifetime cancer risk assumes that exposure to a potential carcinogen is continuous and constant over the lifetime of the individual (EPA 1986c; NAS 1983; OSTP 1985). When administration of the chemical to the animal in the experimental protocol differs from continuous, constant lifetime exposure, EPA (1986c) recommends that unless there is evidence to the contrary the appropriate measure of exposure is the total or cumulative dose of the chemical of concern averaged or prorated over a lifetime, resulting in an average lifetime daily exposure. This assumes that a high dose of a carcinogen received over a short period of time is equivalent in terms of tumor development to a corresponding low dose averaged over a lifetime. Adjustments to the experimental dose to account for short duration of dosing, short duration of observation, changed dosing patterns, or other factors are made to transform or convert experimental administered doses to a continuous lifetime dose for use in the dose-response models (EPA 1986c; EPA 1989j).

When exposure is by the inhalation route, two cases are considered when estimating dose (EPA 1989j). First, the carcinogenic agent is a water soluble gas or aerosol and is absorbed in proportion to the amount of air inhaled (EPA 1989j). Second, the carcinogen is a poorly water soluble gas that reaches an equilibrium between the air breathed and the body compartments, in which case, after equilibrium is reached, the rate of absorption is expected to be proportional to the rate of oxygen consumption that in turn is assumed to be a function of surface area (EPA 1989j).

Animal-to-human extrapolation is accomplished further by assuming that animals and humans are equally susceptible (in terms of extra risk) when the dose is measured in the same units for both species. One method for risk calculation is based on the usual EPA assumption that doses measured in units of weight per surface area per unit time, typically milligrams per square meter per day, give equal risks in animals and humans. This is known as the surface area equivalency. The experimental

doses in milligrams per kilogram per day (mg/kg per day) administered to animals are converted to equivalent human average daily lifetime doses in mg/kg per day, based on the assumed surface area equivalency, by the formula

$$D_H = D_A \left(\frac{BW_A}{BW_H} \right)^{1/3} \qquad (6\text{-}1)$$

Where D_H = human equivalent dose, mg/kg per day
D_A = experimental animal dose, mg/kg per day
BW_A = body weight in animals, kg
BW_H = body weight in humans, assumed to be 70 kg

This calculation is based on the assumed surface area equivalency and the assumption that surface area is proportional to body weight to the 2/3 power. EPA (1992l) recently proposed an animal to human conversion factor (i.e., species scaling factor) of body weight to the 3/4 power. A cancer slope factor derived using this new conversion factor would be approximately two to three times lower using the same data than that derived using the body weight to the 2/3 power. Subsequent risk estimates using the new cancer slope factor would be two to three times less as well. Other evidence suggests that dose equivalence is best achieved by expressing the dose in mg/kg per day (Allen, Crump, and Shipp 1988; Crump, Allen, and Shipp 1990). Assuming that different units, such as mg/kg per day, yield equal susceptibility in animals and humans would result in different estimates of risk-related doses in humans. When the human equivalent doses, D_H, are the ones used as input to the dose response models (discussed below), the output values are directly applicable to humans.

Extrapolation from High to Low Dose

Extrapolation from high to low dose is based on a presumed dose-response relationship, with parameters estimated from the experimental data. Many dose response functions that relate the probability of disease to some overall measure of dose have been proposed. The mathematical form of the dose-response model selected is an important consideration, as different models can provide very different estimates of risk outside the experimental range. The selection of a dose-response model depends on the biological validity of the models, as well as how well different models describe the data. Although little is known about the true shape of the dose-response curves at low doses, as noted earlier two important assumptions are made, both of which are based on an understanding of the cancer process. First, it is assumed that the dose-response relationship is linear, or at least does not exceed linearity, in the low-dose region. Second, it is assumed that any dose, no matter how small, carries with it a probability that cancer may occur.

The dose-response models used most commonly are the multistage model for quantal data (i.e., data indicating only the number of animals with cancer) (Crump, Guess, and Deal 1977; Crump 1984) and its extension, the multistage Weibull model (Krewski et al. 1983) for time-to-tumor data. Both of the models express upper-confidence limits on cancer risk as a linear function of dose in the low-dose range. These models assume that a threshold does not exist (i.e., any dose causes some incremental increase in risk). The multistage model currently has the highest degree of scientific acceptance, with linear upper-confidence limits on risk, and is the one most often used by regulatory agencies. The mathematical form of the multistage model is

$$P(d) = 1 - \exp(-q_0 - q_1 d - \ldots q_k d^k) \quad (6\text{-}2)$$

Where $q_1 \geq$ zero
   d = average lifetime daily dose of the chemical in mg/kg per day
   $P(d)$ = lifetime probability of cancer from the dose level d
   $q_0 \ldots q_k$ = nonnegative parameters estimated by fitting the model to experimental animal carcinogenicity data

The inputs into this model are the dose, the number of animals with the specific tumor, and the number of animals at risk or examined for that specific tumor. This often is referred to as quantal data.

The quantity of principal interest is not the absolute probability of a cancer $P(d)$, but rather the extra lifetime risk of cancer resulting from exposure to a dose d. This risk is defined as

$$R = \frac{P(d) - P(0)}{1 - P(0)} \quad (6\text{-}3)$$

and may be interpreted as the probability of the occurrence of a tumor at a dose of d, given that no tumor would have occurred without the dose.

Parameters (q values) are estimated by fitting the model to experimental animal carcinogenicity data using the maximum likelihood method. In addition to maximum likelihood estimates (MLEs) of model parameters, an upper bound on the dose-response curve is calculated, reflecting the uncertainty of extrapolating the curve to low doses at which human exposures are anticipated to occur. This upper bound can be considered to represent the largest reasonable linear extrapolation to low doses consistent with the data. The method for determining the upper-confidence limits for extra risk and the lower-confidence limits for risk-related doses is based on the largest value for the linear term q_1 that is consistent with the data. This new term is the q_1^*, also referred to as the unit potency estimate, cancer slope factor, or slope factor. The estimated dose-response curve is linear at low doses whenever the estimate of the linear coefficient, q_1, is greater than zero. The upper bound, specifically the q_1^*, is always linear because there always is some model with a positive coefficient (q_1^*) that is consistent with the data.

The multistage Weibull model is expressed as

$$P(d,t) = 1 - \exp[(-q_0 - q_1^d \ldots$$
$$-q_k d^k)\,(t - \alpha)^\beta] \quad (6\text{-}4)$$

Where $q_i \geq 0$ $(i = 0,1, ..., k)$
          $P(d,t)$ = probability of cancer at time t when exposed to dose d

The degree of the polynomial, k, is set equal to the number of dose groups less one and q_0, $q_1, ..., q_k$, α, and β are estimated by maximum likelihood techniques. To apply the multistage Weibull model, the dose received, the time of death of each animal, and whether the animal had the tumor of interest must be known. All animals, including those that died early in the study, are included as input to the multistage Weibull program that is designed to account for intercurrent mortality thereby accounting for survival problems.

The multistage Weibull model can be fit by assuming either that tumors are fatal or that tumors are incidental. Krewski et al. (1983) describe the differences between these two approaches. Because no q_1^* is defined as in the quantal model, a 'potency' parameter for the time-to-tumor model is defined by the ratio of 10^{-6} to the lower limit on the dose corresponding to an extra risk of 10^{-6}. This definition is analogous to the quantal definition because, in that case, q_1^* can be estimated closely by that same ratio (owing to the linearity of the model at low levels of risk). The time-to-tumor potency parameter also is referred to as q_1^*.

Given the estimates of probability defined by the model above, for a given time, t, extra risk is defined by

$$R = \frac{P(d,t) - P(0,t)}{1 - P(0,t)} \quad (6\text{-}5)$$

The output values of q_1^*, the 95 percent statistical upper limit on the linear term q_1, represent the unit risk in units of (mg/kg per day)$^{-1}$ directly applicable to humans, when appropriate 'scaling-up' dose conversions are applied to the experimental data before application of the dose-response models, as discussed above. Similarly, the output MLE and

statistical lower bounds on risk-related doses are applicable directly to humans. At low doses, estimates of the upper bound on extra cancer risk may be obtained by multiplying the unit potency estimate, q_1^* in (mg/kg per day)$^{-1}$, by the estimated human exposure level in mg/kg per day. The unit potency estimate, q_1^*, describes the potential increase in an individual's risk, where the unit of exposure is expressed as milligrams of the chemical per kilogram of body weight per day. Unit risks also can be expressed as either the reciprocal inhaled concentration in the units $(\mu g/m^3)^{-1}$, or as a reciprocal ingestion concentration in the units $(\mu g/l)^{-1}$. Unit risk values and weight of evidence classifications, accepted by EPA at the time of this writing in 1992, for the toxic air pollutants listed in the 1990 Amendments to the Clean Air Act are shown in Table 6-2.

Other dose response models that have been applied to quantal data include the probit, the one-hit model, the Weibull, or the gamma multihit model. All of these models assume that a threshold does not exist. However, not all of these models assume low dose linearity and consequently can provide very different estimates of risk at low doses than those obtained using the multistage model.

Similar dose response models can be applied to epidemiological data, although the forms of these models are different from those used with animal data, owing to the differences in the structure of animal and human carcinogenicity data.

EPA's program entitled Carcinogen Risk Assessment Verification Endeavor (CRAVE) has developed unit potency factors for various chemicals using the methods described in this section. These values are found in EPA's on-line data base, the Integrated Risk Information System (IRIS).[1] IRIS is described in Appendix A. Should readers wish to verify EPA's numbers or derive unit potency estimates using other data or other assumptions, estimates of the parameters in the multistage model and of the confidence limits on risk at specified doses,

---

[1] For information on IRIS, call (513) 569-7254.

Table 6-2   Inhalation carcinogenic criteria for listed toxic air pollutants[a]

Toxic air pollutant	Weight of evidence[b]	Unit risk[c] $(\mu g/m^3)^{-1}$	Source[d]
Acetaldehyde	B2	2.2E-6	IRIS
Acetamide	NA	NA	
Acetonitrile	NA	NA	
Acetophenone	D	NA	
2-Acetylaminofluorene	NA	NA	
Acrolein	C	NA	
Acrylamide	B2	1.3E-3	IRIS
Acrylic acid	NA	NA	
Acrylonitrile	B1	6.8E-5	IRIS
Allyl chloride	NA	NA	
4-Aminobiphenyl	NA	NA	
Aniline	B2[e]	NA	IRIS
o-Anisidine	NA	NA	
Asbestos	A	2.3E-1/fiber/ml	IRIS
Benzene	A	8.3E-6	IRIS
Benzidine	A	6.7E-2	IRIS
Benzotrichloride	B2[e]	NA	IRIS
Benzyl chloride	B2[e]	NA	IRIS
Biphenyl[c]	D	NA	
Bis(chloromethyl)ether	A	6.2E-2	IRIS
Bis(2-ethylhexyl)phthalate (DEHP)	B2[e]	NA	IRIS
Bromoform	B2	1.1E-6	IRIS
1,3-Butadiene	B2	2.8E-4	IRIS
Calcium cyanamide	NA	NA	
Caprolactam	NA	NA	
Captan	NA	NA	
Carbaryl	NA	NA	
Carbon disulfide	NA	NA	
Carbon tetrachloride	B2	1.5E-5	IRIS
Carbonyl sulfide	NA	NA	
Catechol	NA	NA	
Chloramben	NA	NA	
Chlordane	B2	3.7E-4	IRIS
Chlorine	NA	NA	
Chloroacetic acid	NA	NA	
2-Chloroacetophenone	NA	NA	
Chlorobenzene	D	NA	
Chlorobenzilate	NA	NA	
Chloroform	B2	2.3E-5	IRIS
Chloromethyl methyl ether	A	NA	
Chloroprene	NA	NA	
Cresols/cresylic acid (isomers and mixture)	C	NA	
o-Cresol	C	NA	
m-Cresol	C	NA	
p-Cresol	C	NA	
Cumene	NA	NA	
2,4-D, salts and esters	NA	NA	
DDE	B2[e]	NA	IRIS
Diazomethane	NA	NA	
Dibenzofurans	D	NA	
1,2-Dibromo-3-chloropropane	NA	NA	
Dibutylphthalate	D	NA	
1,4-Dichlorobenzene	NA	NA	

Table 6-2   Continued

Toxic air pollutant	Weight of evidence[b]	Unit risk[c] $(\mu g/m^3)^{-1}$	Source[d]
3,3-Dichlorobenzidene	B2[e]	NA	IRIS
Dichloroethyl ether	B2	3.3E-4	IRIS
1,3-Dichloropropene	NA	NA	
Dichlorvos	B2[e]	NA	IRIS
Diethanolamine	NA	NA	
N,N-Diethylaniline	NA	NA	
Diethyl sulfate	NA	NA	
3,3-Dimethoxybenzidene	NA	NA	
Dimethyl aminoazobenzene	NA	NA	
3,3'-Dimethyl benzidene	NA	NA	
Dimethyl carbamoyl chloride	NA	NA	
Dimethyl formamide	NA	NA	
1,1-Dimethyl hydrazine	NA	NA	
Dimethyl phthalate	D	NA	
Dimethyl sulfate	B2	NA	
4,6-Dinitro-o-cresol and salts	NA	NA	
2,4-Dinitrophenol	NA	NA	
2,4-Dinitrotoluene	B2[e]	NA	IRIS
1,4-Dioxane	B2[e]	NA	IRIS
1,2-Diphenylhydrazine	B2	2.2E-4	IRIS
Epichlorohydrin	B2	1.2E-6	IRIS
1,2-Epoxybutane	NA	NA	
Ethyl acrylate	NA	NA	
Ethyl benzene	D	NA	
Ethyl carbamate	NA	NA	
Ethyl chloride	NA	NA	
Ethylene dibromide	B2	2.2E-4	IRIS
Ethylene dichloride	B2	2.6E-5	IRIS
Ethylene glycol	NA	NA	
Ethylene imine	NA	NA	
Ethylene oxide	NA	NA	
Ethylene thiourea	NA	NA	
Ethylidine dichloride	C	NA	
Formaldehyde	B1	1.3E-5	IRIS
Heptachlor	B2	1.3E-3	IRIS
Hexachlorobenzene	B2	4.6E-4	IRIS
Hexachlorobutadiene	C	2.2E-5	IRIS
Hexachlorocyclopentadiene	D	NA	
Hexachloroethane	C	4.0E-6	IRIS
Hexamethylene-1,6-diisocyanate	NA	NA	
Hexamethylphosphoramide	NA	NA	
Hexane	NA	NA	
Hydrazine	B2	4.9E-3	IRIS
Hydrochloric acid	NA	NA	
Hydrogen fluoride	NA	NA	
Hydroquinone	NA	NA	
Isophorone	C[e]	NA	IRIS
Lindane	NA	NA	
Maleic anhydride	NA	NA	
Methanol	NA	NA	
Methoxychlor	D	NA	
Methyl bromide	D	NA	
Methyl chloride	NA	NA	

Table 6-2   Continued

Toxic air pollutant	Weight of evidence[b]	Unit risk[c] $(\mu g/m^3)^{-1}$	Source[d]
Methyl chloroform	D	NA	
Methyl ethyl ketone	D	NA	
Methyl hydrazine	NA[e]	NA	
Methyl iodide	NA	NA	
Methyl isobutyl ketone	NA	NA	
Methyl isocyanate	NA	NA	
Methyl methacrylate	NA	NA	
Methyl-t-butyl ether	NA	NA	
Methylene chloride	NA	NA	
Methylene diphenyl diisocyanate (MDI)	NA	NA	
4,4-Methylene bis (2-chloroaniline)	NA	NA	
4,4'-Methylenedianiline	NA	NA	
Naphthalene	D	NA	
Nitrobenzene	D	NA	
4-Nitrobiphenyl	NA	NA	
4-Nitrophenol	NA	NA	
2-Nitropropane	NA	NA	
N-Nitroso-N-methylurea	NA	NA	
N-Nitrosodimethylamine	B2	1.4E-2	IRIS
N-Nitrosomorpholine	NA	NA	
Parathion	C	NA	
Pentachloronitrobenzene	NA	NA	
Pentachlorophenol	B2[e]	NA	IRIS
Phenol	D	NA	
p-Phenylenediamine	NA	NA	
Phosgene	NA	NA	
Phosphine	D	NA	
Phosphorus	D	NA	
Phthalic anhydride	NA	NA	
PCBs	B2[e]	NA	IRIS
1,3-Propane sultone	NA	NA	
β-Propiolactone	NA	NA	
Propionaldehyde	NA	NA	
Propoxur (Baygon)	NA	NA	
Propylene dichloride	NA	NA	
Propylene oxide	B2	3.7E-6	IRIS
1,2-Propyleneimine	NA	NA	
Quinoline	NA	NA	
Quinone	NA	NA	
Styrene	NA	NA	
Styrene oxide	NA	NA	
2,3,7,8-Tetrachlorodibenzo-p-dioxin	NA	NA	
1,1,2,2-Tetrachloroethane	NA	NA	
Tetrachloroethylene	NA	NA	
Titanium tetrachloride	NA	NA	
Toluene	D	NA	
2,4-Toluene diamine	NA	NA	
2,4-Toluene diisocyanate	NA	NA	
o-Toluidine	NA	NA	
Toxaphene	B2	3.2E-4	IRIS
1,2,4-Trichlorobenzene	D	NA	
1,1,2-Trichloroethane	C	1.6E-5	IRIS
Trichloroethylene	NA	NA	

Table 6-2   Continued

Toxic air pollutant	Weight of evidence[b]	Unit risk[c] $(\mu g/m^3)^{-1}$	Source[d]
2,4,5-Trichlorophenol	NA	NA	
2,4,6-Trichlorophenol	B2	3.1E-6	IRIS
Triethylamine	NA	NA	
Trifluralin	C[e]	NA	IRIS
2,2,4-Trimethylpentane	NA	NA	
Vinyl acetate	NA	NA	
Vinyl bromide	NA	NA	
Vinyl chloride	NA	NA	
Vinylidene chloride	C	5.0E-5	IRIS
Xylenes (isomers and mixtures)	D	NA	
o-, m-, p-Xylene	NA	NA	
Antimony compounds	NA	NA	
Arsenic compounds	A	4.3E-3	IRIS
Beryllium compounds	B2	2.4E-3	IRIS
Cadmium compounds	B1	1.8E-3	IRIS
Chromium compounds[f]	A	1.2E-2	IRIS
Cobalt compounds	NA	NA	
Coke oven emissions	A	6.2E-4	IRIS
Cyanide compounds[g]	NA	NA	
Glycol ethers	NA	NA	
Lead compounds	B2	NA	
Manganese compounds	D	NA	
Mercury	D	NA	
Fine mineral fibers	NA	NA	
Nickel compounds[h]	A	4.8E-4	IRIS
Polycyclic organic compounds[i]	NA	NA	
Radionuclides	A	NA	
Selenium compounds[j]	B2	NA	

[a] Information from EPA data bases only (November 1992).

[b] Weight of Evidence: A = Human carcinogen (sufficient evidence on humans); B = Probable human carcinogen (B1 = limited evidence in humans; B2 = sufficient evidence in animals with inadequate or lack of evidence in humans); C = Possible human carcinogen (limited evidence in animals and inadequate or lack of human data); D = Not classifiable as to human carcinogenicity (inadequate or no evidence); E = Evidence of no carcinogenicity in humans (no evidence in adequate studies).

[c] Units unless otherwise noted.

[d] IRIS = Integrated Risk Information System.

[e] Evidence of carcinogenicity in oral studies.

[f] Chromium (III) – NA.

[g] Free cyanide Class D weight of evidence.

[h] Nickel subsulfide.

[i] Benzo[a]pyrene.

[j] Selenium sulfide.

or on doses that correspond to specified levels of extra risk, can be obtained using the computer program GLOBAL82 (Howe and Crump 1982); a software package, TOX_RISK, also is available that incorporates route and dose conversions (Crump et al. 1992).[2] It allows for application of a variety of statistical tests on the animal data and can apply several of these dose response models, including the multistage, probit, log-normal, and Weibull, to the data in sequence, reporting the results in separate tabular form.

---

[2] For more information, call (318) 255-4800.

Impact of Selected Assumptions on Unit Potency Estimates

As noted in the above discussions, assumptions made concerning data selection, species extrapolation, or low-dose extrapolation can have a significant impact on estimates of the unit potency estimate and, therefore, a significant impact on estimates of extra lifetime cancer risk from exposure to chemicals in the environment. Using hypothetical animal bioassay data, the impact on risk for the following assumptions was determined (see Tables 6-3 and 6-4).

- Dose is assumed to be equivalent across species (A).
- Only malignant tumors or the combination of benign and malignant tumors in the same organ or tissue are selected (B).
- The multistage model or the multistage Weibull model is selected for data sets in which survival is not a problem and in which intercurrent mortality was high (C).
- Risk is expressed as that above background (extra) or including background (additional) (D).
- Different dose-response models are compared for selected data sets (E).

Table 6-3   Impact of selected assumptions on estimates of cancer potency factors

Unit potency estimates, q_1^* (mg/kg per day)$^{-1}$		
A – EQUIVALENCY UNITS		
Species scaling factor	mg/kg BW	mg/kg $BW^{2/3}$	mg/kg $BW^{3/4}$
Mouse, liver tumors	2.6×10^{-2}	3.4×10^{-1}	1.8×10^{-1}

B – MALIGNANT VERSUS MALIGNANT AND BENIGN TUMORS		
Type of tumor		Malignant	Malignant + benign
Mouse, liver tumors		2.04×10^{-3}	1.55×10^{-2}
Data set (mg/kg per day)			
0		0/49	5/49
250		1/46	13/46
500		6/48	30/48

C – MULTISTAGE MODEL VERSUS MULTISTAGE TIME-TO-TUMOR MODEL		
Type of model		Multistage	Multistage Weibull
Tumor data set without survival problems		1.5×10^{-4}	1.9×10^{-4}
Tumor data set with survival problems		1.6×10^{-4}	6.06×10^{-2}

D – EXTRA VERSUS ADDITIONAL RISK		
Type of risk		Extra risk	Additional risk
Tumor data set with low background rate		3.52×10^{-2}	3.52×10^{-2}
Tumor data set with high background rate		1.22×10^{-1}	4.05×10^{-4}

E – COMPARISON OF DIFFERENT DOSE-RESPONSE MODELS		
Model[a]	Data set 1	Data set 2	Data set 3
Multistage	2.58×10^{-2}	1.06×10^{-2}	3.43×10^{-2}
Log-normal	1.94×10^{-5}	2.37×10^{-6}	1.15×10^{-5}
Mantel Bryan	2.69×10^{-3}	1.84×10^{-3}	2.43×10^{-3}
Weibull	2.58×10^{-2}	2.55×10^{-4}	3.09×10^{-2}

[a] Data sets defined in Table 6-4.

Table 6-4   Impact of selected assumptions on estimates of cancer potency factors, definition of data sets used in Table 6-3

	Dose	Incidence
Data set 1: Monotonic increase	0	5/49
	250	13/46
	500	30/48
Data set 2: Zero in the low dose group then monotonic rise	0	0/49
	100	0/46
	250	13/46
	500	30/48
Data set 3: Responses plateau	0	0/49
	100	1/50
	250	25/49
	500	27/50

REDUCING UNCERTAINTY IN RISK ASSESSMENT

Less than Lifetime Exposure

Some of the difficulties in risk assessment, whether bioassay- or epidemiology-related, arise when exposures are intermittent or less than the expected lifetime. As noted earlier, when exposure situations of concern differ from continuous, constant lifetime exposure, EPA (1986c) recommends that, unless there is evidence to the contrary, the appropriate measure of exposure is the total or cumulative dose of the chemical of concern averaged or prorated over a lifetime, resulting in an average lifetime daily exposure. This assumes that a high dose of a carcinogen received over a short period of time is equivalent to a corresponding low dose averaged over a lifetime in terms of extra cancer risk. The supporting rationale for this relies on the underlying assumption of the carcinogenic process – risk is linearly related to dose, in particular in the low-dose region. This implies dose-response linearity for any level of exposure or duration of exposure. As such, the effective dose, or the amount of chemical present at the target tissue at a given time, is not considered. Differences in the sensitivity of target cells at different ages also are not incorporated into currently used methods.

As exposures become more intense but less frequent, this approach becomes problematic, especially when the agent has demonstrated dose-rate effects (EPA 1986c). Dose-rate effects are a different degree or type of response that may occur with different dose patterns even when the total dose is the same for these dosing patterns. For example, recent information suggests that both age and duration of exposure are important factors in vinyl chloride carcinogenesis, and that lifetime risks may be mostly attributable to exposure occurring early in life (Parker, Cogliano, and Pepelko 1990). This analysis and other studies suggest that prorating exposures over a lifetime may not provide an appropriate measure of cancer risk in some cases. The use of pharmacokinetic modeling to adjust for difference in temporal exposure patterns is discussed below.

One set of programs (ADOLL) is available that consider time-dependent dosing patterns when estimating risk (Crump and Howe 1984). These models are based on the multistage theory of cancer suggested by Armitage and Doll (1961). Dose patterns can be approximated by these models and the risks appropriately related to dose patterns. The requirements for ADOLL are knowledge (or estimation) of the total number of stages in the cancer process and determination (or estimation) of the stages that are susceptible to the action of the carcinogen. Using ADOLL and bioassay data for ethylene dibromide, Crump and Howe (1984) compared the extra risk of cancer by age seventy from a partial lifetime exposure at a given level to the extra risk by age seventy from lifetime exposure at the same level. The results indicated that the extra risk from exposure depends quantitatively on the number of stages, which stage is affected by the carcinogen, the duration of exposure, and the age at first exposure. If the first stage is dose-related, early exposures are more dangerous than later ones. According to these estimates, exposure for the first thirty-five years of life gives almost the same risk as if the exposures continued throughout life, provided there are at least three stages.

Another avenue of investigation and dose-response model development is related to the two-stage models of cancer that recently have been proposed (Chen, Kodell, and Gaylor 1988). These models include consideration of the proliferation of transformed cells. However, no program using these models currently is available to account for variable dosing patterns. The extension of such models along these lines, and along lines suggested by further examination of specific chemicals, provides another potentially useful approach to the estimation of the risk associated with partial lifetime exposures.

Finally, the development of pharmacokinetic models, as discussed below, incorporating the latest advances in the understanding of the cancer process, provide additional hope of deriving relevant and meaningful risk estimates for partial lifetime exposures.

The Use of Pharmacokinetic Data in Risk Assessment

A fundamental assumption concerning the toxic actions of chemical agents and the estimates of risk to exposed populations concerns the relationship of dose to response. Although at first this relationship may appear to be straightforward, such often is not the case in the course of

a risk assessment. For example, questions such as the following may arise:

- Is the active compound the administered compound or one or more metabolite?
- What is the magnitude of the effective dose relative to the administered dose?
- Is the response directly related to the particular interaction being measured?

Underlying any risk assessment is the assumption of a dose-response relationship that assumes that the toxic response results from the interaction of the reactive moiety with a molecular or receptor site. The extent of the response is related to the amount of the toxicant at the target site; the amount of toxicant is related to the magnitude of the administered dose.

Pharmacokinetic models, by describing the kinetics of biodistribution, may be useful in the determination of the rate, amount, and identity of the portion of the administered dose that reaches the target tissues, thereby accounting for species differences due to metabolism, distribution, and elimination of a chemical. Other species differences that may affect risk include differences in total number of cells that could be targets, cell turnover rates, longevity, and repair mechanisms and efficiency. All of these factors exert their effect through the intracellular events that take place after delivery of the active compound to the cell. Pharmacokinetic models may be used to estimate the delivered dose, or the amount of chemical reaching a target organ. Therefore, they may be useful in reducing the uncertainty in several of the extrapolations that often are required in risk assessment.

Perhaps the greatest source of uncertainty in cancer risk assessment is that due to the uncertainty in the shape of the dose-response curve at low doses. Although regulatory agencies frequently assume a linear dose-response, experimental data often exhibit a very nonlinear curve shape. Use of pharmacokinetic data allows estimation of the dose to the organ site (i.e., the measure of dose assumed by mathematical dose-response models). A very important aspect is to determine to what extent nonlinearities in dose-response data are due to nonlinearities in pharmacokinetic processes. If this is the case, obtaining pharmacokinetic data and refining methods for use of this data may lead to a considerable reduction in uncertainty in estimates of risk from low exposure levels.

Species differences in absorption, distribution, and metabolism result in different delivered doses even when exposures – whether measured in mg/kg body weight per day, mg/m^2 surface area per day, ppm, or other terms – are identical. It is possible that much of the variability in responses to various carcinogens exhibited by different animal species is due to differences in metabolism and distribution. This issue could be explored systematically by developing risk estimates in various species, including both animals and humans, both with and without incorporating pharmacokinetic methods. The differences in the concordances between these estimates would allow assessment of the extent to which differences in pharmacokinetic parameters are responsible for the divergence in risk estimates in different species. This information then can be used to adjust for the residual differences in risk estimates between species after pharmacokinetic methods have been applied. For example, if after adjusting for pharmacokinetic differences between species, species A is consistently more susceptible than species B to a particular class of chemicals, such differences can be accounted for quantitatively when performing risk assessments on chemicals from this class.

The route of administration and the type of vehicle used can greatly affect the bioavailability of a chemical, or the relative rate and extent at which the administered dose reaches the systemic circulation. The same applied dose may produce different delivered doses or different metabolic products (e.g., first-pass effect in the liver) depending on the route of administration and the physicochemical characteristics of the chemical. Pharmacokinetic data could be used to predict responses in animals from one route of exposure using data from studies employing a second route, and to compare results to those developed from experiments involving the second route.

It frequently is necessary in risk assessment to predict risk in humans from a temporal exposure pattern that is very different from that experienced by the animals in the experiment on which the risk assessment is based. In a typical case, humans are exposed briefly or intermittently whereas the animals are exposed more or less constantly throughout their lifetime. The approach normally used to account for such differences assumes that risk depends on total dose irrespective of timing of dose. However, dose rate can affect uptake, distribution, metabolism, storage, and elimination of a chemical. As a consequence, it can have an important effect on risk apart from total dose. Pharmacokinetic data can play an important role in determining differences in the relationship between applied dose and concentrations of the active carcinogen in critical tissues caused by the differences in exposure patterns. Not all pharmacokinetic modeling approaches are equally suitable for risk assessment purposes. Classical compartmental pharmacokinetic models are defined for specific exposure situations with little or no capability for prediction or extrapolation to other routes of exposure, to other levels of exposure, or to other species. The compartments included in a classical model are hypothetical and do not readily correlate with recognizable organs or tissues; they are used merely to provide an adequate description of observed concentration time-courses in one or more tissues.

Physiologically based pharmacokinetic (PBPK) models, on the other hand, are ideally designed for prediction and extrapolation. They can be used to reduce the uncertainties associated with many of the extrapolations that are necessary in risk assessment contexts. PBPK models are defined in terms of compartments that represent actual organs or tissue groups and have parameters with meaning independent of the model itself. Biotransformation, absorption, elimination, binding, and diffusion are all processes that can be modeled explicitly in the PBPK approach. Thus, differences across species or differences due to different routes of exposure or exposure levels can be predicted (and used for extrapolation) by changing the parameters that describe the compartments and the processes. As opposed to the curve-fitting exercise that is classical compartmental modeling, PBPK modeling allows predictions to be made and either verified or utilized in risk calculations.

A typical PBPK model (Figure 6-1) has compartments identified with the liver, the lung, and other organs, all of which have specified volumes and perfusion rates (Anderson et al. 1987). Metabolism may be expressed in terms of first-order or saturable pathways with parameters that relate to observable phenomena. Because the parameters of a PBPK model can be estimated independently of the model, route-to-route and species-to-species extrapolations are

facilitated. In fact, the explicit consideration of species differences in model parameters and of differences due to alternate routes of exposure logically lead to extrapolations that account for relevant factors better than extrapolations as 'traditionally' performed (e.g., based on administered doses expressed in terms of mg/kg per day and assuming 100 percent absorption).

For inhalation exposures to chemicals that are not in the vapor phase, additional modeling of the lung may be desirable. In this case, the lung can be divided into its major regions (nasopharyngeal, tracheobronchial, and pulmonary) and the deposition and clearance properties of those regions modeled separately. The processes by which inhaled particles are deposited vary from

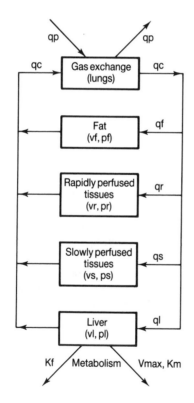

qp = Alveolar ventilation rate
qc = Cardiac output rate
qi = Blood flow rate to compartment i (i = f, r, s, l)
vi = Volume of compartment i (i = f, r, s, l)
pi = Tissue/blood partition coefficient for compartment i (i = f, r, s, l)
Kf, Vmax, Km = Constants describing metabolism

Figure 6-1.   Typical PBPK model. Adapted from Andersen et al. (1987).

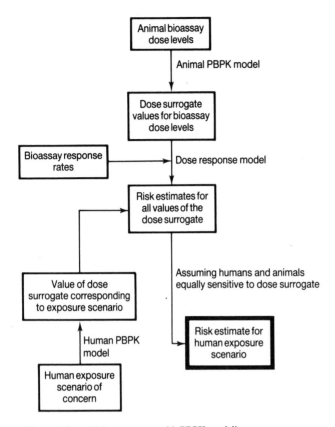

Figure 6-2.   Risk assessment with PBPK modeling.

region to region, the range of particle sizes affected by those processes vary, and the clearance mechanisms of the regions are different.

In general, pharmacokinetic/pharmacodynamic considerations are important when cross-species extrapolations are necessary. Qualitatively, the determination of a species most appropriate as the basis for an extrapolation to humans should consider the differences in pharmacokinetics/pharmacodynamics among species. For example, test species that exhibit pathways of metabolism or mechanisms of action not seen in humans may be inappropriate as the basis of an extrapolation. Moreover, quantitative differences always exist among species (even among those with the same pathways and mechanisms of actions) with respect to the parameter values determining absorption, distribution, and metabolism. PBPK

modeling and calculation of delivered doses permits adjustment for such differences.

Use of delivered dose rather than exposure dose as the independent variable in a risk model can be valuable in each of these extrapolations and may well reduce the overall uncertainty in quantitative risk assessment. One approach to linking pharmacokinetic data in risk assessment is depicted in Figure 6-2. Average values, such as the average under a concentration–time curve and average daily amount metabolized, for the delivered doses corresponding to exposures in an experiment are estimated using PBPK modeling. The average surrogate values are used as the dose variable in the dose-response model and the risks extrapolated. It then is assumed that any exposures that lead to the same average value in humans are equivalent with respect to risk, within and across species. This approach has been applied in risk assessments that

Table 6-5   Impact of incorporating pharmacokinetic data on estimates of lifetime cancer risk

	Extra lifetime cancer risk[a]	Extra lifetime cancer risk[b]
Male mice – hepatocellular carcinomas or adenoma	1.05×10^{-5}	1.95×10^{-6}
Female mice – hepatocellular carcinoma or adenoma	5.26×10^{-6}	1.86×10^{-6}
Male mice – alveolar/bronchiolar carcinoma or adenoma	5.00×10^{-6}	3.77×10^{-7}

[a] Based on experimental dose.
[b] Based on surrogate dose:active metabolite at the target tissue.

incorporate PBPK modeling. It is expected that the use of a pharmacokinetic model to estimate delivered doses will produce greater agreement between the biological data and the assumptions that make such models reasonable. Consequently, the estimated risks produced by fitting the carcinogenesis models should have a greater probability of approximating the true risks. One example of the difference in risk outcome when pharmacokinetic data are incorporated into estimates of risk is given in Table 6-5.

New Methods for Quantitative Cancer Risk Assessment

The practice of quantitative cancer risk assessment is in a constant state of evolution. As additional experimentation reveals more and more about the production of toxic endpoints in humans and animals and as computational capabilities are improved, risk assessment procedures can and should be adapted to incorporate those changes. Moreover, experience gained in applying methods may be as valuable as the acquisition of new data. The goal of any cancer risk assessment is to utilize the most relevant data and quantitative techniques to derive the best possible estimates of risks to humans. For that to be possible, it is necessary to have a thorough understanding of the underlying principles on which the methods are based and the ability to recognize the significance of new data. An experienced team of scientists from many disciplines can accomplish the development of novel methods or adaptation of existing methods suggested by new data, new quantitative techniques, or additional experience.

Ideally, dose-response modeling should be closely tied to the mechanisms assumed to be responsible for the development of a given endpoint. If that is the case, then the choice of a dose-response model is dictated to a large degree by type of endpoint and, if known, the processes responsible for the manifestation of the endpoint. For cancer endpoints, the standard dose-response model applied to bioassay data is the multistage model (Armitage and Doll 1961). The biological assumptions relevant to that model are that cancer is produced when a cell passes through a series of irreversible stages and that exposure to the chemical of interest (if it is carcinogenic) increases the likelihood of one or more of those stages occurring. Furthermore, the occurrence of the sequence of stages in one cell is sufficient to produce a tumor. The multistage model appears to be particularly relevant when the action of a chemical (or its reactive metabolite) is genotoxic and the cancer in question arises as a result of several mutations or other irreversible processes involving DNA. Then the stages represent steps in the transformation of a normal cell to a tumor cell, where cells in a particular stage (before the tumor-cell stage) have experienced some of the mutational events. Lacking data indicating that a chemical produces tumors by nongenotoxic mechanisms, the multistage model is used as the dose-response model for cancer endpoints.

There is growing evidence that the action of some chemicals may not be genotoxic, yet they may still cause cancer. For example, a chemical that selectively enhances the mitotic rate of preneoplastic cells would increase the rate of cancers observed in animals that had preneoplastic cells. Alternatively, a nonspecific cytotoxicant by inducing restorative hyperplasia may enhance the production of tumors if the increased rate of cell production reduces the efficiency of DNA repair processes or if preneoplastic cells proliferate at a greater rate than normal cells.

Processes like those in the preceding paragraph are not well described by the multistage model. Alternate models, such as the MKV model, have been proposed (Moolgavkar and Knudson, 1981) that include consideration of cell turnover rates and other nongenotoxic events. The MKV model postulates the existence of certain steps that a cell must undergo as it is transformed from a normal cell to a cancer cell. It assumes that two irreversible 'transitions' must take place and also considers the birth and death rates of normal and 'precursor' cells. The effect of exposure to a chemical can be modeled as an effect on several of the parameters of the model, such as the rate constant for transition from a normal to a precursor cell, the rate constant for transition from a precursor to a cancer cell, the birth rate constant from precursor cells, and others. The model estimates the probability of the occurrence of a tumor, where the estimate depends on the mechanism believed to be affected by exposure. The price of that degree of specificity is that one must have estimates of several parameters not required by the multistage model as routinely employed. Such parameters include background mutation rates and rates of birth and death rates of normal, precursor, and cancer cells, An approximate solution to the MKV model also is found in the software, TOX_RISK; however, estimates of risk based on this model have not yet been accepted by regulatory agencies.

References

Allen B., K. Crump, and A. Shipp. 1988. Correlation between carcinogenesis potency of chemicals in animals and humans. *Risk Analysis* 8:531–544.

Anderson M., H. Clewell, M. Gargas, F.A. Smith and R.H. Reitz 1987. Physiologically based pharmacokinetics and the risk assessment process for methylene chloride. *Toxicology and Applied Pharmacology* 87:185–205.

Armitage P. 1955. Tests for linear trends in proportions and frequencies. *Biometrics* 11:375–386.

Armitage P. and R. Doll. 1954. The age distribution of cancer and a multistage theory of carcinogenesis. *British Journal of Cancer* 8:1–12.

Armitage P. and R. Doll. 1961. Stochastic models for carcinogenesis. In: *Proceedings of the Fourth Berkeley Symposium on Mathematical Statistics and Probability* 4:19–38. Berkeley, CA: University of California Press.

Bickel P. and K. Doksum. 1977. *Mathematical Statistics: Basic Ideas and Selected Topics*. San Francisco: Holden-Day.

Capen, C.C. and S.L. Martin. 1989. The effects of xenobiotics on the structure and function of thyroid follicular and *C*-cells. *Toxicology and Pathology* 17:266–293.

Chen J., R. Kodell, and D. Gaylor. 1988. Using the biological two-stage model to assess risk from short-term exposure. *Risk Analysis* 8:223–230.

Crump, K. 1984. An improved procedure for low-dose carcinogenic risk assessment from animal data. *Journal of Environmental Pathology and Toxicology* 5:339–348.

Crump K. and R. Howe. 1984. The multistage model with a time-dependent dose pattern: Applications to carcinogenic risk assessment. *Risk Analysis* 4:163–176.

Crump, K.S., H.A. Guess, and I.L. Deal. 1977. Confidence intervals and test of hypotheses concerning dose-response relations inferred from animal carcinogenicity data. *Biometrics* 33:437.

Crump K., B. Allen, and A. Shipp. 1990. An investigation of how well human carcinogenic risk from chemical exposure can be predicted by animal data, with emphasis upon selection of dose measure for extrapolation from animals to humans. *Health Physics* 57:387–393.

Crump K., R. Howe, C. Van Landingham, and W. Fuller. 1990. TOX_RISK. Toxicology Risk Assessment Program. Developed under contract to Electric Power Research Institute.

Crump K., R. Howe, C. Van Landingham, and W. Fuller. 1992. TOX_RISK. Toxicology Risk Assessment Program. Version 3.1. Developed under contract to Electric Power Research Institute.

Farber E. 1982. Sequential events in chemical

carcinogenesis. In: Becker F., ed. *Cancer: A Comprehensive Treatise.* 2nd Edition. Volume 1. Etiology: Chemical and Physical Carcinogenesis. New York: Plenum Press.

Gilbert E., J. Park, and R. Buschbom. 1989. Time-related factors in the study of risks in animals and humans. *Health Physics* 57:379–385.

Hart R. and A. Turturro. 1988. Current views of the biology of cancer. In: Travis C., ed. *Carcinogen Risk Assessment.* New York: Plenum Press. pp. 19–33.

Haseman J.K. 1984. Statistical issues in the design, analysis and interpretation of animal carcinogenicity studies. *Environmental Health Perspectives* 58:385–392.

Howe R. and K. Crump. 1982. GLOBAL82: A computer program to extrapolate quantal animal toxicity data to low doses. Prepared under Contract 41USC252C3 for the Office of Carcinogen Standards, Occupational Safety and Health Administration. Washington, D.C.

Kanerva, R.L., G.M. Ridder, L.C. Stone, and C.L. Alden. 1987. Characterization of spontaneous and decaline-induced hyaline droplets in kidneys of adult male rats. *Food and Chemical Toxicology* 25:63–82

Krewski D., K. Crump, J. Farmer, D. Gaylor, R. Howe, C. Portier, D. Salsburg, R. Sielken, and J. Van Ryzin. 1983. A comparison of statistical methods for low-dose extrapolation utilizing time-to-tumour data. *Fundamental and Applied Toxicology* 3:140–160.

Lutz W. 1990. Endogenous genotoxic agents and processes as a basis of spontaneous carcinogenesis. *Mutation Research* 238:287–295.

McClain, R.M. 1989. The significance of hepatic microsomal enzyme induction and altered thyroid function in rats: Implications for thyroid gland neoplasia. *Toxicologic Pathology* 17:294–306

Miller E. and J. Miller. 1976. The metabolism of chemical carcinogens to reactive electrophiles and their possible mechanism of action in carcinogenesis. In: Searle C., ed. *Chemical Carcinogens.*

American Chemical Society Monographs 173:737–762.

Moolgavkar, S.H. and A.G. Knudson. 1981. Mutation and cancer: A model for human carcinogenesis. *Journal of the National Cancer Institute* 66:1037.

National Academy of Sciences (NAS). 1983. *Risk Assessment in the Federal Government: Managing the Process.* Prepared by the Committee on the Institutional Means for Assessment of Risk to Public Health, Commission on Life Sciences. Washington, D.C. National Academy Press.

Office of Science and Technology Policy (OSTP). 1985. Chemical carcinogens: Review of the science and its associated principles. 50 FR 10372, March 14, 1985.

Parker J., V. Cogliano, and W. Pepelko. 1990. Vinyl chloride: another look. *The Toxicologist* 10:349.

Paynter, O.E., G.J. Burin, R.B. Jaeger, and C.A. Gregoria. 1988. Goiterogens and thyroid follicular cell neoplasia: Evidence for a threshold process. *Regulatory Toxicology and Pharmacology* 31:1506–1512

U.S. Environmental Protection Agency (EPA). 1986c. Guidelines for carcinogen risk assessment. 51 FR 33992, September 24, 1986.

U.S. Environmental Protection Agency (EPA). 1987f. National Primary Drinking Water Regulations: Synthetic Organic Chemicals, Monitoring for Unregulated Contaminants. 40 CFR Parts 141 and 142.

U.S. Environmental Protection Agency (EPA). 1989j. Biological data for pharmacokinetic modeling and risk assessment. Report of a workshop convened by the U.S. Environmental Protection Agency and ILSI Risk Science Institute. EPA 600/3-90-019. Office of Health and Environmental Assessment. Washington, D.C.

U.S. Environmental Protection Agency (EPA). 1992l. A cross-species scaling factor for carcinogen risk assessment based on equivalence of $mg/kg^{3/4}$/day. Notice of draft report. 57 FR 24152. June 5, 1992.

7

Noncancer Toxicity of Inhaled Toxic Air Pollutants: Available Approaches for Risk Assessment and Risk Management

Annie M. Jarabek and Sharon A. Segal

INTRODUCTION

Noncancer health effects refer to toxic endpoints other than cancer and gene mutations that result from the effects of environmental agents on the structure or function of various organ systems. For toxic air pollutants, these effects may manifest themselves as toxicity to the respiratory tract, as a result of direct interaction with these tissues as the portal of entry for inhaled agents, or to a site remote from its entry via the respiratory tract, as a result of absorption and distribution of the toxicant in the body.

This chapter begins with a brief historical perspective on how the regulatory process for air toxics at the federal level resulted in the development of diverse approaches for both noncancer health risk estimation and risk management. An overview of the current approaches available for the health risk assessment and risk management of noncancer health effects caused by inhaled agents for different exposure duration assumptions then is presented. Limitations of various applications also are discussed and a brief summary of research needs is provided.

HISTORICAL PERSPECTIVE

Although federal air pollution legislation was passed in 1955 granting the federal government the authority to conduct research, and provide training and technical assistance programs, the concept of national health criteria for environmental air pollutants was not mandated until the Clean Air Act Amendments of 1970. Under the 1970 Amendments, the newly formed Environmental Protection Agency (EPA) was required to develop and promulgate National Ambient Air Quality Standards (NAAQS) for substances identified as the most common and widespread pollutants. These substances, constituting concern at a national level because of their known health effects, extent of use, ubiquitous occurrence, and nature of release, were known as 'criteria pollutants' and today include: carbon monoxide, lead, nitrogen dioxide, particulate matter (less than 10 microns), ozone, and sulfur dioxide. Primary standards are designed to protect the public health, and secondary standards are designed to protect the public welfare (Code of Federal Regulations, Title 40, Part 50). The primary NAAQS are set by the EPA at a concentration and an averaging time that are adequate to protect the public, including sensitive subpopulations, against adverse health effects. Thus, the primary NAAQS define allowable pollutant concentrations that can be present in the atmosphere without causing adverse health effects and represent a complete health risk characterization according to the NAS risk assessment and risk management framework. Although beyond the scope of this chapter, it is important to emphasize that the research and data evaluation efforts that have been undertaken to support these NAAQS represent the benchmark of health risk estimation procedures for noncancer agents. Many of the concepts and analysis approaches that have been developed for the criteria pollutants only now are beginning to be applied to the types of data available for the chemicals categorized as air toxics. For additional detail, the reader is referred to the NAAQS published in 40 CFR 50, the supporting documentation for the NAAQS (EPA 1979c; EPA 1982e; EPA 1982f; EPA 1982g; EPA 1982h; EPA 1982i; EPA 1984c; EPA 1984d; EPA 1986o; EPA 1986p; EPA 1986q; EPA 1986r; EPA 1986v; EPA 1987h; EPA 1990m; EPA 1990n; EPA 1992h; EPA 1992i), and an overview article describing the NAAQS development process by Padgett and Richmond (1983).

To curtail pollution that was more localized at a point of origin and associated with more serious health effects, National Emission Standards for Hazardous Air Pollutants (NESHAPs) were required for new and existing sources of airborne toxic substances. Listing of pollutants under section 112 required a showing that they 'cause, or contribute to, air pollution which may reasonably be anticipated to result in an increase in mortality or an increase in serious irreversible, or incapacitating reversible, illness.' Between 1970 and 1990, NESHAPs were promulgated for only seven hazardous air pollutants: arsenic, asbestos, benzene, beryllium, mercury, radionuclides, and vinyl chloride.

As described in Chapter 1, EPA's difficulty in promulgating NESHAPs led many state and local air pollution control agencies to attempt to deal with air toxics on their own. However, with minimal federal guidance and often limited resources, the scope and approaches to air toxics regulation varied widely among state and local agencies. The variations in the regulatory approaches included the pollutants to be regulated, the use of emissions standards or ambient air guidelines, and the application of standards or ambient air guidelines to existing or modified sources of a given pollutant (Calabrese and Kenyon 1991).

Similar variability existed in approaches for the development of ambient air levels[1] (AALs) that some state and local agencies use as health-based criteria as part of these regulatory control programs. For noncancer health effects, two approaches most often have been employed. The first is an approach based on the selection of a no-observed-adverse-effect-level (NOAEL) or lowest-observed-adverse-effect-level (LOAEL) from available laboratory animal and human health effects data. The AAL is derived from the available effects data by application of various factors that account for areas of concern such as the sensitivity of high-risk individuals or for other considerations such as the averaging time of the exposure scenario for which it is intended. Although this general approach, under the name 'acceptable daily intake' (ADI), had been used by federal agencies for the development of health-based guidelines for noncarcinogenic substances in a variety of media (EPA 1980f; NRC 1977; NRC 1980a), the lack of guidance on the choice and consistent application of uncertainty factors, the lack of explicit data base criteria, and the lack of appropriate consideration of the dynamics of inhaled agents remain noted deficits of such approaches.

The second approach commonly used to derive AALs was to use occupational exposure

---

[1] Ambient air level (AAL) is a generic term used to describe a guideline or standard concentration used to evaluate environmental exposures to toxic air pollutants. Used as such, the term AAL has an implicit connotation of 'acceptable risk' attached and essentially represents risk management values. Chapter 17 provides a more detailed discussion.

limits[2] (OELs) as the basis for derivation. Occupational exposure limits were a likely choice by the state and local agencies because OELs for nearly 700 pollutants have been developed, mainly by three organizations: the American Conference of Governmental Industrial Hygienists (ACGIH), the National Institute for Occupational Safety and Health (NIOSH), and the Occupational Safety and Health Administration (OSHA). Occupational exposure limits generally are time-weighted average concentrations of airborne substances to which a healthy worker can be exposed during defined work periods and under specific work conditions throughout a working lifetime, without material impairment of health. Another underlying assumption of most OELs is a workplace setting in which industrial hygienists are able to control the environment. Thus, the OEL can represent, in part, a risk management decision that considers nonhealth issues such as the technological feasibility of control measures and analytical detection limits. The appropriateness of some of these assumptions and extenuating considerations to the application of deriving an AAL are discussed below in the section on use of OELs.

Title III of the Clean Air Act Amendments of 1990 requires the EPA to develop standards based on the maximum achievable control technology (MACT) for major sources of 189 listed hazardous air pollutants. Evaluation of residual risk after MACT then will be required to determine if additional regulation is needed to provide 'an ample margin of safety to protect the public health.' Section 112(r) requires EPA to take steps to prevent accidental releases which might cause serious health effects. In addition, consideration of hazard and/or risk is implicit in decisions regarding additions to or deletions from the hazardous air pollutant list (section 112(b)), delisting from the source category list (section 112(b)), and additions to the accidental release list (section 112(r)). Thus, the implementation of Title III requires risk assessment methods and management decisions to address potential toxicity due to various exposure scenarios including single acute, intermittent, and chronic ambient exposures. The remainder of this chapter will discuss the risk assessment and risk management approaches that currently are available to aid this implementation.

CURRENT NONCANCER HEALTH RISK ASSESSMENT PRACTICES

Risk Assessment Versus Risk Management

In 1983, the National Academy of Sciences (NAS) published its ground-breaking report on the use of risk assessment by the federal government (NAS 1983). The NAS had been charged with evaluating the process of risk assessment as performed at the federal level in order to determine the 'mechanisms to ensure that government regulation rests on the best available scientific knowledge and to preserve the integrity of scientific data and judgments' so

that controversial decisions regulating chronic health hazards could be avoided.

The NAS recommended that the scientific aspects of risk assessment be separated explicitly from the policy aspects of risk management. Risk assessment was defined as the characterization of the potential adverse health effects of human exposures to environmental agents. It contained the following four steps:

1. Hazard identification – the determination of whether a chemical is or is not causally linked to a particular health effect.

---

[2] Occupational exposure limit (OEL) is a generic term used to denote a variety of standards that usually reflect a documented body of toxicological, epidemiological, and clinical information pertaining to effects observed upon human occupational exposures to airborne contaminants. Due to their derivation methods, attendant assumptions and intended application, they represent risk management values.

2. Dose-response assessment – the estimation in humans of the relation between the magnitude of exposure and the occurrence of the health effects in question.
3. Exposure assessment – the determination of the extent of human exposure.
4. Risk characterization – the description of the nature and often the magnitude of human risk, including associated uncertainty.

Dose-Response

The importance of understanding the relationship between concentration (i.e., applied dose) and response has been established in the theory and practice of toxicology and pharmacology. Empirical observation generally reveals that as the exposure concentration of the toxicant is increased, the toxic response also increases. For the purposes of this presentation, response is the degree or severity of an effect in an individual or population. A distinction sometimes is made between response and effect as qualitatively different measurements. Effects are graded and measured, whereas responses are quantal and counted (O'Flaherty 1981). Although the distinction is necessary in order to determine an appropriate mathematical model for analysis, for practical and sound conceptual reasons responses and effects can be considered to be identical (Klaassen 1986). In other words, in a qualitative sense when trying to ascertain if a toxic agent exerts an adverse influence, the distinction is unimportant. It is in this context that the term 'response' is used here.

Classic toxicology texts and the NAS framework for risk assessment refer to dose-response assessment as the process of estimating an expected response at various exposure levels (i.e., the response at various applied dose levels or exposure concentrations). Because the tissue dose of the putative toxic moiety for a given response is not always proportional to the applied dose of a compound, emphasis recently has been placed on the need to distinguish clearly between exposure concentration and dose to critical target tissues. The terminology 'exposure-dose-response assessment' has been recommended as more accurate

and comprehensive (Andersen et al. 1992). Characterization of the exposure-dose-response continuum requires that mechanistic determinants of chemical disposition (i.e., deposition, absorption, distribution, metabolism, and elimination), toxicant–target interactions, and tissue responses be integrated into an overall model of pathogenesis. Elucidation of this continuum ensures a more accurate model of pathogenesis and results in reduced uncertainty in the resultant estimates of health risk. The approaches discussed in the following sections address the exposure-dose-response continuum to differing degrees. The convention of the NAS paradigm is used (i.e., in the qualitative sense, the term dose encompasses exposure concentration, delivered dose, or target tissue dose).

Dose-response behavior is exemplified by the following types of data.

1. Quantal responses (dichotomous), in which the number of responding individuals in a population increase (e.g., number of animals in each exposure concentration with a specified health effect).
2. Count responses, in which the number of events is measured (e.g., number of lesion foci in tissue).
3. Dose-graded responses (ordered categorically), in which the severity of the toxic response within an individual or system increases with dose (e.g., pathology graded from mild to severe).
4. Continuous responses, in which changes in a biological parameter (e.g., organ weight or nerve conduction velocity) vary with dose.

Current risk assessment procedures must be developed to describe accurately effect measures expressed using these different types of data and to translate these data to an indication of toxic response.

Threshold versus Nonthreshold Effects

Most chemicals that produce noncancer toxicity do not cause a similar degree of toxicity in all organs, but usually demonstrate major toxicity to one or two organs. These are referred to as the target organs of toxicity for that chemical (Klaassen 1986). Generally, based on understanding of homeostatic, compensatory, and adaptive mechanisms, most risk assessment procedures operationally approach noncarcinogenic health effects as though there is an identifiable threshold (both for the individual and for the population) below which effects are not observable. For an individual, the threshold concept presumes that a range of exposures from zero to some finite value can be tolerated by the organism without adverse effects. For example, there could be a large number of cells that perform the same or similar function whose population must be significantly depleted before the effect is seen. This threshold will vary from one individual to another, so that there will be a distribution of thresholds in the population. Because sensitive subpopulations (i.e., those individuals with low thresholds) frequently are of concern in setting standards, risk assessment efforts are aimed at estimating levels at which these sensitive individuals would not be expected to respond.

The operational identification of a threshold distinguishes approaches for noncancer toxicity assessment from those for carcinogenic and mutagenic endpoints that risk assessment procedures currently approach as resulting from nonthreshold (stochastic) processes. However, it is recognized that there are inherent difficulties in the identification of population thresholds (Gaylor 1985). Identification of a threshold may be the result of the extensive nature of the data and the investigation of the effects in identified sensitive populations. For example, the LOAEL determination for the criteria pollutants (carbon monoxide, lead, nitrogen dioxide, particulate matter and sulfur dioxide) is based on an extensive body of data rather than an explicit identification of a LOAEL or NOAEL. Light microscopy was originally used to characterize morphological changes in the respiratory tract for ozone and that putative effect level was significantly higher than today when the effect level determination is based on electron microscopy. Thus, the operational identification of a threshold depends on the available data and frequently may be revised downward as more

information (e.g., studies encompassing additional endpoints) or more sensitive indicators of toxicity (e.g., effect measures) are developed and evaluated. It should be noted that as the exposure-dose-response continuum described above is characterized better for both certain carcinogens and noncarcinogens, knowledge of the mechanistic determinant may blur this distinction between approaches for noncancer toxicity and carcinogenicity.

Comprehensive Endpoint Array

As mentioned earlier, the operational identification of a threshold depends on the available data and frequently may be revised as more information, such as studies encompassing other endpoints than those previously assessed, becomes available. One of the major challenges to performing dose-response assessment for noncancer endpoints is that it requires the evaluation of effects measured in a number of different tissues. Often different endpoints are investigated in different studies, in different species, and at various concentrations. The effects measured may represent different degrees of severity or adversity within disease continuums. The available information must be synthesized into an assessment of the dose-response for noncancer toxicity based on the entire array of data.

In order to promote technical quality and consistency in risk assessment, guidelines have been published for evaluating toxicity data for a number of different endpoints, including both cancer and noncancer, and within certain issue areas such as for evaluating mixtures (EPA 1987g) or performing an exposure assessment (EPA 1992a). Guidelines also have been published for evaluating developmental toxicity (EPA 1991r) and proposed for evaluating female and male reproductive toxicity. At the time of this writing in 1992, additional guidelines were under development for other noncancer endpoints including neurotoxicity, immunotoxicity, and respiratory tract effects. The historical and conceptual development of the guidelines and their role in the EPA have been discussed elsewhere (EPA 1987g; Jarabek and Farland 1990). Within the context of risk assessment, these guidelines present key considerations and approaches to the evaluation of data within an individual endpoint to arrive at a dose-response estimate.

Although the available guidelines provide insight on how to think about and organize data on individual endpoints, assessment of noncancer toxicity requires determination of whether a suitable array of toxicity endpoints has been evaluated. This often is a difficult determination and requires sophisticated toxicological judgment. For example, if the available chemical disposition data indicate distribution to potential target tissues remote to the portal of entry or if disposition data are unavailable, then the potential for remote effects such as hepatic or renal toxicity, or reproductive and developmental toxicity, cannot be ruled out. Data from a bioassay may rule out the liver and kidney as target tissues, but unless two-generation reproductive and developmental studies also are available, questions remain regarding these important endpoints as potentially critical targets. Chemicals with specific mechanisms of action (e.g., neurotoxicity) also may present specific concerns that should be evaluated explicitly with the appropriate tests. Any methodology attempting to address noncancer toxicity should attempt to account for data gaps across all possible and plausible endpoints.

Temporal Relationships of Toxicity

Experimental exposures to animals usually are divided into four categories: acute, subacute, subchronic, and chronic. Acute exposure is defined as an exposure to a chemical for less than twenty-four hours. Although usually for a single administration (e.g., four hours), repeated exposures sometimes are given within the twenty-four-hour period. Subacute exposure refers to repeated exposure to a chemical for one month or less (e.g., a fourteen-day range finding study). Subchronic exposure refers to repeated exposure for one to three months, usually a ninety-day study. Finally, chronic exposure refers to repeated exposures for longer than three months, most commonly a two-year bioassay in rodents.

Depending on the disposition (i.e., deposition, absorption, distribution, metabolism, and elimination) of a chemical, acute toxicity may or may not manifest itself after prolonged repeated exposures. For many chemicals, the toxic effects following a single exposure are different from those produced by repeated exposure. For example, the acute toxic manifestation of benzene exposure is central nervous system depression, with chronic exposures resulting in blood dyscrasias and leukemia; however, acute exposure also can produce delayed toxicity. Conversely, chronic exposure to a toxic agent may produce some immediate (acute) effects after each exposure in addition to the long-term chronic effects.

The other temporal aspect of toxicity is the frequency of the exposures. In general, fractionation of the dose reduces the effect. If detoxifying biotransformation or elimination occurs between successive doses, or if the damage produced is repaired between successive doses, then a single dose may produce more toxicity than the same amount fractionated into many smaller doses given at intervals. Chronic effects occur if the chemical accumulates, if it produces irreversible effects, or if there is insufficient time for the target tissue to recover from the damage within the exposure frequency interval. Thus, to characterize the toxicity of a specific chemical, information is needed not only on acute and chronic effects but also for exposures of intermediate duration. Further, the appropriate measure of 'dose' must be defined by the nature of the pathogenesis process (i.e., defined according to the mechanism of action for the effect under consideration). Approaches for both 'acute' and 'chronic' duration data are presented here.

The above is important to consider when interpreting data for application in risk assessment and risk management scenarios. Currently, attempts are made to derive health benchmarks based on experimental exposure duration and frequency data that match the anticipated human exposure scenario; for example, one-hour laboratory animal data are used to extrapolate to one-hour human dose-response estimates. Often exposure concentrations associated with effects are prorated linearly to derive estimates for exposures of durations and frequency different from the experimental regimens. The rationale for this proration adjustment is that the resultant human exposure concentration should be the concentration times time ($C \times T$) equivalent of the experimental animal exposure level. This assumption of exposure equivalency is tenuous because steady-state conditions may not have been reached under some acute exposure conditions and is not consistent across different toxicity mechanisms (e.g., an effect mediated by peak blood concentration versus integrated tissue dose). Thus, depending on the mechanism of action, such adjustments may be inappropriate. In other cases, this default assumption has been

shown to be erroneous. However, when $C \times T$ data are lacking, there needs to be a convention for adjusting the exposure duration from the experimental regimen to the scenario of interest. The $C \times T$ proration adjustment, albeit imperfect, provides that convention. An attempt always should be made to take into account the mechanisms of toxic action as related to the temporal parameters of duration and frequency. When these relationships are unknown, greater uncertainty in the derivation is imparted. It should be further noted that this extrapolation is based on the assumption that the temporal relationships of toxicity are the same in laboratory animals and humans (e.g., a two-year 'lifetime' bioassay is equivalent to a seventy-year exposure to humans). Although common practice, further research is needed to substantiate these assumptions.

APPROACHES FOR CHRONIC DURATION DATA

Inhalation Reference Concentration (RfC) Methodology

EPA's interim methodology for development of inhalation reference concentrations is an approach for the dose-response component of risk assessment as originally proposed in the NAS framework. The inhalation reference concentration (RfC) is defined as an estimate (with uncertainty spanning perhaps a factor of ten) of a daily inhalation exposure to the human population that is likely to be without an appreciable risk of deleterious noncancer effects during a lifetime.

The inhalation RfC methodology provides guidance on the evaluation of different types of noncancer toxicity data (e.g., quantal, count, dose-graded, continuous), the designation of effect levels, and the choice of appropriate laboratory animal models in order to discern the critical endpoint and study that is representative of the threshold region for the entire data array of effects. The methodology also defines a minimum data base that is required for derivation.

Generally, when evaluating toxicity data for noncarcinogenic toxicity, the risk assessor must

decide on the critical endpoint to measure as a response and the correct measure of dose. In the simplest terms, a no-observed-adverse-effect-level (NOAEL) and a lowest-observed-adverse-effect-level (LOAEL) are determined for the specified adverse effect from the exposure levels of a given individual study on the various species tested. The NOAEL is the highest level tested at which the specified adverse effect is not produced and is, by definition, a subthreshold level (Klaassen 1986). It also depends on the exposure levels used in the experimental design and, thus, does not necessarily reflect the 'true' biological threshold.

The inhalation RfC methodology has departed from the oral risk reference dose (RfD) paradigm by the incorporation of dosimetric adjustments to scale the NOAELs and LOAELs observed in laboratory animals or in human studies to human equivalent concentrations (HECs) for ambient exposure conditions. These conditions currently are assumed to be twenty-four hours per day for a lifetime of seventy years. The dosimetric conversion to an HEC is necessary before the different adverse effects in the data array can be evaluated and compared.

Although it is preferable to use human studies as the basis for the dose-response derivation, adequate human data are not always available, forcing reliance on laboratory animal data. When presented with data from several laboratory animal studies, the animal model is sought that is most relevant to humans, based on the most defensible biological rationale. For example, a species with comparable pharmacokinetic and pharmacodynamic parameters for metabolizing the administered compound may be chosen. Lacking a clearly 'most relevant' species, however, the 'most sensitive' species (i.e, the species showing a toxic effect at the lowest administered dose) is used by EPA as a matter of scientific policy.

The critical toxic effect used in the dose-response assessment generally is the one characterized by the lowest NOAEL(HEC) that is representative of the threshold region for the available data array. The objective is to select a prominent toxic effect that is pertinent to the chemical's mechanism of action. This NOAEL is the key datum gleaned from an evaluation of all the studies, or from the data array, that summarize the dose-response relationship. This approach is based, in part, on the assumption that if the critical toxic effect is prevented, then all toxic effects are prevented.

This reliance on an individual effect level as the key datum gleaned from the toxicity profile and dose-response of a chemical has been

Table 7-1   The use of uncertainty factors in deriving reference dose (RfD) or reference concentration (RfC)

Standard uncertainty factors (UFs)	Processes considered
H = Human to sensitive human Extrapolation of valid experimental results from studies using prolonged exposure to average healthy humans. Intended to account for the variation in sensitivity among the members of the human population.	Pharmacokinetics, pharmacodynamics, sensitivity, Differences in mass (i.e., children, obese), concomitant exposures, activity pattern, no accounting for idiosyncracies
A = Animal to human Extrapolation from valid results of long-term studies on laboratory animals when results of studies of human exposure are not available or are inadequate. Intended to account for the uncertainty in extrapolating laboratory animal data to the case of average healthy humans.	Pharmacokinetics, pharmacodynamics, relevance of laboratory animal model, species sensitivity
S = Subchronic to chronic Extrapolation from less than chronic exposure results on laboratory animals or humans when there are no useful long-term human data. Intended to account for the uncertainty in extrapolating from less than chronic NOAELs to chronic NOAELs.	Accumulation/cumulative damage, pharmacokinetics/pharmacodynamics, severity of effect, recovery, duration of study, consistency of effect with duration
L = LOAEL to NOAEL Derivation from a LOAEL, instead of a NOAEL. Intended to account for the uncertainty in extrapolating from LOAELs to NOAELs.	Severity, pharmacokinetics/pharmacodynamics, slope of dose-response curve, trend, consistency of effect, relationship of endpoints, functional versus histopathological evidence, exposure uncertainties
D = Incomplete to complete data Extrapolation from valid results in laboratory animals when the data are 'incomplete'. Intended to account for the inability of any single laboratory animal study to adequately address all possible adverse outcomes in humans.	Quality of critical study, data gaps, power of critical study/supporting studies, exposure uncertainties

recognized as a significant limitation of the RfD/RfC approach (EPA 1990k; Crump 1984; Brown and Erdreich 1989). Alternate approaches that address these deficiencies are discussed below.

The inhalation RfC is a benchmark estimate that is operationally derived from the NOAEL-(HEC) for the critical effect by consistent application of uncertainty factors (UFs). The UFs are applied to account for recognized uncertainties in the extrapolations from the experimental data conditions to an estimate appropriate to the assumed human scenario. The standard UFs applied (as required) are listed in Table 7-1. Table 7-1 also lists the processes thought to be encompassed under each factor. The UFs generally are a factor of ten, although incorporation of dosimetry adjustments or other mechanistic data routinely has resulted in the use of reduced uncertainty factors. The composite UF applied to an RfC will vary in magnitude depending on the number of extrapolations required. An RfC will not be derived when use of the data involve greater than four areas of extrapolation. When four or more areas of uncertainty are involved in a given operational derivation, the overall UF applied is restricted to 3,000 in recognition of the lack of independence of these factors. An additional modifying factor (MF) also may be applied when scientific uncertainties in the study are not explicitly addressed by the standard UFs. For example, an MF might be applied to account for the poor statistical power of, or for marginal exposure characterization in, the chosen study.

It should be noted that the basis for the UFs is empirical and has been derived from oral data (Dourson and Stara 1983). Support for the UFs was based on analysis of the ratios of effect levels. For example, analysis of the ratios of NOAELs from oral ninety-day studies compared to NOAELs from chronic studies was used to support the ten-fold factor to account for subchronic to chronic extrapolation. Because the different types of toxicity (portal of entry versus remote) reflected in the different dosimetry adjustments may have different relationships to exposure concentration, the appropriate magnitude for these uncertainty factors when

using inhalation data is a topic of ongoing research at the EPA. Estimation procedures that are not sensitive to the spacing of exposure concentrations, such as the benchmark and Bayesian approaches discussed in the alternate approaches section below, are being explored.

Consideration of Comparative Inhaled Dose

As mentioned above, the inhalation RfC methodology incorporates dosimetric adjustments to scale the exposure concentration associated with an observed effect in laboratory animals or human occupational studies to a human equivalent concentration (HEC) for ambient exposures. Because it is recognized that the various species used in inhalation toxicology studies do not receive identical doses in comparable respiratory tract regions when they receive identical exposures, an attempt was made to account for the dynamics of the respiratory system as the portal of entry and for its interaction with the types of agents (particles and gases) to which it is exposed. The reader is referred elsewhere for documentation of the concepts outlined in this section (Jarabek et al. 1990; Jarabek et al. 1989).

The biological endpoint or health effect may be related more directly to the quantitative pattern of deposition within the respiratory tract than to the exposure concentration. This is because the regional deposition pattern determines not only the initial lung tissue dose but also the specific pathways and rates by which the inhaled material is cleared and redistributed (Schlesinger 1985). Particles initially are deposited by such mechanisms as inertial impaction, sedimentation, diffusion, interception, and electrostatic precipitation. Gas transport and uptake are governed mainly by convection, diffusion, chemical reactivity, solubility and perfusion (Overton 1984). Differences in airway size and branching pattern between species result in significantly different patterns of particle deposition and gas transport due to the effect of these geometric variations on airflow patterns (Snipes 1989). Clearance of the initially deposited dose also varies across species and with the respiratory tract region (i.e., extrathoracic, tracheobronchial, or pulmonary). Interspecies variations in cell morphology, numbers, types, distributions, and functional capabilities (e.g., metabolic enzymes) contribute to variations in clearance (Mercer and Crapo 1987; St. George et al. 1988).

The physicochemical characteristics of the inhaled agent also influence the deposition and retention in the respiratory tract, translocation within the respiratory system, distribution to other tissues, and ultimately the toxic effect (Bond 1989 and Overton 1984). The size and shape of the particles influence aerodynamic behavior and, thus, deposition (RAABE 1979; EPA 1982e; EPA 1982f). For a given aerosol, the two most important parameters determining deposition are the mean aerodynamic diameter and the distribution of the particle diameters about the mean. Information about particle size distribution aids in the evaluation of the effective inhaled dose (Hofmann 1982). The deposition site and rate of uptake of a volatile agent are determined by its reactivity and solubility characteristics.

The physicochemical characteristics of the inhaled agent also interact with physiologic parameters such as pulmonary ventilation, cardiac output (perfusion), metabolic pathways, tissue volumes, and excretion capacities. The pharmacokinetics of gases are governed by the rate of transfer from the environment to the tissue, the capacity of the body to retain the material, and the elimination of the parent compound and metabolites by chemical reaction, metabolism, exhalation, and excretion (Fiserova-Bergerova 1983; Overton 1984; EPA 1986q; Overton and Miller 1988; Bono 1989). The relative contributions or interactions of these processes with the physicochemical characteristics of an agent are, in turn, affected by the exposure conditions (concentration and duration).

Integration of these various physicochemical characteristics and physiologic processes is necessary for estimating the deposition (on airway surfaces) and absorbed dose in order to assess respiratory and extrarespiratory toxicity, respectively. Thus, in order to evaluate the different toxic responses to an inhaled agent observed across species and in order to extrapolate laboratory animal data dosimetrically to humans, a risk assessment methodology should utilize the available detailed data on the differing relative contributions of various deposition and absorption mechanisms for a given lung region. Information on regional differences in clearance and uptake pathways and possible target cell populations should be considered as well. The physicochemical characteristics of an agent should be integrated with this information in an attempt to reconstruct a dynamic picture for assessment of the observed toxicity. The dosimetric adjustments described in the next section were incorporated into the RfC methodology to account for some of these considerations.

Dosimetric Adjustment to Human Equivalent Concentration

Because most risk assessments are based on laboratory animal data, extrapolation from laboratory animals to humans is critical. This extrapolation requires integration of the various factors described in the previous section to estimate the 'dose' (i.e., agent mass deposited per unit tissue volume) delivered to specific target sites in the respiratory tract or made available to uptake and metabolic processes for extrarespiratory distribution (Martonen and Miller 1986). To this end, physiologically-based pharmacokinetic and empirically-based mathematical dosimetry models have evolved into particularly useful tools for predicting disposition differences for risk assessment (Miller et al. 1987). Their use is predicated on providing realistic estimates of target tissue dosimetry across species as the appropriate metric with which to gauge toxicological response rather than exposure concentration or administered dose. It is recognized that in some cases complementary, biologically-based models of the mechanism of action that link target tissue dose and effect also are needed.

The dosimetric adjustment factors applied in the inhalation RfC methodology represent default equations that have been derived from more sophisticated modeling approaches as applications appropriate to less extensive chemical-specific input data. These default adjustments are proposed because it was recognized that the supporting data base on most air toxics was not as robust as for some chemicals (e.g., ozone) that have undergone more sophisticated extrapolation modeling.

Dosimetric adjustment factors are applied to duration-adjusted NOAELs or LOAELs to estimate a concentration that would result in an equivalent exposure to humans. NOAELs or LOAELs are duration-adjusted by the number of hours per day and number of days per week of the laboratory animal exposure regimen for toxicity other than developmental endpoints, which are not duration-adjusted (EPA 1991r). The rationale for this proration-duration adjustment was discussed earlier in the section on temporal relationships. This duration-adjusted NOAEL is then dosimetrically extrapolated to the HEC. The HEC then is used as the basis for identification of the 'most sensitive' species and choice of the critical effect and study. For example, comparison of a mouse NOAEL versus a rat NOAEL may suggest that the mouse is more sensitive because the NOAEL for the mouse appears to be smaller. However, when the comparison is made using the HEC values, this relationship may be reversed and the rat becomes the more sensitive (Jarabek et al. 1990).

The dosimetric adjustments used to derive HECs are calculated according to different scenarios, based on the type of agent (particle or gas) and on the observed toxic effect (respiratory tract or remote toxicity).

1. Particle : Respiratory Effect
2. Particle : Remote Effect
3. Gas : Respiratory Effect

4. Gas : Remote Effect
    a. Optimal Physiologically-Based Pharmacokinetic (PBPK) Approach[3]
    b. Default Method
5. Human Occupational Scenario

Detailed equations for the dosimetric adjustments and major assumptions for each of the above scenarios and illustrative case examples are provided in Appendix A. Review of the appendix is encouraged for an appreciation of the key parameters and for the rationale of data array evaluation involved.

The HEC values are the basis for the operational derivation of the inhalation RfC as follows:

$$RfC = \frac{NOAEL\ (HEC)}{UF \times MF} \qquad (7\text{-}1)$$

The UF and MF are applied as described above. It must be emphasized that the inhalation RfC as a quantitative dose-response estimate is not numeric alone. As risk assessments have become a more prevalent basis for regulatory decision-making, their scientific quality and clarity have gained unprecedented importance (ACGIH 1989). Because of the complexity of many risk assessments, desirable attributes include the explicit treatment of all relevant information and the expression of uncertainty in each element (i.e., hazard identification, dose-response assessment, exposure assessment, and risk characterization). Any dose-response assessment such as the RfC contains uncertainty and imprecision because the process requires some subjective scientific judgment, use of default assumptions, and data extrapolations. A complete dose-response evaluation should include communication of the rationale for data selection, the strengths and weaknesses of the data base, key assumptions, and resultant uncertainties (Habicht 1992a; ACGIH 1989).

---

[3] The PBPK approach is listed for remote effects of gases because a number of peer-reviewed PBPK models for different chemicals exhibiting these types of effects are available; however, it is considered the optimal approach to use a more sophisticated dosimetry or PBPK model for any of the above scenarios whenever available rather than the default dosimetry adjustment.

Each inhalation RfC also has an associated level of confidence described as high, medium, or low. The confidence ascribed to the RfC estimate depends on both the confidence in the quality of the study and confidence in the completeness of the supporting data base. The rationale for the choice of the data from which the RfC is derived, a discussion of data gaps, and the resultant confidence in the RfC all are outlined in the summary of the inhalation RfCs entered in EPA's Integrated Risk Information System (IRIS). A discussion and rationale for the uncertainty factors used in the RfC derivation also are provided. This information is an integral part of the RfC and must be considered when evaluating the RfC as a dose-response estimate and, along with assumptions and resultant uncertainties inherent in the exposure assessment, when attempting to integrate the assessments into a risk characterization for risk assessment.

The minimum data base requirement for derivation of an inhalation RfC with low confidence is a well-conducted subchronic inhalation bioassay that evaluated a comprehensive array of endpoints, including portal of entry effects, and that established an unequivocal NOAEL and LOAEL. For a higher confidence RfC, chronic bioassay data, two-generation reproductive studies, and developmental studies usually are required in two different mammalian species. The requirements for the reproductive and developmental data may be mitigated by data indicating distribution of the chemical is unlikely to sites remote from the portal of entry. For some chemicals, the data base is so weak that the derivation of even a low confidence RfC is not possible. In such cases, the data base supporting an RfC for a chemical is designated as 'not-verifiable' by the EPA. This status is reevaluated upon the availability of new data.

Extrapolation of oral data is not recommended for deriving RfCs because the differences in dosimetry via each administration can drastically influence the disposition and resultant toxicity. In particular, EPA notes that oral data should not be used in the following four instances.

1. For groups of chemicals that are expected to have different toxicity by the two routes (e.g., metals, irritants and sensitizers).
2. When a first-pass effect is expected by the liver, or when the respiratory system was not adequately studied in the oral studies.
3. When a respiratory tract effect is established but dosimetry comparison cannot be clearly established between the two routes.
4. When short-term inhalation studies or in vitro studies indicate potential portal of entry effects at the respiratory tract, but the studies themselves are not adequate for an RfC derivation.

Important parameters to consider when modeling each administration route and a decision matrix to serve as guidance on route-to-route extrapolation has been published by Gerrity and Henry (1990).

Table 7-2 provides a list of the inhalation RfCs available at the time of writing in 1992. IRIS should be consulted for more contemporary information (see Appendix A for information on how to access IRIS).

Alternate Approaches to Dose-Response Assessment

Concentration-Response Modeling
Concentration-response modeling, recently referred to as the 'benchmark dose' approach, has been proposed as an improvement to the NOAEL/UF approach (Crump 1984). The benchmark approach is defined here as the use of a specific mathematical model (e.g., Weibull, polynomial) to determine a concentration (applied dose), and its lower confidence bound, associated with a predefined effect measure (e.g., 10 percent response of a dichotomous outcome) as the benchmark. Figure 7-1 illustrates the concentration–response modeling approach as applied to developmental data from laboratory animals. A mathematical model is superimposed (fitted) on the experimental effects data to estimate a maximum likelihood estimate (MLE) or dose–response function (solid line). The 95 percent confidence limit (dashed line) is calculated using information on

Table 7-2   Noncarcinogenic inhalation criteria for listed toxic air pollutants[a]

Toxic air pollutant	Inhalation RfC (mg/m^3)[b]	Uncertainty factor	Source[c]
Acetaldehyde	9E-3	1,000	IRIS
Acetamide	NA	NA	
Acetonitrile	NA	NA	
Acetophenone	NA	NA	
2-Acetylaminofluorene	NA	NA	
Acrolein	2E-5	1,000	IRIS
Acrylamide	NA	NA	
Acrylic acid	3E-4	1,000	IRIS
Acrylonitrile	2E-3	1,000	IRIS
Allyl chloride	1E-3	3,000	IRIS
4-Aminobiphenyl	NA	NA	
Aniline	1E-3	3,000	IRIS
o-Anisidine	NA	NA	
Asbestos	NA	NA	
Benzene	NA	NA	
Benzidine	NA	NA	
Benzotrichloride	NA	NA	
Benzyl chloride	NA	NA	
Biphenyl	NA	NA	
Bis(chloromethyl)ether	NA	NA	
Bis(2-ethylhexyl) phthalate	NA	NA	
Bromoform	NA	NA	
1,3-Butadiene	NA	NA	
Calcium cyanamide	NA	NA	
Caprolactam	NA	NA	
Captan	NA	NA	
Carbaryl	NA	NA	
Carbon disulfide	NA	NA	
Carbon tetrachloride	NA	NA	
Carbonyl sulfide	NA	NA	
Catechol	NA	NA	
Chloramben	NA	NA	
Chlordane	NA	NA	
Chlorine	NA	NA	
Chloroacetic acid	NA	NA	
2-Chloroacetophenone	3E-5	1,000	IRIS
Chlorobenzene	NA	NA	
Chlorobenzilate	NA	NA	
Chloroform	NA	NA	
Chloromethyl methyl ether	NA	NA	
Chloroprene	NA	NA	
Cresols/cresylic acid (isomers and mixture)	NA	NA	
o-Cresol	NA	NA	
m-Cresol	NA	NA	
p-Cresol	NA	NA	
Cumene	NA	NA	
2,4-D, salts and esters	NA	NA	
DDE	NA	NA	
Diazomethane	NA	NA	
Dibenzofurans	NA	NA	
1,2-Dibromo-3-chloropropane	2E-4	1,000	IRIS
Dibutylphthalate	NA	NA	
1,4-Dichlorobenzene	NA	NA	

Table 7-2   Continued

Toxic air pollutant	Inhalation RfC (mg/m³)[b]	Uncertainty factor	Source[c]
3,3-Dichlorobenzidene	NA	NA	
Dichloroethyl ether	NA	NA	
1,3-Dichloropropene	2E-2	30	IRIS
Dichlorvos	NA	NA	
Diethanolamine	NA	NA	
N,N-Diethylaniline	NA	NA	
Diethyl sulfate	NA	NA	
3,3-Dimethoxybenzidene	NA	NA	
Dimethyl aminoazobenzene	NA	NA	
3,3'-Dimethyl benzidene	NA	NA	
Dimethyl carbamoyl chloride	NA	NA	
Dimethyl formamide	3E-2	300	IRIS
1,1-Dimethyl hydrazine	NA	NA	
Dimethyl phthalate	NA	NA	
Dimethyl sulfate	NA	NA	
4,6-Dinitro-o-cresol and salts	NA	NA	
2,4-Dinitrophenol	NA	NA	
2,4-Dinitrotoluene	NA	NA	
1,4-Dioxane	NA	NA	
1,2-Diphenylhydrazine	NA	NA	
Epichlorohydrin	1E-3	300	IRIS
1,2-Epoxybutane	2E-2	300	IRIS
Ethyl acrylate	NA	NA	
Ethyl benzene	1E+0	300	IRIS
Ethyl carbamate	NA	NA	
Ethyl chloride	1E+1	300	IRIS
Ethylene dibromide	NA	NA	
Ethylene dichloride	NA	NA	
Ethylene glycol	NA	NA	
Ethylene imine	NA	NA	
Ethylene oxide	NA	NA	
Ethylene thiourea	NA	NA	
Ethylidine dichloride	NA	NA	
Formaldehyde	NA	NA	
Heptachlor	NA	NA	
Hexachlorobenzene	NA	NA	
Hexachlorobutadiene	NA	NA	
Hexachlorocyclopentadiene	NA	NA	
Hexachloroethane	NA	NA	
Hexamethylene-1,6-diisocyanate	NA	NA	
Hexamethylphosphoramide	NA	NA	
Hexane	2E-1	300	IRIS
Hydrazine	NA	NA	
Hydrochloric acid	7E-3	1,000	IRIS
Hydrogen fluoride	NA	NA	
Hydroquinone	NA	NA	
Isophorone	NA	NA	
Lindane	NA	NA	
Maleic anhydride	NA	NA	
Methanol	NA	NA	
Methoxychlor	NA	NA	
Methyl bromide	5E-3	100	IRIS
Methyl chloride	NA	NA	

Table 7-2   Continued

Toxic air pollutant	Inhalation RfC (mg/m^3)[b]	Uncertainty factor	Source[c]
Methyl chloroform	NA	NA	
Methyl ethyl ketone	1E+0	1,000	IRIS
Methyl hydrazine	NA	NA	
Methyl iodide	NA	NA	
Methyl isobutyl ketone	NA	NA	
Methyl isocyanate	NA	NA	
Methyl methacrylate	NA	NA	
Methyl-t-butyl ether	5E-1	1,000	IRIS
Methylene chloride	NA	NA	
Methylene diphenyl diisocyanate (MDI)	NA	NA	
4,4-Methylene bis(2-chloroaniline)	NA	NA	
4,4'-Methylenedianiline	NA	NA	
Naphthalene	NA	NA	
Nitrobenzene	NA	NA	
4-Nitrobiphenyl	NA	NA	
4-Nitrophenol	NA	NA	
2-Nitropropane	2E-2	1,000	IRIS
N-Nitroso-N-methylurea	NA	NA	
N-Nitrosodimethylamine	NA	NA	
N-Nitrosomorpholine	NA	NA	
Parathion	NA	NA	
Pentachloronitrobenzene	NA	NA	
Pentachlorophenol	NA	NA	
Phenol	NA	NA	
p-Phenylenediamine	NA	NA	
Phosgene	NA	NA	
Phosphine	NA	NA	
Phosphorus	NA	NA	
Phthalic anhydride	NA	NA	
PCBs	NA	NA	
1,3-Propane sultone	NA	NA	
β-Propiolactone	NA	NA	
Propionaldehyde	NA	NA	
Propoxur (Baygon)	NA	NA	
Propylene dichloride	4E-3	300	IRIS
Propylene oxide	3E-2	100	IRIS
1,2-Propyleneimine	NA	NA	
Quinoline	NA	NA	
Quinone	NA	NA	
Styrene	1	30	IRIS
Styrene oxide	NA	NA	
2,3,7,8-Tetrachlorodibenzo-p-dioxin	NA	NA	
1,1,2,2-Tetrachloroethane	NA	NA	
Tetrachloroethylene	NA	NA	
Titanium tetrachloride	NA	NA	
Toluene	4E-1	300	IRIS
2,4-Toluene diamine	NA	NA	
2,4-Toluene diisocyanate	NA	NA	
o-Toluidine	NA	NA	
Toxaphene	NA	NA	
1,2,4-Trichlorobenzene	NA	NA	
1,1,2-Trichloroethane	NA	NA	
Trichloroethylene	NA	NA	

Table 7-2   Continued

Toxic air pollutant	Inhalation RfC (mg/m^3)[b]	Uncertainty factor	Source[c]
2,4,5-Trichlorophenol	NA	NA	
2,4,6-Trichlorophenol	NA	NA	
Triethylamine	7E-3	3,000	IRIS
Trifluralin	NA	NA	
2,2,4-Trimethylpentane	NA	NA	
Vinyl acetate	2E-1	30	IRIS
Vinyl bromide	NA	NA	
Vinyl chloride	NA	NA	
Vinylidene chloride	NA	NA	
Xylenes (isomers and mixtures)	NA	NA	
o-, m-, p-Xylene	NA	NA	
Antimony compounds	NA	NA	
Arsenic compounds	NA	NA	
Beryllium compounds	NA	NA	
Cadmium compounds	NA	NA	
Chromium compounds	NA	NA	
Cobalt compounds	NA	NA	
Coke oven emissions	NA	NA	
Cyanide compounds	NA	NA	
Glycol ethers	NA	NA	
Lead compounds	NA	NA	
Manganese compounds	4E-4	300	IRIS
Mercury	NA	NA	
Fine mineral fibers	NA	NA	
Nickel compounds	NA	NA	
Polycyclic organic compounds	NA	NA	
Radionuclides	NA	NA	
Selenium compounds	NA	NA	

[a] Information from EPA data bases only (November 1992).

[b] RfC = Risk reference concentration.

[c] IRIS = Integrated Risk Information System.

sample size and variance. It has been recommended that limits based on the distribution of the likelihood ratio statistic be used as the method of choice for this calculation (Crump, 1984). The possible analogues to a NOAEL then can be estimated. For example, the 10th percentile of an effect level could be designated as synonymous to 'no adversity' and the concentration responding to the MLE of that effect level used as the 'effective concentration' (EC_{10}). The lower confidence bound on the EC_{10} could also be used and is shown as the LEC_{10}.

Concentration-response modeling has the advantages that it utilizes more information from the dose-response curve, is less influenced by experimental design (e.g., exposure level spacing), and is sensitive to the influence of sample size. It should be noted that this approach is sensitive to sample size only when the benchmark is defined as the lower-confidence bound. The maximum likelihood estimate alone is not influenced by sample size. Application of this approach has been proposed for developmental endpoints (Kimmel and Gaylor 1988) but it has not yet been applied widely to other noncancer outcomes.

Application of this approach to the myriad of endpoints that can constitute noncancer toxicity will require significantly greater effort directed at modeling continuous data. Guidance also must be developed on the choice of model

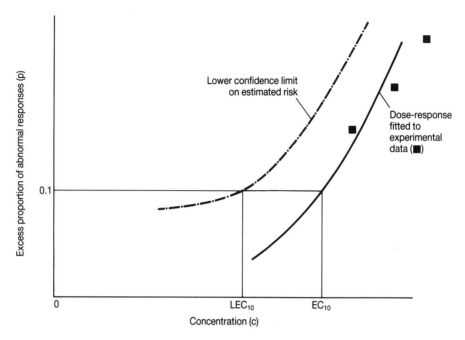

Figure 7-1.   Graphical illustration of concentration-response modeling for developmental toxicity. Adapted from Kimmel and Gaylor (1988).

structures and on goodness-of-fit criteria for models, especially whether it is appropriate to superimpose model structures on data that only have one dose group associated with a nonzero response (relative to control or background).

Despite the advantages of this approach for deriving an estimate for an individual endpoint, the choice of the critical study from the overall data array remains a key assessment decision. In part, this decision is influenced by what is designated as the specified health effect (i.e., the modeled dose-response estimate will be different if, for example, an incidence of 10 percent or 30 percent was used as a specified health effect). Further, if 10 percent incidence is used as a criterion as has been suggested (Kimmel and Gaylor 1988), one could develop a dose-response estimate based on either 10 percent incidence of nasal lesions or 10 percent liver necrosis. Thus, severity of the endpoint becomes an issue and the designation of a specified health effect requires intimate knowledge of the spectrum of severity within a

pathogenesis continuum for an individual endpoint. One must be aware of how the resultant model estimate is influenced by that choice. Previous use of the benchmark approach avoided this controversy because developmental endpoints do not distinguish degrees of severity to a large extent.

Derivation of a dose-response estimate by the benchmark also does not preclude evaluation of the data base for completeness. A comprehensive array of endpoints must be evaluated to identify potential hazards for various target tissues regardless of the way individual endpoints may be modeled. Use of the benchmark approach also does not preclude the necessity for dosimetric adjustment to an HEC and for application of uncertainty factors to account for extrapolations. Once the individual specified health effects are decided, determinations of the appropriate species and critical effect representative of the threshold for the overall data array must be evaluated as described for the RfC methodology.

Application of Bayesian Statistics

The analysis of noncancer toxicity data often involves evaluation of data and a synthesis of information together in order to determine a representative level for the threshold region of the data array. For example, sometimes a NOAEL from one study may be used in conjunction with a LOAEL from another. The advantage to such a synthesis is the utilization of more information rather than the reduction of data to a single effect level that is recognized as a significant limitation to the RfC approach described above.

The Bayesian statistical approach described by Jarabek and Hasselblad (1991) statistically incorporates the attributes of the benchmark approach (incorporates influence of sample size and shape of the dose-response curve) and offers the following three advantages.

1. Visual display and description of the uncertainty in the risk estimates.
2. Explicit synthesis of risk estimates together when determined appropriately.
3. Explicit incorporation of uncertainty in the exposure characterization (proposed).

The proposed approach has been published under the title of the Confidence Profile Method (Eddy, Hasselblad, and Shachter 1992). It combines the standard classical and Bayesian statistical methods to produce likelihood functions and posterior distributions for parameters of interest. Although the likelihood functions and posterior distributions have very different interpretations, their shape usually is highly similar. The likelihood function can be used to compute confidence intervals. The posterior distribution is a continuous plot describing belief about the location of the parameter of interest (i.e., for dose-response estimation purposes), about the dose associated with a specified health effect. The basic formula of Bayesian statistics is

$$p'(\theta) = L(\theta|\text{data})\,p(\theta) \qquad (7\text{-}2)$$

Where

θ = parameter of interest
$p(\theta)$ = prior distribution for θ
$L(\theta|\text{data})$ = likelihood for θ given new data
$p'(\theta)$ = posterior distribution for θ (because $p'(\theta)$ will become the prior for the next experiment, it is denoted by the same letter)

Using a continuous effect measure as an example case, assume the parameter of interest is x_0, the exposure concentration associated with a specified health effect. Because x_0 is not defined for $\beta \leq 0$, it is reasonable to choose the prior for β as $p(\beta) = 1$ for $\beta > 0$, otherwise $p(\beta) = 0$ (because we usually do not believe exposure to a toxic chemical is beneficial). This prior is the horizontal dashed line in Figure 7-2. Assume that an experiment to determine information about θ was conducted, resulting in the likelihood, $L(\theta)$, shown as a dotted line in Figure 7-2. Note that the likelihood is positive for values of β less than 0. The posterior distribution, $p'(\beta)$, is the product of the two prior distributions (properly normalized to be a probability distribution) and is shown as a solid line in Figure 7-2. Note that this distribution has the same general shape as the likelihood function, except that it has no mass below zero. This kind of distribution often is referred to as a truncated distribution. The posterior distribution of x_0 can be calculated from the posterior distribution of β.

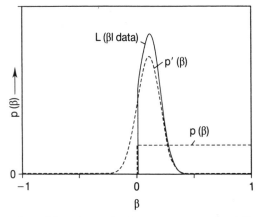

Figure 7-2. Schematic of computing a posterior $p'(\beta)$ from a likelihood function $L(\beta|\text{data})$ and a prior distribution $p(\beta)$. Source: Jarabek and Hasselblad (1991).

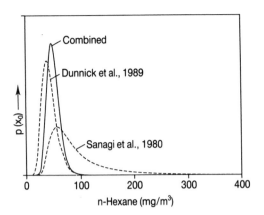

Figure 7-3. Posterior distribution for the concentration of *n*-hexane associated with the specified health effects from the combined evidence of Sanagi et al. (1980) and Dunnick et al. (1989). Source: Jarabek and Hasselblad (1991).

The posterior distribution, $p'(\beta)$, can be used as prior if another experiment is conducted giving additional information about x_0 and β, and the application of Bayes' formula repeated. This application is illustrated in Figure 7-3 when the posterior distribution of concentrations associated with specified health effects for n-hexane are statistically synthesized together (Jarabek and Hasselblad 1991).

Analysis of Figure 7-3 illustrates the advantages of the Bayesian approach. Although experimental details are provided elsewhere (Jarabek and Hasselblad 1991), the two data sets represent both continuous and dichotomous effect measures illustrating the ability of the Bayesian approach to address different outcomes. It should be noted that the mathematical modeling of these data for these effect measures was not different from that achieved using a benchmark approach, but the expression as a normalized posterior distribution is the difference that provides for visual inspection and statistical combination of data. Visual inspection of the posterior distribution concurs with the variability of the data and provides much information about the usefulness of the health effects data for risk evaluation. The shape of the posterior distribution for the data of Dunnick et al. (1989), in contrast to that of Sanagi et al.

(1980), easily highlights that these data were generated from an investigation with an adequate number of animals and test concentrations with a resultant tighter distribution and reduced variance.

The Bayesian approach is the only approach that offers the ability to combine statistically the evidence from different investigations. Figure 7-3 presents the statistical combination of data with different endpoints – neurotoxicity (Sanagi et al. 1980) and respiratory tract effects (Dunnick et al. 1989). The resultant posterior distribution for the combined evidence of different endpoints was not drastically different relative to the individual distributions from which it was derived. This may be due to the fact that both studies investigated very sensitive endpoints (i.e., near the threshold or subthreshold region). Perhaps when data are not comparable with respect to assayed endpoints, but represent very sensitive endpoints, then the combination of these data provide a more likely estimate of the concentration of concern. Future development of this approach will develop guidance on limitations for data combination.

Categorical Regression: Use of Dose-Graded Data

As mentioned in the dose-response section, not all data are expressed as quantal or continuous data that are readily amenable to available standard dose-response models. Results often are reported as 'categorical' (i.e., descriptive or severity-graded results). For example, a particular dose group might exhibit 'mild' toxicity. However, other studies that are not designed explicitly to examine dose-response relationships, such as single-dose studies or mechanistic studies, may provide useful data that should be incorporated into the data array analysis.

An analysis method that allows the combination of quantal data with categorical data and models the relationship between the severity of the effect and the concentration and duration of inhalation exposures has been presented for the analysis of short-term exposures and is described in the following sections (Guth et al. 1992). It should be noted that such an approach

Table 7-3 Comparison of occupational exposure limits

Organization/ exposure limit	Exposure assumption scenario	Concentration duration assumption	Effect severity	Use of UF	Population	Derivation
ACGIH/TLV-TWA	Routine occupational exposure	8 hours/day; 40 hours/week for a working lifetime (40 years)	No adverse effect	Minimal UF used; no systematic application	Nearly all workers; personal protective equipment may be factored	Based on best available information from industrial experience, experimental human and animal studies (human preferred). No systematic basis; derived by expert committee
ACGIH/TLV-STEL	Routine occupational exposure	15 minutes time-weighted average exposure that should not be repeated more than 4 times per day	Protect against irritation, chronic or irreversible tissue damage, or narcosis of sufficient degree to increase chances of accidental injury, impair self-rescue or reduce work efficiency	Same as above	Healthy worker	Same as above
NIOSH/PEL	Routine occupational exposure	Up to 10 hours/day; 40 hours/week; undefined working lifetime duration; appropriate control and surveillance methods	Same as above	Same as above	Same as above	Same as above
OSHA/PEL	Routine occupational exposure	8 hours/day; 40 hours/week; 45 year working lifetime duration; appropriate control and surveillance methods	Protect worker against a wide variety of health effects that could cause material impairment of health or functional capacity	Same as above	Same as above	Same as above. In addition, technological feasibility is considered in establishing a PEL

Table 7-4   AAL values for acetaldehyde in various states

State	OEL	UF	Averaging time (hours)	Conc. (mg/m^3)
Connecticut	TLV	50	8	3.6
Indiana	TLV	200	8	18.0
North Carolina	TLV	10	0.25	27.0
North Dakota	TLV	100	8	1.8
North Dakota	TLV	100	1	2.7
Nevada	TLV	42	8	4.29
New York	TLV	300	1 year	0.6
Virginia	TLV	60	24	3.0
Vermont	TLV	100	8	1.8

Source: NATICH, 1989.

also could be used to analyze data for chronic exposures.

Occupational Exposure Limits (OELs)

Although OELs have made an important contribution to risk management for worker safety, differences in the philosophy, legal mandate, and objectives of the sponsoring organization often result in disparate values for the same chemical. OELs typically are derived by three organizations: the American Conference of Governmental Industrial Hygienists (ACGIH), the National Institute of Occupational Safety and Health (NIOSH), and the Occupational Safety and Health Administration (OSHA). A description and the mandates of these organizations are summarized below.

1. ACGIH
   a. Professional organization
   b. Recommends guidelines
   c. No legal mandate
2. NIOSH
   a. Research institution
   b. Recommends standards to OSHA
   c. No legal mandate
3. OSHA
   a. Regulatory agency
   b. Develops standards[4]

Differences in mandate, operative assumptions, and derivation procedures for the devel-

opment of OELs by the three organizations are shown in Table 7-3.

Compounding the differences in the OEL values themselves, state and local agencies often have different approaches for using the same (or different) OELs to derive an AAL. Table 7-4 shows the range in resultant AALs for a single chemical (acetaldehyde) resulting from different approaches in using the same TLV. Differences across the approaches result principally from the lack of a consistent basis for the application of uncertainty factors and different assumptions for averaging time.

Although OELs have been used to derive AALs for community health protection, most OEL documentation contains the following type of disclaimer.

These limits are intended for use in the practice of industrial hygiene as guidelines or recommendations in the control of potential health hazards and for no other use, e.g., in the evaluation or control of community air pollution nuisances, in estimating the toxic potential of continuous, uninterrupted exposures ... (ACGIH 1989)

Careful consideration of the organizational mandates, operational assumptions, derivation, and populations intended for protection is necessary when evaluating the appropriateness of OELs for application in risk management programs. The use of OELs to derive AALs is discussed in more detail in Chapter 17.

---

[4] As defined by OSHA, a 'standard ... requires conditions, or the adoption or use of one or more practices, means, methods, operations or processes, necessary or appropriate to provide safe or healthful employment and places of employment.' (29 CFR 1910.2(f))

APPROACHES FOR LESS-THAN-LIFETIME EXPOSURE

Historically, EPA's concern with short-term airborne exposures to air toxics has centered primarily around accidental releases. This led to the establishment of EPA's Chemical Emergency Preparedness Program (CEPP) as part of the Air Toxics Strategy released in 1985. This program had two goals: (1) to increase community awareness of chemical hazards and (2) to enhance state and local emergency planning for dealing with accidental airborne releases of acutely toxic chemicals (now known as extremely hazardous substances, or EHS). The original CEPP served as a basis for the present requirements mandated by the Emergency Planning and Community Right-to-Know Act (EPCRA, enacted as Title III of the 1986 Superfund Amendment and Reauthorization Act). Under EPCRA, federal, state, and local governments as well as private industry must comply with numerous requirements with respect to emergency planning, community right-to-know, hazardous emissions reporting, and emergency notification. Title III required EPA to publish a list of EHS and threshold planning quantities (TPQ) for each of the EHS. The TPQ are the levels of emissions that cannot be exceeded under any circumstances and are based on acute toxicity, physical and chemical properties, and production volume. The Toxic Release Inventory (TRI) that resulted from section 313 of EPCRA is discussed in more detail in Chapter 26.

As our understanding of the exposure-dose-response continuum is refined and the temporal aspects of the pathogenesis mechanisms are elucidated, dose-response benchmarks for health risk estimation may be able to address intermediate periodic exposure scenarios with greater accuracy. As discussed in the section on temporal relationships of toxicity, however, current practices to address 'acute' or 'short-term' exposures usually rely on health effects data in which the experimental exposures were also of 'acute' duration and similar frequency. It must be emphasized that the three health risk assessment procedures discussed above – the RfC approach, the benchmark approach, and the Bayesian approach – all also are applicable to short-term data.

Categorical Regression: Use of Dose-Graded Data

Assessment of the health risk of noncancer effects from exposure durations ranging from minutes to hours is complicated by the critical influence of exposure duration on toxicity. When plotted on a concentration versus duration chart, toxicological data generally show a hyperbolic form that can be described as concentration times duration equals a constant or concentration raised to a power times duration equals a constant. These curves indicate that response rate is highly dependent on both exposure concentration and duration. In attempting to develop a general approach for health assessment for short-term exposures, a further complication is the fact that it is difficult to generalize regarding the exposure duration that is most relevant to human health risk because this will be dependent on the specific source and scenario of the exposure as well as on the nature of a chemical's pathogenesis or mechanism of action with respect to duration.

Guth et al. (1992) proposed a regression analysis method that provides for incorporation of both quantal and dose-graded data and for data across different durations. The method has been proposed in order to utilize as much of the available data possible for the evaluation of short-term inhalation exposures defined as less than or equal to twenty-four hours in duration.

A categorization scheme is used in quantitative exposure severity analysis, with severity category as the dependent variable and with concentration and exposure duration as independent variables. The severity scheme consists of three categories representing NOAELs, adverse effect levels (AELs), and lethality. More complicated severity ranking schemes can be applied but become contentious due to the difficulty in equating severity of effect measures across target organs, endpoints, and species (Guth et al. 1992).

The form of the model for regression analysis is

$$\ln \frac{p}{1-p} = A_i + B_1 \ln \text{(concentration)}$$

$$+ B_2 \text{(duration)} \qquad (7\text{-}3)$$

Where p is the probability that, at a given concentration and duration of exposure, severity will be less than or equal to the severity category with rank i, A and B are estimated model parameters.

The model is solved for $P = 1-p$, or the probability that, at a given concentration and duration, the severity will be greater than the severity category with rank i. The regression analysis assumes constant slope parameters, hence the values of B_1 and B_2 are constant across severity categories. The order or rank of the categories is used instead of the numerical values.

The model output is interpreted readily in the context of risk assessment. Figure 7-4 illustrates the method applied to categorical data for exposures of less than eight hours in duration. The maximum likelihood model fit is shown by the line representing the model prediction of $p = 0.1$ that severity is greater than the NOAEL category (i.e., that the predicted effect would be in the 'adverse' range or higher) at the corresponding exposure concentration and duration.

This categorical regression analysis approach appears promising as it addresses many of the problems inherent in dose-response assessment for short-term exposures. A similar approach is used by the European Chemical Industry Ecology and Toxicology Centre (ECETOC) to derive emergency exposure indices (EEIs) for chemicals. The approach is under development by EPA and plans for refinement include development of guidance for model application, guidance on whether to aggregate or disaggregate data on different endpoints for the analysis, and data base requirements (Guth et al. 1991). As with the other alternate approaches discussed, application of interspecies dosimetry adjustments and UFs for data extrapolation is warranted.

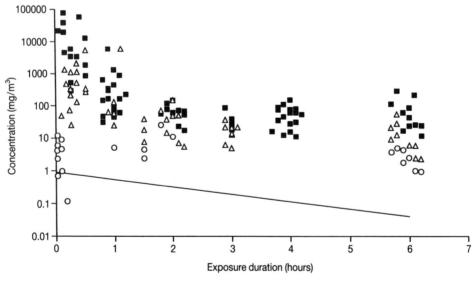

Figure 7-4.   Categorical data from published results on methyl isocyanate for exposures of less than 8 h in duration and shown as NOAEL (O), AEL (△), or lethality (■). The maximum likelihood model fit is shown by the line representing the model prediction ($p = 0.1$) that severity is greater than the NOAEL category at the corresponding exposure concentration and duration. Source: Guth et al. (1993).

Other Measures

EEGLs, SPEGLs, and CEGLs

At the request of the Department of Defense (DOD), the National Research Council's Committee on Toxicology (COT) developed exposure scenario-specific guidance levels for military personnel operating under emergency conditions whose circumstances are peculiar to military operations and for which regulatory agencies have not set relevant standards (NAS 1986). These guidance levels are intended exclusively for use by DOD for its particular exposure scenarios, and are not to be interpreted as synonymous with standards promulgated by regulatory agencies for generic exposure conditions. As such, they represent risk management, rather than risk assessment, decisions. These guidance levels include the Emergency Exposure Guidance Level (EEGL), the Short-Term Public Emergency Guidance Level (SPEGL), and the Continuous Exposure Guidance Level (CEGL). Although it is assumed that exposures to concentrations above these guidance levels may result in adverse effects in some people, these levels do not connote a distinction between safe and unsafe concentrations. Conversely, there may be some individuals that experience adverse health effects when exposed to concentrations of a substance below these guidance levels.

The EEGL (previously known as the Emergency Exposure Limit, or EEL) is defined as follows.

[A] ceiling guidance level for single emergency exposure, usually lasting from 1 hour to 24 hours – an occurrence expected to be infrequent in the lifetime of a person. 'Emergency' connotes a rare and unexpected situation with potential for significant loss of life, property or mission accomplishment if not controlled. (NAS 1986)

An EEGL is intended for use only in short-term emergency situations to prevent irreversible toxicity, and it is expected that some risk or discomfort (e.g., increased respiratory rate from increased carbon dioxide, headache or mild central nervous system effects, or respiratory or eye irritation) that does not impair judgment and does not interfere with proper responses to the situation will occur in order to prevent greater risks (e.g., fire, explosion, or massive release). The EEGL differs from the Short-Term Exposure Limit (STEL) recommended by OSHA or ACGIH in that the STEL is intended to protect workers exposed daily to chemicals for fifteen-minute durations over a lifetime and the EEGL is intended for one-time only emergency situations. Furthermore, it is assumed when deriving an EEGL that there will be complete recovery from any adverse effects. In the case of carcinogens, however, additional guidance is provided so that exposure concentrations will not result in a cancer risk of greater than 1 in 10,000 (1×10^{-4}).

The CEGL (previously known as the Continuous Exposure Limit, or CEL), is defined as follows.

[T]he ceiling concentration designed to avoid adverse health effects, either immediate or delayed, of more prolonged exposures and to avoid degradation in crew performance that might endanger the objectives of a particular mission as a consequence of continuous exposure for up to 90 days. In contrast with EEGLs, which are intended to guide exposures during emergencies (exposures that, although not acceptable under normal operating conditions, should not cause serious or permanent effects), CEGLs are intended to provide guidance for operations lasting up to 90 days in an environment like that of a submarine. (NAS 1986)

When deriving a CEGL, that is designed to be protective of military personnel involved in normal long-term operations, pharmacokinetic processes such as accumulation, biotransformation, and excretion are taken into account (when data are available) and the CEGL is adjusted accordingly. For example, if a substance is known to accumulate in the body with repeated exposure, then the CEGL is adjusted downward.

Inasmuch as EEGLs and CEGLs are not intended for the general civilian population, certain assumptions are inherent in their development. For example, it is assumed that military personnel are generally healthy, young, and homogenous. However, some allowance is made for the interindividual variability in the sensitivity of response to chemicals. In addition, because women are in the military, the potential for the occurrence of developmental and reproductive toxicity also is considered. Although it is assumed that the population to be protected by these guidance levels will have appropriate protective equipment and have planned emergency escape routes, the EEGLs are not based on the availability of these measures.

The SPEGL (previously known as the Short-Term Public Emergency Limit, or SPEL) is defined as follows.

[A] suitable concentration for unpredicted, single, short-term, emergency exposure of the general public. In contrast to the EEGL, the SPEGL takes into account the wide range of susceptibility of the general public. This includes sensitive populations – such as children, the aged, and persons with serious debilitating diseases. Effects of exposure on the fetus and on reproductive capacity of both men and women also should be considered. (NAS 1986)

EEGLs are expressed in units of ppm or mg/m^3 and are derived using acute toxicity data as the primary basis. All health effect endpoints (including reproductive in both sexes, developmental, carcinogenic, neurotoxic, respiratory, and other organ specific toxicities) for both immediate and delayed effects are considered. The most sensitive endpoint (NOAEL or LOAEL), as determined by the available data, is chosen as the basis for the EEGL because it is assumed that a level that protects against the occurrence of this effect also will protect against the occurrence of other, less sensitive effects.

Human data are the preferred basis for an EEGL, but extrapolation from animal data is done lacking reliable quantitative human data. The COT recommends that several default parameters, such as the breathing rate for the experimental animal being used and the breathing rate for a seventy kilogram human, be used to extrapolate from animal data following inhalation exposure to calculate an EEGL for humans. If the EEGL must be extrapolated from an oral animal NOAEL or LOAEL, then it is

assumed that the oral doses are equivalent, and the human oral dose is converted to an inhalation concentration by multiplying the oral dose (in milligrams per kilogram) by seventy kilograms and dividing the product by the appropriate breathing rate in cubic meters to arrive at an EEGL.

Safety factors are not applied routinely in the calculation of EEGLs unless confidence in the available data is low. For example, if the EEGL is based on a short-term animal NOAEL or LOAEL, or if the route of exposure for the selected NOAEL or LOAEL differs from that of the expected exposure scenario for which the EEGL is to apply, then a safety factor of one to ten generally is applied to the NOAEL or LOAEL to arrive at the EEGL. If the chemical is a carcinogen, safety factors are applied to reduce the risk of cancer to 1 in 10,000.

EEGLs generally are calculated first for the shortest exposure duration anticipated (i.e., ten to fifteen minutes) and then longer-duration EEGLs (e.g., one hour and twenty-four hour) are calculated based on this number. In order to calculate the EEGLs for different durations, it is generally assumed that Haber's Law again applies (i.e., concentration times time is a constant) and is appropriate for all short-term exposure situations. In other words, if it is known that exposure to one hundred parts per million of a chemical causes a particular effect in one minute, then a ten-minute exposure to that chemical should produce the same effect at a concentration of ten parts per million (i.e., $100 \times 1 = 10 \times 10$). This assumption may be inappropriate for chemicals where it is known that greater toxicity will result from exposure to high concentrations over a short period as compared to exposure to lower concentrations over a longer period; thus, the application of Haber's Law must be considered on a case-by-case basis.

SPEGLs generally are calculated by dividing the EEGL by a factor of two to ten. A safety factor of two (i.e., divide the EEGL in half) may be applied to protect more sensitive groups in the general civilian population such as children or the elderly, and a safety factor of ten (i.e., divide the EEGL by 10) may be considered more appropriate to protect fetuses or newborns.

CEGLs generally are calculated by dividing the EEGL by a factor of ten to one hundred. A safety factor of ten may be applied to the EEGL to derive a CEGL if it is known that the chemical is substantially and rapidly biotransformed to a nontoxic substance. A safety factor of one hundred is more appropriate if it is known that the detoxification process is slow, or if detoxification does not occur. If it is known that the substance accumulates in the body, a higher safety factor may be applied. Chronic data also may be used to derive a CEGL, with the application of additional safety factors if deemed necessary.

ERPGs

The incident in Bhopal, India, illustrated the need for chemical manufacturers to evaluate plant design and community planning measures and to have standardized emergency response planning guidelines to be used in the event of an accidental chemical release. ERPGs have been developed under the guidance of the Organization Resources Counselors, Inc. Emergency Response Planning Guidelines Task Force (ERPG Task Force).[5] The ERPGs are intended to provide emergency planning guidelines for limits on short-term (one-hour) exposures of the neighboring community occurring as a result of an accidental chemical release caused by process or human failures. The ERPGs are formally reviewed by a technical committee established by the American Industrial Hygiene Association (AIHA) before publication.

As described in AIHA (1989), the assumptions underlying the ERPGs are as follows.

1. The numbers are useful primarily for emergency planning and response.
2. The numbers are suitable for protection from health effects due to short-term exposures. They are not suitable for effects due to

repeated exposures, nor as ambient air quality guidelines.
3. The numbers are guidelines. They are not absolute levels demarcating safe from hazardous conditions.
4. The numbers do not necessarily indicate levels at which specific actions must be taken.
5. The numbers are only one element of the planning activities needed to develop a program to protect the neighboring community.
6. The selection of chemicals needing emergency planning guidelines generally should be based on volatility, toxicity, and releasable quantities.

The ERPGs are calculated for three levels, defined as follows.

ERPG-3: The maximum airborne concentration below which it is believed that nearly all individuals could be exposed for up to one hour without experiencing or developing life-threatening health effects.

The ERPG-3 level is a worst-case planning level above which there is the possibility that some members of the community may develop life-threatening health effects. This guidance level could be used to determine if a potentially releasable quantity of a chemical could reach this level in the community, thus demonstrating the need for steps to mitigate the potential for such a release.

ERPG-2: The maximum airborne concentration below which it is believed that nearly all individuals could be exposed for up to one hour without experiencing or developing irreversible or other serious health effects or symptoms that could impair an individual's ability to take protective action.

Above ERPG-2, for some members of the community, there may be significant adverse health effects or symptoms that could impair an individual's ability to take protective action. These symptoms might

[5]The ERPG Task Force is made up of chemical manufacturers. Their mandate is to recommend ERPGs for use in the event of accidental chemical releases. The resulting ERPGs are reviewed by a technical committee under the auspices of the American Industrial Hygiene Association (AIHA).

Table 7-5 Comparison of less than lifetime exposure limits

Organization/ exposure limit	Exposure assumption scenario	Concentration duration assumption	Effect severity	Use of UF	Population	Derivation
COT/EEGL	Emergency exposure planning	1 and 24 hour exposure	Revesible effects acceptable, e.g., headache, irritation, CNS effects	Generally no (unless confidence in data base is low or chemical is carcinogen)	Military personnel, assumed to be healthy and relatively homogeneous	Based on most sensitive end-point (NOAEL or LOAEL) from human or animal toxicity data (acute toxicity data preferred). All endpoints considered
COT/SPEGL	Emergency exposure planning	1 and 24 hour exposure	Reversible effects acceptable, e.g., headache, irritation, CNS effects	UF of 2–10 applied to EEGL to protect more sensitive subpopulations (UF=2) or fetuses or newborns (UF=10)	General population	EEGL divided by a factor of 2–10 to protect more sensitive subpopulations
COT/CEGL	Emergency exposure planning	90 day exposure	Reversible effects acceptable, e.g., headache, irritation, CNS effects	UF of 10–100 applied to EEGL based on pharmacokinetics (i.e., ability to be rapidly biotransformed or to bioaccumulate)	General population	EEGL divided by a factor of 10–100 to account for pharmacokinetic considerations
ERPG Task Force/ERPG-3	Emergency exposure planning	1 hour exposure	Protect against life-threatening effects	No	General population living in immediate area of release	Acute toxicity data preferred. Based upon most sensitive endpoint from human or animal data. All endpoints considered. Methods vary on a case-to-case basis
ERPG Task Force/ERPG-2	Emergency exposure planning	1 hour exposure	Protect against irreversible or other serious health effects that could impair ability to take protective action	No	General population living in immediate area of release	Same as ERPG-3
ERPG Task Force/ERPG-1	Emergency exposure planning	1 hour exposure	Protect against mild, transient adverse health effects	No	General population living in immediate area of release	Same as ERPG-3

include severe eye or respiratory irritation or muscular weakness.

ERPG-1: The maximum airborne concentration below which it is believed that nearly all individuals could be exposed for up to one hour without experiencing other than mild, transient adverse health effects or without perceiving a clearly defined objectionable effect.

The ERPG-1 identifies a level that does not pose a health risk to the community but that may be noticeable due to slight odor or mild irritation. In the event that a small, non-threatening release has occurred, the community could be notified that they may notice an odor or slight irritation, but that concentrations are below those that could cause health effects. For some materials, because of their properties, there may not be an ERPG-1. Such cases would include substances for which sensory perception levels are higher than the ERPG-2 level. In such cases no ERPG-1 level would be recommended.

As for the EEGLs discussed above, data on the health effects of short-term inhalation exposure in humans are the preferred basis of the ERPGs. Lacking reliable quantitative human data, data from acute inhalation toxicity studies in animals may serve as the basis for the ERPG. All possible endpoints of toxicity, as well as immediate and delayed toxicity, are considered. The NOAEL or LOAEL for the most sensitive endpoint is chosen as the basis for the ERPG. ERPGs are calculated on a case-by-case basis. The precise methods used depend on such factors as the shape of the dose-response, the nature and severity of the endpoint serving as the basis of the ERPG, the type and quality of the data, and the uncertainty arising from the need to perform interspecies or route-to-route extrapolations to arrive at one-hour exposure concentrations for humans. However, even though the ERPG Task Force and AIHA recognize that there may be sensitive subpopulations that would adversely respond to exposure to levels lower than the ERPG, no safety factors are employed. AIHA (1989) provides the following justification:

... since these values have been derived as planning and emergency response guidelines, not as exposure

guidelines, they do not contain the safety factors normally incorporated into exposure guidelines. Instead, they are estimates, by the committee, of the thresholds above which there would be an unacceptable likelihood of observing the defined effects.

The ERPGs, like the EEGLs, SPEGLs, and CEGLS, are risk management tools for specific exposure scenarios, and should not be construed as quantitative risk assessment values or a definitive between safe and unsafe levels. Together with data on volatility and storage volumes, ERPGs can aid in the development of emergency action plans. These plans are situation-specific and should take into account such factors as demographics, terrain, weather conditions, and the nature of the release. The differences in mandate, operative assumptions, and review process for the development of these risk management levels are summarized in Table 7-5.

References

American Conference of Governmental Industrial Hygienists (ACGIH). 1989. Threshold Limit Values and Biological Exposure Indices for 1989–1990. Cincinnati, OH.

American Industrial Hygiene Association (AIHA). 1989. ERPG Committee. ERPGs: Concepts and procedures for the development of emergency response planning guidelines (ERPGs).

Andersen, M.E., K. Krishnan, R.B. Conolly, and R.O. McClellan. 1992. Mechanistic toxicology research and biologically-based modeling: Partners for improving quantitative risk assessments. CIIT Activities. 12(1):1–7.

Bond, J.A. 1989. Factors modifying the disposition of inhaled organic compounds. In: McClellan, R.O. and R.F. Henderson, eds. *Concepts in Inhalation Toxicology.* New York, NY: Hemisphere Publishing Company; pp. 249–270.

Brown, K.G. and L.S. Erdreich. 1989. Statistical uncertainty in the No-Observed-Adverse-Effect-Level. *Fundamental and Applied Toxicology* 13:235–244.

Calabrese, E.J. and E.M. Kenyon. 1991. Air toxics and risk assessment. Chelsea, MI: Lewis Publishers, Inc.

Crump, K. 1984. A new method for determining allowable daily intakes. *Fundamental and Applied Toxicology* 4:854–871.

Dourson, M.L. and J.F. Stara. 1983. Regulatory history and experimental support of uncertainty (safety) factors. *Regulatory Toxicology and Pharmacology* 3:224–238.

Dunnick, J.K., D.G. Graham, R.S.H. Yang, S.B. Haber, and H.R. Brown. 1989. Thirteen-week toxicity study of *n*-hexane in B6C3F1 mice after inhalation exposure. *Toxicology* 57:163–172.

Eddy, D.M., V. Hasselblad, and R. Shachter. 1992. *Meta-analysis by the Confidence Profile Method: The Statistical Synthesis of Evidence.* New York: Academic Press, Inc.

Fiserova-Bergerova, V. 1983. *Modeling of Inhalation Exposure to Vapors: Uptake, Distribution, and Elimination: Volumes I and II.* Boca Raton, FL: CRC Press, Inc.

Gaylor, D.W. 1985. The question of the existence of thresholds: extrapolation from high to low dose. In: Flamm, W.G. and R.J. Lorentzen, eds. *Mechanism in Toxicity of Chemical Carcinogens and Mutagens.* Princeton, NJ: Princeton Scientific Publishing Co., Inc.; pp. 249–260. (Advances in Modern Environmental Toxicology: Vol. 12)

Gerrity, T.R. and C.J. Henry, eds. 1990. Summary report of the workshops on principles of route-to-route extrapolation for risk assessment. In: Principles of route-to-route extrapolation for risk assessment, proceedings of the workshops, March 19–21 and July 10–11: Hilton Head, SC and Durham, NC. New York: Elsevier Science Publishing Co., Inc., pp. 1–12.

Guth, D.J., A.M. Jarabek, L. Wymer, and R.C. Hertzberg. 1991. Evaluation of risk assessment methods for short-term inhalation exposure. Paper No. 91-173.2. Presented at the 84th Annual Meeting of the Air & Waste Management Association. Vancouver, BC.

Guth, D.J., R.C. Hertzberg, and A.M. Jarabek. 1993. (In press). Exposure-response analysis: Modeling severity against concentration and duration. *Fundamental and Applied Toxicology.*

Habicht, F.H. 1992a. Guidance on risk characterization for risk managers and risk assessors. Memorandum from the Deputy Administrator to Assistant Administrators and Regional Administrators. U.S. Environmental Protection Agency. Washington, D.C. February 26, 1992.

Hofmann, W. 1982. The effect of polydispersiveness of natural radioactive aerosols on tracheobronchial deposition. *Radiation Protection Dosimetry* 3:97–101.

Jarabek, A.M. and W.H. Farland. 1990. The U.S. Environmental Protection Agency's risk assessment guidelines. *Toxicology and Industrial Health* 6(5):199–216.

Jarabek, A.M. and V. Hasselblad. 1991. Inhalation reference concentration methodology: Impact of dosimetric adjustments and future directions using the confidence profile method. Paper No. 91-173.3. Presented at the 84th Annual Meeting of the Air & Waste Management Association. Vancouver, BC.

Jarabek, A.M., M.G. Menache, M.L. Dourson, J.H. Overton, Jr., and F.J. Miller. 1989. Inhalation reference dose (RfD): An application of interspecies dosimetry modeling for risk assessment of insoluble particles. *Health Physics* 57(1): 177–183.

Jarabek, A.M., M.G. Menache, M.L. Dourson, J.H. Overton, Jr., and F.J. Miller. 1990. The U.S. Environmental Protection Agency's inhalation RfD methodology: Risk assessment for air toxics. *Toxicology and Industrial Health* 6(5):279–301.

Kimmel, C. and D. Gaylor. 1988. Issues in qualitative and quantitative risk analysis for developmental toxicology. *Risk Analysis* 8:15–20.

Klaassen, C.D. 1986. Principles of toxicology. In: Klaassen, C.D., M. Amdur, J. Doull, eds. *Casarett and Doull's Toxicology: The Basic Science of Poisons.* 3rd Edition. New York: Macmillan Publishing Co.

Martonen, T.B. and F.J. Miller. 1986. Dosimetry and species sensitivity: Key factors in hazard evaluation using animal exposure data. *Journal of Aerosol Science* 17:316–319.

Mercer, R.R. and J.D. Crapo. 1987. Three-dimensional reconstruction of the rat acinus. *Journal of Applied Physiology* 63:785–795.

Miller, F.J., J.H. Overton Jr., E.D. Smolko, R.C. Graham, and D.B. Menzel. 1987. Hazard assessment using an integrated physiologically-based dosimetry modeling approach: Ozone. In: *Pharmacokinetics in Risk Assessment, Drinking Water and Health,* Vol. 8. Washington, D.C.: National Academy Press.

National Academy of Sciences (NAS). 1983. Risk assessment in the federal government: Managing the Process. Committee on the institutional means for assessment of risk to public health. Commission on Life Sciences, and National Research Council. Washington, D.C.: National Academy Press.

National Academy of Science (NAS). 1986. Criteria and methods for preparing emergency exposure guidance level (EEGL), short-term public emergency guidance level (SPEGL), and continuous exposure guidance level (CEGL) documents. Committee on Toxicology. Board on Environmental studies and Toxicology. National Research Council. Washington, D.C.: National Academy Press.

National Air Toxics Information Clearinghouse (NATICH). 1989. Data base report on state, local and federal air toxics activities. EPA 450/5–89–008. U.S. Environmental Protection Agency. Office of Air Quality Planning and Standards. Research Triangle Park, NC.

National Research Council (NRC). 1977. Drinking water and health. Washington, D.C.: National Academy Press.

National Research Council (NRC). 1980a. Drinking water and health, Vol. 2. Washington D.C.: National Academy Press.

O'Flaherty, E.J. 1981. *Toxicants and Drugs: Kinetics and Dynamics.* New York: John Wiley and Sons.

Overton, J.H. Jr. 1984. Physicochemical processes and the formulation of dosimetry models. In: Miller, F.J. and D.B. Menzel, eds. *Fundamentals of Extrapolation Modeling of Inhaled Toxicants: Ozone and Nitrogen Dioxide.* New York, NY: Hemisphere Publishing Company; pp. 93–114.

Overton, J.H. and F.H. Miller. 1988. Absorption of inhaled reactive gases. In: Gardner, D.E., J.D. Crapo, and E.J. Massaro, eds. *Target Organ Toxicology Series: Toxicology of the Lung.* New York, NY: Raven Press, Ltd.; pp. 477-507.

Padgett, J. and H. Richmond. 1983. The process of establishing and revising national ambient air quality standards. *Journal of the Air Pollution Control Association* 33:13–16.

Sanagi, S., Y. Seki, K. Sugimoto, and M. Hirata. 1980. Peripheral nervous system functions of workers exposed to *n*-hexane at a low level. *International Archives of Occupational and Environmental Health* 47:69–79.

Schlesinger, R.B. 1985. Comparative deposition of inhaled aerosols in experimental animals and humans: A review. *Journal of Toxicology and Environmental Health* 15:197–214.

Snipes, M.B. 1989. Long-term retention and clearance of particles inhaled by mammalian species. *Critical Reviews of Toxicology* 20:175–211.

St. George, J.A., J.R. Harkema, D.M. Hyde, and C.G. Plopper. 1988. Cell populations and structure function relationships of cells in the airways. In: Gardner, D.E., J.D. Crapo, and E.J. Massaro, eds. *Target Organ Toxicology Series: Toxicology of the Lung.* New York, NY: Raven Press, Ltd.; pp. 71–102.

U.S. Environmental Protection Agency (EPA). 1979c. Air quality criteria for carbon monoxide. EPA 600/8–79–022. Office of Health and Environmental Assessment, Environmental Criteria and Assessment Office. Research Triangle Park, NC.

U.S. Environmental Protection Agency (EPA). 1980f. Guidelines and methodology used in the preparation of health effects assessment chapters of the consent decree water criteria documents. 45 FR 79347–79357, November 28, 1980.

U.S. Environmental Protection Agency (EPA). 1982e. Air quality criteria for particulate matter and sulfur oxides. EPA 600/8–82–029aF-cF. Office of Health and Environmental Assessment, Environmental Criteria and Assessment Office. Research Triangle Park, NC.

U.S. Environmental Protection Agency (EPA). 1982f. Review of the national ambient air quality standard for particulate matter: Assessment of scientific and technical information. EPA 450/5–82–001. Office of Air Quality Planning and Standards. Research Triangle Park, NC.

U.S. Environmental Protection Agency (EPA). 1982g. Review of the national ambient air quality standard for sulfur oxides: Assessment of scientific and technical information. EPA 450/5–82–007. Office of Air Quality Planning and Standards. Research Triangle Park, NC.

U.S. Environmental Protection Agency (EPA). 1982h. Air quality criteria for oxides of nitrogen. EPA 600/8–82–026. Office of Health and Environmental Assessment, Environmental Criteria and Assessment Office. Research Triangle Park, NC.

U.S. Environmental Protection Agency (EPA). 1982i. Review of the national ambient air quality standard for nitrogen oxides: Assessment of scientific and technical information. EPA 450/5–82–002. Office of Air Quality Planning and Standards. Research Triangle Park, NC.

U.S. Environmental Protection Agency (EPA). 1984c. Revised evaluation of health effects associated with carbon monoxide exposure. EPA 600/8–83–033F. Office of Health and Environmental Assessment, Environmental Criteria and Assessment Office. Research Triangle Park, NC.

U.S. Environmental Protection Agency (EPA). 1984d. Review of the national ambient air quality standard for carbon monoxide: Reassessment of scientific and technical information. EPA 450/5–84–004. Office of Air Quality Planning and Standards. Research Triangle Park, NC.

U.S. Environmental Protection Agency (EPA). 1986o. Second addendum to the air quality criteria for particulate matter and sulfur oxides (1982): Assessment of newly available health effects information. EPA 600/8–86–020F. Office of Health and Environmental Assessment, Environmental Criteria and Assessment Office. Research Triangle Park, NC.

U.S. Environmental Protection Agency (EPA). 1986p. Air quality criteria for lead. EPA 600/8–83–028aF-dF. Office of Health and Environmental Assessment, Environmental Criteria and Assessment Office. Research Triangle Park, NC.

U.S. Environmental Protection Agency (EPA). 1986q. Air quality criteria for ozone and other photochemical oxidants. EPA 600/8–84–020aF-eF. Office of Health and Environmental Assessment, Environmental Criteria and Assessment Office. Research Triangle Park, NC.

U.S. Environmental Protection Agency (EPA). 1986r. Review of the national ambient air quality standard for sulfur oxides: Updated assessment of scientific and technical information. EPA 450/5–86–013. Office of Air Quality Planning and Standards. Research Triangle Park, NC.

U.S. Environmental Protection Agency (EPA). 1986v. Lead effects on cardiovascular function, early development, and stature: An addendum to the Air Quality Criteria for Lead (1986). In: Air quality criteria for lead, v. 1. EPA 600/8–83–028aF. Office of Health and Environmental Assessment, Environmental Criteria and Assessment Office. Research Triangle Park, NC.

U.S. Environmental Protection Agency (EPA). 1987g. The risk assessment guidelines of 1986. EPA 600/8–87/045. Office of Research and Development, Office of Health and Environmental Assessment. Washington, D.C.

U.S. Environmental Protection Agency (EPA). 1987h. Addendum to the air quality criteria for lead. EPA 600/8–87–028B. Office of Research and Development. Research Triangle Park, NC.

U.S. Environmental Protection Agency (EPA). 1990k. Science Advisory Board's comments on the use of uncertainty and modifying factors in establishing reference dose levels. EPA/SAB/EHC/90–005. Letter from L.R. Loehr and A. Upton to W. Reilly. January 17, 1990.

U.S. Environmental Protection Agency (EPA). 1990m. Air quality criteria for lead: Supplement to the 1986 addendum. EPA 600/8–90–049F. Office of Health and Environmental Assessment, Environmental Criteria and Assessment Office. Research Triangle Park, NC.

U.S. Environmental Protection Agency (EPA). 1990n. Review of the national ambient air quality standards for lead: Assessment of scientific and technical information. Staff Paper. EPA 450/2–89–022. Office of Air Quality Planning and Standards. Research Triangle Park, NC.

U.S. Environmental Protection Agency (EPA). 1991r. Guidelines for developmental toxicity risk assessment. 56 FR 63798–63826, December 5, 1991.

U.S. Environmental Protection Agency (EPA). 1992a. Guidelines for exposure assessment. 57 FR 22888–22936, May 29, 1992.

U.S. Environmental Protection Agency (EPA). 1992h. Air quality criteria for carbon monoxide. EPA 600/8–90–020aF-eF. Office of Health and Environmental Assessment, Environmental Criteria and Assessment Office. Research Triangle Park, NC.

U.S. Environmental Protection Agency (EPA). 1992i. Summary of selected new information on effects of ozone on health and vegetation: Supplement to 1986 air quality criteria for ozone and other photochemical oxidants. EPA-600/8–88–105F. Office of Health and Environmental Assessment, Environmental Criteria and Assessment Office. Research Triangle Park, NC.

III

Exposure Assessment

8

Exposure Assessment Principles

Paul C. Chrostowski

INTRODUCTION

Exposure assessment is the formal process whereby the contact of a receptor with an agent is evaluated. In this context, the receptor is most often an individual or population of humans; however, it also may refer to other organisms, aggregates of organisms such as an ecosystem, or a natural or cultural resource. The agent typically is a chemical or mixture of chemicals; however, it also can be a physical agent such as ionizing radiation or a biological agent such as a virus. Exposure assessments are conducted to estimate health risks in parallel with the dose-response quantification step of a risk assessment, to permit the evaluation of causation or association in epidemiological studies, or to estimate ambient concentrations of agents at a particular location or time for comparison to regulatory criteria or standards.

A typical exposure assessment usually starts with a description of the source and a quantification of the rate of emissions of a pollutant from the source. It then continues with a description of the behavior of the emitted pollutants in the environment; this is known as environmental fate (for chemical processes) and transport (for physical processes). This is followed by a description of the receptor populations that are (or potentially are) exposed and a

detailed description of the activity patterns engaged in by the receptors that determine their exposure. Last, the assessment concludes with a quantification of chemical concentrations either at a point external to the receptors (exposure point concentrations) or at a target organ of the receptors, and a statement of the uncertainty associated with the quantification. The steps in the exposure assessment process are shown in Figure 8-1. Table 8-1 presents definitions of important exposure assessment terms presented by the National Academy of Sciences (NAS 1991b); Table 8-2 presents additional exposure assessment definitions from the Environmental Protection Agency (EPA).

The purpose of this chapter is to present an overview of the exposure assessment process as it relates to toxic air pollutants and to serve as a guide to more detailed information on many important aspects of the topic. Subsequent chapters discuss various aspects of exposure assessment in more detail and are referred to throughout this chapter to give this handbook greatest utility.

The tools of exposure assessment generally include environmental measurements, biological monitoring, dosimetry, and mathematical modeling. Most large exposure assessments include

Table 8-1   Exposure assessment definitions

Term	Definition
Biological marker	Indicators of changes or events in human biological systems. Biological markers of exposure refer to cellular, biochemical, or molecular measures that are obtained from biological media such as human tissues, cells, or fluids and are indicative of exposure to environmental contaminants.
Biologically effective dose	The amount of the deposited or absorbed contaminant that reaches the cells or target site where an adverse effect occurs or where an interaction of that contaminant with a membrane surface occurs.
Dose	The amount of a contaminant that is absorbed or deposited in the body of an exposed organism for an increment of time – usually from a single medium. Total dose is the sum of doses received by a person from a contaminant in a given interval resulting from interaction with all environmental media that contain the contaminant. Units of dose and total dose (mass) are often converted to units of mass per volume of physiological fluid or mass tissue.
Environment	Comprises air, water, food, and soil media. Regarding air, it refers to all indoor and outdoor microenvironments, including residential and occupational settings.
Exposure	An event that occurs when there is contact at a boundary between a human and the environment with a contaminant of a specific concentration for an interval of time. Units of exposure are concentration times time.
Exposure assessment	Involves numerous techniques to identify the contaminant, contaminant sources, environmental media of exposure, transport through each medium, chemical and physical transformations, routes of entry to the body, intensity and frequency of contact, and spatial and temporal concentration patterns of the contaminant. An array of techniques can be employed, ranging from an estimation of the number of people exposed and contaminant concentrations to a sophisticated methodology employing contaminant monitoring, modeling, and human biological marker measurement.
Internal dose	Refers to the amount of the environmental contaminant absorbed in body tissue or interacting with an organ's membrane surface.
Microenvironment	A three-dimensional space with a volume in which contaminant concentration is spatially uniform during some specific interval.
Nuisance effect	A subjectively unpleasant effect (e.g., headache) that occurs as a consequence of exposure to a contaminant; it may be associated with some physiological response, but is not permanent.
Potential dose	An exposure value multiplied by a contact rate (e.g., rates of inhalation, ingestion, or absorption through the skin) and assumes total absorption of the contaminant.
Total human exposure	Accounts for all exposures a person has to a specific contaminant, regardless of environmental medium or route of entry (inhalation, ingestion, and dermal absorption). Sometimes total exposure is used incorrectly to refer to exposure to all pollutants in an environment. Total exposure to more than one pollutant should be stated explicitly as such.

Source: NAS 1991b.

combinations of these techniques. The assessment must be constructed to answer relevant questions posed by the end-user (e.g., toxicologist or engineer) as to the magnitude, frequency, and duration of exposure. In order to foster consistency and assist in answering these questions, EPA has developed general guidance for exposure assessment (EPA 1992a), and specific guidance for programs such as pesticides control (EPA 1987c) and Superfund (EPA 1989f). No EPA guidance documents are currently available for exposure assessments conducted to support programs under the Clean Air Act; however,

there is a coalescence of exposure assessment methodologies which makes the programs virtually interchangeable. EPA's recent series of documents (EPA 1989b; EPA 1989c; EPA 1989d; EPA 1989e) that analyze air releases at Superfund sites, for example, are useful to those dealing with toxic air pollutants.

Few organizations other than EPA publish guidelines for exposure assessment. The American Industrial Hygiene Association (Hawkins, Norwood, and Rock 1991) published a strategy for occupational exposure assessment that is useful for the workplace and the indoor air

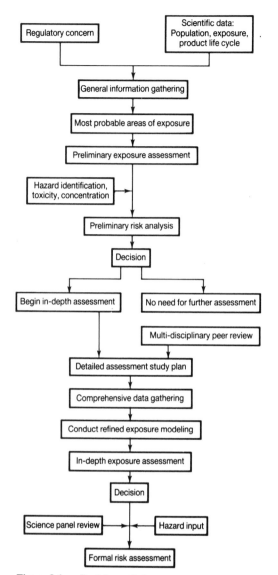

Figure 8-1.   Decision path for exposure assessment.

support the conclusions within well defined bounds of uncertainty. The eight GEAP, listed below, are principles that deserve the attention of all exposure assessors.

1. Every exposure assessment should have a protocol written before its performance.
2. The organization sponsoring the assessment should accept the protocol, have the commitment to carry it out, and draw only those conclusions warranted by the resulting data and its analysis. The assessors should have the education, experience, certification, and facilities to accomplish the work.
3. Any model used in the exposure assessment should be described, defined with respect to parameterization, uncertainty, validation, and assumptions.
4. All statistical procedures, data collection methods, sampling techniques, and analytical methods should be stated and be adequate to support the possible conclusions.
5. A quality assurance plan should be developed and followed.
6. The results and conclusions should be accompanied by an analysis of uncertainty.
7. The protocol, data, reports, and other records should be archived so that they are retrievable for a specified period.
8. The identity of specific individuals participating in an exposure study should be kept confidential. Results involving individuals should be reported on a strict need-to-know basis.

From the scientific or risk assessment standpoint, the ultimate goal of the exposure assessment is to achieve accuracy, precision, and sensitivity. From a regulatory standpoint, however, elements of risk management or policy often are superimposed over the purely scientific exposure assessment process. Good examples of this intrusion are the use of the maximum exposed individual (MEI) in EPA's air program and the reasonable maximum exposure (RME) in EPA's Superfund program. These concepts often assume, for policy purposes, that an individual is practicing extreme behavior that exaggerates his or her exposure. Hawkins (1991) and Patrick (1992b) analyzed

environment in general. At the time of writing this chapter in 1992, the American Society for Testing and Materials (ASTM) also was in the process of developing standards for risk assessment that include exposure assessment. Recently, Hawkins, Jaycock, and Lynch (1992) proposed a set of guidelines, called 'Good Exposure Assessment Practices' (GEAP), that are analogous to Good Laboratory Practices for analytical measurements. The GEAP are designed to assure that the methods and data that are used in the exposure assessment clearly

Table 8-2   Explanation of exposure and dose terms

Term	Refers to	Generic units	Specific example units
Exposure	Contract of chemical with outer boundary of a person (e.g., skin, nose, mouth)	Concentration × time	*Dermal*: (mg chem./l water) × (hours of contact); (mg chem/kg soil) × (hours of contract) *Respiratory*: (ppm chem. in air) × (hours of contact); (μg/m^3 air) × (days of contact) *Oral*: (mg chem./kg soil) × (minutes of contact); (mg chem./kg food) (minutes of contact)
Potential dose	Amount of a chemical contained in material ingested, air breathed, or bulk of material applied to skin	Mass of chemical	*Dermal*: (mg chem./kg soil) × (minutes of contact); (mg chem/kg food) (minutes of contact) *Dose rate* is mass of chem./time; dose rate is sometimes normalized to body weight mass of chem./unit body weight × time *Respiratory*: (μg chem./m^3 air) × (m^3 air breathed/minute) × (minutes exposed) = μg chem in air breathed *Oral*: (mg chem./l water) × (l water consumed/day) × (days exposed) = mg chem ingested in water (also dose rate in mg/day)
Applied dose	Amount of chemical in contact with primary absorption boundaries (e.g., skin, lungs, gastrointestinal tract) and available for absorption	Mass of chemical	*Dermal*: (mg chem./kg soil) × (kg soil directly touching skin) × (1 percent chem. actually touching lung) = mg chem. actually touching lung absorpiton barrier *Oral*: (mg chem./kg food) × (kg food consumed/day) × (1 percent chem. touching GI tract) = mg chem. actually touching GI tract absorption barrier (also absorbed dose rate in mg/day) chemical available to organ or cell (dose rate in mg chem. available to organ/day)
Internal (absorbed) dose	Amount of chemical penetrating across an absorption barrier or exchange boundary via either physical or biological processes	Mass of chemical	*Dermal*: mg chem. available through skin *Respiratory*: mg chem. absorbed via lung *Oral*: mg chem. absorbed via GI tract (dose rate in mg chem. absorbed/day or mg/kg × day) mg chem. available to organ or cell (dose rate in mg chem. available to organ/day)
Delivered dose	Amount of chemical available for interaction with any particular organ or cell	Mass of chemical	mg chem. available to organ or cell (dose rate in mg chem. available to organ/day)

Source: EPA 1992a.

the degree of conservatism that is associated with MEI-type exposure assessments. Although these conservative concepts are useful tools from a regulatory perspective, the assessor always should keep in mind that their use does not necessarily lead to an accurate estimate of exposure.

Exposure is linked intimately with both chemical concentrations and effects. For this reason, the statistical properties of an exposure assessment must be compatible with both the input data and the function that it serves in a risk assessment. Most exposure assessments currently are conducted according to an average daily dose paradigm that was developed by EPA and is described below. This is a deterministic, steady-state concept that averages exposure over a lifetime for carcinogens or over a specified exposure period for noncarcinogens. Because EPA's mandate is to be health-protective, exposures calculated by their methods inevitably exceed actual exposures. More accurate exposure assessment techniques involve the use of stochastic and time-variant models that incorporate pharmacokinetics. For example, Monte Carlo simulation (Thompson, Burmaster, and Crouch 1992; Chrostowski, Foster, and Dolan 1991) allows for the calculation of the probability distribution of exposures instead of a single average value. The assessor then can use any given point on the probability distribution of exposures as input to a risk assessment. Similarly, time-variant models allow for the elimination of chemicals with short biological half-lives. Ott (1980) discusses various models of exposure to air pollution with special attention paid to averaging periods and time constructs; however, he does not integrate this information with pharmacokinetic data.

There are numerous overviews of exposure assessment available in the literature. Severn (1987) presents a brief introduction to the topic within a risk assessment context. Neely and Blau (1984) describe modeling of exposures both in general and in specific situations. Quackenboss and Lebowitz (1989) discuss requirements for exposure assessment, modeling, and validation for human health effects studies with a focus on inhalation. The National Academy of Sciences (NAS) recently published two important documents related to exposure assessment. One of these (NAS 1991a) reports on a symposium concerning frontiers in assessing human exposures to environmental toxicants. The other (NAS 1991b) is a monograph specifically related to human exposure assessment for airborne pollutants. EPA also published several documents concerning exposure assessment including the Superfund Exposure Assessment Manual (EPA 1988b) and the Exposure Factors Handbook (EPA 1989a).

Most exposure assessments rely heavily on data that are collected specifically for the purpose of performing the assessment or that have been collected for another purpose but are useful in performing the assessment. The primary use of data in exposure assessments is to infer more general information about chemical concentrations, contact times, exposures, or doses. For example, measured concentrations of particulate matter at an industrial property boundary may be used to infer concentrations at a residential receptor several kilometers removed. In all cases, the exposure assessor should have a clear picture of the relationship of the data to what is being characterized in the assessment. This often is known as the conceptual model of the problem that is the subject of the assessment. As the assessment proceeds, elements of the conceptual model are replaced by the corresponding elements of a quantitative model until a quantitative measure of uncertainty is reached. Many of the topics that are treated in detail in this chapter are specifically designed to assist the assessor in forming and using conceptual and quantitative models.

REQUIRED EXPERTISE AND RESOURCES

The nature of exposure assessment is multidisciplinary, meaning it requires the application of a number of scientific disciplines. For example, a typical exposure problem may be solved by: (1) a statistician who designs a study plan, (2) a medical doctor who obtains a tissue sample, (3) an analytical chemist who measures concentrations of chemicals in the sample, or (4) an environmental scientist who integrates a range of information. Most exposure assessments require team efforts involving individuals with training in various academic disciplines and at different levels or expertise. Table 8-3 lists important academic disciplines and their corresponding relationship to exposure assessment. The professional exposure assessor typically has a graduate degree in the physical sciences, biological sciences, or engineering, has several years experience in numerical analysis, is computer literate, and communicates clearly.

Physical resources for exposure assessments run the gamut from field sampling equipment through analytical and experimental laboratories to computer systems. Other chapters in this section describe some of these resources. Access to appropriate information is critical to successfully conducting exposure assessments. EPA (1987d) is a particularly useful source of general exposure assessment information.

Table 8-3   Academic disciplines and corresponding areas of exposure assessment

Discipline	Exposure assessment area
Medicine	Tissue sampling, medical monitoring
Chemistry	Analytical methods, fate and transport
Sociology Psychology Anthropology	Behavior and activity patterns
Statistics	Study design, data analysis
Chemical engineering Hydrogeology Hydrology Meteorology	Mathematical modeling
Biology	Ecological exposure assessment
Chemical engineering Pharmacology Toxicology	Pharmacokinetics

EXPOSURE ASSESSMENT METHODS

Exposure assessments typically involve a number of tools, including environmental and biological measurements, dosimetry, and mathematical (more rarely physical) modeling. Each of these are discussed below. There is a general hierarchy of preference of measurements that is related to the proximity of the measurement to the target organ and its ability to account for variations in magnitude, frequency, and duration of exposure. Figure 8-2 presents a block diagram of this hierarchy. The top of the hierarchy consists of direct measurements of exposure at the target tissue; this generally is considered to represent exposures most accurately because no extrapolations are involved. These measurements, however, are rare owing to several factors, including lack of methods and the extreme invasiveness of some procedures required to obtain the data. For example, liver biopsy to evaluate chemical concentrations of a liver carcinogen is a surgical procedure involving both expense and discomfort on the part of the patient. Direct target

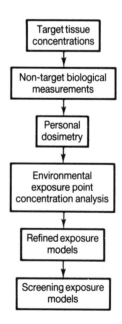

Figure 8-2.   Hierarchy of exposure measurements.

organ measurements are rare and are used primarily in conjunction with toxic tort lawsuits that involve exposure to chemicals (e.g., determination of asbestos levels and types in the lungs).

The second level of the hierarchy is far more common. However, it involves uncertainty in that inferences must be made concerning concentrations at the target organ from concentration measured elsewhere. This usually is done using pharmacokinetic models and, thus, is associated with modeling uncertainties. For example, inferences have been made about developmental effects of mercury exposure by the analysis of mercury in hair (Cox et al. 1989). This is predicated on an exposure pathway in which mercury is ingested and transported to the target tissue (i.e., the developing central nervous system) and the tissue where measurements are made (i.e., hair). Recent tests (Lasorda 1992), however, dispute the validity of this pathway.

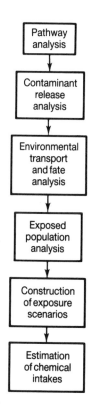

Figure 8-3.   Exposure assessment process: overview.

Proceeding to the bottom of the hierarchy adds more uncertainty and detracts from the accuracy of an exposure assessment. The screening models at the bottom of the hierarchy may be associated with inaccuracies of several factors of ten. Screening models used in exposure and risk assessments for EPA's Superfund program may be useful for regulatory purposes; however, the user always should understand the uncertainties associated with their use.

Several levels of the hierarchy often are combined to formulate an integrated exposure assessment. For example, the exposure tools used to assess lead exposures from windblown mine tailings may include biological monitoring, environmental monitoring, dosimetry, and both screening level and refined models. Although the hierarchy emphasizes chemical measurements, a large portion of the exposure assessment process involves the analysis of data from measurements. As usually practiced, the components of exposure assessment that are independent of chemical measurements may be expressed in the six-step process illustrated in Figure 8-3.

Biological Monitoring

Biological monitoring is defined as the routine analysis of human tissues or excreta for direct or indirect evidence of exposure to chemical substances. Typically, biomonitoring includes three types of measurements.

1. The concentration of the chemical of concern in various biological media such as blood, urine, and expired air.
2. The concentration of metabolites of the chemical of concern in the same media.
3. Determination of nonadverse biological changes resulting from the reaction of the organism to exposure.

Figure 8-4 is a schematic showing the role of biological monitoring in the exposure assessment process. Biological monitoring may take place at the target tissue (e.g., lead in blood) or in an unrelated tissue (arsenic in hair, where the target is the skin). Measurements of target

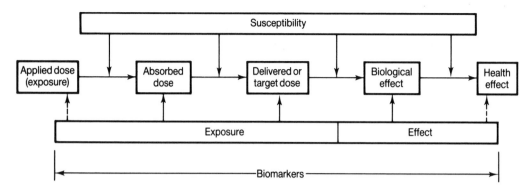

Figure 8-4.   Relationship between biomarkers of susceptibility exposure and effect.

organs are limited and may be difficult to relate to the toxicological endpoint of interest. For example, although the blood is a target tissue for lead, we usually are more interested in its effects on the central nervous system. The link between the measurements and the effects is difficult to make unless the mechanism of action and fate of chemical in the body are known. For chemicals with a long biological half-life (e.g., dioxin in lipids), biological measurements integrate over long exposure periods; those with a short half-life (e.g., mercury in urine) often are reflective only of a particular duration of exposure. Table 8-4 lists selected biological methods. The reader is referred to the appropriate chemical-specific Toxicological Profiles published by the Agency for Toxic Substances Disease Registry (ATSDR) for further details.[1] Standard works

on biological monitoring include Baselt (1980) and Lauwerys (1983). Large biological measurement studies of particular methodological interest include the NHANES program and the Missouri dioxin surveillance program (Patterson et al. 1986).

In addition to measurements of chemicals in a biological system, a new technique with considerable potential is the measurement of biomarkers. Biomarkers are indicators of variation in cellular or physiological components or processes, structures, or functions that are measurable in a biological system or sample. In contrast to biological monitoring for chemicals, biomarker analysis focuses on the organism. Figure 8-5 presents a schematic view of the fate of a xenobiotic compound in a biological organism. As used in this chapter, biological monitoring takes place in Zone A in Figure 8-5

Table 8-4   Selected biomonitoring methods

Chemical	Monitoring technique
Polyhalogenated aromatics (PCBs, PCDD/PCDFs)	Serum lipid chemical concentrations
Organophosphates/carbamates	Cholinesterase
Aromatic nitro compounds	Methemoglobin
Polyurethanes	5-Hydroxyindolic acid
Arsenic	Urinary concentrations
Cadmium	Blood acid phosphatase
Lead	Blood or bone concentrations, porphyrins
Mercury	Hair concentrations
Cyanide	Blood cyanide or thiocyanate concentrations

---

[1] For information call (404) 639–6000.

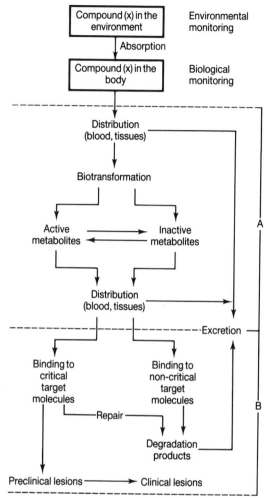

```
Compound (x) in the        Environmental
    environment             monitoring
        │ Absorption
Compound (x) in the          Biological
       body                 monitoring
- - - - - - - - - - - - - - - - - - - - - - - -
    Distribution
   (blood, tissues)
        │
   Biotransformation
                                              A
   Active    ──────▶   Inactive
 metabolites  ◀──────  metabolites

    Distribution ──────────────────────▶
   (blood, tissues)
- - - - - - - - - - - - - - - - - - Excretion ─
 Binding to            Binding to
  critical             non-critical
   target                target
  molecules            molecules
        └──Repair──┘        │       B
                       Degradation
                        products
 Preclinical lesions ──────▶ Clinical lesions
- - - - - - - - - - - - - - - - - - - - - - - -
```

Note: Depending on the biological parameter selected, and the relationship of time of sampling to that of exposure, different information may be obtained; data may reflect the amount of the chemical recently absorbed, the amount already stored in one or more compartments of the organism (body burden), or the amount of the active chemical species bound to the sites of action.

Figure 8-5.   Fate of xenobiotics in the organism.

and biomarker analysis takes place in Zone B. Hattis (1988) lists the advantages of biomarker analysis including better modeling of dose-response relationships, better interspecies extrapolation of doses, improved evaluation of past doses, and improved insight into the matter of intraspecies variability. Although biomarker analysis currently is not used on a routine basis,

numerous techniques are at varying stages of development. EPA (1984b) evaluated biomonitoring for improvements to assessment of human genetic risk with emphasis placed on chemical and immunological analysis of genetic material such as DNA and chromosomes. EPA does not distinguish between biological monitoring and biomarker analysis. More recently, EPA (1990f) assessed the state of the art for biomonitoring and developed a research agenda. According to that report, it will be three to five years before biomonitoring techniques reach the stage where they can be used on a routine research basis. The International Agency for Research on Cancer (IARC 1984b) has reviewed methods for monitoring human exposure to carcinogenic and mutagenic agents. IARC concluded that there is an urgent need to develop biomonitoring methods to assess exposure to carcinogens and mutagens and to establish a set of criteria by which the methods currently or potentially available can be validated. According to IARC, factors to be considered in setting such criteria should include the following list.

- Appropriate for exposure assessment and health effect assessment
- Results valid for individuals and groups
- Reproducibility within and between laboratories
- Accuracy (i.e., specificity, recovery)
- Detection limit
- Interindividual and intraindividual variations in nonexposed reference populations (e.g., due to race, sex, and age)
- Effects of possible interfering factors (e.g., diet, smoking, and alcohol)
- Absence of background levels
- Simplicity
- Possibility of sample storage

IARC also concluded that the criteria should be refined and expanded and then applied to evaluate the methods considered to be most promising at present. This would make possible a common understanding of the significance, current stage of development, and potential use of these methods. IARC indicated that the following methods appear to be those most suited, with appropriate precautions, for development and limited use.

- Determination of chemicals and their metabolites in biological fluids and tissues
- Determination of thioethers in urine
- Detection of mutagenic activity in urine
- Detection of chromosomal aberrations
- Detection of sister chromatid exchange
- Testing for micronuclei in lymphocytes or epithelial cells
- Determination of sperm morphology (in selected situations)

IARC noted that a number of methods are promising but require extensive development and validation before they can be used. These include:

- Detection of protein and DNA adducts
- Detection of protein variants in blood
- Detection of point mutations in blood cells
- Investigation of DNA repair in somatic cells
- Detection of tumor markers

Comparative programs and collaborative studies at the international level are encouraged by IARC in order to improve the techniques and thus to reduce variability resulting from methodology. Once a method has been properly validated from a technical point of view, baseline reference values should be established, keeping in mind the special problems of selecting control groups.

Personal Dosimetry

Personal dosimetry usually involves taking measurements at the exposure point using measurement devices that are attached to the receptor. The measuring device or dosimeter integrates the receptor's exposure over all activities. This technique has been used extensively to measure exposure to ionizing radiation (Attix 1991); however, it has had limited use for monitoring chemicals in the environment, primarily because of problems associated with often cumbersome measuring equipment and high detection limits relative to the environmental levels of chemicals of concern. Standard methods for personal dosimetry are described in the literature (Wallace and Ott 1982; Ott et al. 1986). One recent and intriguing application of personal dosimetry in exposure measurements is the passive diffusion monitor for nicotine from environmental tobacco reported by Hammond and Leaderer (1987).

Environmental Measurements

Environmental measurements are the most common source of chemical exposure information. The first step in obtaining measurements is to develop a conceptual model of how exposure occurs. This enables the exposure assessor to design the sampling and analysis study. Important questions that are answered by the conceptual model include identification of chemicals of concern, definition of environmental media to be sampled, and location of sampling activities. For example, a conceptual

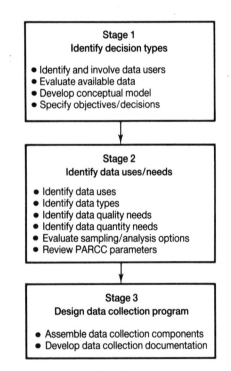

Figure 8-6.   Data quality objectives, three-stage process.

Table 8-5   Data quality objectives terminology

Term	Meaning
Precision	Precision measures the reproducibility of measurements under a given set of conditions. Specifically, it is a quantitative measure of the variability of a group of measurements compared to their average value. Precision is usually stated in terms of standard deviation but other estimates such as the coefficient of variation (relative standard deviation), range (maximum value minus minimum value), and relative range are common.
Accuracy	Accuracy measures the bias in a measurement system; it is difficult to measure for the entire data collection activity. Sources of error are the sampling process, field contamination, preservation, handling, sample matrix, sample preparation and analysis techniques. Sampling accuracy may be assessed by evaluating the results of field/trip blanks, analytical accuracy may be assessed through use of known and unknown QC samples and matrix spikes.
Representativeness	Representativeness expresses the degree to which sample data accurately and precisely represent a characteristic of a population, parameter variations at a sampling point, or an environmental condition. Representativeness is a quality parameter which is most concerned with the proper design of the sampling program. The representativeness criterion is best satisfied by making certain that sampling locations are selected properly and a sufficient number of samples are collected. Representativeness is addressed by describing sampling techniques and the rationale used to select sampling locations. Sampling locations can be biased (based on existing data, instrument surveys, observations, etc.) or unbiased (completely random or stratified-random approaches). Either way, the rationale used to determine sampling locations must be explicitly explained. If a sampling grid is being utilized, it should be shown on a map of the site. The type of sample, such as a grab or composite sample, as well as the relevant standard operating procedure (SOP) for sample collection, should be specified.
Completeness	Completeness is defined as the percentage of measurements made which are judged to be valid measurements. The completeness goal is essentially the same for all data uses: that a sufficient amount of valid data be generated. It is important that critical samples are identified and plans made to achieve valid data for them.
Comparability	Comparability is a qualitative parameter expressing the confidence with which one data set can be compared with another. Sample data should be comparable with other measurement data for similar samples and sample conditions. This goal is achieved although using standard techniques to collect and analyze representative samples and reporting analytical results in appropriate units. Comparability is limited to the other terms listed here because only when precision and accuracy are known can data sets be compared with confidence.

model of the potential impact on the natural environment of mercury emitted from an industrial facility could show that measurements of inorganic mercury at the source (i.e., in the stack) and in the ambient water, and methyl mercury measurements in a top predator, could be adequate to provide input to a risk assessment.

Once the conceptual model is formulated, the next stage of the process is definition of data quality objectives (DQOs) for the study. The use of DQOs is a concept developed by EPA (1987e) to formalize principles concerning precision, accuracy, representativeness, completeness, and comparability[2] for a particular study. The sensitivity of a measurement,

although not formally acknowledged by EPA, also is an important DQO. The output of the process of identifying DQOs is information that leads to selection of sampling and analytical techniques in addition to the design of the study. The DQO process is shown in Figure 8-6 and important DQO terms are defined in Table 8-5. It is important to involve data users in each stage of the DQO process. In the context here, the data user is the exposure analyst.

The study design derives from the output of the DQO process. Study design components consist of identifying the number, location, and type of samples to be obtained in each environmental medium. In this context, type of sample refers to composite, continuous, or discrete

---

[2] Sometimes referred to as PARCC.

sampling. Most complex environmental studies involve some type of statistical sampling plan (Gilbert 1987). Three types of sampling methods typically are used.

- Judgment sampling is subjective selection of sampling locations based on prior knowledge or anticipation. For example, knowledge that an area contained a leaking tank could lead to sampling in that vicinity.
- Probability sampling refers to the use of a specific method of random selection of sample points. Common types of probability sampling include simple random, stratified, two-stage, cluster, systematic, compositing three-stage, and double sampling. Probability sampling is most useful for calculating average values or total amounts of a chemical in a medium.
- Search sampling is conducted to locate chemical hot spots (maximum values) or sources of contamination.

In a typical study, all three types of sampling may be used. Before embarking on the development of a sampling program, however, it is important to define how the data will be statistically analyzed and how uncertainty may be propagated through the exposure assessment process. For example, it makes little sense to specify a sampling density that leads to conclusions at 95 percent confidence when the chemical analytical techniques used are associated with conclusions at 80 percent confidence.

Many types of probability sampling depend on prior knowledge of the distribution of chemicals at the site. Most often, a knowledge of the chemical concentration variance is required as input to the equations for developing the number of samples or sample density. If there is no prior knowledge, it may be necessary to conduct a pilot study. Pilot studies are limited sampling programs that can yield information about the heterogeneity of the chemical distribution in the media of concern and then can be used to help select the appropriate method of probability sampling; they also can yield an estimate of the sample standard deviation to be used in subsequent calculations. In many cases, simple field instrumentation can be used to

conduct pilot studies, thus avoiding the cost and time associated with laboratory analysis.

There are numerous sampling and chemical analysis techniques available for obtaining environmental measurements. These techniques range from real-time, unsophisticated field measurement devices to analytical methods that push the frontiers of chemical analysis. The choice of a particular method depends on the DQOs in addition to practical factors such as cost and time. Chapter 9 and the bibliography contain some important references of analytical techniques that may be important to the exposure assessor.

Once data are generated in the analytical laboratory, it should not be used in an exposure assessment until validated. Validation is the process whereby data are evaluated for quality assurance/quality control (QA/QC) and to assure that the study DQOs are met. Validation can range from a cursory inspection of the data package by a technician using a checklist to an in-depth review of individual chromatogram or mass spectra by a graduate-level chemist. Information often evaluated during validation includes the following:

- Sample holding times before analysis
- Spectrometer tuning information
- Surrogate spike results (a measure of accuracy)
- Matrix spike results (an additional measure of accuracy)
- Field and laboratory precision
- Blank analysis
- Initial and continuing calibration
- Presence of chemicals that interfere with the measurement
- Information about detection limits
- Information about how the samples were handled in the laboratory

Formal data validation is a new concept in environmental analysis; thus, there is not a substantial amount of literature on the topic. EPA (1991g, 1991h) published guidelines for the validation of data in Superfund's Contract Laboratory Program (CLP). These guidelines, however, should be used with caution outside of Superfund because other measurement programs may require more or less validation. For example, exposure assessments that are being conducted to support toxic tort litigation require a higher degree of validation than routine Superfund projects. The validation process can result in the rejection of some data and the qualification of other data. Qualified data (e.g., a statement that a data point is outside a calibration range or a blank is contaminated with trace amounts of an analyte) can be used in exposure and risk assessments as long as the uncertainties associated with the qualification are stated explicitly in the assessments and taken into account by the assessors. EPA's guidance for use of data in Superfund risk assessments is a useful reference for determining the utility of qualified data.

In many cases, the exposure assessor is faced with what is known as a censored data set. This is the case when a large number of samples in a data set are below the limits of detection or quantification. In this case, the assessor often is faced with a choice among several different methods for inputing values to the area below

the detection limit. The manner in which this is done is significant because many environmental situations have a majority of measurements that are below the detection limit. If these measurements are not used in subsequent calculations, exposures could be significantly over-estimated.

Mathematical Modeling

Most exposure assessments are conducted using a combination of environmental measurements and mathematical modeling. For example, exposure assessments conducted for incinerator emissions usually involve measuring stack gas concentrations of the chemicals of concern, performing dispersion modeling, and conducting human uptake or dose modeling. Modeling often is used to predict exposure point concentrations where direct measurements are impractical or at points in time or space that differ from those where measurements were conducted. Some definitions pertinent to modeling are given in Table 8-6. Advantages and

Table 8-6   Definitions of terms relevant to exposure modeling

Term	Meaning
Analytical solution	Method for solving differential equations in a model using classical tools of algebra and calculus.
Calibration	The process of using a set of observed data to adjust the structure and/or internal coefficients of a model such that the output values are accurate with respect to a known value.
Causation	The independent change in the value of one variable causes a predictable change in the value of dependent variables.
Deterministic model	One in which the variations in the variables do not include a random component – there is one output for each set of inputs.
Equilibrium	A system which is not exchanging energy or mass with its surrounding is in equilibrium with the surroundings.
Model	A theoretical construct, together with assignments of numerical values to model parameters, and incorporating prior observation, which relates external inputs to system variable responses.
Numerical solution	Method for solving differential equations in a model using numerical approximation techniques.
Sensitivity analysis	The investigation of changes in output values resulting from changes in values of independent variables and in the posited relationships among variables.
Stability	The ability of a numerical integration method to iterate to a solution.
Steady state	The case where input to a system is balanced by output. A model in which no variables change for the time period under consideration.
Stochastic model	One in which the variation in one or more variables includes a random component.
Verification	Testing of a model following initial calibration to evaluate mode validity.

disadvantages of mathematical models are shown below:

Advantages of mathematical models

- Allow spatial and temporal extrapolation
- Cost effective and rapid
- Uncertainties can be precisely defined
- Avoid measurement uncertainties
- Can be calibrated if reliable data are available
- Allow formal assembly and integration of large bodies of data

Disadvantages of mathematical models

- Models often are influenced by risk management/policy decisions
- Contingent on operator skill, input data quality
- Uncalibrated models often are used without regard to loss of accuracy

More than one type of model generally is used in an exposure assessment. A source model can define the rate, magnitude, and duration of the emission of material from a specific source into an environmental medium. The output of a source model then may be input to an environmental fate or transport model, such as an air dispersion or particle deposition model, that is used to calculate the magnitude of exposure at the exposure point. Intake models that take population activity patterns into account are used to calculate the average daily intake of a chemical from a specific route of exposure. Pharmacokinetic models are used to calculate the concentration of a chemical of concern at a target tissue. Finally, propagation of error models may be used to define the uncertainty associated with the exposure assessment. Many exposure models integrate information from different areas of exposure assessment. For example, McKone and Layton (1986) published multimedia exposure models that are useful for

calculating exposure from various environmental media. The SHEAR model (Anderson and Lundberg 1983) and its successors are examples of an integrative exposure model that links airborne exposures with human activity patterns; other examples are described in Appendix G. A useful integrated pharmacokinetic model was developed by Baskin and Falco (1989) for assessing human exposure to gaseous pollutants. This model simulates connective and diffusive transport of gases from the ambient environment into the human body by way of the respiratory and circulatory systems. The model has been validated by comparison to measured data from oxygen and halothane.

Air dispersion models are discussed in Chapter 11. A detailed discussion of other types of models is beyond the scope of this handbook, although numerous sources of information on other types of models are provided in the bibliography.

Integrated Approaches

Some recent major studies integrated various exposure assessment techniques to evaluate total exposure. The Total Human Environmental Exposure Study (THEES) was designed to assess multimedia exposure to benzo[a]pyrene in an urban environment (Lioy and Daisey 1987). EPA's Total Exposure Assessment Methodology (TEAM) studies (Wallace 1987; Wallace et al. 1983; Wallace et al. 1986; Wallace et al. 1987; Wallace et al. 1988) are the most comprehensive integrated studies to date. These studies used personal monitors, outdoor monitors, and drinking water and breath samples for a large number of volatile organic chemicals.

ENVIRONMENTAL FATE AND TRANSPORT

Environmental fate and transport is the link between the release of a chemical from a source and its contact with a receptor at the exposure point (Figure 8-3). The analysis of environmental fate and transport involves obtaining the distribution of chemical concentrations at the source and following it to the receptor, taking

into account all of the physical and chemical changes that a chemical can undergo. Figure 8-7 is a schematic of a conceptual model of environmental fate at an inactive waste site. This illustration emphasizes the interrelationships between exposure pathways and underscores the need to conduct multimedia exposure

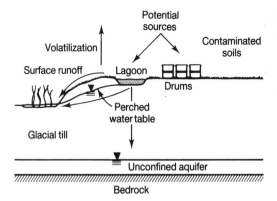

Figure 8-7.   Example conceptual model illustration.

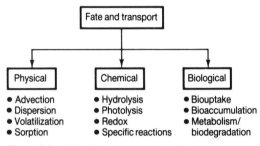

Figure 8-8.   Dominant environmental fate processes.

assessments to arrive at a comprehensive estimate of exposure for a particular population. In terms of exposure assessment, 'fate' usually refers to a process in which a chemical undergoes a chemical transformation either through chemical or biological mediation. 'Transport,' on the other hand, refers to processes that transfer a chemical from one location to another with or without a change in concentration, but with no change in chemical identity. Figure 8-8 shows dominant environmental fate and transport processes. Several important texts are available that deal with environmental fate and transport, including Tinsley (1979), Neely (1980), and Mills et al. (1985).

Pathway analysis is a concept that is linked strongly to environmental fate and transport. The exposure pathway is the course that a chemical takes from its source to the exposed receptor. An exposure pathway describes a unique mechanism by which an individual or population is exposed to chemicals at, or originating from, a source or group of sources.

Four criteria normally are necessary and sufficient for a complete exposure pathway.

1. A source and mechanism of chemical release to the environment
2. An environmental transport medium (e.g., air) for the released chemical and a mechanism of transfer of the chemical from one medium to another (e.g., deposition of particles onto soil)
3. A point of potential contact (i.e., the exposure point) of humans or biota with the contaminated medium
4. A route of exposure (e.g., inhalation) at the exposure point

If one of these conditions is absent, the pathway is incomplete and exposure cannot occur. Figure 8-9 illustrates the concept of exposure pathways. Both direct and indirect exposure may occur. Direct exposure is when only one medium intervenes between the source and the receptor. An example is inhalation of chemicals that have been released from an industrial stack or vent. Indirect exposure occurs when more than one medium intervenes between the source and the receptor. For example, if a chemical is released from a stack in the form of fly ash, it can be transported through the air (the first medium), deposited on the earth's surface and absorbed onto a soil particle (the second medium), and washed off into a water body (the third medium), where it contacts a fish (the receptor). Exposure pathways that should be considered in performing risk assessments of larger stationary air sources include the following:

■ Inhalation during normal operating conditions
■ Inhalation during off-normal operating conditions
■ Ingestion of incidental soil
■ Ingestion of indoor dust
■ Ingestion of homegrown produce
■ Ingestion of commercial produce
■ Ingestion of freshwater fish
■ Ingestion of saltwater fish
■ Ingestion of shellfish
■ Ingestion of locally raised beef

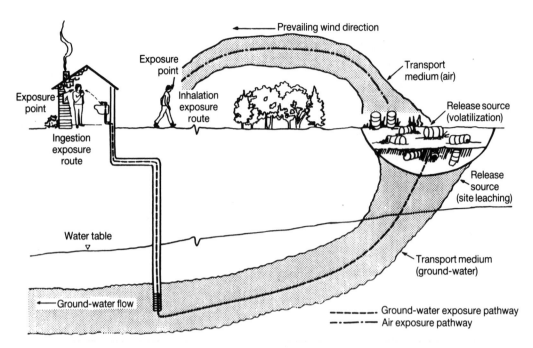

Figure 8-9.   Illustration of exposure pathways.

- Ingestion of locally produced dairy milk
- Ingestion of venison
- Ingestion of locally produced eggs
- Ingestion of human breast milk
- Ingestion of drinking water
- Incidental ingestion of water while swimming
- Dermal absorption from water while swimming
- Dermal absorption from contacted soil
- Whole body immersion (radiation only)

Advection, Dispersion, and Direct Pathways

The group of phenomena characterized by advection and dispersion are particularly significant in analyzing exposure along direct pathways. These processes are common to air, surface water, and groundwater and may be used to evaluate the transport of pure, dissolved, or sorbed chemical species. A typical application is the emission of a gas from a stack, or a liquid from a diffuser, where the receptor is a downwind person or downstream aquatic organism. Advection is the process where a

chemical is transported from one location to another, usually without a change in concentration. Pure advection rarely occurs in the environment, although a close approximation occurs when an immiscible chemical is spilled into a low-velocity water stream. The chemical is advected downstream in a plug-flow mode with minimal subsequent dilution. In reality, advection usually is accompanied by dispersion that involves transport with an accompanying dilution of chemical concentration. Dispersion may be caused either by properties of the chemical or properties of the environment.

If dispersion is caused by properties of the chemical, it is known as diffusion. Diffusion is transport along a gradient from a location of high concentration to a location of lower concentration. Diffusion typically is a slow process and only assumes importance in environmental media where there are weak external forces. For example, a stagnant pond, quiescent air, or slowly flowing groundwater are suitable media for chemical diffusion to be important. Chemical diffusion is well known from both theoretical and empirical standpoints and often can be predicted from first principles.

Environmental dispersion, however, largely is a change in concentration brought about by turbulent mixing. Typically, dispersion and advection occur together. For example, when a plume is released from a stack, it is advected downwind to the receptor while undergoing dilution by dispersion. Dispersion is not well understood from the theoretical standpoint and the assessor usually must rely on empirical methods to predict dispersion. For example, Pasquill's coefficients are empirically based dispersion coefficients that depend on path length (i.e., distance from source to receptor) and atmospheric conditions related to turbulence (e.g., wind speed and stability) for their application. Air dispersion is discussed in Chapter 11. The output of a dispersion model is the exposure point concentration. It may be used alone to calculate exposure by direct pathways or in conjunction with other parameters to calculate exposure by indirect pathways.

Deposition and Indirect Pathways

In many cases, materials are emitted in the form of particulate matter or dense gases that tend to settle to the surface of the earth or a water body where the materials can become available for exposure through indirect pathways. Recent risk assessments of incinerators that include both direct and indirect pathways indicate that the majority of the risk usually is associated with the indirect pathways – the majority of modeling uncertainty also is associated with these pathways. Deposition can take place both in dry and wet forms. Dry deposition usually is a consequence of gravitational settling of particles heavier than air. As they approach the surface of the earth, particles become subject to complex forces in the boundary layer that may accelerate or decelerate the deposition rates. Several mathematical models have been developed to assess dry deposition. Key deposition models include: (1) an empirical model developed by Sehmel and Hodgson (1978) based on wind tunnel experiments and appropriate for calculating size-specific deposition velocities to surfaces that are relatively smooth (e.g., soils,

short grasses, low crops); (2) an empirical model developed by Scire et al. (1987) designed for application to dry deposition of both particles and gases; and (3) a mathematical model developed for calculating size-specific deposition velocities to forest canopies.

Wet deposition usually is the consequence of washout and rainout. Washout results from the scavenging action of precipitation falling through air containing a chemical of concern. Rainout is the removal of particles that serve as condensation points for the formation of raindrops. The turbulent nature of the atmosphere suggests that dry deposition and settling is a minor component of the deposition process; however, wet deposition processes are less well understood than dry deposition processes. Because of this lack of understanding, there is little consensus regarding wet deposition models. Methods have been developed whereby wet deposition can be incorporated into EPA's Industrial Source Complex (ISC) model. For example, COMPDEP is a modification of EPA's Complex I model (Complex I is a Gaussian plume complex terrain model) to account for both wet and dry deposition. The algorithms in the model allow for calculation of deposition velocity based on particle size and atmospheric conditions including the ability to estimate deposition during storm events. RTDMDEP is a modification of RTDM that also accounts for wet and dry deposition. The primary difference between COMPDEP and RTDMDEP is in the ways in which the models handle complex terrain; deposition is dealt with in an identical fashion in both models. The development and application of COMPDEP and RTDMDEP are discussed in detail in EPA (1990g).

The output of a deposition model is the deposition velocity (V_d) that is used to calculate deposition flux (F_d) by the following simple relationship.

$$F_d = V_d \times C_a \qquad (8\text{-}1)$$

Where F_d = deposition flux (mass chemical/ area per unit time)
V_d = deposition velocity (length/time)
C_a = air concentration (mass/volume)

The air concentration often is obtained using a dispersion model; thus, the uncertainty associated with deposition flux combines uncertainties in modeling both deposition velocity and airborne concentrations. Deposition fluxes are used in indirect pathway analyses to calculate the concentration of a chemical of concern in an exposure medium, as

$$C_s = \frac{F_d \times [1 - e^{-k_sT}]}{Z \times BD \times k_s} \qquad (8\text{-}2)$$

Where C_s = soil concentration (mass chemical/mass soil)
k_s = soil loss constant (time^{-1})
T = time for deposition
Z = soil depth (length)
BD = soil bulk density (mass/volume)

These concentrations then are used directly in exposure calculations or as input to food chain models.

Physicochemical Processes and Chemical Transformations

Chemicals can undergo many types of transformations while they are being transported from a source to a receptor. Many of these changes are oxidation-reduction changes related to the oxidation state of most environmental media. For example, many chemicals are oxidized by reactive oxygen species such as ozone or hydroxyl radicals in air or water. Other chemicals are reduced by chemical or microbial actions in anoxic sediments. In some cases, the parent chemical is detoxified as a consequence of the chemical reaction, whereas in other cases its toxicity is increased. In almost all cases, the products of chemical transformations have different physicochemical properties than the parent chemical that then needs to be accounted for in exposure intake and pharmacokinetic models as well as indirect pathway analyses. Chapter 12 of this handbook discusses atmospheric transformation processes in more detail. The remainder of this section is devoted to an overview of transformations in the water and soil that may be relevant to the analysis of indirect exposure pathways.

Environmental fate and transport in aquatic and soil compartments is considerably more complex than in air primarily due to the presence of biotic components that can interact with chemicals through both biotransformation and bioconcentration mechanisms. In addition to biological processes, aquatic media are the location of many purely chemical or physical processes including acid–base equilibria, hydrolysis, photolysis, volatilization, precipitation–dissolution, adsorption–desorption, oxidation–reduction, and specific chemical reactions. These processes also occur in soil and sediment. Photolysis, for example, occurs only at the soil surface that is accessible to sunlight. The dominant aquatic and soil physicochemical processes are shown in Figure 8-10. These processes are categorized and discussed individually below.

Speciation processes are related to the specific form of a chemical that exists in the environment. Acid–base speciation relates to the fraction of an acid or base that is in the neutral or ionized state. The extent of acid–base speciation is controlled by pH. Typically, neutral molecules are more readily taken up into biological systems, volatilized, and sorbed. Most hydrophobic organic molecules are capable of sorption to particulate matter. The extent of sorption is determined by hydrophobicity and molecular surface area. The fate of sorbed material is determined largely by the fate of the particles onto which the chemicals are sorbed. Speciation reactions can cause a chemical to exist as a particular salt or ion pair that has unique properties. These reactions are most relevant to metals and anions. Slightly soluble species may precipitate onto sediments. Additionally, solubility limits of both organic and inorganic chemicals can be exceeded causing a pure chemical phase to form that has properties of its own.

Volatilization is a physical process whereby a chemical enters the air from a pure, sorbed, or dissolved phase. Volatilization behavior is extremely important from surface and subsurface spills.

Biological processes include biotransformation, bioconcentration, and bioaccumulation.

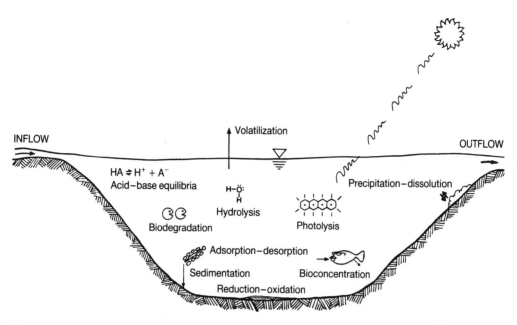

Figure 8-10. Speciation, transport, and transformation processes in the aquatic environment.

Many organisms are capable of transforming organic chemicals through a variety of metabolic processes. Even chemicals that generally are considered to be stable may be biotransformed under appropriate conditions. One of the most interesting series of biotransformations in the environment is the reductive dehalohydrogenation reactions of chlorinated aliphatics that continue until the stable monochloro members of the chemical families are reached. Bioconcentration refers to the uptake of a chemical by an organism by passive means (i.e., through lungs, gills, or dermal absorption). Bioaccumulation refers to the uptake of a chemical by an organism by active means such as ingestion of food or sediment.

Photolysis is a chemical process in which a chemical is transformed by its interaction with sunlight. Hydrolysis is the reaction of a chemical with water that usually results in smaller, less complex products. Both inorganics and organics can undergo a wealth of oxidation–reduction reactions that are characterized by the addition and substraction of electrons. Many biological transformations are oxidation–reduction reactions. Complex biogeochemical cycles involving natural materials, such as the nitrogen cycle, involve several sequential oxidation–reduction reactions.

In the natural environment, chemicals are subjected to numerous physicochemical processes, often occurring simultaneously. For example, mercury can be emitted from a source to the air as a metallic vapor sorbed onto particles. If the particles deposit onto water, a fraction of the mercury may be transformed into dimethylmercury by a microbially mediated reduction reaction. The dimethylmercury then may volatilize or be bioconcentrated by fish.

There are several references containing data or methods for evaluating environmental fate and transport processes. These include Mills et al. (1985), Lyman et al. (1982), and Bodek et al. (1988).

ROUTES OF EXPOSURE AND THEIR QUANTITATIVE EXPRESSION

The ultimate outcome of an exposure assessment is a calculation of a quantitative expression of exposure, dose, intake, or uptake that can be used in conjunction with a numerical expression of toxicity (i.e., the dose-response parameter) or regulatory standard to evaluate health risk or compliance. The calculation of exposure is route-specific where the term 'route' refers to inhalation, ingestion, or dermal absorption (most other routes, such as intravenous, are unimportant for assessing environmental exposure). In general, exposure is the product of the concentration of a chemical at the exposure boundary (e.g. lung) and the duration for which the concentration remains at the boundary. The derivation that follows is based on that provided in NAS (1991a). A critical element in understanding this derivation is the concept of the microenvironment. The microenvironment is defined as a location in time and space within which chemical concentrations can be assumed to be homogeneous (Duan 1982). In this context, total human exposure can be considered to be a time-weighted average or the product of the chemical concentration in the microenvironment multiplied by the time spent in that microenvironment divided by the total time.

The mathematical relationship for the exposure to an airborne chemical by a receptor in a single microenvironment (e.g., residence) is described by

$$\Delta E = C \times \Delta t \qquad (8\text{-}3)$$

Where ΔE = the exposure of a person
C = the concentration
t = the time

The units of exposure are concentration multiplied by time (e.g., $\mu g/m^3 \times h$). This is represented in the integral form as

$$E = \int_{t_1}^{t_2} C(t)\, dt \qquad (8\text{-}4)$$

$C(t)$ represents the relationship of concentration as a function of time for an interval t_1 through t_2. The time can be instantaneous or it can represent longer contact periods.

An operational form of the above equation that delineates the exposures of an individual for different microenvironments (e.g., residence, workplace, recreation area) in which that individual spends time is given by:

$$\Delta E_{j,k} = C_{j,k}(\Delta t) \times \Delta t_{j,k} \qquad (8\text{-}5)$$

Where $\Delta E_{j,k}$ = the exposure of person k to a given pollutant during time interval Δt as a result of that person's activities in microenvironment j
$C_{j,k}(\Delta t)$ = the average concentration to which person k is exposed during the time Δt while in microenvironment j
$\Delta t_{j,k}$ = the time spent by person k in microenvironment j; for acute health effects, t must have a short enough interval to assure that the average reflects the peak concentration that can cause an effect.

This equation can be written to include exposures of multiple contaminants to various persons in diverse microenvironments as

$$E_{i,j,k} = \sum_{i=1}^{I} C_{i,j}\, \Delta t_{j,k} \qquad (8\text{-}6)$$

The kth person is exposed to a concentration C_{ij} of the ith chemical in the jth microenvironment. The concentration C_{ij} is considered to be constant for that location during the interval $\Delta t_{j,k}$. The integrated exposure, E^T, then can be calculated for many different situations. For the exposure of the kth person to the ith contaminant, the time-integrated exposure for the kth person is the sum of the individual exposures to the ith contaminant over all of the possible microenvironments, or

$$E_{i,k}^T = \sum_{j=1}^{J} C_{i,j}\, \Delta t_{j,k} \qquad (8\text{-}7)$$

Alternatively, population exposure can be calculated for a group of persons in contact with the ith contaminant in a single microenvironment, such as for all people at a workstation, as

$$E_{i,j}^T = \sum_{k=1}^{K} C_{i,j} \, \Delta t_{j,k} \qquad (8\text{-}8)$$

A time-integrated population exposure to a single chemical as shown below, also could be defined for a population for a series of diverse microenvironments

$$E_i^T = \sum_{j=1}^{J} \sum_{k=1}^{K} C_{i,j} \, \Delta t_{j,k} \qquad (8\text{-}9)$$

Total airborne exposure to all chemicals can be defined by summing this equation for chemicals $i = 1$ to I. Thus, a quantitative exposure assessment requires accounting of the time spent by each person in the presence of each different concentration of every different contaminant of ultimate biological significance.

Because a human must be in an air environment at all times, some researchers use the concept of 'time-weighted average exposure for air exposure.' The equation below is the same as that for integrated exposure, except that $(\Delta t)_{i,j,k}$ refers to the proportion of T (time-weighted average associated with exposure) that a person k spends in microenvironment j, with the ith contaminant

$$T_k = \sum_{j=1}^{J} (\Delta t)_{i,j,k} \qquad (8\text{-}10)$$

In the summation over exposure, Δt was not explicitly written because the person k 'brings' the time of the microenvironment. The microenvironment and the time are defined by the person being studied; an exposure occurs in a microenvironment only if person k is in that microenvironment at a specific time

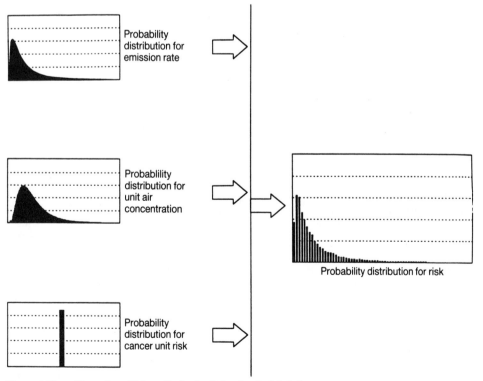

Figure 8-11.   Illustration of Monte Carlo simulation for the inhalation exposure pathway. During each iteration, a randomly selected value from each input parameter distribution (left) is combined to estimate the risk (right).

The dose of a chemical incorporating intake is the amount of the chemical deposited in biological tissue over an increment of time

$$D = \int_{t_1}^{t_2} C(t)I(t)\, dt \qquad (8\text{-}11)$$

or in conventional integrated form:

$$D = E \times I \times A \times t \qquad (8\text{-}12)$$

Where D = dose (may be expressed per unit time or per body weight)
E = exposure as above
I = intake rate of the medium containing the chemical of concern
A = absorption fraction
t = duration of exposure

Risk assessments conducted by EPA often use an expression of exposure known as the average daily intake. Two intake terms that appear often in the regulatory literature are the average daily dose (ADD) and the chronic daily intake (CDI)

$$CDI = ADD = \frac{C \times I \times ED}{BW \times AT} \qquad (8\text{-}13)$$

Where ED = exposure duration
BW = body weight
AT = averaging time

This type of characterization may be referred to as a time-weighted average daily intake. Intake may be converted to chronic daily uptake (CDU) by application of an absorption factor A:

$$CDU = CDI \times A \qquad (8\text{-}14)$$

Chapter 13 of this handbook presents equations for calculating CDIs for various exposure pathways and gives values for significant input variables. This methodology for the calculation of exposure and dose is deterministic in that it yields only one output value. If the output value reflects inputs that correspond to a 'reasonable maximum' exposure, the output will reflect greater than the 95th-percentile of exposure. If average values are used in the calculation, the output will reflect an average exposure. Stochastic simulation methods, in addition to deterministic methods, are highly useful for exposure and risk assessments. They have the ability to present the uncertainty in the assessment in addition to deriving the entire spectrum of exposures. An example of an output from a stochastic simulation that illustrates this technique is shown in Figure 8-11.

PHARMACOKINETIC AND PHYSIOLOGICAL FACTORS

Considerable emphasis is placed in this handbook on the properties of the environment that are important to the calculation of exposure. In addition to these properties, there are numerous properties of the receptor that must be taken into account in an exposure assessment. These properties may be incorporated into an exposure assessment as single points on a distribution or as a probability distribution. They include:

■ Intake rates (i.e., how much of a medium containing a chemical is taken in by a receptor)
■ Absorption and bioavailability (i.e., how much of the chemical of concern will be released from the medium when it is taken in)

■ Physiological and anatomic factors that describe the behavior of the organism

Intake Rates

The most important intake rates for exposure assessments concern the rate of inhalation of ambient air and the rate of ingestion of food and contaminated nonfood materials such as soil and water. EPA's Exposure Factors Handbook (EPA 1989a) presents analyses of ingestion exposure from drinking water, consumption of various foodstuffs, consumption of soil, and values for inhalation and dermal uptake. This information is arranged, where possible, into categories by age and sex so that specific calculations may be made. The reader should

Table 8-7   Exposure parameters used to estimate doses for multiple pathways

Exposure pathway	Age period (years)	Average body weight over period (kg)	Assumed duration of exposure (years)	Assumed frequency of exposure (days/year)	Exposure rate per exposed day
Inhalation of ambient air	1–70	70	70	365	20 m^3
Incidental ingestion of outdoor soil	1–6	16	6	150	62 mg
	7–17	44	11	100	24 mg
	18+	70	53	100	24 mg
Ingestion of fresh produce					
Vine crops	1–8	20	8	49	80 g
	9+	70	62	49	126 g
Root crops	1–8	20	8	61	61 g
	9+	70	62	61	76 g
Leafy crops	1–8	20	8	54	62 g
	9+	70	62	54	107 g
Ingestion of beef	1–8	20	8	156	92 g
	9–18	52	10	156	152 g
	19+	70	52	156	166 g
Ingestion of milk from local dairy cows	1–8	20	8	365	460 g
	9–18	52	10	365	524 g
	19+	70	52	260	326 g
Ingestion of venison	1–70	70	70	NA	22 kg/yr
Ingestion of eggs	1–8	20	8	104	67 g
	9+	70	62	104	86 g
Ingestion of fish	1–8	20	8	90	87 g
	9–18	52	10	90	123 g
	19+	70	52	90	163 g
Ingestion of drinking water	1–10	22	9	365	0.71
	11–19	57	10	365	0.91
	20+	70	51	365	1.31
Ingestion of human breast milk	0–1	9	1	365	0.8 kg
Incidental ingestion of household dust	1–6	17	6	183	62 mg

NA = Not applicable.

exercise caution in using the Exposure Factors Handbook, however, because significant amounts of exposure assessment research have taken place since publication of the document and many of the uptake rates reported in the document do not reflect current information. Table 8-7 lists some typical age-specific intake rates that are useful in exposure assessment.

Bioavailability and Absorption

In general terms, bioavailability is the amount of a chemical taken into the body that is available to be taken up into the systemic circulation. Bioavailability usually is broken down into two components – the desorption of the chemical of concern from its matrix and the absorption of the chemical across membranes. Obviously, bioavailability can have a major impact on exposure. For example, studies show that gastrointestinal absorption of inorganic arsenic varies from 30 percent to 80 percent depending on chemical speciation and the medium from which it is absorbed. Some forms of a chemical (e.g. elemental lead) are not absorbed whereas other forms (e.g. lead nitrate) are absorbed almost totally. A hierarchical approach is useful for performing and evaluating bioavailability studies. An example of a hierarchy is shown in Figure 8-12. This hierarchy is designed to select biological models that most closely resemble the human situation and that also meet certain

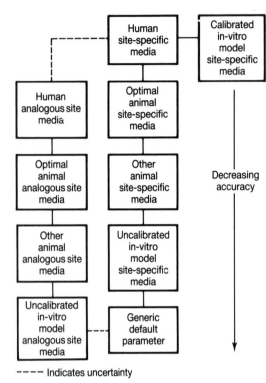

---- Indicates uncertainty

Figure 8-12.   Hierarchy of bioavailability models.

requirements for quality assurance and good laboratory practice.

Physiological and Anatomic Factors

Physiologic and anatomic factors are required to calculate intake or uptake of a chemical. The number and use depends on the pharmaco-kinetic complexity of the exposure models. For the simplest models, information often is limited to readily obtained values such as body weight or percent lipids; more sophisticated models require intercompartment transfer coefficients that are highly dependent on physiology. EPA's Exposure Factors Handbook is a useful source of data for human physiological and anatomic factors.

Use of Pharmacokinetic Models in Exposure Assessment

Pharmacokinetics is the body of scientific knowledge dealing with the absorption, distribution, metabolism, and elimination of chemicals within an organism. Pharmacokinetic models can be used to calculate the fraction of an administered dose that can ultimately reach a target organ and cause an adverse health effect. The application of these models can result in accurate calculations of exposure. For example, a pharmacokinetic model developed by Paterson and Mackay (1987) can be used to show that only about 1 percent of an administered dose of a particular chlorinated biphenyl is transported to the liver, which is taken to be the target organ. The reader is referred to Chapters 6 and 7 of this handbook, as well as standard works (Ritschel 1986, Krishna and Klotz 1990) for general discussions of the application of pharmacokinetics to risk assessment. One application is particularly important to current risk assessment practice – EPA's Integrated Uptake/Biokinetic Model (U/BK) for evaluating exposure to lead (Chrostowski and Wheeler 1992). This model allows the user to calculate blood (or other tissue) lead levels as a function of exposure. The blood lead levels then are compared to levels of toxicological concern.

CHARACTERIZATION OF THE EXPOSED POPULATION

The object of this component of the exposure assessment is to identify the individuals or populations who could potentially be exposed and the profile of contact with the chemicals of concern based on behavior, location as a function of time, and characteristics of the individuals. Population characterization is substantially more complex for ecological exposure assessments than for human exposure assess-ments. A forest, for example, may contain thousands of species, some of which are resident and some migratory. Some of these species can spend particularly sensitive portions of their lifetime (e.g., an early life stage) within an exposure study area. Often the behavior of a particular species is dependent on that of other species or on environmental factors which may or may not be related to the chemicals of concern being studied in the exposure assessment. A classic example is the avoidance behavior that some fish show to highly turbulent aquatic discharges that may prevent their exposure to chemicals of concern in those discharges. Typically, ecological exposure assessments characterize exposure to only a few of the potential receptor species. The chosen receptors are known as indicator species. The indicators may be chosen because they are taxonomic analogs (e.g., domestic cat is analogous to wild cats), because they occupy a critical niche in a food chain, because they are related to standard test organisms (e.g., Ceriodaphnia or fathead minnows), or because they are endangered, threatened, or of particular commercial value. Ecological assessment is discussed in Chapter 15. Useful references related to ecological risk assessments are EPA (1991i) and EPA (1988f).

Static and Dynamic Population Location

The concept of location is intimately tied to the concept of microenvironments. In theory, a large number of microenvironments are possible. It is up to the exposure assessor to define and limit the number that are used in an assessment. The simplest type of microenvironment analysis is represented by the case where the exposure parameter of interest is a spatial average air concentration over a large area and the potential receptors never leave the area.

Demographic Factors

Demographic factors are related to the end-use of the exposure data. In a typical assessment for exposure to a major stationary source, age and sex distributions normally are considered to be most important. Additional information on socioeconomic status or race can help define sensitive subgroup populations. Sophisticated exposure analyses often integrate demographic and anatomic data (e.g., by calculating the average body weight associated with a particular population). Demographic factors associated with land use also are important

determinants of behavior patters. For example, typical individuals exhibit different behavior on land that is primarily residential-single family, residential-multiple family, agricultural, commercial, industrial, or recreational. Useful sources of demographic information are the U.S. Bureau of the Census, state or local zoning or land use agencies, zoning maps, tax or insurance maps, topographic, land use or housing maps, sequential aerial photography, and commercially available geographic information systems (GIS). A recent example of the development of demographic data for incorporation into exposure assessment can be found in the paper by Israeli and Nelson (1992) who researched the distribution and expected time of residence for U.S. households.

Activity Patterns

The definition of activity patterns is necessary to determine the amount of time a person spends at a particular exposure point and what the significant routes are at that point. The current literature is summarized in Chapter 14 and EPA (1989a). Recent advances in this area stem from the application of surveys from the social sciences (Robinson 1988). Activity diaries in which a study participant logs daily activities and their duration also are useful tools (Quackenboss and Lebowitz 1989). Some studies can become quite specific. For example, a recent study of exposure to lead (Bornshein et al. 1990) asked parents to note the frequency of the ingestion of snow, paper, dirt, cigarettes, plaster, and paint by their children. A recent example of an activity pattern study can be found in Schwab et al. (1990). This study used diary techniques to determine the amounts of time which different population subgroups spent in seven microenvironments. Interestingly, these authors concluded that overall activity patterns exhibited low variation across the population. Table 8-8 is a list of questions that are useful for characterizing activity patterns for inhalation and soil ingestion pathways. Similar lists can be developed for other exposure pathways.

Table 8-8   Behavior and activity pattern questionnaire

Category	Question
General Information (personal)	■ Address/location of individual ■ Number of individuals in household ■ Length of time individuals in household have been at this residence ■ Age, sex, body weight, height, and occupation of all individuals in household ■ Location of workplaces or schools of individuals in household ■ Amount of time spent commuting to/from work or school ■ Name of the stores/markets where groceries are purchased

Incidental ingestion and dermal absorption of surface soil	■ Size of property on which individuals live
	■ Nature of property
	  ● How is property vegetated (i.e., grass, shrubs, trees, etc.)?
	  ● What percentage, if any, is unvegetated?
	  ● Are there any buildings other than the home itself?
	■ How many hours per day are spent at home (this may differ on weekends against weekdays)?
	■ Of the time spent at home, what proportion of time is spent indoors against outdoors?
	■ What activities are performed outdoors and how much time is spent at each?
	  ● Gardening
	  ● Lawn care
	  ● Home maintenance
	  ● Automobile maintenance
	  ● Recreational activities/playing
	  ● Other
	■ What kind of clothes are worn outdoors when performing these activities?
	■ Is there play equipment outdoors at home (e.g., swings, basketball court, sandbox, etc.)?
	■ Does individual's job entail outdoor work? If so, what type(s) of work is conducted?
	■ At work, what portion of time is spent indoors against outdoors (if applicable)?
	■ If there are children in the household, how much time is spent outside at school?
	■ Have any children in the household been diagnosed as exhibiting pica?
	■ Do children in the household incidentally eat soil, paper, fabric, or other materials?

Inhalation of outdoor air	■ Questions for this exposure pathway are covered by the questions asked for soil ingestion.

Inhalation of indoor air	■ Of the time spent indoors, what proportion of time is spent in the following levels of physical activity:
	  ● Resting (e.g., watching television, reading, sleeping)
	  ● Light activity (e.g., level walking, meal cleanup, care of laundry and clothes, domestic work, hobbies, minor indoor repairs)
	  ● Moderate activity (e.g., climbing stairs, scrubbing surfaces and other heavy cleanup, major indoor repairs and alterations)
	  ● Heavy activity (vigorous physical exercise such as weight lifting, dancing, riding exercise bike)
	■ Is there air conditioning? How often is it used? What type?
	■ What type of heating system is used for the house? How often is it used?
	■ How often are windows open in the house? Are there screens?
	■ Are there any air purifiers or humidifiers?
	■ What is the smoking activity of members of the household?

Sensitive Subgroups

Certain components of the population may require particular attention in an exposure or risk assessment owing to special sensitivities. These can be exposure based (e.g., a subsistence hunter) or toxicologically based (e.g., fetuses with their particular sensitivity to the effects of lead). People who may be at greater risk from the toxicological standpoint include children, pregnant and lactating women, the elderly, and people with preexisting illnesses, such as hospitalized

populations. From an ecological exposure standpoint, endangered and threatened species, or species that occupy a critical niche in a particular ecosystem, constitute sensitive subgroups. Behavior patterns that can define a sensitive subgroup include children who ingest nonfood items, people who depend largely on a limited source of food, and people who are exposed to chemicals from a variety of sources (i.e., occupationally, at home, and during recreation). Local demographic studies usually are the best source of information on sensitive subgroups.

TYPICAL AIR TOXICS EXPOSURE SCENARIOS

Inhalation Exposure to Routine Emissions from Chemical Manufacturing

This is one of the simpler types of scenarios that may confront the exposure assessor. A typical example concerns emissions that are reported under section 313 of the Emergency Response and Community Right-to-Know Act. In the example, a manufacturer collects and reports emissions data to EPA; however, the manufacturer is concerned as to the nature of the risks that could be associated with these emissions. The answer to this question is determined through a screening exposure assessment.

A screening exposure assessment involves definition of the source, definition of the receptor, and performance of air dispersion modeling. Because the emission rates already have been defined, the exposure assessor needs only to define other characteristics of the source such as its height above sea level, its geometry (i.e., point, line, area, or volume source), the velocity of the release of the chemicals of concern, and whether one or multiple sources are involved. Two receptors commonly are chosen for this type of evaluation. One is the classic maximum exposed individual (MEI) and the other is a hypothetical individual located at the property fenceline. If population risks are to be calculated, the population that is affected by an average concentration from the facility also must be obtained. This information can be used later to draw risk contours associated with the

emissions. Air dispersion modeling can be performed using either a discrete air model such as ISCLT (see Chapter 11) or an integrated dispersion-exposure model such as SHEAR (Anderson and Lundberg 1983); other integrated exposure-dose models are described in Appendix G. If this assessment yields risks that are considered to be unacceptable, the usual next step is to undertake a refined assessment that calculates average, instead of maximum, exposures and risks.

Indirect Exposure to Windblown Mine Tailings

The western United States has been extensively mined for over 100 years. Most of the major mining sites have accumulated large quantities of mine tailings that are the residue remaining after economically important metals have been extracted from ore. In many cases, these tailings consist of small particles that are deposited as a waterborne slurry. When the water evaporates, the particles become exposed to the environment. They often may be entrained by wind, especially in the presence of activities that disturb the tailings. In addition to tailings from mining operations, these sites may contain tailings or other residuals from smelters and processors or mills of various types. All of these activities can contribute to large reservoirs of small particles that may be entrained readily and that often contain significant quantities of potentially toxic metals. At a Superfund site in

Utah, for example, there were thousands of cubic yards of tailing materials that contained significant quantities of arsenic, lead, and cadmium. Exposure assessors observed the entrainment and redistribution of particles from the surface of the tailings piles during wind events. In this case, the particles were transported to a residential area adjacent to the site and deposited.

At such a site, several exposure pathways are important. First is the air pathway, both as a source of direct exposure and as a means of transporting particles for subsequent deposition. Important indirect pathways related to deposited particles include incidental ingestion of soil and ingestion of homegrown produce. The exposure assessment methodology for complex tailings sites usually combines environmental measurements, biological measurements, and mathematical modeling.

Environmental measurements are obtained both at the source (i.e., tailings piles) and at the exposure point. In addition to the normal analysis for total metal concentrations, information on geochemical speciation and mineralogy often is important for subsequent parts of the exposure assessment as are measurements of particle size distribution. Chemical speciation is important for subsequent determination of toxicity and bioavailability, whereas particle size is important for use in transport models and determining bioavailability. Biomonitoring procedures are available for most common chemicals of concern that are found at mining sites. For example, lead is commonly measured in blood or bone and arsenic is measured in urine or hair. It is important to match the biomonitoring method to the exposure of interest. Lead measurements in bone are more appropriate as an indication of chronic exposure whereas lead measurements in blood are more indicative of subchronic or acute exposure. A combination of environmental and biological measurements can be used to develop exposure pathways. For example, if lead is measured in indoor dust,

outdoor soil, food, water, in wipe samples on walls, in indoor and outdoor air, and in blood, statistical techniques such as structural equation modeling (Buncher, Succop, and Dietrich 1991) can be used to create exposure pathways.

Mathematical modeling is useful for the characterization of exposures at mining sites. Air dispersion and particle deposition models are used to evaluate the transport of heavy metals from the source to the receptor. A model such as EPA's Fugitive Dust Model, described in Chapter 11, is useful in this regard. Exposure models are useful for calculating uptake of chemicals of concern from foodstuffs. Finally, pharmacokinetic models are useful for calculating concentrations of chemicals of concern at target organs. For example, EPA's UBK model for lead can be calibrated with site-specific environmental and biological measurements after which it can be used to simulate other situations such as exposure during the implementation of remedial alternatives or exposure at locations that are currently undeveloped or that could be developed in the future.

Exposure to an Emergency Release

Exposure assessors often are called on to provide rapid assessments of exposure that results from emergency releases. These releases can come from many sources including industrial manufacturing facilities, pipelines, and transportation spills. The prediction and evaluation of emergency releases constitutes an entire branch of risk assessment that is beyond the scope of this handbook. The reader is referred to standard references such as the work of Greenberg and Cramer (1991) for further details.

For purposes of this example, a railway accident results in the release of benzene from a tank car[3] The method of analysis consists of determining the total amount of benzene that was discharged, calculating the radius of the resulting pool, and performing simple air

---

[3] This example has been abstracted from materials for handling benzene spills developed by the Department of the Environment, Conservation, and Protection, Canada, which should be consulted for further details (Phone 819–997–1575).

dispersion modeling to define the extent of a hazard zone. In this case, the hazard zone is an area that should not be entered by workers without respiratory protection. In this example, a tank car spills twenty metric tons of benzene. The spill takes five minutes and occurs during the night when the temperature is 20° C and the wind speed is 7.5 km/h. The first step in the process is to determine the radius of the spill, either from direct observations or using a nomograph technique described in the manual. A second nomograph allows for the calculation of the benzene vapor emission rate as a function of the spill radius. These data then are used in a virtual point source screening level air dispersion model in conjunction with the permissible exposure limit for benzene to determine the area of the hazard zone. In this particular case, the hazard zone is 0.52 km wide by 14 km long.

Multiple-Pathway Risk Assessment for a Waste-to-Energy Plant

Incinerators[4] have been the subject of more exposure and risk assessments than any other stationary source category. These assessments are required by permitting agencies, used by facility proponents as educational tools for informing the public about potential health effects, and are used to decide among alternate sites, design configurations, or pollution control technologies. EPA has published guidance for performing exposure and risk assessment for indirect pathways associated with emissions from WTE facilities (EPA 1990g). This guidance should be used with caution, however, because postpublication peer review has revealed several inaccuracies in the methodology. Numerous state-level regulatory guidance documents also have been published for various types of incinerator risk and exposure assessments. These documents contain considerable information of general utility in addition to state-specific requirements.

The steps in a typical incinerator risk assessment include the following:

■ Selection of a list of chemicals of potential concern
■ Development of emission rates
■ Performance of dispersion and deposition modeling
■ Calculation of exposures via indirect pathways
■ Calculation of risks
■ Analysis of uncertainties

The selection of chemicals of concern and calculation of risks are beyond the scope of this chapter and are not considered further. Emission rates usually are obtained from quality assured and validated measurements of comparable facilities or calculated. Care should be taken to match the feedstock, type of combustors, and pollution control equipment for facilities being considered. Emission rates may be input to the exposure assessment as single values or probability distributions.

Air dispersion modeling usually is performed using EPA or state approved models such as ISCST or VALLEY, described in Chapter 11. Because most incinerators require air quality permits (both construction and operating), that also require modeling, it usually is more efficient to use the same models in both the exposure assessment and permit preparation. Dry deposition modeling requires particle size distributions, generally obtained from comparable facilities, and site-specific factors such as roughness height. The decision to use wet deposition modeling typically is made by balancing the accuracy of existing models with the objectives of the assessment and the presence of climatological factors conducive to wet deposition.

Numerous indirect pathways often are evaluated in incinerator risk assessments based on site-specific conditions. Normally, the assessment contains an analysis of ingestion of various foodstuffs, incidental ingestion of soil,

---

[4] Including waste-to-energy (WTE) facilities for municipal solid waste, biohazardous waste incinerators, sewage sludge incinerators, and hazardous waste incinerators.

Table 8-9   Sample qualitative uncertainty analysis

Assumption	Magnitude of effect on risk[a]	Direction of effect on risk
Stack source parameters were based on the proposed facility design.	Low	May over- or underestimate risk
The ISCST model was used for long-term and short-term air concentrations.	Low	May over- or underestimate risk
For long-term dispersion and deposition modeling, five years of local meteorological data were used.	Low	May over- or underestimate risk
Annual average air concentrations were predicted assuming no prior deposition of particles from the plume.	Low	May overestimate risk
Chemical removal processes such as photolysis not included in air dispersion or deposition modeling.	Moderate[b] Low[c]	May overestimate risk
The deposition modeling only took dry deposition into account (i.e., wet deposition not included in the model).	Low to moderate	May underestimate risk
Average surface roughness lengths were assumed for both land and water in deposition modeling.	Low	May over- or underestimate risk
Particle resuspension not included in deposition modeling.	Low	May over- or underestimate risk
Deposition fluxes were estimated using the Wurzburg, Germany Waste-to-Energy Facility particle size distribution data.	Low	May over- or underestimate risk
Plume depletion was not considered in the inhalation estimates.	Low	May overestimate risk
Maximum emission rates were LAER or BACT technology-based.	Low to moderate	May overestimate risk
Mean emission rates were based on the geometric mean of emissions data from operating MSW facilities.	Low	May over- or underestimate risk
Emission rates during upset conditions were based on pollution control technology removal efficiencies or on data from operating MSW facilities.	Low	May over- or underestimate risk
Rural dispersion coefficients were used instead of dispersion coefficients.	Low	May underestimate urban risk

[a] Low means ≤1 order of magnitude effect.
Moderate means 1–2 orders of magnitude effect.
High means >2 orders of magnitude effect.
[b] Reactive organics.
[c] Other chemicals.

and the transmission of chemicals to the newborn through mother's milk. Analysis of numerous assessments strongly suggests that dermal exposure to emissions from incinerators is not significant; therefore, it usually does not require evaluation.

The uncertainty analysis should indicate the location of the exposure on the probability distribution of all possible exposures (e.g.,

median, mean, 95th-percentile, and maximum). In addition, the uncertainty analysis should identify all assumptions used in the assessment and evaluate their impact on the results. Table 8-9 is an example of this type of uncertainty analysis. Quantitative uncertainty analyses involving Monte Carlo simulations or propagation of error also are useful tools in this part of the assessment.

References[5]

Anderson, G.E. and G.W. Lundberg. 1983. User's Manual for SHEAR. Publication No. SYSAPP-83/124. Prepared for U.S. Environmental Protection Agency, Office of Air Quality Planning and Standards. Research Triangle Park, NC.

Attix, F.H. 1991. *Introduction to Radiological Physics and Radiation Dosimetry.* New York: Wiley Interscience.

Baselt, R.C. 1980. *Biological Monitoring Methods for Industrial Chemicals.* Davis, CA: Biomedical Publications.

Baskin, L.B. and J.W. Falco. 1989. Assessment of human exposure to gaseous pollutants. *Risk Analysis* 9(3):365–375.

Bodek, I., W.J. Lyman, W.F. Reehl, and D.H. Rosenblatt. 1988. *Environmental Inorganic Chemistry.* New York: Pergamon Press.

Bornshein, R., S. Clark, W. Pan, and P. Succop. 1990. Midvale community lead study. University of Cincinnati, Cincinnati, Ohio.

Buncher, C.R., P.A. Succop, and K.N. Dietrich. 1991. Structural equation modeling in environmental risk assessment. *Environmental Health Perspectives* 90:209–213.

Chrostowski, P.C., S.A. Foster, and D. Dolan. 1991. Monte Carlo analysis of the reasonable maximum exposure (RME) concept. Presented at the Hazardous Materials Control '91 Conference. December 3–5, 1991.

Chrostowski, P.C. and J.A. Wheeler. 1992. A comparison of the integrated uptake biokinetic model to traditional risk assessment approaches for environmental lead. Superfund Risk Assessment, In: *Soil Contamination Studies*, ASTM STP 1158. K.B. Hoddinott, G. D. Knowles, eds. American Society for Testing and Materials, Philadelphia, PA.

Cox, C., T.W. Clarkson, D.O. March, L. Amin-Zaki, S. Tikriti, and G.G. Myers. 1989. Dose-response analysis of infants prenatally exposed to methyl mercury: An application of a single compartmental model to single strand hair analysis. *Environmental Research* 49:318–332.

Duan, N. 1982. Models for human exposure to air pollution. *Environmental International* 8:305–315.

Gilbert, R.O. 1987. *Statistical Methods for Environmental Pollution Monitoring.* New York: Van Nostrand Reinhold.

Greenberg, H.R. and Cramer, J.J. 1991. *Risk Assessment and Risk Management for the Chemical Process Industry.* New York: Van Nostrand Reinhold.

Hammond, S.K. and B.P. Leaderer. 1987. A diffusion monitor to measure exposure to passive smoking. *Environmental Science and Technology* 27:494–497.

Hattis, D. 1988. The use of biological markers in risk assessment. *Statistical Science* 3(3):358–366.

Hawkins, N.C. 1991. Evaluating conservatism in maximally exposed individual (MEI) predictive exposure assessments: A first-cut analysis. *Regulatory Toxicology and Pharmacology* 14:107–117.

Hawkins, N.C., S.K. Norwood, and J.C. Rock. 1991. A strategy for occupational exposure assessment. American Industrial Hygiene Association. Akron, Ohio.

Hawkins, N.L., M.A. Jaycock, and J. Lynch. 1992. A rationale and framework for establishing the quality of human exposure assessments. *American Industrial Hygiene Association Journal* 53:34–41.

International Agency for Research on Cancer (IARC). 1984b. Monitoring Human Exposure to Carcinogenic and Mutagenic Agents. IARC Scientific Publication No. 59. World Health Organization. Lyon, France.

Israeli, M. and C.B. Nelson. 1992. Distribution and expected time of residence for U.S. households. *Risk Analysis.* 12:65–72.

Krishna, D.R. and U. Klotz. 1990. *Clinical Pharmacokinetics.* New York: Springer Verlag.

Lasorda, B. 1992. Trends in mercury concentrations in the hair of women of Nome, Alaska: Evidence of seafood consumption or abiotic absorption. International Conference on Mercury as a Global Pollutant. Monterey, CA. May 1992.

Lauwerys, R.P. 1983. *Industrial Chemical Exposure: Guidelines for Biological Monitoring.* Davis, CA: Biomedical Publications.

Lioy, P.J. and J.M. Daisey. 1987. *Toxic Air Pollution.* Chelsea, MI: Lewis Publishers.

Lyman, W.J., W.F. Reehl, and D.H. Rosenblatt. 1982. *Handbook of Chemical Property Estimation Methods.* New York: McGraw-Hill.

McKone, T.E. and D.W. Layton. 1986. Screening the potential risks of toxic substances using a multimedia compartment model. Report No. UCRL-94548. Lawrence Livermore Laboratory. Berkeley, CA.

---

[5] Additional useful reference material is listed in the bibliography.

Mills, W.B., D.B. Porcella, M.J. Ungs, S.A. Gherini, K.V. Summers, L. Mok, G.L. Rupp, and G.L. Bowie. 1985. Water Quality Assessment: A screening procedure for toxic and conventional pollutants in surface and ground water. EPA 600/6–85–002a. U.S. Environmental Protection Agency, Office of Research and Development. Athens, GA.

National Academy of Sciences (NAS). 1991a. *Frontiers in Assessing Human Exposures to Environmental Toxicants*. Washington, D.C. National Academy Press.

National Academy of Sciences (NAS). 1991b. *Human Exposure Assessment for Airborne Pollutants – Advances and Opportunities*. Washington, D.C. National Academy Press.

Neely, W.B. 1980. Chemicals in the Environment. Distribution – Transport Fate – Analysis. In: P.N. Cheremisinoff, ed. Volume 13 – *Pollution Engineering and Technology Series*. New York: Marcel Dekker.

Neely, W.B. and G.E. Blau. 1984. *Environmental Exposure From Chemicals*. Two Volumes. Boca Raton, FL: CRC Press.

Ott, W.R. 1980. Models of human exposure to air pollution. Technical Report #32. Department of Statistics. Stanford University.

Ott, W.R., L.E. Rodes, R.J. Drago., et al. 1986. Automated data-logging personal exposure monitors for carbon monoxide. *Journal of the Air Pollution Control Association* 36:883.

Patrick, D.R. 1992b. The impact of exposure assessment assumptions and procedures on estimates of risk associated with exposure to toxic air pollutants. Paper 92–95.02. Presented at the 85th Annual Meeting of the Air & Waste Management Association. Kansas City, MO.

Paterson, G. and D. MacKay. 1987. A steady-state fugacity-based pharmacokinetic model with simultaneous multiple exposure routes. *Toxicology and Environmental Chemistry* 6:395–408.

Patterson, D.G., R.E. Hoffman, L.L. Needham, D.W. Roberts, J.R. Bagby, J.L. Pirkle, H. Falk, E.J. Sampson, and V.N. Houk. 1986. 2,3,7,8-Tetrachlorodibenzo-*p*-dioxin levels in adipose tissue of exposed and control persons in Missouri. An Interim Report. *Journal of the Air and Waste Management Association* 256(19):2683–2686.

Quackenboss, J.J., and M.D. Lebowitz. 1989. The utility of time-activity data for exposure assessment: Summary of procedures and research needs. Paper No. 89–100.7. Presented at the 82nd Annual Meeting of the Air & Waste Management Association. Anaheim, CA.

Ritschel, W.A. 1986. *Handbook of Basic Pharmacokinetics – Including Clinical Applications*. Third Edition. Hamilton, IL: Drug Intelligence Publications, Inc.

Robinson, J.P. 1988. Time-diary research and human exposure assessment: Some methodological considerations. *Atmospheric Environment* 22: 2085–2090.

Schwab, M., S.D. Colome, J.D. Spengler, P.B. Ryan, and I.H. Billick. 1990. Activity patterns applied to pollutant exposure assessment: Data from a personal monitoring study in Los Angeles. *Toxicology and Environmental Health* 6:517–532.

Scire, J., R. Yamartino, D. Strimaitis, and S. Hanna. 1987. Design for a non-steady state air quality modeling system.

Sehmel, G.A. and W.J. Hodgson. 1978. A model for predicting dry deposition of particles and gases to environmental surfaces. PNL-SA-6721. Prepared for the U.S. Department of Energy by Battelle Pacific Northwest Laboratory. Richland, WA.

Severn, D.J. 1987. Exposure assessment. *Environmental Science and Technology* 21:1159–1163.

Thompson, K.M., D.E. Burmaster, and E.A.C. Crouch. 1992. Monte Carlo techniques for quantitative uncertainty analysis in public health risk assessments. *Risk Analysis* 12:53–63.

Tinsley, I.J. 1979. *Chemical Concepts in Pollutant Behavior*. New York: John Wiley & Sons.

U.S. Environmental Protection Agency (EPA). 1984b. Approaches for improving the assessment of human genetic risk – Human biomonitoring. EPA 600/9–84–016. Office of Health and Environmental Assessment. Washington, D.C.

U.S. Environmental Protection Agency (EPA). 1987c. Pesticide assessment guidelines: Applicator exposure. EPA 540/9–87–127. Office of Pesticides Programs. Washington, D.C.

U.S. Environmental Protection Agency (EPA). 1987d. Risk assessment, management, and communication – A guide to selected sources. EAP/IMSD/ 87–002a. Office of Toxic Substances. Washington, D.C.

U.S. Environmental Protection Agency (EPA). 1987e. Data quality objectives for remedial response activities. Development process. EPA 540/6–87–003. Office of Emergency and Remedial Response and Office of Waste Programs Enforcement. Washington, D.C.

U.S. Environmental Protection Agency (EPA). 1988b. Superfund exposure assessment manual. EPA 540/1–88–001. OSWER Directive 9285.5–1. Office of Emergency and Remedial Response. Washington, D.C.

U.S. Environmental Protection Agency (EPA). 1988f. Review of ecological risk assessment methods. EPA 230/10–88–041. Office of Policy Analysis. Washington, D.C.

U.S. Environmental Protection Agency (EPA). 1989a. Exposure factors handbook. EPA 600/8–89–043. Office of Health and Environmental Assessment. Washington, D.C.

U.S. Environmental Protection Agency (EPA). 1989b. Air Superfund national technical guidance series. Volume I: Application of air pathway analyses for Superfund activities. Interim Final Report. EPA 450/1–89–001. Office of Air Quality Planning and Standards. Research Triangle Park, NC.

U.S. Environmental Protection Agency (EPA). 1989c. Air Superfund national technical guidance series. Volume II: Estimation of baseline air emissions at Superfund sites. Interim Final Report. EPA 450/1–89–002. Office of Air Quality Planning and Standards. Research Triangle Park, NC.

U.S. Environmental Protection Agency (EPA). 1989d. Air Superfund national technical guidance series. Volume III: Estimation of air emissions from cleanup activities. Interim Final Report. EPA 450/1–89–003. Office of Air Quality Planning and Standards. Research Triangle Park, NC.

U.S. Environmental Protection Agency (EPA). 1989e. Air Superfund national technical guidance series. Volume IV: Procedures for dispersion modeling and air monitoring for Superfund air pathway analysis. Interim Final Report. EPA 450/1–89–004. Office of Air Quality Planning and Standards. Research Triangle Park, NC.

U.S. Environmental Protection Agency (EPA). 1989f. Risk assessment guidance for Superfund. Volume 1. Human health evaluation manual. Part A. EPA 540/1–89–002. Office of Emergency and Remedial Response. Washington, D.C.

U.S. Environmental Protection Agency (EPA). 1990f. ORD health biomarker research program. A strategy for the future. Science Advisory Board. Washington, D.C.

U.S. Environmental Protection Agency (EPA). 1990g. Methodology for assessing health risks associated with indirect exposure to combustor emissions. Interim Final. EPA 600/6–90–003. Office of Health and Environmental Assessment. Washington, D.C.

U.S. Environmental Protection Agency (EPA). 1991g. EPA Contract Laboratory Program. State of work for organics analysis. Multi-media, multi-concen-tration. Doc. No. OLM01.5 (April 1991 revision). Washington, D.C.

U.S. Environmental Protection Agency (EPA). 1991h. EPA Contract Laboratory Program. State of work for inorganics analysis. Multi-media, multi-concentration. Doc. No. ILM01.0. Washington, D.C.

U.S. Environmental Protection Agency (EPA). 1991i. Summary report on issues in ecological risk assessment. EPA 625/3–91–018. Office of Research and Development. Research Triangle Park, NC.

U.S. Environmental Protection Agency (EPA) 1992a. Guidelines for exposure assessment. 57 FR 22888–22938, May 29, 1992.

Wallace, L. 1987. The total exposure assessment methodology (TEAM) study: Summary and analysis, Volume 1. EPA 600/6–87–002a. U.S. Environmental Protection Agency, Office of Research and Development. Washington, D.C.

Wallace, L.A. and W.R. Ott. 1982. Personal monitors. A state-of-the-art survey. *Journal of the Air Pollution Control Association* 32:602.

Wallace, L., E. Pellizzari, T. Hartwell, C. Sparacino, and H. Zelon. 1983. Personal exposures to volatile organics and other compounds indoors and outdoors – the TEAM Study. NTIS PB83–121357. U.S. Environmental Protection Agency, Office of Research and Development. Washington, D.C.

Wallace, L., E. Pellizzari, T. Hartwell, R. Whitmore, C. Sparacino, and H. Zelon. 1986. Total exposure assessment methodology (TEAM) study: Personal exposures, indoor–outdoor relationships and breath levels of volatile organic compounds in New Jersey. *Environment International* 12:369–387.

Wallace, L., E. Pellizzari, T. Hartwell, C. Sparacino, R. Whitmore, L. Sheldon, H. Zelon, and R. Perritt. 1987. The TEAM (Total Exposure Assessment Methodology) study: Personal exposures to toxic substances in air, drinking water and breath of 400 residents of New Jersey, North Carolina and North Dakota. *Environmental Research* 43:290–307.

Wallace, L., E. Pellizzari, T. Hartwell, R. Whitmore, H. Zelon, R. Perritt, and L. Sheldon. 1988. California TEAM study: Breath concentrations and personal air exposures to 26 volatile compounds in air and drinking water of 188 residents of Los Angeles, Antioch and Pittsburgh, California. *Atmospheric Environment* 22: 2141–2163.

9

Source Sampling and Analysis

Robin R. Segall and Peter R. Westlin

INTRODUCTION

Title III of the 1990 Clean Air Act Amendments (CAAA) requires establishment of emission standards for 189 hazardous air pollutants (also called air toxics) listed in section 112(b)(1). Source emission testing may be required both to determine whether an emission standard applies to a particular source and, if it does, whether the source then complies with the standard. In addition, source emission testing also may be used by industry to collect emission data to demonstrate early reduction under section 112(i)(5) of the CAAA to defer application of maximum achievable control technology (MACT).

The majority of the 189 HAPS are volatile and semivolatile organic compounds; most of the remainder are metal compounds. At the time of writing this handbook in 1992, there were not validated emission test methods for sampling and analyzing all of these pollutants. In fact, fully validated methods were available for less than a quarter of these pollutants. Further, even where validated methods are available for one source, they may not be applicable to other emission sources. The Environmental Protection Agency (EPA) recently published a field validation protocol applicable to emission test methods – Proposed Method 301 (EPA 1991o;

EPA 1991c). This method provides procedures for determining the performance of a method applied to source-specific emissions in terms of systematic error (bias) and random error (precision). Once a specified level of performance is documented, the method is 'validated' for that specific application; validated methods do not automatically extend to substances other than the ones specifically covered by the validation testing or to sources other than the one where the validation was performed.

There are several purposes for which emission measurements are routinely conducted:

1. Compliance with a Standard – emission tests are conducted to determine compliance with an applicable emission standard. Compliance testing includes data collected for application for early reduction as well as after application of MACT.
2. Standard Setting – emission data are obtained to investigate options for, and to support the establishment of, air pollution-related standards.
3. Emission Factors – emission data are used to calculate emission factors, such as those in the EPA publication AP-42 (1985a), and that routinely are used for emission inventories,

dispersion modeling, and other tools employed in making air quality management decisions.[1]

4. Screening Measurements – screening emission tests are conducted to provide a preliminary indication of the presence or levels of a pollutant. This type of testing often is conducted to provide direction for subsequent test programs or to guide air quality management.

Different levels of data quality are required from the source emission tests conducted for these four purposes. Compliance determination requires the most rigorous data, including defined data quality objectives (DQO) and documented sample integrity; emission testing for screening purposes generally does not necessitate such exacting measurements. A validated method, when conducted according to the prescribed procedures, should yield data of known quality. Data quality typically is expressed in terms of the bias (i.e., systematic error in the measurement) and precision (i.e., random error in the measurement). Generally, only a validated method should be used to collect emission data for the first three purposes. A validated method, however, may not be necessary to collect screening data.

Although there are not validated test methods for the majority of the listed air toxics, a number of methods that may be applicable already are validated for similar substances at one or more sources. In some cases, test data substantiate, at least in part, performance of an analytical methodology, but there is no corresponding information available on an appropriate sampling technique. Of the currently validated and potentially applicable emission sampling and analytical methods, the majority involve state-of-the-art measurement techniques, and are normally complex and comparatively expensive.

EPA has ongoing programs to develop and validate methodologies for the most critical listed air toxics. Industry is expected to develop and validate test methods for use in demonstrating early reduction or compliance with existing or potential regulations. In addition, EPA is funding studies to explore emerging measurement technologies including Fourier transform infrared (FTIR) spectroscopy, ultraviolet (UV) spectroscopy, and multiphoton ionization (MPI). Without test methods, quantification of emissions of air toxics generally have to rely on alternate techniques including mass balances, engineering estimates, and emission factors.

The primary objective of this chapter is to provide an overview of the available techniques for sampling and analysis of air toxics from stationary sources. Primary goals are to present information on current approaches to measurement including potentially suitable methodologies for sampling and analysis, as well as to discuss some of the principal concerns in choosing and applying measurement methodology. The focus of the chapter is on methodology designed for sampling ducted emissions, although some mention will be made of techniques that may be suitable for measurement of fugitive emissions. Fugitive emissions are discussed in more detail in Chapter 22.

An important secondary objective of this chapter is to provide the reader with direction to reference materials and other available sources of relevant information on emission measurement technology. This field will change rapidly over the next decade and it is imperative that those conducting or using emission tests have access to the most up-to-date technical information. This information is available not only from the traditional sources, such as technical and trade journals, conference proceedings, and books and reports, but also through technical telephone hot lines and electronic bulletin boards.

---

[1] Emission factors are used to estimate typical emissions from a source type; the emission factor normally is expressed as an average rate at which a pollutant is released to the atmosphere as a function of the activity causing the release (see Chapter 10). Emission inventories are compilations of emissions data and other information relevant to sources of a pollutant or groups of pollutants. Dispersion (or diffusion) modeling is a mathematical technique for calculating the atmospheric distribution of air pollutants based on emissions data and meteorological data for an area (see Chapter 11).

AVAILABLE TECHNIQUES FOR MEASURING AIR TOXICS EMISSIONS

At present, there are a number of accepted emission measurement methods that already are applicable or may prove applicable for measuring a significant number of the listed air toxics. In this section, each of these methods is described in terms of applicability, principles of operation, advantages and disadvantages, and relationship to specific environmental regulations. Most of these methods are of the manual type where an integrated sample is collected over a specified time in one or more media and analyzed at a later time using one or more analytical techniques. The remainder are continuous or instrumental methods, where sample gas is analyzed continuously or semicontinuously using instrumental techniques.

The emphasis of these discussions is on methods designed to measure air toxics emissions confined to stacks, ducts, and vents. However, there is a brief discussion in the final section of methodologies potentially applicable to fugitive emissions of the specified air toxics. Because of the potential for significant cost savings in conducting emission test programs for multiple air toxics, the discussions favor those methods that allow simultaneous measurement of several air toxics.

Manual Methodologies with Multipollutant Applicability

There are a number of manual methodologies applicable to three or more of the listed hazardous air pollutants, each of which is described in the following sections. These methodologies measure target analytes (i.e., the substance(s) to be analyzed) within the same chemical family (e.g., metals, semivolatile organics, sulfur compounds, halogens, and halides). Multipollutant applicability is most valuable in cases when the project objectives require simultaneous measurement of multiple air toxics, particularly chemically similar substances. Measurement of multiple air toxics is useful when screening source emissions for the presence of the listed air toxics and cost-effective where multiple air toxics in the same chemical family have been selected as the target analytes.

Method 0010 and Associated Analytical Techniques (Semivolatile Organics)

Method 0010 (EPA 1986k), also known as Modified Method 5 (MM5) and the semivolatile organic sampling train (semi-VOST), was developed under the Resource Conservation and Recovery Act (RCRA) for determining the destruction and removal efficiency (DRE) of hazardous waste incinerators for the semivolatile principal organic hazardous compounds (POHCs) in the hazardous waste feed. For the purposes of the method, semivolatile organic species are defined as compounds with boiling points greater than 100°C.

Gaseous and particulate pollutants are withdrawn from the source at representative points[2] in the emission gas stream at an isokinetic sampling rate.[3] The sample is collected in a

---

[2] When particulate matter is involved, representative sampling involves traversing the duct with the sampling train probe and nozzle assembly. EPA Method 1 (40 CFR Part 60, Appendix A) provides the requirements for designing the sampling pattern. This practice compensates for stratification of the particulate matter in the duct.

[3] Isokinetic sampling is important to ensure a representative sample when particulate matter is involved. When the velocity of the sample gas drawn into the sampling train closely matches that of the sample gas passing by the nozzle, the sampling is said to be 'isokinetic.' A mismatch of these two velocities results in anisokinetic sampling. Inertial effects during anisokinetic sampling may result in over- or undercollection of particulate matter depending on the direction of the mismatch and the size and mass of the particulate. This effect is generally of marginal importance for gaseous pollutants and particulate material less than 2 μm in diameter.

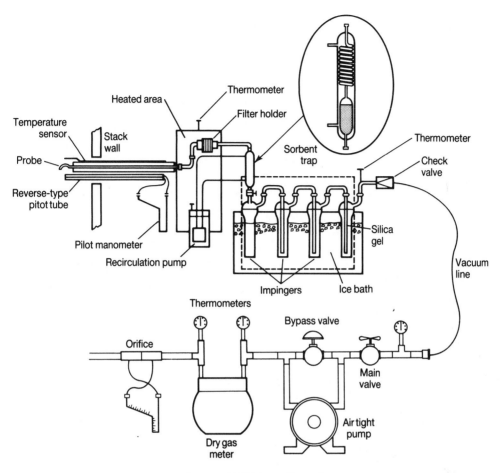

Figure 9-1.   Modified Method 5 (MM5), or semi-VOST, sampling train.

multicomponent sampling train (see Figure 9-1). A high-efficiency glass- or quartz-fiber filter is used to collect organic-laden particulate materials, a packed bed of porous polymeric resin (XAD-2™) serves to adsorb semivolatile organic species, and a series of water-filled impingers may collect some semivolatile organics that pass through the filter and sorbent. The sampling train configuration is adapted from EPA Method 5, a methodology routinely used to determine the particulate emissions from stationary sources. As such, most of the sampling equipment is readily available and familiar to the stack sampling community. The Method 0010 sampling train components not described in Method 5 are the condenser coil and the sorbent module (see Figure 9-1).

The collected sample fractions (XAD-2™ resin, filter, aqueous impinger solution, and solvent rinses) typically are extracted and analyzed by gas chromatography/mass spectrometry (GC/MS) (EPA 1990c). High-performance liquid chromatography (HPLC) and other analytical methods may be used for analysis, particularly of compounds not amenable to GC/MS. The sample fraction collected on the high-efficiency filter may be analyzed gravimetrically for a determination of particulate emissions before the extraction; however, this procedure is not acceptable in the determination of DRE under RCRA regulations and generally is discouraged except for use in making screening measurements, because it could result in loss of organics from the sample.

Table 9-1 HAP applicability of Method 0010 and analytical methods[a]

Compound	Analytical methods	Reference for analytical methods
	Analytical method demonstrated[b]	
Acetophenone	8270	EPA 1986
Aniline	8270	EPA 1986
Benzidine	8270	EPA 1986
Bis (2-ethylhexyl) phthalate	8270	EPA 1986
Chlordane	8270	EPA 1986
Chlorobenzene	8270	EPA 1986
o-Cresol	8270	EPA 1986
m-Cresol	8270	EPA 1986
p-Cresol	8270	EPA 1986
Cresylic acid	8270	EPA 1986
DDE	8270	EPA 1986
Dibenzofurans	8290	EPA 1986
Dibutyl phthalate	8270	EPA 1986
1,4-Dichlorobenzene	8270	EPA 1986
Dimethylaminoazobenzene	8270	EPA 1986
Dimethyl phthalate	8270	EPA 1986
2,4-Dinitrophenol	8270	EPA 1986
2,4-Dinitrotoluene	8270	EPA 1986
1,4-Dioxane	8270	EPA 1986
1,2-Diphenylhydrazine	8270	EPA 1986
Heptachlor	8270	EPA 1986
Hexachlorobenzene	8270	EPA 1986
Hexachlorobutadiene	8270	EPA 1986
Hexachlorocyclopentadiene	8270	EPA 1986
Hexachloroethane	8270	EPA 1986
Hexamethylene-1,6-diisocyanate	8270	EPA 1986
Isophorone	8270	EPA 1986
Lindane	8270	EPA 1986
Methoxychlor	8270	EPA 1986
Naphthalene	8270	EPA 1986
Nitrobenzene	8270	EPA 1986
4-Nitrophenol	8270	EPA 1986
N-Nitrosodimethylamine	8270	EPA 1986
Pentachloronitrobenzene	8270	EPA 1986
Pentachlorophenol	8270	EPA 1986
Phenol	8270	EPA 1986
2,3,7,8-Tetrachloro-dibenzodioxin	8280/Draft 8290	EPA 1986
1,1,2,2-Tetrachloroethane	8270	EPA 1986
Toluene	8270	EPA 1986
Toxaphene	8270	EPA 1986
1,2,4-Trichlorobenzene	8270	EPA 1986
2,4,5-Trichlorophenol	8270	EPA 1986
2,4,6-Trichlorophenol	8270	EPA 1986
	Potentially suitable methods[c]	
Acetamide	8270	EPA 1986
2-Acetylaminoflourene	8270	EPA 1986
Acrylamide	8270	EPA 1986
Acrylic acid	8270 (derivatized)	EPA 1986
4-Aminobiphenyl	8270	EPA 1986

Table 9-1 Continued

Compound	Analytical methods	Reference for analytical methods
	Potentially suitable methods[c] (Continued)	
o-Anisidine	8270	EPA 1986
Benzotrichloride	8270	EPA 1986
Benzyl chloride	8270	EPA 1986
Biphenyl	8270/8310	EPA 1986
Bromoform	8270	EPA 1986
Caprolactam	632/8270	EPA 1985b/EPA 1986
Captan	8270	EPA 1986
Carbaryl	8318/632	EPA 1986/EPA 1985b
Catechol	8270	EPA 1986
Chloramben	515/615	EPA 1988/EPA 1982
Chloroacetic acid	8270 (derivatized)	EPA 1986
2-Chloroacetophenone	8270	EPA 1986
Chlorobenzilate	8270	EPA 1986
Cumene	8270	EPA 1986
2,4-D salts and esters	515/615/8270	EPA 1988/EPA 1982/
		EPA 1986
1,2-Dibromo-3-chloropropane	8270	EPA 1986
3,3'-Dichlorobenzidene	8270	EPA 1986
Dichloroethyl ether	8270	EPA 1986
Dichlorvos	8270	EPA 1986
Diethanolamine	8270	EPA 1986
N,N-Diethylaniline	8270	EPA 1986
Diethyl sulfate	8270	EPA 1986
3,3'-Dimethoxybenzidine	8270	EPA 1986
3,3'-Dimethylbenzidine	8270	EPA 1986
Dimethyl carbamoyl chloride	531	EPA 1988
Dimethyl formamide	8270	EPA 1986
Dimethyl sulfate	8270	EPA 1986
4,6-Dinitro-o-cresol and	8270/515/615	EPA 1986/EPA 1988
salts		EPA 1982
Epichlorohydrin	8270	EPA 1986
Ethyl benzene	8270	EPA 1986
Ethyl carbamate	8270/632	EPA 1986/EPA 1985b
Ethyl dibromide	8270	EPA 1986
Ethylene glycol	8270	EPA 1986
Ethylene thiourea	632	EPA 1985b
Glycol ethers	8270/632	EPA 1986/EPA 1985b
Hexamethylphosphoramide	632	EPA 1985b
Hydroquinone	8270	EPA 1986
Maleic anhydride	8270	EPA 1986
Methyl t-butyl ether	8270	EPA 1986
4,4'-Methylene bis(2-chloroaniline)	8270	EPA 1986
Methylene diphenyl diisocyanate	8270/632	EPA 1986/EPA 1985b
4,4'-Methylenedianiline	8270	EPA 1986
Naphthalene	8310	EPA 1986
4-Nitrobiphenyl	8270/8310	EPA 1986
2-Nitropropane	8270	EPA 1986
N-Nitroso-N-methylurea	8270/632	EPA 1986/EPA 1985b
N-Nitrosomorpholine	8270/632	EPA 1986/EPA 1985b
Parathion	8270	EPA 1986
p-Phenylenediamine	8270	EPA 1986

Table 9-1   Continued

Compound	Analytical methods	Reference for analytical methods
	Potentially suitable methods[c] (Continued)	
Phthalic anhydride	8270	EPA 1986
PCB	680	EPA 1985b
Polycyclic organic matter	CARB 429/8270/8310	EPA 1986/CARB 1989
1,3-Propane sultone	8270	EPA 1986
β-Propiolactone	8270	EPA 1986
Propoxur	8270/632	EPA 1986/EPA 1985b
Quinoline	8270	EPA 1986
Styrene	8270	EPA 1986
Styrene oxide	8270	EPA 1986
Tetrachloroethylene	8270	EPA 1986
2,4-Toluene diamine	8270/632	EPA 1986/EPA 1985b
2,4-Toluene diisocyanate	8270/632	EPA 1986/EPA 1985b
o-Toluidine	8270	EPA 1986
1,1,2-Trichloroethane	8270	EPA 1986
Trifluralin	8270	EPA 1986
Xylenes(isomers and mixtures)	8270	EPA 1986
o-Xylene	8270	EPA 1986
m-Xylene	8270	EPA 1986
p-Xylene	8270	EPA 1986

[a] Adapted from Bursey 1991.
[b] Analytical method performance has been demonstrated for these compounds; in most cases, the precision and bias (method performance) of an overall sampling and analytical methodology have not been established.
[c] Neither sampling nor analytical method performance have been established.

Various methods compatible with Method 0010 for sample preparation and analysis of specific semivolatile organic compounds and compound classes are presented in EPA Publication No. SW-846 (EPA 1986k); additional suitable methods have been published by EPA and others for analysis of drinking water and wastewater (40 CFR Part 136, Appendix A; EPA 1979b; Rand, Greenberg, and Taras 1975). Air toxic analytes to which Method 0010 applies, or potentially applies, are listed in Table 9-1 together with reference to the appropriate or potentially appropriate analytical method. To maintain simplicity, the appropriate sample preparation methods are not included in the table. More information on the methods and modifications that must be made to the analytical methodologies for analysis of new analytes can be found in the Bursey (1991).

The sample preparation and analytical techniques required, as well as the need for special sampling techniques for certain compounds, determine the analytes that can be measured using a single sampling train. However, if the analytes are similar (e.g., nonpolar semivolatile organic compounds), on the order of ten to one hundred might be measured using a single train. Also, if the extract volume is sufficient, multiple analyses may be conducted using different techniques; GC/MS and HPLC are a powerful complementary pair. Clearly, Method 0010 should prove to be an extremely powerful and versatile methodology in the measurement of air toxics.

Method 0030 and Methods 5040 and 5041 (Volatile Organics)

Method 0030 (EPA 1986k) is known as the Volatile Organic Sampling Train (VOST). Like Method 0010, Method 0030 was developed under RCRA for use in determining the DRE of hazardous waste incinerators for POHCs in the hazardous waste feed. The VOST is designed to collect organic compounds more volatile than the semivolatile compounds collected by the MM5 train. For the purposes of definition, volatile organic compounds are those with boiling points less than 100°C. In practice, both the MM5 and the VOST may collect organic compounds with boiling points in the range of approximately 100°C to 120°C. If the boiling point of an organic compound of interest is less than 30°C, the sorbed compound may migrate through and off the VOST sorbent (volumetric breakthrough) under typical VOST sampling conditions.

Under Method 0030, a twenty-liter sample of effluent gas is withdrawn from the source at a constant sampling rate using a glass-lined probe and the VOST (see Figure 9-2). The sample gas is cooled to 20°C by passage through a water-cooled condenser and volatile organic compounds are collected on a pair of sorbent resin traps. Liquid condensate is collected in an impinger placed between the two resin traps. The first resin trap (front trap) contains Tenax-GC™ and the second contains Tenax-GC™ followed by petroleum-based charcoal.

The normal VOST sampling rate is one liter per minute (l/min) (FAST-VOST). The pairs of traps are replaced every twenty minutes over a two-hour sampling period to yield a total of six trap pairs, each representing a twenty liter sample. An alternate procedure involves collecting a sample volume of twenty liters or less at a reduced sampling rate (SLO-VOST). This procedure has been used to collect five liters of sample per trap pair (at 0.25 l/min for twenty minutes, the lowest acceptable rate) or twenty liters of sample on each trap pair (0.5 l/min for forty minutes). Smaller sample volumes collected at lower flow rates are recommended when the boiling points of the organic compounds of interest are less than 30°C. Because detection limit is inversely proportional to the

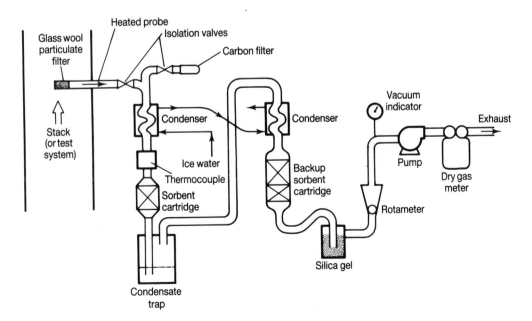

Figure 9-2.   Volatile organic sampling train (VOST).

Table 9-2   HAP applicability of Method 0030 and analytical methods[a]

Compound	Analytical methods	Reference for analytical methods
Analytical method demonstrated[b]		
Benzene	5040/Draft 5041	EPA 1986
Carbon tetrachloride	5040/Draft 5041	EPA 1986
Chlorobenzene	5040/Draft 5041	EPA 1986
Chloroform	5040/Draft 5041	EPA 1986
Methyl chloroform	5040/Draft 5041	EPA 1986
Tetrachloroethylene	5040/Draft 5041	EPA 1986
Toluene	5040/Draft 5041	EPA 1986
1,1,2-Trichloroethane	5040/Draft 5041	EPA 1986
Trichloroethylene	5040/Draft 5041	EPA 1986
Vinyl chloride[c]	5040/Draft 5041	EPA 1986
Potentially suitable methods[d]		
Acetaldehyde	5040/Draft 5041	EPA 1986
Acrylonitrile	5040/Draft 5041	EPA 1986
Allyl chloride	5040/Draft 5041	EPA 1986
Carbon disulfide	5040/Draft 5041	EPA 1986
Chloro methyl ether	5040/Draft 5041	EPA 1986
Chloroprene	5040/Draft 5041	EPA 1986
1,3-Dichloropropene	5040/Draft 5041	EPA 1986
1,1-Dimethylhydrazine	5040/Draft 5041	EPA 1986
1,2-Epoxybutane	5040/Draft 5041	EPA 1986
Ethyl acrylate	5040/Draft 5041	EPA 1986
Ethyl chloride	5040/Draft 5041	EPA 1986
Ethylene dichloride	5040/Draft 5041	EPA 1986
Ethylene imine	5040/Draft 5041	EPA 1986
Ethylidene dichloride	5040/Draft 5041	EPA 1986
Hexane	5040/Draft 5041	EPA 1986
Methyl bromide	5040/Draft 5041	EPA 1986
Methyl ethyl ketone	5040/Draft 5041	EPA 1986
Methyl hydrazine	5040/Draft 5041	EPA 1986
Methyl iodide	5040/Draft 5041	EPA 1986
Methyl methacrylate	5040/Draft 5041	EPA 1986
Methylene chloride	5040/Draft 5041	EPA 1986
Propylene dichloride	5040/Draft 5041	EPA 1986
Propylene oxide	5040/Draft 5041	EPA 1986
1,2-Propyleneimine	5040/Draft 5041	EPA 1986
Triethylamine	5040/Draft 5041	EPA 1986
2,2,4-Trimethylpentane	5040/Draft 5041	EPA 1986
Vinyl acetate	5040/Draft 5041	EPA 1986
Vinyl bromide	5040/Draft 5041	EPA 1986
Vinylidene chloride	5040/Draft 5041	EPA 1986

[a] Adapted from Bursey 1991.

[b] Analytical method performance has been demonstrated for these compounds; in most cases, the precision and bias (method performance) of an overall sampling and analytical methodology have not been established.

[c] Vinyl chloride's boiling point is outside the normal limits for Method 0030. With good quality control procedures, however, Method 0030 can be used to measure this compound, but Methods 106 and 18 are preferred.

[d] Neither sampling nor analytical method performance have been established.

volume of gas sampled, the VOST procedures collecting twenty liters provide more detection ability.

Analysis of the resin traps is carried out by thermal desorption purge-and-trap GC/MS. This procedure is described in detail in SW-846 Method 5040 (EPA 1986k) and prospective SW-846 Method 5041. The sampling train was designed to use six pairs of sorbent traps sequentially. The first set of traps often is analyzed as a range finder. If an adequate amount of the compound(s) of interest is found, then the other five pairs of traps can be analyzed in the same manner. If the first analysis detects no compound(s) of interest, the other five pairs are desorbed and collected onto a single analytical trap that allows a five-fold increase in the detection limit. Separate analysis of the Tenax and Tenax/charcoal traps allows evaluation of breakthrough. The purge-and-trap heat desorption procedure is limiting in that it is a one-shot analysis and offers no opportunity for reanalysis or multiple analytical finishes.

Air toxic analytes to which Method 0030 is or potentially is applicable are listed in Table 9-2 together with reference to the appropriate or potentially appropriate sample preparation and analytical methods. Suggested modifications to the analytical methodology for analysis of new analytes also are included in the table (Bursey 1991). An EPA study currently is under way to validate the VOST and MM5 for measurement of halogenated organic compounds.

Method 0011 and Method 0011A (Aldehydes and Ketones)

Method 0011 (EPA 1991p; EPA 1990d) also was issued under RCRA in conjunction with standards regulating boilers and industrial furnaces (BIF) burning hazardous waste (EPA 1991n). Method 0011 is the sample collection

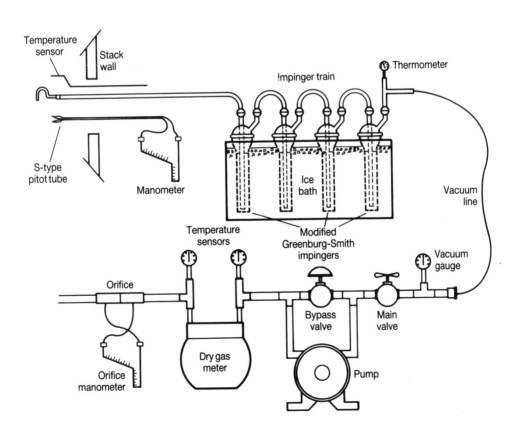

Figure 9-3. Method 0011 sampling train.

Table 9-3   HAP applicability of Methods 0011/0011A

Compound	Validation status
Acetaldehyde	Sampling/analytical methods partially validated; further evaluation in progress
Acetophenone	Evaluation of sampling/analytical methods in progress
Acrolein	Evaluation of sampling/analytical methods in progress
2-Chloroacetophenone	Evaluation of sampling/analytical methods in progress
Formaldehyde	Sampling/analytical methods partially validated; further evaluation in progress
Isophorone	Evaluation of sampling/analytical methods in progress
Methyl ethyl ketone	Evaluation of sampling/analytical methods in progress
Methyl isobutyl ketone	Evaluation of sampling/analytical methods in progress
Propionaldehyde	Evaluation of sampling/analytical methods in progress
Quinone	Evaluation of sampling/analytical methods in progress

technique for determining the DRE of BIF for formaldehyde. The methodology has been applied specifically to formaldehyde, but many laboratories have extended the application to other aldehydes and ketones. Extension of the methodology should prove useful because ten aldehydes and ketones are included among the 189 listed air toxics.

Under Method 0011, gaseous and particulate pollutants are withdrawn isokinetically from an emission source and collected in an aqueous acidic solution of dinitrophenylhydrazine (DNPH). The sampling train containing the DNPH impinger solution is similar to that specified in EPA Method 5, except that all of the sample-exposed surfaces are glass or quartz and the heated filter is eliminated (see Figure 9-3). The formaldehyde present in the emission sample reacts with the DNPH to form a formaldehyde-dinitrophenylhydrazone deriva-tive; other aldehydes and ketones present in the emissions should form derivatives in a similar manner and, thus, can be collected using the same sampling train.

The dinitrophenylhydrazone derivatives are extracted with methylene chloride, solvent-exchanged, concentrated, and then analyzed by high performance liquid chromatography according to Method 0011A (EPA 1991p; EPA 1990d). This method was optimized for the determination of formaldehyde and acetalde-

hyde in aqueous environmental matrices and leachates of solid samples and emission sam-ples collected using Method 0011. An EPA-sponsored study is currently under way to validate the use of these methods for measure-ment of additional aldehydes and ketones. The air toxics listed in the 1990 CAAA to which Methods 0010/0011A are applicable or are potentially applicable are listed in Table 9-3.

Method 0012 (Metals)

Method 0012 (EPA 1991p; EPA 1990d), com-monly known as the multiple metals method, was developed for the determination of multiple trace metals in emissions from hazardous waste incinerators and similar combustion sources. Method 0012 recently was issued under RCRA in conjunction with the standards for BIF burning hazardous waste. Under certain condi-tions, the method also may be used for the determination of particulate emissions.

Method 0012 sampling is conducted using a Method 5 sampling train modified to include a low metals background quartz-fiber filter, a Teflon™ frit, glass nozzle and probe, extra impingers, and special absorbing reagents (see Figure 9-4). The gaseous and particulate pollu-tants are withdrawn isokinetically from the source, with particulate metals collected in the probe and on the heated filter and gaseous metals collected in the back-half[4] in chilled

---

[4.] The front-half fraction includes the filter and rinses of all glassware in the sampling train prior to the filter. The back-half fraction includes the impinger reagents and rinses of all glassware in the portions of the sampling train subsequent to the filter.

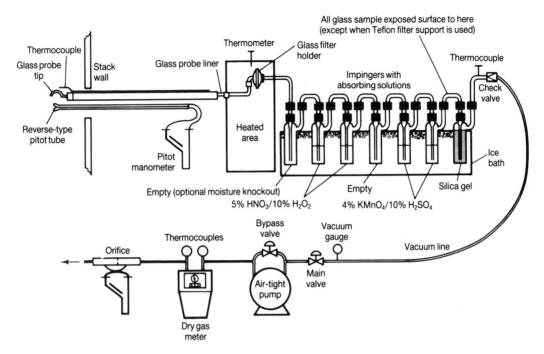

Figure 9-4. Method 0012 sampling train.

impingers containing an aqueous solution of dilute nitric acid combined with dilute hydrogen peroxide followed by impingers filled with acidic potassium permanganate.

Sampling train components are recovered and digested in separate front- and back-half fractions. Materials collected in the sampling train are digested with acid solutions using conventional (including Parr[R] Bomb) or microwave techniques. Aliquots of the front-half fraction, the nitric acid/hydrogen peroxide solution, the potassium permanganate solution, and a hydrochloric acid rinse of the permanganate impingers are analyzed for mercury using cold vapor atomic absorption spectroscopy (CVAAS). Aliquots of the front-half fraction and the nitric acid/hydrogen peroxide solution are analyzed for all the remaining target metals using inductively coupled argon plasma emission spectroscopy (ICP) or atomic absorption spectroscopy (AAS). Graphite furnace atomic absorption spectroscopy (GFAAS) may be used to improve the analytical sensitivity for certain

Table 9-4   HAP applicability of Method 0012

Metal/compound	Method performance established[a]
Antimony compounds	Yes
Arsenic compounds	Yes
Beryllium compounds	Yes
Cadmium compounds	Yes
Chromium compounds	Yes
Cobalt compounds	No
Lead compounds	Yes
Manganese compounds	Yes
Mercury compounds	Yes
Nickel compounds	Yes
Phosphorus	Yes
Selenium compounds	Yes
Titanium tetrachloride	No

[a] Yes: combined sampling/analytical method performance established for these metals in emissions from municipal waste combustors, hazardous waste incinerators, and/or sewage sludge incinerators.
No: combined sampling/analytical method performance not established, but method is potentially suitable for measuring total amount (without regard to valence state) of the metals in these compounds.

metals. An important consideration in the use of this method is that it does not allow for speciation of the various metal compounds or valence states. The metallic air toxics listed in the 1990 CAAA to which Method 0012 is applicable or is potentially applicable are listed in Table 9-4.

Methods 0050, 0051, and 9057/9056 (Halogens and Halides)

Methods 0050 and 0051 (EPA 1991p; EPA 1990d) also were recently issued under RCRA in conjunction with the standards for BIF. Both methods are applicable to the determination of hydrogen chloride (HCl) and chlorine (Cl_2) emissions from hazardous waste incinerators, municipal waste combustors, and BIF. Method 9057 (EPA 1991p; EPA 1990d) is the companion analytical method for analyzing the samples collected using Methods 0050 and 0051; however, it likely will be replaced by prospective Method 9056 in the near future.

Method 0051 is based on constant rate sampling using a midget impinger sampling train (see Figure 9-5); Method 0050 utilizes a Method 5-type train (see Figure 9-6) coupled with isokinetic sampling and, therefore, is particularly suited for sampling at sources such as those controlled by wet scrubbers that emit acid particulate matter (e.g., HCl dissolved in water droplets). Method 0050 also may be used for the determination of particulate emissions following the additional procedures described. Method 26 (40 CFR Part 60, Appendix A) is nearly identical to the combination of Methods 0051 and 9057/9056; it was developed in conjunction with regulations controlling emissions from municipal waste combustors (EPA 1991m). Draft Method 26A is a prospective method for measuring multiple hydrogen halide and halogen compounds emitted from synthetic organic chemical manufacturing industry (SOCMI) reactor, air oxidation, and distillation process vents. Draft Method 26A parallels Methods 0050 and 9057/9056, but includes additional procedures to expand the applicability to measurement of hydrogen bromide, hydrogen fluoride, and bromine. Draft Method 26A, along with comparable revisions to Method 26, will be proposed in conjunction

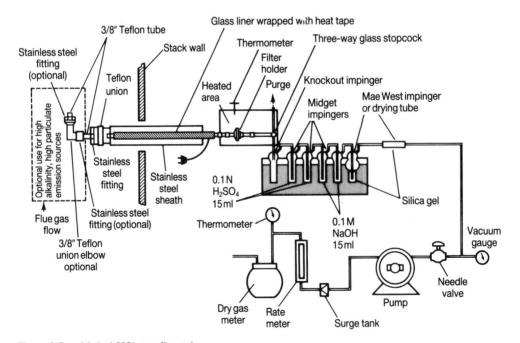

Figure 9-5.   Method 0051 sampling train.

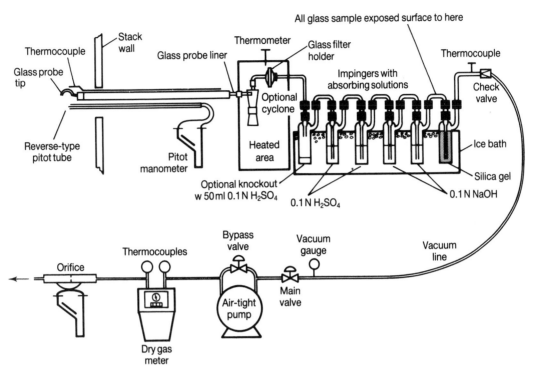

Figure 9-6.   Method 0050 sampling train.

with chemical industry regulations to be published by EPA.

Method 0051 (and Method 26) sampling involves collection at a constant rate of an integrated gas sample that passes through a Teflon™ filter, acidified water, and finally through an alkaline solution (see sampling train in Figure 9-5). The filter serves to remove particulate matter, such as chloride salts, which potentially could react and form analytes in the absorbing solutions. In the acidic absorbing solution, the HCl gas is solubilized and forms chloride ions (Cl^-). Chlorine gas present in the emissions has a very low solubility in the acidified water and passes through to the alkaline absorbing solution to undergo hydrolysis to form a proton, chloride ion, and hypochlorous acid. Method 0050 (Draft Method 26A) sampling is conducted isokinetically using a Method 5-type train with a Teflon™ or quartz filter (see Figure 9-6). The principle of collection is the same as that for the midget impinger sampling train of Methods 0051 and 26. The

chloride ions in the separate acidic and alkaline absorbing solutions from the midget impinger and Method 5-type trains are measured by ion chromatography (IC) (Methods 9057/9056, 26, and 26A). Air toxics to which these methods are applicable are listed in Table 9-5.

Method 18 (Gaseous Organics)

Method 18 (40 CFR Part 60, Appendix A) is a generic method for measuring gaseous organic compounds and was developed as a test method to be used in association with New Source Performance Standards (NSPS) (40 CFR Part 60). Method 18 is based on separating the major gaseous organic components of an emission gas stream by gas chromatography (GC) and measuring the separated components with a suitable detector. The emission gas samples are analyzed immediately as taken from the emission source (direct interface option) or within a set period of time after being collected in a container (e.g., Tedlar™ bag) or on an adsorption tube. Generalized schematics of three

Table 9-5   HAP applicability of Methods 0050/0051 and Methods 9057/9056

Compound	Method performance established[a]
Chlorine	Yes
Hydrogen chloride (hydrochloric acid)	Yes
Hydrogen fluoride	No

[a] Yes: combined sampling/analytical method performance established for these compounds in emissions from incinerators. No: combined sampling/analytical method performance not established, but method is potentially suitable.

principal types of sampling trains used to collect, and in the case of the direct interface option, also analyze Method 18 samples are shown in Figures 9-7, 9-8, and 9-9.

To identify and quantify the major components in a Method 18 sample, the retention times of each separated component are compared with those of known compounds under identical chromatographic conditions. The method requires that the identity and approximate concentrations of the target components are known beforehand so that calibration

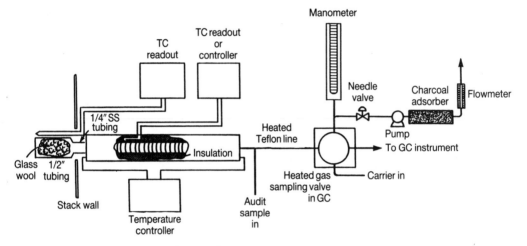

Figure 9-7.   Method 18 direct interface sampling system.

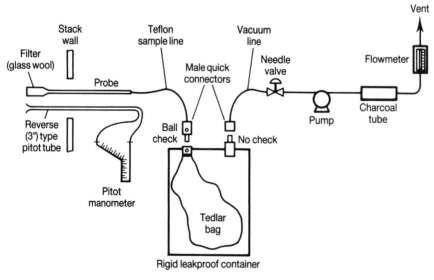

Figure 9-8.   Method 18 integrated bag sampling system.

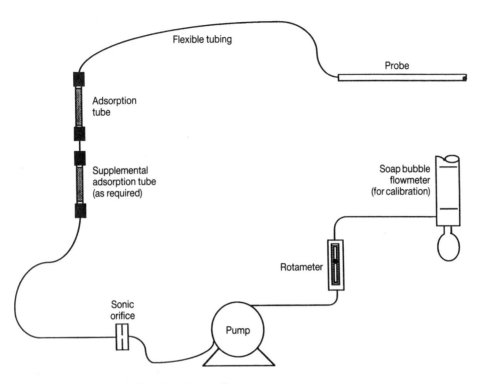

Figure 9-9.   Method 18 absorption tube sampling system.

standards can be mixed or purchased, if commercially available. Also critical is prior information on potential sample matrix interferents, components, and conditions requiring sampling modifications (e.g., particulate matter requiring filtration and gas temperature or moisture content necessitating particular sampling approaches).

Method 18 is typically applicable to pollutants in concentrations greater than one ppm. Target compounds at concentrations less than one ppm generally are measured best using the MM5 train (Method 0010) or the VOST (Method 0030) coupled with the appropriate analytical techniques. Method 18 also is not applicable to organic compounds that are reactive, polymeric (i.e., high molecular weights), polymerize before analysis, or have very low vapor pressures at stack or instrument conditions. The air toxics to which Method 18 is applicable or may be applicable are listed in Table 9-6. It is noteworthy that Method 18

specifies procedures that define analytical method performance and result in partial validation of the method each time it is applied (see section below on method validation).

Method 23 (Dioxins/Furans)

Method 23 (40 CFR Part 60, Appendix A) is applicable to the determination of emissions of polychlorinated dibenzo-*p*-dioxins (PCDDs) and polychlorinated dibenzofurans (PCDFs) from stationary sources. Method 23 was published in association with NSPS for municipal waste combustors (EPA 1991m).

Method 23 includes both the sampling and analytical procedures for isomer-specific measurement of PCDDs/PCDFs. The sampling train employed is essentially identical to that of Method 0010 (MM5) (see Figure 9-10); however, sorbent spiking and sample recovery differ from the MM5 procedures. Gaseous and particulate PCDDs/PCDFs are withdrawn isokinetically from the source and collected in the

Table 9-6   HAP compounds to which Method 18 is potentially applicable[a]

Compound	Method performance established[b]
Allyl chloride	No
Benzene	Yes
Bis(chloromethyl)ether[c]	No
Bromoform[d]	No
1,3-Butadiene	Yes
Carbon tetrachloride	Yes
Chlorobenzene[d]	No
Chloroform	Yes
Chloromethyl methyl ether	No
Chloroprene	No
Cumene[d]	No
1,2-Dibromo-3-chloropropane[d]	No
1,3-Dichloropropene	No
1,1-Dimethylhydrazine	No
1,2-Epoxybutane	No
Ethyl acrylate	No
Ethyl chloride	No
Ethylene dibromide[d]	No
Ethylene dichloride	Yes
Ethylene imine	No
Ethylene oxide	Yes
Ethylidene dichloride	No
Hexane	No
Methanol	No
Methyl bromide	No
Methyl chloride	No
Methyl hydrazine	No
Methyl isobutyl ketone	No
Methyl iodide	No
Methyl methacrylate	No
Methyl t-butyl ether	No
Methylene chloride	Yes
Phosphine	No
Propylene dichloride	No
Propylene oxide	No
1,2-Propyleneimine	No
Styrene[d]	No
Tetrachloroethylene	Yes
Toluene	No
Trichloroethylene	Yes
2,2,4-Trimethylpentane	No
Vinyl acetate	No
Vinyl bromide	No (see Method 106)
Vinyl chloride	Yes (see Method 106)
Vinylidene chloride	No
o-Xylenes[d]	No
p-Xylenes[d]	No
m-Xylenes[d]	No

[a] Compiled from Bursey 1991 and Section 3.16 of EPA 1977.
[b] Yes: certain method performance parameters established for combined sampling/analytical procedures for these compounds. No: combined sampling/analytical method performance unknown for these compounds.
[c] Highly reactive with water; Method 18 would only be applicable when the emissions do not contain moisture.
[d] These compounds have relatively high boiling points (>120°C). Special care must be taken in sampling them using containers (e.g., Tedlar® bags) or adsorbent tubes, including comprehensive quality control procedures such as sampling system matrix spikes, duplicate samples, etc. Particulate matter contained in the emission gas stream may cause a negative bias in measurements through association with the gases of interest.

sampling probe, on a glass- or quartz-fiber filter, and a packed bed of adsorbent resin (XAD-2™).

The collected sample is solvent-extracted, cleaned using various columns, separated by high resolution gas chromatography (HRGC), and measured by high resolution mass spectrometry (HRMS). Isotopically labelled internal standards and ion-specific monitoring are used in the HRGC/HRMS analysis to quantify the individual PCDD/PCDF isomers. The sample preparation and analytical techniques are similar to Method 8280 (EPA 1986k) and prospective successor Method 8290. Method 23 is applicable to 2,3,7,8-tetrachlorodibenzo-p-dioxin and dibenzofurans, both listed air toxics.

Method 15 (Sulfide Compounds)
Method 15 (40 CFR Part 60, Appendix A) was developed to measure hydrogen sulfide (H_2S), carbonyl sulfide (COS), and carbon disulfide (CS_2) emissions from tail gas control units of sulfur recovery plants. Carbonyl sulfide and carbon disulfide are listed air toxics. The method originally was published in association with NSPS for petroleum refineries (40 CFR Part 60).

In applying Method 15, a gas sample is extracted from the emission source, passed through a filter and citrate buffer scrubber to remove particulate matter and sulfur dioxide, and diluted with clean dry air. A schematic of the sampling apparatus is shown in Figure 9-11. An aliquot of the diluted sample then is analyzed for the three target analytes by GC separation and flame photometric detection (FPD). Permeation tubes are used to make the calibration gas mixtures necessary to identify and quantify the target analytes.

CARB Method 429 (Polycyclic Organic Matter)
California Air Resources Board (CARB) Method 429 (CARB 1989a) is applicable to measurement of polycyclic aromatic hydrocarbon (PAH) emissions from stationary sources. PAH are a subset of polycyclic organic matter (POM), which is listed in the 1990

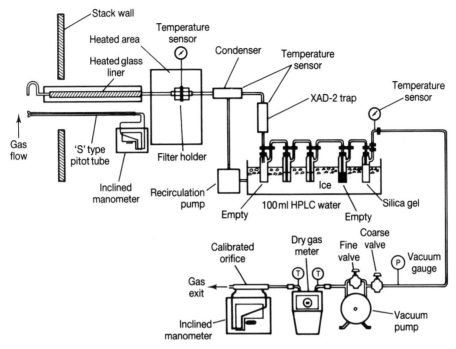

Figure 9-10. Method 23 sampling train.

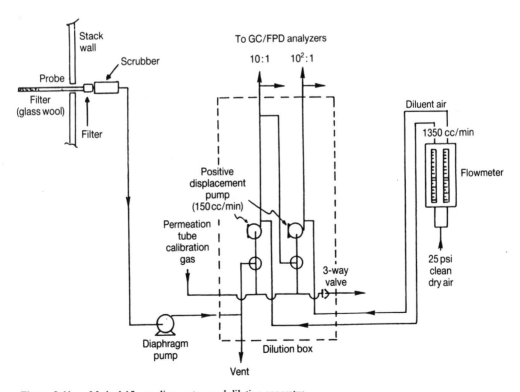

Figure 9-11. Method 15 sampling system and dilution apparatus.

CAAA. The sampling and analytical procedures for this method are much like those for Methods 0010 and 8270 in EPA Publication No. SW-846. Particulate and gaseous phase PAH are withdrawn isokinetically from the duct using an MM5-type sampling train and collected on a filter, using XAD-2™ resin, and in impingers. The sample is composed of the filter, front-half rinses, and impinger contents and rinses. The analytical method involves addition of internal standards in known quantities to each sample, followed by matrix-specific extraction with appropriate solvents, preliminary fractionation and cleanup of extracts (if necessary), and analysis of the processed extract for PAH using HRGC coupled with low resolution mass spectroscopy (LRMS) or HRMS. The specific target compounds to which this method is applicable are: naphthalene, acenaphthylene, acenaphthene, fluorene, phenanthrene, anthracene, fluoranthene, pyrene, benz[a]anthracene, chrysene, benzo[b]fluoranthene, benzo[k]fluoran-

thene, benzo[a]pyrene, benzo[ghi]perylene, dibenz[a,h]anthracene, and indeno[1,2,3-cd]pyrene.

Manual Methodologies with Single Pollutant Applicability

Certain manual methodologies are applicable to only one (or two) pollutants on the air toxics list. These manual methodologies may not be as useful or cost effective in testing situations where multiple air toxics are to be measured, but may be valuable when only a single air toxic must be measured or when none of the multiple pollutant methods are applicable to the analyte of interest. Each of these methodologies is discussed below.

Method 12 (Lead)

Method 12 (40 CFR Part 60, Appendix A) is applicable to the determination of inorganic lead emissions and was developed in conjunction

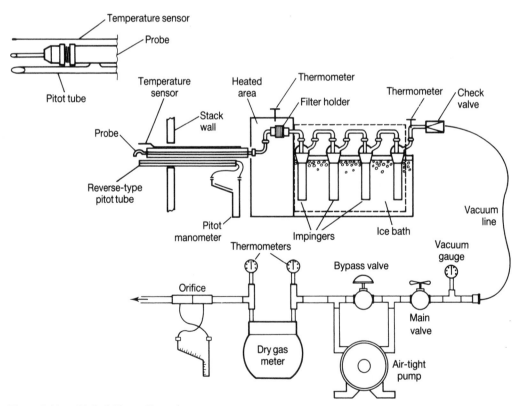

Figure 9-12.   Method 12 sampling train.

with the NSPS for lead-acid battery manufacturing plants (40 CFR Part 60). Like Method 0012, Method 12 does not allow for speciation of lead compounds or determination of lead valence states. Method 12 sampling is conducted using a Method 5-type sampling train with a low lead background filter and a dilute nitric acid absorbing reagent in the impingers (see Figure 9-12). Particulate and gaseous lead emissions are withdrawn isokinetically from the source and collected in the probe, on the heated filter, and in the chilled impingers containing the dilute nitric acid. The collected samples are recovered and digested with concentrated nitric acid. The resulting extract is analyzed for lead by AAS using an air/acetylene flame.

Methods 13A, 13B, and 14 (Fluorides)

Methods 13A, 13B, and 14 (40 CFR Part 60, Appendix A) all are applicable to measurement of fluoride emissions. These methods were developed in conjunction with NSPS for primary aluminum reduction plants, wet-process phosphoric acid plants, superphosphoric acid plants, diammonium phosphate plants, triple superphosphate plants, and granular triple superphosphate storage facilities (40 CFR Part 60). Methods 13A, 13B, and 14 provide for measurement of total fluorides and do not allow speciation. These methods also are applicable to the measurement of hydrogen fluoride, a listed air toxic, but only when no other fluorides are present in the emission gas stream.

Method 13A sampling is conducted using a Method 5-type sampling train; the filter may be located between the third and fourth impingers or before the first impinger (see Figure 9-13). Gaseous and particulate fluoride emissions are withdrawn isokinetically from the source and collected in deionized distilled water and on a

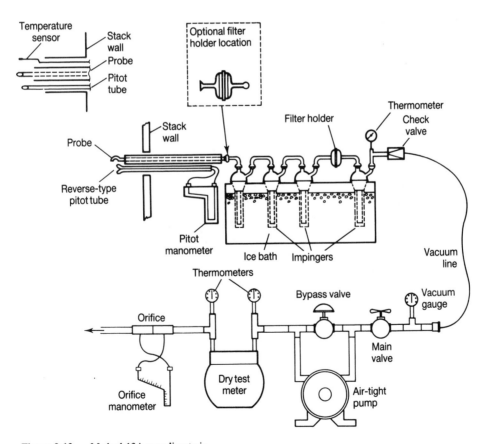

Figure 9-13.   Method 13A sampling train.

filter. The collected samples are recovered, distilled, and analyzed for total fluorides by the SPADNS Zirconium Lake colorimetric method (Bellack and Schouboe 1958).

The Method 13B sampling procedure is identical to that of Method 13A. The recovered samples are distilled as for Method 13A, but the total fluorides are determined using a specific ion electrode. Method 14 provides specifications for a sampling manifold to be used in application of Methods 13A and 13B to fluoride emissions from potroom roof monitors at primary aluminum plants.

Methods 101, 101A, and 102 (Mercury)

Methods 101, 101A, and 102 (40 CFR Part 61, Appendix B) are applicable to measurement of mercury emissions. Method 101 is designed for sources, such as chlor-alkali plants, where the carrier gas in the duct is principally air; Method 101A was developed for sewage sludge incinerators. Method 102 provides procedures for sampling in hydrogen gas streams in chlor-

alkali plants. The three methods were developed in association with National Emission Standards for Hazardous Air Pollutants (NESHAPs) for mercury ore processing facilities, chlor-alkali plants, and sewage sludge incineration and drying (40 CFR Part 61). They all measure total mercury and do not allow speciation.

Method 101 sampling is conducted using a Method 5-type sampling train without the filter (see Figure 9-14). Gaseous and particulate mercury emissions are withdrawn isokinetically from the source and collected in acidic iodine monochloride (ICl) absorbing solution in the chilled impingers. The mercury collected (in the mercuric form) is reduced to elemental mercury, that then is aerated from the solution into an optical cell and measured by cold vapor atomic absorption spectroscopy (CVAAS). Method 101A is similar to Method 101, except acidic potassium permanganate solution is used instead of acidic ICl for sample collection. The collected mercury sample is reduced and analyzed by CVAAS.

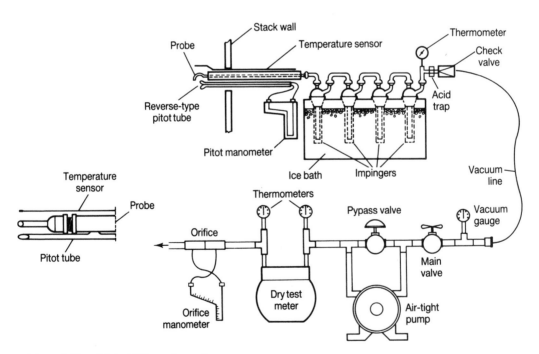

Figure 9-14.   Method 101 sampling train.

Method 102 is identical to Method 101 except for safety modifications required to sample in gas streams containing hydrogen. These modifications include elimination of probe heating and the glass-fiber filter, use of no other electrical equipment besides the vacuum pump, sealing the port, venting the sampled gas at least ten feet from the train, and calibration of the meter box orifice using hydrogen or other gas with a similar Reynolds number.

Methods 103 and 104 (Beryllium)

Methods 103 and 104 (40 CFR Part 60, Appendix B) are designed for the measurement of beryllium emissions from stationary sources and were developed in association with NESHAPs for beryllium sources and for beryllium rocket motor firing. These standards cover beryllium ore extraction plants, ceramic plants, incinerators, propellant plants processing various beryllium-containing material, and rocket motor test sites (40 CFR Part 61). Both methods measure total beryllium and do not allow speciation.

Method 103 is referred to as the 'beryllium screening method,' but is acceptable for compliance determination under the previously mentioned NESHAP regulations. The Method 103 sampling train includes a stainless steel nozzle, glass probe, Millipore™ filter, and gas metering system (see Figure 9-15). Sampling is conducted isokinetically at three points in the duct. The filter and probe rinse samples are recovered, and prepared for and analyzed by AAS, spectrography, fluorometry, chromatography, or other appropriate technique.

Method 104 sampling is conducted using a Method 5 sampling train with a Millipore™ filter and deionized distilled water in the impingers. Gaseous and particulate beryllium emissions are withdrawn isokinetically from the source and collected in the probe, on the filter, and in the chilled impinger water. The filter and impinger solution/probe rinse samples are recovered, digested with nitric and perchloric acid, and analyzed for beryllium using AAS with a nitrous oxide/acetylene flame. This method is not intended to apply to gas streams other than those emitted directly to the atmosphere without further processing.

Method 106 (Vinyl Chloride)

Method 106 (40 CFR Part 61, Appendix B) is applicable to the measurement of vinyl chloride in emissions from ethylene dichloride, vinyl chloride, and polyvinyl chloride manufacturing processes. Method 106 was developed in association with the NESHAPs for vinyl chloride (40 CFR Part 61). The method does not measure vinyl chloride contained in particulate matter.

Method 106 is similar to the bag sampling option of Method 18. An integrated sample of the emission gas stream containing vinyl chloride is collected in a Tedlar™ bag (see Figure 9-16). This sample is analyzed by GC/FID.

Method 108 (Arsenic)

Method 108 (40 CFR Part 61, Appendix B) applies to the determination of inorganic arsenic emissions from stationary sources. Method 108 was developed in association with the NESHAPs for furnaces at glass manufacturing plants (40 CFR Part 61). Method 108 does not speciate the arsenic compounds nor provide information on the arsenic valence states.

Method 108 sampling is conducted using a Method 5 sampling train with deionized distilled water in the impingers. Gaseous and particulate arsenic emissions are withdrawn isokinetically from the source and collected in the probe, on the filter, and in the chilled impinger water. The filter, impinger, and probe rinse samples are recovered, digested, and analyzed for arsenic by AAS.

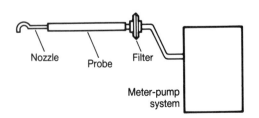

Figure 9-15.   Method 103 sampling train.

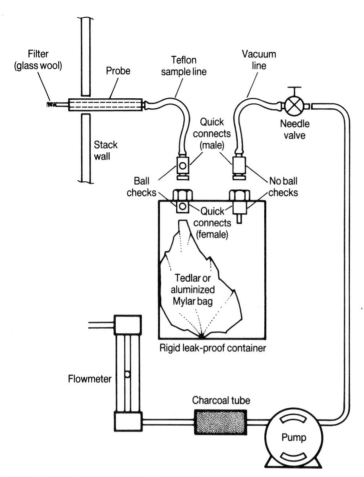

Figure 9-16.   Method 106 sampling train.

Method 111 (Polonium-210)

Method 111 (40 CFR Part 61, Appendix B) is applicable to the determination of polonium-210, a specific radionuclide, in particulate emission samples and was developed in association with the NESHAPs for radionuclide emissions from elemental phosphorus plants (40 CFR Part 61). In performing Method 111, a particulate sample is collected isokinetically from the emission gas stream using a Method 5 sampling train. The recovered filter and acetone probe rinse are acid digested; the polonium-210 in the solution is deposited on a silver disc and the radioactive disintegration rate measured using an alpha spectrometry system. Polonium in acid solution spontaneously deposits on surfaces of metals that are more electropositive than polonium.

Method 114 (Radionuclides)

Method 114 (40 CFR Part 61, Appendix B) is a generic stationary emission source test method that provides the monitoring, sample collection, analytical, and quality assurance requirements appropriate for radionuclides. Method 114 was developed in conjunction with the NESHAPs for radionuclide emissions from Department of Energy (DOE) facilities, facilities licensed by the Nuclear Regulatory Commission, and Federal facilities other than DOE facilities (40 CFR Part 61).

The possible radionuclide emissions to be measured differ in their chemical and physical forms, half-lives, and types of radiation (i.e., alpha, beta, and gamma) emitted. The appropriate combination of sample extraction, collection, and analysis techniques for an individual radionuclide is dependent on many interrelated factors including the mixture of other radionuclides present. Because of the wide range of potential conditions, no single method for monitoring or sample collection and analysis of a radionuclide is applicable to all types of facilities. Method 114 presents a series of techniques based on 'principles of measurement' for monitoring, sample collection, and analysis applicable to the measurement of radionuclides found in the exhaust streams of stationary sources. Flexibility is provided to choose the most appropriate combination of these techniques for the particular effluent characteristics. Method 114 also provides a list of approved methods for specific radionuclides as well as quality assurance methods that must be applied in conjunction with the radionuclide emission measurements.

Method 115 (Radon-222)
Method 115 (40 CFR Part 61, Appendix B) provides several procedures for determining radon-222 emissions from underground uranium mine vents, uranium mill tailings piles, phosphogypsum stacks, and other piles of waste material emitting radon. This method was developed in association with the NESHAP covering these emission sources (40 CFR Part 61). The procedure applicable to underground uranium mine vents employs Methods 1 and 2 and either continuous radon monitoring or alpha track detector techniques from Method 114. The two procedures applicable to uranium mill tailings piles, phosphogypsum stacks, and other piles of waste material emitting radon both involve measurement and calculation of radon flux (i.e., emission rate per area). The flux calculation is based on results determined for numerous activated charcoal radon collectors placed on the surface of the emission source according to method specifications.

CARB Method 427 (Asbestos)
California Air Resources Board (CARB) Method 427 (CARB 1989a) is applicable to measurement of asbestos emissions from stationary sources. CARB Method 427 also may be applicable to measurement of mineral fibers,

another air toxic listed in the 1990 CAAA. A particulate emission sample is collected isokinetically using a sampling train consisting of a probe, a filter placed behind the probe within the stack, a series of impingers or other condenser, and a calibrated dry gas meter. The sample is composed of the filter and the dried rinses of the probe, probe nozzle, and the front half of the filter holder; the bulk of the particulate sample is collected on the filter. Transmission electron microscopy (TEM) can be used to analyze the collected sample. TEM is capable of classifying the collected fibers as chrysotile, amphibole, or nonasbestos, and describing the aggregation of asbestos into fibers, bundles, or mats. Analysis by phase contrast microscopy (PCM) also is permitted by CARB Method 427 but does not allow for differentiation between asbestos and other mineral fibers.

CARB Method 431 (Ethylene Oxide)
CARB Method 431 (CARB 1989a) was developed to determine ethylene oxide (EtO) emissions from sterilization chambers. The volumetric flow of the sample gas stream is monitored during purging of the sterilization chamber. Concurrently, samples of the gas stream are analyzed one after the other by gas chromatography. Total emissions of ethylene oxide during the sterilization cycle are calculated from curves of flow and concentration over time.

Method for Hexavalent Chromium (Cr^{+6})
A method (EPA 1991p; EPA 1991c; Carver 1991), known as Method Cr^{+6}, has been developed for the determination of hexavalent chromium emissions from hazardous waste incinerators, municipal waste combustors, sewage sludge incinerators, and BIF burning hazardous waste. Method Cr^{+6} recently was issued in conjunction with the standards regulating BIF. Optionally this method also may be used on approval of EPA for the determination of total chromium emissions.

For incinerators and combustors, the hexavalent chromium emissions are collected isokinetically from the emission source. To

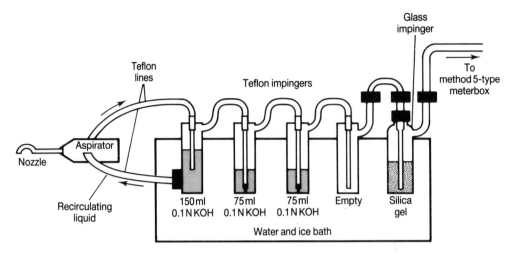

Figure 9-17.   Hexavalent chromium recirculatory impinger train with aspirator assembly.

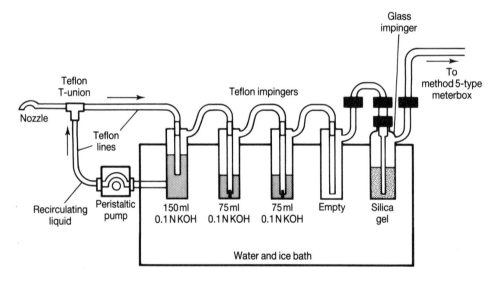

Figure 9-18.   Hexavalent chromium recirculatory impinger train with pump/sprayer assembly.

eliminate the possibility of reduction of hexavalent chromium in the sampling train between the nozzle and the impingers, the emissions are collected with a recirculatory train (see Figures 9-17 and 9-18) where the alkaline impinger reagent is continuously circulated to the noz-zle.[5] Otherwise, the sampling train is based on a Method 5 train without a filter. The components of the train are constructed principally of Teflon™ to minimize the interaction of the hexavalent chromium with glass; other portions of the train that come in contact with the sample

---

[5] The alkaline impinger reagent (a solution of potassium hydroxide) provides the elevated pH conditions (pH 7.4 or higher) necessary to avoid conversion of the hexavalent chromium in the emissions to another valence state.

are constructed of glass, quartz, or Tygon™ tubing. The gas metering system specified is identical to that required by Method 5.

Sample recovery procedures include: (1) a postsampling nitrogen purge of the impinger train as a safeguard against the conversion of hexavalent chromium to another valence state, and (2) filtration of the liquid sample to remove insoluble matter including the majority of the trivalent chromium potentially present. When total chromium will be determined, the filter residue is retained for analysis; otherwise, it is discarded. The impinger train samples are analyzed for hexavalent chromium by using an ion chromatograph equipped with a postcolumn reactor (IC/PCR) and a visible wavelength detector. The IC/PCR separates the hexavalent chromium as chromate from the other components in the sample matrices that may interfere with hexavalent chromium-specific diphenyl-carbizide reaction that occurs in the postcolumn reactor. To increase sensitivity for trace levels of hexavalent chromium, a preconcentration system may be used in conjunction with the IC/PCR.

Instrumental and Continuous Emission Measurement Methodologies

The gaseous toxic emission measurement methods discussed up to this point have involved extraction of a known quantity of sample, absorption or adsorption of the components of interest on a solid material or in a liquid solution, and generally complex laboratory analysis. Such a measurement approach is called manual sampling and is useful for short-term compliance demonstrations and emission screening.

Some instrumental analytical techniques have been applied to source emission measurements including some air toxics. Many of these instrumental methods require sample extraction much like a manual method but the analysis is conducted on-site with a compound-specific detector and an electronic data reduction device. Sampling techniques appropriate to the sample matrix and proper calibration of the instrument are crucial to obtaining valid data. Instrument

sensitivity, interference factors, and calibration linearity must be established within acceptable levels before and during the test. Such instrumental method performance checks are included in 40 CFR Part 60 (Appendix A), Methods 3A for O_2 and CO_2, Method 6C for SO_2, Methods 7E and 20 for NO_x, Method 10 for CO, and Methods 25A and 25B for total volatile organic compounds. These instrumental methods are intended for short-term (i.e., less than twenty-four hours) operation in a manner similar to that for the manual methods.

Instrumental measurement techniques can be modified to operate on a 'continuous' basis to collect emission data for long term data averaging. Continuous emission monitoring systems (CEMSs) have been applied to emissions from major stationary sources for over twenty years. Two types of CEMSs are commonly applied. Extractive CEMSs draw a sample from the stack, condition the sample gas (i.e., remove particulate matter and dry the gas), and analyze for the specific compounds of interest. In situ CEMSs are designed to provide a measure of target compounds in the stack without sample extraction or conditioning. The components of in situ CEMSs commonly include a light or radiation source, a detector, and a data reduction device mounted on the stack. In situ CEMSs allow analysis of the sample in its normally occurring condition and, thus, may be more representative than analysis of an extracted, conditioned sample.

Data from CEMSs are used for compliance determinations for long-term or rolling averaging periods (e.g., thirty days) in several current regulations. CEMSs for SO_2 and NO_x emissions monitoring are applied at large fossil-fuel-fired utility boilers. SO_2 and H_2S CEMSs are applied at petroleum refineries for emissions from catalytic cracking units. Some state agencies require continuous monitoring of H_2S, CO, and HCl for hazardous waste and medical waste incinerators. CEMS applied to combustion sources usually include an analyzer for O_2 or CO_2 to provide data for correcting the pollutant concentration measurements for dilution air.

Performance criteria for CEMSs are different from the short-term instrumental methods described above. Appendix B of 40 CFR Part 60 contains performance specifications for CEMSs used on stationary sources. The performance of a CEMS is evaluated by comparing the results of the CEMS output with the results of a number of test runs conducted concurrently with a manual or instrumental reference test method. The reference method for the relative accuracy test is usually the method used for the initial compliance demonstration and is specified in the applicable regulation. The relative accuracy test consists of at least nine simultaneous comparison test runs. The relative accuracy (RA) is determined as the sum of the mean difference (DIFF) between the CEMS and the reference method results and a confidence coefficient (CC) divided by the mean of the reference method value (RM). In equation form, the relative accuracy is

$$RA = \frac{DIFF + CC}{RM} \times 100 \text{ percent} \quad (9\text{-}1)$$

The performance specifications also include checks and limits on CEMS calibration drift on a daily basis.

The relative accuracy test procedures and acceptance limits in Performance Specification 2 of 40 CFR Part 60, Appendix B are similar to those that will be applied to CEMSs required for air toxics emissions monitoring. These limits are twenty percent of the mean reference method value or ten percent of the applicable emission limit, whichever is less stringent. The purpose of the dual relative accuracy limits is to allow the CEMSs applied to very low emission sources more opportunity to pass the relative accuracy test. With the conditions specified, CEMSs applied to sources with emissions less than one-half the regulatory limit may pass the relative accuracy test using the ten percent of the emission limit criterion.

Continuously recorded emission data will be used in the future not only for establishing compliance, but also for market-based or emission trading regulations. Acid rain regulations under Title IV of the 1990 CAAA (see Chapter 26) are an example of emission trading (EPA 1991q). For this regulation, sources will measure SO_2 and NO_x emission rates to establish compliance with an allowance based on a proportion of a national emission limit. A source that exceeds the applicable allowance of emissions may purchase credits equal to the excess emissions from another source that emitted an amount less than its allowance. The price of these credits will be established by the market. The objective of the program is to reduce the overall acid gas emission impact in an economically sound manner.

Similar continuous emission monitoring and trading regulations are being considered for air toxics sources. Although the primary purpose of the continuous emission monitoring requirements in these regulations would be to monitor and maintain compliance with the emission limits, the regulations may allow sources to trade emission credits as described above. Regulation most likely would include a hierarchy of air toxics with assigned risk weights that must be taken into account in trading air toxics emissions. For example, dioxin may have a weight of ten thousand units and hydrogen chloride (HCl) a weight of one unit. Thus, a source of dioxin emissions would have to purchase ten thousand units of HCl emissions for each dioxin unit of emission over the limit to meet its overall air toxics emission limit.

Implementation of long-term compliance or emissions trading regulations requires that CEMSs be subject to continuous and long-term quality assurance and quality control (QA/QC) procedures and acceptance criteria. Appendix F of 40 CFR Part 60 describes procedures and format for preparing QA/QC plans for CEMSs, daily CEMS drift checks, periodic relative accuracy tests, and acceptance criteria. Appendix F, the form and format of which would be applied to CEMSs used for air toxics emission monitoring, also defines procedures to follow during periods of out-of-control CEMS operation. Such procedures may include manual emission testing to meet minimum data reporting requirements.

The following discussion describes the currently available technology for continuously monitoring air toxics. At present, demonstrated CEMS performance information is available only for volatile organic compounds (VOC). The section on emerging technologies later in this chapter discusses some of the potentially applicable instrumental measurement methods and CEMSs for air toxics emissions being developed (A&WMA 1990; Draves, Dayton, and Bursey 1991).

VOC Continuous Emission Monitoring Systems

A majority of the listed air toxics are VOC. Analyzers are available that detect relative VOC concentrations using methods such as flame ionization detection (FID), photoionization detection (PID), nondispersive infrared (NDIR) absorption, or Hall detection. These VOC analyzers do not speciate the VOCs nor do they respond equally to all VOCs; thus, VOC analyzers provide only a measure of the relative concentration level of a mixture of compounds. Although VOC analyzers can be used for air toxics measurements when consistent proportions of multiple air toxics are in the exhaust stream and they generate equal instrumental responses, the best application for a VOC CEMS for air toxics determination is for an exhaust gas stream that contains a single air toxic and no other interfering substances.

The effect that sample conditioning can have on air toxics VOC measurement with a CEMS can be significant and must be considered in the design of the system. A CEMS sample conditioning system that includes a condenser to reduce the sample gas moisture content to below one percent can protect the analyzer from damage and prolong the CEMS life. On the other hand, the condenser also can remove some or all of the air toxic(s) of interest and cause the CEMS to have a low measurement bias. A conditioning system that maintains the sample gas temperature above the moisture dew point up to injection into the analyzer is less likely to 'lose' the target air toxic(s) before analysis; however, sample conditioning equipment and analyzer maintenance problems can become significant.

The planned Appendix A to 40 CFR Part 64 will include performance specifications (e.g., Performance Specification 101) for installing and evaluating VOC CEMSs. The CEMS performance test includes a measure of calibration error, CEMS response time, and a performance audit test. The calibration gas must be a mixture that includes the air toxic(s) of interest and an appropriate dilution gas (e.g., nitrogen or purified air) for a compound-specific application of a VOC CEMS. Performance Specification 101 calls for a performance audit gas, when available, or a Protocol 1 (EPA 1977a) gas to be introduced to the VOC CEMS as a QA check of the entire VOC CEMS including the gas conditioning system. Procedures for evaluating the CEMS response time and twenty-four-hour analyzer calibration drift over a minimum seven-day period are included.

Methodologies Applicable to Fugitive Emissions

Stationary sources of emissions that are not confined to stacks, ducts, or vents often are referred to as fugitive emission sources. Examples of fugitive emission sources include leaks from furnace and oven doors; leaks from process equipment at valves, seals, flanges, pumps, and compressors; emissions from open tanks such as plating baths and degreasers; uncontrolled emissions from spraying operations; uncontrolled emissions from coating operations; emissions from storage vessels such as storage tanks and drums; emissions from material transfer operations; emissions from storage and waste piles; and emissions from waste treatment facilities such as landfills and lagoons. Although fugitive emissions are discussed in more detail in Chapter 22, sampling and analysis issues are summarized here.

In many cases, the methodologies discussed previously may be applicable to fugitive emission sources. Methods developed for the sampling and analysis of ambient air samples also may be applicable, usually with some adaptation or modification. When ambient sampling methodologies are applied, proper precautions must be taken to avoid exceeding the capacity of the method. Ambient concentrations typically are in the part per billion (ppb) range and ambient air methodologies usually are designed

Table 9-7   Ambient air methods with potential applicability to HAPS[a]

Compound	Potentially applicable methods[b]
Acetaldehyde	TO-5, -11; NIOSH 3507; OSHA 68
Acetamide	TO-13
Acetonitrile	TO-14; NIOSH 1606
Acetophenone	TO-5, -13
2-Acetylaminofluorene	TO-13
Acrolein	TO-5, -11; OSHA 52
Acrylamide	TO-13; OSHA 21
Acrylic acid	TO-13; OSHA 28
Acrylonitrile	TO-1, -14; NIOSH 1604; OSHA 37
Allyl chloride	TO-1, -2, -14; NIOSH 1000
4-Aminobiphenyl	TO-13
Aniline	TO-13; NIOSH 2002
o-Anisidine	TO-13; NIOSH 5013
Asbestos	NIOSH 7400, 9001
Benzene	TO-1, -2, -14; NIOSH 1501; OSHA 12
Benzidine	TO-13; NIOSH 5013, 5509; OSHA 65
Benzotrichloride	TO-1, -13
Benzyl chloride	TO-1, -13; NIOSH 1003
Biphenyl	TO-13
Bis(2-ethylhexyl)phthalate	TO-13
Bis(chloromethyl)ether	TO-1, -14; OSHA 10
Bromoform	TO-1, -14; NIOSH 1003
1,3-Butadiene	TO-1, -2, -14; NIOSH 1024; OSHA 56
Calcium cyanamide	
Caprolactum	TO-13
Captan	TO-4, -10
Carbaryl	TO-13; OSHA 63
Carbon disulfide	TO-1; NIOSH 1600
Carbon tetrachloride	TO-1, -2, -14; NIOSH 1003
Carbonyl sulfide	TO-1, -14
Catechol	TO-13
Chloramben	TO-13
Chlordane	TO-4, -10; NIOSH 5510; OSHA 67
Chlorine	OSHA ID-101
Chloroacetic acid	TO-13
2-Chloroacetophenone	TO-5, -11, -13
Chlorobenzene	TO-1, -4; NIOSH 1003
Chlorobenzilate	TO-13
Chloroform	TO-1, -2, -14; NIOSH 1003; OSHA 5
Chloromethyl methyl ether	TO-1, -14; OSHA 10
Chloroprene	TO-1, -14; NIOSH 1002
Cresols/cresylic acid	TO-8, -13; OSHA 32
o-Cresol	TO-8, -13; OSHA 32
m-Cresol	TO-8, -13; OSHA 32
p-Cresol	TO-8, -13; OSHA 32
Cumene	TO-1, -13, -14; NIOSH 1501
2,4-D, salts and esters	TO-10
DDE	TO-4, -10
Diazomethane	NIOSH 2515
Dibenzofurans	TO-9
1,2-Dibromo-3-chloropropane	TO-13
Dibutylphthalate	TO-13; NIOSH 5020
1,4-Dichlorobenzene	TO-1, -13, -14
3,3-Dichlorobenzidene	TO-13; NIOSH 5509; OSHA 65

Table 9-7   Continued

Compound	Potentially applicable methods[b]
Dichloroethyl ether	TO-13
1,3-Dichloropropene	TO-1, -14
Dichlorvos	TO-13; OSHA 62
Diethanolamine	TO-13
N,N-Diethylaniline	TO-13; NIOSH 2002
Diethyl sulfate	TO-13
3,3'-Dimethoxybenzidine	TO-13
Dimethyl aminoazobenzene	TO-13
3,3'-Dimethyl benzidine	TO-13
Dimethyl carbamoyl chloride	TO-13
Dimethyl formamide	TO-13; NIOSH 2004
1,1-Dimethylhydrazine	TO-1, -14
Dimethyl phthalate	TO-13
Dimethyl sulfate	TO-13
4,6-Dinitro-o-cresol and salts	TO-13
2,4-Dinitrophenol	TO-13
2,4-Dinitrotoluene	TO-13; OSHA 44
1,4-Dioxane	TO-1, -14
1,2-Diphenylhydrazine	TO-13
Epichlorohydrin	TO-13; NIOSH 1010
1,2-Epoxybutane	TO-1, -14
Ethyl acrylate	TO-1, -14
Ethyl benzene	TO-1, -14; NIOSH 1501
Ethyl carbamate	TO-13
Ethyl chloride	TO-1, -2, -14
Ethylene dibromide	TO-1, -13, -14; NIOSH 1008; OSHA 2
Ethylene dichloride	TO-1, -2, -14; NIOSH 1003; OSHA 3
Ethylene glycol	TO-13; NIOSH 5500
Ethylene imine	TO-1, -14
Ethylene oxide	TO-14; NIOSH 1614, 1607; OSHA 50
Ethylene thiourea	TO-13; NIOSH 5011
Ethylidene dichloride	TO-1, -14
Formaldehyde	TO-5, -11; NIOSH 3501, 2541, 3500; OSHA 52
Heptachlor	TO-4, -10
Hexachlorobenzene	TO-13
Hexachlorobutadiene	TO-1, -13, -14
Hexachlorocyclopentadiene	TO-13
Hexachloroethane	TO-1, -13, -14; NIOSH 1003
Hexamethylene-1,6–diisocyanate	TO-13; OSHA 42
Hexamethylphosphoramide	TO-13
Hexane	TO-1, -14
Hydrazine	NIOSH 3503; OSHA 20
Hydrochloric acid	
Hydrogen fluoride	
Hydrogen sulfide	
Hydroquinone	TO-13; NIOSH 5004
Isophorone	TO-5, -11, -13
Lindane	TO-4, -10
Maleic anhydride	TO-13; OSHA 25
Methanol	TO-14; NIOSH 2000
Methoxychlor	TO-4, -10, -13
Methyl bromide	TO-1, -2, -14
Methyl chloride	TO-1, -2, -14; NIOSH 1001
Methyl chloroform	TO-1, -2, -14; NIOSH 1003

195

Table 9-7 Continued

Compound	Potentially applicable methods[b]
Methyl ethyl ketone	TO-1, -5, -11, -14
Methyl hydrazine	TO-1, -14
Methyl iodide	TO-1, -14; NIOSH 1014
Methyl isobutyl ketone	TO-1, -5, -11, -14
Methyl isocyanate	TO-1, -14; OSHA 54
Methyl methacrylate	TO-1, -14
Methyl *t*-butyl ether	TO-1, -5, -11, -14
4,4'-Methylene bis(2-chloroaniline)	TO-13
Methylene chloride	TO-1, -2, -14; NIOSH 1005; OSHA 80
Methylene diphenyl diisocyanate	TO-13; OSHA 47
4,4'-Methylenedianiline	TO-13; NIOSH 5029; OSHA 57
Naphthalene	TO-13; NIOSH 1501; OSHA 35
Nitrobenzene	TO-1, -14; NIOSH 2005
4-Nitrobiphenyl	TO-13
4-Nitrophenol	TO-13
2-Nitropropane	TO-1, -13, -14; OSHA 46
N-Nitroso-*N*-methylurea	TO-13
N-Nitrosodimethylamine	TO-1, -7, -13; OSHA 27
N-Nitrosomorpholine	TO-7, -13; OSHA 27
Parathion	TO-4, -10; OSHA 62
Pentachloronitrobenzene	TO-13
Pentachlorophenol	TO-13; NIOSH 5512; OSHA 39
Phenol	TO-13; NIOSH 3502; OSHA 32
p-Phenylenediamine	TO-13
Phosgene	TO-6; OSHA 61
Phosphine	TO-13; NIOSH 7905
Phosphorus	OSHA ID-180
Phthalic anhydride	TO-13
PCB	TO-13
1,3-Propane sultone	TO-13
Polycyclic organic matter	TO-13
β-Propiolactone	TO-13
Propionaldehyde	TO-5, -11
Propoxur	TO-13
Propylene dichloride	TO-1, -14
Propylene oxide	TO-1, -14; NIOSH 1612
1,2-Propyleneimine	TO-1, -14
Quinoline	TO-13
Quinone	TO-5, -11
Styrene	TO-1, -14; NIOSH 1501; OSHA 9
Styrene oxide	TO-13
2,3,7,8-Tetrachlorodibenzo-*p*-dioxin	TO-9
1,1,2,2-Tetrachloroethane	TO-1, -14; NIOSH 1019
Tetrachloroethylene	TO-1, -14; NIOSH 1003
Titanium tetrachloride	
Toluene	TO-1, -2, -14; NIOSH 1501, 4000
2,4-Toluene diamine	TO-13; NIOSH 5516; OSHA 65
2,4-Toluene diisocyanate	TO-13; OSHA 42
o-Toluidine	TO-13; NIOSH 2002, 5013; OSHA 73
Toxaphene	TO-4, -10
1,2,4-Trichlorobenzene	TO-1, -13; NIOSH 5517
1,1,2-Trichloroethane	TO-1, -14; NIOSH 1003; OSHA 11
Trichloroethylene	TO-1, -2, -14; NIOSH 1022
2,4,5-Trichlorophenol	TO-13

Table 9-7   Continued

Compound	Potentially applicable methods[b]
2,4,6-Trichlorophenol	TO-13
Triethylamine	TO-13
Trifluralin	TO-4, -10
2,2,4-Trimethylpentane	TO-1, -14
Vinyl acetate	TO-1, -14; OSHA 51
Vinyl bromide	TO-1, -2, -14; NIOSH 1009; OSHA 8
Vinyl chloride	TO-1, -2, -14; NIOSH 1007; OSHA 75
Vinylidene chloride	TO-1, -2, -14; NIOSH 1015; OSHA 19
Xylenes (isomers and mixtures)	TO-1, -14; NIOSH 1501
o-Xylenes	TO-1, -14
p-Xylenes	TO-1, -14
m-Xylenes	TO-1, -14
Antimony compounds	
Arsenic compounds	OSHA ID-105
Beryllium compounds	NIOSH 7102
Cadmium compounds	NIOSH 7048, 7200
Chromium compounds	NIOSH 7200, 7024; OSHA ID-103
Cobalt compounds	NIOSH 7027
Coke oven emissions	OSHA 58
Cyanide compounds	NIOSH 7904, 9010, 9012; OSHA ID-120
Glycol ethers	
Lead compounds	NIOSH 7082
Manganese compounds	NIOSH 7200
Mercury compounds	OSHA ID-121
Fine mineral fibers	NIOSH 7400
Nickel compounds	NIOSH 7200
Radionuclides	
Selenium compounds	

[a] Adapted from Bursey 1991.

[b] TO-methods from Winberry, Murphy, and Riggin 1990; NIOSH methods from NIOSH 1985; OSHA methods from OSHA 1990.

to measure in that range. In contrast, stationary source concentrations may exceed many parts per million (ppm). Methods for ambient air based on the use of sorbents particularly are susceptible to exceeding their collection capacity when applied to source emission measurement. Table 9-7 presents a number of potentially applicable ambient methods developed by EPA, the Occupational Health and Safety Administration (OSHA), and the National Institute of Occupational Safety and Health (NIOSH) (Winberry, Murphy, and Riggin 1990; NIOSH 1985b; OSHA 1990; Bursey 1991). Bursey (1991) provides suggested modifications to adapt these methods.

Probably the most difficult problem in testing a fugitive emission source is that of obtaining a representative sample. Some examples of techniques that have been used in sampling fugitive emissions are discussed below.

Enclosure with a Bag

The bag collection technique has been used to measure emissions from equipment leaks at sources such as petroleum refineries and synthetic organic chemical manufacturing facilities. The leaking equipment (e.g., a flange) is enclosed in a bag of nonreactive material such as Tedlar™. The atmosphere inside the bag can then be sampled to yield a value corresponding to the total emissions within the bag over a specific period of time. One procedure uses a portable FID to sample directly from the bag and analyze the collected sample (CMA 1989). In another procedure, a continuously operating personnel sampling pump is used to circulate the atmosphere from the bag, through adsorbent tubes or an impinger absorbing solution where the sample is collected, and back into the bag (Smith, Grosshandler, and Mastrianni 1990). This procedure can offer improved detection limits and accuracy, as well as the ability to measure specific compounds.

Temporary Total Enclosure

A temporary total enclosure is an enclosure installed to surround a source of emissions such that all the emissions can be contained for discharge through ducts that allow for accurate measurement (Edgerton, Kempen, and Lapp 1991). Temporary total enclosures have been applied in the coating industry to measure fugitive emissions of VOC. Considerations in their application include avoiding disruption of the normal air flow patterns around the fugitive emission source and ensuring the evacuation of potentially toxic or explosive atmospheres.

Temporary Hoods

Temporary hooding arrangements also have been used in the capture and measurement of fugitive emissions (Kolnsberg 1976). Temporary hoods are most manageable in situations where the area of the emissions is small. For example, in one case movable temporary hoods were used in sampling the particulate and gaseous emissions from door leaks on coke ovens (Barrett et al. 1977).

Sampler Arrays

Use of a vertical, horizontal, or two-dimensional array of samplers is a classic approach for sampling emissions from fugitive sources such as landfills, lagoons, Superfund sites, and waste piles. Typically, the samplers are positioned downwind and upwind of the source, with the upwind measurements made to determine the background levels of the target analyte(s) (Kolnsberg 1976; Lewis et al. 1985). Depending on the analyte(s), the samplers may include pump and filter combinations, pump and impinger combinations, pump and sorbent tube combinations, or passive collectors such as charcoal canisters.

As an example, Method 115, discussed earlier for measurement of radon-222 emissions from waste piles, involves placement of numerous large area charcoal collectors to yield individual radon flux values (rate per area) that subsequently are used to calculate the radon emissions for the total pile.

Mobile Integrating Sampler

A mobile integrating sampler has been proposed by Wisner and Davis (1989) to avoid problems inherent with the stationary sampler array approach. These problems include difficulty in maintaining the samplers within the emission 'plume' throughout the sampling period when there are changes in wind direction, as well as the related problem of obtaining a sufficiently large sample to meet the detection threshold while within the plume. The mobile sampler system is affixed to a vehicle that traverses the plume at a constant rate in a crosswind direction while one or more samplers are operated continuously. Weather balloons also have been used as a technique to introduce the vertical component into the mobile sampler approach.

Flux Chambers

The flux chamber method is an isolation chamber method that has been used to measure fugitive emissions from landfills, liquid impoundments (e.g., lagoons, overflow basins, storage tanks), and contaminated soils (Kienbusch 1986; Gholson et al. 1987; Dupont 1985; Eklund, Balfour, and Schmidt 1985). The principle of the flux chamber involves enclosure of a representative area of the source surface with the chamber. A controlled flow of clean sweep gas is added to the chamber, allowed to mix, and then released to the chamber exit(s). Concentrations of the target species are measured in the exit gas using an appropriate method. Use of multiple flux chambers for sampling also is a specific case of sampler arrays.

CHOICE OF AVAILABLE TECHNIQUES

Several factors must be considered in selecting the proper air toxics measurement technique. Some of these are readily apparent, others may not be so obvious. Important factors are described below.

Target Air Toxics

The first consideration is the air toxics to be measured. One approach is to determine the 189 listed air toxics that are emitted from the subject emission source. To conduct a test program to satisfy this objective, it often is prudent to select methods with multiple pollutant applicability. This approach, however, can be problematic. Despite using multipollutant methods, numerous sampling and analytical techniques must be applied; this results in extremely costly testing. Further, some of these techniques may not have been validated for certain of the target air toxic or for the emission source; this results in qualitative data. Qualitative or unvalidated data should be used only for screening purposes, to indicate the presence or absence of the air toxic and possibly its relative level in the emission gas stream.

Screening results indicating the absence of a target compound are not, however, always reliable. For example, use of a multiple pollutant methodology instead of a specialized methodology could result in significantly higher detection limits for certain air toxics of interest and the risk that trace quantities might not be measured. Once specific compounds, for which the applied methods are not validated, are identified at significant levels, sampling and analysis for these compounds must be repeated using validated methods to provide data sufficiently reliable to be used for compliance determination, standard setting, or development of emission factors.

It is preferable to narrow the scope of the air toxics test program. Narrowing of the scope typically is accomplished by applying knowledge of the process and physical and chemical phenomena to determine the potential air toxics in the gas stream. Even for a shortened list of air toxics, use of multiple pollutant methods can reduce testing costs. Also, with a narrower scope of target pollutants, the test program may be designed to validate methodologies for those air toxics or sample matrices for which the methods are not already validated.

A decision tree that may prove helpful in making initial determinations of potentially applicable methods is presented in Figure 9-19 (Bursey 1991). The decision tree does not address all listed air toxics nor all potentially applicable methods; nor is information provided on the status of method validation or potential sampling and analytical problems with certain target analytes. Some of these topics are discussed in later sections of this chapter. Additional pertinent information is provided in Bursey (1991), Johnson et al. (1989), and Johnson and Merrill (1983).

Air Toxics Without Candidate Methodology

At the time this chapter was written in 1992, there were several listed air toxics that may not be readily measured by the candidate methods presented thus far. One is diazomethane, an explosive and highly reactive compound that may not even exist under stack or atmospheric conditions.

Another is phosgene, an extremely toxic ketone that also is highly reactive. OSHA Method 61 (OSHA 1990) provides for measurement of phosgene in workplace air by collection in a solution of 4,4'-nitrobenzyl pyridine in diethyl phthalate that then is analyzed spectrophotometrically. Another method for measuring phosgene in ambient air, Method TO-6 (Winberry, Murphy, and Riggin 1990), collects the analyte in an aniline-toluene mixture with analysis by HPLC. Analysis by HPLC would be preferable, but this method has not been shown to be suitable for source emission measurement. A third approach (Hendershott 1986) uses XAD-2™ resin treated with di-*n*-butylamine for sample collection and GC/FID for analysis. This method appears promising for source

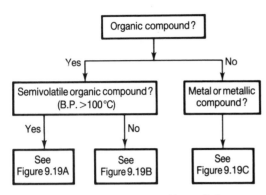

emission testing if an HPLC or GC/MS analytical technique does not prove compatible with an effective sampling procedure.

Acetonitrile emissions also are difficult to measure in the presence of moisture because acetonitrile is highly soluble in water. When moisture is present, neither the VOST nor the Method 18 container sampling techniques are suitable. Previous techniques for direct injection GC analysis of acetonitrile did not offer sufficiently low detection limits to allow collection of the analyte in water. However, EPA recently completed development of a GC/nitrogen phosphorus detector (NPD) method for analysis of

Figure 9-19.   Decision tree for making preliminary method selections. Adapted from Bursey (1991).

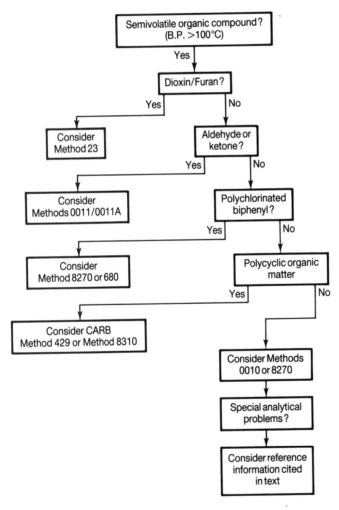

Figure 9-19A.   Decision tree for making preliminary method selections. Adapted from Bursey (1991).

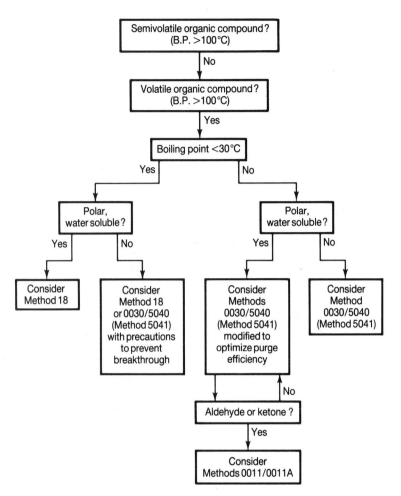

Figure 9-19B. Decision tree for making preliminary method selections. Adapted from Bursey (1991).

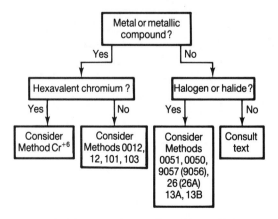

Figure 9-19C. Decision tree for making preliminary method selections. Adapted from Bursey (1991).

acetonitrile in water with significantly improved detection limits; it is slated for publication in SW-846 (EPA 1986k) as analytical Method 8033. A potential sample collection technique for use with this analytical method is isokinetic sampling using a Method 0011-type sampling train with water in the impingers.

The four listed isocyanates (i.e., hexamethylene-1,6-diisocyanate, methyl isocyanate, methylene diphenyl diisocyanate, and toluene diisocyanate) also may prove difficult to measure. Isocyanates react readily with water to form species not easily measured. For this reason, they likely will require derivatization upon collection. A NIOSH sampling method for isocyanates (NIOSH 1985b) uses this principle, but may not have the capacity for source emission measurement. Measurement of urea derivatives by HPLC or GC/MS may prove useful (Dunlap, Sandridge, and Keler 1976). A Method 0011-type impinger train most likely could be used for sample collection.

There is no method that measures polycyclic organic matter (POM) as a compound class. Available methods measure individual PAH, which are a subset of POM.

There are two other air toxics for which a recommended method will yield only a 'worst case' value; these are titanium tetrachloride and phosphine. Both potentially can be measured using Method 0012; however, Method 0012 will yield values only for total phosphorus and total titanium. If other phosphorus or titanium compounds are present in the sample matrix, the results will be biased high. It may be possible to measure phosphine more specifically using Method 18.

Measurement Objectives

As discussed earlier, the objectives of an emission measurement or test program must be considered in selecting the test methods. Some methods may be suitable only for providing qualitative data for screening purposes, and should not be used to collect data for development of emission factors, standard setting purposes, or compliance determination where data of known quality are essential. Other methods may be suitable for providing qualitative data, but with modification and validation could provide quantitative data of known quality. Lastly, for certain analytes emitted from certain sources, validated methods will provide data of known quality without modification.

Detection and Quantification Limits

Levels of the detection and quantification limits must always be considered in applying an emission measurement method. Method detection limit is defined as 'the minimum concentration of a substance that can be measured and

reported with 99 percent confidence that the analyte concentration is greater than zero and is determined from analysis of a sample in a given matrix containing the analyte' (40 CFR 136, Appendix B). The American Chemical Society Subcommittees on Environmental Improvement and Environmental Analytical Chemistry use a more general definition: 'limit of detection is the lowest concentration of an analyte that the analytical process can reliably detect' (MacDougall et al. 1980). For instrumental methods, a detection limit often is calculated based on the relationship between the analyte signal, the field blank signal, and the variability of the field blank signal (MacDougall et al. 1980; Long and Winefordner 1983; EPA 1986k). The limit of quantification for a method is the level at which the results for an analyte can be quantified with some degree of certainty and is above the limit of detection. The limit of quantification often is calculated as some multiple of the detection limit (MacDougall et al. 1980; EPA 1986k). Both the detection and the quantifiable limits sometimes are expressed on a sample-specific basis.

Before testing, the detection and quantification limits of the candidate method should be compared to (1) the expected source emission levels, if known, and (2) the level above which it is important to know whether the analyte of interest is present and its concentration. An important distinction is the difference between an analytical detection limit and an 'in-stack' detection limit. The analytical detection limit describes the limit of detection for the sample matrix (e.g., solution or extract) submitted to the analytical technique. The in-stack detection limit defines the limits of measurement of the analyte within the sample gas stream, taking into account the sample gas volume, sample recovery procedures, and sample preparation procedures, together with the analytical detection limit. It is difficult to quantify accurately the 'in-stack' detection limit because of unknown and unquantifiable sample matrix effects. However, estimates of the 'in-stack' detection limit based on the analytical detection limit and the factors listed above are very important in planning a test program.

Method Validation

Another key consideration in selection of a methodology for measurement of a pollutant of interest is the validation status of the methodology for that pollutant as emitted from the emission source in question. Intimately associated with the validation status of the methodology are the DQO of the test program.

Rigorous validation of an emission measurement methodology involves determining and documenting the quality of the data generated (i.e., bias and precision) at the measured concentrations of an analyte of interest in the emission sample matrix. In other words, the performance of the combined sampling and analytical procedure (or emission measurement methodology) is defined under actual field conditions. The bias and precision determined take into account both the target analyte and measurement of that analyte within the specific source emission matrix.

Method 301 (EPA 1991c) provides a standardized method validation procedure to be used whenever an unvalidated emission test method is proposed to meet an EPA requirement. The protocol involves introducing a known concentration of an analyte (this is known as spiking) during the sampling procedure and carrying it through the entire sampling and analytical procedure to assess the bias of the proposed method. Alternatively, when possible, the proposed method is compared against a previously validated method to assess the bias. In all cases, multiple colocated simultaneous samples must be collected to determine the precision of the test method.

Method 301 includes procedures for spiking and specifies the number of spiked samples necessary to allow statistical determination of the validity of the proposed test method. If the bias and precision meet the requirements of Method 301, the methodology is considered validated for measurement of a particular analyte in the emissions from a particular type of source. Method 301 also includes optional procedures that may be used to expand the applicability of a proposed method. These include: 'ruggedness testing,' which is a laboratory evaluation demonstrating the sensitivity of the method to various parameters; a procedure for assessing sample recovery and analysis times; and a procedure for the determination of the practical limit of quantitation for the method.

As previously noted, currently there are not validated emission measurement methodologies for the majority of the 189 listed air toxics. Methods validated for certain analytes may not be valid for all possible applications and sample matrices. In the earlier section of this chapter covering 'Available Techniques,' the methodologies discussed were said to be applicable to or validated for certain analytes. Such methods are considered validated for those analytes in emissions from one or more specific source types, but not necessarily validated for measurement of that analyte from every type of emission source. For example, Method 8270 includes a list of 'target compounds' for which the analytical method performance has been established, and through the RCRA regulations the applicability of the combination of Methods 0010 and 8270 has been established for measurement of the target compound emissions from hazardous waste incinerators. However, the validity of applying this combined sampling and analytical methodology to other emission sources has not been established.

Information on method performance, such as analytical precision and bias, constitutes what is termed 'partial validation' of a method. Method 301 allows for the situation where partial validation of a method may be adequate for certain applications. For example, if the proposed test method is validated for a 'similar' source, the requirements of Method 301 may be waived or partially waived provided the requester can demonstrate to the satisfaction of the EPA Administrator that the emission characteristics (e.g., sample matrices and sample conditions) of the two sources are 'similar.' The method's applicability to the 'similar' source may be demonstrated by conducting a ruggedness test. Section 12 of Method 301 provides specific information concerning waivers of the method requirements.

Reference Information on Potentially Applicable Methods

The next decade should be a time of rapid expansion and change in the area of air toxics emissions measurement. For this reason, this chapter can serve only as an introduction to the subject and will rapidly be outdated. The following are resources that may prove useful for obtaining current information regarding the sampling and analysis of air toxics in the future:

1. U.S. EPA Emission Measurement Technical Information Center (EMTIC) Computer Bulletin Board System (BBS), one of twelve BBSs on the Office of Air Quality Planning and Standards' Technology Transfer Network (TTN) located in Research Triangle Park, NC. The telephone number is (919) 541-5742. A voice helpline is available at (919) 541-5384.

2. Problem POHC Reference Directory, a database of physical, chemical, toxicological, and sampling/analytical information on the air toxics and toxic compounds listed in Appendices 8 and 9 of RCRA. Available from the National Technical Information Service (NTIS) as Publications PB 91-507749 and PB 91-201061. Also available in the EMTIC BBS.

3. 'Test Methods for Evaluating Solid Waste, Physical/Chemical Methods, Third Edition,' EPA Publication No. SW-846, November 1986, EPA Office of Solid Waste and Emergency Response, Washington, D.C. Available from U.S. Government Printing Office, Washington, D.C., Document No. 955-001-0000001.

4. Stationary Source Sampling and Analysis Directory, a database of potential methods for measuring 1990 CAAA air toxics including validation status information and based on the Probe POHC Reference Directory described above. This database will be available on the EMTIC BBS in the summer of 1993.

5. Environmental Monitoring Methods Index (EMMI), a database compiled by EPA to facilitate selection of methods for determination of analytes in environmental media (water, waste, air). This database is updated regularly and eventually will be available for use by the public. Consult the EMTIC BBS for further information.

SAMPLING AND ANALYSIS PRECAUTIONS

The complexity and extent of the currently available techniques for sampling and analysis of air toxics preclude an exhaustive discussion of the techniques and concerns inherent in their application; however, the authors believe it beneficial to address the general techniques and major areas of consideration in employing existing emission measurement methodologies.

Sampling Concerns

The initial, and probably most critical, element in air emission testing is collection of the sample. The important considerations in sample collection include obtaining a sample representative of the emission gas, adequate and known collection efficiency of the sampling media, reducing sample contamination, and ensuring sample integrity is maintained through collection, sample recovery and storage, and analysis.

Collecting a Representative Emission Sample

To ensure a representative emission sample, the sampling probe often is moved from point to point in the duct according to a specified pattern, called traversing. Traversing is intended to compensate for stratification of the sample material in the duct that can be caused mechanically or thermally. EPA Method 1 (40 CFR Part 60, Appendix A) provides specifications for designing the sampling pattern. Stratification generally is of more concern in sampling for particulate pollutants; sampling for gaseous pollutants is routinely conducted at one to three points including the centroid of the duct.

Also of importance in obtaining representative samples of particulate pollutants is isokinetic sampling. Sampling is isokinetic when the velocity of the sample gas entering the sampling train matches that of the sample gas passing the nozzle. A mismatch of the velocities results in anisokinetic sampling and over- or undercollection of particulate matter (see Footnote 2). This effect generally is not significant for gases or particulate matter less than 2 μm in diameter.

Representative sampling for gaseous species requires a sample collection rate in constant proportion to the total gas flow rate in the duct. Steady state or constant stack flow rate conditions necessitate a constant sampling flow rate. For a nonsteady state process, the sampling flow rate must be varied in proportion to the changing duct flow rate. The duct flow rate is directly proportional to the sample gas velocity that, in turn, is related linearly to the square root of the velocity pressure (typically measured near the sampling point).

Another consideration in collecting an emission sample is the conditioning required to get the analyte successfully to the sampling media. Conditioning often includes some combination of heating, filtration, cooling, or condensation.

Care also must be taken in design of the sampling system and procedures to prevent contamination of the sample. For example, the sample-exposed components of the sampling system should not be constructed of metal when measuring trace metal emissions and halogenated volatile organics and sealants must not be used in assembling sampling systems used to measure trace organic compounds. Procedures for preparation and cleaning of the sampling system before sampling should be directed toward eliminating the presence of the analyte(s) and any interferents in the sampling system. Sampling reagents must be chosen to avoid introduction of the analyte or interferents into the sample. Sampling procedures must ensure that nothing except the sample gas enters the sampling system. For instance, if improperly handled the VOST discussed earlier may be extremely susceptible to contamination from organic compounds in the atmosphere at the sampling location.

Blank samples, including reagent blanks and field blanks that are addressed in more detail in the section on QA/QC, often are used to assess the levels of sample contamination. Blanks cannot indicate all types of contamination, such as that resulting from interaction of the sample matrix with the sample-exposed components of the sampling system.

The sampling system also must be capable of collecting the target analyte(s) with reasonable efficiency. Ideally, collection efficiency would be nominally 100 percent; however, for a number of reasons, 100 percent collection efficiency is not always possible. Evaluation of actual collection efficiency is important. In the case of an analyte collected in a series of impingers, the relative level of the analyte collected in the last impinger of the series provides a measure of the sampling system collection efficiency. A relative level of 1 percent indicates good collection; less than 10 percent often is considered acceptable.

Sorbents used for sample collection should have adequate affinity for the target analyte under sampling conditions, the sorbent should have sufficient weight capacity for the analyte, and volumetric breakthrough should not occur under sampling conditions. Some of the sorbents currently in use for air toxics sampling include XAD-2™, Tenax GC™, charcoal, silica gel, and Florisil™; each has its strengths and weaknesses. For example, charcoal, silica gel, and Florisil™ are sensitive to high moisture conditions (Johnson and Merrill 1983). Weight capacity is related to saturation of the resin with sorbed material. Volumetric breakthrough involves the migration of sorbed material through unsaturated sorbent beds (Johnson and Merrill 1983) and may be assessed through measurement of the analyte captured by 'back up' sorbent modules added to the sampling system during field testing or determination of recoveries for compounds spiked on the sorbent before sampling. Many sorbents require cleaning before use and all should be analyzed for blank levels of interfering compounds.

Collection of air toxics samples in plastic bags (e.g., Tedlar™) or passivated stainless steel canisters (e.g., SUMMA™) allows for multiple analyses and dilutions, as well as storage and analysis of the sample gas in its original state. However, stability characteristics of the specific compound may limit applications. For example, polar organic compounds, such as alcohols, generally show poor stability in sampling bags due to adsorption on the bag walls. Sample gases containing acid gases can react with the metal walls of the SUMMA™ canister and be lost. High-temperature and high-moisture gas streams should not be sampled using these containers because of the high potential for condensation.

Artifact formation is one of the more difficult sampling problems. Reaction of components of the sample gas stream with elements of the sampling system or with each other may produce compounds not emitted from the source; further, compounds emitted from the source may be destroyed by unwanted reactions in the sampling system. There are several published accounts of sample reactions on filter substrates and some work has been done to define the sensitivities of sorbent resins (Gallant et al. 1978; Piecewicz, Harris, and Lewis 1979; Harris, Miseo, and Piecewicz 1982). The possibility of artifact formation always should be considered in applying emission measurement methods.

Sample Recovery/Storage Concerns

Following collection of an acceptably representative sample, consideration must be given to: (1) quantitative recovery of the analyte(s) from the sampling system, and (2) storage conditions necessary to preserve sample integrity. Filter samples requiring gravimetric analysis must be recovered to include the entire tare-weighed filter and the particulate matter captured thereon. The sample recovery protocol should be designed so that the sample collection reagents, rinse solvents, and sample train cleaning procedures are adequate to remove the analyte of interest from the sampling system. In particular, the recovery properties of specific sorbent/solvent/analyte combinations must be assessed. Interference from components of the sample gas stream always should be considered in the course of determining sample recovery characteristics.

Critical elements of sample storage include container selection and storage conditions. Sample containers should be chosen and prepared to prevent both sample contamination and degradation. Glass containers typically are a good choice for both organic and metals samples because glass is inert and easy to clean. However, sometimes low concentrations of metals may either sorb to glass or leach out of glass; in this case, very clean inert plastics may be preferable. For example, Method 0012 specifies polyethylene containers for certain types of samples. Teflon™ is often the choice for the container seals that must be secure to avoid sample loss and contamination. Organic samples that are light-sensitive are best stored in amber glass, placed in a dark area, and in many cases, also refrigerated. Sorbent sample storage must protect against sample migration (and possible loss). Samples must not be stored in the vicinity of significant quantities of the target analyte(s) or corresponding interferents. Data should be obtained regarding acceptable storage times in relation to sample stability. Section 8 of Method 301 offers guidance in this area.

Analysis

Sample Preparation Concerns

Sample preparation includes those procedures necessary to prepare a sample for analysis and typically is dependent on the sample matrix and the target analytes. Sample preparation generally involves some combination of physical operations (e.g., grinding, blending, sieving, filtering, drying, and phase separation) and chemical operations (e.g., dissolution, digestion, pH adjustment, extraction, fractionation, derivatization, and addition of preservatives). Evaluation and modification of sample preparation procedures are particularly important in cases where an analytical methodology applicable to waste, water, or other media samples is being adapted for analysis of air emission samples.

Although the instrumental analysis of standard solutions in the majority of cases will yield data of known quality, the preparation of the sample before analysis is an additional critical element that can greatly affect the quality of the data. Potential effects of sample preparation include contamination (e.g., from glassware, reagents, and in handling), physical losses (e.g., in glassware or other containers such as sieves, inadequate recovery from sampling media, during transfer operations, and from volatilization), chemical losses (e.g., analyte decomposition, reaction with containers, and reaction with other components of sample matrix or with reagents), and inadequate separations. Such effects may result in positive or negative biases, and increased variance. These effects must be considered carefully in the development and application of an analytical protocol. Additionally, each step of the protocol should be documented. Quality assurance samples including blanks should be subjected to the same preparation steps as the field samples for use in identifying and assessing these effects if they should occur.

Sample Analysis Concerns

From the discussion of available techniques for measuring air toxics emissions, there are several analytical techniques that have extensive application. These analytical techniques are addressed briefly below, along with some general concerns in emission sample analysis. The major emphasis is on instrumental instead of wet chemical techniques because instrumental analytical techniques tend to be more universal and are much more sensitive. Additional information on these techniques can be found in the bibliography.

The most prevalent techniques for trace organic analysis are gas chromatography (GC), liquid chromatography (LC), and mass spectrometry (MS); for trace inorganic analysis, the prevalent techniques include atomic absorption spectroscopy (AAS), inductively coupled plasma emission spectroscopy (ICP), and ion chromatography (IC). Combinations of these techniques include GC/MS, LC/MS, and ICP/MS.

Gas chromatography gained widespread use in the late 1950s and is used widely for analysis of volatile and semivolatile organics in a gaseous, liquid, or solid state. However, high molecular weight compounds, certain ionic compounds, highly polar compounds, and thermally unstable compounds cannot be directly analyzed by GC (Johnson and James 1988). The sample is injected into the GC, vaporized, and moved by a carrier gas (the mobile phase) into a column. The column contains the stationary phase, either an active packing material (i.e., packed column chromatography) or a liquid coating on a solid support or the column walls (i.e., capillary column chromatography). The packing or liquid coating has different affinities for the components of the sample causing them to pass through the column at different rates, related to retention times. The various components exiting the column are measured by a detector. Gas chromatography detectors include thermal conductivity (TCD), flame ionization (FID), electron capture (ECD), flame photometric (FPD), Hall electrolytic conductivity (HECD), nitrogen phosphorus (NPD), photo-ionization (PID), FTIR, and MS.

Because GC identifies compounds only by retention time and more than one compound may have the same retention time, particularly in complex sample matrices, GC often is used in combination with MS. MS can accommodate the same types of compounds as GC and provides information on the molecular structure of the compounds in a sample. Mass spectrometry is a newer technique, gaining wide application in the early 1960s. In the mass spectrometer, the sample is vaporized under a high vacuum, then bombarded by a beam of electrons accelerated from a filament that ionizes the molecules of vapor by removal of electrons, forming positively charged molecular ions. Other ionization techniques such as chemical ionization are used less frequently. The molecular ions are accelerated into an analyzer where they are separated according to their mass-to-charge ratios. The mass-to-charge ratio coupled with the relative abundance of each ion is used to deduce the molecular structure of the compounds in a sample. In total ion current mode, MS can be used as a scanning technique to identify components of a sample; in selected ion monitoring mode, MS is used for quantitation of specific components at extremely low concentrations. The most common application of MS techniques uses GC as the sample introduction system.

High performance liquid chromatography (HPLC) often is used for analysis of the low volatility or labile organic compounds that cannot be measured by GC. The mobile phase is a liquid, which carries the sample through the column containing the stationary phase, a solid. Separated sample components exit the column in inverse order of their retention on the column and enter the detector where they are quantified. Detectors include ultraviolet (UV), fluorescence, and electrochemical, with the UV detector being the most universally applicable. Mass spectrometry sometimes is used in combination with LC offering advantages parallel to GC/MS.

Atomic absorption spectroscopy (AAS) is the most widely used technique for quantitative analysis of trace metals. Species in a sample solution are quantified based on the absorption of specific wavelengths of light produced by a special lamp and measured by a photomultiplier detector. In this method, the analyte(s) in the sample solution is introduced into the optical path of the spectrophotometer often in atomized form. Sample introduction/activation systems include flame (FAAS), graphite furnace (GFAAS), hydride (HAAS), and cold vapor (CVAAS); the latter can be made highly selective for mercury.

Inductively coupled argon plasma emission spectroscopy (ICP), a technique introduced in the 1970s, also is applicable to trace metals, along with most nonmetal elements. In ICP, a liquid sample is introduced into the center of a gaseous plasma (usually an argon plasma) where the sample is dissociated into its atomic form and excited to high energy levels. The excited species emit characteristic light radiation as they drop to the atomic and ground states. The emitted light is collected and transferred to a wavelength selector that simultaneously or sequentially directs the selected wavelengths to the photomultiplier detector(s). ICP generally has replaced the older arc or spark activated emission spectroscopy tech-

niques. ICP is usually more cost-effective than AAS for analysis of samples for more than three elements, but GFAAS is capable of lower detection limits.

Ion chromatography (IC) utilizes anion and cation exchange columns for separation of inorganic and organic sample ions. The general application of this technique began in the mid-1970s. For anion analysis, a liquid sample is injected and carried onto an anion exchange column by a basic eluent. In the column, the anions in the sample are retained by displacing the eluent ions from the exchange resin. The retained ions are later displaced; the time required for this displacement depends on the ability of the retained ions to compete with the anions of the eluent for the resin exchange sites. After their release from the column, the separated anions are measured by a detector. Although the conductivity detector is the most widely used, other detectors are available including spectrophotometric and fluorescence. Many IC systems include a postcolumn device for chemical suppression of the background noise from the eluent; chemical suppression can increase the sensitivity at least a factor of ten. Postcolumn reactor systems have been used to alter the detected species for purposes of increasing either sensitivity or selectivity. Preconcentration techniques used before injection onto the column can increase sensitivity by up to three factors of ten. Cations are determined in a manner similar to anions.

Radioactivity in emission samples is measured by counting either alpha, beta, or gamma radiation. Although there are a number of methods for detecting these radiations, almost all are based on the concept of capturing the energy released through radioactive decay and converting it to a current flow or pulse that can be recorded. The instruments used to measure radiation make use of its ionizing properties by providing a medium with which the radiation interacts and a means of detecting this interaction. These instruments generally are classified by the type of medium: gas ionization detectors, scintillation detectors, and solid state detectors. Gas ionization detectors measure on the basis of electrical changes occurring in two collecting electrodes exposed to an enclosed volume of gas that is ionized by the radiation. Scintillation detectors are based on the fact that ionizing radiation produces flashes of light or scintillations in phosphors that can be detected by a photomultiplier tube. Solid state detectors employ the property of some insulators to become conductive when struck by radiation. One unique technique used for measurement of radon-222 utilizes microscopic analysis of sub-micron damage tracks made by alpha particles on a plastic strip.

Selection of the appropriate measurement technique for radionuclides depends on the type of radiation to be measured, the radionuclide composition of the sample, the energy of the radiation, the condition of the sample, the activity level of the sample, and the desired accuracy. Particulate radionuclides may be collected on filter media and subsequently analyzed; gaseous species may be measured using continuous techniques or collected from an effluent sample by sorption, condensation, dissolution, or oxidation, and subsequently analyzed. Because of the low activity of most radionuclide emission samples, minimization of the effects of background radiation such as cosmic rays is very important.

General areas of sample analysis concern include analyte decomposition during analysis, analyte reaction during analysis, improper application of analytical techniques because of unknown analyte properties, positive or negative interferences, cross-contamination from analyzing high- and low-level samples in succession, contamination introduced by the mobile phase during liquid or gas chromatography, and inadequate detection limits or method performance for DQO.

Importance of QA/QC

Use of appropriate QA/QC procedures is imperative in assessing and documenting data quality. QA/QC procedures also are beneficial in identifying and correcting sampling and analytical problems as they occur. Quality control typically is defined as the system of

activities designed to provide a high-quality product, whereas quality assurance is the system of activities designed to provide assurance that the quality control system is performing adequately (EPA 1976b).

Preparation of, and adherence to, a quality assurance project plan (QAPjP) can be instrumental in achieving an effective system of QA/QC activities for a particular emission measurement exercise. QAPjP development promotes early definition of DQO and compels consideration of how the QA/QC activities will act to ensure that these objectives are met. A typical QAPjP for an emission test project (EPA 1980b) addresses project objectives; project and QA organization and responsibilities; data quality objectives in terms of accuracy, precision, completeness, representativeness, and comparability; test site selection and sampling procedures; sample custody; analytical procedures; calibration procedures including frequency; data reduction, validation, and reporting; internal QC checks; performance and systems audits; preventive maintenance; calculation of data quality indicators; corrective action; and QC reports to management. Some of the principal QA/QC activities typically associated with air emission sampling and analysis are discussed by Jackson and Midgett (1991) and Segall et al. (1991), and are summarized below.

The sampling equipment typically is calibrated before testing; the equipment includes gas metering devices, temperature sensors, pitot tubes, nozzle diameters, barometers, and balances. QC limits for calibration tolerances can be specified in established methods or set forth in the QAPjP. To avoid compromising emission results through sample contamination, sample containers and sample-exposed surfaces of the sampling equipment should be cleaned in a manner compatible with the target analyte(s) and sampling reagents checked for background levels of the analyte(s). A check for contamination introduced through these combined pathways can be conducted by analyzing proof rinse samples; the sampling system is charged with the reagents, the reagents are recovered and stored in the sample containers, and they subsequently are analyzed for the target analytes. In general, the blank analyte levels should be below the analytical detection limit.

The major QC activities conducted during the sampling portion of an emission test project are collection of reagent blanks, collection of field blanks, spiking of the sampling system, performance and systems audits, and calibration of instruments to be used on site. Reagent blanks collected in the field are analyzed to determine background contamination of the reagents and solvents used in sampling and sample recovery. Field blanks typically consist of sampling trains that are charged with reagents, assembled, transported to and from the sampling location with the other trains (but not used for sample collection), then recovered and analyzed to assess contamination introduced in handling during testing. Field spiking of the sampling system may be done to assess sample recovery under field conditions. When continuous instrumental measurements are conducted, calibrations and other system checks are conducted using certified calibration gases.

Systems audits are a qualitative evaluation of all components of a measurement system to determine proper selection and use. Components of a measurement system include facilities, equipment, calibration procedures, operating procedures, maintenance procedures, record keeping, data validation, reporting, and internal QC checks. Performance evaluation audits quantitatively evaluate the accuracy of a measurement system. Performance audits conducted during the sampling phase of an emission measurement project include audits of dry gas meters, barometers, and temperature sensors, as well as cylinder gas audits of any continuous monitoring instrumentation.

Key QA/QC activities during the analytical phase of an emission measurement project are instrument calibration; analysis of instrument calibration check samples, interference check samples, matrix spike samples, duplicate samples, and method blanks; addition of internal, surrogate, or recovery standards; and performance and systems audits. Calibration proce-

dures are specific to the particular analysis and instrument; however, as a rule the calibration curve must be in a range to bracket the sample analysis. Calibration check samples are analyzed at regular intervals to confirm the calibration. Interference check samples are analyzed in cases where a species has a potential to interfere with analysis of the target analyte. Spikes of the sample matrix are conducted to assess matrix effects on the analysis of the target analyte at the sample level. Analysis of duplicate samples provides an assessment of analytical precision. Method blanks are prepared from the laboratory reagents used and analyzed to determine possible laboratory contamination. GC/MS analyses for trace organic compounds often include addition of isotopically labeled or internal standards to the samples before extraction. The recoveries of these compounds can be used to assess the efficacy of sample extraction and analytical procedures; in some cases, the recoveries of the internal standards are used to adjust results for the native species in the samples. Isotopically labeled recovery standards, for example, are used in the GC/MS analysis of dioxins and furans and are added to the sample extracts immediately before their injection into the instrument; recoveries of these compounds are used in the calculation of the internal standard recoveries. Whenever appropriate audit samples are available, it is beneficial to analyze them along with the field samples to assess the analytical accuracy.

QA/QC in data reduction and reporting involves systems audits of data recording, transcription, and calculations; data validation; and calculation of data quality indicators. The systems audits involve qualitative evaluation of each step of the data reduction/reporting process. Data validation includes evaluation of data for consistency and reasonableness based on knowledge of the site characteristics and reasonable data ranges, examination of QC data in relation to control limits, and evaluation of performance audit results. Lastly, calculation of data quality indicators involves determining, where possible, the accuracy, precision, and completeness of the emissions data.

EMERGING MEASUREMENT TECHNOLOGIES

Traditional laboratory instrumental techniques for identifying and quantifying specific air toxics are under evaluation by EPA, as well as other instrumental methods or CEMS that may be applied under stack conditions in response to the need for air toxics compliance testing methods, for continuous compliance determination methods, and in support of potential air toxics emissions trading regulations. The following are brief descriptions of the principles of operation for some of the air toxics instrumental methods and CEMS being investigated (A&WMA 1990; Draves, Dayton, and Bursey 1991).

Gas Chromatography

Gas chromatography (GC) may be applied to a VOC CEMS to allow separation of specific organic air toxics before detection. Chromatographic separation results from the differential transport rate of compounds through a medium or column that adsorbs or absorbs the compounds of interest. Typical columns include DB-624, DB-5, Megabore™, fused silica, or VOCOL™ up to thirty meters in length. These and other chromatographic columns can adequately separate on the order of five VOC air toxics and allow compound-specific analysis and quantification on a repetitive basis. Detection devices used in combination with separation columns include FIDs, PIDs, mass spectrometers, Fourier transform infrared (FTIR) analyzers, and ECDs. Gas chromatography CEMS are extractive by design as gas conditioning and column separation are necessary. Appendix A of 40 CFR Part 64 will include performance specifications (Performance Specification 102) for installing and evaluating CEMSs employing gas chromatographic principles.

Mass Spectrometry

Mass spectrometry (MS) may be used to identify and quantify individual species based on mass and characteristic patterns resulting from ion decomposition reactions. Gaseous molecules are ionized through exposure to an electron beam and are separated by mass. Most compounds have distinctive resulting mass spectra. Separation and identification of mass spectra can be enhanced through the use of chromatographic columns in conjunction with mass spectrometry (GC/MS).

Further refinement of MS includes the combination of mass spectrometry/mass spectrometry (MS/MS) in which ions are formed and focused in an initial mass spectrometer. The ions of a specific mass then are reacted with a neutral gas in a collision cell to form a series of specific product ions to be monitored by a second mass analyzer. Ion trap detector technology also may be applied to generate the ion-molecule reactions that are characteristic of MS/MS. In this technology, radio frequency voltages are applied to a ring electrode around an ion trap chamber which traps stable ions for several seconds. A conventional mass spectrometer then can scan the chamber for the ion of interest. MS/MS technology is applicable only for extractive CEMSs because of the sample conditioning required. The MS/MS techniques offer separation advantages over GC/MS or high-resolution GC/MS in species separation and detection.

Infrared Spectroscopy

The principle of infrared (IR) spectroscopy involves excitation of the vibrational modes of molecules. Infrared techniques are sensitive to molecules only and will not detect metals or homonuclear diatomic compounds (e.g., O_2, N_2, or Cl_2). The strengths of excitation in the IR range, about 200 to 4,000 cm^{-1}, generally are weaker than electronic absorbance signals in the ultraviolet range. IR methods require a spectral source, a detector, and, in most cases, a method for wavelength selection. Following are descriptions of specific IR techniques.

Fourier Transform Infrared Spectroscopy
The Fourier transform infrared (FTIR) approach provides for collection of the entire IR spectrum in a few seconds allowing selection of analysis

regions with low or no interfering spectra. FTIR equipment includes a standard interferometer, associated optical support, and a computer system. A typical extractive FTIR system includes a path length of about twelve meters achieved using multiple IR beam reflections in a contained sample cell. Path length for in-situ CEMSs without multiplication through reflective techniques is limited to the width of the stack. Sensitivity at path lengths across a stack is limited to about one to five ppm and may not be sufficient for some air toxics measurement situations. The frequency of measurement at about 1,800 measurements per hour is sufficient for CEMS applications. Reference spectra for air toxics measurable with FTIR and standard protocols for their measurement are under development.

Sensitivity of the FTIR extractive technique may be enhanced through combination with matrix isolation (MI/FTIR). In MI/FTIR, the extracted sample is frozen (at about 4K) in a matrix of nitrogen gas that narrows the inherent spectral features of the species. Narrowed spectra allow for significant interference reduction and extended observation time producing better detectability. Detection levels can be less than 1 ng/m^3 with measurement frequencies of about 300 per hour.

A further extension of the MI/FTIR technique is a combination of gas chromatography and MI/FTIR. Addition of GC provides species separation before quantification and removes potential spectroscopic interferences. Longer scan times are possible than with MI/FTIR, further improving signal-to-noise ratios. The GC/MI/FTIR has a measurement frequency of about 6 per hour and, at present, requires extensive maintenance for the GC separation equipment, the matrix isolation system, and the computer timing and analysis considerations.

IR Laser Absorption

Carbon dioxide and diode lasers that operate in the IR region of the spectrum provide advantages over FTIR. Laser absorption provides a high degree of spectral resolution allowing for more accurate identification over a smaller spectral range and a larger signal-to-noise ratio

because of the laser light source intensity. A diode laser produces a tunable spectral beam over about 10 cm^{-1}; the carbon dioxide laser produces discrete spectral lines that are not easily shifted.

The laser absorption technique is not yet well developed. No vapor phase spectra exist and identification of appropriate laser wavelengths is difficult. Diode lasers are limited in range requiring several lasers to be used in tandem to scan a significant portion of the IR range. Frequency of measurement should be about 60 per hour.

Gas Filter Correlation

Gas filter correlation (GFC) is a nondispersive IR technique that provides a concentration measure based on characteristic absorbance frequency modulation. A vapor phase reference sample of the compound of interest is necessary. Infrared light passed through the reference gas is diminished at the wavelengths that coincide with the absorbencies of the molecules of this compound. Exposing the IR light to a sample containing the same compound as the reference sample further diminishes the light intensities at these wavelengths. A detector measures the light intensity changes as the IR beam is cycled through reference and sample cells.

More than one compound can be monitored with the same light source and detector with multiple reference and heated sample cells. Detection limitations for GFC techniques can provide only low-ppm-level measurements.

Ultraviolet Absorbance Spectroscopy

Ultraviolet (UV) spectroscopy is a well-developed technology that uses a broadband light source (e.g., xenon arc or quartz iodide lamps) or a discrete-line source (e.g., high-pressure mercury arc lamp) to produce electronic excitation of atoms or molecules in the 200 to 400 nm spectral range. Laser approaches to producing specific UV absorbance spectra also are under investigation. Ultraviolet spectroscopy applied to some air toxics compounds provides better sensitivity than infrared technology because of stronger band absorbance; however, not all air toxics absorb in the UV region.

Water and carbon dioxide that can interfere significantly with infrared techniques provide negligible interference in the UV region. Other stack gas constituents can produce interferences. These stack gas constituents include polycyclic aromatic hydrocarbons, dioxins and furans, and compounds attached to the surface of particulate matter. Ultraviolet spectrometers can provide quantification of specific compounds in about one minute and detection limits are below $100\,ng/m^3$ for most air toxics that absorb in the UV region.

Fluorescence techniques provide a measure of the excitation of the molecule of interest from an excited state produced through photon absorption. The photon is produced from a broadband light source at the same or lower frequency and the photon emission or fluorescence is detected using a spectrophotometer. Measurement considerations include path length, fluorescence volume, and detector angle. Gas phase fluorescence spectra are not widely available. No significant water or CO_2 interferences exist with the fluorescence technique. Measurement frequency is about 60 per hour and sensitivity is less than $100\,ng/m^3$.

Enhanced molecule excitation and specific compound identification is possible through laser induced fluorescence (LIF). A laser light source can be more selective regarding the species that are excited and the species will be excited in a narrow band making emission spectra easier to interpret. The laser light source can be an exciter-pumped dye laser or an argon laser, and a photomultiplier tube is used for a detector.

Fluorescence and LIF techniques are in the development stage at this time. Application to CEMS will depend on development of durable equipment and reference spectra.

Other New CEMS Technologies

Several laboratory analytical techniques are being considered for air toxics CEMS applica-

tion. Development work is focusing on ruggedness testing and preparing necessary individual air toxics spectra or species-specific reference materials. These technologies are described in brief.

Photoacoustic spectroscopy involves excitation of the molecules of interest with a light source, usually a laser, of known absorption frequency. The absorbed energy converts to heat and subsequently raises the pressure in the closed sample cell. A sensitive microphone or density detecting interferometer then senses the pressure change proportional to the compound concentration.

Multiphoton ionization (MPI) provides a measure of the ionization of the compound of interest through absorption of multiple laser-induced photons. An electrometer detects ions instead of light absorbance or fluorescence.

Shpol'skii spectroscopy relies on an effect that is the result of a geometric structure and an electronic correlation between the molecule of interest and a specific solvent. For air toxics measurement, the sample is subjected to GC separation and a sample containing the molecule of interest is mixed with a carefully chosen alkane and the mixture is frozen to liquid nitrogen temperatures. The frozen mixture is exposed to laser light and the absorption or emissions are monitored with a photomultiplier tube. The resulting detection limits can be less than $10\,ng/m^3$ making the approach attractive for very low concentration air toxics sources.

References[6]

Air & Waste Management Association (A&WMA). 1990. Continuous emission monitoring: Present and future applications. Proceedings of the Air & Waste Management Association Specialty Conference, Chicago, IL. November 1989. Publication No. SP-71, 1990.

Barrett, R.E., W.L. Margard, J.B. Purdy, and P.E. Strup. 1977. Sampling and analysis of coke-oven door emissions. Publication No. EPA 600/2-22-213. Prepared by Battelle-Columbus Laboratories for U.S. Environmental Protection Agency, Office of Research and Development. Research Triangle Park, NC.

[6.] Additional useful reference material is listed in the bibliography.

Bellack, E.A. and P.J. Schoube. 1958. Rapid photo-metric determination of fluoride in water. *Analytical Chemistry* 30:2032.

Bursey, J.T. 1991. Screening methods for the development of air toxics emission factors. EPA 450/4-91-021. U.S. Environmental Protection Agency, Office of Air Quality Planning and Standards. Research Triangle Park, NC.

California Air Resources Board (CARB). 1989a. Stationary source test methods, Volume III: Methods for determining emissions of toxic air contaminants from stationary sources. Monitoring and Laboratory Division. Sacramento, CA.

Chemical Manufacturers Association (CMA). 1989. Improving air quality: Guidance for estimating fugitive emissions from equipment. Washington, D.C.

Draves, J.A., D. Dayton, and J.T. Bursey. 1991. Innovative sensing techniques for monitoring and measuring selected dioxins, furans, and polycyclic aromatic hydrocarbons in stack gas. Report prepared by Radian Corporation under Contract No. 68-D1-0010. U.S. Environmental Protection Agency. Office of Research and Development. Research Triangle Park, NC.

Dunlap, K.L., R.L. Sandridge, and J. Keler. 1976. Determination of isocyanates in working atmospheres by high speed liquid chromatography. *Analytical Chemistry* 48(3):497.

Dupont, R.R. 1985. Measurement of volatile hazardous organic emissions from land treatment facilities. *Journal of the Air Pollution Control Association* 37:168.

Edgerton, S.W., J. Kempen, and T.W. Lapp. 1991. The measurement solution: Using a temporary total enclosure for capture efficiency testing. EPA 450/4-91-020. U.S. Environmental Protection Agency, Office of Air Quality Planning and Standards. Research Triangle Park, NC.

Eklund, B.M., W.D. Balfour, and C.E. Schmidt. 1985. Measurement of fugitive volatile organic emission rates. *Environmental Progress* 4(3):199.

Gallant, R.F., J.W. King, P.L. Levins, and J.F. Piecewicz. 1978. Characterization of sorbent resins for use in environmental sampling. EPA 600/7-78-054, NTIS No. PB 284-347. U.S. Environmental Protection Agency, Office of Research and Development. Washington, D.C.

Gholson, A.R., J.R. Albritton, R.K.M. Jayanty, J.E. Knoll, and M.R. Midgett. 1987. Field evaluation of flux chamber method for measuring volatile organic emissions from hazardous waste surface

impoundments. In: Proceedings of the 1989 EPA/A&WMA International Symposium Measurement of Toxic and Related Air Pollutants. Publication No. VIP-13. EPA 600/9-89-060, p. 358. Published by A&WMA, Pittsburgh, PA.

Harris, J.C., E.V. Miseo, and J.F. Piecewicz. 1982. Further characterization of sorbents for environmental sampling – II. EPA 600/7-82-052. U.S. Environmental Protection Agency, Office of Research and Development. Washington, D.C.

Hendershott, J.P. 1986. The simultaneous determination of chloroformates and phosgene at low concentrations in air using a solid sorbent sampling - Gas chromatography procedure. *American Industrial Hygiene Association Journal* 47(12):742.

Jackson, M.D. and M.R. Midgett. 1991. Importance of quality for collection of environmental samples – The role of quality assurance in minimizing errors in stationary source field sampling. In: Proceedings of the Eighteenth Annual National Energy Division Conference, Danvers, MA, p. 5D-3.1.

Johnson, L.D. and R.H. James. 1988. Sampling and analysis of hazardous wastes. In: Standard Handbook for Hazardous Waste Treatment and Disposal. H.M. Freeman, ed. New York: McGraw-Hill. 1988.

Johnson, L.D. and R.G. Merrill. 1983. Stack sampling for organic emissions. *Toxicology and Environmental Chemistry* 6:109.

Johnson, L.D., M.R. Midgett, R.H. James, M.M. Thomason, and M.L. Manier. 1989. Screening approach for principal organic hazardous constituents and products of incomplete combustion. *Journal of the Air Pollution Control Association* 39(5):709.

Kienbusch, M.R. 1986. Measurement of gaseous emission rates from land surfaces using an emission isolation flux chamber: User's guide. EPA 600/8-86/008. U.S. Environmental Protection Agency, Office of Research and Development. Las Vegas, NV.

Kolnsberg, H.J. 1976. Technical manual for the measurement of fugitive emissions: Upwind/downwind sampling method for industrial fugitive emissions. EPA 600/2-76-089a. U.S. Environmental Protection Agency, Office of Research and Development. Washington, D.C.

Lewis. R.G., B.E. Martin, D.L. Sgontz, and J.E. Howes. 1985. Measurement of fugitive atmospheric emissions of polychlorinated biphenyls from hazardous waste landfills. *Environmental Science and Technology* 19(10):986.

Long, G.L. and J.D. Winefordner. 1983. Limit of detection: A closer look at the IUPAC definition. *Analytical Chemistry* 55(7):712A.

MacDougall, D. et al. 1980. Guidelines for data acquisition and data quality evaluation in environmental chemistry. *Analytical Chemistry* 52(14):2242.

National Institute for Occupational Safety and Health (NIOSH). 1985b. *NIOSH Manual of Analytical Methods*, Third Edition. U.S. Department of Health, Education, and Welfare. Document No. SN-917-011-000001, available from U.S. Government Printing Office, Washington, D.C.

Occupational Safety and Health Administration (OSHA). 1990. *OSHA Analytical Methods Manual*, Second Edition. Catalog Nos. 4542, 4543, and 4544. OSHA Analytical Laboratory, Salt Lake City, UT. Published by American Conference of Governmental Industrial Hygienists, Cincinnati, OH.

Piecewicz, J.F., J.C. Harris, and P.L. Lewis. 1979. Further characterization of sorbents for environmental sampling. EPA 600/7-79-216, NTIS No. PB 80-118763. U.S. Environmental Protection Agency, Office of Research and Development. Washington, D.C.

Rand, M.C., A.E. Greenburg, and M.J. Taras. 1975. *Standard Methods for the Examination of Water and Wastewater*, Fourteenth Edition. American Public Health Association. Washington, D.C.

Segall, R.R., S.A. Shanklin, A.L. Cone, and C.E. Riley. 1991. External QA/QC for medical waste incinerator emission testing. In: Municipal Waste Combustion, Papers and Abstracts from the Second International Conference, published by Air & Waste Management Association, Publication No. VIP-19, p. 485, April 1991.

Smith, W.S., T.M. Grosshandler, and A.L. Mastrianni. 1990. Determination of fugitive VOC emissions for SARA 313 reporting requirements. Paper presented at Air & Waste Management Association Florida Regional Section Annual Meeting.

U.S. Environmental Protection Agency (EPA). 1976b. Quality Assurance Handbook for Air Pollution Measurement Systems: Volume I. Principles. Section No. 1.3. EPA 600/9-76-005. Office of Research and Development. Research Triangle Park, NC.

U.S. Environmental Protection Agency (EPA).

1977a. Quality Assurance Handbook for Air Pollution Measurement Systems: Volume III. Stationary Source Specific Methods. Section 3.0.4. EPA 600/4-77-027b. Office of Research and Development. Research Triangle Park, NC.

U.S. Environmental Protection Agency (EPA) 1979b. Methods for chemical analysis of water and wastes. EPA 625/6-74-003. National Environmental Research Center. Cincinnati, OH.

U.S. Environmental Protection Agency (EPA). 1980b. Interim guidelines and specifications for preparing quality assurance project plans. EPA/QAMS 005-870. Office of Monitoring Systems and Quality Assurance, Office of Research and Development. Washington, D.C.

U.S. Environmental Protection Agency (EPA). 1985a. Compilation of air pollutant emission factors. Volume 1. Stationary point and area sources, Fourth Edition. Publication AP-42. Supplement A (October 1986) and Supplement B (September 1988). Office of Air Quality Planning and Standards. Research Triangle Park, NC.

U.S. Environmental Protection Agency (EPA). 1986k. Test Methods for Evaluating Solid Waste, Physical/Chemical Methods, Third Edition. Publication No. SW-846, Document No. 955-001-0000001, available from U.S. Government Printing Office. Office of Solid Waste and Emergency Response. Washington, D.C.

U.S. Environmental Protection Agency (EPA). 1990d. Methods manual for compliance with the BIF regulations. Publication No. EPA 530/SW-91-010, NTIS No. PB-91-120-006, Office of Solid Waste and Emergency Response. Washington, D.C.

U.S. Environmental Protection Agency (EPA). 1991c. Protocol for the field validation of emission concentrations from stationary sources. EPA 450/4-90-015. Office of Air Quality Planning and Standards and Office of Research and Development. Research Triangle Park, NC.

U.S. Environmental Protection Agency (EPA). 1991m. Regulations for municipal waste combusters. 56 FR 5488. February 11, 1991.

U.S. Environmental Protection Agency (EPA). 1991n. Regulations for boilers and industrial furnaces. 56 FR 7135. February 21, 1991.

U.S. Environmental Protection Agency (EPA). 1991o. Requirements for early reductions in hazardous air pollutants. 56 FR 27370. June 13, 1991.

U.S. Environmental Protection Agency (EPA).

1991p. Regulations for boilers and industrial furnaces. 56 FR 32688. July 17, 1991.

U.S. Environmental Protection Agency (EPA). 1991q. Acid rain regulations. 56 FR 63002. December 3, 1991.

Winberry, W.T., N.T. Murphy, and R.M. Riggin. 1990. Compendium of methods for the determination of toxic organic compounds in ambient air. Publication No. EPA 600/4-89-017, NTIS No. PB 90-116989/AS (also available from Noyes Data Corporation, Parkridge, NJ). Prepared by Engi-neering-Science, Inc. for U.S. Environmental Protection Agency, Office of Research and Development. Research Triangle Park, NC.

Wisner, C.E. and T.W. Davis. 1989. Characterization of fugitive emission rates using a mobile integrating sampler. In: Proceedings of the 1989 EPA/AWMA International Symposium Measurement of Toxic and Related Air Pollutants. Published by Air & Waste Management Association, Pittsburgh, PA. Publication No. VIP-13, EPA 600/9-89-060, p. 155.

10

Emissions Estimation

David R. Patrick

INTRODUCTION

Knowledge of emissions to the air is an essential component of most steps in the assessment and control of toxic air pollutants. Emissions ultimately determine community exposures and public health risks. Emissions information also is required for many federal, state, and local regulatory and information gathering programs. For example, information on emissions of regulated pollutants is a fundamental requirement in all construction and operating permit programs; estimated increases in emissions determine the applicability of regulatory programs that regulate processes and process modifications; and emissions of most of the hazardous air pollutants listed in the 1990 Amendments must be reported under section 313 of the Emergency Response and Community Right-to-Know Act.

Ideally, emissions are determined by actual measurement, as discussed in Chapter 9, using extractive stack sampling procedures or continuous emissions monitoring equipment. However, stack sampling and continuous monitors are expensive, technically demanding, and only applicable to operating sources. Thus, before sources are constructed or modified, where emissions are difficult or impossible to measure accurately, and in many other situations, emissions estimation techniques are required. The purpose of this chapter is to describe typical methods for emissions estimation and to identify the primary resources, both published and computerized, of emissions estimation information.

SOURCES OF TOXIC AIR POLLUTANT EMISSIONS

Sources of air pollutants generally are categorized as 'point' or 'fugitive.' Examples of point sources are stacks, vents, or ducts through which emissions usually are intentionally directed and released. Fugitive sources are nondirected sources such as open doors, windows, storage piles, and equipment leaks. Other common release points include secondary sources and storage, handling, and loading and unloading processes. Secondary sources include wastewater treatment and cooling towers, and are called secondary because the air pollutants are released as a part of a supporting process. In many instances, secondary sources are fugitive

Table 10-1   Source categories for which emission factors have been developed by EPA

1. *External combustion sources*
   1.1 Bituminous coal combustion
   1.2 Anthracite coal combustion
   1.3 Fuel oil combustion
   1.4 Natural gas combustion
   1.5 LPG combustion
   1.6 Wood waste combustion in boilers
   1.7 Lignite combustion
   1.8 Bagasse combustion
   1.9 Residential fireplaces
   1.10 Wood stoves
   1.11 Waste oil combustion

2. *Solid waste disposal*
   2.1 Refuse incinerators
   2.2 Automobile body incinerators
   2.3 Conical burners
   2.4 Open burning
   2.5 Sewage sludge incinerators

3. *Internal combustion engines*
   3.1 Highway vehicles
   3.2 Off-highway mobile sources
   3.3 Off-highway stationary sources

4. *Evaporation losses*
   4.1 Dry cleaning
   4.2 Surface coating
   4.3 Storage of organic liquids
   4.4 Petroleum transport/marketing
   4.5 Asphalt/asphalt cement
   4.6 Solvent degreasing
   4.7 Waste solvent reclamation
   4.8 Tank and drum cleaning
   4.9 Graphic arts
   4.10 Commercial/consumer solvent
   4.11 Textile printing

5. *Chemical process industry*
   5.1 Adipic acid
   5.2 Synthetic ammonia
   5.3 Carbon black
   5.4 Charcoal
   5.5 Chlor-alkali
   5.6 Explosives
   5.7 Hydrochloric acid
   5.8 Hydrofluoric acid
   5.9 Nitric acid
   5.10 Paint and varnish
   5.11 Phosphoric acid
   5.12 Phthalic anhydride

   5.13 Plastics
   5.14 Printing ink
   5.15 Soap and detergents
   5.16 Sodium carbonate
   5.17 Sulfuric acid
   5.18 Sulfur recovery
   5.19 Synthetic fibers
   5.20 Synthetic rubber
   5.21 Terephthalic acid
   5.22 Lead alkyl
   5.23 Pharmaceutical production
   5.24 Maleic anhydride

6. *Food and agriculture*
   6.1 Alfalfa dehydrating
   6.2 Coffee roasting
   6.3 Cotton ginning
   6.4 Feed/grain mills/elevators
   6.5 Fermentation
   6.6 Fish processing
   6.7 Meat smokehouses
   6.8 Ammonium nitrate fertilizers
   6.9 Orchard heaters
   6.10 Phosphate fertilizers
   6.11 Starch manufacturing
   6.12 Sugar cane processing
   6.13 Bread baking
   6.14 Urea
   6.15 Beef cattle feedlots
   6.16 Cotton harvesting
   6.17 Harvesting of grain
   6.18 Ammonium sulfate

7. *Metallurgical industry*
   7.1 Primary aluminum production
   7.2 Coke production
   7.3 Primary copper smelting
   7.4 Ferroalloy production
   7.5 Iron and steel production
   7.6 Primary lead smelting
   7.7 Zinc smelting
   7.8 Secondary aluminum operations
   7.9 Secondary copper operations
   7.10 Gray iron foundries
   7.11 Secondary lead smelting
   7.12 Secondary magnesium smelting
   7.13 Steel foundries
   7.14 Secondary zinc processing
   7.15 Storage battery production
   7.16 Lead oxide/pigment production
   7.17 Miscellaneous lead products
   7.18 Lead-bearing ore crushing and grinding

Table 10-1   Continued

8. *Mineral products industry*	8.21 Coal conversion
8.1 Asphaltic concrete plants	8.22 Taconite ore processing
8.2 Asphalt roofing	8.23 Metallic minerals processing
8.3 Bricks and related clay products	8.24 Western surface coal mining
8.4 Calcium carbide manufacturing	
8.5 Castable refractories	9. *Petroleum industry*
8.6 Portland cement manufacturing	9.1 Petroleum refining
8.7 Ceramic clay manufacturing	9.2 Natural gas processing
8.8 Clay and fly ash sintering	
8.9 Coal cleaning	10. *Wood products industry*
8.10 Concrete batching	10.1 Chemical wood pulping
8.11 Glass fiber manufacturing	10.2 Pulpboard
8.12 Frit manufacturing	10.3 Plywood veneer and layout operations
8.13 Glass manufacturing	10.4 Woodworking waste collection operations
8.14 Gypsum manufacturing	
8.15 Lime manufacturing	
8.16 Mineral wool manufacturing	
8.17 Perlite manufacturing	11. *Miscellaneous sources*
8.18 Phosphate rock process	11.1 Forest wildfires
8.19 Construction aggregate processing	11.2 Fugitive dust sources
8.20 Reserved	11.3 Explosives detonation

Source: AP-42 (EPA 1985a).

sources. Storage, handling, and loading and unloading processes can be either point (i.e., vents from a storage tank or loading dock) or fugitive (i.e., leaks). Examples of source categories for common releases to the air are shown in Table 10-1.

Over the past twenty or more years, air pollution regulatory agencies have developed accurate measurement techniques for many point source emissions. These methods have been published by EPA in 40 CFR 60 and 61; many are described in Chapter 9. Fugitive emissions measurements are less well defined owing both to the difficulty in measuring a diffuse source of air pollutants and the extreme variability of fugitive emission sources. Techniques that are available to measure fugitive emissions are described in Chapters 9 and 22.

As discussed in EPA (1987b), process vents are a primary source of air emissions in manufacturing and processing operations under normal conditions. Process vents are intended to release excess gas, heat, and waste materials from the process. Process vents also include emergency venting devices, including pressure relief valves. Point source emissions often are released through tall stacks, have a higher temperature than the ambient, and are forced through use of fans. Each of these factors leads to increased height and dispersion in the atmosphere, directing pollutants away from the nearby population.

Fugitive emissions, on the other hand, typically are emitted close to the ground and nearby population can be exposed before the emissions significantly disperse. Although the total fugitive emissions from a source often are less than the point source emissions, they often can be more significant from a public health standpoint. Fugitive emissions sources include process leaks, evaporation from open processes and spills, chemical loading and unloading losses, and windblown losses from materials storage piles. Fugitive emissions are particularly difficult to measure because of their diffuse nature. EPA (1989i), developed for treatment, storage, and disposal (TSDF) facilities, is useful for estimating emissions from container loading, storage, and cleaning; waste treatment and disposal operations; and equipment leaks in the synthetic organic chemical manufacturing industry.[1]

Another useful method for estimating process fugitive emissions (EPA 1987b) is plant air measurement data. In many instances, health and safety regulations require routine measurement of regulated air concentrations. The lists of pollutants regulated in the work place (ACGIH 1989) contains many of the 189 hazardous air pollutants listed in the 1990 Amendments.

Important sources of gaseous and particulate matter emissions are material handling, storage, loading, and unloading operations. These sources can be either vented or fugitive. Typical vented emissions are breathing and working losses. Breathing losses arise from the expansion and contraction of the liquid owing to temperature change; working losses arise when liquid is withdrawn or added to the tank. When the liquid level in the vessel drops, liquid evaporates to maintain equilibrium between the liquid and vapor. When the liquid level rises, vapor is forced out of the vessel through a vent. The same principles apply to loading and unloading of liquids (i.e., loading liquid into a vessel forces vapor out and withdrawing liquid from a vessel allows more liquid to evaporate).

Leaks from vessels, pipes, valves, and other equipment have been studied since the late 1970s. The search for routine leaks rarely was conducted before that time, but the significant increase in petrochemical costs after the 1973 oil embargo led to research aimed at conserving raw materials; many leaks were discovered where none had been suspected. Classically, process engineers believed that if they did not see or smell a leak, then it could not be significant. The research of the late 1970s proved that premise wrong. In fact, equipment frequently leaked without visible evidence and the total amounts of the leaks were found in many cases to be highly significant. Considerable research has gone into the subject since that time. The results are discussed in detail in Chapter 22.

As noted above, secondary emissions arise from waste treatment and handling. In particular, chemical contaminants in wastewater frequently volatilize during treatment processes. Many processes (e.g., aeration to facilitate biological degradation) enhance the volatilization. Estimating releases from secondary sources can be complex and substantial research has been conducted to measure these emissions. A number of analytical models have been developed, and are described in the TSDF report referenced above. For example, models are described for the following.

■ Nonaerated impoundments, including quiescent surface impoundments and open-top tanks
■ Aerated impoundments, including aerated surface impoundments and aerated tanks
■ Disposal impoundments, including nonaerated disposal impoundments
■ Land treatment
■ Landfills

EPA (1987b) notes that computerized methods are under development. One integrated spreadsheet, CHEMDAT4, allows the user to calculate the partitioning of volatile compounds among different pathways based on the parameters of the facility of interest.

EMISSIONS ESTIMATION TECHNIQUES

There are four principal techniques typically used to estimate emissions to the air: emission factors, engineering calculation, mass balance, and predictive emissions modeling. The choice of the technique to be used is based on the estimator's knowledge of the process, his or her educational background, the quantity of readily available data, the complexity and size of the process, and many other factors. Each of these techniques is discussed below.

---

[1] For information, call (919) 541-5373.

Emission Factors

Using emission factors is desirable as a first step in emission estimation because it requires only nominal understanding of the process in question and few technical resources are required. An emission factor is a tool for estimating typical emissions and is defined as 'an estimate of [the average] of the rate at which a pollutant is released to the atmosphere as a result of some activity (such as combustion or industrial production) divided by the level of that activity' (EPA 1985a). In most cases, these factors are used without information on process parameters (e.g., temperature and reactant concentrations). However, in a few cases, including estimation of volatile organic emissions from petroleum storage tanks, it has been possible to develop empirical formulas that relate emissions to such variables as tank diameter, liquid storage temperature, and wind velocity.

An emission factor is only as good as the data or other information on which it is based. Emission factors typically are based on source test data, process data, engineering estimates, or a combination of these. The number of data points, the variability of the process and resultant emissions, along with the quality of the data used in calculating the emission factor must all be considered in determining if an emission factor is adequate for the intended application. In some cases, published emission factors include associated ratings to provide information on the suitability of the factors for a specific purpose. Emission factors ordinarily are developed to describe typical emission sources in a category and may not yield accurate emission estimates for a particular facility. In general, emission factors are most appropriate for use in air quality management applications such as diffusion models, to estimate the impact of proposed or new emission sources, and for community or nationwide air pollutant emission estimates.

Emission factors typically are given in units of weight of pollutant per weight of product produced, weight of pollutant per unit of time, or weight of pollutant per volume of air exhausted. Emission factors have been developed for many years by EPA and are published in AP-42, 'Compilation of Air Pollution Emission Factors' (EPA 1985a). AP-42 is the most comprehensive emission factor source available and is updated occasionally by EPA. AP-42 and other emission factor activities are directed by EPA's Emission Factor and Methodologies Section, Emission Inventory Branch in Durham, NC.[2] Table 10-1 lists the processes for which EPA has developed emission factors and that are provided in AP-42.

Other emission factor documents of potential use to those concerned with toxic air pollutants are found in EPA (1986l), EPA (1980d), and Carl et al. (1984). In addition, EPA began in 1984 and continues to publish documents on specific toxic air pollutants and source categories. Table 10-2 lists many of those reports, along with the EPA report numbers and National Technical Information Service (NTIS) order numbers. These reports identify source categories for which emissions have been characterized and the air toxics emissions from specific source categories. The reports include general process descriptions with potential release points and emission factors.

More recently, EPA is providing emission factor information through computerized services.[3] These information data bases are described in a brochure published by EPA, entitled 'Tools for Estimating Air Emissions of Criteria and Toxic Pollutants,' and are summarized below. These data bases can be accessed via modem by anyone experienced in on-line information retrieval. Information on air emission inventories and emission factors is provided through EPA's Clearinghouse for Inventories and Emission Factors (CHIEF). This bulletin board system provides access to

[2]Contact the Info Line: (919) 541-5285.

[3]EPA's information services typically are free to government agencies and nonprofit organizations; however, fees may be charged for other users.

Table 10-2   Locating and estimating emissions of specific substances and source categories

Substance/source category	EPA report number	NTIS order number	Date
Acrylonitrile	EPA-450/4-84-007a	PB-84-200609	1984
Carbon tetrachloride	EPA-450/4-84-007b	PB-84-200625	1984
Chloroform	EPA-450/4-84-007c	PB-84-200617	1984
Ethylene dichloride	EPA-450/4-84-007d	PB-84-239193	1984
Nickel	EPA-450/4-84-007f	PB-84-210988	1984
Chromium	EPA-450/4-84-007g	PB-85-106474	1985
Chromium (supplement)	EPA-450/2-89-002	PB-90-103243	1989
Manganese	EPA-450/4-84-007h	PB-86-117587	1986
Phosgene	EPA-450/4-84-007i	PB-86-117595	1986
Epichlorohydrin	EPA-450/4-84-007j	PB-86-117603	1986
Vinylidene chloride	EPA-450/4-84-007k	PB-86-117611	1986
Ethylene oxide	EPA-450/4-84-007l	PB-87-113973	1987
Chlorobenzenes	EPA-450/4-84-007m	PB-87-189841	1987
Polychlorinated biphenyls (PCBs)	EPA-450/2-84-007n	PB-87-209540	1987
Polycyclic organic matter (POM)	EPA-450/4-84-007p	PB-88-149059	1988
Benzene	EPA-450/4-84-007q	PB-88-196175	1988
Organic liquid storage tanks	EPA-450/4-88-004	PB-89-129019	1988
Coal and oil combustion sources	EPA-450/2-89-001	PB-89-194229	1989
Municipal waste combusters	EPA-450/2-89-006	PB-89-195226	1989
Perchloroethylene and trichloroethylene	EPA-450/2-89-013	PB-89-235501	1989
1,3-Butadiene	EPA-450/2-89-021	PB-90-160003	1989
Sewage sludge incinerators	EPA-450/2-90-009	PB-90-226614	1990
Formaldehyde	EPA-450/4-91-012	PB-91-2181242	1991
Styrene (interim)	EPA-450/4-91-029	PB-92-126788	1992
Medical waste	In Draft in CHIEF		1992

For more information on these and other documents in preparation, call EPA's Info Line, (919) 541-5285.

several tools, described below. To access the CHIEF bulletin board, a personal computer (PC), modem, and communication package is required capable of communicating at 300, 1,200, 2,400, or 9,600 baud. For CHIEF registration information, contact EPA's Info Line at (919) 541-5285.[4]

Through CHIEF, the user can access and download several emission factor and source data bases, including all of AP-42 and EPA's Aerometric Information Retrieval System (AIRS). AIRS is an integrated data base system developed by EPA that replaces all previous data bases, files, and software for storing and retrieving ambient air quality data, stationary source emissions data, and compliance data. A five-volume users manual is available and a quarterly newsletter is published by EPA.

Air CHIEF CD

Air CHIEF CD is a compact disk-read only memory (CD-ROM) system containing reports and data bases (e.g., AP-42, locating and estimating reports, XATEF, and SPECIATE). Air CHIEF CD is updated annually and allows rapid search and retrieval of data by pollutant or source description. A user's manual accompanies the CD. Users need an IBM compatible PC that runs MS-DOS 3.0 (or later), 640 kilobytes (KB) of random access memory (RAM), and a CD-ROM disc drive. The Air CHIEF CD also is available from the

---

[4]CHIEF access numbers are (919) 541-5742 (300, 1,200, and 2,400 baud), and (919) 541-1447 (9,600 baud).

Government Printing Office (Order No. S-N055-000-00398-0, cost $15.00).[5]

XATEF

The Crosswalk/Air Toxic Emission Factor Data Base Management System (XATEF) provides toxic air pollutant/source crosswalk information and air toxic emission factors. XATEF is updated annually. A user's manual accompanies the 5.25-inch high-density diskette. Users need an IBM compatible PC that runs MS-DOS 3.0 (or later), with a fixed disk with at least 20 megabytes (MB) of storage, 640 KB of RAM, and a high-density 5.25-inch external disk drive.

XATEF-Crosswalk

This companion system provides a link between toxic air pollutants and sources, using pollutant names, Chemical Abstract Service (CAS) numbers, Standard Industrial Classification (SIC) Codes, and EPA Source Classification Codes (SCCs). The crosswalk does not contain emission factor information.

XATEF Emission Factors

Emission factors for selected air toxics are available and are identified by pollutant name, CAS number, process description, SIC Codes, emission source description, and SCCs. Brief descriptions of the derivation of the factor, notes on control measures associated with emission factors, and references also are included.

SPECIATE

EPA maintains a clearinghouse, designated SPECIATE, of speciation factors for volatile organic compounds (VOC) and particulate matter (PM). Speciation factors are used to derive emission factor estimates of individual species from factors or estimates of total VOC or PM. SPECIATE presents data by source category and by SCC. The data base is updated annually. EPA (1991e) accompanies the diskette and explains the use of the data management system. SPECIATE is distributed on a 5.25-inch diskette and requires an IBM compatible PC, MS-DOS 3.0 (or later), with a fixed disk with at least 8 MB of storage, 640 KB free RAM, and a high-density 5.25-inch external disk drive.

SPECIATE VOC

This portion of the data base contains (as of the 1991 update) 275 profiles representing twenty-eight source categories. The profiles are listed by SCC and profile name.

SPECIATE PM

This portion of the 1991 update contains 343 profiles representing twenty source categories. Particle size information is provided.

Engineering Calculation

Engineering calculation relies on engineering evaluation of a process and estimation of emissions using standard engineering procedures or engineering equations developed specifically for the purpose. For example, significant effort has been expended over the past fifteen years to develop equations that accurately estimate volatile organic compound (VOC) emissions from petroleum and chemical storage, loading, and unloading operations, as well as fugitive emissions from petroleum and chemical processes. These sources in particular often are too large or diffuse to measure emissions directly. Accurate engineering calculation requires knowledgeable engineers who understand the physical and chemical processes involved and the underlying technologies. The 'Chemical Engineer's Handbook' (Perry and Green 1984) is one of the most important resources for these engineering calculations.

Classical chemical engineering generally is divided into 'unit operations' and 'unit processes.' As described by Shreve (1956), a unit operation involves a physical change connected with the handling of chemicals and allied materials whereas a unit process involves a

---

[5] See footnote 1.

chemical change. Unit operations include fluid flow, evaporation, distillation, drying, mixing, size reduction, and material handling. Most of these operations can be associated with the generation of gases or dusty solids and the release of these materials to the ambient environment. Unit processes include combustion, oxidation, and a wide range of chemical conversion processes; these also can be associated with the generation of gases or solids and their release to the environment. The economic success of today's chemical and allied industries depends to a great extent on the understanding by engineers of the unit operations and unit processes underlying each industry.

Mass Balance

Mass balance involves the quantification of total materials into and out of a process and the assumption that the difference has been lost to the environment through the air, water, and waste. Mass balance is particularly useful when the input and output streams can be quantified; this most often is the case for small processes and operations. For larger industrial facilities or operations, mass balance is far less useful because of the large potential error that can arise in calculating inputs and outputs. In other words, in large processes the calculation of materials 'in process' is so difficult and the amount so potentially large, that an accurate emission estimate is almost impossible to obtain. In many cases, the calculation errors are larger than the potential emissions.

One way to improve the usefulness of mass balances is to reduce the size of the process under consideration. For example, each unit operation or unit process can be viewed as a single entity and a mass balance conducted around it. This reduces the error potential but the input and output values often are more difficult to estimate.

Mass balance techniques and engineering estimates are best used where there is a system with prescribed inputs, defined internal conditions, and known outlets. Mass balance techniques frequently have been applied to volatile organic compound emissions from spray and coating operations, although with varying results. The emission values calculated are only as good as the values used in performing the calculations. For example, small errors in estimating conditions internal to the system (e.g., hood capture efficiency, transfer efficiency, temperature, pressure) can result in larger errors in the final emission estimates. When sampling of input or output materials is required to conduct a mass balance or other engineering calculations, lack of a representative sample also will contribute to uncertainty in the results. In some cases, the combined uncertainty in the mass balance/engineering estimate may be quantifiable; this is useful in determining if the values are suitable for their intended use. In other cases, such as where the representativeness of a sample is not known, the uncertainty of the calculation cannot be fully defined.

Predictive Emissions Models

Emissions models are available that predict emission rates for fugitive releases, landfills, lagoons, open dumps, waste piles, land treatment operations, and other area sources. These are used both in screening studies and to make more in-depth estimates. When used for screening purposes, the models generally utilize data that can be obtained or calculated from information in the literature or that can be assumed with some level of confidence. When used for in-depth estimates, the models require site-specific site and waste characterization data such as vapor diffusion or mass transfer coefficients, physical area of the source, physical parameters of any covering medium, physical and chemical parameters of the waste, and atmospheric conditions.

Some important predictive models are discussed in Chapter 22. Additional information regarding these alternatives to direct emission measurement can be found in several references in the bibliography.

Control Device Efficiency

Accurate estimation of emissions requires an understanding of the efficiency of control devices in place. Air pollution control devices generally operate in one of four ways: (1) contaminants are transferred to another medium where removal is easier (e.g., absorption); (2) contaminants are captured on an adsorbing media (e.g., carbon); (3) contaminants are destroyed or converting to a less harmful state (e.g., oxidation); and (4) contaminants are mechanically collected (e.g., filtration). Principles of control of toxic air pollutants are discussed in detail in Chapter 21.

The efficiency of various control devices can be measured in some cases (e.g., by conducting a stack test before and after the device). In some cases, however, tests are not possible and a mass balance or engineering calculation is required. Under ideal conditions, the most commonly used control devices can achieve control efficiencies exceeding 95 percent, often exceeding 99 percent. Efficiency tends to increase as the concentration of the contaminant increases. Also, the cost of control generally increases as the concentration of the contaminant decreases.

References

American Conference of Governmental Industrial Hygienists (ACGIH). 1989. *Threshold Limit Values and Biological Exposure Indices for 1989–1990*. Cincinnati: ACGIH.

Carl, J.E., et al. 1984. Receptor model source composition library. EPA 450/4-85-002. U.S. Environmental Protection Agency, Office of Air Quality Planning and Standards. Research Triangle Park, NC.

Perry, R.H. and D.W. Green. 1984. *Chemical Engineer's Handbook*. Sixth Edition. New York: McGraw-Hill.

Shreve, R.N. 1956. *The Chemical Process Industries*. New York: McGraw-Hill.

U.S. Environmental Protection Agency (EPA). 1980d. Volatile Organic Compound (VOC) Species Data Manual. Second Edition. EPA 450/4-80-015. Office of Air Quality Planning and Standards. Research Triangle Park, NC.

U.S. Environmental Protection Agency (EPA). 1985a. Compilation of air pollutant emission factors. Volume 1. Stationary point and area sources, Fourth Edition. Publication AP-42. Supplement A (October 1986) and Supplement B (September 1988). Office of Air Quality Planning and Standards. Research Triangle Park, NC.

U.S. Environmental Protection Agency (EPA). 1986l. VOC emission factors for NAPAP emission inventory. EPA 600/7-86-052. Office of Research and Development. Research Triangle Park, NC.

U.S. Environmental Protection Agency (EPA). 1987b. Estimating releases and waste treatment efficiencies for the toxic chemical release inventory form. EPA 560/4-88-002, Office of Pesticides and Toxic Substances. Washington, D.C.

U.S. Environmental Protection Agency (EPA). 1989i. Hazardous Waste Treatment, Storage, and Disposal Facilities (TSDF) – Air Emission Models. EPA 450/3-87-026. Emission Standard Division. Office of Air Quality Planning and Standards. Research Triangle Park, NC.

U.S. Environmental Protection Agency (EPA). 1991e. VOC/PM speciation data system documentation and user's guide, Version 1.4. EPA 450/4-92-027. Office of Air Quality Planning and Standards. Research Triangle Park, NC.

11

Air Dispersion and Deposition Models

James F. Lape

INTRODUCTION

Analysis of the impacts of air pollutant emissions requires determination of the resulting ambient air concentrations and in some cases the rate of pollutant deposition. This information may be obtained through either field monitoring or modeling studies. The use of modeling studies to predict ambient air concentrations and deposition rates has long been recognized as an acceptable and accurate means of evaluating the impacts of air pollutants. The spatial and temporal requirements for analysis of air pollutant emissions can limit severely the use of field monitoring studies; monitoring also can be quite expensive. Modeling studies, on the other hand, provide the only means of analysis for proposed new emission sources and pollution reduction techniques, and for developing emergency response plans for accidental releases. In most cases, modeling provides a highly cost-effective alternative to field monitoring studies.

The purpose of this chapter is to provide a general understanding of air quality modeling and provide guidance and resources for selection and execution of appropriate air models. The guidance follows current Environmental Protection Agency (EPA) recommendations on air quality modeling. First, the nature of air models and air modeling are discussed. Next, the expertise and resources required for the range of air models are discussed. Finally, currently available air models are identified and guidance provided on their selection.

THE NATURE OF AIR QUALITY MODELING

In general, air quality modeling can be conducted using either a physical model or computational techniques. Physical modeling makes use of wind tunnels or water tanks to provide scaled versions of the atmospheric processes. Because of the level of technical expertise and the specialized facilities necessary for physical modeling, this chapter is limited to discussion of computational techniques of air quality modeling.

Based on the computational techniques used, air quality models can be empirical or deterministic. Empirical models employ measurements of gases and particles at both the emission source and receptor locations to provide a quantitative estimate of the contribution of a source to a receptor concentration. Determi-nistic models use mathematical descriptions of the dynamics in the atmosphere along with emission source characteristics to predict the associated air concentrations and deposition rates. The focus of this chapter is the determi-nistic type of air model, and the remaining usage of the term air model implies a determi-nistic model.

As described above, an air model is com-posed of numerical schemes used to approx-imate the physical processes in the atmosphere. Typical processes considered in these models include transport of emissions via advection by the wind, dilution through turbulent mixing in the atmosphere, removal processes such as wet and dry deposition, and chemical transforma-tions. The complexity of a model depends in large part on the number of processes that are simulated and the sophistication of the numer-ical approximations. This chapter considers air dispersion models ranging from screening mod-els that represent only a limited number of atmospheric processes using simple approxima-tions to refined models that represent a more sophisticated treatment and a broader range of atmospheric phenomena. Owing to differences in sources, climate, terrain, and availability of air model input data, there is considerable diversity in the level of sophistication of air models and the expertise and resources required for application. As such, there is no one air model that is appropriate for all air quality evaluations.

Air models have been developed for a wide range of temporal and spatial considerations. The time scales for model predictions, known as averaging times, range from instantaneous to annual. The model receptor locations, or dis-tances at which model predictions are made, range from tens of meters to hundreds of kilometers (km).

An emission released into the atmosphere has defined boundaries within which the mass of the emission is contained. This cloud commonly is known as a plume in the case of continuous releases, and a puff in the case of instantaneous releases. The plume or puff of material is dispersed in the atmosphere primarily through the mechanism of turbulent mixing. There are two types of turbulence in the atmosphere – mechanical turbulence and convective turbu-lence. Mechanical turbulence results as the atmosphere flows over the rough surface of the earth. The larger the roughness elements the greater the mechanical turbulence. Convective turbulence results from thermal differences in the layers of the atmosphere. An example of convective turbulence is when the earth's surface warms through solar radiation and conducts heat into the overlying air. The air at the surface is now warmer and less dense than air above it, rises due to the positive buoyancy forces, and is replaced with cooler air from above. Turbulence in the atmosphere causes the plume to expand or disperse as the adjacent layers of the atmosphere are mixed into the plume.

The mechanical and convective turbulence effects take place in the lowest layer of the atmosphere. This is referred to as the mixing layer because it is the region of the atmosphere where emissions typically mix and dilute. The depth of the mixing layer varies depending on the time of day, wind speed, cloud cover, and terrain. This layer can be thought of as a lid that blocks the vertical migration of emissions. Mixing heights during daylight hours are typi-cally greater than at nighttime and, thus, can have greater dispersion potential. Mixing heights are determined from observations of the ambient temperature as a function of height. Holzworth (1972) provides a listing of after-noon and morning mixing heights for the continental United States. Mixing depths also can be obtained from the National Oceanic and Atmospheric Administration, National Climatic Data Center in Asheville, North Carolina.[1]

Air models can be distinguished by the scheme used to approximate the turbulent mixing or dispersion of the plume or puff within the atmosphere. The most elementary technique

---

[1] For information, call (704) CLI-MATE (254–6283).

is the box model. A box model assumes that emissions are mixed uniformly into a single volume or box within the atmosphere. Horizontal and vertical dimensions of the box are defined and the mixing of emissions is limited to the air within these boundaries. The resulting air concentrations are a mass balance of the source emission rate with advection out of the box via the transport wind speed. The simplified box model provides a reasonable approximation for conditions of steady and uniform emissions and wind, and sufficient volume so that lateral (i.e., perpendicular to wind flow) and vertical dispersion out of the box can be ignored. Box models have been employed for many years to estimate pollutant concentrations over long averaging times (months to annual) within urban areas (tens of kilometers) with reasonable accuracy.

The most commonly used technique for dispersion calculations is the Gaussian model. The Gaussian model approximation is based on the assumption that concentrations in the atmosphere take on a normal or bell-shaped curve when the distribution of values is plotted. Assumption of a Gaussian distribution in the atmosphere makes physical sense because the concentrations at the edges of the plume or puff are decreased through mixing with adjacent air. The normal or Gaussian distribution is defined by the height of the curve that is the maximum value in the distribution, and the standard deviation that is the square root of the variance and defines the width of the curve. In the Gaussian model, dispersion coefficients are used to define the standard deviation of the concentration distribution.

The dispersion coefficients used in the Gaussian model are a measure of the turbulence in the atmosphere and, therefore, vary depending on the meteorological conditions, surface characteristics, and distance from the source. Numerous schemes have been developed to relate atmospheric turbulence to easily measured meteorological conditions. The most common method is called the Pasquill-Gifford stability classification that uses observations of wind speed at approximately ten meters with solar radiation intensity (during the day) and

cloud cover or vertical temperature differences (during the night) to classify the atmospheric turbulence. Stability classifications are designated by one of six letters from A through F. Stability class A, the most unstable class, has the greatest dispersion potential; F stability, the most stable class, represents conditions that tend to dampen turbulence, thereby reducing dispersion. Classes A through C are limited to daytime whereas E and F are nighttime conditions only. A neutral stability classification, D, can occur during day or nighttime periods. In some cases a seventh stability class, denoted G stability, is used to represent very stable conditions with low wind speeds.

Having determined the stability classification, the appropriate dispersion coefficients can be specified for use in the Gaussian model. For the Gaussian plume model, two dispersion coefficients, symbolized as σ_y and σ_z, typically are required. The σ_y value represents diffusion in a horizontal plane perpendicular to the plume trajectory; σ_z represents diffusion in the vertical plane. Gaussian puff models also include a σ_x value to represent dispersion along the axis parallel to plume trajectory. Quantitative values for the dispersion coefficients have been developed from experimental results of dispersion in the atmosphere and theoretical concepts. Most refined air models include an option, typically called buoyancy induced dispersion, to modify the dispersion parameters during the initial stages after release when velocity differences between the plume or puff and the ambient air increases turbulence at the edges. Computer-based models contain algorithms to calculate the appropriate values based on user inputs. Many air models allow the user to specify the use of urban or rural terrain dispersion coefficients. EPA (1986i) provides recommendations for determining if the terrain is rural or urban. Screening techniques using desktop calculations typically provide tables or equations to determine the dispersion coefficients for the stability and downwind distance.

Although a Gaussian concentration distribution is fundamental to all Gaussian models, there are a number of specific formulations of this concept. Some Gaussian plume models include consideration of removal processes, such as dry deposition of particles, and chemical transformation assuming exponential decay. The model dry deposition process attempts to simulate the removal and subsequent dilution of the plume due to gravitational settling of particles to the surface. The removal of particles during precipitation events, known as wet deposition, requires considerable expertise to model. The process is discussed in Chapter 12.

The most sophisticated air dispersion models are numerical models that use one of several numerical techniques (e.g., finite difference, particle in cell) to define the transport and dispersion of emissions in the atmosphere. The numerical technique is used to solve the conservation of mass equation at various locations within a two- or three-dimensional grid of the atmosphere. Currently, the most extensive use of numerical models is to simulate the fate of reactive emissions in the atmosphere; however, these models require considerable input data, computer resources, and technical expertise.

Critical to any modeling exercise is the characterization of the emission source, as this determines not only what models may be appropriate but also how the emissions are treated in the model. A distinction previously was drawn between continuous and instantaneous emission sources in relation to the terms plume and puff. Emission sources can be classified further, based on their physical characteristics into one of four source types: point, line, area, and volume. The most common type of emission source is the point source. Examples of point sources are stacks and vents. Emissions of dust or vehicle exhaust from a highway are characterized as a line source. Sources with similar lengths and widths, such as surface impoundments or a landfill surface, are characterized as area sources. A volume source is used to describe emissions from a storage pile where the source has initial dimensions both laterally and vertically.

Another important aspect of any source is the initial buoyancy of the emissions. Gaseous emissions that are heavier than air, either due to molecular weight or storage conditions (e.g.,

low temperature or high pressure), are known as 'dense gas' releases. Dense gases have an initial negative buoyancy and tend to sink to the surface and slump along the ground. In almost all cases the transport and dispersion of these emissions must be evaluated using specialized dense gas models. Modeling dense gas releases is beyond the scope of this chapter and is not discussed further. Readers interested in modeling dense gas releases are referred to guidance provided in EPA (1991b).

The need for meteorological data in air models has been discussed in terms of specifying the dispersion coefficients. Meteorological data also are used to determine the direction and rate of transport of an emission, and if the emission will rise in the atmosphere through buoyancy forces. Most screening models make use of a generic set of meteorological data, that are not site-specific but are useful in providing a rough estimate of the air concentrations. The generic data may represent the worst-case meteorological conditions (i.e., the combination of wind speed and atmospheric stability that results in the highest predicted air concentrations for the modeled scenario) or they may represent a matrix of all possible combinations of wind speed and atmospheric stabilities. Refined air models require input of meteorological data that represent the conditions within the area of interest. These data are typically supplied in the form of hourly observations (known as RAMMET data) or joint frequency distributions of stability, wind speed, and direction (known as STAR data). The required meteorological data can be obtained from measurements on site or from nearby sources (e.g., National Weather Service, military bases, power plants, and universities). Both EPA and the Nuclear Regulatory Commission have established rigorous guidelines for collection and analysis of meteorological data. The National Climatic Data Center is the repository of meteorological data collected at various locations throughout the nation, including National Weather Service stations, and can provide the data in the format necessary for most air models (see footnote 1).

When selecting meteorological data from off-site locations for use in an air modeling project, a number of issues must be considered. EPA recommends five years of data from an off-site location be used in the air modeling, compared to one year of observations required for on-site data. The proximity of the station and similarity of terrain are essential features to be evaluated. For instance, meteorological data collected from coastal areas should not be used for locations far inland. Difficulties also arise in hilly or mountainous terrain where observed wind speeds and directions can differ substantially within short distances. Considerable expertise often is required to determine the representativeness of meteorological data from off-site locations.

The goal of any air modeling project is to predict ambient air concentrations at a particular location or receptor. Model predictions may be required for a predetermined location, such as the facility boundaries, or for an initially unknown location such as the point of maximum concentration. Some screening models provide only the location and value for the maximum concentration. Refined models provide the user with the ability to specify the location of receptors in the form of a polar or Cartesian coordinate system. When specifying an input receptor grid to identify maximum concentrations, both the distance from the source and the spacing between receptors must be considered. Screening models or procedures can be used to identify the distance from the source to the maximum concentration and this information then can be used to select the receptor locations and grid resolution in a refined model. A coarse grid (e.g., five hundred meter spacing) can be used in a refined model initially to locate areas of high concentrations for which a fine grid (e.g., fifty meter spacing) can be used in a second model run. The concentrations should decrease in value in all directions from the location of the maximum value. If concentrations are not decreasing with distance toward the downwind edge of the grid, then it is likely that the maximum value has not been located.

Under ideal conditions of an isolated source in relatively flat terrain, it is possible to predict the meteorological conditions that result in the maximum ground-level air concentrations for a short-term (i.e., one hour or less) averaging time. Elevated point source releases exhibit a maximum concentration during unstable conditions, as the large scale turbulence eddies bring the plume into contact with the ground close to the release point. However, depending on the elevation of the point source release, the difference between maximum concentrations under unstable and stable conditions can be within a factor of two, with the stable maximum occurring at a much greater distance from the source. For ground-level sources, the maximum concentration occurs during stable conditions because turbulent mixing is restricted. Because the source is at ground level, the maximum occurs at the nearest downwind receptor.

REQUIRED EXPERTISE

Regardless of the level of sophistication built into an air quality model, the accuracy of model predictions ultimately depends on the inputs and modeling decisions by the user. Most of today's computer based air models are user friendly and it is not difficult for an untrained individual to obtain model predictions from even the most refined air models. However, this ease of use can mask the actual level of expertise required to execute and interpret the model results correctly. This section discusses the required expertise to conduct air modeling exercises for various situations and model types.

The previous section defined air models in a broad sense as either screening or refined depending on the level of sophistication incorporated into the model. In general, screening models require less expertise than refined models due to reduced demands for model input and fewer model options. Most screening models employ a default set of meteorological data and are configured to provide conservative estimates of the maximum ground-level concentrations from a source. Typically, the primary expertise required for screening models is the ability to depict the physical nature of the emission source correctly.

On the other hand, the level of expertise required to execute a refined air model is dependent on both the model and the modeling scenario. Refined models require meteorological data that are both representative of the area and compiled in a specific format. Refined models typically provide the user with a variety of options that are designed to provide a more realistic simulation of transport and dispersion of emissions, but may require expert judgment for appropriate selection. The nature of the terrain surrounding the source also affects the level of expertise required to model emissions correctly. The complexity of the emission source also dictates the level of expertise required to conduct the modeling exercise. Each of these issues is addressed below in order to provide insight into the required skills and resources or guidance available.

The selection of representative meteorological data for use in a refined air model is critical. By definition, representative meteorological data depict the transport and dispersion conditions in the area to be modeled. The proximity of the source location to the site where the meteorological data were collected must be considered. However, proximity alone does not assure a representative database as the nature of the intervening terrain can be influential. For example, the presence of hills or mountains can affect atmospheric stabilities, wind speeds, and wind directions. Similarly, large water bodies can influence these same parameters. In less than ideal conditions, where ideal is defined as flat terrain and no large water bodies, the selection of representative meteorological data requires expert judgment. When modeling is conducted for regulatory purposes, experts within the reviewing agency should provide guidance regarding the adequacy of the input meteorological data.

Once a representative set of meteorological data are identified, it must be compiled in a specific format for use in a refined air model. The two most common formats for Gaussian models are RAMMET and STAR. The RAMMET format provides hourly averages of model required meteorological data based on concurrent observations of both surface and upper air conditions. The STAR data format is a joint frequency distribution of atmospheric stability, wind speed, and wind direction over a period of months to years. The STAR data cannot be used to predict short-term air concentrations because the data represent long-term average conditions. Programs are available for converting hourly observations from meteorological monitoring stations to the RAMMET and STAR data formats. However, these programs require that there are no periods of missing data in the hourly observations; expert judgment again is required to fill in the missing data. RAMMET and STAR data can be purchased from many private vendors as well as the National Climatic Data Center.

Through the use of a variety of options, refined models allow the user considerable flexibility in the simulation and evaluation of emissions. Some options have no effect on the model calculations but simply allow the user to specify the format of the model output. However, other options can have profound effects on the model predictions, and it is critical that the user make the appropriate selection. EPA's guidelines (EPA 1986i) and in some cases the model user's guide provide guidance for selection of options that influence model calculations.

The nature of the terrain in the area surrounding the source determines the type of air model that must be used and, thus, the level of expertise required to conduct and interpret the results. For the purposes of air modeling, terrain is classified as either simple or complex. EPA (1986i) defines complex terrain as terrain exceeding the height of the stack being modeled. Complex terrain can become a critical issue during stable conditions when elevated plumes undergo limited dispersion before contacting elevated terrain. Simple terrain models assume flat terrain or reduce complex terrain elevations to stack top elevation so that plume impaction on elevated terrain is ignored. Complex terrain models allow terrain elevations above stack top elevation and contain techniques for evaluating plume impaction. The presence of complex terrain in the vicinity of the stack requires the use of both complex and simple terrain models. EPA (1986i) establishes a series of screening techniques for conducting complex terrain evaluations. At the time of this writing in 1992, a supplement to EPA's guidelines was planned to include refined modeling techniques for complex terrain. Evaluation of complex terrain requires considerable expertise even for screening techniques.

For elevated point sources located near large water bodies, fumigation events can result in elevated short-term concentrations. Fumigation occurs as a plume emitted into a stable layer is advected into a less stable region of the atmosphere and the plume is rapidly mixed to the ground. This phenomenon occurs near large water bodies because the land warms faster than the water from the input of solar radiation. A circulation pattern then can develop with warm air over the land rising aloft to be replaced by cooler stable air moving over the land at low levels from the adjacent water body. This stable layer is eroded upwards from the surface as the air moves over the surface. Plumes emitted into the stable air undergo limited mixing until the convective turbulence reaches the plume height then rapidly mixes the plume down to the surface.

Fumigation events are not restricted to areas along land–water interfaces. During cloudless nights with very low winds, the land can rapidly cool the overlying air from the surface upwards, creating a stable layer in the lower atmosphere. Following sunrise the land once again heats up and erodes the stable layer from the surface upwards by convective turbulence. Plumes emitted into the stable layer can mix rapidly to the surface as the stable layer decays from below. Fumigation events of this type typically last for a period of thirty minutes or less. Fumigation events along land–water interfaces, however, can last for hours because the circulation pattern continues to bring stable marine air over the land surface.

Evaluation of fumigation events requires a high degree of technical expertise. EPA (1986i)

recommends screening techniques for evaluating fumigation events, but cautions the user to exercise care when using the results from these procedures. As part of the planned supplement to the current air quality modeling guidelines, EPA states that the selection of the appropriate model for evaluation of shoreline fumigation for regulatory purposes should be determined in consultation with the responsible EPA Regional Office.

The complexity of the emission source or sources also affects the level of expertise required to model the resulting transport and dispersion. Sources located on or near large buildings may be subject to the effects of aerodynamic downwash. As the wind blows over and around the building the flow is disturbed and several regions of increased turbulence and distorted flow patterns develop. The most critical of these regions is the cavity region located along the roof and sides of the building, which is effectively cut off from the mean transport wind flow. The cavity region can extend up to approximately three building heights downwind from the lee building face (depending on the building dimensions), with typical lateral and vertical dimensions of no more than 1.5 times the width and height of the building, respectively. Within the cavity region the chaotic airflow tends to recirculate so that emissions within the cavity region are carried back upwind. Thus, emissions released into the cavity can result in very high concentrations as mixing with air outside the cavity zone is limited, and the emissions can be brought quickly to the ground with little or no dilution. Above the cavity zone, a second region of disturbed flow called the wake region also can mix emissions quickly to the ground downwind of the cavity region. This same phenomena also can occur on the lee side of hills.

Modeling source emissions subject to aerodynamic building downwash requires varying levels of expertise. For elevated point sources, a good engineering practice (GEP) stack height analysis should be used to determine if building downwash affects the emissions. For regulatory purposes the GEP stack height is defined as the height necessary to insure that emissions from a stack do not result in excessive concentrations in the immediate vicinity of the source as a result of atmospheric downwash caused by the source itself or nearby structures and obstacles. EPA (1985d) provides technical guidance for determination of GEP stack height for various building configurations and terrain considerations. Most air models contain some method for assessing the effects of building downwash on emissions based on user input building dimensions. However, determining the relevant or controlling building dimensions for input to the model can require significant expertise when sources are located in a complex industrial setting with multiple buildings of varying heights, irregular geometry, or numerous tiers. Additionally, to take advantage of refined air modeling of building downwash it becomes necessary to specify building dimensions as a function of the wind direction. Depending on the method used, this results in

either sixteen (i.e., every 22.5 degrees) or thirty-six (i.e., every 10 degrees) separate specifications of building dimensions for a given stack.

A second type of aerodynamic effect that must be considered for stack sources is stacktip downwash. Similar to the building downwash condition, stacktip downwash occurs as the mean flow pattern becomes distorted upon encountering a solid obstacle, in this case the stack itself. A low pressure region develops in the lee of the stack that can draw the stack emissions down below the release point. This has the effect of lowering the emission height and changing the ambient concentrations. It generally is recognized that downwash will not occur if the velocity of the emissions at the stack exit is at least 1.5 times greater than the wind speed. The effects of stacktip downwash can be considered easily both in desktop and in computer based air models.

AVAILABLE MODELS AND MODELING GUIDANCE

A large number of air models and modeling guidance have been developed for application in EPA regulatory programs. This section presents a detailed discussion of the most commonly used and available guidance and models. The focus of this section is on EPA recommended models and guidance. As such, it is by no means a complete listing of all the air models or guidance available today. The models and guidance discussed here, however, meet the great majority of modeling tasks for regulatory or other purposes.

Screening Analysis

A valuable resource for screening analysis of air toxics is EPA (1988i). This publication provides the user with methods to estimate steady-state emission rates and associated short-term ambient air concentrations for eighteen scenarios representing some of the most common release mechanisms of air toxics. This workbook is not recommended for use in evaluating accidental releases, and does not consider complex terrain effects. Based on the release scenario, the user

calculates ambient air concentrations using either a continuous Gaussian model, an instantaneous Gaussian puff model, or a dense gas model. The workbook is highly structured and provides sufficient background information to guide the user through the selection and application of the appropriate screening technique.

A computerized version of the screening techniques used in the workbook is available as TSCREEN. TSCREEN provides a series of menus displayed on the computer screen to allow the user to: (1) select which of the eighteen scenarios are to be evaluated, and (2) select input data. The model automatically links the emission rate to the appropriate air dispersion model to determine the maximum concentrations using worst-case meteorological conditions. However, the model has no provisions to calculate dense gas dispersion. TSCREEN allows the user to select averaging times ranging from fifteen minutes to twenty-four hours for continuous plumes, and instantaneous to fifteen minutes for puff model calculations. Help screens for each menu option are provided to assist the user in selecting the appropriate scenario and input of required parameters. The model also contains a chemical database of relevant physicochemical parameters for emissions calculations, which the user may edit and expand. There also are options for creating graphs of the calculated air concentrations as a function of distance. TSCREEN is an IBM-PC based system that requires 500 kilobytes (KB) of random access memory (RAM). The TSCREEN model has a user manual, but it is recommended that the workbook be used when running TSCREEN.

Refined Models

EPA (1991b) is an extension to the workbook on screening techniques. This publication provides guidance on the use of refined air models for both dense gas and neutrally buoyant releases of air toxics. The publication includes information on the following: (1) general model input considerations for air toxic releases, including release characteristics, meteorological data, and averaging times; (2) specific guidance for three dense gas models, including sample problems with step-by-step discussions of the model application; and (3) interpretations of the model results. The model output for the dense gas applications is provided in an appendix. EPA (1991b) is an excellent reference for information on critical issues in executing and interpreting the results from refined air models applied to air toxic releases, but it is not intended to replace the specific model user's guide.

Air Toxic Models

AFTOX is a refined PC-based Gaussian puff model for use with continuous or instantaneous releases. AFTOX can model releases from both point and area sources but does not evaluate dense gas releases or complex terrain. The AFTOX model has the ability to calculate emission rates for liquid spills and the resulting air concentrations. Detailed source characteristics and meteorological data must be supplied by the user for model execution. The model output includes concentrations at specified

locations and times, the maximum concentration at a specified elevation and time, and a plot of concentration isopleths. The model is not recommended for averaging times of greater than one hour. AFTOX does not consider chemical transformation or deposition.

Guidance for both modeling and assessing the human health risks for air toxic releases is presented in EPA (1992c). Guidance is provided for analysis of single and multiple area and point sources of one or more pollutants. The tiered modeling approach begins with a simple screening approach that provides worst-case estimates of one hour and annual average air concentrations. The worst-case air concentrations are determined from a table of values that lists air concentration as a function of distance from the source and the height of emission for point sources or the maximum horizontal length for area sources. The second tier involves the use of a screening air dispersion model using generic meteorological data and additional information regarding the source characteristics. The third tier requires detailed information regarding the source characteristics, and site-specific meteorological data in refined air dispersion models. For each tier, the document provides guidance on calculating the acute and chronic inhalation risks for the resulting air concentrations. The tiered approach allows a user to begin with a simple procedure for analysis and increase in complexity if the results of the analysis are not acceptable. The air models used in the second and third tiers, SCREEN and ISC2, respectively, are discussed in detail below. EPA (1992c) is intended for use with the user manual for the air models used in the second and third tiers. It does not provide guidance for modeling dense gas releases or releases in complex terrain.

The preceding guidance documents were developed specifically to address modeling of air toxic releases. EPA also provides several guidance documents for air modeling to support permitting under the Clean Air Act. A brief description of several of these documents is provided below. Although these documents were not developed specifically for air toxics analysis, they provide useful, and in some cases the sole, guidance for air modeling issues.

In response to the 1977 Clean Air Act requirements for such analyses, EPA published a guidance document (EPA 1977b) that contained screening procedures for evaluating the air quality impacts of new stationary sources. This document was updated and revised in EPA (1988j). At the time of this writing in 1992, EPA (1988j) currently is in draft form for public comment with the intention of producing a final version for inclusion as a recommended screening procedure in a supplement to EPA (1986i). In the interim, EPA encourages and accepts the use of the document for screening analysis, as do many state agencies. Detailed screening procedures are provided for calculating short-term and annual average concentrations for both point and area sources of continuous emissions using Gaussian formulations. Screening methodologies also are provided for evaluating sources affected by plume impaction on elevated terrain and fumigation. A calculator is sufficient to carry out most of the screening procedures presented in this document.

In conjunction with the publication of the revised screening procedures document described above, EPA also makes available the SCREEN model that provides a computerized version of the procedures for analysis of short-term impacts on ambient air quality. A user's guide for the SCREEN model is included as an appendix in the guidance document. SCREEN is an interactive program which prompts the user for keyboard input of all information required to execute the model. Both point and area sources can be evaluated in either rural or urban terrain. The SCREEN model can be run with a user input wind speed and stability, or with a matrix of wind speed and stability conditions contained in the model. Building downwash, buoyancy induced dispersion, fumigation, and plume impaction on elevated terrain can be treated. The model calculates and displays to the screen the location and value of the maximum one-hour concentration for a source, as well as the maximum concentration at user-specified locations. At the conclusion of the model run, the user is presented with the

option to produce a hard copy of the inputs, model option selections, and model results. The primary hardware requirements for SCREEN are an IBM-PC or compatible PC with at least 256 KB of RAM.

Preferred Air Models

EPA (1986i) provides recommendations for the preferred air models for screening and refined modeling in both simple and complex terrain, and provides procedures for their appropriate application. These models are designated as preferred for a particular application based on either superior performance or, in the case of no clearly superior model, based on the modeling community's familiarity and past use of a model and the requirements to obtain and execute the model. EPA (1986i) is a valuable resource for EPA recommendations on air modeling issues. As stated earlier, EPA planned to publish Supplement B to the guideline. The draft version of the supplement (EPA 1990h) indicates that the inventory of preferred models is to be revised to include several additional models. These models are identified and discussed below along with preferred models in the current guideline that can be used in evaluating air toxic releases.

CRSTER

The CRSTER model is the preferred simple terrain model for evaluation of a single point source located in a rural area, for short-term and annual average concentrations. The model also is considered appropriate for use in urban areas. CRSTER is a Gaussian plume model that treats up to nineteen individual point sources in a single run. However, the model assumes that all sources are located at the same point relative to the receptor grid. Source data input consists of emission rate, stack height, stack gas exit velocity, stack inside diameter, and stack gas exit temperature. Required meteorological input is one year of RAMMET data, and the actual height of wind speed measurements (i.e., anemometer height). The model receptor grid is a series of five receptor rings located at user specified distances from the source, with receptors spaced at 10-degree increments around each ring, for a total of 180 receptors. The user may input topographic elevation for receptors; however, all receptor elevations must be below the stack top or the model will abort the run. The model automatically provides the location and value of the daily maximum one-hour and twenty-four hour concentrations for the receptor grid, the highest and second highest concentrations during the year for three short-term (one, eight, and twenty-four hours) and annual averaging times at each receptor. Additional short-term averaging times may be selected by the user (i.e., two, four, six, or twelve hours). When multiple stacks are considered, a model option allows the user to obtain output showing the contribution of each source to the concentrations for selected receptors. All concentrations are calculated at ground level. Chemical transformations can be treated by CRSTER using an exponential decay, based on a user input half-life. CRSTER does not treat fumigation, building downwash, or particle removal via deposition. CRSTER contains a regulatory default option that the user may select to switch relevant model options to the appropriate settings for regulatory applications. The model is available in IBM-PC format.

RAM

For single or multiple point and area sources in an urban environment, the RAM model is the preferred model for both short-term and long-term concentrations. RAM is a Gaussian plume dispersion model requiring user input of source, meteorological, and receptor data. User input for point sources consists of source location, emission rate, stack height, stack gas exit velocity, stack inside diameter, and stack gas exit temperature. For area sources the user inputs the location and size of the area source, the emission rate, and the height of emissions. Meteorological data are input in RAMMET format along with the actual anemometer height. Receptor locations are specified by the user; however, actual terrain elevation cannot be entered as the model assumes flat terrain. Unlike CRSTER, which assumes that all sources are colocated when calculating

concentrations, the RAM model applies the user-specified locations for all sources and receptors to calculate concentrations based on actual source–receptor separation. Model output is specified by the user, with options for short-term concentrations at each receptor for averaging times from one hour to twenty-four hours, annual average concentrations at each receptor, individual source contribution to calculated concentrations, and the highest through fifth highest concentrations for the period at each receptor. RAM does not evaluate building downwash, fumigation, or removal via deposition. Chemical transformation can be evaluated using exponential decay based on a user input half-life. RAM contains a regulatory default option that the user may select to switch relevant model options to the appropriate settings for regulatory applications. The model is available in IBM-PC format.

CDM 2.0

When calculating long-term (i.e., seasonal or annual) concentrations for more than a 'few' sources in an urban environment, the Climatological Dispersion Model (CDM 2.0) is preferred. The meteorological data input for CDM 2.0 is the STAR data, the average mixing height and wind speed for each stability class, and the average air temperature. An anemometer height of ten meters is assumed by the model. Source data input requirements for a maximum of two hundred point sources are source location, emission rate, stack height, stack gas exit velocity, stack inside diameter, and stack gas exit temperature. For a maximum of 2,500 area sources, the user inputs the location and size of the area source, the emission rate, and the height of emissions. The model receptor grid is based on user input coordinates for each receptor. Terrain elevations for individual receptors cannot be input; however, CDM 2.0 allows the user to input a single height at or above ground level at which concentrations are calculated for all receptors. The model output is the average concentration for the period of the STAR data (e.g., monthly, seasonal, or annual), and an optional histogram for point and area concentrations as a function of stability for each receptor. CDM 2.0 does not evaluate building downwash, fumigation, or particle deposition. Chemical transformation can be evaluated using exponential decay based on a user input half-life. CDM 2.0 contains a regulatory default option that the user may select to switch relevant model options to the appropriate settings for regulatory applications. The model is available in IBM-PC format.

MPTER

The Multiple Point Gaussian Dispersion Algorithm With Terrain Adjustment (MPTER) model is the preferred multiple point source model for calculating short- and long-term concentrations in a rural environment. MPTER also is a recommended model for multiple point sources in urban areas. As the name implies, MPTER is a Gaussian plume model for point sources. The optional terrain adjustment feature of MPTER allows the model to adjust the plume height upwards to simulate flow over receptors on elevated terrain. The terrain adjustment feature is limited to receptor elevations less than or equal to the lowest input stack top elevation (i.e., physical stack height plus ground level elevation). The model requires user input of source, meteorological, and receptor information. The meteorological data input is a RAMMET data set and the actual anemometer height. Source input data, for a maximum of 250 sources, consist of stack location, emission rate, stack height, stack gas exit velocity, stack inside diameter, and stack gas exit temperature. Optionally, the user may specify the topographic elevation for the stack location for use with the terrain adjustment feature. The model allows a maximum of 180 receptor locations. The user may specify the model receptor grid by entering five downwind distances and allowing the model to construct a series of five receptor rings, located at the user specified distances, with receptors spaced at 10-degree increments around each ring, for a total of 180 receptors. Optionally the model receptor grid is constructed from user input coordinates for each receptor. When the terrain adjustment feature of MPTER is selected, the user must input terrain elevations for each receptor.

MPTER contains numerous options that allow the user to specify model output. Concentration output for each receptor can be specified for short-term averaging times from one to twenty-four hours, including highest through fifth highest concentrations for the period, annual average concentrations, and source contributions (up to a maximum of the twenty-five most significant sources). MPTER does not treat fumigation, building downwash, or particle deposition. Chemical transformation processes can be treated using an exponential decay based on the user input half-life. A regulatory default option allows the user to switch relevant model options to the appropriate settings for regulatory applications. MPTER is available in IBM-PC format.

ISC2

The Industrial Source Complex (ISC2) dispersion models arguably are the most commonly used refined air models for simple terrain applications. The popularity of the ISC2 models is due in large measure to their ability to treat a great number of simple modeling scenarios. The ISC2 models do not evaluate fumigation, but can treat building downwash conditions and the removal of particles in the plume via dry deposition. Chemical transformation is treated assuming exponential decay based on a user input half-life. The ISC2 models contain a regulatory default option to select and configure relevant model parameters and options to the appropriate setting for regulatory applications. The two models that share the ISC prefix are ISCST2 and ISCLT2. The ISCST2 is a Gaussian plume model designed to calculate short-term impacts (i.e., one to twenty-four hours) but it can be used to determine annual average concentrations. The ISCLT2 is a climatological Gaussian plume model designed to calculate monthly, seasonal, or annual average impacts. The ISC2 models are available in IBM-PC format.

The ISC2 models are the preferred models for evaluations of complicated sources in rural and urban areas. Complicated sources are defined loosely as those sources for which the refined models described above are not adequate. This would include sources subject to building downwash effects, emissions of particulate matter, and emissions characterized as volume sources. The ISC2 models are user friendly; however, their extensive flexibility requires a detailed understanding of the model requirements and capabilities. A brief description focusing on the highlights of the ISC2 dispersion models follows. The reader is directed to Volumes I and II of the ISC2 User's Guide (EPA 1992d) for more detailed information.

Multiple point, area, and volume sources can be treated in the ISC2 models. Input source locations are used in the models to calculate concentrations based on the actual separation from receptor locations. Emission rates for input sources can be constant or varied as functions of time, wind speed, and stability. When evaluating multiple sources in the ISC2 models, the user has the option to evaluate each source individually, specify all sources as a single source group, or subdivide the sources into several source groups based on user-defined combinations. The contribution from each source in a group is combined by the model to determine the total concentration at a receptor. The model output reports the total receptor concentration for a source group and the contribution from each source in the group.

For particulate matter emissions, the user can input up to twenty separate gravitational settling velocities for each source along with the corresponding mass fraction and the surface reflection coefficient. The surface reflection coefficient defines the percentage of particles that do not deposit at the surface upon impaction. Volume II of the ISC2 User's Guide provides techniques for calculating the settling velocity and surface reflection coefficient. The user specifies whether the model is to calculate ambient air concentrations or the deposition values.

For evaluation of building downwash conditions, the ISC2 models require user input of building height and width. These dimensions are direction specific, in that they are the building dimensions relative to the source in the direction toward which the wind is blowing. For the ISCST2 model, the wind direction is broken down into thirty-six, 10-degree sectors requiring building dimensions for each sector. The ISCLT2 model requires only sixteen direction-specific building dimensions because it uses 22.5-degree sectors. The building downwash calculations are used only for point source emissions.

The meteorological input requirements are different for the two ISC2 models. The ISCST2 model requires hourly observations of meteorological data. The user has a number of options for the format of the input meteorological data including RAMMET format. The ISCLT2 model requires a user input STAR data and the average mixing height and air temperature for each stability. Both models require user input of the actual anemometer height. The ISC2 models also contain an option that allows the user to rotate the input wind direction data.

The receptor grid options in the ISC2 models provide the user with many useful alternatives. The initial program configuration allows the user to specify up to five separate receptor grids for a single run. With this option the user could specify a coarse receptor grid for the entire modeling domain, with fine receptor grids located at various points of interest within the coarse grid. The number of simultaneous receptor grids that can be used in the ISC2 models can be increased by the user. In addition to the receptor grids, the user also may input discrete receptor locations (e.g., at property boundaries). The user also may input the elevation of each receptor including elevations above ground level. However, elevations above the release height being evaluated are reduced to the source release height during model calculations.

The output options for the ISC2 models are extensive. The ISC2 models provide a detailed listing of the user specification for model options, and inputs for source, meteorological, and receptor data. The ISCST2 model output can include summaries of the highest and second highest values for each receptor, averaging period and source group combination, a listing of the locations and values of the fifty highest values across a receptor grid for each averaging period and source group combination, and daily tables of concurrent values by receptor for each source group and averaging period for each day evaluated. The ISCLT2 provides a tabular listing of the values for each receptor evaluated and the maximum values within the receptor grid. This listing can be developed for a source group and individual sources, with the option to print individual source contributions for the maximum value table. The ISC2 models also produce output files that can be used in graphics packages to produce contour plots of model output values.

BLP

In some industrial processes (e.g., aluminum reduction plants), buoyant emissions are vented through the roof, forming an effective line source of emissions. The Buoyant Line and Point Source (BLP) model was designed expressly and is the preferred model for emissions from buoyant line sources in rural areas for short- and long-term evaluations. BLP is a Gaussian plume model with the capability to treat building downwash effects. Source input data for a maximum of ten parallel line sources consist of emission rates, release height, coordinates for the endpoints of the line source, average line source width, average line source buoyancy, average building width, and average spacing between buildings. BLP has the ability to model a maximum of fifty point sources using standard input characteristics. BLP requires a user-input RAMMET meteorological data set. The user may specify the individual locations and elevations for receptors, specify the location and size for a receptor grid, or allow the model to generate the receptor grid. The BLP model provides a number of model output options including: the five highest concentrations at each receptor for averaging times of one, three, and twenty-four hours; the fifty highest concentrations across the receptor grid for one-, three- and twenty-four-hour averaging times; long-term (annual or period modeled) concentrations for each receptor; and total concentration or individual source contribution at each receptor. BLP treats chemical transformation assuming a linear decay based on the

user input decay rate. Fumigation and particle deposition via gravitational settling are not treated in BLP. The BLP model is available in IBM-PC format.

FDM

The Fugitive Dust Model (FDM) is a refined model specially designed to deal with particulate matter emissions from fugitive sources in urban or rural areas with simple terrain. Examples of fugitive dust sources that could be modeled in FDM include wind erosion of storage piles or sparsely vegetated surfaces, vehicle traffic on paved or unpaved roads, surface mining activities, and materials handling (e.g., excavation, dumping, and grading). FDM is a Gaussian plume model that calculates both short- and long-term values of air concentration and particle deposition. Input meteorological data can be either RAMMET or STAR data format. Source characteristics including emission rate and source location are required model input. Multiple sources may be input and emission rates may be varied as a function of wind speed. The user may specify up to twenty particle size classes for dry deposition calculations. Air concentrations or deposition rates are calculated for user-specified receptor locations using actual source–receptor spacing. Model output contains the user specification for model options, the source input data, and the receptor locations and elevations. Optional model output includes meteorological data, concentration and deposition values at each receptor for the averaging times selected, and a file containing the coordinate location of receptors and the corresponding concentration/deposition values for use in plotting programs. The FDM model is available in IBM-PC format.

Complex Terrains

To determine air concentrations at receptors with elevations greater than the release height, a complex terrain model must be used. Because of the substantial expertise required to model complex terrain dispersion properly, only a limited discussion of these models is presented. EPA (1986i) provides guidance for the selection of complex terrain models and modeling procedures. The guidelines list four screening techniques for evaluation of concentrations due to plume impaction during stable conditions. The initial screening technique involves the use of the VALLEY air model with default worst-case assumptions for meteorological data. This screening technique for the VALLEY model has been incorporated into the SCREEN model described previously. The remaining screening techniques require input meteorological data with a full year of on-site data as the preferred input. The proposed changes to EPA's current guidelines are to include a fifth screening technique using the CTSCREEN model and the first refined air model, CTDMPLUS. The CTSCREEN model uses a fixed matrix of meteorological data to determine the maximum one-hour concentration.

EPA Model Bulletin Board

A final resource that should prove valuable to persons wishing to conduct air modeling is EPA's Support Center for Regulatory Air Models (SCRAM) bulletin board. SCRAM is part of the EPA's Technology Transfer Network (TNN) and can be accessed via modem. SCRAM provides IBM-PC versions of regulatory air models and related computer programs, meteorological data, bulletins regarding model status, modeling guidance, and state and federal contacts for air modeling issues. Information on the SCRAM bulletin board can be obtained by calling (919) 541-5742.

References

Holzworth, G.C. 1972. Mixing heights, wind speeds, and potential for urban air pollution throughout the contiguous United States. AP-101. U.S. Environmental Protection Agency, Office of Air Programs. Research Triangle Park, NC.

U.S. Environmental Protection Agency (EPA). 1977b. Guidelines for air quality maintenance planning and analysis. Volume 10, Revised: Procedures for evaluating air quality impacts for new stationary sources. EPA 450/4-77-001 (OAQPS Number 1.2-029R). Office of Air Quality Planning and Analysis. Research Triangle Park, NC.

U.S. Environmental Protection Agency (EPA). 1985d. Guideline for determination of good engineering practice stack height (Technical support document for stack height regulations). (Revised). EPA 450/4-80-023R. Office of Air Quality Planning and Standards. Research Triangle Park, NC.

U.S. Environmental Protection Agency (EPA). 1986i. Guideline on air quality models (Revised) and Supplement A (1987). EPA 450/2-78-027R. (NTIS PB 86-245248). Office of Air Quality Planning and Standards, Research Triangle Park, NC.

U.S. Environmental Protection Agency (EPA) 1988i. A workbook of screening techniques for assessing impacts of toxic air pollutants. EPA 450/4-88-009. Office of Air Quality Planning and Standards. Research Triangle Park, NC.

U.S. Environmental Protection Agency (EPA) 1988j. Screening procedures for estimating the air quality impacts of stationary sources (Draft for public comment). EPA 450/4-88-010. Office of Air Quality Planning and Standards. Research Triangle Park, NC.

U.S. Environmental Protection Agency (EPA). 1990h. Supplement B to the guideline on air quality models (Revised) – Draft. Office of Air Quality Planning and Standards. Research Triangle Park, NC.

U.S. Environmental Protection Agency (EPA). 1991b. Guidance on the application of refined dispersion models for air toxics releases. EPA 450/4-91-007. Office of Air Quality Planning and Standards. Research Triangle Park, NC.

U.S. Environmental Protection Agency (EPA). 1992c. A tiered approach for assessing the risks due to sources of hazardous air pollutants. EPA 450/4-92-001. Office of Air Quality Planning and Standards. Research Triangle Park, NC.

U.S. Environmental Protection Agency (EPA). 1992d. User's guide for the industrial source complex (ISC2) dispersion models, Volumes I, II, and III. EPA 450/4-92-008a. Office of Air Quality Planning and Standards. Research Triangle Park, NC.

12

Atmospheric Transformation and Removal of Air Toxics

Mary P. Ligocki

INTRODUCTION

In assessing the potential impacts of emissions of toxic species into the atmosphere, it is important to have an understanding of their atmospheric persistence. Species which persist for long periods of time can accumulate to high concentrations during stagnation periods and can be transported further from their sources than species that are destroyed rapidly by atmospheric reactions.

A variety of atmospheric transformation and removal processes of importance to air toxics can occur in the atmosphere. Chemical processes include reaction with atmospheric oxidants and photolysis. Physical processes include wet and dry deposition. The term 'atmospheric transformation' also can apply to the atmospheric formation of secondary toxic species. Finally,

phase distribution processes that result in partitioning of toxics between the gas phase, aqueous droplets, and solid aerosol particles are atmospheric transformation processes. The phase in which an atmospheric pollutant resides (i.e., gas, aqueous, or solid) will have a major impact on the removal processes that are operative.

Among the hazardous air pollutants listed in section 112(b)(1) of the 1990 Clean Air Act Amendments, atmospheric reactions are most important for volatile organic compounds (VOCs). The listed metals generally will not be treated here although some can and do undergo atmospheric reactions; however, these reactions typically are much slower (e.g., involving phase changes altering the rate of physical removal from the atmosphere).

CHEMICAL REACTIVITY

In the gas phase, toxic air pollutants are destroyed by oxidation. The oxidant of most importance on a global scale is the hydroxyl radical (OH), which is produced photolytically everywhere in the atmosphere and reacts with nearly every organic substance. In urban atmospheres, ozone (O_3) also can be an

important oxidant. At night, OH concentrations drop significantly because little OH is produced in the absence of sunlight, but concentrations of the nitrate radical (NO_3) can increase to high levels when high concentrations of nitrogen dioxide (NO_2) and ozone are present. Other atmospheric oxidants are the hydroperoxyl radical (HO$_2$), the oxygen atom (O^{3P}), and the chlorine atom (Cl), which may be important under some circumstances.

Toxic air pollutants differ widely in their rate of reaction in the atmosphere. The most reactive, such as 1,3-butadiene (C$_4$H$_6$) and aniline (C$_6$H$_5$NH$_2$), react with OH at rates 100 thousand times as fast as the least reactive, such as carbon tetrachloride (CCl$_4$). In order to be able to predict the reactivity of a given toxic species in the atmosphere, some information on its chemical structure must be known. Procedures for estimating OH rate constants for many organic species based on their structures are given by Atkinson (1986). Periodic reviews of rate constants (Atkinson et al. 1989; DeMore et al. 1987) and tropospheric chemistry (Atkinson 1990) do not emphasize toxics, but include those toxics that also are important to atmospheric chemistry (such as the aldehydes and xylenes). Reaction rates for polychlorinated biphenyls, dioxins, and furans are discussed by Atkinson (1987). A comprehensive electronic database of gas phase rate constants, including OH, O$_3$, and NO$_3$ reactions, also has been developed by the National Institute of Standards and Technology and is updated annually (NIST 1991).

In general, organic compounds containing double bonds (termed alkenes, olefins, or unsaturated hydrocarbons) are highly reactive. In addition to reacting with OH, these compounds react rapidly with ozone and NO$_3$. However, as chlorine atoms are added to a molecule, the reactivity decreases. Hence, tri- and tetrachloroethene are less reactive than ethene, and hexachlorobutadiene is much less reactive than butadiene.

Another category of reactive organic compounds is the aromatic species, which are characterized by the presence of a six-membered planar ring structure. The simplest aromatic species, benzene (C$_6$H$_6$), is not very reactive. However, when substituent groups are added to the benzene ring structure the reactivity increases. Toluene (methylbenzene) and ethylbenzene are more reactive than benzene; xylenes (dimethylbenzenes), trimethylbenzenes, and phenols (hydroxybenzenes) are quite

reactive. As with the olefins, the addition of chlorine causes a decrease in reactivity. Polycyclic organic matter (POM) species contain two or more fused benzene rings. The lower molecular weight POM species (two to four rings) exist primarily in the gas phase; they also are extremely reactive. The higher molecular weight POM species (five rings or more) generally are adsorbed to particles, where their reactivity is a strong function of the particle matrix.

A third category of reactive species is the aldehydes. Formaldehyde (HCHO), acetaldehyde (CH$_3$CHO), acrolein (C$_2$H$_3$CHO), and propionaldehyde (C$_2$H$_5$CHO) are listed in section 112(b)(1); all are quite reactive. Not only do aldehydes react with OH, they also photolyze (i.e., decompose in the presence of sunlight). With the exception of acetaldehyde, photolysis is of comparable importance to reaction with OH. In addition, photolysis of aldehydes is important to atmospheric chemistry because it produces radicals. Although aldehydes are emitted by many sources, their major source is from atmospheric oxidation of other organic species. Aldehydes are produced naturally from the oxidation of biogenic hydrocarbons. In remote environments, concentrations of formaldehyde and acetaldehyde may be as high as 1 ppb (NRC 1981). In urban atmospheres, aldehydes are intermediate products of the ozone cycle. They are formed in the initial oxidations of olefins, and then photolyze to produce radicals and fuel the ozone production process.

In general, acids are not reactive in the gas phase. This applies to inorganic as well as organic acids. Thus, although hydrofluoric and hydrochloric acids are extremely reactive in liquid form, they are essentially inert in the gas phase.

It is important to note that atmospheric reactions of air toxics may not result in complete removal of toxicity. One must consider the products of the atmospheric reaction. For example, the atmospheric degradation of 1,3-butadiene produces formaldehyde and acrolein, two species that also are listed in the 1990 amendments. Atmospheric oxidations of POM can produce nitro-POM (Arey et al. 1986), which are among the most mutagenic species known. Often, the products formed in atmospheric oxidations are more reactive than the parent compound, in which case the initial oxidation can be considered the rate-limiting step.

Many atmospheric species react rapidly in the aqueous phase of clouds, fog, and aqueous aerosols. For some toxic air pollutants, this can be a major atmospheric transformation pathway. For metals, virtually all atmospheric chemistry will take place in the aqueous phase. In the atmospheric aqueous phase, metals can be oxidized or reduced. They also can form complexes with organic ligands. For some metals, such as mercury, atmospheric aqueous phase reactions can significantly alter their toxicity.

For VOCs, the importance of the aqueous transformation pathway depends not only on the aqueous reaction rate, but also on the Henry's Law constant H of the species, that is a measure of the partitioning of a species between the gas and aqueous phases. All polar organic species, including aldehydes, alcohols (including phenols and cresols), acids, and ketones will partition readily into the aqueous phase; nonpolar organics such as hydrocarbons will not.

In the atmospheric aqueous phase, OH is again the most important oxidant. However, the importance of the HO_2 radical and its conjugate base, O_2^-, has recently been recognized (Gunz and Hoffmann 1990). Compilations of aqueous phase rate constants for metals and organic compounds are given by Buxton et al. (1988) and Bielski et al. (1985). A comprehensive review of Henry's Law constants is given by Mackay and Shiu (1981). Procedures for estimating H are given by Lyman, Reehl, and Rosenblatt (1982).

GAS-PARTICLE PARTITIONING

Atmospheric transformation also can include the condensation of semivolatile vapors onto atmospheric aerosols. This process can be of major importance for semivolatile organic compounds (SVOCs) such as pesticides, PCBs, and POM, as well as for metals such as mercury and selenium. For most of these species, physical removal from the atmosphere will be more rapid for the particulate-phase component (Bidleman 1988); for some, chemical removal will be more rapid for the gas-phase component.

Gas-particle partitioning has been shown to be a function of the vapor pressure of the species and the amount of aerosol present in the atmosphere (Yamasaki, Kuwata, and Miyamoto 1982). Both theoretical (Junge 1977; Pankow 1987) and experimental (Bidleman, Billings, and Foreman 1986; Ligocki and Pankow 1989) evidence indicates that adsorption to particles is negligible for species with vapor pressures greater than 10^{-4} torr. Because vapor pressure is a strong function of temperature, species with vapor pressures near 10^{-6} torr may be found predominantly in the gas phase in summer and daytime, yet be found predominantly in the particulate phase in the winter and nighttime. Benzo[a]pyrene (B[a]P) is an example of a toxic air pollutant that is associated nearly exclusively with particles in urban atmospheres, yet in more remote environments where atmospheric particle loadings are low, a substantial fraction of B[a]P is found in the gas phase (Baker and Eisenreich 1990).

PHYSICAL REMOVAL PROCESSES

The physical processes of wet and dry deposition also can be significant removal routes for some atmospheric pollutants. Wet deposition refers to the capture and removal of species by hydrometeors (e.g., rain, snow, and hail). Dry deposition refers to the loss of atmospheric species to surfaces by diffusion, sedimentation, and impaction.

For particulate species, the rate of removal by wet and dry deposition depends primarily on the particle size distribution. Large particles are removed rapidly from the atmosphere by inertial processes such as sedimentation and impaction. Smaller particles do not contain sufficient mass for inertial processes to be efficient, but diffuse much more rapidly than do large particles. As a result, removal rates of atmospheric particles are governed by the competition between these two types of processes, and generally reach a minimum somewhere in the range 0.1 to 1 micron (μm). This size range often is referred to as the accumulation mode, because particles in this size range tend to persist and, hence, accumulate.

For gaseous species, the rate of removal by wet deposition will depend on the Henry's Law constant (Ligocki, Leuenberger, and Pankow 1985). Dry deposition rates of gaseous species are less well understood; most work has focused on understanding air–sea exchange (e.g. Slinn et al. 1978). Even for land surfaces, dry deposition of gaseous species has been parameterized in terms of the Henry's Law constant (Tucker and Preston 1984; Wesely 1989) because many relevant deposition surfaces, such as leaf stomata, are aqueous media.

ATMOSPHERIC LIFETIMES

In order to compare the persistence of various species in the atmosphere, the concept of atmospheric lifetime τ often is used. The lifetime represents the time required for a fixed concentration of a species to decay to a certain percentage of its initial concentration. Atmospheric lifetimes of toxic air pollutants can vary from minutes to years.

It is important to recognize the limitations of characterizing atmospheric persistence in terms of lifetimes. Lifetime calculations do not incorporate chemical production rates for secondary species. Thus, lifetime calculations may indicate that a species such as formaldehyde is removed rapidly during the daytime, when formaldehyde is being produced more rapidly than it is being removed. Also, lifetime calculations consider atmospheric reactions as destruction processes and do not consider the possible transformation of toxic species into equally toxic products. Despite these limitations, atmospheric lifetime calculations can be valuable when viewed in context with these other issues.

All atmospheric removal processes can be represented by mathematical equations relating the removal rate R of the species to its atmospheric concentration C. Removal processes for which the removal rate is directly proportional to the concentration are termed first-order, or linear processes, and

$$R = kC \qquad (12\text{-}1)$$

The constant k depends on factors such as temperature and time of day, but is considered to be constant under a given set of conditions. This assumption of linearity, or near-linearity, is necessary for the calculation of lifetimes.

The solution to Equation 12-1 is:

$$C(t) = C_0 e^{-kt} \qquad (12\text{-}2)$$

Where $C(t)$ represents the concentration at time t

C_0 represents the initial concentration.

The lifetime, τ, is defined to be equal to $1/k$ and, thus, represents the time at which the product kt is equal to 1. At that time, the concentration has been reduced to 37 percent of its original value.

Each atmospheric removal process will have a characteristic lifetime associated with it for a given set of conditions. The overall lifetime for a toxic species under a given set of conditions is obtained by summing the individual lifetimes in a manner analogous to the summation of parallel electrical resistances

$$\tau = \left(\Sigma_i \, \tau_i^{-1} \right)^{-1} \qquad (12\text{-}3)$$

Where τ_i represents the lifetime due to the individual processes.

For example, if it has been determined that a certain toxic substance has a chemical lifetime of one day, and a lifetime due to dry deposition of three days, its overall atmospheric lifetime is

$$\tau = \frac{1}{1/1 + 1/3} = 0.75 \text{ day} \quad (12\text{-}4)$$

Atmospheric lifetimes due to gas-phase chemical reaction τ_g can be calculated

$$\tau_g = (k_g[x]_g)^{-1} \quad (12\text{-}5)$$

Where k_g is the reaction rate constant
$[x]_g$ represents the concentration of the appropriate reactant, such as OH and O_3.

Thus, in addition to rate constants, typical atmospheric oxidant concentrations are required for lifetime calculations. Concentrations of OH radicals are of particular importance, because chemical lifetimes for many atmospheric species are determined by their rate of reaction with the OH radical. Few reliable measurements of ambient OH concentrations exist; for estimating atmospheric lifetimes it is probably adequate to use OH concentrations calculated from a simple model as a function of latitude, time of day, and month of the year, such as that presented by Altshuller (1979).

A good first approximation for daytime average ambient OH concentrations is 10^{-7} ppm. For the most reactive toxics (e.g., butadiene, styrene, aniline), this translates to atmospheric lifetimes on the order of a couple of hours. Summer, noontime OH concentrations can reach 10^{-6} ppm. Under these conditions, the most reactive toxics have lifetimes on the order of minutes.

Atmospheric lifetimes for the representative toxic species benzene, butadiene, formaldehyde, acetaldehyde, and POM have been estimated under a variety of conditions, including day/night, clear/cloudy, and summer/winter (Ligocki et al. 1991; Ligocki and Whitten 1991). These calculations show for most species that seasonal and day/night differences in atmospheric lifetime can be large, with much shorter lifetimes in summer and in the daytime than in winter or at night.

For species that partition between the gas and particulate phases, the expression on the right-

hand side of Equation (12-5) must be multiplied by the gas-phase fraction of the species concentration f_g. An expression analogous to Equation (12-5) for the atmospheric lifetime due to particulate-phase chemical reaction τ_p can be written as

$$\tau_p = (k_p[x]_g f_p)^{-1} \quad (12\text{-}6)$$

Where k_p is the particulate-phase rate constant
f_p is the particulate-phase fraction.

However, very little is known about reactivity in the particulate phase. Despite years of research on the subject, reported rates of reactivity are often contradictory (Grosjean, Fung, and Harrison 1983; Van Vaeck and Van Cauwenberghe 1984) or so dependent on particle composition (Korfmacher et al. 1980) that generalizations cannot be made for a typical ambient aerosol.

Nonprecipitating clouds affect the lifetime of atmospheric pollutants in two major ways. First, clouds affect the solar ultraviolet (UV) radiation at ground level. Dense clouds will decrease UV, slowing photolysis rates and decreasing radical concentrations, but thin or broken clouds may enhance UV at ground level. Second, clouds are a reactive medium in which rapid chemical transformation will take place. The role of clouds in the atmospheric production of acids has been known for years; the importance of clouds to other aspects of atmospheric chemistry is just beginning to be investigated (Lelieveld and Crutzen 1990). The presence of clouds has been estimated to increase formaldehyde lifetimes during the day but decrease them at night (Ligocki et al. 1991).

The atmospheric lifetime due to in-cloud chemical destruction τ_{aq} is given by

$$\tau_{ag} = (k_{aq}[x]_{aq} f_{aq})^{-1} = \left(\frac{k_{ag}[x]_{ag}}{1 + a/H^*} \right)^{-1}$$
$$(12\text{-}7)$$

Where k_{aq} is the in-cloud reaction rate
$[x]_{aq}$ is the in-cloud oxidant (e.g., OH) concentration

f_{aq} is the fraction of the total concentration that is present in the aqueous phase

The fraction f_{aq} depends on H^*, the effective Henry's Law constant. Thus, even if in-cloud destruction of a particular substance is rapid, it will not lead to a short lifetime unless a large fraction of the total atmospheric concentration is present in the aqueous phase.

The overall expression for chemical lifetime τ_c includes the contributions from gas-phase chemical reaction, particulate-phase chemical reaction, and aqueous phase chemical reaction

$$\tau_c = (\tau_{c,g}^{-1} + \tau_{c,p}^{-1} + \tau_{c,aq}^{-1})^{-1} \quad (12\text{-}8)$$
$$= (k_g[x]_g f_g + k_p[x]_g f_p$$
$$+ k_{aq}[x]_{aq} f_{aq})^{-1}$$

Precipitating clouds clearly have a more dramatic effect on atmospheric lifetimes than nonprecipitating clouds. Soluble gases and particles are removed efficiently from the atmosphere during rainstorms in a matter of hours. For gases, a lifetime due to wet deposition $\tau_{w,g}$ can be calculated as

$$\tau_{w,g} = \left(\frac{RTH^*r}{z_w} \right)^{-1} \quad (12\text{-}9)$$

Where R is the gas constant
   T is temperature
   r is the rainfall rate
   z_w is the thickness of the layer through which rain is falling.

For 'typical' conditions, $\tau_{w,g}$ for soluble toxics such as formaldehyde and cresol is on the order of a few hours.

For toxics that are associated with particles, the lifetime due to wet deposition $\tau_{w,p}$ depends on the particle diameter d_p, the hygroscopic/hydrophobic nature of the particle surface, and the type of storm and other meteorological and cloud physics parameters. The two methods most commonly used to estimate wet deposition of particle-associated species involve the scavenging coefficient Λ and the scavenging ratio

W_p. Theoretical values of the scavenging coefficient have been determined as a function of rainfall rate and the geometric mean and geometric standard deviation of the particle size distribution (Dana and Hales 1976). For size distributions appropriate to many toxics, and a rainfall rate of 5 millimeters per hour (mm/h), the value of Λ is about $2\,h^{-1}$, corresponding to a $\tau_{w,p}$ value of $0.5\,h$. Measured scavenging ratios for semivolatile organics have been reviewed by Bidleman (1988), who suggests that W_p for SVOCs may typically fall in the range 10^5 to 10^6. This approach also leads to lifetime estimates of about one hour.

For species that exist in both gas and particulate phases, the overall lifetime due to wet deposition τ_w is obtained from

$$\tau_w = \left(\frac{f_g}{\tau_{w,g}} + \frac{f_p}{\tau_{w,p}} \right)^{-1} \quad (12\text{-}10)$$

Where f_g and f_p are the gaseous and particulate fractions of the concentration, respectively.

Wet deposition is clearly an efficient removal mechanism, but it is episodic. On average, dry deposition is often more important because, although slow, it is continuous. The most common measure of dry deposition rate is the deposition velocity v_d. Lifetimes due to dry deposition τ_d are obtained from deposition velocities from

$$\tau_d = \left(\frac{v_d}{z_d} \right)^{-1} \quad (12\text{-}11)$$

Where z_d is the thickness of the mixed layer.

For particles, the dry deposition velocity depends on wind speed, atmospheric stability, and particle diameter and density (Slinn et al. 1978). Toxic air pollutants often are associated with submicron sized particles; deposition velocities for particles in that size range are on the order of 0.1 to 1 cm/s, leading to lifetimes on the order of days. For gaseous toxics, deposition velocities are not well known. Dry deposition velocities for several chlorinated pesticides (Bidleman and Christensen 1979) and other hydrocarbons (Farmer and Wade 1986) have been reported. The measured values generally fall in the range 0.01 to 0.1 centimeters per second (cm/s). These values translate to atmospheric lifetimes on the order of weeks. For many gas-phase toxics, deposition is a much slower removal process than chemical reaction.

References

Altshuller, A.P. 1979. Model predictions of the rate of homogeneous oxidation of sulfur dioxide to sulfate in the troposphere. *Atmospheric Environment* 13(12):1653–1662.

Arey, J., B. Zielinska, R. Atkinson, A.M. Winer, T. Ramdahl, and J.N. Pitts, Jr. 1986. The formation of nitro-PAH from the gas-phase reactions of fluoranthene and pyrene with the OH radical in the presence of NO_x. *Atmospheric Environment* 20:2339–2345.

Atkinson, R. 1986. Kinetics and mechanisms of the gas-phase reactions of hydroxyl radical with organic compounds under atmospheric conditions. *Chemical Review* 86:69–201.

Atkinson, R. 1987. Estimation of OH radical reaction rate constants and atmospheric lifetimes for polychlorobiphenyls, dibenzo-*p*-dioxins and dibenzofurans. *Environmental Science and Technology* 21:305–307.

Atkinson, R. 1990. Gas-phase tropospheric chemistry of organic compounds: a review. *Atmospheric Environment* 24A(1):1–43.

Atkinson, R., D.L. Baulch, R.A. Cox, R.F. Hampson, Jr., J.A. Kerr, and J. Troe. 1989. Evaluated kinetic and photochemical data for atmospheric chemistry: Supplement III. *Journal of Physical Chemistry Reference Data* 18:881–1097.

Baker, J.E. and S.J. Eisenreich. 1990. Concentrations and fluxes of polycyclic aromatic hydrocarbons and polychlorinated biphenyls across the air–water interface of Lake Superior. *Environmental Science and Technology* 24:342–352.

Bidleman, T.F. 1988. Atmospheric processes. *Environmental Science and Technology* 22:361–367.

Bidleman, T.F. and E.J. Christensen. 1979. Atmospheric removal processes for high molecular weight organochlorines. *Journal of Geophysical Research* 84:7857–7862.

Bidleman, T.F., W.N. Billings, and W.T. Foreman. 1986. Vapor-particle partitioning of semi-volatile organic compounds: Estimates from field collections. *Environmental Science and Technology* 20:1038–1043.

Bielski, B.H.J., D.E. Cabelli, R.L. Arudi, and A.B. Ross. 1985. Reactivity of HO_2/O_2^- radicals in aqueous solution. *Journal of Physical Chemistry Reference Data* 14:1041–1100.

Buxton, G.V., C. L. Greenstock, W.P. Helman, and A.B. Ross. 1988. Critical review of rate constants for reactions of hydrated electrons, hydrogen atoms, and hydroxyl radicals in aqueous solution. *Journal of Physical Chemistry Reference Data* 17:513–886.

Dana, M.T. and J.M. Hales. 1976. Statistical aspects of the washout of polydisperse aerosols. *Atmospheric Environment* 10:45–50.

DeMore, W.B., D.M. Golden, R.F. Hampson, C.J. Howard, M.J. Kurylo, M.J. Molina, A.R. Ravishankara, and S.P. Sander. 1987. Chemical kinetics and photochemical data for use in stratospheric modeling. Evaluation #8. JPL Publ. 87–41. Pasadena, CA.

Farmer, C.T. and T.L. Wade. 1986. Relationship of ambient atmospheric hydrocarbon (C_{12}–C_{32}) concentrations to deposition. *Water, Air, and Soil Pollution* 29:439–452.

Grosjean, D., K. Fung, and J. Harrison. 1983. Interactions of polycyclic aromatic hydrocarbons with atmospheric pollutants. *Environmental Science and Technology* 17:673–679.

Gunz, D.W. and M.R. Hoffmann. 1990. Atmospheric chemistry of peroxides: A review. *Atmospheric Environment* 24A:1601–1634.

Junge, C.E. 1977. Basic considerations about trace constituents in the atmosphere as related to the fate of global pollutants. In: *Fate of Pollutants in the Air and Water Environment*, I. H. Suffet, ed. New York: John Wiley & Sons.

Korfmacher, W.A., E.L. Wehry, G. Mamantov, and D.F.S. Natusch. 1980. Resistance to photochemical decomposition of polycyclic aromatic hydrocarbons vapor-adsorbed on coal fly ash. *Environmental Science and Technology* 14:1094–1099.

Lelieveld, J. and P.J. Crutzen. 1990. Influences of cloud photochemical processes on tropospheric ozone. *Nature* 343:227–233.

Ligocki, M.P., C. Leuenberger, and J.F. Pankow. 1985. Trace organic compounds in rain. II. Gas scavenging of neutral organic compounds. *Atmospheric Environment* 19:1609–1617.

Ligocki, M.P. and J.F. Pankow. 1989. Measurements of the gas/particle distributions of atmospheric organic compounds. *Environmental Science and Technology* 23:75–83.

Ligocki, M.P. and G.Z. Whitten. 1991. Atmospheric transformations of air toxics: acetaldehyde and polycyclic organic matter (POM). SYSAPP-91/113. Systems Applications International, San Rafael, CA.

Ligocki, M.P., G.Z. Whitten, R.R. Schulhof, M.C. Causley, and G.M. Smylie. 1991. Atmospheric transformations of air toxics: benzene, 1,3-butadiene, and formaldehyde. SYSAPP-91/106. Systems Applications International, San Rafael, CA.

Lyman, W.J., W.F. Reehl, and D.H. Rosenblatt. 1982. *Handbook of Chemical Property Estimation Methods.* New York: McGraw-Hill.

Mackay, D. and W.Y. Shiu. 1981. A critical review of Henry's Law constants for chemicals of environmental interest. *Journal of Physical and Chemical Reference Data* 10:1175–1199.

National Institute of Standards and Technology (NIST). 1991. *Chemical Kinetics Database* (Electronic version). Gaithersburg, MD.

National Research Council (NRC). 1981. *Formaldehyde and Other Aldehydes.* Washington, D.C. National Academy Press.

Pankow, J.P. 1987. Review and comparative analysis of the theories on partitioning between the gas and aerosol particulate phases in the atmosphere. *Atmospheric Environment* 21:2275–2283.

Slinn, W.G., L. Hasse, B.B. Hicks, A.W. Hogan, D. Lal, P.S. Liss, K.O. Munnich, G.A. Sehmel, and O. Vittori. 1978. Some aspects of the transfer of atmospheric trace constituents past the air–sea interface. *Atmospheric Environment* 12:2055–2087.

Tucker, W.A. and A.L. Preston. 1984. Procedures for estimating atmospheric deposition properties of organic chemicals. *Water, Air, and Soil Pollution* 21:247–260.

Van Vaeck, L. and K. Van Cauwenberghe. 1984. Conversion of polycyclic aromatic hydrocarbons on diesel particulate matter upon exposure to ppm levels of ozone. *Atmospheric Environment* 18:323–328.

Wesely, M.L. 1989. Parameterization of surface resistances to gaseous dry deposition in regional-scale numerical models. *Atmospheric Environment* 23:1293–1304.

Yamasaki, H., K. Kuwata, and H. Miyamoto. 1982. Effects of ambient temperature on aspects of airborne polycyclic aromatic hydrocarbons. *Environmental Science and Technology* 16:189–194.

13

Human Intake

Sarah A. Foster

INTRODUCTION

A crucial step in evaluating exposure and calculating potential risks is the prediction of chemical intakes into the human body. Estimation of the intake of chemicals for a risk assessment can be required as part of the permit process, to provide information to the public, and to assist in the siting of facilities.

The methods for calculating human intakes described in this chapter focus on practical applications of exposure assessment in which the exposure is estimated from environmental concentration data (e.g., air concentrations, soil concentrations). As discussed in Chapter 8, refined methods, such as biochemical markers and personal breath samples, are available that can be used to predict exposures more accurately. However, these methods are not yet applied widely in evaluating sources of air toxics, and they can involve more significant time and resources than the methods described here. If more refined estimates of exposure are required, the reader should consider the approaches described in NAS (1991a).

The terminology used in this chapter for calculating human intakes conforms to exposure assessment guidance provided in EPA (1992a). Following this guidance, calculated exposures are referred to as 'doses' rather than 'intakes.' The methods used to calculate doses for application in an exposure assessment generally are similar, although the meaning of the calculated doses and the specific terms used to describe them can differ.

The five major steps involved in estimating human doses for air toxics are listed below and then discussed in more detail to provide a general understanding of the processes involved.

1. Identify relevant exposure pathways
2. Compile algorithms to calculate chemical doses
3. Identify the parameters needed to calculate doses
4. Develop exposure scenarios addressing these parameters
5. Calculate chemical doses

IDENTIFY EXPOSURE PATHWAYS

The number and variety of exposure pathways evaluated in an exposure assessment can range from a straightforward, single, direct inhalation pathway to a much more complex and time-consuming analysis of a dozen varied pathways. To identify air toxics exposure pathways

249

properly, the specific problem being addressed needs to be conceptualized with respect to both the source under consideration and the level of technical complexity required. In some cases, a screening analysis involving only the inhalation pathway may be sufficient to address the issue under consideration. For more site-specific assessments, a refined analysis may be required involving evaluation of multiple exposure pathways. The key to success begins with this conceptualization, as it directs the identification of exposure pathways and ultimately provides the foundation for calculating doses.

For toxic air pollutants, inhalation is not the only exposure pathway of possible concern. Indirect pathways involving ingestion and dermal absorption through the skin also can be of interest. Over the past several years, there has been increased interest in addressing indirect exposure pathways in addition to direct inhalation for sources of air toxics. In rural areas especially, public concerns often focus on the potential chemical doses associated with ingestion of locally produced agricultural crops, beef and dairy products, and fish. In these cases, indirect pathways are potentially important because chemicals present in air emissions can be deposited from an exhaust plume onto the earth's surface and become available for uptake into a variety of flora and fauna.

The conceptual model of an exposure assessment considers several factors. First, the type of air toxics source and its configuration are important. The pathways that are relevant for a continuous emissions source of particulate matter from an exhaust stack differ from those of potential concern for a release of vapors due to a chemical spill. The location of the receptor or receptors (e.g., humans) of concern also should be considered. There is a difference in the degree of complexity of an evaluation that focuses on a single hypothetical receptor (i.e., a person at the maximum impact point) compared to one that focuses on many receptors within a large area surrounding a source. For complex sources, especially those located in densely developed areas (e.g., hospital incinerators), exposures at elevated receptor locations, such as building air intakes, may be of concern. Further,

INPUTS

Factors	Examples
Characterize the source	Stack, fugitive, spill, area
Characterize the time frame of the emissions	Continuous, intermittent, short-term incident
Characterize the type of release	Vapor, particulate matter
Identify receptors of concern	Maximum impact Single/multiple Current/potential Ground level, elevated

OUTPUTS

Factors	Examples
Identify type of pathways	Direct inhalation (vapor and particulate) Indirect ingestion/ dermal (particulate)
Identify type of exposure	Chronic, long-term (continuous emission) Acute, short-term (continuous, intermittent, and short-term incident)

Figure 13-1.   Factors to consider in developing a conceptual model for exposure assessment dose calculations.

it is important to consider whether doses are to be calculated for current land use conditions in the source area, or whether potential future land uses are to be considered.

Figure 13-1 shows some of the factors (i.e., inputs) that must be considered in developing a conceptual model for an exposure assessment, that in turn dictate what types of dose calculations (i.e., outputs) should be made. Figure 13-2 highlights the wide array of direct and indirect pathways that may need to be evaluated for an air toxics source. Some sources of air toxics, such as incinerators, receive considerable public attention. As a result, assessments for these facilities involve the evaluation of many of the exposure pathways noted in this figure. In the future, it is likely that other air toxics sources will come under increased public scrutiny, leading to similarly refined assessments.

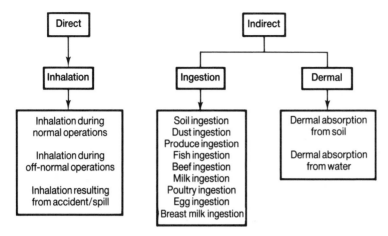

Figure 13-2.   Exposure pathways relevant for emissions from air toxics sources.

HUMAN DOSE CALCULATIONS

This section focuses on the computational procedures used to calculate doses for use in a quantitative exposure and risk assessment. The results of these calculations can be combined with toxicity data (i.e., dose-response information) to predict potential risks.

The approach commonly applied in exposure assessment for most chemicals (N.B., one special exception is lead) involves calculating doses in units of milligrams of chemical per kilogram of body weight per day of exposure (mg/kg per day). A similar approach is used to calculate doses for inhalation, ingestion, and dermal pathways although the meaning and terms used to describe them differ (EPA 1992a). For oral and inhalation pathways of exposure, the doses are referred to as 'potential doses.' The potential dose is the amount of chemical ingested or inhaled, and is analogous to the 'administered dose' in dose-response studies. For the dermal absorption pathway, the dose is referred to as an 'internal dose.' It reflects the amount of chemical that has been absorbed and is available for interaction with biologically important tissues. To evaluate chronic, long-term exposures, these doses are calculated as average daily doses, averaged over the duration of exposure (i.e., to evaluate noncancer effects) or averaged over a lifetime (i.e., to evaluate cancer risks). For the inhalation pathway, exposures also can be estimated in air concentration units of milligrams per cubic meter (mg/m^3), often referred as inhalation exposure concentrations, or IECs.[1]

The calculation of exposures as average daily doses or IECs is designed for compatibility with the dose-response data used to quantify risks. The most readily available sources of verified dose-response data are EPA's Integrated Risk Information System (IRIS) and Health Effects Assessment Summary Tables (HEASTs). These publicly available data bases contain, among other information, the following four critical types of dose-response toxicity values.

---

[1] The choice of approach for estimating inhalation exposures depends upon the availability of compatible toxicity data, the desired level of refinement of the exposure assessment, and the preferred method by the reviewing regulatory agency (if applicable). Different EPA Regional Offices, for example, may recommend using dose units or air concentration units in exposure assessment. The air concentration method does not, however, allow for consideration of age-specific or activity-specific variations (e.g., body weight, inhalation rate) and, thus, generally is appropriate only for a screening level assessment.

1. Oral and inhalation cancer slope factors in (mg/kg per day)$^{-1}$ for direct combination with lifetime average daily doses, to evaluate cancer risks.
2. Unit risk factors in (mg/m^3)$^{-1}$ for direct combination with IECs, to evaluate inhalation cancer risks.
3. Oral reference doses in (mg/kg per day) for direct combination with average daily doses, to evaluate noncancer effects.
4. Reference air concentrations in (mg/m^3) for direct combination with IECs, to evaluate noncancer effects.

Calculation of human doses compatible with available dose-response data is critical. This means that doses should match dose-response data in two major ways: (1) the units should be the same, and (2) the average daily doses should be calculated as either an internal dose (i.e., into the body) or a potential dose (i.e., at the exposure/body interface) depending on the type of dose that formed the basis of the dose-response data. For almost all chemicals, exposures should be calculated as potential doses because dose-response values typically are based on these.

In general, consistent with EPA's guidelines (EPA 1986g; EPA 1989j), IECs and average daily doses are averaged over a seventy-year lifetime for carcinogens and over the duration of exposure for noncarcinogens. For regulatory purposes, the overall exposure period for a continuous air toxics source often is assumed to be seventy years. For some pathways or sources, however, shorter exposure durations may be appropriate (e.g., an infant exposed via human breast-milk ingestion may conservatively be assumed to be exposed for one year).

Example calculation procedures are provided and briefly discussed below for several pathways. This is not intended to be an exhaustive presentation but, instead, one that is illustrative of the methods used to calculate exposures in an applied risk assessment. For more information, the reader should refer to EPA (1989j) and EPA (1992a).

Inhalation of Ambient Air

Inhalation exposure concentrations (IECs) for exposures to chemicals in ambient air are calculated using

$$\text{IEC} = C_a \times \frac{ET}{24} \times \frac{EF}{365} \times \frac{ED}{70} \times \text{BIO} \quad \text{(13-1)}$$

Where IEC = inhalation exposure concentration (mg/m^3)

C_a = chemical concentration in air (mg/m^3)

ET = exposure time (hours/day)

EF = frequency of exposure (days/year)

ED = duration of exposure (years)

BIO = relative inhalation bioavailability factor, which adjusts for differences in chemical bioavailability if applicable (unitless)

Inhalation exposures also can be calculated in dose units as

$$\text{ADD, LADD} = \frac{C_a \times \text{BIO}}{AT \times \text{Days}}$$

$$\times \sum_{i=1}^{n} \frac{HR_i \times EF_i \times ED_i}{BW_i}$$

$$\text{(13-2)}$$

Where ADD = average daily dose for non-cancer effects (mg/kg per day)

LADD = lifetime average daily dose for cancer effects (mg/kg per day)

C_a = chemical concentration in air (mg/m^3)

BIO = relative inhalation bioavailability factor, which adjusts for differences in chemical bioavailability if applicable (unitless)

HR_i = inhalation rate in age period i (m^3/day)

EF_i = frequency of exposure in age period i (days/year)

ED_i = duration of exposure in age period i (years)

BW_i = average body weight in age period i (kg)

AT = averaging time (70 years for carcinogens, duration of exposure for noncarcinogens)

Days = conversion factor (365 days/year)

These equations can be modified to incorporate toxic equivalency factors (TEFs) when dealing with complex chemical mixtures such as polychlorinated dibenzo-*p*-dioxins (PCDDs) and polychlorinated dibenzofurans (PCDFs), for example

$$IEC = C_a \times \frac{ET}{24} \times \frac{EF}{365} \times \frac{ED}{70} \times BIO \times TEF$$

(13-3)

Where TEF = toxic equivalency factor (for calculating IECs for PCDDs/PCDFs)

Exposures to PCDD/PCDF congeners are calculated as 2,3,7,8-TCDD toxic equivalents because 2,3,7,8-TCDD has been shown to be the most potent carcinogen member of the class of PCDDs/PCDFs. The TEFs used to derive IECs, ADDs, and LADDs for the various congeners are specified by EPA (1989a) and are presented in Table 13-1.

Another factor affecting the IEC, ADD, and LADD calculation is a chemical's bioavailability. Bioavailability reflects both the extent to which a chemical becomes available to be absorbed (i.e., the extent to which a chemical is desorbed from an inhaled or ingested fly ash or particulate matter matrix), as well as its subsequent absorption into the body. The absorption is the extent to which a chemical is transported across, or moves through, the lining of the lung or gut into the body.

In some cases, a chemical's bioavailability from one environmental medium (e.g., fly ash or soil) differs from its bioavailability from the vehicle used in toxicity studies (e.g., vapors, solvent, or solution) from which risk assessment

Table 13-1   Toxic equivalency factors (TEFs) for PCDD/PCDF congeners

Toxic equivalency congener	Factor (TEF)[a]
2,3,7,8-TCDD	1
Other TCDDs	0
2,3,7,8-PeCDD	0.5
Other PeCDDs	0
2,3,7,8-HxCDD	0.1
Other HxCDDs	0
2,3,7,8-HpCDD	0.01
Other HpCDDs	0
OCDD	0.001
2,3,7,8-TCDF	0.1
Other TCDFs	0
1,2,3,7,8-PeCDF	0.05
2,3,4,7,8-PeCDF	0.5
Other PeCDFs	0
2,3,7,8-HxCDF	0.1
Other HxCDFs	0
2,3,7,8-HpCDF	0.01
Other HpCDFs	0
OCDF	0.001

[a] TEFs represent a consensus of North Atlantic Treaty Organization (NATO) members and are compatible with EPA (1989a) recommendations for calculating exposures to PCDDs/PCDFs.

toxicity criteria are derived. For example, most chemicals emitted from a combustion source are adsorbed to fly ash particles or exist as gases at ambient temperatures and pressures. Most chemicals released as gaseous fugitive emissions travel freely as vapors or aerosols. Chemicals inhaled in a fly ash or particulate matter matrix may be less bioavailable than chemicals inhaled in a gaseous or aerosol form (Fraser and Lum 1983; Poiger and Schlatter 1980). In these cases, the IEC, ADD, or LADD can be multiplied by a relative bioavailability factor (BIO) to account for the chemical's reduced bioavailability from fly ash compared to the matrix in which it was tested in a toxicity study. This unitless factor represents the ratio of the chemical's bioavailability from an environmental exposure medium (e.g., fly ash) to its bioavailability from the vehicle used in the relevant toxicity study. A relative bioavailability factor of 1.0 indicates that the chemical is equally bioavailable from both the environmental exposure medium and the relevant toxicity study vehicle.

In most assessments, inhalation exposures usually are not adjusted to reflect reduced chemical bioavailability. Thus, a default assumption that BIO = 1.0 may overestimate exposures and risks, especially for chemicals with systemic effects (i.e., effects at target organs other than the lung).

Incidental Ingestion of Soil or Dust

Exposures resulting from incidental ingestion of soil or dust are estimated by calculating average daily doses

$$\text{ADD, LADD} = \frac{C_{s/d} \times \text{BIO} \times CF}{AT \times \text{Days}}$$

$$\times \sum_{i=1}^{n} \frac{IR_{s,d,i} \times EF_i \times ED_i}{BW_i}$$

(13-4)

Where ADD = average daily dose for non-cancer effects (mg/kg per day)

LADD = lifetime average daily dose for cancer effects (mg/kg per day)

$C_{s/d}$ = chemical concentration in soil or dust (mg/kg)

BIO = relative oral bioavailability factor, which adjusts for differences in chemical bioavailablilty if applicable (unitless)

CF = conversion factor (kg/ 10^6 mg)

$IR_{s/d,i}$ = soil or dust ingestion rate in age period i (mg/day)

EF_i = frequency of exposure in age period i (days/year)

ED_i = duration of exposure in age period i (years)

BW_i = average body weight in age period i (kg)

AT = averaging time (70 years for carcinogens, duration of exposure for carcinogens)

Days = conversion factor (365 days/ year)

As noted above, this dose equation also can be multiplied by a TEF parameter when dealing with complex chemical mixtures such as PCDDs/PCDFs.

The soil and dust ingestion dose equation also includes a bioavailability factor (BIO). This is because, as noted above, chemicals ingested in a fly ash or soil matrix may be less bioavailable than when ingested in food, solvent, or the solution used in toxicity studies. This reduction in bioavailability likely is most important for highly hydrophobic and halogenated chemicals (e.g., those with large octanol-water coefficients, K_{ow}) and for metals. Thus, relative oral bioavailability factors can be used to adjust for this difference in bioavailability for the soil or dust ingestion pathway. The relative oral bioavailability factor is the ratio of a chemical's bioavailability (i.e., ability to be absorbed and potentially exert an effect) when administered in a fly ash or soil matrix compared to its bioavailability when administered in the vehicle used in the toxicity study (e.g., solvent or solution).

Relative oral bioavailability factors can be derived from studies on the gastrointestinal absorption of 2,3,7,8-TCDD adsorbed onto soil and fly ash (Poiger and Schlatter 1980; McConnell et al. 1984; Lucier et al. 1986; Van den Berg et al. 1986; Van den Berg, Sinke, and Wever 1987; Wendling et al. 1989). The experimental evidence indicates that the oral bioavailability of 2,3,7,8-TCDD from a fly ash or soil matrix may range from 7 to 50 percent when compared to the bioavailability of 2,3,7,8-TCDD from a solution matrix (e.g., corn oil). The fraction that becomes bioavailable is dependent in part on the composition of the soil or particulate matrix (e.g., amount of organic carbon) and the length of contact between the soil and the chemical. An average TCDD relative oral bioavailability factor derived from several experimental studies on 2,3,7,8-TCDD is 21 percent. Because the toxicity data for 2,3,7,8-TCDD are based on a dietary feeding study, and the relative bioavailability of 2,3,7,8-TCDD from feed versus soil or fly ash is likely to be greater than from solution, an average value for its relative oral bioavailability factor likely is much higher than 21 percent.

For large aromatic chemicals that are highly hydrophobic (e.g., chemicals with a log $K_{ow} > 3$) and that have three or more chlorine atoms, a reasonable assumption is that their adsorptive properties will approach those of 2,3,7,8-TCDD, because they have similar physicochemical properties. For example, the effect of a fly ash, particle, or soil matrix on the gastrointestinal bioavailability of 1,2,4-trichlorobenzene (a large aromatic hydrophobic, chlorinated compound with a log $K_{ow} \simeq 4$) is expected to resemble that for 2,3,7,8-TCDD. For those chemicals that are hydrophilic (chemicals with a log $K_{ow} < 3$), or that have two or less chlorine atoms and are not aromatic compounds, a default relative oral bioavailability factor of 1.0 should be used unless specific experimental data are available. This conservatively assumes that there is no difference in bioavailability between chemicals incorporated in a fly ash, particulate matter, or soil matrix, and chemicals in the vehicle from which the toxicity criteria were derived in the animal study.

For metals, only a few experimental studies on relative bioavailability have been conducted; these have focused primarily on lead and arsenic bioavailability from soil contaminated by mine tailings (Johnson, Freeman, and Killinger 1989; Davis, Ruby, and Bergstrom 1992). Some studies also have shown that bioavailability of metals such as lead is strongly influenced by particle size (Healy et al. 1982). Lacking experimental bioavailability data for fly ash or emitted particulate matter, a rough estimate of the relative bioavailability of metals can be obtained from fly ash leaching studies that are readily available. For example, Fraser and Lum (1983) used fly ash obtained from the Hamilton (Ontario) municipal combustion facility to obtain the fractions of various chemicals that could be leached from ash following several intensive extraction procedures at a low pH. These fractions can be assumed to be representative of relative oral bioavailability factors based on the assumption that after a chemical is leached from a fly ash matrix, it is as available for absorption as it would be if administered in food, solvent, or solution.

A reasonable approach for the dust ingestion pathway is to evaluate one age period representative of a child (e.g., one to six years). Because indoor dust may be composed primarily of outdoor soil, exposures to chemicals from household dust ingestion can be adjusted by the same relative oral bioavailability factors developed for chemicals in soil.

Ingestion of Produce

Exposures resulting from ingestion of fresh produce (e.g., vine, leafy, and below-ground root crops) are estimated by

$$ADD, LADD = \frac{C_{lvr} \times CF \times BIO \times Prep}{AT \times Days}$$

$$\times \sum_{i=1}^{n} \frac{IR_{lvr,i} \times EF_i \times ED_i}{BW_i}$$

(13-5)

Where ADD = average daily dose for noncancer effects (mg/kg per day)

LADD = lifetime average daily dose for cancer effects (mg/kg per day)

C_{lvr} = chemical concentration in leafy, vine, or root crop (mg/kg)

CF = conversion factor (kg/10^3 g)

BIO = relative oral bioavailability factor, which adjusts for differences in chemical bioavailability, if applicable (unitless)

Prep = reductions in concentration due to food preparation (unitless)

$IR_{lvr,i}$ = ingestion rate for leafy, vine, or root crops in age period i (g/day)

EF_i = exposure frequency in age period i (days/year)

ED_i = duration of exposure in age period i (years)

BW_i = average body weight in age period i (kg)

AT = averaging time (70 years for carcinogens, duration of exposure for noncarcinogens)

Days = conversion factor (365 days/year)

Doses from ingestion of above-ground vine and leafy crops contaminated by direct deposition of particles onto exposed surfaces can be modified by the relative oral bioavailability factors used to calculate soil ingestion ADDs or LADDs. This is because chemicals deposited onto produce surfaces are assumed to be bound in fly ash or other particulate matter and, therefore, would not be as bioavailable as in the food or solvent matrix used in a toxicity study. For intakes associated with concentrations due to translocation of chemicals from the soil, a relative oral bioavailability factor of 1.0 typically is used because chemicals are assumed to be bound in the plant tissues (e.g., the food or solvent matrix, as in the toxicity study).

A conservative assumption that can be applied is that chemical concentrations present in produce would not be decreased in any way before ingestion, for example as a result of peeling, washing, or cooking (i.e., Prep = 1.0). Results from relevant experimental studies can be used to modify this food preparation parameter.

Ingestion of Milk and Beef

Exposures for ingestion of beef and milk are estimated using

$$ADD, LADD = \frac{C_{b/m} \times CF \times BIO \times FL_{b/m} \times Prep}{AT \times Days}$$

$$\times \sum_{i=1}^{n} \frac{IR_{b/m,i} \times EF_i \times ED_i}{BW_i}$$

(13-6)

Where ADD = average daily dose for non-cancer effects (mg/kg per day)

LADD = lifetime average daily dose for cancer effects (mg/kg per day)

$C_{b/m}$ = chemical concentration in beef or milk (mg/kg)

CF = conversion factor (kg/10^3 g)

BIO = relative oral bioavailability factor, which adjusts for differences in chemical bioavailability if applicable (unitless)

$FL_{b/m}$ = fraction of beef or milk consumed that is locally raised (unitless)

Prep = reductions in concentrations due to food preparation (unitless)

$IR_{b/m,i}$ = ingestion rate for beef or milk in age period i (g/day)

EF_i = exposure frequency in age period i (days/year)

ED_i = duration of exposure in age period i (years)

BW_i = average body weight over period i (kg)

AT = averaging time (70 years for carcinogens, duration of exposure for noncarcinogens)

Days = conversion factor (365 days/year)

For the beef and milk pathways, no reduction in concentration resulting from preparation may be conservatively assumed (i.e., Prep = 1.0), unless data are available to indicate otherwise. In addition, because doses via these pathways are of chemicals present in food matrices (beef or milk), the relative oral bioavailability factor should be assumed to be 1.0. Based on nationally averaged data by EPA (1989a), it can be assumed that 44 percent of the total beef eaten by an individual and 40 percent of the total milk that an individual consumes are produced locally (i.e., $FL_b = 0.44$ and $FL_m = 0.40$).

CHEMICAL-SPECIFIC CONSIDERATIONS

It is important to consider chemical-specific factors in the calculation of exposures, because not all chemicals being evaluated in an assessment should be carried through every exposure pathway. For example, exposures via indirect pathways (e.g., soil ingestion) need not be evaluated for acid gases and highly volatile organic chemicals. Once emitted, these chemicals most likely remain in the air rather than adhere to soil particles or other surface materials.

To determine the organic chemicals likely to remain in the air, rather than be deposited and retained in other environmental media, the screening approach described below can be used. This approach is conducted using a soil:air partition coefficient K_{as}, which combines a chemical's affinity for organic carbon with its affinity for air, as follows:

$$K_{as} = \frac{K_d}{H} \qquad (13\text{-}7)$$

Where K_{as} = soil:air   partition   coefficient (unitless)

K_d = soil:water partition coefficient (unitless)

H = Henry's Law constant (unitless)

Those chemicals with $K_{as} < 10$ can be considered unlikely to persist in soil, or on surfaces in general, owing to a strong tendency to partition into air. Using this method, examples of organic chemicals that would not be carried through indirect pathways in an air toxics exposure assessment include benzene, chloroform, 1,2-dichloroethane, 1,1,1-trichloroethane, trichloroethylene, and vinyl chloride.

Chemical-specific information also can be used to identify those organic chemicals most likely to be of concern with respect to food chain pathways. Chemicals with $\log K_{ow} < 3$ do not tend to partition significantly into fat (EPA 1982a); therefore, accumulation in animal tissue from exposure to these chemicals would be unlikely to contribute significantly to exposures.

INPUT PARAMETERS

As demonstrated in the above equations, a wide variety of input data are required to calculate doses. There are many reference sources for input parameter values; however, the choice of values in part is determined by the purpose of the exposure and risk assessment to be conducted. The intake parameter values for a risk assessment being conducted for regulatory compliance may be specified by the regulatory agency. Typically this type of exposure assessment focuses on the hypothetical maximum exposed individual (MEI). The MEI for the inhalation pathway generally is assumed to be exposed for 24 hours/day, 365 days/year, and 70 years at the single location where the maximum annual average air concentrations are predicted to occur. Although these assumptions do not reflect realistic conditions, and consequently overestimate exposure, this method does allow for a consistent basis for comparison among air toxics sources and exposure conditions.

Default exposure parameter values also have been identified for other exposure pathways, primarily under EPA's Superfund program (EPA 1989a; EPA 1989f). These often even further overestimate exposure than for the inhalation pathway, owing to the scarce experimental database from which the parameters may be obtained (e.g., soil ingestion rates for adults, bioavailability factors) and to the greater impact of personal activity patterns on the potential for exposure (e.g., individuals may or may not engage in activities that result in soil contact and incidental ingestion exposures). In general, the use of default exposure parameter values constitutes a screening level analysis whose results are highly likely to be overestimated but unlikely to be underestimated. The advantages in this approach are that the ready availability of these parameters in reference documents (e.g., EPA 1989f) can save time, and the risk result typically reflects an upper-bound estimate. If the screening level results are below the range of interest, the assessment then is sufficient to document the potential upper-bound risks of an air toxics source.

The use of standard default values, however, also can result in a decrease in accuracy in intake calculations. Both site-specific factors and advances in scientific understanding of activity patterns affecting exposures are important in developing reasonable exposure estimates. More refined exposure assessments can rely on a wider array of primary and secondary sources of data to develop exposure parameter values. These include reports that provide detailed information on age- and sex-specific body weights, ventilation rates, skin surface areas and food ingestion rates (EPA 1985b; USDA 1983) and experimental studies providing estimates of soil and dust ingestion rates (Calabrese et al. 1989; Calabrese et al. 1990) and chemical bioavailabilities (see earlier references in this chapter). Site-specific climate information is compiled and available from the National Oceanic and Atmospheric Administration, National Climatic Center in Asheville, North Carolina.[2]

The discussion thus far has focused on deterministic estimation of chemical doses using single values for each input parameter. An alternate exposure assessment method that is gaining wider acceptance in applied risk assessment makes use of stochastic simulation to derive probability distributions of possible exposures and risks (Habicht 1992a). These results can identify different percentiles of risk (e.g., the 95th percentile), providing more useful and accurate information on risks associated with an air toxics source than is achieved in a deterministic evaluation. The simulation applies the standard ADD and LADD equations for calculating deterministic risks but uses as inputs distributions for the environmental concentrations and other exposure assumptions. Statistics from the resulting dose probability distribution are combined with dose-response values, that can be a single point or distributions as well, to calculate corresponding risks. Thus, instead of using a single-value, and often upper-bound, estimate for each exposure parameter, distributions characterize the magnitude and likelihood of the complete range of possible intake parameter values.

EXPOSURE SCENARIOS

The next step in the human exposure assessment is the development of scenarios for each pathway selected for evaluation. The level of desired complexity in these scenarios should be determined at the conceptual stage of the assessment. A screening assessment may evaluate a single, simple scenario, whereas a multipathway analysis may be more complex.

An example of the exposure parameters used in a moderately refined exposure scenario for the soil ingestion pathway is shown in Table 13-2. This example was obtained from a Superfund hazardous waste site risk assessment where the focus was on a 'reasonable maximum exposure' (RME) case, a term that has been defined in EPA (1989f) specifically for use in the Superfund program. In this

assessment, three receptor populations were evaluated (child/teenage trespasser, child/teenage recreational user, and resident) for which exposure scenarios were developed. Some of the exposure parameters in this table are default values specified by EPA (e.g., soil ingestion rate), whereas others were developed based on site-specific conditions (e.g., exposure frequency).

More refined exposure assessments examining many exposure pathways require development of many exposure scenarios. Table 8-7 provides examples of scenarios that are identified for a multipollutant, multipathway risk assessment. In this instance, key primary or secondary scientific sources were used (EPA 1985b; Calabrese et al. 1990; USDA 1983) and

---

[2]For more information, call (704) 254-6283.

Table 13-2   Exposure parameters used to estimate doses for incidental ingestion of soil

	EPA reasonable maximum exposure (RME)[a]		
Case parameters	Child/teenage trespasser	Child/teenage recreational user	Resident
Age period (years)	6–16	6–16	0–30
Exposure frequency (days/year)[b]	52	100	140
Exposure rate (years)	10	10	30[c]
Soil ingestion rate (mg/day)[d]	110	110	120
Fraction ingested (dimensionless)[e]	0.5	0.5	0.5
Body weight (kg)[f]	40	40	48
Averaging time (years)			
Carcinogenic[g]	70	70	70
Noncarcinogenic	10	10	30

[a] RME is a term developed by EPA's Office of Solid Waste for use in evaluating potential risks associated with Superfund hazardous waste sites (EPA 1989a).

[b] For recreational user, assumes that children and teenagers will play at the playground area five days per week during the warmer twenty weeks of the year when temperature exceeds 60°F (May through September) (NOAA 1989) for a total of one hundred exposure events per year. For residents, assumes exposure will occur seven days per week during the warmer months (May through September) for a total of 140 exposure events year. For site trespassers, assumes exposure occurs three days per week during the summer months (June, July, and August), and one day per week during the spring and fall (April/May and September/October).

[c] Based on national upper-bound time at one residence (EPA 1991a, EPA 1989a).

[d] Value for child/teenage trespasser and recreational user is a weighted-average ingestion rate assuming six-year-olds ingest soil at a rate of 200 mg/day and older children/teenagers ingest soil at 100 mg/day (EPA 1991a). Value for 0 to 30-year-old is the weighted-average, assuming six years at 200 mg/day and twenty-four years at 100 mg/day (EPA 1991a, EPA 1989a).

[e] Fraction of total daily soil intake from a contaminated source. Value for child/teenage trespasser, recreational user, and 0 to 30-year-old resident is based on direction by EPA, assuming one-half of daily intake will result from activities occurring at the site, at the recreational area, or on residential property, respectively.

[f] A time-weighted average for a 6 to 16-year-old and 0 to 30-year-old (EPA 1989b).

[g] EPA standard assumption for lifetime (EPA 1991a, EPA 1989a).

site-specific conditions were considered for derivation of many parameter values. Finally, for most pathways age-specific variations in physiology and activity patterns are taken explicitly into account through the evaluation of several age periods.

CALCULATION OF CHEMICAL DOSES

The final step in the quantitative exposure assessment is the calculation of average daily dose (ADD or LADD). The ADD or LADD provides an estimate of the daily intake of a chemical per unit of body weight, averaged over many years. The doses ultimately are combined with toxicity criteria to calculate potential human health risks.

A relative comparison of ADDs and LADDs can provide an indication of which pathways and chemicals are most important. However, because some chemicals have exposure route specific toxicity criteria (e.g., benzo[a]pyrene has different cancer slope factors for ingestion and inhalation), the relative difference between oral and inhalation doses may not resemble the relative difference in risks.

Figures 13-3, 13-4, and 13-5 illustrate the differences in calculated doses across pathways and for several chemicals for three different air

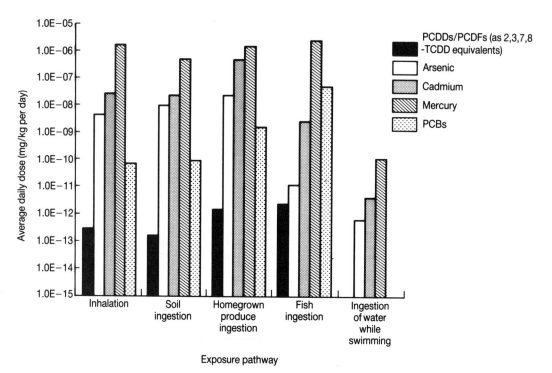

Figure 13-3. Calculated average daily doses for a municipal solid waste facility risk assessment.

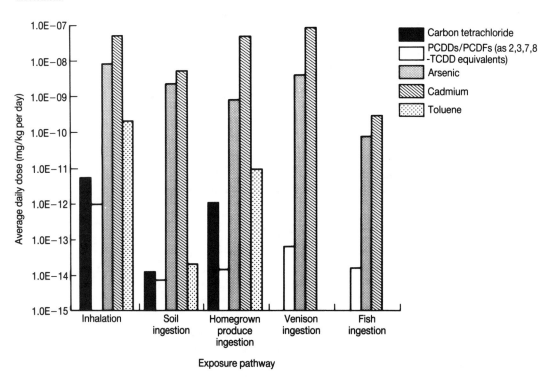

Figure 13-4. Calculated average daily doses for a hazardous waste incinerator risk assessment.

260

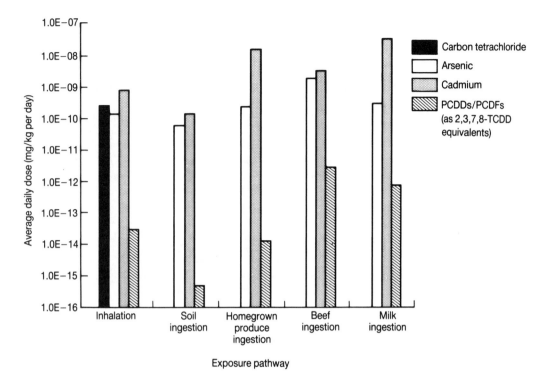

Figure 13-5. Calculated average daily doses for a cement kiln burning coal supplemented with hazardous waste fuel.

toxics sources: municipal solid waste incinerator, hazardous waste incinerator, and cement kiln burning liquid hazardous waste fuel. The doses for the municipal solid waste and hazardous waste incinerators were calculated for a hypothetical maximum case, involving combining maximum emission rates with maximum air dispersion modeling results and conservative exposure assumptions. The doses that result from combining these types of upper-bound assumptions do not reflect plausible exposure situations; instead, they provide an indication of the upper-bound of potential exposures and are likely to overestimate actual exposures greatly. The doses for the cement kiln reflect a plausible maximum case, involving combining average emission rates with average air dispersion modeling results and conservative exposure assumptions.

Table 13-3   Monte Carlo simulation of possible average daily doses for soil ingestion associated with resource recovery facilities

Percentile (≤ value shown)	Soil ingestion chronic daily intake (mg/kg per day)[a]	
	PCDDs/PCDFs[b]	Arsenic
0	1.0E-20	2.0E-15
10	2.0E-18	2.0E-12
20	5.0E-18	5.0E-12
30	9.0E-18	1.0E-11
40	1.4E-17	2.0E-11
50	2.2E-17	4.0E-11
60	3.6E-17	7.0E-11
70	6.1E-17	1.2E-10
80	1.1E-16	2.5E-10
90	2.6E-16	7.3E-10
100	3.9E-14	1.4E-07

[a] Values shown in scientific notation.
[b] PCDDs/PCDFs as 2,3,7,8-TCDD toxic equivalents.

As discussed earlier in this chapter, stochastic simulation can provide much more information regarding potential exposures than deterministic calculations such as those summarized above. Table 13-3 illustrates the results that are obtained from a stochastic simulation for a municipal solid waste incinerator. The table shows the doses calculated for a range of different percentiles for PCDDs/PCDFs and arsenic for the soil ingestion pathway. The input distributions are emission rates, air concentrations, soil ingestion rates, body weights, and exposure frequency.

References

Calabrese, E.J., R. Barnes, E.J. Starek, H. Pastideo, C.E. Gilbert, P. Veneman, X. Wang, A. Lasztity, and P.T. Kostecki. 1989. How much soil do young children ingest: An epidemiological study. *Regulatory Toxicology and Pharmacology* 10:123–137.

Calabrese, E.J., E.J. Starek, C.E. Gilbert, and R.M. Barnes. 1990. Preliminary adult soil ingestion estimates: Results of a pilot study. *Regulatory Toxicology and Pharmacology* 12:88–95.

Davis, A., M.V. Ruby, and P.D. Bergstrom. 1992. Bioavailability of arsenic and lead in soils from the Butte, Montana mining district. *Environmental Science and Technology* 26:461–468.

Fraser, J.L. and K.R. Lum. 1983. Availability of elements of environmental importance in incinerated sludge ash. *Environmental Science and Technology* 17:52–54.

Habicht, F.H. 1992a. Guidance on risk characterization for risk managers and risk assessors. Memorandum from the Deputy Administrator to Assistant Administrators and Regional Administrators. U.S. Environmental Protection Agency. Washington, D.C. February 26, 1992

Healy, M, P. Harrison, M. Aslam, S. Davis, and C. Wilson. 1982. Lead sulfide and traditional preparations: Routes for ingestion, and solubility and reactions in gastric fluid. *Journal of Clinical and Hospital Pharmacy* 7:164–173.

Johnson, J.D., G.B. Freeman, and J.M. Killinger. 1989. Bioavailability study of lead and arsenic in soil following oral administration to rabbits. Performed by Battelle Columbus Laboratory (Columbus, OH) for ARCO Coal Company.

Lucier, G.W., R.C. Rumbaugh, Z. McCoy, R. Hass, D. Harvan, and K. Albro. 1986. Ingestion of soil contaminated with 2,3,7,8-tetrachlorodibenzo-*p*-dioxin (TCDD) alters hepatic enzyme activities in rats. *Fundamental and Applied Toxicology* 6:364–481.

McConnell, E.E., G.W. Lucier, R.C. Rumbaugh, P.W. Albro, D.J. Harvan, J.R. Hass, and M.W. Harris. 1984. Dioxin in soil: bioavailability after ingestion by rats and guinea pigs. *Science* 223:1077–1079.

National Academy of Sciences (NAS). 1991a. *Frontiers in Assessing Human Exposures to Environmental Toxicants*. Washington, D.C. National Academy Press.

Poiger, H. and C. Schlatter. 1980. Influence of solvents and adsorbents on dermal and intestinal absorption of TCDD. *Food and Chemical Toxicology* 18:477–481.

U.S. Department of Agriculture (USDA). 1983. Food Intakes: Individuals in the 48 States, Year 1977–78. NFCS 1977–78 Report no. I-1. Hyattsville, MD.

U.S. Environmental Protection Agency (EPA). 1982a. Pesticide assessment guidelines. Subdivision E. Hazard evaluation: Wildlife and aquatic organisms. Ecological Effects Branch, Hazard Evaluation Division, Office of Pesticides Programs. Washington, D.C.

U.S. Environmental Protection Agency (EPA). 1985b. Development of statistical distributions or ranges of statistical factors used in exposure assessments. OHEA-E-161. Final Report. Office of Health and Environmental Assessment. Washington, D.C.

U.S. Environmental Protection Agency (EPA). 1986g. Guidelines for exposure assessment. 51 FR 34042, September 24, 1986.

U.S. Environmental Protection Agency (EPA). 1989a. Exposure factors handbook. EPA 600/8-89-043. Office of Health and Environmental Assessment. Washington, D.C.

U.S. Environmental Protection Agency (EPA). 1989f. Risk assessment guidance for Superfund. Volume 1. Human health evaluation manual. Part A. EPA 540/1-89-002. Office of Emergency and Remedial Response. Washington, D.C.

U.S. Environmental Protection Agency (EPA). 1989j. Biological data for pharmacokinetic modeling and risk assessment. Report of a workshop convened by the U.S. Environmental Protection Agency and ILSI Risk Science Institute. EPA 600/3-90-019. Office of Health and Environmental Assessment. Washington, D.C.

U.S. Environmental Protection Agency (EPA). 1992a. Guidelines for exposure assessment. 57 FR 22888. May 29, 1992.

Van den Berg, M., M. Van Greevenbroek, K. Olie, and O. Hutzinger. 1986. Bioavailability of PCDDs and PCDFs on fly ash after semi-chronic ingestion by rat. *Chemosphere* 15:509.

Van den Berg, M., M. Sinke, and H. Wever. 1987. Vehicle dependent bioavailability of polychlorinated dibenzo-*p*-dioxins (PCDDs) and dibenzofurans (PCDFs) in the rat. *Chemosphere* 8:1193–1203.

Wendling, J., F. Hileman, R. Orth, T. Umbreit, E. Hesse, and M. Gallo. 1989. An analytical assessment of the bioavailability of dioxin contaminated soils to animals. *Chemosphere* 18:925–932.

14

Population and Activity Analysis

Arlene S. Rosenbaum

INTRODUCTION

Human exposure to air pollution may be characterized as contact between a human and air with a specific pollutant concentration for a specific period of time. The most direct route of exposure is inhalation, although human exposure also may occur through indirect pathways, such as ingestion of plant material or soil contaminated through pollutant deposition. Because air concentrations of pollutants vary from one location to another the activity pattern of an individual may influence his or her inhalation exposure. Moreover, patterns of activities, such as ingestion rates of various food types, may influence exposure through noninhalation pathways.

Air pollutant exposure assessment is carried out in at least two different contexts. The first is for epidemiological studies that attempt to evaluate the association or relationship between exposure and health effects. These studies may be of a descriptive type, where the unit of analysis is a group; a cohort type, where individuals are included on the basis of their exposure; or a case-control type, where individuals are included on the basis of their health effect status. The second context for exposure assessment is that of an environmental assessment, where exposure is used as a surrogate

measure for health effects or to combine with a predetermined exposure-health response relationship.

In both contexts, exposure sometimes is inferred on the basis of demographic characteristics, such as residential location, occupation, or the presence of certain appliances in the home. For example, some simple carcinogenic health risk models assume that an individual's lifetime inhalation exposure to an air pollutant, and consequent increment to his probability of developing cancer, may be estimated from the long-term average ambient concentration in the vicinity of his or her home. The model thus implies that the individual spends seventy years outdoors at that location. In some epidemiological studies, the presence or absence of an association between occupation and health status is evaluated. Sources of some potentially useful demographic data are discussed in the next section.

Exposure of individuals may be measured directly with personal monitors. However, such monitoring is expensive and available monitors often lack sensitivity or do not exist for many pollutants of interest. Moreover, such data by themselves are limited to the population actually measured, unless the sample is selected properly to represent a larger population. Without additional information about demographic factors that are expected to covary with exposure, sound extrapolation to other individuals or populations is limited. Furthermore, past conditions or future emission control scenarios cannot be evaluated directly.

Another alternative is the use of biomarkers, increasingly employed in industrial hygiene for work place exposure monitoring (ACGIH 1986; ACGIH 1989). Examples of potential biomarkers of exposure include concentrations of contaminants in exhaled air, blood, urine, breast milk, or semen. In addition, for many carcinogens and reproductive toxins that are metabolically activated to electrophilic metabolites that covalently bind to DNA, the DNA adducts in blood and other tissues may serve as biological markers of exposure. NAS (1991b, Chapter 4) provides a useful introduction to the subject of biomarkers in exposure assessment. The use of biomarkers alone, however, suffers from some of the same limitations with respect to extrapolation as does direct monitoring.

A third method is the indirect estimate of exposure through combining concentration data, either measured or estimated with models, with time-activity data. An activity profile defines a schedule of movements among specified locations (e.g., indoors at home, outdoors at a neighborhood park) and activities (e.g., sleeping, walking the dog) for an individual over a period of time. The specified locations where the activities take place are referred to as 'microenvironments.' A microenvironment is a location within which the pollutant concentration is assumed to be uniform at any point in time, although it may vary over time and may vary with the associated activity. When the focus is on exposure to air pollutants, the locations and activities specified should include those associated with other pollutants of interest (e.g., smoking or proximity to gas appliances). This indirect approach lends itself more easily to extrapolation than the others. For example, if the participants in a time-activity study constitute a random probability sample with known demographic characteristics, the resulting data base may be generalizable to other populations, perhaps with adjustments for the relative size of constituent demographic groups. The exposure of the second population then may be estimated by combining their projected activity patterns with microenvironment concentrations appropriate for their location. The required fixed-site monitoring generally is easier and less expensive to conduct than personal monitoring of individuals. To estimate exposure for a future control scenario, only the microenvironment concentrations need to be forecast if activity patterns are assumed to remain the same. The development, sources, and use of time-activity data also are described below.

LOCATION OF POPULATION

Because ambient air pollutant concentrations vary geographically, the location of the population is of primary importance in assessing population exposure. For many toxic air pollutants and emission scenarios, ambient concentrations may change rapidly in the vicinity of the emission source. Very fine resolution of the geographical distribution of population is desirable in such cases.

For many analyses the residential population distribution is used as a surrogate for actual distributions, that change over the course of a day as people travel to and from their homes for work, school, shopping, and other activities. In other studies the residential population distribution may be used as a starting point from which to estimate diurnal variations on the basis of patterns of mobility. Major sources of population location data, crosstabulated with other important demographic factors, are described in this section.

Sources of Data

The most complete source of data on population location and demographic factors cross tabulated with location for the United States is the Bureau of the Census.[1] Two sets of available data are of particular value in exposure assessment. The first is the *Census of Population and Housing, 1990: Summary Tape File 1 (STF 1)*, available on tape and compact disc – read only memory (CD-ROM). This data set contains summary statistics of all persons and housing units in the United States. It is 100 percent data (i.e., it is not a sample). With the exception of version STF 1C, it is available on a state-by-state basis. Population items of potential interest include age, race, sex, and Hispanic origin. Geographic breakdown varies according to the file version. STF 1A provides data for states and subareas down to the block group level.[2] The geographical hierarchy is state, county, place, census tract/block numbering area (BNA), block group. Also included are state portions of American Indian and Alaska Native areas, as well as county portions. STF 1B provides the same hierarchy as STF 1A but with additional subarea breakdowns of census blocks,[3] metropolitan areas, urban/rural areas, and urbanized areas. (Note: STF 1B is available on extract fiche as well.) STF 1C is a national, rather than a state, file. Its geographical hierarchy goes down to place subareas with a population of 10,000 or more only. STF 1D provides data by congressional district within a state. A number of summary reports also are available from the U.S. Government Printing Office; an example is *1990 CPH-1, Summary Population and Housing Characteristics*.

A second data set of special interest is the *Census of Population and Housing, 1990: Summary Tape File 3 (STF 3)*, also available on tape and CD-ROM. This file contains socioeconomic data from a 17 percent sample taken concurrently with the 100 percent census data.

The data are weighted to represent the total population. In addition, it contains 100 percent counts and unweighted sample counts for total persons and total housing units. Population items of potential interest include age, sex, race, Hispanic origin, education, farm and nonfarm population, income in 1989, work status in 1989, occupation, industry, place of work, means of transportation to work, travel time to work, time leaving home to go to work, poverty status in 1989, and residence in 1985. Housing items include heating fuel and water source. Geographic coverage again varies according to version. STF 3A has a subarea set much like STF 1A (i.e., to block group level). It is available in microfiche as well. STF 3B provides ZIP Code data for the five-digit ZIP Codes within each state including county portions of the ZIP Codes. STF 3C and 3D have geographic coverage as in STF 1C and 1D.

The location of potentially sensitive populations, such as children and elderly and infirm persons, may be of special interest. The sites of schools and hospitals generally are indicated on U.S. Geological Survey maps, and often on commercially produced maps. Childcare and nursing home facility information also often can be obtained from state and local agencies. For example, counties in California generally have a County Childcare Council. Yellow page listings of telephone books also are useful sources of information on these types of sites, as well as retirement communities.

Use of Data

Each specified geographical area in the census files contains the latitude and longitude of an internal point, generally approximating the geographic centroid. However, if the shape of

---

[1] The Bureau of the Census is in Washington, D.C. It both produces and distributes these files. They are accompanied by technical documentation with user instructions and detailed information on the content and formats. The Customer Service number is (301) 763-4100.

[2] A census block group is a cluster of census blocks, usually containing from 250 to 550 housing units. There are approximately 350,000 block groups in the U.S.

[3] A census block is a small area bounded on all sides by visible features such as streets, and invisible political boundaries, such as city limits. There are more than seven million census blocks in the U.S.

the specified area causes the geographic centroid to fall outside the area or in a water area the designated internal point will differ from the centroid. An approximate population centroid may be constructed for a geographical area by calculating the population-weighted average of the internal point parameters of its constituent subareas (e.g., the constituent blocks of a block group).

Once the location of the centroid is known, an ambient concentration may be assigned to it either from monitored or modeled data. For example, EPA's Human Exposure Model (HEM) (Anderson 1983) uses Gaussian dispersion modeling to estimate ambient concentrations (or increments to ambient concentrations) at a set of grid receptors. It then interpolates the concentrations from the grid receptors to the block group centroids encompassed by the modeling region. HEM further assumes that the ambient concentration at the centroid is uniform throughout the block group for purposes of estimating exposure, and that the ambient concentration is the exposure concentration for all residents of the block group.

More advanced models, such as the South Coast Risk and Exposure Assessment Model – Version 2 (SCREAM-II) (Rosenbaum, Anderson, and Lundberg 1993), use a similar procedure for assigning ambient concentrations to population centroids, but then estimate exposure to resident population groups on the basis of time-activity patterns and derived indoor, as well

as outdoor, concentrations. In this application, some of the supplementary data from the census file may be useful in assigning subgroups of the resident population to time-activity patterns (e.g., age or work status), or for estimating indoor concentrations of some pollutants in the home (e.g., heating fuel). The data concerning travel to work may be useful in estimating the mobility patterns for the population.

Demographic data also may be important to stratify the population exposure estimates according to the demographic characteristics of subpopulations, if those characteristics are expected to influence health response. For example, sensitivity to reproductive toxins may be related to gender and sensitivity to lead may be related to age. Moreover, the presence of certain occupations in an area, such as fishing or agriculture, may indicate the need to investigate specific exposure pathways like ingestion of particular foods. The development and use of time-activity and geographic mobility patterns are discussed in more detail below.

In statistical models of personal exposure, supplementary census data, such as age, sex, and occupation, may be useful independent variables in addition to pollutant concentrations. For example, in a study of personal exposure to nitrogen dioxide that included personal monitoring (Quackenboss et al. 1986), regression models were tested that included household characteristics and individual characteristics as predictor variables.

HEALTH STATUS

Data bases of health status may be used in a number of different ways. Epidemiologists investigate associations between particular diseases and specific occupations. Similarly, they may be used to investigate associations between particular diseases and specific locations. Both kinds serve as leads for further analysis of possible exposure factors.

An individual's health status may influence his or her sensitivity to particular pollutants or the adverse consequences of a given response. For example, asthmatics are believed to be more sensitive to the effects of sulfur dioxide. On the

other hand, no evidence has been found that asthmatics are more sensitive to tropospheric ozone with respect to pulmonary function, but a given decrement in function may have more serious consequences for an asthmatic than for others. Moreover, health status may influence activity patterns especially with respect to exercise and, therefore, may constitute a factor by which the target population of a time-activity study should be stratified.

This section describes some of the publicly available health data bases that contain geographic breakdowns of information. Specific studies may require investigation of state or local data sources or development of original data through survey techniques.

Mortality

Vital Statistics of the U.S., Volume II: Mortality is compiled annually by the National Center for Health Statistics (NCHS) from original death certificates for all deaths registered in the U.S. It contains deaths and death rates with demographic breakdowns (e.g., age, sex, and race) by detailed cause of death. Causes of death are classified and numbered according to the *Ninth Revision, International Classification of Diseases, 1975*. In Part A, geographic breakdown is limited primarily to states. Part B contains a geographic breakdown for total deaths and selected causes by state, county, Standard Metropolitan Statistical Area (SMSA), towns of ten thousand or more, and metropolitan–nonmetropolitan areas. There also is some cross-tabulation with race, sex, and age. At the time of this writing, 1988 was the most recent year for which this publication was available. NCHS also publishes monthly reports with data corresponding to Part A.

Mortality data also are available on tape for individual years from 1968 to 1988, with geographic breakdowns by state, county and city. For years from 1968 through 1978 cities of 250,000 or more are included. For subsequent years the files contain data for cities of one hundred thousand or more. Data more recent than 1988 may be obtained by special request.

The NCHS *Catalog of Publications* and *Catalog of Electronic Data Products* may be obtained from the Department of Health and Human Services, Public Health Service, Centers for Disease Control, Hyattsville, Maryland.[4]

Morbidity

Also available on tape or CD-ROM from NCHS through its *Catalog of Electronic Data Products*

is data from its National Health Interview Survey. The survey has been conducted continuously since 1957 on a nationwide probability sample of the U.S. civilian noninstitutionalized population. Each week a sample of households is interviewed to obtain information about health and other characteristics of each living member of the household. During a year the sample typically is composed of 36,000 to 47,000 households, including 92,000 to 125,000 persons. Data on the number of restricted-activity days, bed days, work- or school-loss days, and all physician visits during the two-week period before the interview are recorded. Data on acute and chronic conditions responsible for these days or visits also are obtained, as well as information on all hospital episodes during the previous twelve months. Additional material on current health topics, that generally change each year, is included. For example, in 1987 cancer was one of the special topics covered. The geographic breakdown of the data is limited to the SMSA level. Data from 1987 also are available on CD-ROM with special search software available to extract data and build tables.

Another data set available from NCHS is the National Ambulatory Medical Care Survey (NAMCS). Its findings are based on a sample of representative ambulatory office visits to physicians in the U.S. engaged in patient care in an office setting. The data include demographic characteristics of patients, such as age, sex, race, and ethnicity, as well as clinical aspects of visits. The geographic breakdown is limited to one of the nine census regions and to SMSAs.

Sources of cancer data are tumor registries, where available. For example, the Los Angeles County Cancer Surveillance Program has been collecting information on new cases in the county since 1972 (Hisserich, Preston-Martin, and Henderson 1975). The data base includes age, sex, race, address, religion, occupation, and industry. The state of California currently is planning a statewide registry.

---

[4]The Customer Service number is (301) 436-8500.

POPULATION ACTIVITY PATTERNS

As noted above, an important indirect method for estimating exposure is to combine time-activity data with information on microenvironment concentrations, either measured or estimated with models. A microenvironment is a location within which the pollutant concentration is assumed to be uniform at any point in time, although it may vary over time. For this type of procedure microenvironment concentrations may be measured independently or concurrently with time-activity data. Examples of each approach are discussed below.

Activity patterns generally are defined with respect to a general set of locations and, possibly, activities. In reference to inhalation exposure, Thomas and Behar (1989) suggest that 'modeling of exposure to virtually any pollutant can be achieved if we consider all activities in all microenvironments in relation to seven basic activity/location groups.' The suggested groups are comprised of three major categories: outdoor, indoor, and in-transit. The in-transit category is subdivided according to combustion- and noncombustion-powered modes whereas the indoor category is subdivided into sleeping in residence, other in residence, office, and other nonresidence. Thomas and Behar also note that adjustments or additions to these categories may be required for specific pollutants with respect to occupational exposures, other pollution intensive activities or locations, seasonal variations, or geographic location.

In addition to descriptions and locations, another important characteristic of activities with respect to inhalation exposure is breathing rate. For a given exposure concentration a higher ventilation rate may result in enhancement of the dose of a pollutant delivered to body tissues, both because more contaminated air enters the body and because more air enters through the mouth, which lacks some of the screening mechanisms of the nasal cavities. In some studies, breathing rate is inferred from the reported activity, but an explicit estimate is preferable. Precise ventilation rate measurements normally are made with a cumbersome

apparatus that is impractical for use outside of the laboratory. As a result, most time-activity studies that do track this variable utilize a rough, subjective scale (e.g., low, medium, high). A recent study by Shamoo et al. (1991) investigated the use of more convenient heart rate monitors, precalibrated to an individual's breathing rate, for use in time-activity studies.

This section describes methods for estimating time-activity patterns, some examples of time-activity studies, and examples of how time-activity data have been used in exposure assessment.

Overview of Approach for Inhalation Exposure

For exposure to pollutants like carcinogens, where the long-term average exposure concentration is of concern, the general approach is to estimate individual exposure as

$$E = \frac{\sum_{i=1}^{I} C_i T_i}{\sum_{i=1}^{I} T_i} \qquad (14\text{-}1)$$

Where E = the long-term average individual exposure concentration
i = a microenvironment
C_i = the concentration in the microenvironment
T_i = the time spent in the microenvironment

If the concentration for any of the microenvironments is time varying, then the time of occurrence of activity/location combinations must be tracked as well as their durations. This is particularly important when significant pollutant exposure occurs outdoors, where concentrations at a fixed site may vary with time (e.g., diurnally, daily, or seasonally) owing to variations in pollutant emissions or to temporal meteorological patterns even with uniform emissions. Population exposure is estimated by weighting the individual exposures according to the number of people who are expected to follow each activity pattern.

For exposure to pollutants for which short-term exposures are of concern and a threshold effect is expected, such as some reproductive toxins, individual exposure may be estimated as

$$E = \sum_{i=1}^{I} \Delta_i \qquad (14\text{-}2)$$

Where E = the number of time units of exposure to concentrations exceeding c^*

c^* = the threshold concentration of interest

i = a microenvironment

Δ_i = T_i when c_i exceeds c^*, and equals 0 otherwise

T_i = duration of time spent in the microenvironment

c_i = the concentration in the microenvironment

Again, time-varying microenvironment concentrations require tracking times of occurrence of activity/locations; and population exposure is estimated by weighting of individual exposure estimates. A variation of this type of measure weights each time unit by either the corresponding concentration, c_i, or the difference between the concentration and the threshold, $(c_i - c^*)$.

Methods of Collecting Time-Activity Data

Time-activity information generally is obtained through survey research methods, such as questionnaires and time-activity diaries. NAS (1991b, Chapter 5) provides a useful introduction to the subject of survey research methods for exposure assessment. A summary of discussions on methodological issues concerning these instruments, held in 1988 by the Environmental Inventory Working Group of the Air Pollution Control Association (now the Air &

Waste Management Association), is contained in Quackenboss and Lebowitz (1989). Detailed summaries of the protocols followed in specific studies are contained in the references noted in the discussion of example time-activity studies below.

The primary distinction between time-activity questionnaires and diaries is that questionnaires may solicit information about the frequency and duration of various activities and locations but not the scheduling of such events (i.e., questionnaires commonly require recall over longer periods of time than diaries). Questionnaire data, therefore, may be inadequate for situations where pollutant concentrations are expected to vary over time (e.g., diurnally or seasonally). Moreover, the reliability of the data may be subject to the increased uncertainties of fading memory. The advantage of the questionnaire, however, is that it is usually less burdensome to the respondent than a diary. In addition, a questionnaire concerning demographic characteristics of the respondent and other pertinent information may be a useful supplement to data obtained from time-activity diary studies.

The recent development of a standard Environmental Inventory Questionnaire (Lebowitz et al. 1989) is an attempt to permit comparisons of results from different studies. The survey instrument contains sections on housing characteristics, demographic characteristics of household occupants, smoking in the home, household appliances, radon, and organic pollutants.

In diary studies the sequence and locations of activities for a given period of time are described. This may be done in a number of different ways. Observers may be used to monitor participants or information may be self-reported. If self-reported, the data may be recalled from a previous time period, usually a day, or may be recorded concurrently with the activities. Some studies also include concurrent measurement of concentrations, through either personal or fixed-site monitoring within the relevant microenvironments. Some investigations of the reliability of self-reported activity data have had favorable results (Robinson 1985;

Juster 1985), whereas others (Stock and Morandi 1989) have reported high discrepancy rates between self-reports and independent observations.

One important methodological issue in the design of such a study is the sample selection. First, the target population or sampling frame must be chosen. For example, the target population may be the entire residential population of a metropolitan area. Alternatively, it may be limited to some subset, such as nonsmokers or asthmatics. A scientifically generalizable sample requires a suitable sample size and adequate response rate in addition to random probability methods of selection. A small sample size results in low precision estimates whereas a low response rate may indicate nonresponse bias and, therefore, would warrant careful investigation.

Another important issue in study design is the length of time covered for each respondent. The reliability of predicting exposures over extended periods on the basis on a single day's activities is unknown. However, the quality of the respondent's self-reports generally will diminish over time when multiday periods are

involved. Because activities on weekends tend to differ significantly from those on weekdays, acquiring data for both day types is desirable.

Examples of Time-Activity Studies

Recent U.S. time-activity studies developed primarily for exposure assessment are highlighted below and are summarized in Table 14-1. Summaries of previous studies undertaken for other purposes can be found in Roddin, Ellis, and Siddiqee (1979) and Ott (1989). The latter report also shows some national and international comparisons of activity patterns. Studies undertaken without a focus on pollutant exposure often neglect to specify locations and activities that influence exposure, such as pumping gasoline or applying pesticides.

EPA's Total Exposure Assessment Methodology (TEAM) program includes an extensive series of exposure studies conducted over the period 1981 to 1985 in cities of four states: New Jersey, North Carolina, North Dakota, and California (Wallace 1987; Wallace et al. 1983; Wallace et al. 1986; Wallace et al. 1987; Wallace et al. 1988; Pellizzari et al. 1987a;

Table 14-1   Time-activity studies conducted in the U.S. for air pollutant exposure assessment

Study area	Probability sample	Pollutant monitoring	Recall or concurrent	References
NJ, ND, NC, CA (TEAM)	Yes	Yes	Recall	Wallace (1987); Pellizzari et al. (1987a, 1987b)
Denver, CO	Yes	Yes	Concurrent	Johnson (1984a, 1984b); Akland et al. (1985)
Washington, D.C.	Yes	Yes	Concurrent	Hartwell et al. (1984); Akland et al. (1985)
Cincinnati, OH	Yes	No	Concurrent	Johnson (1987)
California	Yes	No	Recall	Wiley et al. (1991); Jenkins, Phillips, and Mulberg (1990)
WI, OH, KS		Yes	Both	Ferris et al. (1979); Adair and Spengler (1989)
West Virginia		Yes	Concurrent	Schwab et al. (1989); Schwab et al. (1991)
MA, OH	No	Yes	Concurrent	Dockery and Spengler (1981)
Waterbury, VT	No	Yes	Concurrent	Sexton, Spengler, and Treitman (1984)
New Jersey (THEES)	No	Yes	Recall	Freeman, Waldman, and Lioy (1991)
Tennessee	Yes	Yes	Concurrent	Spengler et al. (1985)
Portage, WI	Yes	Yes	Concurrent	Quackenboss et al. (1986)
Los Angeles, CA	Yes	Yes	Concurrent	Schwab, Spengler, and Ryan (1989)
Tucson, AZ	Yes	Yes		Quackenboss, Krzyzanowski, and Lebowitz (1991)
Boise, ID		Yes		Glen, Highsmith, and Cupitt (1991)

Pellizzari et al. 1987b). Twenty chemicals were targeted for analysis and a total of six hundred subjects were selected with random probability methods. Personal air samples over two consecutive twelve-hour periods, two drinking water samples, and a sample of exhaled air were collected for each subject. Concurrent ambient air samples were collected near some participants' homes. The study design included consideration of seasonal variation and studies of population subgroups of special concern, such as nursing mothers. At the end of the sampling period a twenty-four-hour recall diary about activities was completed that specified the duration, but not the schedule, of activities.

Two related time-activity studies were conducted during the winter of 1982 to 1983 concurrently with personal monitoring of carbon monoxide (CO) exposure (Johnson 1984a; Johnson 1984b; Hartwell et al. 1984; Akland et al. 1985). The target populations of these studies were noninstitutionalized, nonsmoking, adults (ages eighteen to seventy) in urbanized portions of the Denver, CO, and Washington, D.C., metropolitan areas, with populations of approximately 245,000 and 1.2 million, respectively. Both studies used random probability sampling methods for participant selection. More than eight hundred valid person-days of data were gathered in Denver (two per subject) and more than seven hundred in Washington (one per subject) from November through February. Because the primary interest of the study was exposure to carbon monoxide, the location/activity specifications highlighted proximity to mobile sources and unvented combustion sources, attached garages, and smoking activities. Thirty-four activity codes and thirty-three location codes were combined in subsequent processing to yield twenty-two microenvironments relevant to CO exposure.

Analysis of the Washington data (Schwab 1989) empirically demonstrated that differences in activity patterns among demographic groups lead to differences in exposure patterns. The analysis showed that travel is an important contributor to CO exposure for those who work outside the home, especially men, and the home is the major site of CO exposure for others,

especially women. Comparison of the activity pattern results from the two different locations show some differences, such as longer time spent in office and in transit in Washington D.C., but a notable overall similarity (Thomas and Behar 1989). This finding suggests the feasibility of extrapolating activity patterns from one geographic area to another.

In 1985, a large time-activity study also was carried out in Cincinnati, OH (Johnson 1987). Concurrent air concentration monitoring was not included. The target population were all residents of Hamilton County, OH, Clermont County, OH, and Kenton County, KY, including both urban and rural areas. Random probability methods were used for selection. Children were included by having their parents provide information. More than nine hundred subjects provided self-reported activity diaries, recorded concurrently with activities, for three consecutive days, either in March or August. The diary entries consisted of the new activity/location initiation time, one of forty-five activity codes, one of forty-four location codes, one of three breathing rate or breathing problem codes, and one of three smoking activity codes. For indoor locations open windows and operating fireplace or wood-burning stove also were reported. A detailed questionnaire was administered to each participant for demographic group determination. It included items on age, sex, smoking status, asthma status, occupation, commuting habits, leisure activities, education, health, indoor pollutant sources, estimates of time spent indoors and outdoors, and estimates of time spent in high physical exertion. Summary analysis of the resulting data showed that time spent in transit and time spent outdoors varied with age, gender, and season, ranging from 2 to 9 percent and from 2 to 20 percent, respectively.

A large time-activity study also recently was completed for the target population of all English-speaking California residents over the age of twelve who live in households with telephones (Wiley et al. 1991; Jenkins, Phillips, and Mulberg 1990). The target population, stratified into three geographic regions, was sampled by telephone with random probability methods over four seasonal study periods from October 1987 through September 1988. Telephone interviews were conducted with 1,762 respondents over six- to eight-week periods within each season. The information gathered consisted of twenty-four-hour recall diaries of activities and locations, and questions about proximity to and use of potential pollutant sources, including cigarettes, consumer products, combustion appliances, and vehicular exhaust. Interviews were conducted with a computer-assisted system that allowed open-ended responses, that were coded into forty-four location codes and ninety-nine activity codes. There were separate interview protocols for adolescents (ages twelve to seventeen) and adults.

The results showed that Californians spend an average of 87 percent of time indoors. Roughly fifteen hours per day of that time occurs at home and six hours indoors elsewhere. An average of two hours per day (8 percent) is spent in transit, and one hour (5 percent) outdoors. Californians average 72 percent of their time in sleeping, working, using electronics (TV or computer), traveling, and eating. More women than men use consumer products that constitute pollutant sources, but more men than women engaged in activities near gasoline sources, such as pumping gasoline. Only 22 percent of Californians are active tobacco smokers, but 62 percent are exposed to environmental tobacco smoke (ETS) on any given day. These types of data analyses aid in identification of potential exposures for various population subgroups.

A time-activity study of children was included as part of the Harvard Indoor Air Pollution Health Study (Ferris et al. 1979; Adair and Spengler 1989). The project was a prospective epidemiological study involving roughly twenty thousand people, including 1,800 children, in six communities. The primary concern was indoor air pollution and respiratory effects. The children were surveyed for respiratory symptoms and monitored for exposure to nitrogen dioxide and respirable particles. In three of the communities, children completed a technician administered time-activity diary for

three summer and three winter days. The protocol included a twenty-four-hour to thirty-six-hour recall portion with the remainder completed by the subject or parent. Eight microenvironments were distinguished. The time-activity data were combined with information about the home environment (e.g., cooking fuel, presence of smokers) to determine proper weights to apply to microenvironment concentrations from fixed monitors when estimating exposure.

Another time-activity study of children was included as part of the Kanawha County Health Study (Schwab et al. 1989a; Schwab et al. 1991; Schwab, Spengler, and Ozkaynak 1990). The project, initiated in 1986 in a large center of chemical manufacturing in West Virginia, was a three-phase epidemiological study of the health of the population. Longitudinal data were collected on children's time-activity patterns from a sample of one hundred children in grades four to six over two-week periods in both summer and fall. The sample was stratified by gender and respiratory health status. The diary entries specified geographic location, whether indoors or outdoors, and whether traveling in a vehicle, and one of three exertion levels.

Results showed that the children spent most of their time at home, averaging eighteen to twenty-one hours per day depending on the season and day of week. Although the great majority of time was spent indoors, the children averaged four hours per day outdoors, which is roughly twice as much as previous studies have shown for adults. A subsequent analysis of the self-reported exertion levels highlights some of the problems of consistency in collecting this type of data, especially for children (Schwab et al. 1991).

Several other projects have included time-activity data collection along with personal monitoring on small sample populations (Dockery and Spengler 1981; Sexton, Spengler, and Treitman 1984; Freeman, Waldman, and Lioy 1991) and on larger sample populations (Spengler et al. 1985; Quackenboss et al. 1986; Schwab et al. 1989b; Schwab et al. 1990). Others have combined time-activity data collection with fixed-site monitoring (Quack-enboss, Krzyzanowski, and Lebowitz 1991; Glen, Highsmith, and Cupitt, 1991).

Use of Time-Activity Data in Exposure Models

Time-activity data collected concurrently with personal monitoring data may be used to develop a statistical model of the relationship between exposure, activities, and locations. An example of this approach is reported for partic-ulate exposure of nonsmoking adults in rural Tennessee (Spengler et al. 1985). Independent variables used as predictors of average exposure concentrations in regression models included travel time and work time.

Time-activity data also have been incorpo-rated into physical inhalation exposure models in a number of different ways. Early versions of EPA's National Ambient Air Quality Standards Exposure Model (NEM) (Johnson and Paul 1981; Johnson and Paul 1982) and the South Coast Risk and Exposure Assessment Model – Version 2 (SCREAM-II) (Rosenbaum, Ander-son, and Lundberg 1993) incorporate data from a national time-activity study (Juster et al. 1979). This data base was used to develop prototype activity patterns for twelve major groupings of age-occupation groups, further divided into fifty-six subgroups (Roddin, Ellis, and Siddiqee 1979; Roddin and Lieberman 1979). For each subgroup, a prototypical activ-ity profile is specified for each of three day types (weekday, Saturday, Sunday). Because these models typically are used to analyze exposure to pollutants with time-varying con-centrations, the timing of activities, as well as duration, is specified. The activity profile defines a geographic location (i.e., vicinity of home or work), one of six microenvironment locations, and one of three activity levels for each hour of the day. The models also incorpo-rate site-specific information on the demo-graphic composition of the population to assign activity profiles, and mobility patterns (see below) to track geographic locations of the population over the course of a day. The geographic distribution of ambient and

microenvironment concentrations, evaluated separately in the models, are combined with the resulting geographic distributions of population groups to estimate the frequency distribution of exposures.

A weakness of this approach is that the limited number of activity patterns may not cover the full range of variation in exposure, even if the central tendency is well represented. This problem is less serious for estimating population exposure and resulting health effects for pollutants that are assumed to have health effects in proportion to exposure concentrations in the range of interest, such as carcinogens. Although the full variation in individual risk may not be captured by this approach, the overall average exposure for the population, from which the expected number of excess cancer cases is estimated, will be.

In order to capture more of the individual variability in exposure, a stochastic approach to development of activity profiles from time-activity data is used in the Activity Pattern Simulator (APS) (Johnson, Wijnberg, and Mersch 1987). The APS utilizes data from the Cincinnati time-activity study described above (Johnson 1987). The population is divided into nineteen demographic groups on the basis of such factors as age, sex, work status, work shift, school status, commute time, time spent out-doors, and health. Five exposure states are defined by combining three microenvironments with breathing rate options. The time-activity data base then is used to develop event probability tables for transition from one expo-sure state to another and duration of exposure states. A separate set of tables is developed for each population group/day of the week/season of the year combination. A set of tables consists of an initial event table and a table for each time of day/vacated state combination.

A similar approach was taken in the Simula-tion of Human Air Pollutant Exposure (SHAPE) model (Ott 1981) and the Regional Human Exposure Model (REHEX) (Winer et al. 1989). To achieve consistency with this approach, careful attention must be paid to the specifica-tion of microenvironments or of sampling protocols. For example, if more than one mode of travel is available a simulated activity pattern might define travel from home and travel to home by different modes. The APS, SHAPE, and REHEX models are able to avoid such problems by maintaining a very small number of microenvironments. For example, REHEX specifies only indoor, outdoor, and in-transit. Expansion to a more extensive set of microenvironments in order to capture variability in concentrations requires a more complex sampling protocol.

An alternative stochastic approach designed to address these issues of consistency is utilized in the current probabilistic version of NEM (p-NEM) (Johnson et al. 1990; McCurdy et al. 1991; Johnson et al. 1992). With this approach, each person-day of diary data for the time-activity study is coded into an event sequence containing the microenvironment definitions of interest and the diary days are grouped according to demographic factors, season, temperature range, and day type (weekday or weekend). The local population under study is categorized into cohorts by demographic group, geographic location of home, geographic location of workplace, and residential air-conditioning status. Demographic groups are defined with reference to age, work status, and probability of working outdoors (for ozone application) or simply age and work status (for CO application). In order to construct a cohort event sequence for an extended period of time (e.g. a year) an entire person-day of coded diary data is sampled from the category appropriate for each calendar day. Note that the sampled unit for this approach is a consistent twenty-four-hour sequence so that a detailed set of microenvironments may be specified. However, because the sampling comes from a pool of diary days describing many individuals, consistency from day to day is dependent on the variables used for grouping.

Geographic Mobility Patterns

Because of the geographic variability of pollutant concentrations, another important type of activity information for exposure assessment is geographic mobility patterns (i.e., how the population distribution shifts over time).

Ideally, the mobility patterns would be known with respect to each demographic group of interest, and for each day type (weekday or weekend), season, and any other relevant variable. Part of the required information may be contained in time-activity data, specifying the initiation and duration of time in transit (e.g., driving to work). Some of this data also are available from the Bureau of the Census, as described above. However, stipulation of the destination with respect to local geography also is required in order to assess exposure (e.g., in what part of the modeling region is the workplace located). Many local transportation agencies have information on travel patterns that may be useful in this regard. For example, California's South Coast Association of Governments (SCAG) uses the Los Angeles Regional Transportation Simulation (LARTS) model for transportation planning in the South Coast Air Basin. The model estimates the number of trips between different parts of the basin on weekdays, distinguishing between those commuting to and from the workplace from others.

Although transportation planning is concerned primarily with the total volume of traffic, in a special project SCAG also combined geographic transportation data with demographic data to estimate the composition of trips by demographic group (Hayes, Austin, and Rosenbaum 1988). This was done by distributing the total home-to-work trips for each region-to-region combination among the working population living in the home region in proportion to their demographic composition. For example, if there are one hundred home-to-work trips per day from region one to region two, and if 25 percent of the working population residing in region one are in demographic group five, then it is estimated that twenty-five of the home-to-work trips from region one to region two are taken by members of demographic group five.

Food Ingestion and Other Activities

Other types of population activities that may be important for assessment of exposure to air pollutants from noninhalation pathways include ingestion of various food types, drinking water, soil, and household dust, and dermal absorption. A comprehensive data base on food consumption patterns in the continental U.S. is published by the U.S. Department of Agriculture (Pao et al. 1982; USDA 1983). The information is based on a national survey of more than 36,000 individuals for three-day food intakes conducted in 1977 to 1978. Data are reported by gender, age, race, income, urbanization of residence, and season.

For most exposure assessments, however, the portion of consumed food that is grown locally is the primary interest. EPA (1989a) addresses ingestion rates of homegrown food, as well as ingestion rates of recreationally caught fish, ingestion rates of soil, and rates of dermal loading. The fraction of consumed produce that is grown in home gardens was estimated on the basis of information from the USDA survey. The data are reported by food type and urbanization of residence. Data also are presented on the fraction of households that have gardens by region of the country, by urbanization of residence, and by household size. Homegrown beef and dairy products also are discussed.

Another comprehensive reference on ingestion rates and dermal absorption rates is Clement Associates, Inc. (1988). Regulatory guidance on recommended values for use in California is contained in CAPCOA (1991).

References

Adair, J.H. and J.D. Spengler. 1989. Assessing activity patterns for air pollution exposure research. In: Proceedings of the Research Planning Conference on Human Activity Patterns. Office of Research and Development. U.S. Environmental Protection Agency. Las Vegas, Nevada.

Akland, G.G., T.D. Hartwell, T.R. Johnson, and R.W. Whitmore. 1985. Measuring human exposure to carbon monoxide in Washington D.C. and Denver, Colorado, during the winter of 1982–1983. *Environmental Science and Technology* 19:911–918.

American Conference of Governmental Industrial Hygienists (ACGIH). 1986. *Documentation of the Biological Exposure Indices.* Cincinnati, OH.

American Conference of Governmental Industrial Hygienists (ACGIH). 1989. *Threshold Limit Values and Biological Exposure Indices for 1989–1990.* Cincinnati, OH.

Anderson, G.E. 1983. Human exposure to atmospheric concentrations of selected chemicals, Volume 1. NTIS PB84-102540. U.S. Environmental Protection Agency, Office of Air Quality Planning and Standards. Research Triangle Park, NC.

California Air Pollution Control Officers Association (CAPCOA). 1991. Air Toxics 'Hot Spots' Program Risk Assessment Guidelines.

Clement Associates, Inc. 1988. Multi-pathway health risk assessment input parameters guidance document. Prepared for South Coast Air Quality Management District. Fairfax, VA.

Dockery, D.W. and J.D. Spengler. 1981. Personal exposure to respirable particulates and sulfates. *Journal of the Air Pollution Control Association* 31:153–159.

Ferris, B.G., F.E. Speizer, J.D. Spengler, D.W. Dockery, Y.M. Bishop, M. Wolfson, and C. Humble. 1979. Effects of sulphur oxides and respirable particles on human health: Methodology and demography of populations in study. *American Review of Respiratory Disease* 120:769–779.

Freeman, N.C., J.M. Waldman, and P.J. Lioy. 1991. Design and evaluation of a location and activity log used for assessing personal exposure to air pollutants. *Journal of Exposure Analysis and Environmental Epidemiology* 1:327–338.

Glen, W.G., V.R. Highsmith, and L.T. Cupitt. 1991. Development of an exposure model for application to wintertime Boise. Paper No. 91-131.7, presented at the 84th Annual Meeting of the Air & Waste Management Association. Vancouver, BC.

Hartwell, T.D., C.A. Clayton, R.M. Michie, R.W. Whitmore, H.S. Zelon, S.M. Jones, and D.A. Whitehurst. 1984. Study of carbon monoxide exposure to residents of Washington DC. and Denver, Colorado. EPA 600/S4-84-031, NTIS PB 84-183516. U.S. Environmental Protection Agency, Office of Research and Development. Research Triangle Park, NC.

Hayes, S.R., B.S. Austin, and A.S. Rosenbaum. 1988. A technique for assessing the effects of ROG and NO_x reductions on ozone exposure and health risk in the South Coast Air Basin. SYSAPP-88/007. Systems Applications International, San Rafael, California.

Hisserich, J.C., S. Preston-Martin, and B.E. Anderson. 1975. An areawide cancer reporting network. *Public Health Reports* 90:15–17.

Jenkins, P.L., T.J. Phillips, and E.J. Mulberg. 1990. Activity patterns of Californians: Use of and proximity to indoor pollutant sources. In: Proceedings of the 5th International Conference on Indoor Air Quality and Climate. Toronto. pp 465–470.

Johnson, T. 1987. A study of human activity patterns in Cincinnati, OH. Electric Power Research Institute. Palo Alto, CA.

Johnson, T.R. 1984a. A study of personal exposure to carbon monoxide in Denver Colorado. Paper No. 121.3, presented at the 77th Annual Meeting of the Air Pollution Control Association. San Francisco, CA.

Johnson, T.R. 1984b. A study of personal exposure to carbon monoxide in Denver, CO. EPA 600/4-84-014, NTIS PB-84-146125. U.S. Environmental Protection Agency, Office of Research and Development. Research Triangle Park, NC.

Johnson, T.R. and R.A. Paul. 1981. The NAAQS Exposure Model (NEM) and its application to nitrogen dioxide. PEDCo Environmental. Durham, NC.

Johnson, T.R. and R.A. Paul. 1982. The NAAQS Exposure Model (NEM) applied to carbon monoxide. PEDCo Environmental. Durham, NC.

Johnson, T.R., L. Wijnberg, and R. Mersch. 1987. A probabilistic model for simulating human activity patterns. PEI Associates. Durham, NC.

Johnson, T.R., R.A. Paul, J.E. Capel, and T. McCurdy. 1990. Estimation of ozone exposure in Houston using a probabilistic version of NEM. Paper No. 90-150.1, presented at the 83rd Annual Meeting of the Air & Waste Management Association. Pittsburgh, PA.

Johnson, T.R., R.A. Paul, and J.E. Capel. 1992. Application of the hazardous air pollutant exposure model (HAPEM) to mobile source pollutants. IT Corporation, Air Quality Services. Draft report.

Juster, F.T. 1985. The validity and quality of time use estimates obtained from recall diaries. In: *Time, Goods and Well-Being*, F.T. Juster and F.P. Stafford, eds. Ann Arbor, Mich.: University of Michigan Survey Research Center.

Juster, F.T., P. Courant, G. Duncan, J. Robinson, and F. Stafford. 1979. Time use in economic and social accounts. Survey Research Center, Institute for Social Research, University of Michigan. Ann Arbor.

Lebowitz, M.D., J.J. Quackenboss, M.L. Soczek, M. Kollander, and S.D. Colome. 1989. The new standard environmental inventory questionnaire for estimation of indoor concentrations. *Journal of the Air Pollution Control Association* 39:1411–1419.

McCurdy, T., J.E. Capel, R.A. Paul, and T.R. Johnson. 1991. Preliminary analysis of ozone exposures in Houston using NEM/O3. Paper No. 91-141.1, presented at the 84th Annual Meeting of the Air & Waste Management Association. Vancouver, BC.

National Academy of Sciences (NAS). 1991b. *Human Exposure Assessment for Airborne Pollutants – Advances and Opportunities*. Washington, D.C. National Academy Press.

Ott, W.R. 1981. Computer simulation of human air pollution exposures to carbon monoxide. Paper No. 81-57.6, presented at the 74th Annual Meeting of the Air Pollution Control Association. Philadelphia, PA.

Ott, W.R. 1989. Human activity patterns: a review of literature for estimating time spent indoors, outdoors and in transit. In: Proceedings of the Research Planning Conference on Human Activity Patterns. T.H. Starks, ed. U.S. Environmental Protection Agency, Office of Research and Development. Las Vegas, NV.

Pao, E.M., K.H. Fleming, P.M. Guenther, and S.J. Mickle. 1982. Foods commonly eaten by individuals: Amounts per day per eating occasion. Home Economics Research Report No. 44. U.S. Department of Agriculture. Washington, D.C.

Pellizzari, E.D., K. Perritt, T.D. Hartwell, L.C. Micheal, R. Whitmore, R.W. Handy, D. Smith, and H. Zelon. 1987a. The total exposure assessment methodology (TEAM) study: Elizabeth and Bayonne, NJ, Devil's Lake, ND, and Greensboro, NC, Volume II. EPA 600/6-87/002b. U.S. Environmental Protection Agency, Office of Research and Development. Washington, D.C.

Pellizzari, E.D., K. Perritt, T.D. Hartwell, L.C. Micheal, R. Whitmore, R.W. Handy, D. Smith, and H. Zelon. 1987b. The total exposure assessment methodology (TEAM) study: Selected communities in Northern and Southern California, Volume III. EPA 600/6-87/002c. U.S. Environmental Protection Agency, Office of Research and Development. Washington, D.C.

Quackenboss, J.J., M. Krzyzanowski, and M.D. Lebowitz. 1991. Exposure assessment approaches to evaluate respiratory health effects of particulate matter and nitrogen dioxide. *Journal of Exposure*

Analysis and Environmental Epidemiology 1:83–107.

Quackenboss, J.J. and M.D. Lebowitz. 1989. The utility of time-activity data for exposure assessment: Summary of procedures and research needs. Paper No. 89-100.7. Presented at the 82nd Annual Meeting of the Air & Waste Management Association. Anaheim, CA.

Quackenboss, J.J., J.D. Spengler, M.S. Kanarek, R. Letz, and C.P. Duffy. 1986. Personal exposure to nitrogen dioxide: relationship to indoor/outdoor air quality and activity patterns. *Environmental Science and Technology* 20:775–783.

Robinson, J.P. 1985. The validity and reliability of diaries versus alternative time use measures. In: *Time, Goods and Well-Being*, F.T. Juster and F.P. Stafford, eds. Ann Arbor, Michigan: University of Michigan Survey Research Center.

Roddin, M.F., H.T. Ellis, and M.W. Siddiqee. 1979. Background data for human activity patterns, Volume 1. SRI International. Menlo Park, CA.

Roddin, M.F., and R. Lieberman. 1979. Background data for human activity patterns, Volume 2. SRI International. Menlo Park, CA.

Rosenbaum, A.S., G.E. Anderson, and G.W. Lundberg. 1993. User's guide to the South Coast Risk and Exposure Assessment Model – Version 2 (SCREAM-II), for the personal computer. Systems Applications International. San Rafael, CA.

Schwab, M. 1989. The influence of daily activity patterns on differential exposure to carbon monoxide among social groups. In: Proceedings of the Research Planning Conference on Human Activity Patterns. U.S. Environmental Protection Agency, Office of Research and Development. Las Vegas, NV.

Schwab, M., J. Spengler, and H. Ozkaynak. 1990. Using longitudinal data to understand children's activity patterns in an exposure context: Data from the Kanawha County Health Study. In: Proceedings of the 5th International Conference on Indoor Air Quality and Climate. Toronto. pp 471–475.

Schwab, M., J. Spengler, H. Ozkaynak, and P. Terblanche. 1989a. The time/activity component of the Kanawha County Health Study. In: Proceedings of the EPA/A&WMA Specialty Conference on Total Exposure Assessment Methodology. Las Vegas, NV.

Schwab, M., J.D. Spengler, P.B. Ryan, S.D. Colome, A.L. Wilson, E. Becker, and I.H. Billick. 1989b. Describing activity patterns for use in exposure assessment: Data from the Los Angeles personal monitoring study. Paper No. 89-100.4 presented at the 82nd Annual Meeting of the Air & Waste Management Association. Anaheim, CA.

Schwab, M., S.D. Colome, J.D. Spengler, P.B. Ryan, and I.H. Billick. 1990. Activity patterns applied to pollutant exposure assessment: Data from a personal monitoring study in Los Angeles. *Toxicology and Environmental Health* 6:517–532.

Schwab, M., J.D. Spengler, J.H. Ware, H. Ozkaynak, G.Wagner, and J.M. Samet. 1991. The Kanawha County Health Study: A multi-phase investigation of children's respiratory health and exposure to chemical industry emissions. Presented at the 84th Annual Meeting of the Air & Waste Management Association. Vancouver, BC.

Sexton, K., J.D. Spengler, and R.D. Treitman. 1984. Personal exposure to respirable particles: A case study in Waterbury, Vermont. *Atmospheric Environment* 18:1385–1398.

Shamoo, D.A., T.R. Johnson, S.C. Trim, D.E. Little, W.S. Linn, and J.D. Hackney. 1991. Activity patterns in a panel of outdoor workers exposed to oxidant pollution. *Journal of Exposure Analysis and Environmental Epidemiology* 1:423–438.

Spengler, J.D., R.D. Treitman, T.D. Tosteson, D.T. Mage, and M.L. Soczek. 1985. Personal exposures to respirable particulates and implications for air pollution epidemiology. *Environmental Science and Technology* 19:700–707.

Stock, T.H. and M.T. Morandi. 1989. A comparative evaluation of self-reported and independently-observed activity patterns in an air pollution health effects study. In: Proceedings of the Research Planning Conference on Human Activity Patterns. T.H. Starks, ed. U.S. Environmental Protection Agency, Office of Research and Development. Las Vegas, NV.

Thomas, J. and J. Behar. 1989. Basic activity patterns structure for modeling pollution exposure. In: Proceedings of the Research Planning Conference on Human Activity Patterns. U.S. Environmental Protection Agency, Office of Research and Development. Las Vegas, NV.

U.S. Department of Agriculture (USDA). 1983. Food Intakes: Individuals in the 48 States, Year 1977–78. NFCS 1977–78 Report no. I-1. Hyattsville, MD.

U.S. Environmental Protection Agency (EPA). 1989a. Exposure factors handbook. EPA 600/8-89-043. Office of Health and Environmental Assessment. Washington, D.C.

Wallace, L. 1987. The total exposure assessment methodology (TEAM) study: Summary and analysis, Volume 1. EPA 600/6-87-002a. U.S. Environmental Protection Agency, Office of Research and Development. Washington, D.C.

Wallace, L., E. Pellizzari, T. Hartwell, C. Sparacino, and H. Zelon. 1983. Personal exposures to volatile organics and other compounds indoors and outdoors – the TEAM Study. NTIS PB83-121357. U.S. Environmental Protection Agency, Office of Research and Development. Washington, D.C.

Wallace, L., E. Pellizzari, T. Hartwell, R. Whitmore, C. Sparacino, and H. Zelon. 1986. Total exposure assessment methodology (TEAM) study: Personal exposures, indoor–outdoor relationships and breath levels of volatile organic compounds in New Jersey. *Environment International* 12:369–387.

Wallace, L., E. Pellizzari, T. Hartwell, C. Sparacino, R. Whitmore, L. Sheldon, H. Zelon, and R. Perritt. 1987. The TEAM (Total Exposure Assessment Methodology) study: Personal exposures to toxic substances in air, drinking water and breath of 400 residents of New Jersey, North Carolina and North Dakota. *Environmental Research* 43:290–307.

Wallace, L., E. Pellizzari, T. Hartwell, R. Whitmore, H. Zelon, R. Perritt, and L. Sheldon. 1988. California TEAM study: Breath concentrations and personal air exposures to 26 volatile compounds in air and drinking water of 188 residents of Los Angles, Antioch and Pittsburgh, California. *Atmospheric Environment* 22:2141–2163.

Wiley, J.A., J. Robinson, T. Piazza, K. Garrett, K. Cinksena, Y. Cheng, and G. Martin. 1991. Activity patterns of California residents. State of California, Air Resources Board. Sacramento, California.

Winer, A.M., F.W. Lurmann, L.A. Coyner, S.D. Colome, and M.P. Poe. 1989. Characterization of air pollutant exposures in the California South Coast Air Basin: Application of a new Regional Human Exposure (REHEX) Model. Statewide Air Pollution Research Center, University of California. Riverside, CA.

15

Ecological Risk Assessment

Judi L. Durda

INTRODUCTION

Section 112(b)(2) of the 1990 Clean Air Act Amendments defines hazardous air pollutant in terms both of adverse human health effects and adverse environmental effects. This is in contrast to the hazardous air pollutant provisions of the 1970 Clean Air Act Amendments that focused solely on serious adverse health effects to humans. Environmental risk assessment, more generally referred to as ecological risk assessment, in the context of this handbook deals with the assessment of the impacts of toxic air pollutant emissions on plants and animal life.[1] As a science, it is less well developed than that dealing with human exposure to toxic air pollutants both because human health effects are of more personal concern and because of the enormous diversity of plant and animal life. As such, few regulatory actions have been taken for the sole purpose of preventing ecological damage resulting from hazardous or toxic air pollutants.[2] Notwith-

standing, ecological risk assessment is becoming an increasingly important tool in the management of environmental problems. Through the last half of the 1980s, the Environmental Protection Agency (EPA) made a concerted effort to begin integrating and standardizing ecological risk assessment in its programs (EPA 1992k; Norton et al. 1992).

The purpose of this chapter is to provide an overview of the methods currently used to assess and characterize ecological risks. First, the basic paradigm and approaches for ecological risk assessment are discussed and then individual steps in the ecological risk assessment process are described. This discussion is provided in the exposure assessment section of this handbook, even though it contains aspects both of effects and exposure, because at this time there is more information and general understanding of ecological exposure than there is of ecological effects.

---

[1] Other environmental effects, such as global climate change, also are of concern but are beyond the scope of this handbook.

[2] A notable exception is EPA's Superfund program that frequently considers ecological effects in the remediation of hazardous waste sites.

BASIC PARADIGM AND APPROACHES

The approach to ecological risk assessment is conceptually similar to that of human health risk assessment and is based on the basic paradigm presented by the National Academy of Sciences (NAS 1983), that has been widely adopted by governmental agencies for human health assessment. Under this approach, ecological risk assessments are conducted in four basic steps.

1. Hazard assessment identifies the receptors, exposure pathways, and chemicals or other stressors of concern and identifies data needs for further study (as appropriate).
2. Exposure assessment describes the extent of environmental exposures to a chemical or other stressor.
3. Dose-response assessment defines the relationship between the magnitude of exposure to a chemical or other stressor and the magnitude or probability of response.
4. Risk characterization integrates information on hazard, exposure, and dose-response to define the potential for, and type and magnitude of, ecological impacts in the exposed system, along with the attendant uncertainties.

Within this basic paradigm, there are three general types of assessments that can be used to evaluate ecological risks associated with chemical pollutants: (1) epidemiological assessments, (2) toxicity assessments, and (3) model-based assessments.

Epidemiological assessments use direct field measurements to identify adverse effects and to determine their significance, as well as to establish whether there is a relationship between these effects and the suspected chemical contaminants or other stressors. Epidemiological assessments focus on measurement of structural and functional characteristics of biological populations and communities. Typically, these types of assessments are most useful for evaluating large-scale pollution problems of sufficient potential scope to affect the structure, function, or interactions of biological populations,

communities, and ecosystems. Epidemiological assessments can be used to establish a correlation between a chemical stress and an effect, although they cannot be used to define cause-effect relationships. These types of assessments generally are time- and resource-intensive, however, and must be conducted over a number of years and a number of seasons using a variety of reference locations before data can be used to infer a pollutant-related effect. EPA's Environmental Monitoring and Assessment Program (EMAP) is a type of epidemiological assessment that is geared toward monitoring status and trends in the condition of representative ecosystems within the U.S. (Kutz et al. 1992). This program encompasses major ecosystems within the U.S. and is designed to occur over many years, if not decades.

Toxicity-based assessments are less resource- and time-intensive and can be used if it is impractical to measure effects directly. Toxicity assessments rely on the use of bioassays of environmental media to infer potential impact. Toxicity tests may be conducted in situ or in the laboratory. They are useful for evaluating the extent and severity of chemical contamination, particularly when chemical contamination consists of complex mixtures or chemicals that have not been well characterized with respect to ecological toxicity. Although toxicity assessments can be applied to a large number of potential contaminant problems, standardized test protocols have not been developed for many potential receptor species. Further, the utility of toxicity assessments in ecological risk assessments is limited because they do not provide a direct measure of effects on resident biota or on the populations and communities of which they are a part.

Model-based risk assessments can be used in place of epidemiological or toxicity-based assessments or can be used to interpret data collected using these other approaches. When used in place of other approaches, model-based assessment resembles the predictive assessment approach used to evaluate human health risks, in that potentially exposed populations are identified and then information on exposure and toxicity are combined to estimate risk. When used in conjunction with other approaches, model-based assessment can be used to provide a link between a particular chemical and an observed response, to define more appropriate exposure regimes, or to interpret the results of the assessment in the context of potential population-, community-, or ecosystem-level impacts. As such, model-based assessments are broadly applicable. The use of model-predicted risks to support environmental management decisions might, in some instances, be limited by the degree of uncertainty associated with the predictions.

The particular approach selected to support a given ecological assessment will depend on the nature of the problem being investigated and on the resources and time available to assess ecological risks. Often, a combination of approaches is most useful for evaluating ecological risks. For example, toxicity assessments are more useful when used in combination with another approach. Toxicity assessments can be used in conjunction with epidemiological assessments to define impact and to link that impact to particular chemical stressors. Toxicity assessments also can be used in conjunction with model-based approaches to evaluate potential risks to populations, communities, or ecosystems.

Regardless of the selected approach, all ecological risk assessments follow the basic paradigm to a large degree. The basic steps in the risk assessment process are described below.

HAZARD ASSESSMENT

Hazard assessment is the critical first step in conducting an ecological assessment. It is at this stage that the objectives of the ecological risk assessment are defined, and the data required to achieve those objectives are identified. This step is critical to assuring that the proposed work is adequate to support the evaluation of ecological impacts associated with the environmental problem, as well as regulatory decisions regarding the need for and extent of remedial actions. In essence, the hazard assessment process is a scoping activity. EPA (1992k) has used the term 'problem formulation' to describe this initial step.

The supporting information for defining assessment objectives and data needs is obtained by conducting a preliminary ecological assessment. The principal data sources for the preliminary assessment are existing information concerning chemical hazards (e.g., type, probable distribution, and concentrations) and potentially affected habitats (e.g., type, prevalence, and relative distribution within the region of concern), along with chemical-specific fate and effects information obtained from the literature. The output of this step is a preliminary identification of the principal ecological receptors, exposure pathways, and chemicals of concern, and a preliminary evaluation of the extent and magnitude of potential ecological impacts.

The objectives of the ecological risk assessment should be identified after consideration of these results. Suter (1989; 1990) has introduced the concept of 'assessment endpoint' as a way to define the ultimate goals and objectives of the ecological risk assessment, and this approach has recently been adopted in EPA (1992k). Assessment endpoints are formal expressions of the actual environmental values that are to be protected and, if found to be affected, would indicate a need for some type of corrective action. Generally, assessment endpoints represent ecological attributes that are valued by society, have biological relevance, and are susceptible to the hazard being assessed (Suter 1989). Assessment endpoints are most often defined in terms of attributes of the population (e.g., yield, abundance, and extinction), community (e.g., diversity), or ecosystem (e.g., productive capability). Effects on individual organisms (with the exception of endangered species) are not the focus of ecological assessments as they are not generally valued by society and have little biological relevance. In other words, death of a few animals is not generally of great societal concern and represents a negligible perturbation in the structure and function of populations, communities, or ecosystems.

Once the assessment endpoints have been identified, the way in which they will be evaluated must be defined. Assessment endpoints rarely are measurable directly, either because exposure to the hazardous agent has not yet occurred or because the scale and complexity of the system of interest precludes direct measurement. For example, it is difficult to measure decreased abundance of a sport fish population in the Chesapeake Bay or in a bird population distributed over a given geographic region. In these instances, surrogate endpoints are selected that are related to the assessment endpoint and that can be measured. These are termed 'measurement endpoints.

Once the assessment and measurement endpoints have been selected, sampling and analysis plans can be developed to provide the necessary data for evaluating ecological impacts. If a literature-based and model-based assessment is to be conducted, the assessment and measurement endpoints are used to define the types of data that are required to complete the assessment.

EXPOSURE ASSESSMENT

After the assessment and measurement endpoints have been selected, the type and degree of exposure to the chemicals of concern must be defined. Exposure assessment is generally conducted in two basic steps – exposure pathways identification and exposure quantification.

Exposure pathways describe the way a chemical moves from a source to a potential receptor. Exposure pathways are developed based on considerations of the sources, releases, types of chemicals being evaluated, the likely environmental fate and transport of these chemicals, and the location and activity of receptor populations.

Terrestrial wildlife may be exposed to chemicals in the environment by several routes: (1) ingestion of soil or sediment while foraging or grooming; (2) ingestion of food (plant or animal) that has accumulated chemicals from soil, surface water, or sediment; (3) ingestion of surface water; (4) dermal absorption; and (5) inhalation. Ecological assessments most often focus on potential exposures via ingestion of food and ingestion of surface water. Exposures via ingestion of soil or sediment while foraging or grooming, dermal absorption, or inhalation are less frequently quantified because few data are available to quantify wildlife exposures or risk via these pathways.

Terrestrial plants can be exposed to environmental contaminants via the following: (1) uptake of chemicals from soil, (2) absorption of vapor-phase chemicals, or (3) direct contact with particulate-phase atmospheric contaminants. The pathways evaluated depend to a large degree on the source of the chemicals of concern and also on the availability of phytotoxicity data for the chemicals and species of concern.

Aquatic life may be exposed to chemicals in the environment by the following: (1) respiration, (2) direct contact with water and sediment, (3) ingestion of water or sediment (e.g., in filter feeders), and (4) ingestion of food. The pathways evaluated are dictated principally by the types of aquatic toxicity data that are available. For example, few toxicity data are available to evaluate dietary exposures in aquatic organisms. Typically, exposures are quantified for selected exposure pathways either by estimating chemical concentrations in the exposure medium (e.g., air or water) or by combining these concentration estimates with estimates of intake.

Exposure concentrations can be estimated using environmental monitoring data, if a release has occurred, or environmental fate and transport models, if a release has not yet occurred or if too few monitoring data are available to estimate exposure concentrations. Temporal factors should be considered in characterizing exposure concentrations. Factors to consider include changes in chemical concentrations due to seasonal influences (e.g., increased or decreased precipitation), diurnal patterns (e.g., tidal cycles and changes in wind direction), or episodic events (e.g., accidental releases and storm events).

When estimating exposures for ecological receptors, it also is important to estimate exposure concentrations for the area over which a population or community of organisms could be exposed, rather than for a particular maximum exposure point. This type of approach provides data that are consistent with the assessment endpoints, that most often focus on impacts in the population or community rather than on the individual organism. The area for which exposure concentrations are estimated depends on the known or expected distribution of the contaminant in the different habitats of concern and on the spatial distribution of the receptor species or communities of concern. For example, surface water exposure concentrations might be estimated separately for particular aquatic habitats such as marsh, shoreline, and open water habitats, so that exposures can be evaluated for receptors that differentially use these aquatic habitats. Further, within these habitats, concentrations might be averaged over large areas to evaluate the impacts in wide-ranging species (e.g., some fish species) and over smaller areas to evaluate the impacts in less mobile species (e.g., benthic invertebrates). In all cases, however, species-specific behaviors should be considered in determining the appropriate area over which to estimate concentrations.

If exposures are to be expressed as intakes (i.e., dosage) rather than concentration in the exposure medium, additional estimates are required based on assumption regarding the rate and frequency of media intake and the body weight of the exposed receptor. Factors such as habitat and food preferences and home range should be considered when estimating intakes for terrestrial wildlife species. This information can be obtained from the literature.

In some cases, exposure can be quantified based on measurement of biomarkers in individual organisms. Exposure biomarkers are physiological (e.g., gas exchange in plants), biochemical (e.g., enzyme response in wildlife), or chemical (e.g., tissue residues) indicators of contaminant exposure. These measurements provide the most direct way to document exposure in resident organisms. Although exposure biomarkers are useful for evaluating the degree of exposure that is occurring, they are limited in their utility for evaluating toxic effects in the exposed organism. Consequently, the principal utility of these types of biomarkers is as part of larger studies to support inferences regarding the relationship between contaminant sources and exposure or effects. DiGuilio (1989) provides a review of exposure biomarkers that have been developed to support chemical exposure assessment for ecological receptors.

DOSE-RESPONSE ASSESSMENT

The purpose of the dose-response assessment is to quantify the relationship between the magnitude of exposure to a particular stressor (e.g., a chemical) and the likelihood or magnitude of an adverse effect. Two types of data can be used to support chemical dose-response assessment: (1) direct measures of toxicity derived from site-specific or source-specific bioassays, and (2) toxicity data derived from the literature.

Bioassays are important tools for ecological risk evaluations because they provide a direct measure of stress associated with complex chemical mixtures, including any stress due to synergistic or potentiating interactions among chemicals. If bioassays are to be conducted to support the ecological risk assessment, it is important that the suite of bioassays chosen be representative of the environments, trophic levels, and organisms potentially exposed within the study area. To the extent possible, bioassays should be conducted with test species that are representative of study area receptors. Further, species or endpoints known to be sensitive to environmental toxicants should be selected preferentially over less sensitive species or endpoints so that impacts on potentially sensitive receptors within the study area can be evaluated. In addition, bioassays that measure effects on reproduction, development, or survival should be selected to the extent possible as these endpoints have direct relevance to evaluation of potential population-level impacts. Finally, if possible the bioassay program should evaluate possible variations in toxic response due to temporal changes in contaminant loading (e.g., during storm events or accidental releases) or in the use of the study area by receptor species (e.g., as a spawning or nursery ground). These type of data are important to interpreting the results of the bioassay program in the context of potential population or community-level responses (i.e., the assessment endpoints).

If site-specific or source-specific bioassay data are not available, toxicity data from the literature can be used to describe exposure-response relationships. For literature-based assessments, toxicity data derived from tests with the selected receptor species or taxonomically similar species should be used to the extent possible. If the selected receptor for the assessment is a community, data from studies on similar communities (e.g., from mesocosm studies) should be used.

For population-level assessments, dose-response relationships should be derived from toxicity studies that evaluated effects on reproduction, development, or survival, as these endpoints have direct relevance to evaluation of potential population-level impacts. Less relevant endpoints (e.g., enzyme effects or organ lesions) can be used lacking more appropriate data or if they can be linked directly to reproduction, development, or survival. For example, reductions in acetylcholinesterase (AChE) activity can be linked directly to

survival and, therefore, would be an appropriate measurement endpoint.

Two approaches are available for establishing toxicity criteria under a literature-based approach. Under the first approach, toxicity data are used to define the exposure-response function for a given chemical and a given species. Under the second approach, toxicity data are used to estimate no-effect doses, conceptually similar to human health risk reference doses (RfDs) discussed in Chapter 7. Although the first approach provides the most useful data to support ecological risk assessment, too few data generally exist for the species and endpoints of interest to apply this approach in most assessments. Consequently, literature-based exposure-response assessments most often focus on an estimation of a dose at which no adverse effects are likely to occur in the selected receptor.

Procedures for estimating no-effect doses consist of identifying a no-effect or lowest-effect dosage or concentration and applying uncertainty factors to account for interspecies variability or differences in exposure regimes, or to estimate a no-effect level from a low-effect level. There currently are no explicit guidelines on the development and use of uncertainty factors in ecological risk assessment. Newell et al. (1987) proposed using a weight of evidence approach when selecting an appropriate uncertainty factor. This approach is similar to that used by EPA (1987j) to derive water quality advisories. Others have relied on analysis of dose-response data to define appropriate uncertainty factors. This is the approach originally adopted by EPA for estimating ecological toxicity as part of its pesticide registration program (Urban and Cook 1986).

RISK CHARACTERIZATION

Risk characterization is the final phase of the ecological risk assessment process. During this step, the probability and magnitude of adverse effects occurring as a result of exposure are estimated and the uncertainty associated with these estimates is described. Three general approaches are available for characterizing

ecological risk (EPA 1992k): (1) comparing single-value estimates of toxicity and exposure, (2) comparing the distribution of dose-response and exposure information, and (3) conducting simulation modeling.

Historically, the most common approach for characterizing ecological risk has been a comparison of single-effect toxicity values with predicted or measured levels of a chemical. The ratio or quotient of the exposure value to the toxicity value provides the estimate of risk. The approach has been termed the quotient method (Barnthouse et al. 1986) and is the most simplistic and least data intensive of the available risk characterization approaches. The quotient method, however, provides little information on the probability that a toxic effect will occur or on the magnitude of the toxic effect at varying levels of exposure.

An evaluation of the distribution of dose-response and exposure values can provide greater insight into the magnitude of effects expected at differing levels of exposure. Under this approach, risk is quantified by the degree of overlap between the two distributions, with a greater overlap implying greater risk. This approach is data intensive, however, as it requires sufficient data amenable to statistical treatment to construct exposure and dose-response distributions.

Simulation modeling can be used to integrate the dose-response profile and the exposure profile to provide probabilistic estimates of risk. The two types of simulation modeling used in ecological risk assessment are individual-based models and multispecies models. Individual-based models use measured responses in individual organisms to predict effects at the population level. Multispecies models use measured responses in individuals of a given species along with information on the types of interactions between species to predict effects in the community or ecosystem.

Few models currently exist to support these types of evaluations. For example, few models exist that link measured responses in individual organisms (e.g., as with bioassays) to population-level or higher responses, and those that do exist require a detailed understanding of the life history and population structure and dynamics of the species of concern, as well as an understanding of the dose-response relationship for the chemical mixtures and receptors of concern in the study area. Methods for linking laboratory-derived toxicity test data to fish population models have been published

(Barnthouse et al. 1990; Connolly and Tonelli 1985), but are data intensive. Individual-based models also have been developed for evaluating chemical effects on *Daphnia* populations (Hallam et al. 1990), but again these models are data intensive. Few terrestrial models have been proposed (Emlen 1989). Barnthouse (1992) provides a more detailed discussion of the approaches for modeling population- or community-level responses. Although few models are currently available, this is an active area of research in ecological risk assessment.

The final stage of the risk characterization process is an evaluation of the uncertainties associated with the overall assessment. The uncertainty analysis identifies the sources of uncertainty at each step in the risk assessment process and quantifies this uncertainty to the extent possible. This information should be used by risk managers along with the results of the assessment to support environmental management decisions.

References

Barnthouse, L.W. 1992. The role of models in ecological risk assessment: a 1990s perspective. *Environmental Toxicology and Chemistry* 11:1751–1760.

Barnthouse, L.W., G.W. Suter, and A.E. Rosen. 1990. Risks of toxic contaminants to exploited fish populations: Influence of life history, data uncertainty and exploitation intensity. *Environmental Toxicology and Chemistry* 9:297–311.

Barnthouse, L.W., G.W. Suter, S.M. Bartell, J.J. Beauchamp, R.H. Gardner, E. Linder, R.V. O'Neill, and A.E. Rosen. 1986. User's Manual for Ecological Risk Assessment. Publication No. 2679, ORNL-6251. Environmental Sciences Division, Oak Ridge National Laboratory. Oak Ridge, TN.

Connolly, J. and R. Tonelli. 1985. Modeling kepone in the striped bass food chain of the James River estuary. *Estuarine, Coastal and Shelf Science* 20:-349–366.

DiGuilio, R. 1989. Biomarkers. In: W.W. Hicks, B.R. Parkhurst, and S.S. Baker, eds. Ecological Assessments of Hazardous Waste Sites: A Field and Laboratory Reference. EPA 600/3-89/013. Environmental Protection Agency, Environmental Research Laboratory. Corvallis, OR.

Emlen, J.M. 1989. Terrestrial population models for ecological risk assessment: A state-of-the-art review. *Environmental Toxicology and Chemistry* 8:831–842.

Hallam, T.G., R.E. Lassiter, J. Li, and W. McKinney. 1990. Toxicant induced mortality in models of *Daphnia* populations. *Environmental Toxicology and Chemistry* 9:597–621.

Kutz, F.W., R.A. Linthurst, C. Riordan, M. Slimack, and R. Frederick. 1992. Ecological research at EPA: New directions. *Environmental Science and Technology* 5:860–866.

National Academy of Sciences (NAS). 1983. *Risk Assessment in the Federal Government: Managing the Process.* Prepared by the Committee on the Institutional Means for Assessment of Risk to Public Health, Commission on Life Sciences. Washington, D.C. National Academy Press.

Newell, A.J., D.W. Johnson, and L.K. Allen. 1987. Niagara River Biota Contamination Project: Fish Flesh Criteria for Piscivorous Wildlife. Technical Report 87-3. New York State Department of Environmental Conservation.

Norton, S.B, D.J. Rodier, J.H. Gentile, W.H. Van Der Shalie, and E.P. Wood. 1992. A framework for ecological risk assessment at the EPA. *Environmental Toxicology and Chemistry* 11: 1663–1672.

Suter, G.W. 1989. Ecological endpoints. In: W.W. Hicks, B.R. Parkhurst, and S.S. Baker, eds. Ecological Assessments of Hazardous Waste Sites: A Field and Laboratory Reference. EPA 600/3-89/013. Environmental Protection Agency. Environmental Research Laboratory. Corvallis, OR.

Suter, G.W. 1990. Endpoints for regional ecological risk assessments. *Environmental Management* 14:19–23.

U.S. Environmental Protection Agency (EPA). 1987j. Guidelines for Deriving Ambient Aquatic Life Advisory Concentrations. Office of Water Regulation and Standards, and Office of Research and Development. Duluth, MN.

U.S. Environmental Protection Agency (EPA). 1992k. Framework for Ecological Risk Assessment. EPA 630/R-92/001. Risk Assessment Forum. Washington, D.C.

Urban, D.J. and N.J. Cook. 1986. Standard Evaluation Procedures for Ecological Risk Assessment. Hazard Evaluation Division. Washington, D.C.

IV

Regulatory Strategies

16

Legislative and Legal Considerations

Kathy D. Bailey

INTRODUCTION

The Clean Air Act Amendments of 1990 dramatically changed the federal approach to regulation of air toxics. After thirteen years of debate, section 112 of the Clean Air Act was completely revised. These amendments were designed to spur activity in an area that has been almost moribund due to reevaluations and litigation. Emissions of 189 substances listed as hazardous air pollutants (HAPs or toxic air pollutants) will be reduced significantly in the next decade and beyond. In doing so, American industry will spend billions of dollars to install control equipment

and find new processes that do not emit HAPs.

This complete overhaul of section 112 was the result of frustration on the part of industry, environmentalists, and regulators in dealing with the requirements of the old section 112. To understand how things reached this point, this chapter will review the early attempts by the EPA and some states to control air toxics and the litigation that resulted from those attempts. The chapter also outlines the new approach to air toxics control that results from the 1990 Clean Air Act Amendments.

FEDERAL LEGISLATIVE ACTIVITY BEFORE 1990

Before the 1970 Amendments to Clean Air Act, virtually nothing was done to control air toxics directly. To the extent that some air toxics were emitted from automobile exhaust or were precursors to smoke or smog, they were reduced incidentally as the main regulatory targets were reduced.

In 1970, however, that changed. The 1970 Amendments contained section 112, that was

designed specifically to reduce emissions of HAPs to protect public health. Section 112(a)(1) defined a hazardous air pollutant as:

An air pollutant to which no ambient air quality standard is applicable and which in the judgment of the Administrator causes, or contributes to, air pollution which may reasonably be anticipated to result in an increase in serious irreversible, or incapacitating reversible, illness.[1]

---

[1] 42 USC § 7412(a)(1) (1982).

Once determined to be a HAP, a substance was to be listed and regulated within a very short period of time. The level of control was defined in the statute by the requirement that:

The Administrator shall establish any such standard at the level which in his judgment provides an ample margin of safety to protect the public health from such hazardous air pollutant.[2]

The drafters of this language could never have anticipated the controversy that would surround almost every phrase. They apparently thought it would be simple for EPA to expand to hazardous air pollutants the control program for criteria pollutants that already existed. The clear expectation was that EPA would be able to identify, list, and regulate a significant number of HAPs in a short time. So sure were they of this result, Congress required the Administrator of EPA to publish a list of HAPs within ninety days of enactment. Once listed, the Administrator had 180 days to publish a proposed National Emission Standard for Hazardous Air Pollutants (NESHAP) and schedule a public hearing. After another 180 days, the Administrator was required to publish a final NESHAP or issue a determination that the substance was not a HAP.

Instead, what happened was that EPA managed to regulate only seven substances under the NESHAP program before the Act was amended in 1990. The interval between listing of a HAP and promulgation of standards has varied from two to six years, rather than the one year contemplated by the 1970 Amendments

(see Table 16-1). In many cases, standards were produced only when the agency was compelled to do so by court order. For example, in 1980 EPA listed inorganic arsenic as a HAP. Standards were proposed in 1983, but only after the state of New York successfully sued and the court ordered the agency to produce rules.[3] Radionuclide standards also were proposed in 1983, following a successful suit by the Sierra Club.[4] Getting the agency to issue final radionuclide standards took until 1985, however, after a second Sierra Club suit and a contempt citation.[5] Most recently, the Natural Resources Defense Council sued to force EPA to regulate additional sources of benzene emissions.[6] (A recent list of the cases involved in Section 112 standards is provided in Table 16-2.)

What was behind all this litigation? Why was the agency apparently so slow to list and regulate substances that might threaten public health? The agency had two major problems in developing NESHAPs: (1) how to define what was really a HAP, and (2) whether the standards had to eliminate all risk, regardless of cost or technological feasibility.

Despite the obvious difficulty the agency was having in dealing with these issues, Congress made almost no changes to section 112 as part of the 1977 Amendments to the Clean Air Act. By 1990, however, frustration was high on all sides of the debate. Public concern about air toxics also was growing, whereas simultaneously there was an increasing recognition that nothing in life is risk-free. In the 1990 Amendments, Congress tried to balance these concerns.

THE 1990 CLEAN AIR ACT AMENDMENTS

The approach to regulation of air toxics was changed dramatically by the 1990 Amendments. Instead of putting the burden on the Administrator to identify toxic air pollutants, Congress

included in the statute a list of 172 substances and 17 categories of substances for regulation. The levels to which emissions of those substances are to be controlled were made much

---

[2] 42 USC § 7412(b)(1)(B) (1982).

[3] *New York* v. *Gorsuch*, 554 F.Supp. 1060 (S.D.N.Y. 1983).

[4] *Sierra Club* v. *Gorsuch*, 554 F.Supp. 785 (N.D. Cal. 1982).

[5] *Sierra Club* v. *Ruchelshaus*, 602 F.Supp. 892 (N.D. Cal. 1984).

[6] *Natural Resources Defense Council* v. *U.S. EPA*, Nos. 83-2011, 83-2951 (D.D.C. Feb. 7, 1989).

Table 16-1   Timetable of NESHAPs listing decisions

Pollutant	Date listed	Date of proposed standard	Date of final standard	Date of reproposal
Asbestos	3/31/71 36 FR 5931	12/7/71 36 FR 23239	4/6/73 38 FR 8826	
Benzene	6/8/77 42 FR 29332	1/5/81 46 FR 1165[a]	6/6/84 49 FR 23498[a]	7/28/88 53 FR 28496[b]
Beryllium	3/31/71 36 FR 36560	12/7/71 36 FR 23239	4/6/73 38 FR 8826	
Coke oven emissions	9/18/84 49 FR 36560	4/23/87 52 FR 13586		
Inorganic arsenic	6/15/80 45 FR 37886	7/20/83 48 FR 33112	8/4/86 51 FR 27956	
Mercury	3/31/71 36 FR 76738	12/7/71 36 FR 23239	4/6/73 38 FR 8826	
Radionuclides	12/27/79 44 FR 76738	4/6/83 48 FR 15076	2/6/85 50 FR 5190[c]	3/7/89 54 FR 9612
		2/21/85 50 FR 7280[d]	4/17/85 50 FR 34056[d]	
		2/21/86 51 FR 6382[e]	9/24/86 51 FR 34056[e]	
Vinyl chloride	12/24/75 40 FR 59477	12/24/75 50 FR 59532	10/21/76 41 FR 46560	6/2/77 42 FR 28154[f]
			9/30/86 51 FR 34904[g]	

a Standard for benzene equipment leaks (fugitive emissions sources) only.

b Standards for benzene emissions from maleic anhydride process vents, ethylbenzene/styrene process vents, and benzene storage vessels were proposed in 1980, at 45 FR 26660, 45 FR 83448, and 45 FR 83952, respectively. On June 6, 1984, EPA withdrew these proposed standards (49 FR 23558) and proposed standards for benzene emissions from coke by-product recovery plants (49 FR 23522). On August 3, 1984, the Natural Resources Defense Council filed a petition in the Court of Appeals for the D.C. Circuit seeking review of these withdrawals and the final standard for equipment leaks (*Natural Resources Defense Council, Inc.* v. *Thomas*, No. 84-1387, generally known as the Benzene case). Before a decision was rendered, EPA requested a voluntary remand of its benzene standards for reconsideration in light of the vinyl chloride decision. The remand was approved, and in accordance with a court-ordered deadline, EPA proposed new standards for these sources on July 28, 1988.

c Standards for nonradon sources only.

d Standards for radon emissions from underground uranium mines only.

e Standard for radon emissions from licensed uranium mill tailings only.

f Date of final promulgation: 9/30/86, 51 Fed. Reg. 34904. This standard remains in effect, but the agency was required to propose new standard by decision in *Natural Resources Defense Council* v. *EPA*, 824 F.2d 1146 (D.C. Cir. 1987) – generally known as the vinyl chloride case.

g This standard remains in effect, but the agency was required to propose a new standard by decision in *Natural Resources Defense Council* v. *EPA*, 824 F.2d 1146 (D.C. Cir. 1987) – the vinyl chloride case.

more clear, as were the roles of cost and technological feasibility. The 1990 Amendments recognized that not only the large industrial sources traditionally regulated by EPA should be controlled, but so should smaller, 'area' sources.

Sources to be Regulated

The first step in this new control scheme is to identify all major sources of the listed pollutants. For purposes of section 112, a major source is one that emits 10 tons per year (TPY)

Table 16-2   Recent Court decisions on NESHAP rule-makings

Case	Date	Holding
Sierra Club v. *Gorsuch* 551 F. Supp. 785 (N.D.Cal. 1982)	9/30/82	EPA required to propose radionuclide standards
New York v. *Gorsuch* 554 F. Supp. 1060 (S.D.N.Y. 1983)	1/12/83	EPA required to propose inorganic arsenic standards
Sierra Club v. *Ruckelshaus* 602 F. Supp. 892 (N.D.Cal. 1984)	7/24/84 12/11/84	EPA required to issue final radionuclide standards or find radionuclides not hazardous; Administrator and agency cited for contempt for failure to comply with initial order
Natural Resources Defense Council v. *EPA* 804 F.2d 710 (D.C. Cir. 1986)	11/4/86	EPA allowed to consider economic and technological factors in withdrawing proposed vinyl chloride regulations
Natural Resources Defense Council v. *EPA* 1146 (D.C. Cir. 1987)	7/28/87	Previous decision of court vacated *en banc*: required to set acceptable risk level without considering factors other than health before setting margin of safety which may include consideration of cost and technological factors
Natural Resources Defense Council v. *Thomas* 689 F. Supp. 246 (S.D.N.Y. 1988)		District court did not have jurisdiction to compel EPA to list pollutants on basis of alleged delays; complaint dismissed without prejudice
Natural Resources Defense Council v. *EPA* 695 F. Supp. 48 (D.D.C. 1988)	9/14/88	EPA not required to propose emission standards for every source category of benzene after listing it, but must propose standards or issue determination not to regulate for each source within statutory timetable
Natural Resources Defense Council v. *EPA* Nos. 83-2011 and 83-2951 (D.D.C. Feb. 7, 1989)	2/7/89	Earlier order vacated. EPA required to propose standards for benzene source categories covered in previously proposed standards and later to propose final standards or determination that standards are not necessary for each such source category

or more of a single listed substance or 25 TPY or more of any combination of listed substances. The Administrator also has the authority, that he has indicated that he will use, to designate even smaller sources as major. In addition, the Administrator can designate area sources for control. EPA's list of HAP sources to be regulated was published in mid-1992.[7]

Technology-Based Standards

Once source categories are listed, EPA must develop technology-based standards designed to produce the maximum emission reduction achievable. These standards, generally referred to as requiring maximum achievable control technology (MACT), may be different for new

---

[7] 57 FR 31576, July 16, 1992.

and existing sources. They also may distinguish, within a source category, between individual sources based on age, size, or other factors.

For new sources, standards can be no less stringent than the emission control achieved in practice by the best controlled similar source. For existing sources, standards must be no less stringent than the emission control achieved by the best performing 12 percent of sources in the category or subcategory (or five sources if the category or subcategory has less than thirty sources). Any source that has first installed lowest achievable emission rate (LAER) controls within thirty months before promulgation or eighteen months before proposal of a MACT standard is excluded from the comparison pool. For area sources, the standard may be different, based on generally available control technology (GACT).

The amendments also established a stringent schedule for setting standards. Emissions standards for forty source categories were to be promulgated by November 15, 1992. Twenty-five percent of the listed categories are to be regulated by November 15, 1994; 50 percent are to be regulated by November 15, 1997; and all remaining categories are to be regulated by November 15, 2000. Although this schedule may be unrealistically tight, it was a reflection of Congress' determination to keep the process moving.

All the parties involved with the 1990 Amendments agreed that section 112 should continue to be focused on protection of public health. The technology-based MACT and GACT standards were designed to jump start the system and guarantee significant reductions in air toxics emissions in the short term. Although those reductions are being achieved, Congress also developed an elaborate system that, in the long term, will result in section 112 standards designed to protect public health.

Health-Based Standards

The requirements for health-based standards in the 1990 Amendments comprise two major steps. First, the National Academy of Sciences (NAS) is required to undertake a review of the methodology used by EPA for carcinogenic risk assessment and recommend improvements in such methodology. The academy is specifically directed to consider the following, at a minimum.

■ Techniques for estimating carcinogenic potency to humans.
■ Techniques for estimating exposure (including use of the hypothetical most exposed individual).
■ Methodologies for assessing risks of health effects other than cancer.

The report of the NAS findings is to be submitted to Congress and to the Administrator by May 15, 1993. To assist the NAS, the Administrator may obtain information from private entities and may require private entities to conduct tests and report results. Before promulgating any health-based standards, and after notice and opportunity for comment, the Administrator must publish revised Guidelines for Carcinogenic Risk Assessment. If the recommendations of the NAS are not accepted, the Administrator must publish a detailed explanation.

The second step in the health-based regulatory process is the requirement for the Administrator, after consultation with the Surgeon General, to investigate and report to Congress on the following.

■ Methods for calculating risks to public health after application of MACT.
■ The public health significance of the remaining risks and the technologically and commercially available methods and costs of reducing such risks.
■ The actual health effects to persons in the vicinity of sources, including epidemiological and other health studies, risks from background concentrations, uncertainties in health risk assessment methodologies, and negative health or environmental consequences of efforts to reduce such risks.
■ Legislative recommendations to address any remaining risks.

In the event that, after all this work, Congress does not act on the recommendation submitted by the Administrator, then the Administrator is required to set residual risk standards for listed source categories and subcategories, if such standards are required to protect public health with an ample margin of safety or to prevent adverse environmental effects. Any standards to address environmental effects must consider costs, energy, safety, and other relevant factors. Standards must be promulgated for any source category that contains a source of known, probable, or possible human carcinogen where the lifetime excess cancer risk to the individual most exposed to emissions from the source is more than one in a million.

Standards under this subsection must be promulgated no later than eight years after promulgation of MACT for the category or subcategory, except that for the categories or subcategories for which MACT must be set by two years after enactment, residual risk standards must be set no later than nine years after promulgation of MACT.

REGULATORY ACTIVITY BEFORE 1990

Despite the lack of regulations issued under section 112, emissions of many HAPs still were controlled and were significantly reduced even before the 1990 Amendments. An important factor in this reduction was the use of federal authorities other than section 112. For example, activities of other EPA environmental programs that were useful in reducing air toxics included information gathering requirements (e.g., the Comprehensive Assessment and Information Rule under the Toxic Substances Control Act); toxics release inventory requirements under the Emergency Planning and Community Right-to-Know Act; and information dissemination, health assessments, and regulatory actions under other environmental legislation. Specific examples of regulatory development and assessment activities in other programs that influenced emissions of air toxics include the following.

1. New source performance standards for wood stoves, that significantly reduce the polycyclic organic matter components of wood stove emissions.
2. Regulations for seven types of hazardous waste treatment, storage, and disposal facilities developed under authority of the Resource Conservation and Recovery Act.
3. Regulations under the mobile source provisions of the Clean Air Act to reduce toxic constituents either directly or indirectly, including ongoing activities to reduce the lead content of gasoline, prevent tampering and misfueling, phase in more stringent standards for light- and heavy-duty diesel engines, and regulate methanol-fueled vehicles.
4. Regulations banning carbon tetrachloride in pesticide formulations, eliminated the major emission source of carbon tetrachloride.
5. Superfund sites frequently have a significant air emission component. Increased cleanup activity under the Superfund will further accelerate reduction in air emissions from these sites.

Other Clean Air Act Authorities Reducing Air Toxics

Section 109 – National Ambient Air Quality Standards

Since the early 1970s, EPA and the states have implemented the National Ambient Air Quality Standards program to control the 'criteria' pollutants, including ozone and particulate matter. Although this program focuses on controlling emissions of criteria pollutants, it results in substantial reductions in emissions of toxic air pollutants as well. For example, controls on particulate matter reduce emissions of toxic metals and other pollutants that are entrained on particulates. State Implementation Plan limitations on particulate matter frequently reduce emissions of toxic metals by 80 to 98 percent. Similarly, controls on volatile organic compound emissions to attain the ozone standard can significantly reduce emissions of toxic volatile organic compounds.

Section 111 – New Source Performance Standards (NSPS)

Like the National Air Ambient Air Quality Standards program, new source performance standards promulgated under section 111 of the Clean Air Act have substantially reduced emissions of toxic air pollutants. Section 111 standards, that are source-specific, limit emissions of specific pollutants that may include various air toxics. For example, NSPS for electric utility steam generating units, incinerators, sewage treatment plants, and other similar sources establish limitations on particulate matter that will reduce emissions of toxic metals. Control programs to limit particulate emissions also reduce toxic air pollutants in the form of polycyclic organic matter. Similarly, NSPS for synthetic organic chemicals manufacturing, petroleum refineries, and other similar sources reduce emissions of toxic volatile organic compounds.

Resource Conservation and Recovery Act

The Resource Conservation and Recovery Act (RCRA) has been viewed primarily as a statute to prevent exposure to hazardous chemicals through groundwater and surface water pathways. However, particularly in light of the Hazardous and Solid Waste Amendments of 1984, RCRA also is concerned with protecting human health from exposure to hazardous air emissions. Even before the 1984 Amendments, however, EPA had taken various actions under RCRA to control air emissions from hazardous waste facilities including monitoring and efficiency requirements for incinerators and other thermal treatment devices.

Control of air emissions at hazardous waste facilities has been greatly expanded in the wake of the 1984 Amendments. Section 3004(n) of RCRA directs EPA to 'promulgate such regulations for the monitoring and control of air emissions at hazardous waste treatment, storage and disposal facilities, including, but not limited

to, tanks, surface impoundments and landfills, as may be necessary to protect human health and the environment.' EPA has carried out this provision by proposing an accelerated program for the control of volatile organic emissions at hazardous waste facilities.

Emergency Planning and Community Right-to-Know Act (EPCRA)

Title III of the Emergency Response and Community Right-to-Know Act of 1986 established emergency planning and community right-to-know programs that will have a significant impact on the emission of toxic air pollutants. Subtitle A (sections 301–305) establishes emergency planning and notification requirements designed to alert local communities to actual and potential emergency releases of hazardous chemicals and to provide for the establishment of effective emergency planning and response mechanisms. Subtitle B (sections 311–313) establishes a variety of nonemergency information and reporting requirements. These include: the submission of Material Safety Data Sheets for hazardous chemicals (section 311); the provision of inventory, location, and related information on hazardous chemicals (section 312); and, annual reports on releases of toxic chemicals to the air and other environmental media (section 313).

Comprehensive Environmental Response, Compensation and Liability Act (CERCLA)

Under section 103 of CERCLA, also known as Superfund, anyone in charge of a facility is required to notify immediately the National Response Center of any release of a hazardous substance over the reportable quantity of that substance.

Section 104 establishes the Agency for Toxic Substances and Disease Registry for the purpose of developing a list of hazardous substances and information on the potential health effects of those substances. If any study conducted under this section contains a finding that the exposure concerned presents a significant risk to human health, the President may take whatever steps are necessary to reduce exposures and eliminate or mitigate the significant risk to human health.

Under section 105, a National Contingency Plan is to be developed to establish procedures and standards for responding to the release of hazardous substances, including air emissions.

Occupational Safety and Health Act

Section 5 of this Act requires every employer to furnish to each employee a workplace free from recognized hazards. That mandate has been broadly interpreted and has allowed regulation of air emissions of a number of substances.

Toxic Substances Control Act (TSCA)

Since TSCA's passage in 1976, EPA has completed twenty-two regulatory actions to control five different substances under an unreasonable risk standard contained in section 6. Four of the substances controlled under the regulations also are listed as an air toxic in the 1990 Clean Air Act Amendments: polychlorinated biphenyls (PCBs), asbestos, dioxin, and chromium.

In addition, EPA has taken other actions to control toxic substances under the authority of TSCA that did not involve the use of the section 6 unreasonable risk standard. Nine actions have been taken to control new substances and chemicals under the authority of section 5 of TSCA, and over three hundred orders have been issued that set conditions on the manufacture of new chemicals. Further, information and reports on seven substances have been referred to the Occupational Safety and Health Administration (OSHA) for its consideration in reducing risks to worker health, and nine chemical advisories have been issued warning various workers, manufacturers, and others of potential hazards from substances such as used motor oil and nitropropane.

National Emission Standards for Hazardous Air Pollutants (NESHAP)

After an early flurry of activity that resulted in NESHAP for asbestos, beryllium and mercury, the program ground to a virtual halt. During the Carter administration, EPA developed and proposed an airborne carcinogen policy that would have greatly expanded the scope of section 112.[8] It also would have incorporated economic and technological feasibility questions into the standard-setting process. The proposal called for listing of carcinogens emitted to the air and would have required installation of best available technology (BAT). However, the policy had still not been completed when President Reagan took office. The new administration showed no interest in rejuvenating the NESHAP program. In 1983, when William Ruckelshaus returned to EPA, he pledged to increase activity under section 112. Using new risk assessment and risk management techniques, the agency tried to move more quickly. Twenty-six pollutants were reviewed for health effects (see Table 1-2). The agency issued an intent-to-list notice for nine of the twenty-six. The agency announced that it did not intend to list nine of the twenty-six and reserved judgment on the remaining eight.

The intent-to-list process was not mentioned in the statute, but represented an attempt by EPA to collect data so that if a substance was ultimately listed there might be some chance of meeting the tight statutory deadlines for regulation. The process was criticized by many and resulted in a legal challenge by the NRDC.[9] NRDC sought a court order compelling EPA to list several probable carcinogens from the 'intent-to-list' list. Instead, the court held that EPA's delay in listing these substances was 'supported by sufficient evidence in the record that it would be improper' to conclude that there was bad faith on the agency's part.[10]

Despite the controversies and seemingly endless starts and stops, EPA did manage to

---

[8] 44 FR 58642, October 10, 1979.
[9] *Natural Resources Defense Council* v. *Thomas*, 689 F.Supp. 246 (S.D.N.Y. 1988).
[10] Ibid. at 257.

promulgate several NESHAPs (see Table 16-1). The more significant are described below.

Benzene

The first standards issued after Administrator Ruckelshaus's promise to revive the program were for fugitive emissions of benzene.[11] The standard, promulgated in 1984, required sources in petroleum refining and chemical manufacturing industries that use equipment containing at least a 10 percent concentration of benzene to adopt certain practices to control leaks.[12] The standard allowed no detectable emissions due to leaks from pressure relief devices, required a leak detection and repair program for valves and pumps, and prescribed equipment standards for compressors, sampling connections, open-ended valves, product accumulation vessels and closed-vent systems.[13]

Radionuclides

In 1985, EPA promulgated standards for radionuclides other than radon.[14] The standards established radiation exposure limits for populations around federal and NRC-licensed facilities and emission limits for polonium-210 from elemental phosphorous plants. The emission standards did not require installation of any controls beyond those already in place.

Uranium Mines and Tailings Piles

In 1985, EPA promulgated standards for radon emissions from underground uranium mines.[15] The standards require mine owners and operators to install bulkheads to block off abandoned parts of mines to minimize the escape of radon emissions. Radon emissions from uranium mill tailings were controlled by standards proposed in 1986 and promulgated that year. The standards require that tailings be disposed of either by burial in impoundments no larger than 40 acres or by continuous disposal in which no more than 20 acres of tailings are exposed at any time before burial.

Inorganic Arsenic

Final standards for inorganic arsenic were promulgated in 1986.[16] The standards apply to copper smelters, glass manufacturing plants, and arsenic trioxide and metallic arsenic production facilities. Standards for production facilities are work practice standards, whereas those for the other sources are emission limitations.

Coke Oven Emissions

The last NESHAP activity before vinyl chloride (see below) was the proposal of standards for coke oven emissions in 1987.[17] Issuance of a final standard was deferred pending the vinyl chloride decision and has now been replaced by specific statutory requirements for coke oven emissions in the 1990 Amendments.

Acrylonitrile

EPA reviewed acrylonitrile under the NESHAP program, but ultimately decided not to regulate it at the federal level.[18] Instead, EPA announced a pilot project in which it would negotiate with fourteen state and local agencies, within whose jurisdiction there were significant acrylonitrile sources, to identify approaches to deal with acrylonitrile emissions.[19] Almost immediately, the State and Territorial Air Pollution Program Administrators (STAPPA) and the Association of Local Air Pollution Control Officials (ALAPCO) criticized the proposal. They said it lacked federal enforceability and could lead to inconsistency in state regulation.[20]

[11] 49 FR 23498, June 6, 1984.

[12] 40 C.F.R. § 61.110 et. seq.

[13] Ibid. at § 61.240 et. seq.

[14] 50 FR 5190, February 6, 1985.

[15] 50 FR 34056, April 17, 1985.

[16] 51 FR 27956, August 4, 1986

[17] 52 FR 13586, April 23, 1987.

[18] 50 FR 24319, June 10, 1985.

[19] Ibid.

[20] See [Current Developments] 15 Envir. Rep. 235 (BNA) (June 7, 1985).

Ultimately, the referral program was suspended because of objections from STAPPA/ALAPCO. Those organizations recommended that any referral program be limited to substances that did not pose a national emissions problem, that were produced in no more than five states, and that had an estimated lifetime cancer risk of less than one in one million for the most exposed individual.[21] EPA never identified another substance for the referral program. In many cases, the states went ahead with their own air toxics programs, as described later in this chapter.

Vinyl Chloride

Each of the NESHAPs has a story; vinyl chloride's is particularly colorful. Vinyl chloride was first listed as a hazardous air pollutant in 1975. Emissions of vinyl chloride often result from equipment leaks, thus EPA's original standard proposed to control emissions based on work practices, instead of setting an emissions cap. In 1976, the Supreme Court declared work practice standards to be illegal under section 112.[22] This decision led to one of the few changes to section 112 in the 1977 Amendments; they expressly allowed use of work practice standards. EPA again proposed standards in 1977, but later withdrew the proposal. Environmental groups filed a petition with EPA challenging the withdrawal. Denial of the petition led to the now famous Vinyl Chloride decision that finally established the Court's views of how section 112 was supposed to work.[23]

One of the central issues in the Vinyl Chloride case was the use by EPA of cost-benefit analysis. The Natural Resources Defense Council (NRDC) challenged this consideration, claiming that section 112 was designed to protect public health regardless of cost or feasibility. The Court of Appeals for the D.C. Circuit initially sided with EPA. In an opinion written by Judge Bork, the court held that the language and legislative history of section 112 were vague enough that 'in the absence of congressional direction as to the values the Agency should use in guiding the Administrator's discretion, the Administrator's choice of economic and technological feasibility amounted to a reasonable one.'[24]

The NRDC quickly petitioned the court for a rehearing *en banc*. In July 1987, the full court handed down its opinion, also authored by Judge Bork.[25] The new opinion remanded the vinyl chloride standards to EPA for reconsideration consistent with the procedures set out in the opinion. Important aspects of the decision are the following.

■ Identification of 'safe' level of emissions – the court ordered EPA to set NESHAPs by following a two-step process. First the Administrator is to make a purely health-based decision about the level of acceptable risk:

The Congressional mandate to provide an ample margin of safety 'to protect the public health' requires 'safe'. This determination must be based exclusively upon the Administrator's determination of the risk to health at a particular emission level.[26]

Nonhealth factors are explicitly excluded from this stage of the analysis:

The Administrator cannot consider cost and technological feasibility in determining what is safe.[27]

The Administrator's decision does not require a finding that 'safe' means 'risk-free' ... or a

---

[21] See [Current Developments] 15 Envt. Rep. 1663 (BNA) (December 27, 1985).

[22] *Adamo Wrecking Co.* v. *U.S.*, 98 S.Ct. 566 (1978).

[23] 51 FR 34904, September 30, 1986.

[24] *Natural Resources Defense Council* v. *EPA*, 804 F.2d. 710 at 727 (D.C. Cir. 1986).

[25] 824 F.2d. 1146.

[26] Ibid. at 1164.

[27] Ibid. at 1166.

finding that the determination is free from uncertainty. Instead, we find only that the Administrator's decision must be based upon an expert judgment with regard to the level of emissions that will result in an 'acceptable' level of risk to health ... the Administrator must determine that inferences should be drawn from available scientific data and decide what risks are acceptable in the world in which we live. (See Industrial Union Dept., 448 U.S. at 642, 100 S. Ct. at 2864. 'There are many activities that we engage in every day – such as driving a car or even breathing city air – that entail some risk of accident or material health impairment; nevertheless, few people would consider those activities unsafe').[28]

■ Determination of margin of safety – once the 'safe' level of emissions is determined, the court then requires EPA to consider whether emissions levels should be further reduced to increase the margin of safety.

Because the purpose of the margin of safety is to protect against incompletely understood dangers, the degree of uncertainty in risk estimates can be taken into account. In determining what is an 'ample margin' the Administrator may, and perhaps must, take into account the inherent limitations of risk assessment and the limited scientific knowledge of the effects of exposure to carcinogens at various levels, and may therefore decide to set the level below that previously determined to be 'safe.'[29]

It is only at this point of the regulatory process that the Administrator may set the emission standard at the lowest level that is technologically feasible ... Once 'safety' is assured, the Administrator should be free to diminish as much of the statistically determined risk as possible by setting the standard at the lowest feasible level.[30]

STATE AND LOCAL PROGRAMS

Increasingly frustrated with the slow pace of federal regulation, more state and local governments initiated their own programs to control air toxics. The development of these programs, like that of federal efforts, has often been slowed by controversy. Only nineteen states had air toxics programs in 1984, but by 1989 every state had some type of program (STAPPA/ALAPCO 1989).

The type of programs vary widely. Some address only new sources whereas others include existing sources; some are based on ambient concentration whereas others use emission limitations; and some require risk assessments and others do not. Tables 16-3 and 16-4 summarize the state and local programs, respectively.

Control of air toxics at the state level generally involves both a legislative mandate and administrative rules promulgated by the state air regulatory agency. Six states

(California, Colorado, Illinois, Maine, New Hampshire, and Wisconsin) have legislatively mandated air toxics programs. The other forty-four states rely on general air pollution control legislation for their authority.

States that have developed programs to control air toxics generally have chosen one of two ways to carry out these programs: policy or regulation. States selecting the policy approach are concerned with operating flexibility, both for their own air pollution program and for the regulated community as well. Program flexibility often is a plus given the uncertainties inherent in programs to control air toxics (e.g., the list of air toxics, the populations at risk, and the diversity of sources).

On the other hand, those states selecting the regulation route largely are striving for a pragmatic approach that is unambiguously enforceable. The administrative procedures that serve as a foundation for any regulatory

[28] Ibid. at 1164–65.
[29] Ibid. at 1165.
[30] Ibid. at 1165.

Table 16-3   State toxic air pollution control programs

| Agency | Regulatory status[a] | Control approach[b] | |
		Carcinogens	Noncarcinogens
Alabama			
New sources	Policy	AAL	AAL
Existing sources	Informal/plan	ND	ND
Alaska			
New sources	Informal	ND	ND
Existing sources	Informal/plan	ND	ND
Arizona			
New sources	Informal	AAL,RA	AAL
Existing sources	Informal/plan	AAL,RA	AAL
Arkansas			
New sources	Informal	LAER,AAL,RA	LAER,AAL,RA
Existing sources	Informal/plan	LAER,AAL,RA	LAER,AAL,RA
California			
New sources	Policy	TBACT,RA	ND
Existing sources	Limited reg.	TBACT	ND
Colorado			
New sources	Policy	CT[c],AAL[c]	CT[c],AAL[c]
Existing sources	Informal/plan	CT[c],AAL[c]	CT[c],AAL[c]
Connecticut			
New sources	Regulatory	CT,AAL,RA	CT,AAL,RA
Existing sources	Regulatory	CT,AAL,RA	CT,AAL,RA
Delaware			
New sources	Proposed	TDBACT[c]	TDBACT[c],AAL[c]
Existing sources	Proposed	TDBACT[c]	TDBACT[c],AAL[c]
Florida			
New sources	Policy	CT[c],AAL[c],RA[c]	CT[c],AAL[c]
Existing sources	Limited	CT[c],AAL[c],RA[c]	CT[c],AAL[c]
Georgia			
New sources	Policy	AAL	AAL
Existing sources	Informal/plan	AAL	AAL
Hawaii			
New sources	Informal	AAL[c]	AAL[c]
Existing sources	No activity	No Activity	No Activity
Idaho			
New sources	Policy	BACT,AAL,RA	BACT,AAL,RA
Existing sources	Informal/plan	ND	ND
Illinois			
New sources	Informal	ND	ND
Existing sources	Informal/plan	ND	ND
Indiana			
New sources	Policy	ND	ND
Existing sources	Informal/plan	ND	ND
Iowa			
New sources	Policy	RA	AAL
Existing sources	Informal/plan	ND	ND
Kansas			
New sources	Policy	BACT/LAER,AAL,RA	Same
Existing sources	Informal/plan	ND	ND
Kentucky			
New sources	Regulatory	BACT	AAL
Existing sources	Regulatory	RACT	AAL

Table 16-3   Continued

Agency	Regulatory status[a]	Control approach[b]	
		Carcinogens	Noncarcinogens
Louisiana			
New sources	Policy	AAL,RA	AAL
Existing sources	Informal/plan	AAL,RA	AAL
Maine			
New sources	Informal	ND	ND
Existing sources	Informal/plan	ND	ND
Maryland			
New sources	Regulatory	TBACT,AAL,RA	TBACT,AAL,RA
Existing sources	Regulatory	TBACT,AAL,RA	TBACT,AAL,RA
Massachusetts			
New sources	Informal	CT,AAL,RA	CT,AAL,RA
Existing sources	Informal/plan	CT,AAL,RA	CT,AAL,RA
Michigan			
New sources	Policy	CT,RA	CT,AAL
Existing sources	Informal/plan	ND	ND
Minnesota			
New sources	Policy	TDBACT,RA	AAL
Existing sources	Informal/plan	RA,CT[c]	AAL
Mississippi			
New sources	Policy	CT,RA	AAL
Existing sources	Policy	RA	AAL
Missouri			
New sources	Informal	BACT[c],AAL[c],RA[c]	AAL[c]
Existing sources	Informal/plan	ND	ND
Montana			
New sources	Informal	ND	ND
Existing sources	Informal/plan	ND	ND
Nebraska			
New sources	Regulatory	TDBACT	TDBACT
Existing sources	Informal/plan	ND	ND
Nevada			
New sources	Regulatory	CT,AAL	CT,AAL
Existing sources	Regulatory	CT,AAL	CT,AAL
New Hampshire			
New sources	Policy	AAL,RA[c]	AAL,RA[c]
Existing sources	Informal/plan	RA[c]	RA[c]
New Jersey			
New sources	Regulatory	LAER/TD,RA	LAER/TD,RA[c]
Existing sources	Regulatory	LAER/TD,RA	LAER/TD[c],RA[c]
New Mexico			
New sources	Regulatory	CT,AAL	CT,AAL
Existing sources	Informal/plan	ND	ND
New York			
New sources	Policy	CT,AAL,RA[c]	CT,AAL,RA[c]
Existing sources	Informal/plan	CT,AAL,RA[c]	CT,AAL,RA[c]
North Carolina			
New sources	Proposed	AAL[c],RA[c]	AAL[c]
Existing sources	Proposed	AAL[c],RA[c]	AAL[c]
North Dakota			
New sources	Policy	CT,AAL	AAL
Existing sources	Informal/plan	ND	ND

Table 16-3   Continued

Agency	Regulatory status[a]	Control approach[b]	
		Carcinogens	Noncarcinogens
Ohio			
New sources	Policy	BAT,AAL,RA	BAT,AAL
Existing sources	Limited	CT[c],AAL[c],RA[c]	CT[c]
Oklahoma			
New sources	Regulatory	CT,AAL,RA	CT,AAL,RA
Existing sources	Regulatory	CT,AAL,RA	CT,AAL,RA
Oregon			
New sources	Policy	RA	AAL
Existing sources	Informal/plan	ND	ND
Pennsylvania			
New sources	Policy	BACT,AAL,RA	BACT,AAL,RA
Existing sources	Informal/plan	BACT[c],AAL[c],RA[c]	BACT[c],AAL[c],RA[c]
Rhode Island			
New sources	Regulatory	LAER,AAL,RA	LAER,AAL,RA
Existing sources	Regulatory	LAER,AAL,RA	LAER[c],AAL[c],RA[c]
South Carolina			
New sources	Policy	AAL	AAL
Existing sources	Informal/plan	ND	ND
South Dakota			
New sources	Informal	ND	ND
Existing sources	Informal/plan	ND	ND
Tennessee			
New sources	Policy	CT,AAL	CT,AAL
Existing sources	Informal/plan	ND	ND
Texas			
New sources	Policy	CT,AAL	CT,AAL
Existing sources	Limited	CT,AAL	CT,AAL
Utah			
New sources	Informal	ND	ND
Existing sources	Informal/plan	ND	ND
Vermont			
New sources	Regulatory	AAL,RA	AAL
Existing sources	Regulatory	AAL,RA	AAL
Virginia			
New sources	Regulatory	CT,AAL	CT,AAL
Existing sources	Regulatory	AAL	AAL
Washington			
New sources	Informal	CT[c],AAL[c],RA[c]	CT[c],AAL[c]
Existing sources	Informal/plan	ND	ND
West Virginia			
New sources	Proposed	LAER[c],RA[c]	ND
Existing sources	Proposed	TDBACT[c],RA[c]	ND
Wisconsin			
New sources	Regulatory	CT,RA	AAL
Existing sources	Regulatory	CT,RA	AAL
Wyoming			
New sources	Informal	CT,AAL,RA	CT,AAL
Existing sources	Informal/plan	ND	ND

[a] Policy – Source review based on agency policy; regulatory – comprehensive regulatory program; informal – informal control through new source review; limited – limited source categories regulated; informal/plan – case-by-case control or planning development of more comprehensive program.
[b] AAL – acceptable ambient level; LAER – Lowest achievable emission rate; RA – risk assessment; BACT – best available control technology; TBACT – BACT for toxics; TDBACT – top down BACT; CT – unspecified control technology; RACT – reasonably available control technology; ND – not yet determined.
[c] Proposed.

Source: STAPPA/ALAPCO 1989.

Table 16-4   Selected local toxic air pollution control programs

Agency	Regulatory status[a]	Control approach[b] Carcinogens	Noncarcinogens
Alabama – Jefferson County			
New sources	Policy	AAL	AAL
Existing sources	Policy	AAL	AAL
Arizona – Maricopa			
New sources	Informal/plan	CT[c],RA[c]	CT[c],RA[c]
Existing sources	Informal/plan	CT[c],RA[c]	CT[c],RA[c]
California – Bay Area AQMD			
New sources	Policy	RA	AAL
Existing sources	Informal/plan	LAER/BACT[c]	LAER/BACT[c]
California – South Coast AQMD			
New sources	Policy	TDBACT,RA	AAL
Existing sources	Informal/plan	TDBACT,RA	AAL
Colorado – Denver			
New sources	Informal/plan	ND	ND
Existing sources	Informal/plan	ND	ND
Florida – Palm Beach			
New sources	No activity	ND	ND
Existing sources	No activity	ND	ND
Indiana – Indianapolis			
New sources	No activity	ND	ND
Existing sources	No activity	ND	ND
Kansas – Kansas City/Wyandotte County			
New sources	Policy	CT,RA	AAL
Existing sources	No activity	ND	ND
Kentucky – Jefferson County			
New sources	Regulatory	CT,AAL	CT,AAL
Existing sources	Regulatory	CT,AAL	CT,AAL
Maryland – Baltimore			
New sources	Regulatory	TBACT,AAL,RA	TBACT,AAL,RA
Existing sources	Regulatory	TBACT,AAL,RA	TBACT,AAL,RA
Michigan – Wayne County			
New sources	Policy	CT,AAL,RA	CT,AAL
Existing sources	Informal/plan	CT,AAL,RA	CT,AAL
Missouri – St. Louis			
New sources	No activity	ND	ND
Existing sources	No activity	ND	ND
Nevada – Clark			
New sources	Regulatory	CT	ND
Existing sources	Informal/plan	ND	ND
North Carolina – Mecklenburg County			
New sources	Informal/plan	AAL[c]	AAL[c]
Existing sources	Informal/plan	AAL[c]	AAL[c]
Ohio – Cleveland			
New sources	No activity	ND	ND
Existing sources	No activity	ND	ND
Ohio – Southwestern APCA			
New sources	Policy	ND	ND
Existing sources	Informal/plan	ND	ND
Oregon – Lane Regional APA			
New sources	Informal/plan	CT,RA[c]	CT,RA[c]
Existing sources	Informal/plan	AAL[c]	AAL[c]

Table 16-4   Continued

| Agency | Regulatory status[a] | Control approach[b] | |
		Carcinogens	Noncarcinogens
Pennsylvania – Allegheny County			
New sources	Informal/plan	AAL[c]	AAL[c]
Existing sources	Informal/plan	ND	ND
Pennsylvania – Philadelphia			
New sources	Regulatory	AAL	AAL
Existing sources	Regulatory	AAL	AAL
Tennessee – Hamilton County			
New sources	Informal/plan	CT,RA	CT,RA
Existing sources	Informal/plan	ND	ND
Washington – Puget Sound			
New sources	Limited	CT,AAL	CT,AAL
Existing sources	Limited	CT	CT

[a] Policy – Source review based on agency policy; regulatory – comprehensive regulatory program; informal – informal control through new source review; limited – limited source categories regulated; informal/plan – case-by-case control or planning development of more comprehensive program.

[b] AAL – acceptable ambient level; LAER – lowest achievable emission rate; RA – risk assessment; BACT – best available control technology; TBACT – BACT for toxics; TDBACT – top down BACT; CT – unspecified control technology; RACT – reasonably available control technology; ND – not yet determined.

[c] Proposed.

Source: STAPPA/ALAPCO 1989.

program, such as the opportunity for public involvement in official-decision making, offer the regulated community some sense of partnership in the overall process. Some states have instituted a policy approach as an interim measure while developing a regulatory approach.

Each state program has grappled with the fundamental question of what constitutes a toxic air pollutant. Again, most states have chosen one of two ways to identify what they want to consider an air toxic: a state-specific list or a list developed by some scientific or regulatory body. States that have developed their own lists typically have done so with consideration of the types and quantities of specific toxic substances emitted within their borders. States that have depended on another body to develop a list of toxic substances are relying on the specific expertise of that body to determine what is toxic and what is not.

Ambient Concentration Programs

At least thirty-eight states have programs that are based on ambient concentrations of air toxics.[31] In determining acceptable ambient concentrations, most states take occupational standards and apply a factor that represents the difference in exposure scenarios. The most commonly used occupational standard is the threshold limit value (TLV) developed by the American Conference of Governmental Industrial Hygienists (ACGIH). However, the developers of those standards state explicitly that they are not intended for use as ambient standards. Other states use Occupational Safety and Health Administration (OSHA) or National Institute of Occupational Safety and Health (NIOSH) standards. A few states use toxicity data from the Registry of Toxic Effects of Chemical Substances (RTECS). (See Chapter 17 for more information.)

---

[31] Ibid. at 3–6.

Uncertainty factors are applied to these standards to convert them from workplace to community standards. These uncertainty factors vary widely. For example, the occupational standards were developed to protect workers who generally are exposed to higher concentrations and in an eight-hour day, forty-hour week. Ambient environmental exposures, on the other hand, can be continuous over a lifetime but generally are lower in concentration. In addition to exposure, some states consider the toxicity of the substance in setting safety factors. As a result, uncertainty factors range from 1/7,500 to 1.

Another factor to be considered is averaging time. Different averaging times may be applied depending on the health effects associated with exposure to a particular pollutant. For example, an annual averaging time may be applied to a carcinogen that results in chronic effects, whereas a much shorter averaging time may be applied to pollutants with acute effects.

All of these decisions are arbitrary in the sense that there is no scientifically agreed method of converting protective workplace levels to acceptable community levels. Nevertheless, the system has been effective in requiring emission controls that would otherwise not have been required.

Control Technology Requirements

As of 1989, twenty-one states have air toxics programs that defined specific control requirements.[32] At least seven of these states base requirements on source type, six on the type of pollutant, and five on both source type and pollutant. Eleven of these states require controls on new sources only, whereas seven require controls on both new and existing sources.

Risk Assessment

More than twenty-five states report that they use quantitative risk assessment for determining

control requirements.[33] Risk assessment may be used in two different ways. The first way is to calculate the risk posed by a source's emissions and then to decide if that level of risk is acceptable. If not, the source is required to reduce emissions to the point where risks are acceptable.

The second way in which risk assessment is used is in development of acceptable ambient guidelines. The state may determine the acceptable level of risk and translate that into an acceptable ambient concentration. Sources must control emissions so as not to exceed the acceptable ambient levels.

Selected State Programs

Florida

The Florida Air Toxics Permitting Strategy[34] was developed as a means of controlling toxic air contaminant emissions to protect the public health. This strategy was adopted by the Florida Air Toxics Working Group composed of Florida Department of Environmental Regulation (FDER) and county air toxics staff. The strategy is currently being carried out on the basis of a policy; this provides FDER flexibility and the emitting sources latitude for negotiation in conforming to the strategy.

The substance of the strategy is the comparison of an estimated ambient concentration resulting from a facility's emissions to an acceptable ambient level. There are three elements to this strategy: (1) the identification of what defines a toxic air pollutant subject to review under this strategy; (2) the determination of what is an acceptable amount of that pollutant in the ambient air (i.e., the no threat level, NTL); and (3) the prediction of the ambient impact due to the anticipated emission of that pollutant.

There is no set list of air toxics to which FDER will apply this strategy. All compounds potentially emitted by a new or modified source

---

[32] Ibid. at 3–8.
[33] Ibid. at 3–9.
[34] J. Glunn, Florida Department of Environmental Regulation, Tallahassee, FL, personal communication, 1992.

may be considered to be hazardous. For such evidence, FDER relies most frequently on the ACGIH-TLVs. Other sources of health data include OSHA's list of Permissible Exposure Limits, the EPCRA section 313 reporting list, the National Toxicology Program's list of carcinogens, the International Agency for Research on Cancer's list of carcinogens, and EPA's inhalation reference dose list.

An acceptable amount of an air toxic is a concentration that is not likely to cause appreciable health risks. The selection of a conservative acceptable level will provide an ample margin of safety in order to be health protective. Therefore, FDER selects the health-based TLV for any given compound and incorporates a safety factor in order to define the NTL. A safety factor of 100 is used for sources permitted to operate eight hours per day or less, whereas a safety factor of 420 is used for sources permitted to operate twenty-four hours per day.

The FDER will conduct screening-level dispersion modeling of the air toxic emissions using the actual source characteristics and parameters to estimate the applicable NTL; if the predicted concentration is less than the NTL, the anticipated emissions of the HAP satisfy the strategy and no further evaluation should be required. However, if the predicted concentration is greater than the NTL, then the facility owner/operator has several options: (1) conduct refined dispersion modeling; (2) redesign the process to limit potential emissions or ground-level concentrations; (3) add abatement equipment; (4) restrict the operating schedule; or (5) conduct a health risk assessment to determine an appropriate NTL.

Georgia

The Georgia Environmental Protection Division (GEPD) follows a policy guideline on hazardous air pollutant emissions when reviewing permit applications to construct new sources or to modify or simply continue to operate

sources.[35] The approach requires a dispersion modeling demonstration to predict whether the expected ambient impact of air toxics emissions will be within an acceptable ambient concentration (AAC).

First, the AAC is determined for the air toxic being emitted. The AAC is derived from available literature in the following order of preference: (1) OSHA's permissible exposure limits (PELs); (2) ACGIH-TLVs; and (3) NIOSH's recommended exposure limits (RELs). Other references can be used to support an AAC if no value for a specific pollutant is found in one of these three. The guideline is silent as to how discrepancies between these lists will be addressed. However, once a literature value is identified, it is converted to an exposure-equivalent concentration for the number of hours per week the pollutant is emitted. Next, this value is further reduced by a safety factor of 100 for noncarcinogens, or 300 for carcinogens. This value then becomes the AAC and is so defined as a twenty-four-hour hour averaging time.

Second, the dispersion of the hazardous air pollutant is simulated using a screening model run by the GEPD. Only the off-property maximum ground-level concentration (MGLC) is considered and the averaging time of the concentration estimate must be determined on a twenty-four-hour basis. If the MGLC is greater than the AAC, a refined dispersion model may be run. If the source's emissions still result in exceeding the AAC, the owner/operator may either: (1) reduce the emission rate of the HAP; (2) increase the stack height; or (3) conduct his/her own impact assessment.

The result of the impact assessment that shows compliance with the AAC will be used by the GEPD to establish allowable emission limits for the air toxic of concern. The Georgia Guideline also has provisions for multiple sources of toxics, multiple toxics at the same facility, and additive/synergistic effects from multiple pollutant exposures.

---

[35] D. Yardumiau, Guideline for Ambient Impact Assessment of Toxic Air Pollutant Emissions, Georgia Department of Natural Resources, Atlanta, GA, 1984.

North Carolina

North Carolina's Division of Environmental Management (NCDEM) requires certain sources to obtain a permit to emit hazardous air pollutants.[36] Exemptions from permitting are clearly stated in the regulations as are the alternates if the owner/operator is unable to make a modeling demonstration of compliance with acceptable ambient levels of 105 listed compounds or compound classes.

If the facility is subject to the regulations, the owner/operator must conduct a dispersion modeling study of the emissions. The permit application to emit air toxics should be approvable if the owner/operator can show that the off-property maximum ground-level concentrations are less than the acceptable ambient levels (AALs) listed for the 105 regulated compounds and compound classes. These AALs are based on annual, twenty-four-hour, one-hour, and fifteen-minute averaging times. These AALs are not ambient air standards but are to be used as 'guides' in determining whether to permit the source. If the owner/operator cannot show compliance with the AALs by modeling, there are several options: (1) demonstrate that the modeled air quality will not adversely affect human health; (2) demonstrate that the level of control needed to achieve the AALs is technologically infeasible; (3) demonstrate that compliance with the AALs will result in serious economic hardship; or (4) submit a compliance schedule to accomplish the AALs within three years.

California

The California Air Resources Board (CARB) administers a statewide program to identify and control stationary sources of toxic air contaminants (TACs). Under the Tanner Act, or AB 1807 as it is referred to, CARB and Department of Health Services (DHS) identify a substance as a TAC based on the chemical's health effects and its prevalence in the state. The chemical is then designated as a TAC. After designation,

CARB investigates appropriate measures to limit emissions of the TACs that may be emissions limitations, control technologies, operation and maintenance requirements, closed system engineering, cost, or substitution of compounds. CARB then prepares a report on the appropriate degree of regulation and adopts control measures accordingly. These control measures are the minimum regulations that must be imposed by each of the local air districts in the form of regulations.

Often the local air districts are more restrictive. On the local level, the Air Quality Management Districts (AQMDs) are the most important regulatory agencies. They have the responsibility for enforcing federal regulations, setting regulatory standards, and implementing state regulations. As a special district their authority is very significant and can allow them to dictate operating procedures for industries under their jurisdiction.

To date, fourteen chemicals have been formally designated as TACs (essentially all are listed in the 1990 Clean Air Act Amendments): asbestos, benzene, cadmium, carbon tetrachloride, chlorinated dioxins and dibenzofurans, chloroform, hexavalent chromium, inorganic arsenic, ethylene dibromide, ethylene dichloride, ethylene oxide, methylene chloride, trichloroethylene, and vinyl chloride. Control requirements have been adopted for benzene, hexavalent chromium, cadmium, dioxin, and ethylene oxide. Draft reports also have been issued on nickel and perchloroethylene.

Existing sources of TACs are incorporated into the control program through source registration, operating permit renewal, and emissions inventories. Some existing sources are identified from federal data bases or citizen complaints, and then incorporated through the permit program. New sources are incorporated through permits to construct and operate, and on a case-by-case basis.

The California Air Toxics 'Hot Spots' Information and Assessment Act (AB 2588), has

---

[36] North Carolina Administrative Code, Chapter 2, Regulation 2D.1100, Control of Toxic Air Pollutants and Regulation 2H.0610, Permit Requirements for Toxic Air Pollutants, 1990.

gone a step further and identified 326 toxic substances for which facilities have to inventory their emissions. This list was developed using the TACs list and candidate TACs list, Proposition 65 (another important California legislative action) chemicals, certain hazardous chemicals identified under the California Labor Code, certain substances listed by the National Toxicology Program, and any additional substances recognized by the CARB as presenting a chronic or acute threat to public health.

Using a tiered schedule, all potential sources of listed air toxics (based on those facilities emitting criteria pollutants in certain amounts per year) and facilities that were listed in any toxics use or toxics air emissions survey, have to prepare emissions inventories and update these lists every two years. Facilities must then prepare risk assessments in accordance with a priority list set by each local air pollution control district. On approval of the risk assessment, the facility must then notify all exposed persons if the district determines that there is a significant health risk. These inventories and risk assessments may then be used by CARB or the districts to identify and control additional TACs under programs including AB 1807.

On the local level, the Bay Area Air Quality Management District (BAAQMD) and the South Coast Air Quality Management District (SCAQMD) have released their lists of high priority facilities under AB 2588. CARB also has recently added sixty-six new toxic substances to the (AB 2588) list. SCAQMD also adopted regulations in June 1990 known as New Source Review of Carcinogenic Air Contaminants (Rule 1401). Rule 1401 establishes the requirements for health risk assessments, specific limits for individual cancer risk, describes the procedures to be used to calculate the risk, and lists the carcinogenic air contaminants to be evaluated. This listing was increased in December 1990 from nine to thirty-nine compounds including inorganic arsenic, chloroform, acetaldehyde, formaldehyde, PCBs, trichloroethylene, and vinyl chloride.

SCAQMD intends the rule to apply to both new and existing sources. All sources of the listed chemicals must go through the permitting process and may be denied a permit based on residual risk remaining after installation of best available control technology (BACT).

Texas

The air toxics program in Texas is an integral part of the Texas Air Control Board's (TACB) activities to prevent and control air pollution. The TACB began a new and modified source permit review program in 1972 for all air contaminants, both criteria and noncriteria.[37] This review starts with an engineering evaluation of the process to determine BACT on a case-by-case basis. The resulting emissions are quantified and dispersion modeling is conducted to predict residual ground-level concentrations. Ambient effects are evaluated in the light of best available health effects information. The effects review includes evaluation of pertinent toxicological information, including standards set for exposure in occupational situations, the documented effects levels on which the occupational standards are based, and epidemiological and toxicological information from the literature and from consultation with other federal, state, and private health-related agencies. Generally, the guideline is 1 percent of the occupational limit for short-term (thirty-minute) exposures and 0.1 percent of the occupational limit for long-term (annual) exposures. Control requirements can be adjusted if the need is indicated based on the health effects review. The use of guidelines instead of standards provides needed flexibility in dealing with potentially thousands of materials. It allows the reviewer to consider new toxicity information when it is available. It also allows the reviewer to consider information such as operating schedules, adjacent property, and types of health effects associated with each substance.

The TACB and the Texas Water Commission adopted, in joint rule-making, regulations for

---

[37] Texas Administrative Code, Title 31.

control of air pollution from hazardous waste or solid waste management facilities. These became effective on April 30, 1986. In addition, the TACB and the Texas Department of Health adopted, in joint rule-making, regulations for the control of air pollution from municipal solid waste facilities that became effective on December 17, 1986. Under these programs, facilities are undergoing review to determine the type and quantity of air emissions present and their potential effects. Review of potential air quality impacts from Superfund site cleanup proposals is another type of review conducted in conjunction with another agency's authority. This crossmedia interaction is proving to be a useful method of dealing with the complex environmental problems of handling and disposing of hazardous or potentially hazardous wastes.

Wisconsin
On October 1, 1988, new hazardous air pollution regulations became effective in Wisconsin.[38] For the first time, all wastewater treatment facilities, both publicly owned and private industrial facilities, are required to report on hazardous air pollution emissions from the diffused sources that exist in wastewater treatment processes, such as aeration tanks. Wastewater treatment facilities as a class are one of the largest group of sources that previously had not been affected by air pollution controls but are covered by the new rules. In addition, a survey of facilities conducted by the Wisconsin Department of Natural Resources (WDNR) before rule adoption found that most large manufacturing facilities, a few smaller facilities incinerators, large fossil fuel users, and very large gasoline stations would be affected by the rule. Of the manufacturing facilities, the chemical manufacturing and allied products facilities would be most affected.

Since 1986, emissions of hazardous air pollutants have been prohibited in Wisconsin 'in such quantity, concentration, or duration as to be injurious to human health, plant or animal life unless the purpose of that emission is for the control of plant or animal life.'[39] In addition, mercury, asbestos, beryllium and vinyl chloride were regulated specifically under those existing rules. However, this general prohibition had little or no effect in curtailing the emission of other unspecified hazardous air contaminants. The new rules aim to fill the gap left by the lack of federal regulation and to give meaning to the Wisconsin prohibition against hazardous emissions.

Toward that end, the definition of 'hazardous air contaminant' is broad and includes all pollutants for which no ambient air quality standards previously have been set.[40] Included in the definition are sulfur oxides, particulates, carbon monoxide, ozone, nitrogen dioxide, lead and pollutants that the WDNR determines may 'cause or significantly contribute to an increase in mortality or an increase in serious irreversible or incapacitating reversible illness, or may pose a significant threat to human health or the environment.'[41] Within this class, specific hazardous contaminants are identified.

More than four hundred hazardous contaminants have been included on four lists contained in Chapter NR 445.[42] These lists are not exhaustive, and the WDNR is required to modify the lists as new information becomes available.[43] The lists define three sets of hazardous air contaminants: (1) those with acceptable ambient concentrations (i.e., 'acute' contaminants shown in Tables 1 and 4); (2) those that are pesticides, rodenticides, insecticides, herbicides or fungicides with acceptable ambient concentrations shown in Table 2; and (3) those without acceptable ambient concentrations that are known or suspected carcinogens

---

[38] Wis. Admin. Code §NR 445.
[39] Wis. Admin. Code §NR 445.03.
[40] Wis. Admin. Code §NR 404.4.
[41] Wis. Admin. Code §NR 445.02(6).
[42] Wis. Admin. Code §NR 445.04, Tables 1–4.
[43] Wis. Admin. Code §NR 445.06(3).

shown in Table 3. For contaminants with acceptable ambient concentrations, the tables specify permissible emission rates in pounds per hour. For the known or suspected carcinogens, a 'de minimis emission' rate is stated in pounds per year.

Reference

State and Territorial Air Pollution Program Administrators and the Association of Local Air Pollution Control Officials (STAPPA/ALAPCO). 1989. *Toxic Air Pollutants: State and Local Regulatory Strategies – 1989.* Washington, D.C.

17

Ambient Concentration Limits

David R. Patrick

INTRODUCTION

State and local air pollution control agencies frequently specify an ambient concentration limit (ACL) to achieve their regulatory goals for a toxic air pollutant. ACLs also are called acceptable ambient limits (AALs) and acceptable ambient concentrations (AACs). An ACL is determined by a regulatory agency to be the maximum ambient air concentration to which people may be exposed. ACLs generally are derived from health criteria developed from human or animal studies and normally are presented in units of weight of the pollutant per unit volume of air. A major advantage of the ACL is that it can be translated easily into a maximum allowable release rate for a source of the pollutant. ACLs normally are used only for pollutants, such as noncarcinogens, that are associated with a threshold of effect. ACLs rarely are used for carcinogens because cancer generally is viewed as a health effect of concern at any exposure.

It should be noted that although many state and local agencies have used ACLs directly to regulate toxic air pollutants, EPA generally has not.[1] A primary reason is that the Clean Air Act Amendments of 1970 required EPA to regulate toxic air pollutants through the use of national emissions standards. The 1990 Amendments continued and strengthened this requirement.

There are a number of reasons why state and local agencies use ACLs for regulatory purposes. First, for many air pollutants ACLs can be derived easily and economically from readily available health effects information. This is of particular value to a regulatory agency with limited scientific staff and monetary resources. Second, the maximum emission rate for a source that corresponds to the selected ACL can be determined easily through mathematical calculation (e.g., by using an air dispersion model in reverse). This means that compliance or noncompliance by a source can be determined easily by the regulator. Third, by using ACLs regulators do not have to identify and specify acceptable process or control technologies, that requires considerable technical

---

[1] The exceptions were beryllium and mercury, two noncarcinogens regulated in 1973. In both cases, an acceptable ambient concentration was translated into an acceptable emission rate.

Table 17-1   Use of ACLs to regulate toxic air pollutants

Advantages	Disadvantages
■ ACLs can be derived easily from RfCs and RfDs, occupational limits or other criteria by assuming standardized human uptakes and standard exposure situations. ■ ACLs can be verified through ambient monitoring, although this can be costly and time-consuming. ■ ACLs are easy to revise when new data are developed. ■ ACLs can be derived by any trained scientist, thus improving their utility and timeliness.	■ ACLs may bear no relationship to the exposures that can occur at community. ■ In order to verify compliance with an ACL, ambient monitoring is necessary. This can be expensive and time-consuming. ■ Emission rates derived from ACLs generally based on many assumptions, i.e., meteorology, population number, population distribution. ■ Application of safety factors to derive ACLs must be done with careful evaluation of the test data to ensure that the process is scientifically appropriate. ■ ACLs derived from criteria other than EPA RfCs and RfDs may not have been reviewed by independent scientists.

expertise; they only have to require that sources meet the ACL or the maximum emission rate with which it corresponds.

On the other hand, the use of ACLs has significant shortcomings. The most important is that ACLs frequently are derived from occupational health criteria and there are no widely accepted procedures for converting occupational criteria to community criteria (i.e., from healthy workers exposed forty hours per week to the general population potentially exposed twenty-four hours per day). Another disadvantage is that both animal and occupational exposures from which health criteria are developed typically are at concentrations far greater than normal community exposures. This requires extrapolation from higher to lower dosages and often from animals to man. Some of the advantages and disadvantages of using ACLs are listed in Table 17-1.

In this chapter, the typical derivation and use of ACLs are described. In particular, the advantages and disadvantages of various methods are discussed. Finally, the current use of ACLs by state and local air pollution control agencies is summarized.

DERIVATION OF AMBIENT CONCENTRATION LIMITS

ACLs typically are derived from accepted health criteria for the substance in question. Most frequently, ACLs are expressed in concentration terms such as micrograms per cubic meter ($\mu g/m^3$). Accepted health criteria may have been derived in those terms or in terms of dosage (e.g., weight of pollutant taken into the body per unit of body weight per day). When dosage values are used, these must be translated into concentration units using generally accepted assumptions for average breathing rates, average rates of food and water consumption, and substance specific absorption factors. EPA has described these procedures (EPA 1988b, EPA 1989a).

Other methods of deriving ACLs rely on the concept of an absolute threshold (CMA 1988). These methods set ACLs at some fraction of an observed threshold or established guideline, and a safety margin is added depending on the type and severity of the effect, the quality of the data, and other factors. Still other methods rely on extrapolation from limits established for other but similar purposes.

Currently, the health criterion accepted as the most appropriate for deriving ACLs is the risk reference dose (RfD) established by EPA. EPA has been deriving RfDs for a number of years to provide benchmarks with which to compare doses of environmental concern to the general population (EPA 1986h). RfDs have been developed by EPA both for inhalation and ingestion pathways, although they are available for the ingestion pathway for a greater number of substances.

RfDs must be derived by trained scientists and should be independently reviewed by qualified scientists. Moreover, the derivation process is highly resource intensive and the number of chemicals for which criteria potentially are required is vast. RfDs developed by EPA also are appropriate only for longer-term human health effects. The RfD derivation process is described below along with other techniques used when RfDs are not available. Each of the methods has advantages and disadvantages. Short-term adverse health effects also can be of concern for some toxic air pollutant exposures; other procedures must be used to derive ACLs for short-term exposures.

Use of the Risk Reference Dose

The derivation of RfDs is described in Chapter 7. RfDs currently are the preferred values for evaluating long-term noncarcinogenic toxicity in EPA risk assessments and, thus, serve as the best basis for establishing ACLs. However, RfDs derived by EPA still are approximate numbers and conservative (i.e., health protective). In other words, doses that are less than the RfD are not believed to be of concern; doses that are significantly greater than the RfD are generally viewed as indicating a potential for an adverse effect; and doses that are slightly higher than the RfD have a higher probability of producing an adverse effect, although EPA generally does not intend for these doses to be considered, by definition, as unacceptable. These latter doses generally require further evaluation.

Different RfDs are developed for oral and inhalation exposure routes, that can be quite different. If available, ACLs for air pollutants should be based on inhalation RfDs. EPA

currently is deriving reference values in terms of $\mu g/m^3$ for inhalation health effects; this reference value is called the risk reference concentration, RfC.

RfD and RfC values are available through EPA's Integrated Risk Information System (IRIS).[2] Many state and local regulatory agencies use the EPA-derived RfDs and RfCs to establish ACLs. The resulting ACLs are available from the state and local agencies. They also are generally provided in EPA's National Air Toxics Information Clearinghouse (NATICH),[3] although NATICH may not be up to date in all respects. The advantages and disadvantages of using RfDs to establish ACLs are summarized in Table 17-2.

A value similar to the ACL, called the acceptable daily intake (ADI) level, was used in the past by EPA and others but largely has been supplanted by the RfD. The concept was the same; the ADI was defined as the amount of a chemical to which one can be exposed on a daily basis over an extended period of time (usually a lifetime) without suffering a deleterious effect (EPA 1986h). The term ADI rarely is used any longer primarily because of the desire to avoid absolute terms such as 'acceptable.'

In cases where an inhalation RfD or RfC is not available, ACLs must be derived from other sources. One method is to convert an ingestion RfD, if one is available, to an equivalent inhalation RfD. This is not always scientifically appropriate because absorption and uptake by the lungs and the gastrointestinal system can differ substantially. When used, adjustments must be made to equalize the dosages, as discussed above. These should be undertaken only by appropriately trained scientists. Although actual inhalation rates vary greatly with age, sex, work level, and other factors, regulators frequently conservatively assume that everyone breathes in $20 \, m^3/day$ (an average adult value) and that the inhaled pollutant is totally absorbed.

---

[2] For information on IRIS, call (513) 569-7254.
[3] For information on NATICH, call (919) 541-5332 or (919) 541-0850.

Table 17-2   Use of RfDs to establish ACLs

Advantages	Disadvantages
■ Inhalation RfDs expressed as μg/m³ can be directly used as the ACL. ■ Toxicity characterization typically considers potency, duration, and route specificity. ■ Primary, peer-reviewed sources normally used. Considers data quality and weight of evidence. Typically developed by an EPA work group and undergo a formalized peer review and verification process. ■ Most appropriate for estimating consequences of longer exposures, such as a lifetime. ■ RfDs developed for use with the public. Typically consider effects on sensitive subpopulations such as children, the elderly, and the infirm.	■ RfDs developed only by EPA. Oral RfDs are currently available for more substances than inhalation RfDs, although programs are under way by EPA to develop additional inhalation RfDs. ■ RfDs developed for substances that exhibit threshold effects. Some effects, such as teratogenic effects, are not easily characterized as to whether they exhibit thresholds or not. ■ RfDs not appropriate for evaluating acute toxicity. ■ The derivation of RfDs does not generally distinghish among severity of effects of different substances.

Use of Occupational Exposure Limits

Frequently, RfDs or RfCs are not available and ACLs must be derived from other sources. In addition, other sources must be explored for use in deriving acute exposure ACLs. Because regulatory officials prefer to use accepted, peer-reviewed data, occupational limits often are used to establish both chronic and acute exposure ACLs. The occupational limits typically used are the threshold limit value (TLV) and the permissible exposure limit (PEL). Each of these establishes allowable concentrations and times that a worker may be exposed in the workplace to a wide variety of potential air pollutants.

Threshold limit values have been developed for many years by the American Conference of Governmental Industrial Hygienists (ACGIH). ACGIH periodically publishes workplace inhalation exposure limits for several hundred chemicals (ACGIH 1989). There are three types of TLVs.

■ Time-weighted average (TLV-TWA) – the TWA is the time-weighted average concentration for a normal eight-hour workday and forty-hour workweek to which nearly all workers may be repeatedly exposed, day after day, without adverse effect.

■ Short-term exposure limit (TLV-STEL) – the STEL is a fifteen-minute time-weighted average concentration that should not be exceeded at any time during the workday even if the eight-hour TWA is met. Exposures to the STEL should not be longer than fifteen minutes, they should not be repeated more than four times per day, and there should be at least sixty minutes between successive exposures. STELs also supplement the TWA when there are recognized acute effects from a substance whose toxic effects are primarily of a chronic nature.

■ Ceiling limit (TLV-C) – The TLV-C concentration should not be exceeded during any part of the working exposure. Ceiling limits may supplement other limits or stand alone.

Permissible exposure limits are established by the U.S. Occupational Safety and Health Administration (OSHA) and are defined in much the same way as the TLVs. OSHA adopted ACGIH's values when federal occupational standards were published originally in 1974.[4] More recently, OSHA revised the air contaminant standards[5]; although many concentration limits remained the same as in the past, a number were revised.

---

[4] 39 FR 23502, June 27, 1974.
[5] 54 FR 2332, January 19, 1989.

Both ACGIH and OSHA specifically recommend *against* the direct use of occupational limits to establish community exposure limits because the occupational values are developed specifically for workers and workplace exposures. However, because the ACGIH and OSHA limits represent scientifically peer reviewed health criteria for air pollutants, and because few other extensive, peer reviewed databases are readily available, many state and local agencies use occupational limits to derive ACLs.

To extrapolate a workplace limit to community use, the occupational limit must be adjusted with safety factors. Unfortunately, there are no widely accepted guidelines for doing this. Actual adjustments that have been used by state and local agencies vary significantly. The safety factors typically are applied in the following manner. As noted above, occupational limits apply to normal workday exposures (e.g., eight-hour day, five-day week). For application to the general community, occupational limits of most pollutants are adjusted to apply to exposures generally greater than a workweek, up to and including continuous exposures (i.e., twenty-four-hour day, seven-day week). Occupational limits also are derived for healthy workers; thus, additional safety factors frequently are used when applied to the general population, that can include children, the elderly, and the infirm. Finally, adjustments may be made depending on the quality of the data and the regulatory stringency of the agency.

Adjustments also may be needed depending on the potential health effect that may result from exposure to the substance of concern. For example, some pollutants may not respond linearly with concentration. In these cases, a single concentration limit would not be appropriate. Finally, the use and selection of safety factors also depends on the needs and philosophies of the regulating agency and the characteristics of the specific substance and emission situation. Specific derivations for chronic and acute exposures are shown below. The advantages and disadvantages of using occupational exposure limits to establish ACLs are summarized in Table 17-3.

Chronic Exposures

To develop criteria for chronic exposures, safety factors generally are applied to occupational limits in the following ways.

■ The TWA can be reduced by a factor of 4.2 (168/40) to convert from the typical forty-hour work week to the maximum community exposure time of twenty-four hours per day, seven days per week (i.e., 168 hours). This factor assumes that workers and community residents breathe air at the same rate.

■ The TWA also can be reduced by a factor of 2.8 (20×7)/(10×5). This factor is based on the assumption that community residents inhale an estimated $20\,m^3$ of air each twenty-four-hour weekday and workers inhale an

Table 17-3   Use of occupational limits to establish ACLs

Advantages	Disadvantages
■ ACGIH and OSHA limits available and widely accepted for large number of air contaminants in workplace. Typically derived by trained scientists. ■ ACGIH and OSHA limits based upon adverse effects resulting from inhalation; can be considered toxic air pollutants. ■ Many contaminants regulated in the workplace are listed in 1990 Clean Air Act Amendments. ■ Deriving community exposure limits from occupational limits is simple and does not require significant resources or technical expertise; revising exposure limits when new data are obtained also is simple.	■ TLVs and PELs are established only by ACGIH and OSHA, respectively. ■ Limits are not revised often, they vary in accuracy, and many are not recent. ■ TLVs and PELs derived for healthy workers only, not children, elderly, or infirm. ■ Limits established only for typical workplace exposures (eight-hour day, five-day week). ■ Limits do not always distinguish among severity of effects. ■ There are no scientifically accepted criteria for deriving safety factors.

estimated 10 m^3 of air in an eight-hour workday. This factor assumes that persons at work breathe, on the average, at a higher rate than community residents. Other factors have been used depending on the particular situation and assessment needs.

■ The TWA can be reduced by an additional factor of ten to account for the fact that workers are a relatively healthy group of the population and are exposed only during their adult life, as opposed to the community population that includes children, the elderly, and the infirm, and where lifetime exposure can occur.

■ The TWA also can be reduced by another factor of ten to compensate for the fact that community populations may have varying sensitivities to environmental exposures.

■ Safety factors also may be tailored specifically to the toxic substance and the exposure situation by appropriately trained scientists.

Acute Exposures

To develop ACLs for acute exposures, maximum instead of time-weighted averages generally are used. For example, single safety factors can be applied to STELs based on the assumption that the community population is potentially more sensitive to acute effects than the workforce. Approaches that have been used include the following.

■ The STEL may be reduced by a factor of ten if the adverse effect is generally nonharmful irritation (e.g., eye or skin irritation).

■ The STEL limit may be reduced by a factor of one hundred if the adverse effect is potentially more serious than nonharmful irritation (e.g., liver or kidney damage).

■ Safety factors may be tailored specifically to the toxic substance and the exposure situation by appropriately trained scientists.

Use of Other Approaches

NOAELs and LOAELs

Two experimentally derived health criteria typically are used to derive RfDs and also can be used, if available, to establish ACLs when no RfD has been derived. These criteria are: (1) the no-observed-adverse-effect level (NOAEL) or no-observed-effect level (NOEL), and (2) the lowest-observed-adverse-effect-level (LOAEL) or lowest-observed-effect-level (LOEL). The NOAEL is the experimental exposure level to the substance that is the highest level tested at which no adverse effect is observed. The NOAEL is the key result of animal dose-response studies although it also can be derived from human studies. The NOAEL frequently forms the basis for toxicity criteria such as RfDs because it represents the level of no adverse effect. RfDs are derived from NOAELs through the application of uncertainty factors (UFs) and, in some cases, a modifying factor (MF). The NOEL is the experimental exposure level representing the highest level of a substance tested at which any effect, adverse or not, is observed. The NOEL generally is less useful because some effects may be trivial from a health standpoint.

The LOAEL is the experimental exposure level to the substance that is the lowest level tested at which an adverse health effect is observed. It generally is less useful than the NOAEL because adverse effects may occur at lower levels than those tested; thus, the LOAEL usually is adjusted with an additional safety factor before use in setting human health criteria. Similarly, the LOEL is the experimental exposure level representing the lowest level of a substance tested at which any health effect, adverse or not, is observed. It is even less useful than the LOAEL.

A substance may elicit more than one toxic effect (endpoint), even in one test animal, in tests of the same or different duration (e.g., acute, subchronic, and chronic exposure studies). In general, NOAELs derived for these effects will be different. The critical toxic effect used in the derivation of RfDs generally is the one characterized by the lowest NOAEL after conversion to adjust for species differences. This level is selected based in part on the assumption that if the critical toxic effect is prevented, then all toxic effects are prevented.

The uncertainty factors that are used generally consist of multiples of 10 and each factor represents a specific area of uncertainty inherent in the extrapolation from available animal data to human exposures. Either the NOAEL or the LOAEL is divided by the product of all appropriate uncertainty factors (UF) and an additional modifying factor (MF), if appropriate.

The process of deriving the RfD, or the ACL, can be represented by the following relationship.

$$RfD = NOAEL/(UF_1 \times UF_2 \times \ldots \times MF)$$

(17-1)

Generally, uncertainty factors are applied in the following sequence.

1. An uncertainty factor of ten is used to account for the uncertainty associated with extrapolating from animals to humans (i.e., interspecies variation).
2. An uncertainty factor of ten is used to account for variability in sensitivity among individuals within the human population (i.e., intraspecies variation).
3. An uncertainty factor of ten is used if a NOAEL derived from a subchronic study must be used as the basis for a value that will be used to evaluate chronic exposures (i.e., no lifetime studies available).
4. An uncertainty factor of ten is used if a LOAEL is used instead of a NOAEL.
5. A modifying factor ranging up to ten may be included reflecting a professional assessment

of the overall uncertainties of the study and of the entire data base for the chemical, not explicitly addressed by the above uncertainty factors.

The advantages and disadvantages of using NOAELs and LOAELs in establishing ACLs are summarized in Table 17-4.

Minimal Risk Level (MRL)

The Agency for Toxic Substances and Disease Registry (ATSDR) was formed as a result of the Comprehensive Environmental Response, Compensation and Liability Act of 1980 (CERCLA, also known as Superfund). This law, and the Superfund Amendments and Reauthorization Act of 1986 (SARA), mandated that ATSDR undertake a variety of activities to identify and fill data gaps and to meet the public's need for relevant and understandable toxicological information concerning chemicals found or suspected to be present at uncontrolled hazardous waste sites. A critical provision of CERCLA requires ATSDR to prepare and update toxicological profiles for the hazardous substances most commonly found at facilities on the Superfund National Priorities List (NPL) that pose the greatest significant potential risk to human health.

As part of these toxicological profiles, ATSDR derives minimal risk levels (MRLs) based on human and animal studies. MRLs are derived for both inhalation and oral exposures, with inhalation MRLs expressed as concentrations and oral MRLs expressed as doses. The

Table 17-4 Use of NOAELs/LOAELs to establish ACLs

Advantages	Disadvantages
■ Toxicity criteria can be derived using NOAELs or LOAELs that are similar to RfDs in concept and are developed for use with the general population. ■ NOAELs or LOAELs are available or can be developed from the literature for many substances.	■ Published values frequently are not peer-reviewed or EPA approved. ■ Considerable toxicological expertise is required to evaluate all existing data and to designate studies appropriate for the derivation of NOAELs or LOAELs. ■ The direct application of safety factors without regard to the substance, its effect, and its mechanisms of action can lead to substantial errors in the derived community limits.

Table 17-5   Use of MRLs to establish ACLs

Advantages	Disadvantages
■ MRLs are independently developed and peer-reviewed. ■ MRLs are based upon evaluation of the available literature.	■ MRLs are available only for a limited number of substances on the Clean Air Act list of hazardous air pollutants. ■ The MRL information provided in the toxicological profiles is not easily translated into a single number for regulatory purposes.

Table 17-6   Use of IDLHs, ERPGs, and EEGLs to establish ACLs

Advantages	Disadvantages
■ These are available for a number of substances on the section 112(b) list of hazardous air pollutants.	■ As derived, these values generally are applicable to long-term human exposure.

estimates typically are made for the most sensitive noncarcinogenic endpoint, that includes developmental and reproductive effects, and includes acute, intermediate and chronic exposures. The acute exposure duration is assumed to be one to fourteen days; the intermediate is assumed to be fourteen to 365 days; and the chronic is assumed to be greater than 365 days. MRLs are derived using a procedure that is consistent with that used by EPA to derive RfDs and RfCs, as discussed above. The advantages and disadvantages of using MRLs to establish ACLs are summarized in Table 17-5.

MRLs can be found in the ATSDR Toxicological Profile documents in the section entitled Health Effects Summary.[6] The values are shown on the significant exposure figure. In the latter part of 1991, approximately sixty-two MRLs (twenty-four inhalation and thirty-eight oral) had been derived by ATSDR.

Immediately Dangerous to Life and Health (IDLH)

The National Institute for Occupational Safety and Health (NIOSH) has developed values for approximately 390 chemicals that represent the concentration that is immediately dangerous to life and health, IDLH (NIOSH 1985a). IDLHs were developed principally for respirator selection in the workplace. The IDLH represents the maximum concentration of a substance in air from which healthy male workers can escape without loss of life or without suffering irreversible health effects under conditions of a maximum thirty-minute exposure time. Practically, IDLHs are concentrations above which a highly reliable breathing apparatus is required for escape. IDLH concentrations generally far exceed those for TWAs, STELs, and PELs. Advantages and disadvantages of using IDLHs in establishing ACLs are summarized in Table 17-6.

Emergency Response Planning Guidelines

The American Industrial Hygiene Association (AIHA) derived emergency response planning guidelines (ERPGs) for a number of substances. Advantages and disadvantages of using ERPGs are shown in Table 17-6. As the name implies, these values are for use in planning appropriate

---

[6]For information on ATSDR Toxicological Profiles call (404) 639-6000.

responses to the emergency releases of chemicals. Three levels are derived.

Level 1

The maximum airborne concentration below which most individuals could be exposed for as much as one hour without experiencing other than mild, transient adverse health effects (or without perceiving a clearly defined objectionable odor).

Level 2

The maximum airborne concentration below which most individuals could be exposed for up to one hour without experiencing (or developing) irreversible or other serious health effects, or symptoms that could impair the ability to take protective action.

Level 3

The maximum airborne concentration below which most individuals could be exposed for as much as one hour without experiencing (or developing) life-threatening health effects.

Emergency Exposure Guidance Levels

The National Research Council (NAS 1986) derived for the Department of Defense emergency exposure guidance levels (EEGLs) for use in planning for sudden contamination of air during military and space operations. Specifically, they are used to select protective equipment and response plans after nonroutine occurrences such as line breaks, spills, and fires. An EEGL is defined as a concentration of a substance in air judged to be acceptable for the performance of specific tasks by personnel during emergency conditions lasting one to twenty-four hours. Exposure to EEGLs may produce some transient adverse effects and eye or respiratory irritation but nothing serious enough to prevent proper responses to emergency conditions. Exposure to an EEGL is not considered safe but is acceptable during tasks that are necessary to prevent greater risks, such as fire or explosion. EEGLs are peak levels of exposure considered acceptable for rare situations, but are not to be applied in instances of repeated exposure. Advantages and disadvantages of using EEGLs are shown in Table 17-6.

COMPLIANCE WITH AMBIENT CONCENTRATION LIMITS

ACLs generally are used by requiring sources of the substances to reduce emissions to an amount that assures that the ACL will not be exceeded at the property boundary of the emitting facility. By using mathematical dispersion modeling techniques (discussed in Chapter 11) or other techniques accepted by the regulatory agency, a maximum emission limit can be back-calculated from an accepted ACL. Although effective, this type of calculation can be complicated, particularly when a source has several emission points, the local topography is complex, or meteorological conditions vary significantly.

Compliance with the required maximum emission limit then can be determined by sampling at the point of emission, often called source or stack testing (see Chapter 9). When available, source testing can be conducted promptly and provide reasonably accurate results; however, source test procedures are not widely available for many toxic air pollutants and the procedures typically can be complex and costly. In addition, stack tests only measure in 'real time' and are not practical for estimating past or future concentrations.

Alternatively, ambient monitoring (see Chapter 10) can be undertaken to measure the concentrations of the substance at the property line or at locations in the community where people live and, thus, provide a direct measure of success with meeting the ACL. If the monitoring indicates that the ACL is being exceeded, then emissions must be reduced. However, ambient monitoring has several major drawbacks. One is that ambient test methods have not yet been developed for many toxic air pollutants. When they have been developed the methods typically are expensive, complex, and time-consuming. For example, natural variations in meteorology and topography can require numerous monitors to be used and long (e.g., often one year or more) sampling times to achieve a high probability of measuring the 'maximum' concentration. Ambient tests also measure only in real time.

Although these techniques are described in more detail in Chapters 9, 10, and 11, they are summarized briefly here.

Source and Ambient Sampling

Emissions at the source frequently are sampled and analyzed to identify chemicals that are released and their concentration. Samples can be taken at the point of emission (i.e., source sampling). This is done by sampling the exhaust stacks or vents at the source. Samples also can be taken at a site removed from the point of emission (i.e., ambient sampling). This frequently takes place at the fenceline of the source or at locations where community residents can be exposed, often at those locations believed to represent the highest potential exposure.

In both source and ambient sampling, the pollutants are extracted from the exhaust stream or ambient air and either analyzed immediately using a portable analyzer or the extract is collected and sent to a laboratory for analysis. For sampling purposes, air pollutants generally fall into three categories: solids, semivolatiles, and gases. Source and ambient sampling methods for some toxic air pollutants in these categories have been developed and generally validated by EPA; however, many more remain to be developed and validated.

Source sampling equipment typically consists of probes that are inserted into a stack or vent and collection equipment into which samples are withdrawn and analyzed immediately or contained for later analysis. Because stack and vent conditions can vary widely (e.g., location, dimension, temperature, pressure, velocity, and corrosiveness), the sampling equipment must be able to perform across a wide range of situations and conditions. A minimum of at least three source tests normally are conducted and the results usually averaged.

Ambient sampling equipment generally consists of a suction pump and a collection medium. These are protected from the elements and from vandalism and often are designed to run unattended for long periods of time. Because ambient samplers measure only what they are exposed to, numerous samplers often are required to optimize the chances of measuring the 'maximum' concentration of the pollutant in the air. Several samplers also may be required in order to compensate for variations in wind direction and topography. The cost of ambient sampling is related directly to the number of samplers and the sample time.

Samples that are collected by ambient or source samplers must be analyzed. Because many chemicals can be toxic at very low concentrations, the state-of-the-art in analytical techniques has undergone dramatic improvement in the past few years. Some chemicals now are routinely measured at parts per trillion (ppt) levels. To ensure accuracy and enforceability, analyses must be conducted by persons highly trained in the use of accepted techniques and the laboratories typically must be certified by an appropriate governmental authority. Analytical techniques can range from simple weighing and measuring to use of extremely expensive, automated systems (e.g., gas chromatography/mass spectrometry) for analyzing to very low concentrations.

Air Dispersion Modeling

The concentrations of pollutants to which people in a community are exposed can be estimated by mathematical modeling if emission rates are known or can be estimated. As noted above, source and ambient sampling can be expensive and only measure pollutant concentrations in real time. Mathematical dispersion models, on the other hand, can estimate air concentrations under almost any set of conditions, with accuracy a function only of the knowledge of the source, the emission characteristics, the meteorology, and the surrounding topography. These models can range from simple versions with a large error potential to sophisticated versions with a much reduced

Table 17-7 State and local agency use of ambient concentration limits

State	Derivation of ACL
Alabama	TLV/40 (one-hour), TLV/420 (annual)
Alaska	Case-by-case analysis
Arizona	0.0075 × Lower of TLV or TWA
Arkansas	TLV/100 (twenty-four-hour), $LD_{50}/10,000$
California	Risk assessment used
Colorado	Generally uses risk assessment
Connecticut	TLV/50 low toxicity
	TLV/100 medium toxicity
	TLV/200 high toxicity
Delaware	TLV/100
Florida	Ranges from TLV/50 to TLV/420 depending upon the situation
Georgia	TLV/100 (eight-hour), noncarcinogens
	TLV/300 (eight-hour), carcinogens
Hawaii	TLV/200
Idaho	Case-by-case analysis
	BACT can be required
Illinois	Case-by-case analysis
Indiana	Case-by-case analysis
Iowa	Case-by-case analysis
Kansas	TLV/100 (twenty-four-hour), irritants
	TLV/420 (annual), serious effects
Kentucky	Case-by-case analysis
Louisiana	TLV/42 (one-hour) screening level
Maine	Case-by-case analysis
Maryland	Varies, TLV/100 (eight-hour)
Massachusetts	Health-based program
Michigan	TLV/100 (eight-hour)
Minnesota	TLV/100 (eight-hour)
Mississippi	TLV/100 (ten-minute)
Missouri	TLV/75 to TLV/7500 (eight-hour)
Montana	TLV/42
Nebraska	Case-by-case analysis
Nevada	TLV/42 (eight-hour) and case-by-case analysis
New Hampshire	TLV/100 (twenty-four-hour) low toxicity
	TLV/300 (twenty-four-hour) medium toxicity
	TLV/420 (twenty-four-hour) high toxicity
New Jersey	Case-by-case analysis
New Mexico	TLV/100 (eight-hour)
New York	TLV/50 (eight-hour) low toxicity
	TLV/300 (eight-hour) high toxicity
North Carolina	TLV/10 (one-hour) acute toxicity
	TLV/20 (one-hour) systemic toxicity
	TLV/160 (twenty-four-hour) chronic toxicity
North Dakota	TLV/100 (eight-hour)
Ohio	TLV/42
Oklahoma	TLV/10, TLV/50, TLV/100
Oregon	TLV/50, TLV/300
Pennsylvania	TLV/42, TLV/420, TLV/4200 (one-week)
Rhode Island	Case-by-case analysis
South Carolina	TLV/40 (eight-hour) low toxicity
	TLV/100 (eight-hour) medium toxicity
	TLV/200 (eight-hour) high toxicity
South Dakota	Case-by-case analysis
Tennessee	TLV/25, screening
Texas	TLV/100 (thirty-minute)
	TLV/1000 (annual)
Utah	TLV/100 (twenty-four-hour)
Vermont	TLV/420 (eight-hour)
Virginia	TLV/60 (eight-hour), TLV/100
Washington	TLV/420
West Virginia	Case-by-case analysis
Wisconsin	TLV/42 (twenty-four-hour), screening
Wyoming	TLV/4

error potential. Sophisticated models typically account for variations in source configuration, meteorological conditions, topography, and a range of emission characteristics. Their greatest advantage is their ability to inexpensively (compared to sampling) predict ambient concentrations over a wide range of conditions and time.

Guidelines have been published by EPA that describe available dispersion models and the conditions under which each should be used (EPA 1986i). Models are available for simple and complex terrains, various meteorological conditions, both stack and fugitive releases,

single or multiple sources, and varying times. Meteorological data bases frequently are built directly into these models.

Once pollutants are emitted into the atmosphere, they are transported through the atmosphere. As this occurs, concentrations are reduced because of dilution. Gases generally remain in the air; however, particles tend to settle to the ground. Dispersion models can account for both processes. These are described in more detail in Chapter 11. Again, to ensure accuracy and defensibility, dispersion modeling must be conducted by trained persons using accepted techniques.

CURRENT USAGE OF AMBIENT CONCENTRATION LIMITS

State and local regulators have employed a wide range of safety factors in using ACLs. Table 17-7 shows typical ACLs recently used by state and local agencies.

References

American Conference of Governmental Industrial Hygienists (ACGIH). 1989. *Threshold Limit Values and Biological Exposure Indices for 1989–1990.* Cincinnati, OH.

Chemical Manufacturers Association (CMA). 1988. Chemicals in the community: methods to evaluate airborne chemical levels. Washington, D.C.

National Academy of Science (NAS). 1986. Criteria and methods for preparing emergency exposure guidance level (EEGL), short-term public emergency guidance level (SPEGL), and continuous exposure guidance level (CEGL) documents. Committee on Toxicology. Board on Environmental Studies and Toxicology. Washington, D.C.: National Academy Press.

National Institute of Occupational Safety and Health

Administration (NIOSH). 1985a. *NIOSH Pocket Guide to Chemical Hazards.* U.S. Department of Health and Human Services. Washington, D.C.

U.S. Environmental Protection Agency (EPA). 1986h. Integrated risk information system (IRIS) database. Appendix A. Reference Dose (RfD): Description and use in health risk assessments. Office of Health and Environmental Assessment. Washington, D.C.

U.S. Environmental Protection Agency (EPA). 1986i. Guideline on air quality models (Revised) and Supplement A (1987). EPA 450/2-78-027R. (NTIS PB 86-245248). Office of Air Quality Planning and Standards. Research Triangle Park, NC.

U.S. Environmental Protection Agency (EPA). 1988b. Superfund exposure assessment manual. EPA 540/1-88-001. OSWER Directive 9285.5-1. Office of Emergency and Remedial Response. Washington, D.C.

U.S. Environmental Protection Agency (EPA). 1989a. Exposure factors handbook. EPA 600/8-89-043. Office of Health and Environmental Assessment. Washington, D.C.

18

Technology Standards

Jack R. Farmer

INTRODUCTION

Technology standards are standards that result in the control of point and area sources of toxic air pollutant emissions and are based on process, equipment, and economic consideration. Technology standards are not related to any ambient air quality standard or goal. Allowable emissions are provided for as numerical emission limits, or design, equipment, work practice, or operational procedure requirements.

The emission limits usually are based on the performance of an emission control system and take into account the variation in process and control equipment parameters. An emission control system may include add-on control devices, raw material limitations, manufacturing process requirements, equipment requirements, work practice requirements, operational requirements or the banning of a process, product, or raw material. Air pollution control methods are discussed in Chapters 21 and 22, as well as other handbooks (Danielson 1973, Stern 1977).

Several acronyms are associated with technology standards. The following are the most common and will be discussed later in this chapter: NESHAP, MACT, GACT, NSPS, CTG, RACT, BACT, and LAER.

THE STANDARDS DEVELOPMENT PROCESS

The development of legally enforceable technology standards involves a rigorous engineering and economic investigation. The most comprehensive and effective process is that developed by the U.S. Environmental Protection Agency (EPA) in response to the requirements of the 1970 Clean Air Act Amendments. The development of emission standards by EPA covers three primary phases: (1) screening and evaluation of information availability, (2) data gathering and analysis, and (3) decision-making. The first phase involves a review of the affected source category or subcategory, gathering available information and planning the next phase. The second phase is resource intensive and can require many months to complete. The processes, pollutants, and emission control systems used by facilities in the category are evaluated. Performance of emission control systems is measured; costs of the control systems are developed; and environmental, energy, and economic effects associated with the control systems are evaluated. Several regulatory alternatives are selected and evaluated. During the third and final phase of the process, one of the regulatory alternatives is selected as the basis for the standard and undergoes the procedures for informal rule-making. Under federal informal rule-making procedures, the standard is published in the Federal Register as a proposed rule. A public comment period of thirty to ninety days follows. During the public comment period, oral testimony may be presented at a public hearing or written comments may be submitted. EPA reviews and evaluates all comments received on the proposed rule, decides what if any changes are appropriate, and then publishes the final rule in the Federal Register.

The development of standards is an open process and offers many opportunities for input by the public and affected industry. During the process, EPA consults with federal agencies, advisory committees, and other experts. Before

a standard is proposed in the Federal Register, EPA consults with the National Air Pollution Control Techniques Advisory Committee (NAPCTAC) in a meeting open to the public. When the standard is proposed in the Federal Register, all of the information and analysis that supports the proposal is published in a Background Information Document (BID). EPA maintains a docket for all rule-making. The docket contains all information pertinent to the rule-making and is open to the public.[1]

The Federal Register is a daily publication of federal regulations and other legal documents of the executive branch of the government. For more information contact: Office of the Federal Register, National Archives and Records Administration, Washington, D.C. 20408. Annually, all standards published in the Federal Register are codified in the Code of Federal Regulations (CFR). Copies of the CFR are for sale by the Superintendent of Documents, Government Printing Office, Washington, D.C. 20402.

ELEMENTS OF AN EMISSION STANDARD

Emission standards are legal instruments and must clearly define what sources are subject to the standard and what is required by the standard. Effective and efficient standards contain four main elements: (1) applicability; (2) emission limits; (3) compliance procedures and requirements; and (4) monitoring, reporting, and record keeping requirements.

Applicability

Applicability provisions define who and what are subject to the emission standard requirements. This includes a clear definition of the affected source category or subcategory, the process or equipment included, and any size limitations or exemptions. Definitions of words and terms often are used in clarifying the applicability of a standard. Any distinction among classes, types, and sizes of equipment

within the affected source category also is part of the applicability.

Emission Limits

Emission limits specify the pollutant being regulated and the maximum permissible emissions of that pollutant. In developing the emission limits, the performance, cost, energy, and environmental effects of alternate control systems are evaluated. This evaluation provides the information from which to select the control system that serves as the basis for the standard. An emission limit must be selected that reflects the best performance of the selected control system. Selecting the proper emission limit is important because it insures that the affected sources install control systems similar to the one that serves as a basis for the emission limit, or a different control system

---

[1] For information on the Clean Air Act docket, call (202) 260-7549.

that has equivalent performance in reducing pollutant emissions.

Typically, an emission limit is written in terms of an allowable mass emission rate (i.e., mass of pollutant per unit time) or an allowable concentration (i.e., mass of pollutant per volume of exhaust gas). In some instances, a process weight limit (i.e., weight of pollutant per unit of product or input material) or a minimum percent emission reduction of the pollutant (i.e., control system collection efficiency) is used. All of these types of emission limits require the direct measurement of emissions to determine compliance.

In some cases, a numerical emission limit is not feasible because (1) measurement methodology is not available or practicable due to technological or economic limitations, or (2) the pollutants cannot be emitted through a conveyance designed and constructed to emit or capture the pollutant. In these situations, design, equipment, work practice, or operational standards are used. Restrictions on opacity or visible emissions also may be included as an alternative or as a supplement to a standard for particulate emissions.

Compliance Procedures and Requirements

The compliance procedures and requirements are as important as the emission limits. In fact, the stringency of a numerical emission limit may be established by the compliance procedures and requirements. The emissions standard specifies the condition the affected facility is to operate under during compliance testing, the test methods to be used, and the averaging time for the tests to be conducted. The compliance procedures for emission standards that includes design, equipment, work practice, or operational requirements usually are different from those for numerical emission limits and are tailored to the specific design, equipment, work practice, or operation required.

For regulatory purposes, test methods are classified as reference, equivalent, or alternative. A reference method is the primary method for determining compliance and is published in the CFR. An equivalent method is one that has been demonstrated to have a consistent and quantitatively known relationship to the reference method. An alternative method has been demonstrated to produce results, in specific cases, that are adequate for determining compliance.

The averaging time for an emission standard is important if the emission rate is variable. For a facility with an emission rate that varies on an hourly or daily basis, a monthly averaging time would be less stringent than a daily or hourly averaging time for the same emission limit set on a not to be exceeded basis.

During a compliance test, a facility usually is required to operate under normal conditions. A facility would not be required to operate at greater than the design rate unless such operation was normal, nor would it be allowed to operate at substantially below the design rate.

Monitoring, Reporting, and Record-Keeping

The primary purpose of the monitoring, reporting, and record-keeping requirements of an emission standard is to insure that the owner or operator of a facility is properly operating and maintaining the emission control system. Even though a control system may meet all requirements when it is installed and may pass the initial compliance test, poor performance and excess emissions can occur without proper operation and maintenance.

Continuous emission monitors (CEMs) are available for a wide range of pollutants and applications. Because they give a direct measurement of a facility's emissions for a direct comparison to the emission limit, emission standards require CEMs when they are applicable and available at a reasonable cost. When CEMs are not available, other process parameters that give an indication of whether the owner or operator is performing proper operation and maintenance on the emission control system may be monitored.

Record-keeping requirements usually cover the information and data collected by the required monitoring systems. Although these data and information may be required to be kept at the affected facility, the more critical information or summary of the information usually is required to be reported to the enforcing agency. Typical reporting periods are monthly, semiannually, and annually.

HAZARDOUS AIR POLLUTANT STANDARDS

Section 112 of the 1970 Clean Air Act Amendments directed the EPA Administrator to establish national emission standards for hazardous air pollutants (NESHAP). The 1990 Amendments totally revise section 112 and direct the Administrator to establish standards that require the installation of maximum achievable control technology (MACT). The provisions are described in more detail in Chapter 16. The technology standards developed by EPA under the hazardous air pollutant program are described in this section.

NESHAP

From 1970 to 1990, eight pollutants (asbestos, beryllium, mercury, vinyl chloride, arsenic, benzene, radionuclides, and coke oven emissions) were listed as hazardous air pollutants and NESHAP were promulgated for all but coke oven emissions. A listing of these NESHAP with their Federal Register citations is presented in Table 18-1. With a few exceptions, all of these NESHAP are technology standards. The NESHAP for mercury and beryllium are designed to prevent specified ambient concentrations of mercury or beryllium from being exceeded. Some parts of the NESHAP for benzene and radionuclides also are based on the risk assessment procedure that was issued with the benzene standards published on September 14, 1989.[2] Chapter 19 provides a more detailed discussion of this procedure.

Asbestos
The NESHAP for asbestos was promulgated in 1973 and is a technology standard. In the early 1970s, asbestos was a known carcinogen but it had no known dose-response relationship and

risk assessment tools had not been developed. This presented EPA with a dilemma because the 1970 Amendments required NESHAP to provide an 'ample margin of safety to protect public health.' Because there was no clear method for establishing an 'ample margin of safety,' EPA decided to use a report on asbestos published by the National Academy of Science (NAS 1971) as the basis of the standard. The Academy report concluded that 'Asbestos is too important in our economy for its essential use to be stopped. But, because of the known serious effects of uncontrolled inhalation of asbestos minerals in industry and uncertainty as to the shape and character of the dose-response curve in man, it would be highly imprudent to permit additional contamination of the public environment with asbestos. Continued use at minimal risk to the public requires that the major sources of man-made asbestos emission into the atmosphere be defined and controlled.'

A substantial complication was the unavailability of a test method for measuring asbestos emissions. As a result, the first asbestos NESHAP used limitations on visible emissions with an option in some cases to use a fabric filter or, in some cases, a wet-collection air cleaning device. These requirements were based on what was considered to be the best that could be done on a technological basis; costs did not play a significant role in the first asbestos standard.

Vinyl Chloride
The NESHAP for vinyl chloride was promulgated October 21, 1976. This standard was designed to minimize the health risks associated with vinyl chloride by requiring reasonable control measures, although risk assessment methods still were not available at that time.

---

[2] 54 FR 38044.

Table 18-1 National emission standards for hazardous air pollutants

References	Affected facility	Emission level	Monitoring
	Asbestos (Subpart B)		
Proposed/effective 12/07/71 (36 FR 23239)	Asbestos mills	No visible emissions or meet equipment standards	No requirement
Promulgated 04/06/73 (38 FR 8820)	Roadway surfacing	Contain no asbestos, except temporary use	No requirement
Revised 05/03/74 (39 FR 15396)	Manufacturing	No visible emissions or meet equipment standards	No requirement
10/14/75 (40 FR 48299) 03/02/77 (42 FR 12127) 08/17/77 (42 FR 41424)	Demolition/renovation	Wet friable asbestos or equipment standards and no visible emissions	No requirement
03/03/78 (43 FR 8800) 08/19/78 (43 FR 26372)	Spraying friable asbestos Equipment and machinery	No visible emissions or meet equipment standards	No requirement
	Buildings, structures, etc.	<1 percent asbestos dry weight	No requirement
Background information APTD-1503	Fabricating products	No visible emissions or meet equipment standards	No requirement
EPA-450/2-74-009a	Friable insulation	No asbestos	No requirement
EPA-450/2-77-030	Waste disposal	No visible emissions or meet equipment and work practice requirements	No requirement
	Waste disposal sites	No visible emissions; design and work practice requirements	No requirement
	Beryllium (Subpart C)		
Proposed/effective 12/07/71 (36 FR 23239)	Extraction plants Ceramic plants	1. 10 g/hour, or 2. 0.01 μ/m^3 (thirty-day)	1. Source test 2. Three years CEM[a]
Promulgated 04/06/73 (38 FR 8820)	Foundries Incinerators		
Revised 08/17/77 (42 FR 41424)	Propellant plants Machine shops (Alloy >5 percent by weight		
03/03/78 (43 FR 8800)	beryllium)		
Background information APTD-1503			
	Beryllium (Subpart D)		
Proposed/effective 12/07/71 (36 FR 23239)	Rocket motor test sites Closed tank collection of combustion	75 μg min/m^3 of air within 10 to 60 minutes during two consecutive weeks	Ambient concentration during and after test
Promulgated 04/06/73 (38 FR 8820)	products	2 g/hour, maximum 10 g/day	Continuous sampling during release
Revised 08/17/77 (42 FR 41424) 03/03/78 (43 FR 8800)			
Background information APTD-1503			

Table 18-1   Continued

References	Affected facility	Emission level	Monitoring
		Mercury (Subpart E)	
Proposed/effective 12/07/71 (36 FR 23239) *Promulgated* 04/06/73 (38 FR 8820)	Ore processing Chlor-alkali plants	2300 g/24 hour 2300 g/24 hour	Source test Source test or use approved design, maintenance and housekeeping
Revised 10/14/74 (40 FR 48299) 08/17/77 (42 FR 41424) 03/03/78 (43 FR 8800) 06/08/82 (47 FR 24703) *Background information* APTD-1503 EPA-450/2-74-009a	Sludge dryers and incinerators	3200 g/24 hour	Source test or sludge test
		Vinyl chloride (Subpart F)	
Proposed/effective 12/24/75 (40 FR 59532) *Promulgated* 10/21/76 (41 FR 46560)	Ethylene dichloride (EDC) manufacturing	1. EDC purification: 10 ppm[b] 2. Oxychlorination: 0.2 g/kg of EDC product	Source test/CEM Source test
Revised 12/03/76 (41 FR 53017) 06/07/77 (421 FR 29005) 08/17/77 (42 FR 4124) 03/03/78 (42 FR 8800) 07/10/90 (55 FR 28346)[c]	Vinyl chloride manufacturing Polyvinyl chloride manufacturing Equipment Reactor opening loss Reactor manual vent valve	10 ppm[b] 10 ppm[b] 0.02 g/kg No emission except emergency	Source test/CEM Source test/CEM Source test
Proposed Revisions 06/02/77 (42 FR 28154) *Background information* EPA-450/2-75-009a&b *Reviewed* EPA-450/3-82-003	Sources after stripper	Each calendar day: 1. Strippers – 2000 ppm (PVC disposal resins excluding latex); 400 ppm other	Source test
		2. Others – 2 g/kg (PVC disposal resins excluding latex); 0.4 g/kg other	Source test
	EDC/VC/PVC manufacturing Relief valve discharge Loading/unloading Slip gauges Equipment seals Relief valve leaks Manual venting Equipment opening Sampling (>10 percent by weight VC) LDAR[d] In-process wastewater	None, except emergency 0.0038 m^3 after load/unload or 10 ppm when controlled Emissions to control Dual seals required Rupture disc required Emissions to control Reduce to 2.0 percent VC or 25 gallon Return to process Approved program required 10 ppm VC before discharge	 Source test Source test Approved program Source test

Table 18-1   Continued

References	Affected facility	Emission level	Monitoring
	Inorganic arsenic (Subpart N)		

Proposed/effective
07/20/83 (48 FR 33112)
03/20/84 (49 FR 10278)
 	
Promulgated
08/04/86 (51 FR 28025)
 	
Revised
10/03/86 (51 FR 35355)
	Glass melting furnace	Existing: <2.5 Mg/year^e or
		  85 percent control
		New or modified: <0.4 Mg/
		  year or 85 percent control
			Method 108
			Continuous opacity
			  and temperature
			  monitor for control

Proposed/Effective
07/20/83 (48 FR 33112)
 	
12/16/83 (48 FR 55880)
	Copper converter	Secondary hooding system
		Particle limit 11.6 mg/dscm^f
		
		Approved operating plan
			Methods 5 and 108A
			Continuous opacity for
			  control
			Airflow monitor for
			  secondary hood

Promulgated
08/04/86 (51 FR 28029)
 	
Proposed/effective
07/20/83 (48 FR 33112)
12/16/83 (48 FR 55880)
 	
Promulgated
08/04/86 (51 FR 28033)
 	
Revised
10/03/86 (51 FR 35355)
	Arsenic trioxide and
	  metallic arsenic plants
	  using roasting/
	  condensation process
		Approved plan for control of
		  emissions
			Opacity monitor for
			  control
			Ambient air
			  monitoring

	Benzene (Subpart J)		

Proposed/effective
01/05/81 (46 FR 1165)
 	
Reproposal
07/28/88 (53 FR 28496)
 	
Promulgated
06/06/84 (49 FR 23522)
09/14/89 (54 FR 38044)
	Equipment leaks
	  (Servicing liquid or gas
	  ≥10 percent by weight
	  benzene; facilities
	  handling <1,000 Mg/
	  year and coke oven by-
	  product exempt)
		Leak is 10,000 ppm using
		  Method 21; no detectable
		  emissions^g is 500 ppm
		  using Method 21	

Background information
EPA-450/3-80-032a&b
EPA-450/3-82-010
EPA-450/3-89-31
	Pumps	Monthly LDAR, dual seals,
		  95 percent control or
		  NDE^g
			Test for NDE

	Compressors	Seal with barrier fluid, 95 percent control or NDE	Test for NDE

	Pressure relief devices
	Sampling connection
	  systems
	Open-end valves/lines
	Valves
		NDE or 95 percent control
		Closed purge or closed vent
		
		Cap, plug, or second valve
		Monthly LDAR (quarterly if
		  not leaking for two
		  consecutive months) or
		  NDE
			Test for NDE
			
			Test for NDE

	Pressure relief
	  equipment
	Product accumulators
	Closed-vent systems
	  and control devices
		LDAR
		
		95 percent control
		NDE or 95 percent control
			
			Monitor annually

Table 18-1   Continued

References	Affected facility	Emission level	Monitoring
		Benzene (Subpart L)	
Proposed/effective 06/06/84 (49 FR 23522)	Coke by-product plants Equipment and tanks	Enclose source, recover or destroy. Carbon adsorber or incinerator alternate	Semiannual LDAR, annual maintenance
Revised Proposal 07/28/88 (53 FR 28496) 04/01/91 (56 FR 13368)	Light-oil sumps Naphthalene equipment Equipment leaks (servicing ≥10 percent by weight)	Cover, no venting to sump Zero emissions See 40 CFR 61, Subpart J	Semiannual LDAR
Promulgated 09/14/89 (54 FR 38044) 09/19/91 (56 FR 47404)	Exhausters (≥1 percent by weight)	Quarterly LDAR or 95 percent control or NDE	Test for NDE
Background information EPA-450/3-83-016a&b EPA-450/3-89-31			
		Benzene (Subpart Y)	
Proposed/effective 07/28/88 (53 FR 28496)	Benzene storage vessels Vessels with capacity ≥10,000 gallon	Equip with: 1. Fixed roof with internal floating roof-seals, or	Periodic inspection

Promulgated
09/14/89 (54 FR 38044)
		2. External floating roof
		   with seals, or
		3. Closed vent and 95
		   percent control
			Periodic inspection
			
			Maintenance plan and
			monitoring

Background information EPA-450/3-84-004 EPA-450/3-89-31			
		Benzene (Subpart BB)	
Proposed/effective 09/14/89 (54 FR 38083)	Benzene transfer Producers and terminals (loading ≥1,300,000 1/year)	Vapor collection and 95 percent control	Annual recertification
Promulgated 03/07/90 (55 FR 8292)	Loading racks (marine, rail, truck) Exemptions Facilities loading <70 percent benzene	Load vapor-tight vessels only	Yes
Corrections 04/03/90 (55 FR 12440) 10/31/90 (55 FR 45804)	Facilities loading less than required of ≥70 percent benzene Both of above subject to record-keeping		

Table 18-1 Continued

References	Affected facility	Emission level	Monitoring
		Benzene (Subpart FF)	

Proposed/effective
09/14/89 (54 FR 38083)
12/09/91 (56 FR 64217)1[h]
	Waste Operations
	Chemical manufacturing
	  plants
	Petroleum refineries
	Coke by-product plants
		1. Facilities ≥10 Mg/year in
		  aqueous wastes must
		  control streams ≥10 ppm.
		  Control to 99 percent or
		  <10 ppm
			Monitor control and
			treatment. Also,
			periodically monitor
			certain equipment
			for emissions

Promulgated
03/07/90 (55 FR 8292)
03/05/92 (57 FR 8012)[h]
	TSDF[j] treating wastes
	  from the above three
		2. If >10 ppm in wastewater
		  treatment system:
		    Wastes in <10 ppm
		    Total in <1 Mg/year
		3. >1 Mg/year to <10 Mg/
			>500 ppm and
			inspect equipment
			
			Report annually

Corrections 04/03/90 (55 FR 12440) 05/02/90 (55 FR 18330) 09/10/90 (55 FR 37230)		year 4. <1 mg facilities	One-time report
		Radionuclides (Subpart H)	

Promulgated
12/15/89 (54 FR 51695)
	DOE facilities (radon not
	  included)
		10 mrem/year[k] radionuclides
		  (any member of the
		  public)
			Approved EPA
			computer model and
			Method 114 or direct
			monitoring
			(ANSIN13.1-1969)

		Radionuclides (Subpart I)	

Promulgated
2/15/89 (54 FR 51697)
	NRC licensed facilities
	  and facilities not
	  covered by Subpart H
		10 mrem/year[k] radionuclides
		  (any member of the
		  public)
		3 mrem/year iodine (any
		  member of the public)
			Approved EPA
			computer model or
			Appendix E
			Emissions determined
			by Method 114 or
			direct monitoring
			(ANSIN13.1-1969)

		Radionuclides (Subpart K)	

Promulgated
2/15/89 (54 FR 51699)
	Calciners and nodulizing
	  kilns at elemental
	  phosphorus plants
		2 curies per year (polonium-
		  210)
			Method 111

		Radionuclides (Subpart Q)	

Promulgated
2/15/89 (54 FR 51701)
	Storage and disposal
	  facilities for radium-
	  containing material,
	  owned/operated by
	  DOE
		20 pCi/m^2 per second[m]
		  (radon-222)
			None specified

Table 18-1   Continued

References	Affected facility	Emission level	Monitoring
	Radionuclides (Subpart R)		
Promulgated 2/15/89 (54 FR 51701)	Phosphogypsum stacks (waste from phosphorus fertilizer production)	20 pCi/m^2 per second^m (radon-222)	Method 115
	Radionuclides (Subpart W)		
Promulgated 2/15/89 (54 FR 51703)	Disposal of uranium mill tailings (operational)	20 pCi/m^2 per second^m (radon-222)	Method 115

[a] CEM = continuous emission monitor.
[b] Before opening equipment, VC must be reduced to 2.0 percent (volume) or 25 gallons, whichever is larger.
[c] This revision clarified several provisions of the standard but did not change the standard.
[d] LDAR = leak detection and repair.
[e] Mg/year = megagrams per year.
[f] mg/dscm = milligrams per dry standard cubic meter.
[g] NDE = no detectable emissions.
[h] Stayed effectiveness of Subpart FF.
[j] TSDF = treatment, storage and disposal facilities.
[k] mrem/year = millirems per year (the rem is the unit of effective dose equivalent for radiation exposure).
[m] pCi/m^2 per second = picocuries per square meter per second.

Costs were considered to a very limited extent to assure that the costs of control technology were not grossly disproportionate to the amount of emission reduction achieved. The standard included numerical emissions limits, concentration limits, equipment specifications, and work practice requirements.

Most of the equipment subject to the standard operates under pressure and is equipped with pressure relief valves. Interestingly, while the standard was being developed, a significant number of relief valve discharges occurred releasing large amounts of vinyl chloride to the atmosphere. To correct this situation, the standard allows only 'emergency' discharges (i.e., discharges that cannot be avoided by taking preventive measures). This was the most controversial part of the standard and, after reviewing the standard, EPA proposed on January 9, 1985 (50 FR 1182), to reformat the standard for relief valve discharges to reflect the number of discharges that occur from those plants complying with the format of the current standard. The agency decided on September 30, 1986 (51 FR 34904), not to promulgate the proposed revision in light of public comments and other findings, and retained the original requirements.

Benzene

Benzene was added to the list of hazardous air pollutants on June 8, 1977. The first NESHAP for benzene was proposed April 18, 1980, for maleic anhydride plants. The proposed standard was consistent with the EPA's proposed Policy and Procedures for Identifying, Assessing, and Regulating Airborne Substances Posing a Risk of Cancer (44 FR 58642), See Chapter 16. Proposed standards also were issued for ethylbenzene/styrene (EB/S) on December 18, 1980, benzene storage vessels on December 19, 1980, and equipment leaks on January 5, 1981. On March 6, 1984 (49 FR 8386), however, the agency proposed to withdraw the proposed standards for maleic anhydride, EB/S, and storage vessels because the agency concluded 'that both the benzene health risks (annual leukemia incidence and maximum lifetime risks) to the public from these source categories and the potential reduction in health risks achievable with available control techniques are too small to warrant federal regulatory action under Section 112 of the Clean Air Act.' On June 6, 1984, standards were proposed for benzene emissions from coke by-product recovery plants along with a final standard for benzene equipment leaks based on the application of BAT. On that date, EPA also withdrew the proposed standards for maleic anhydride, EB/S, and storage vessels.

The Natural Resources Defense Council (NRDC) immediately took legal action against EPA because of the three withdrawals and the final standard for equipment leaks. As a result of the court decision in a separate case on vinyl chloride (see Chapter 16), the Agency requested a voluntary remand on the benzene case to reconsider its June 6, 1984, rule-makings. On September 14, 1989, EPA issued final rules for maleic anhydride, EB/S, storage vessels, equipment leaks, and coke by-product recovery plants and proposed standards for chemical manufacturing process vents, industrial solvent use, benzene waste operations, benzene transfer operations, and gasoline marketing system (final rules on these proposals were issued March 7, 1990). These were the last NESHAP issued under section 112 of the 1970 and 1977 Amendments to the Clean Air Act. The problems associated with the benzene rules helped convince Congress that section 112 of the Act needed substantial revision.

Arsenic

The NESHAP for arsenic was proposed on July 20, 1983. In this proposal (48 FR 33116), EPA interpreted section 112 in the following way: 'All source categories of the pollutant that are estimated to result in significant risks should be evaluated. Each such source category should be controlled at least to the level that reflects best available technology (BAT), and to a more stringent level, if in the judgment of the Administrator, it is necessary to prevent unreasonable risks ... By BAT, EPA means the best controls available, considering economic, energy, and environmental impacts.' The final standard for arsenic was promulgated on August 4, 1986, using a risk management strategy as the basis. The Agency explained risk management as follows (51 FR 27958): 'The EPA's strategy for risk management under section 112 first provides for the identification of source categories that may pose significant risks to public health as a result of air emissions. Next,

the Agency conducts an assessment of candidate source categories to evaluate current control levels and associated health risks, future or ongoing emissions reductions from other regulatory activities (e.g., State Implementation Plans (SIPs) and Occupational Safety and Health Administration (OSHA) standards), the availability of more stringent options such as further controls or process modifications, and the costs and economic effects associated with each option. Based on this assessment, the Administrator selects a level of control that, in his judgment, reduces the health risks to the greatest extent that reasonably can be expected after considering the uncertainties in the analysis, the residual risk remaining after the application of the selected control level, the costs of further control, and the societal and other environmental impacts of the regulation.'

MACT/GACT

Section 112, as rewritten in the 1990 Clean Air Act Amendments, directs EPA to require control of major and area sources of 189 hazardous air pollutants (HAP) listed in section 112(b). The control of these substances is to be achieved through the initial promulgation of technology-based emission standards that require major sources to install maximum achievable control technology (MACT) and area sources to install generally available control technologies (GACT). MACT/GACT standards are to be developed to control HAP emissions from both new and existing sources.

MACT/GACT standards must reflect consideration of the cost of achieving the emission reduction, and any nonair quality health and environmental effects, and energy requirements for control levels more stringent than the MACT floor. Section 112(d)(2) specifies that the emission reduction may be accomplished through application of measures, processes, methods, systems or techniques including, but not limited to, measures that:

(A) reduce the volume of, or eliminate emissions of, such pollutants through process changes, substitution of materials or other modifications,

(B) enclose systems or processes to eliminate emissions,

(C) collect, capture or treat such pollutants when released from a process, stack storage or fugitive emissions point,

(D) are design, equipment, work practice, or operational standards (including requirements for operator training or certification) as provided in subsection (h), or

(E) are a combination of the above.

The 1990 Amendments limit the Agency's discretion by establishing a minimum baseline or 'floor' for standards. Section 112(d)(3) states that for new sources, the standards for a source category or subcategory 'shall not be less stringent than the emission control that is achieved in practice by the best controlled similar source as determined by the Administrator.' Existing source standards may be less stringent than new source standards but may be no less stringent than the average emission limitation achieved by the best performing 12 percent of the existing sources (excluding certain sources) for categories and subcategories with thirty or more sources or the best performing five sources for categories or subcategories with fewer than thirty sources.

Once the floor has been determined for new or existing sources for a category or subcategory, the Administrator must set MACT standards that are no less stringent than the floor. Such standards must then be met by all sources within the category or subcategory. In establishing standards, however, the Administrator may distinguish among classes, types and sizes of sources within a category or subcategory. For example, the Administrator could establish two classes of sources within a category or subcategory based on size and establish a different emission standard for each class, provided both standards are at least as stringent as the MACT floor.

EPA must decide whether it is more appropriate to follow the MACT or the GACT approach for regulating an area source category. If all or some portion of the sources in a source category emit less than 10 tons per year (TPY)[3] of any HAP, or less than 25 TPY[4] of total HAPs, then it may be appropriate to define subcategories within the source category and apply a combination MACT/GACT approach (i.e., MACT for major sources and GACT for area sources). In other cases, it may be appropriate to regulate both major and area sources in a source category under MACT.

OTHER TECHNOLOGY STANDARDS

New Source Performance Standards

Section 111 of the 1970 Clean Air Act Amendments first authorized the EPA Administrator to establish new source performance standards (NSPS). Only minor revisions were made to section 111 by the 1990 Amendments. The Administrator is required to establish NSPS for any category of new stationary source of air pollution that causes, or contributes significantly, to air pollution that may reasonably be anticipated to endanger public health or welfare. Section 111(a)(1)(C) specifies that NSPS reflect the degree of emission limitation achievable

through the application of the best system of emission reduction that (taking into account the cost of achieving such reduction and any nonair quality health and environmental impact and energy requirements) the Administrator determines has been adequately demonstrated. NSPS apply only to stationary sources, the construction or modification of which commences after the NSPS are proposed in the Federal Register.

It is important to recognize that many of the NSPS published by EPA will serve as a basis for MACT emission standards under section 112 of the 1990 Amendments. The reason is that

---

[3] 9.1 megagrams per year (Mg/yr).
[4] 22.7 Mg/yr.

Table 18-2 New source performance standards for some sources potentially emitting toxic air pollutants

Source category	Citation[a]	Pollutants regulated[b]
Incinerators (>50 tons/day)	Subpart E	PM
Municipal waste combusters (>250 tons/day)	Subpart Ea	PM, organics, NO_x, acid gases
Portland cement	Subpart F	PM
Asphalt plants	Subpart I	PM
Petroleum refineries	Subpart J	PM, CO, SO_2, VOC
Petroleum storage vessels (>40,000 gallon)	Subpart K, Ka	VOC
Secondary lead smelters	Subpart L	PM
Secondary brass/bronze	Subpart M	PM
Basic oxygen furnaces	Subpart N, Na	PM
Sewage treatment plants	Subpart O	PM
Primary copper smelters	Subpart P	PM, SO_2
Primary zinc smelters	Subpart Q	PM, SO_2
Primary lead smelters	Subpart R	PM, SO_2
Primary aluminum reduction	Subpart S	Fluorides
Phosphoric acid plants	Subpart T	Fluorides
Superphosphate acid plants	Subpart U	Fluorides
Diammonium phosphate plants	Subpart V	Fluorides
Triple superphosphate plants	Subpart W	Fluorides
Triple superphosphate storage	Subpart X	Fluorides
Coal preparation plants	Subpart Y	PM
Ferroalloy production	Subpart Z	PM
Electric arc furnaces	Subpart AA, AAa	PM
Kraft pulp mills	Subpart BB	PM, TRS
Glass manufacturing	Subpart CC	PM
Surface coating – metal furniture	Subpart EE	VOC
Lime manufacturing	Subpart HH	PM
Lead-acid battery manufacturing	Subpart KK	Lead
Metallic minerals	Subpart LL	PM
Surface coating – automobiles and light-duty trucks	Subpart MM	VOC
Phosphate rock plants	Subpart NN	PM
Ammonium sulfate manufacture	Subpart PP	PM
Graphic arts and printing	Subpart QQ	VOC
Surface coating – tapes and labels	Subpart RR	VOC
Surface coating – large appliances	Subpart SS	VOC
Surface coating – metal coils	Subpart TT	VOC
Asphalt processing/roofing	Subpart UU	PM
Equipment leaks – organic chemical manufacturing industry	Subpart VV	VOC
Surface coating – beverage cans	Subpart WW	VOC
Bulk gasoline terminals	Subpart XX	VOC
Residential wood heaters	Subpart AAA	PM
Rubber tire manufacturing	Subpart BBB	VOC
Polymer manufacturing	Subpart DDD	VOC
Flexible vinyl and urethane coating and printing	Subpart FFF	VOC
Equipment leaks in petroleum refineries	Subpart GGG	VOC
Synthetic fiber production	Subpart HHH	VOC
Air oxidation processes – organic chemical manufacturing	Subpart III	VOC
Petroleum dry cleaners (dryer capacity $\geq$ 38 kg)	Subpart JJJ	VOC
Onshore natural gas processing		
Equipment leaks	Subpart KKK	VOC
SO_2 emissions	Subpart LLL	SO_2
Distillation processes – organic chemical manufacturing	Subpart NNN	VOC
Nonmetallic minerals	Subpart OOO	PM
Wool fiberglass insulation manufacturing	Subpart PPP	PM
Petroleum refinery wastewater	Subpart QQQ	VOC
Magnetic tape manufacturing	Subpart SSS	VOC
Surface coating – Plastic parts for business machines	Subpart TTT	VOC

[a] All citations are in the Code of Federal Regulations, Title 40, Part 60.
[b] PM = particulate matter; CO = carbon monoxide; SO_2 = sulfur dioxide; NO_x = nitrogen oxides; VOC = volatile organic compounds; TRS = total reduced sulfur.

section 112(d)(3) defines MACT in terms of those emission controls 'achieved in practice by the best controlled similar source[s].' For new sources, it is the single best source; for existing sources, it is the best-performing 12 percent of existing sources (or best-performing five sources in categories or subcategories with less than thirty sources). In addition, as noted below, many NSPS indirectly control emissions of hazardous or toxic air pollutants. A listing of the NSPS and their appropriate CFR subparts are presented in Table 18-2. These subparts provide useful information on control measures for similar sources anticipating MACT standards.

NSPS by themselves do not guarantee protection of health or welfare because they are not designed to achieve any specific air quality levels. Instead, they are technology standards and are designed only to reflect the degree of emission limitation achievable through application of the best demonstrated technology (BDT) considering costs.

In addition to direct air quality benefits, the NSPS program provides a number of other significant benefits. NSPS help avoid situations where some states may attract industries by relaxing standards in comparison to other states. NSPS also enhance the potential for long-term growth, help achieve long-term cost savings by avoiding the need for more expensive retrofitting when pollution ceilings may be reduced in the future, and create an incentive for improving technology.

NSPS are established for criteria pollutants (i.e., those pollutants for which national ambient air quality standards, NAAQS, apply) and designated pollutants (i.e., those pollutants subject to an NSPS but not subject to NAAQS or NESHAP).[5] When NSPS are promulgated for a designated pollutant under section 111(b) of the 1970 Amendments, section 111(d) requires that states submit plans that establish emission standards for existing sources and provide for implementation and enforcement of emission standards for the designated pollutant. In general, this means that control under section 111(d) is appropriate when the pollutant may cause or contribute to endangerment of public health or welfare but is not known to be 'hazardous' within the meaning of section 112 or is not controlled under sections 108–110. An example of the latter situation would be a pollutant not emitted from 'numerous or diverse' sources, as required by section 108. Pollutants regulated under section 111(d) include sulfuric acid mist, total reduced sulfur compounds, and fluorides.

The NSPS program also results indirectly in the control of some hazardous pollutants and toxic air pollutants. For example, NSPS that limit particulate matter, for which a NAAQS is established, and volatile organic compounds, that are controlled to meet the ozone NAAQS, will control any hazardous pollutants in the affected gas streams.

BACT/LAER

The 1977 Clean Air Act Amendments provided for the Prevention of Significant Deterioration (PSD) program to assure that operators of regulated sources in relatively unpolluted areas would not allow a decline in air quality to the minimum level permitted by the NAAQS. Under this program, no major emitting facility may be constructed or modified without meeting specific requirements, including demonstrating that the proposed facility is subject to the best available control technology (BACT) for each pollutant subject to regulation under the Act. BACT is defined in section 169(3) of the 1977 Amendments as: 'an emission limitation based on the maximum degree of reduction of each pollutant subject to regulation under this Act emitted from or which results from any major emitting facility, which the permitting authority, on a case-by-case basis, taking into account energy, environmental, and economic impacts and other costs, determines is achiev-

---

[5] Section 111(d).

able for such facility through application of production processes and available methods, systems, and techniques, including fuel cleaning, clean fuels, or treatment or innovative fuel combustion techniques for control of each such pollutant. In no event shall application of best available control technology result in emissions of any pollutants that will exceed the emissions allowed by any applicable standard established pursuant to section 111 or 112 of the Act. Emissions from any source utilizing clean fuels, or any other means, to comply with this paragraph shall not be allowed to increase above levels that would have been required under this paragraph as it existed prior to enactment of the Clean Air Act Amendments 1977.'

Each BACT decision is a case-by-case determination based on a careful analysis and informed judgment for each emission unit and affected pollutant. The emission limit must be as stringent as any applicable NSPS or NESHAP and reflect the maximum degree of emissions reduction achievable considering energy, environmental, and economic effects and other costs.

In 1987, EPA began using what is now referred to as the 'top-down' BACT process. The first step in this process is for the affected emission source to determine the most stringent control available for a similar or identical source or source category. If it can be shown that this level of control is technically or economically infeasible for the source, then the next most stringent level of control is determined and similarly evaluated. This process continues until the BACT level under consideration cannot be eliminated by any substantial or unique technical, environmental, or economic objection. The top-down BACT process has been the subject of considerable controversy, including litigation; as such, at the time of this writing its future was uncertain.

The 1977 Amendments also added requirements for state implementation plans relating to attainment and maintenance of NAAQS in nonattainment areas. These requirements, generally known as New Source Review, include permits for the construction and operation of new or modified major stationary sources that require the lowest achievable emission rate (LAER). LAER is defined under section 171(3) of the 1977 Amendments as: 'that rate of emissions which reflects – (A) the most stringent emission limitation which is contained in the implementation plan of any state for such class or category of source, unless the owner or operator of the proposed source demonstrates that such limitations are not achievable, or (B) the most stringent emission limitation which is achieved in practice by such class or category of source, whichever is more stringent. In no event shall the application of this term permit a proposed new or modified source to emit any pollutant in excess of the amount allowable under applicable new source standards of performance.'

The 1990 Amendments did not change the definition of LAER but made a small change to the definition of BACT. The 1990 Amendments extended the applicability of the new source review requirements in ozone nonattainment areas by establishing lower potential emission rates. The base emission rate remains 100 tons per year but is lowered to as little as 10 tons per year for nonattainment areas that exceed the ozone NAAQS by the greatest amounts. Chapter 26 provides a more complete discussion.

Because BACT and LAER are case-by-case determinations, EPA established the BACT/LAER Clearinghouse (see Chapter 1 for access) to assist state and local air pollution control agencies in selecting BACT and LAER in a nationally consistent manner. The goals of the BACT/LAER Clearinghouse are to: (1) provide state and local agencies with current information on case-by-case technology determinations that are made nationwide, and (2) promote communication, cooperation, and sharing of control technology information among permitting agencies.

T-BACT

Before the enactment of the 1990 Amendments, many states developed programs for toxic air pollutants. Some states, such as Maryland and

Michigan, adopted regulations that require new and modified sources of toxic air pollution to minimize emissions using best available control technology for toxics, often called T-BACT. Although the definitions of T-BACT vary among the states, they all reflect the maximum degree of emission reduction achievable considering technical and economic feasibility. The requirements generally are implemented through the state's permit program and T-BACT is determined on a case-by-case basis. The term T-BACT is used to give it separate and distinct identity from BACT.

In order to avoid regulating very small sources that emit a de minimis level of toxic emissions, state programs typically incorporate exemption provisions. Most state permit rules include some exemptions that apply to certain sources. Other sources may be handled through specific exemptions.

States usually require a health assessment analysis after the application of T-BACT to insure that the residual emissions do not adversely affect human health. The approaches used in this analysis vary from state to state and are not discussed here because they vary so widely.

The 1990 Clean Air Act Amendments recognize that these state programs for air toxics were developed before the enactment of the Amendments and contemplate the states playing a significant role in the implementation of hazardous air pollutant programs. Section 112(l) provides that states may develop and submit to EPA programs for the implementation and enhancement of standards issued by EPA under the Amendments. EPA is required to publish guidance that would be useful to the states in developing their programs.

RACT/CTG

The 1977 Amendments required states in which NAAQS are exceeded to adopt and submit revised state implementation plans (SIP) to EPA. These SIP's must require the installation of reasonably available control technology (RACT) for select stationary sources. RACT is defined as: 'The lowest emission limitation that a particular source is capable of meeting by the application of control technology that is reasonably available considering technological and economic feasibility. For a particular industry RACT is determined on a case-by-case basis, considering the technological and economic circumstances of the individual source.' A control techniques guideline (CTG) document is issued by EPA and provides guidance on RACT for control of volatile organic compound (VOC) emissions in nonattainment areas.

The 1990 Amendments require EPA to issue within three years CTGs for eleven categories of stationary sources of VOC emissions for which CTGs had not already been issued by November 15, 1990. The 1990 Amendments also require EPA to review and if necessary, update existing CTGs by November 1993.

References

Danielson, J.A. ed. 1973. *Air Pollution Engineering Manual.* Second Edition. U.S. Environmental Protection Agency and Air Pollution Control District, County of Los Angeles.

National Academy of Sciences (NAS). 1971. *Asbestos (The Need for and Feasibility of Air Pollution Controls.* Washington, D.C. National Academy Press.

Stern, A.C. ed. 1977. *Air Pollution – Engineering Control of Air Pollution. Volume IV.* First Edition. New York: Academic Press.

19

Risk Assessment and Risk Management

David R. Patrick

INTRODUCTION

Humans always have assessed the risks associated with activities in which they were involved and made decisions based on those assessments. It was not until the mid-seventeenth century, however, that Blaise Pascal turned the process into a science to more precisely calculate the probabilities associated with gambling. By this century, actuaries were assessing risks and calculating the probabilities of a wide range of events to establish insurance premiums. With the advent of nuclear power and the serious consequences associated with exposure to radiation, scientists expanded these processes into a science called risk assessment to predict both the likelihood of the occurrence of an undesirable event and the significance of the event on humans and property. As public concerns with exposures to environmental carcinogens grew to new heights in the mid-1970s, regulators and scientists expanded the principles of risk assessment further to address environmental concerns. This type of risk assessment has been particularly difficult to develop and use constructively because of the complexity of the events and a substantial lack of scientific understanding of many of the adverse human health effects resulting from environmental exposures.

Congress and federal regulators have approached the subject of environmental risks with little consistency. One of the earliest attempts to deal with environmental risks was the Delany Amendment to the Food, Drug and Cosmetics Act that banned additives shown to cause cancer in humans or animals. Although politically expedient, this was found later to be unrealistic in the case of saccharine where the benefits to diabetics and dieters of a non-nutritive sweetener overshadowed the small estimated cancer risks. Environmental legislation later required the use of risk assessment or risk-benefit analyses; however, little effective guidance was offered by the legislators or developed by the regulators.

In 1976, scientists at EPA adopted the first policy and guidelines for assessing the risks of exposure to carcinogens (EPA 1976a; Albert, Train, and Anderson 1977). These guidelines were derived largely from earlier experience in assessing the risks associated with radiation exposures. To deal with the many uncertainties of environmentally induced cancer, the guidelines contained very conservative (i.e., health protective) assumptions to be sure that the risk of cancer would not be underestimated. Public officials wanted to be certain that if errors were

341

made in the estimation of cancer risks those errors would be made in favor of health protection.

At the same time, it was becoming obvious to regulators at EPA that suspected carcinogens, that were being identified more frequently as environmental contaminants, could not all be regulated to a zero risk level, that was the goal of the Delany Amendment and the environmental movement of the 1970s. Thus, EPA's guidelines attempted to establish procedures both to help regulators set priorities for regulatory action and to establish appropriate levels of control. EPA's guidelines also were based on the presumption that some residual cancer risk was acceptable as a regulatory policy. This was a highly controversial concept but one that generally has been upheld by the courts.

One stimulus to the development of these early policies and guidelines was the growing capability within the scientific community to evaluate cancer risks in quantitative terms (EPA 1991a). Although the early quantitative techniques were limited to cancer and were insufficiently accurate to define risks in absolute terms, they were useful for comparing the relative risks of similar exposures to carcinogens. This information could be used to help set regulatory priorities; however, there was no widely accepted method for using the information to establish regulatory control levels.

In an attempt to establish a consistent framework for dealing with environmental risks, the National Academy of Sciences published *Risk Assessment in the Federal Government: Managing the Process* (NAS 1983). In it the

Academy presented a number of recommendations for the application of risk assessment by the federal government. The Academy also more precisely defined environmental risk assessment as the scientific activity of evaluating the toxic properties of a substance and the conditions of human exposure to it in order both to ascertain the likelihood that exposed humans would be adversely affected and to characterize the nature of the effects they might experience. The Academy strongly recommended separation of risk assessment and risk management. This was to assure that 'the scientific findings and policy judgments embodied in risk assessments [are] explicitly distinguished from the political, economic, and technical considerations that influence the design and choice of regulatory strategies.' Importantly, EPA Administrator Ruckelshaus embraced these recommendations and stated that the agency would work toward that end and that the public would become more involved in the decision-making process (Ruckelshaus 1983b).

In the remainder of this chapter, EPA's development and application of the techniques of risk assessment and risk management are described. These are separated into the two basic categories used by scientists to evaluate health effects: carcinogens and noncarcinogens. Although this categorization is simplistic, it serves as a useful differentiation between substances that are associated with a threshold of effect (i.e., noncarcinogens) and those that appear to have no threshold of effect (i.e., carcinogens). The importance is that assessment of risks for these two categories must be handled differently.

ENVIRONMENTAL RISK ASSESSMENT

Environmental risk assessment is the process of gathering and evaluating available information related to the potential for adverse human health effects (or environmental effects) to occur as a result of environmental exposures and estimating the magnitude and potential consequences of those exposures. EPA has focused considerable attention on developing processes and procedures for assessing risks resulting from

community exposures to carcinogens. Furthermore, since publication by EPA of *Interim Risk Assessment Guidelines for Suspected Carcinogens* (EPA 1976a), EPA's regulatory decision-making has relied heavily on quantitative risk assessment. EPA's early approach (Anderson et al. 1983) to the assessment of cancer risks in humans consisted of establishing the likelihood that a substance is a human carcinogen and then estimating the magnitude of the public health effects. In assessing the magnitude of cancer risk to the public from exposure to a carcinogen, risks generally were bracketed between a plausible upper bound and a lower bound that approached zero. The upper bound risks generally were estimated in terms of both increased cancer risks to individuals and increased number of annual cancer cases estimated in the total exposed population. Again, early methods were based on the premise that, in the face of scientific uncertainty, risk estimates should err on the side of public safety.

Significant improvements in the risk assessment process have taken place since 1976. Importantly, in their report (NAS 1983) the National Academy of Sciences more clearly defined the nature of environmental risk assessment as consisting of four components.

1. *Hazard identification* – the determination of whether a particular chemical is or is not causally linked to particular health effects.
2. *Dose-response assessment* – the determination of the relation between the magnitude of exposure and the probability of occurrence of the health effects in question.
3. *Exposure assessment* – the determination of the extent of human exposure before and after application of regulatory controls.
4. *Risk characterization* – the description of the nature and often the magnitude of human risk, including attendant uncertainties.

Each of these components has potential uncertainties associated with it. These uncertainties result from: (1) variations in the models used in the analysis, (2) variations in inputs to the models, (3) imprecise knowledge of the underlying science, and (4) natural variability. Uncertainties resulting from the first three of these sources frequently can be reduced through additional research and more detailed analysis. Regardless of the source of the uncertainty, however, assumptions or default values frequently are required. The manner in which these assumptions or default values are selected depends on the purpose of the assessment and the needs of the assessor.

In the *hazard identification* component, a large quantity of information is needed for proper evaluation. This typically includes the following: physical/chemical properties; routes and patterns of exposure; metabolic and pharmacokinetic data; toxicological studies, including short-term tests and long-term animal studies; human studies; and ancillary information, including in vitro studies and structure-activity relationships. The availability and accuracy of this information varies widely depending on the substance and the adverse health effect (i.e., the hazard) being considered. All of the information is taken together to establish the weight of evidence that a substance is capable of causing a particular effect. Importantly, there is at present no accepted means of quantifying the weight of evidence.

Once the potential hazard associated with exposure to a substance is established, the *dose-response* component projects the potential effects on humans at varying exposure concentrations. This component quantifies the adverse effects occurring at test exposure levels and extrapolates the results to a much wider range of exposures. For many substances, there are no human test data and animal test data must be used. Evaluation of animal test data generally entails the development of mathematical models to aid in describing the range of effects. Mathematical extrapolation is required because animals almost always are tested at higher concentrations than the community exposures of concern. This is done both to maximize the potential for the effect to occur and to observe an effect in the lifetime of the animal, that typically is much shorter than that of humans. Depending on the substance and animal species tested, the dose-response relationship may involve a threshold and it may be linear, quadratic, or obey another mathematical relationship.

The *exposure assessment* component involves the process of determining or estimating the magnitude, duration, and route of exposure of an organism to a substance of concern. In the early risk assessment years, this component of the process often received less attention than the other components of the process in terms of understanding its uncertainties and variabilities. Information important to the assessment of exposure to an air contaminant includes: the nature of the source and the releases; the different pathways by which humans can be exposed; transport and fate characteristics of the contaminant; human intake patterns; the number of people potentially exposed; the frequency and duration of the exposures; and the locations and activities of the exposed population.

Finally, in the fourth component of the risk assessment process all of the gathered information and methodologies are used to *charac-terize* (i.e., estimate) the risks associated with exposure to the pollutant. Risk characterization expresses both the weight of evidence and the quantitative estimates of the risks. It also ideally presents the limitations and inherent uncertainties in the calculations.

The procedures and uses of hazard identification and dose-response assessment are described in Chapters 5, 6, and 7. The procedures and uses of exposure assessment are described in Chapters 8 through 15. The remainder of this chapter focuses on describing the risk characterization component of the process and discussing how it typically is translated into regulatory decisions.

RISK CHARACTERIZATION

For risk assessment purposes, pollutants normally are separated into two toxicological categories depending on whether the adverse health effect occurs only above a specific dose, often called the threshold, or whether the adverse effect appears to occur at any dose above zero. Cancer is the most widely studied health effect that appears to be associated, at least for many chemicals, with no threshold of effect. Although simplistic, scientists and regulators tend to classify toxic substances as either carcinogens or noncarcinogens.

For risk assessment purposes, both chronic and acute chemical intake exposures also are considered. Chronic health effects occur from longer-term exposures, up to a human lifetime normally considered to be seventy years. Chronic health effects can result from exposure to chemicals associated with both thresholds and lack of thresholds. Acute health effects generally result from short-term exposures, up to a few days, and they generally are associated with a threshold of effect.

Carcinogens

The Risk Characterization Process
At the time of this writing in 1992, EPA had derived inhalation cancer potency factors and classified the weight of evidence for thirty-eight of the substances listed in section 112(b) of the 1990 Amendments. The information is available through the agency's Integrated Risk Information System (IRIS).[1] EPA and most states use these cancer potency factors to evaluate the acceptability of emissions of carcinogens from various sources. Only a few organizations have scientists with the training and expertise necessary to assess cancer data and develop proper cancer potency values and weights of evidence.

In 1986, EPA published procedures for estimating the cancer risk resulting from exposures to potential human carcinogens.[2] Of critical importance in the procedure is the numerical value, termed the cancer potency factor, that is multiplied by the lifetime average human intake of the potential carcinogen to yield the lifetime cancer risk for the human(s) exposed to that substance at that intake. When there is no other information, a linear dose-response relationship at low concentrations is assumed. This provides the most conservative answer. When assuming linearity, cancer risks at other doses and for other exposure times can be calculated easily. Most regulatory agencies have adopted these procedures for assessing the risks of carcinogens.

---

[1] For information on IRIS, call (513) 569-7254.
[2] 51 FR 33992, September 24, 1986.

Cancer is considered by health scientists to be a long-term (or chronic) disease. Thus, cancer risk estimates normally are based on an assumed human lifetime of seventy years of exposure. Cancer potency factors are established using this lifetime assumption.

The determination that a substance is a carcinogen and the estimation of its cancer potency factor are based largely on the results of human occupational or long-term animal studies. Animal studies normally are conducted at high doses in order to observe an effect during the short lifetime of the animal; occupational exposures also typically are much higher than those experienced in the community. Because community exposures normally are lower than these test exposures, the occupational or animal data must be extrapolated to lower doses using mathematical models in order to estimate the potential response by those in the community. When based on experimental animal studies, the cancer potency factors are estimates of the statistical 95 percent upper confidence limits on the dose-response association for a chemical carcinogen and, as a result, are conservative. When based on human epidemiological studies, the cancer potency factors are maximum likelihood estimates. Because the dose-response association for human epidemiological studies rarely can be measured, it typically is assumed to be linear at low doses. Thus, these cancer potency factors also are considered to be conservative. The term used by EPA to describe these cancer potency factors is upper bound because of their conservative nature. These upper bound cancer potency factors are unlikely to underestimate cancer risks but may greatly overestimate cancer risks. Only a limited number of substances have been shown to be definitely carcinogenic in humans based on occupational or epidemiological studies; thus, most cancer potency factors are derived from animal studies.

There also are varying degrees of confidence in the strength or weight of the scientific evidence that a substance is a carcinogen. EPA uses a system that characterizes the overall quality and quantity of the evidence based on a careful evaluation of all of the animal, human and other supportive data. The weight of evidence classification is an attempt to establish the likelihood that an agent is a human carcinogen and, thus, provides a confidence level in the potential health risk estimates. As noted earlier, there is no way at present to consider quantitatively the weight of evidence in the risk estimation process. As such, it is important to provide the weight of evidence categorization for a substance along with each cancer potency estimate. This allows the user to modify an action indicated by a specific risk estimate based on a stronger or weaker weight of evidence. Cancer risk estimates for substances with less certain weight of evidence often are treated by regulators the same as those for substances with more certain weight of evidence. This is another instance where the procedures typically used to estimate cancer risks are conservative.

The EPA weight of evidence classification system contains five categories.

Group A – human carcinogen: this category indicates that there is sufficient evidence from epidemiological studies to support a causal association between an agent and cancer.

Group B – probable human carcinogen: this category generally indicates that there is at least limited evidence from epidemiological studies of carcinogenicity to humans (Group B1) or that, lacking adequate data on humans, there is sufficient evidence of carcinogenicity in animals (Group B2).

Group C – possible human carcinogen: this category indicates that there is limited evidence of carcinogenicity in animals and an absence of data on humans.

Group D – not classified: this category indicates that the evidence for carcinogenicity in animals is inadequate.

Group E – no evidence of carcinogenicity to humans: this category indicates that there is no evidence for carcinogenicity in at least two adequate animal tests in different species or in both epidemiological and animal studies.

Calculation of Cancer Risks

The lifetime cancer risk associated with a particular level of exposure is calculated by multiplying the cancer potency factor by the estimated human chronic intake dose. Two types of risk generally are calculated. The first measure is individual risk, that is the cancer risk estimated to be experienced by an individual from a lifetime of exposure at a specified level. The maximum individual risk (MIR) is commonly used. Generally, the exposure component in the MIR is calculated from the maximum annual average air concentration as estimated using an appropriate dispersion model. The modeled maximum 'annual' average exposed concentration normally is assumed to represent the maximum 'lifetime' average exposed concentration.

The second measure is total population risk, that is the cancer risk estimated to be experienced by the entire population exposed to the carcinogen. It can be presented in several ways. The most common is the aggregate population risk, that is defined as the average number of excess cancer cases expected in the exposed population residing in the vicinity of the source, usually expressed as annual estimated cancer cases. Individual risk obviously diminishes because of dispersion as distance increases from the source of the emissions. Because mathematical dispersion models vary as to their level of sophistication and accuracy, total population risk should be estimated only within the geographical area for which the modeled air concentrations are valid. One of the more widely used air dispersion models, EPA's Industrial Source Complex (ISC) model, generally is assumed to be valid to about fifty kilometers (roughly thirty miles) from the source.

The advantages and disadvantages of using each of these measures of risk are summarized below.

Advantages of using individual risk

- Focuses attention on the risk to individuals, often a very few individuals that may experience the highest exposures.
- Not highly influenced by the presence of other emission sources, population distribution or dispersion.
- Easier to estimate than total population risks.

Disadvantages of using individual risk

- Precludes consideration of potentially large segments of the population that are exposed to lower concentrations of the emitted carcinogen.
- Generally overpredicts risk because conservative assumptions are used (e.g., most exposed individual often assumed to be at one location for a seventy-year lifetime).

Advantages of using total population risk

- Provides a more realistic estimate of the risk experienced by the majority of the exposed population.
- Applies to all of the population that can be exposed, not just a few highly exposed individuals.

Disadvantages of using total population risk

- Deemphasizes risks estimated to be experienced by the individuals that may be the most exposed (e.g., such as persons living close to the source).
- More complicated to estimate because it requires considerable data on the distribution and characteristics of the exposed population.

Cancer risk is calculated by multiplying the chronic daily intake by the cancer potency value. For the inhalation pathway, the chronic daily intake is typically calculated using the following equation (EPA 1989g).

$$CDI = \frac{CA \times IR \times ED \times EF \times L}{BW \times AT \times 365}$$

(19-1)

Where CDI = chronic daily intake (mg/kg per day)

CA = contaminant concentration in air (mg/m^3)

IR = inhalation rate (m^3/hour)

ED = exposure duration (hours/week)

EF = exposure frequency (weeks/year)

L = length of exposure (years)

BW = body weight (kg)

AT = averaging time (period over which exposure is averaged – usually seventy years lifetime for carcinogens and one year for noncarcinogens)

365 = days per year

The cancer risk is calculated as

$$R = \text{CDI} \times P \qquad (19\text{-}2)$$

Where R = cancer risk (unitless)

CDI = chronic daily intake (mg/kg per day), see Equation (19-1)

P = cancer potency, $(\text{mg/kg per day})^{-1}$

For simultaneous exposures to several carcinogens, regulatory agencies generally assume that the cancer risks are additive. This is another conservative assumption because different carcinogens can cause different types of cancer that may or may not be comparable. Still, the scientific understanding of carcinogenesis generally is not yet at a level that allows individual carcinogens to be considered separately.

Acceptability of Cancer Risks

The acceptability of cancer risks varies depending on the needs and wishes of the regulating agency or concerned party. Until recently, EPA had not published specific guidance for its hazardous air pollutant regulatory program for use in defining 'acceptable' and 'unacceptable' cancer risks. As a matter of practice, EPA's program offices generally have considered regulation where maximum individual risks ranged from 1×10^{-7} (a probability of one excess cancer death in a population of ten million exposed for a lifetime) to 1×10^{-4} (a probability of 1 in 10,000).

However, in response to a court ruling[3] EPA published in 1989 a decision rule[4] for protecting public health from cancer risks resulting from exposure to hazardous air pollutants. EPA's stated approach is to '(i) protect the greatest number of persons possible to an individual lifetime cancer risk level no higher than approximately 1×10^{-6} and (ii) limit to no higher than approximately 1×10^{-4} the estimated cancer risk that a person living near a source would have if he or she were exposed to the maximum pollutant concentrations for 70 years.'

Most state and local air pollution control agencies regulate carcinogens from new stationary sources when estimated maximum individual risks are in the range of 1×10^{-5} to 1×10^{-6} or greater. Table 19-1 summarizes acceptable risk levels used by many states and the manner in which they are used. Specific regulation of carcinogenic emissions is much less frequent for existing stationary sources. Public interest groups generally support maximum individual lifetime cancer risks of 1×10^{-6}.

Noncarcinogens

The Risk Characterization Process

For substances exhibiting effects other than cancer, the mechanism of action is considered by most authorities to involve a physiological reserve capacity within organisms that must be exceeded by some critical concentration before clinical disease results. This threshold view holds that a range of exposures from just above zero to some higher value can be tolerated by organisms for long periods without an appreciable risk of adverse effects.

Health criteria for noncarcinogens are developed that represent the maximum accepted intake or dose for a specific period. Processes that are used are described in detail in Chapter 7 and are summarized here. The health criterion currently used most widely is the risk reference dose (RfD) established by EPA. The RfD is defined as an estimate of the daily dosage to a substance that is likely to be without appreciable risk of deleterious effect during a lifetime of exposure to a person (EPA 1986h). RfDs generally are derived from human studies involving workplace exposures or from animal studies. In both cases, for application to general community exposures the occupational or animal test results normally are adjusted using uncertainty factors. These are used to compensate for such factors as the higher pollutant concentrations in the tests, the extrapolation from animal to man, the healthy worker effect,

---

[3] *Natural Resources Defense Council, Inc.* v. *EPA*, 824 F.2d 1146[1987].

[4] 54 FR 38044, September 14, 1989.

Table 19-1   Use of risk assessment by state agencies

State	Acceptable risk	How used
Arizona	1×10^{-6}	To determine level of control required to meet ambient standards
California	1×10^{-6}	To set priorities and determine benefits of control
Colorado	1×10^{-6}	
Connecticut	Varies	To establish ambient standards
Florida	1×10^{-6}	Generally case by case
Idaho	1×10^{-5}	
Iowa	1×10^{-6}	
Kansas	1×10^{-6}	
Louisiana	1×10^{-4}	
Maryland	1×10^{-5}	Screening analysis and case-by-case risk management
Massachusetts	1×10^{-6}	To set AALs
Michigan	1×10^{-6}	
Minnesota	1×10^{-5}	
Mississippi	1×10^{-6}	
New Jersey	1×10^{-6}	Residual risk analysis
New York	1×10^{-6}	To set AALs for carcinogens
North Carolina	1×10^{-6}	
North Dakota	1×10^{-6}	
Ohio	1×10^{-5}	
Oklahoma	1×10^{-6}	
Oregon	1×10^{-5}	To assess health impact and judge control needs
Rhode Island	1×10^{-5}	To determine BACT for PSD and to set AALs
Vermont	1×10^{-6}	To set ambient levels and evaluate impacts
Washington	1×10^{-6}	To limit health impact
Wisconsin	1×10^{-6}	To serve as de minimis emission level
Wyoming	1×10^{-6}	

Source: STAPPA/ALAPCO 1989.

and others. RfDs also typically are developed for specific routes of entry (e.g., inhalation and ingestion).

The purpose of the RfD is to provide a benchmark with which other route-specific doses (e.g., those projected from human exposure under various environmental conditions) might be compared. Doses that are less than the RfD are not believed to be of concern. Doses that are greater than the RfD generally are viewed as indicating an increased probability for an adverse effect, or at least that an inadequate margin of safety could exist for exposure to that substance. As developed by EPA, the RfD is an approximate number with uncertainty spanning perhaps an order of magnitude (EPA 1986h). Although doses higher than the RfD have a higher probability of producing an adverse effect, the RfD generally is not considered by EPA to represent a level that is 'unacceptable.' Instead, RfDs represent

generally conservative estimates of maximum acceptable doses.

At the time of this writing in 1992, more RfDs were available for the ingestion route than for inhalation. However, a program is under way in EPA's Environmental Criteria and Assessment Office, Office of Research and Development, to develop additional inhalation RfDs for hazardous and toxic air pollutants. Inhalation RfDs usually are converted from an exposure 'dose' to a corresponding exposure 'concentration' using generally accepted average human intake values. These are referred to as risk reference concentrations (RfCs).

In cases where inhalation RfDs or RfCs have not been developed by EPA, and this is currently the case for many hazardous or toxic air pollutants, an inhalation RfD may be derived by trained scientists from other appropriate data. In some instances, an RfD developed for a different route of exposure can be used. This is not normally desired because absorption frequently differs significantly with different routes of exposure. When used, the conversions are made by equalizing dosages. For example, inhalation health criteria normally are presented in units of micrograms per cubic meter ($\mu g/m^3$) and ingestion health criteria normally are presented in units of milligrams per kilogram of body weight per day (mg/kg per day). To convert an ingestion RfD to an inhalation RfD, assumptions are required about average human breathing rates, average daily rates of food and water consumption, and substance and route-specific absorption factors. EPA has published standard assumptions for these physiological parameters (EPA 1985b, EPA 1988b, and EPA 1989a).

When RfDs or other applicable noncarcinogenic health criteria are not available for air pollutants, regulatory agencies often have used occupational (workplace) standards developed by the American Conference of Governmental Industrial Hygienists (ACGIH) or the U.S. Occupational Safety and Health Administration (OSHA) to establish health criteria for community exposures. Two occupational standards commonly are used: (1) the time-weighted average (TWA), that is defined as the average concentration for a normal eight-hour workday and a forty-hour workweek to which nearly all workers may be repeatedly exposed, day after day, without adverse effect, and (2) the short-term exposure limit (STEL), that is defined as the concentration to which workers can be exposed continuously for a short period of time without suffering from irritation, chronic or irreversible tissue damage, or narcosis of sufficient degree to increase the likelihood of accidental injury, impair self-rescue, or materially reduce work efficiency. All workers are assumed to be able to withstand up to four exposures per day of concentrations as high as the STEL with no ill effects, if the TWA also is not exceeded. STELs are generally used to supplement the TWA when there are recognized short-term effects from a substance whose toxic effects are primarily of a long-term nature.

ACGIH derives a TWA value called the threshold limit value (TLV). These are available for over four hundred substances and periodically are updated by ACGIH. OSHA has adopted a similar value, called the permissible exposure limit (PEL). OSHA revised its PEL list in 1989.[5]

It is important to recognize that both ACGIH and OSHA recommend *against* direct use of occupational limits for community exposure protection. The major reason is that occupational limits are developed specifically for workers and workplace exposures and are not developed to protect the public that, under certain circumstances, can be more sensitive or more susceptible to adverse health effects. Another is that TWAs are designed for persons intermittently exposed for only about forty hours out of the 168 total hours in the week, and community exposures can be much longer, even continuous. However, few health criteria data bases as extensive are available for chemicals that are common air pollutants. Thus, occupational values are used widely as an interim measure.

Other Techniques

In cases where there is no useful information from which to derive noncarcinogenic health criteria for a particular substance, comparisons to substances whose chemical structure is similar, and for which health criteria have been derived, have been used. These structure-activity relationships (SARs) assume that toxicity is related to specific chemical structures; SARs generally are accepted only as a means for providing 'ballpark' estimates.

As noted above, in some cases short-term effects are of concern for noncarcinogens. Because RfDs are developed by EPA specifically for longer term, noncarcinogenic human health effects, short-term health criteria generally must be developed from occupational criteria, that frequently are based on shorter-term effects. Of course, the STEL specifically is a short-term exposure limit.

---

[5] 54 FR 2923, January 19, 1989.

Calculation of Noncarcinogenic Risks

For noncarcinogens, potential risks are assessed in three steps. In the first step, the estimated maximum daily intake is divided by the accepted health criterion, that is the maximum acceptable intake, for each substance of concern. This ratio often is called the hazard quotient. When the hazard quotient exceeds one, i.e, when the estimated daily intake exceeds the maximum acceptable intake, it is indicative of a potential for an adverse effect to occur. This process can be represented as

$$\text{Hazard quotient} = \frac{\text{MDI}}{\text{RfD}} \quad (19\text{-}3)$$

Where MDI = maximum daily intake (mg/kg per day)

RfD = risk reference dose (mg/kg per day)

Second, when there is exposure to a mixture of more than one substance of concern, the potential effect of the mixture usually is evaluated by adding the hazard quotients for each substance in the mixture. When this sum, that often is called the hazard index, exceeds one, it also is indicative of a potential for an adverse effect. This process can be represented as

$$\text{Hazard index} = \frac{\text{MDI}_1}{\text{RfD}_1} + \frac{\text{MDI}_2}{\text{RfD}_2} + \cdots + \frac{\text{MDI}_i}{\text{RfD}_i}$$

$$(19\text{-}4)$$

Where MDI_i = maximum daily intake (mg/kg per day) of the ith component

RfD_i = risk reference dose (mg/kg per day) of the ith component

This procedure is conservative (i.e., health protective) because the different substances in the mixture frequently are associated with different health effects, that often are not additive. The third step in the process compensates for this conservatism by more closely

evaluating the individual substances. The closer evaluation frequently takes the form of grouping all substances associated with specific health endpoints (e.g., eye irritation or kidney effects) or grouping all chemicals associated with a specific intake route, and again summing the ratios of the estimated intakes to the maximum acceptable intakes. This still assumes that all substances in each group act additively, that again likely is conservative.

The procedures for using these ratios and sums were developed by EPA and are described in *Guidelines for Health Risk Assessment of Chemical Mixtures*, published in the Federal Register.[6] These procedures currently are applied in EPA's Superfund cleanup program (EPA 1989c; EPA 1989d; EPA 1989e; EPA 1989f).

The intakes of substances by potentially exposed populations typically are estimated assuming specified exposure conditions (such as frequency and duration of exposure) and typical intake parameters; these also have been developed by EPA (1989a). These are discussed more thoroughly in Chapter 13. Frequently, an average exposure case and a plausible maximum exposure case are considered. The average exposure case uses what are considered to be most likely (but generally conservative) exposure conditions. The plausible maximum exposure case uses high estimates of the range of potential exposure parameters. The plausible maximum case, although considered theoretically possible, is likely to apply, if at all, to only a very small percentage of the potentially exposed populations. These exposures are evaluated in order to place a plausible upper bound on the estimated risks.

Noncancer Risk Acceptability

As noted above, the purpose of the RfD is to provide a benchmark with which other doses (e.g., those projected from human exposure under various environmental conditions) might be compared. Doses that are less than the RfD are not believed to be of concern. Doses that are significantly greater than the RfD are generally

---

[6] 51 FR 34014, September 24, 1986.

viewed as indicating a potential for an adverse effect. As developed by EPA, the RfD is an approximate number with uncertainty spanning perhaps an order of magnitude (EPA 1986h); in other words, although doses higher than the RfD have a higher probability of producing an adverse effect, they generally are not considered by EPA to represent levels that are necessarily 'unacceptable' or 'of concern.' Thus, RfDs represent generally conservative estimates of maximum acceptable doses. Another way in which to view a noncancer risk value greater than one is that it indicates a reduced margin of safety and the need for further evaluation.

Although the above caveats are broadly understood, noncancer risk acceptability frequently is established by regulators explicitly at a level of one. In other words, if a target hazard quotient or hazard index exceeds one, regulatory agencies in many cases will require further reduction in emissions by the source. The caveats frequently are ignored because time and resource limitations and a general lack of guidance restrict the ability of agencies to evaluate the uncertainties and act accordingly. Dialogue between regulators and the parties being regulated is essential to identify and consider more reasonable approaches.

CURRENT RISK ASSESSMENT RESEARCH

Researchers are actively investigating many areas of uncertainty associated with conducting risk assessments. Models are being developed that more accurately represent biological and physical processes and, as a result, produce more accurate and more plausible results. Input data for a variety of models for which assumptions were used in the past also are being gathered and developed, again to produce more accurate and more plausible results. Finally, the underlying sciences are being studied more intensively to understand better the chemical, physical, and biological processes at work. A number of current risk assessment research areas are described below (Patrick and Anderson 1990).

Hazard Identification

Work is under way to categorize cancer weight of evidence more correctly not only with respect to the outcomes of animal studies (i.e., tumor formation) and from short-term in vitro and in vivo tests (i.e., potential genotoxic activity), but also with respect to the relevance of high-dose tumor formation to lower-dose environmental exposures. An example of this is ethylene thiourea (ETU) that produces rat liver tumors through a threshold mechanism by virtue of thyroid activity and also produces effects in the mouse liver tumor system. The question is whether these two species should be counted

together in categorizing ETU as a suspect human carcinogen if environmental exposure levels do not exceed the threshold required for thyroid tumor response. If they are not combined, ETU would be classified as a possible human carcinogen instead of a probable human carcinogen. In short, the weight of evidence classification could, although it does not currently, reflect whether biological mechanisms operative at high doses are also operative at low doses and whether some tumor types occurring at high doses also might occur at low doses.

A similar question is whether or not cancers of varying types or severities should be given equal weight. For example, one of the effects of ingesting arsenic can be skin cancer that is far more treatable than the lung cancer produced when certain forms of arsenic are inhaled. It can be argued that these effects should be considered separately, although this is not normally done.

The extraordinary potency of some members of the dioxin family has led to several advances in risk assessment state of the art. First, EPA proposed lowering the cancer potency of dioxins based on several considerations including use of a biologically based two-stage model of carcinogenesis where the promoting activity of dioxin has formed the basis for the model (EPA 1986b). This also has implications for other chemicals, particularly polychlorinated biphenyls (PCBs) and polycyclic aromatic hydrocarbons (PAHs). For example, PCBs respond only in the mouse liver tumor system, indicating the possibility that a biologically based two-stage model could be applied to the mouse liver tumor data to provide more accurate estimation of the dose-response curve. This probably would result in risks that are considerably lower than those estimated from the more commonly used linear nonthreshold upper bound potency factor.

Second, EPA developed new procedures for compensating for the widely varying cancer potencies of the different members of the dioxin family. These methods (see Chapter 13 for more detail) are based on the fact that different suspected carcinogens of a family of compounds generally exhibit different cancer potency values; in fact, some members of the family may not be carcinogenic. When considered individually, this can result in a significant reduction in the estimated risks associated with exposure to dioxin mixtures.

This technique also can be used for PCBs. The toxicity of PCBs frequently is assessed by applying EPA's cancer potency factor for a specific mixture of PCBs known as Aroclor 1260. There is considerable scientific opinion, however, that suggests that only certain individual PCB chemicals are either bioaccumulated or toxic. Short-term tests, such as those that indicate the induction of certain enzymes, may be useful in developing toxicity equivalents for the individual PCB chemicals. For this refined approach to be used in risk assessment, chemical-specific analysis of the PCBs would be required.

For PAHs, a relative potency method is available (Thorslund 1990). In the past, regulators generally assumed that all emissions of PAHs had cancer potencies equivalent to that of benzo[a]pyrene (B[a]P), one of the more potent and best-studied carcinogens of the family. However, it is now known that the relative potencies of complex PAH mixtures estimated directly from animal bioassay experiments are much less than that of B[a]P. Similarly, research demonstrates widely varying potencies of the individual suspect carcinogens in typical PAH mixtures. These illustrate the importance of identifying and measuring specific mixtures of PAHs. If measurement data are available to identify the components, risks can be calculated using the potencies of the individual PAHs. These typically result in estimated risks lower by several factors of ten than the risks estimated based on B[a]P equivalence.

Dose-Response Assessment

Dose-response modeling generally follows the concept of establishing a linear nonthreshold dose response curve at low doses to define 'plausible upper bounds on the risk,' meaning the actual risks are unlikely to be higher than the predicted risks but could be considerably lower (Crump, Guess, and Deal 1977; Crump and Watson 1979; Crump 1981). In 1985, EPA's Risk Assessment Forum formed a subcommittee to investigate the possibility of finding more accurate ways to characterize risk in addition to 'plausible upper bounds.' The outcome of the work was the publication of a generic model that relies on biological information in a two-step mechanism to describe the likely course of disease (Moolgavkar and Knudson 1981; Thorslund, Charnley, and Anderson 1986; Thorslund, Brown, and Charnley 1987). The effect of this model generally is to reduce estimated risks well below the upper bound; the amount of reduction varies up to many factors of ten depending on the system and biological information used.

Recent interest also is focusing on the fact that exposure to a contaminant is not synonymous with intake dose, and intake dose is not synonymous with the effective dose at the site of injury (i.e., target organ). To more precisely establish the dose delivered to an organ in the body that might be damaged by exposure to the toxic chemical, scientists are paying increasing attention to tracking the distribution of chemicals using pharmacokinetic approaches that include consideration of bioavailability, laboratory bioassays of actual environmental materials, and mathematical modeling. These results currently are available for only a few specific chemicals, but show conclusively that estimates of risks are decreased under most circumstances because smaller quantities of chemicals usually are delivered to target organs than are present at the points of human contact.

Because many exposures are episodic in nature, another use of pharmacokinetics is to consider the half-life of chemicals in the body. This can significantly affect the risk assessment outcome. For example, the biological half-life for mercury is only about two weeks, whereas that for dioxin is several years. If an individual receives all of his or her exposure in a short period, mercury will clear rapidly from the body thus providing less overall exposure than is the case from episodic exposures. This would not be true in the case of dioxin.

Exposure Assessment

Exposure assessments are becoming much more site-specific, relying on more accurate and more relevant information concerning local conditions. For example, in the case of the ASARCO copper smelter in Tacoma, Washington, EPA first used generalized dispersion models to define exposure to the general population and later refined these estimates by using local air monitoring data and dispersion parameters specific to the locality (Patrick and Peters 1985). The effect of this work was to lower the estimated risks by well over a factor of ten.

Population location assumptions also can influence exposure and risks to a considerable degree. Analysts frequently have used worst-case population assumptions, sometimes known as the 'fencepost' person (i.e., a person living on a fencepost outside his or her home, at the point of maximum estimated exposure, for a lifetime of seventy years) (EPA 1988b).[7] Although this may be appropriate for some screening purposes, it is unlikely that anyone lives under these conditions and these assumptions result in unrealistically elevated exposure and risk estimates. More recently, computerized techniques are being used that directly compare Bureau of Census population data with modeled community air concentrations. In these assessments, pollutant concentration distributions around sources are overlaid with census population data available for over three hundred thousand nationwide block groups/enumeration districts. This provides a much more accurate estimate of the number of people exposed to specific estimated pollutant concentration ranges.

In the past, population life-styles and activities also rarely were included in exposure assessments. Instead, the exposed population generally was assumed to be equally and continuously exposed to contaminants. Recent research, however, provides a much better ability to account for widely varying human life-styles and activities in exposure and risk assessments. For example, time-use studies describe different human activity patterns, including work, personal care, shopping, education, entertainment, and leisure. The average amounts of time spent by average males and females, and children, are categorized and tabulated in great detail. Additional work has studied geographical and mobility differences (EPA 1988b, 1989a). Because many of these activities and life-styles lead persons to be away from home, that is the primary point of exposure, more accurate consideration of these factors can result in more accurate and generally lower risks. On the other hand, more is being learned about indoor air exposures and the fact that indoor concentrations of some toxic air pollutants frequently are higher than outdoor concentrations, resulting in greater exposures and risks (Wallace et al. 1983, Wallace et al. 1986; Wallace et al. 1987; Wallace et al. 1988).

Bioavailability also is becoming an important issue in exposure assessment. Bioavailability is the fraction of a chemical in an environmental matrix (e.g., soil) that can be released from the matrix and absorbed by the organism of concern and, thus, is available to elicit a biological

---

[7] A humorous term is 'Mr. Max,' who might be described as the person residing at the point of maximum average pollutant concentration, sitting on his front porch for seventy years, wearing no clothes, eating food grown only on his property, and drinking water from his own shallow well.

effect. To illustrate, dioxins originally were assumed to be 100 percent biologically available from inhaled particulate matter. Recent studies, however, have found that this is not correct; instead, dioxins in fly ash and soil are biologically available from 7 to 50 percent (Van den Berg et al. 1986).

Recent risk assessments also are focusing on more accurate exposure pathway analyses. For example, increasingly sophisticated models are being developed to estimate exposures through many indirect pathways involving bioaccumulation through the food chain. More sophisticated particle deposition models are available to predict the amounts of emitted chemicals that may be deposited around a hazardous waste site (Sehmel and Hodgson 1979). In addition, many organic chemicals are capable of reacting with sunlight or other chemicals to form other substances, that can be more or less toxic. Furthermore, metals are capable of being mineralized and integrated into natural biogeochemical cycles. Application of these concepts is in its developmental stages; however, preliminary results indicate that each potentially can result in estimates of risks reduced by a factor of ten or more.

Risk Characterization

Because risk assessment methodologies initially were designed to be very conservative, erring on the side of human health protection, the effects of collecting and applying more scientific and site-specific data to risk assessments for the most part lowers the magnitude of predicted risks. In some cases, however, use of improved information in a risk assessment can increase predicted risks. For example, newer deposition models described above generally predict higher rates of chemical deposition and some chemicals produce, through environmental degradation, more potent by-products than their parents. For example, trichloroethylene can anaerobically degrade to the more toxic vinyl chloride (Cline and Viste 1984; Parsons, Wood, and Demarco 1984).

As a result of new scientific methods, procedures, and information being developed, risk assessments are now possible that are more accurate and credible than in the past, and the capabilities are improving. These risk estimates also are more scientifically defensible and potentially more acceptable to all parties involved. These new methods and procedures generally result in lower estimated risks.

UNCERTAINTIES IN RISK ASSESSMENT

All risk assessments involve the use of assumptions, judgment, and imperfect data to varying degrees. This results in uncertainty in the final estimates of risk. Uncertainty in a risk assessment may arise from, and be magnified through a combination of, sources including:

- Analytical or statistical variation in measured data.
- Misidentification or failure to identify all hazards to which the person is exposed.
- Inappropriate choice of models or evaluation of toxicological data in dose-response quantification.
- Inappropriate choice of models or input parameters in exposure assessment and fate and transport modeling.
- Inappropriate selection of assumptions concerning exposure scenarios and population distributions.

The technique commonly employed to compensate for the uncertainty is to bias the assessment in the direction of overestimation of risk. This is often termed the 'worst case' or 'conservative' analysis, and assures that if the risk estimate is in error, it errs on the side of human health protection. The net effect of combining numerous conservative assumptions obviously is that the final estimates of risk may be greatly overestimated.

Various types of uncertainties are described below and tables are presented with examples of the assumptions that typically are made and the effect of the assumptions on the ultimate risk estimates. It is important to recognize that the vast majority of assumptions lead to over-estimates of risk.

Uncertainties in Emissions Estimates and Air Dispersion Modeling

Several uncertainties affecting emissions estimation and air dispersion modeling are listed in Table 19-2. Emission rates are critical inputs into the dispersion models that calculate the exposure concentrations. Ideally, the emission rates should be based on source test data, that provides the best source of such data in a risk assessment. However, source test data can be difficult and costly to generate, there are variations in the type and quality of the emission tests, and the tests only provide real-time emissions data (i.e., at the time of the test).

Selection of the appropriate air dispersion model depends on the sources, the geography, and the use of the results. For risk assessments in which the results are for screening or priority setting, the models need not be rigorous. However, where the risk results are to be used to support regulation or where significant expenditures of resources might result, more refined models should be used. Typically, the Industrial Source Complex (ISC) is the model of choice by EPA (EPA 1986i) for facilities in urban and rural areas, areas with flat or rolling terrain, transport distances less than fifty kilo-

meters, and one hour to annual averaging times. For elevated terrains, other models should be used. The ISC model does not take into account chemical removal or transformation processes in estimating air concentrations.

Uncertainties in Hazard Identification

In almost all risk assessments, one of the largest sources of uncertainty is the accepted health criteria. Table 19-3 indicates how uncertainties in health criteria can affect the magnitude of the risk estimates. Health criteria used to evaluate long-term exposures (e.g., RfDs and cancer potency factors) are based on concepts and assumptions that bias the evaluation in the direction of overestimation of health risk. As EPA noted in its Guidelines for Carcinogenic Risk Assessment (EPA 1986c):

There are major uncertainties in extrapolating both from animals to humans and from high to low doses. There are important species differences in uptake, metabolism, and organ distribution of carcinogens, as well as species and strain differences in target site susceptibility. Human populations are variable with respect to geometric constitution, diet, occupational and home environment, activity patterns, and other cultural factors.

These uncertainties typically are compensated for by using upper bound potency factors for carcinogens and safety factors for non-

Table 19-2   Typical uncertainties in emissions and air dispersion modeling

Assumption	Magnitude of effect[a]	Direction of effect
Source parameters based on source	Low	May over- or underestimate risk
Emissions based on stack test	Low	May over- or underestimate risk
Five years of meteorology data are used	Low	May over- or underestimate risk
Atmospheric chemical removal ignored	Moderate	May overestimate risk
Particles assumed to be largely inhalable and unlikely to deposit	Low to moderate	May over- or underestimate risk

[a] Low means ≤ one order of magnitude; moderate means > one and ≤ two orders of magnitude; high means > two orders of magnitude.

Table 19-3   Typical uncertainties in hazard identification

Assumption	Magnitude of effect[a]	Direction of effect
Conservatively derived health criteria	Low to moderate	May overestimate risk
Cancer potency factors derived from multistage linearized model	Moderate to high	May overestimate risk
Oral RfDs used where inhalation RfDs unavailable	Moderate	May over- or underestimate risk
Cancer risks assumed to act additively	Moderate	May over- or underestimate risk
Cancer weight of evidence assumed to be equivalent	Moderate	May overestimate risk
Noncancer effects assumed to act additively	Moderate	May overestimate risk

[a] Low means ≤ one order of magnitude; moderate means > one and ≤ two orders of magnitude; high means > two orders of magnitude.

carcinogens. At best, the assumptions provide a rough but plausible estimate of the upper limit of risk (i.e., it is not likely that the true risk would be much more than the estimated risk but it could very well be considerably lower, even approaching zero).

In addition, there are varying degrees of confidence in the weight of evidence for carcinogenicity of a given chemical. EPA's weight of evidence classification provides information that can indicate the level of confidence or uncertainty in the data obtained from studies in humans or experimental animals. However, it cannot be used quantitatively in the estimate of risk.

Uncertainties in Exposure Assessment

There are two major areas of uncertainty in the exposure assessment: (1) the choice of exposure models and input parameters used to estimate exposure point concentrations, and (2) the

methods and parameters used to calculate chemical intakes. Table 19-4 summarizes many of the specific areas of uncertainty involved in estimating exposure point concentrations. Most of the assumptions shown are expected to overestimate exposure point concentrations and, thus, estimated risks (Patrick 1992b).

Table 19-5 summarizes some of the uncertainties affecting the estimation of chemical intakes. These uncertainties are associated with the parameters used to characterize maximum case exposures, including the assumed points of exposure, the contact rates with exposure media, and the frequency and duration of exposure. Again, these uncertainties typically are compensated for by making conservative assumptions. For example, the contact rates by inhalation are normally assumed to be constant throughout the exposure period within each age category. Exposures also typically are estimated assuming no residents migrate from the area during a seventy-year period.

INTERPRETATION OF RISK NUMBERS

Many everyday activities undertaken by people involve risk in some form. Working, driving a car, smoking a cigarette, crossing a street, and living at high altitudes are associated with a possible negative outcome about the situation in which the activity does not occur. Table 19-6 lists some common occurrences and their statis-

tical risk. Table 19-7 lists calculated risks of death associated with various occupations. Individuals often make decisions about the probability of an occurrence and the risk involved before undertaking an activity or an occupation. The basic purpose of an environmental risk assessment is to provide sufficient information

Table 19-4   Typical uncertainties in exposure point concentrations

Assumption	Magnitude of effect[a]	Direction of effect
Facility lifetime assumed to be seventy years	Low	May overestimate risk
Concentrations based on modeling not sampling	Low to moderate	May over- or underestimate risk
Deposition ignored	Low	May underestimate risk
Accumulation in plants, fish, soil and water assumed minimal	Low to moderate	May underestimate risk
Deposited emissions assumed to mix in top 1 cm of soil	Moderate	May overestimate risk
Food uptakes based upon chemical property data direct measurement	Low to moderate	May over- or underestimate risk

[a] Low means ≤ one order of magnitude; moderate means > one and ≤ two orders of magnitude; high means > two orders of magnitude.

Table 19-5   Typical uncertainties in estimation of chemical intakes

Assumption	Magnitude of effect[a]	Direction of effect
Inhalation rate assumed constant at every stage	Low	May overestimate risk
Residents assumed to live in study area for seventy years	Low	May overestimate risk
Maximum exposure combined maximum air and deposition	Moderate	May overestimate risk
Indoor/outdoor concentrations assumed to be the same	Low	May overestimate risk
Ingestion exposure assumed constant over seventy years	Low	May overestimate risk
Bioavailability assumed to be low	Low to moderate	May underestimate risk
Uptake from plants, fish and soil ignored	Low to moderate	May underestimate risk

[a] Low means ≤ one order of magnitude; moderate means > one and ≤ two orders of magnitude; high means > two orders of magnitude.

about the possible health effects of an action or situation so that a decision can be made as to its acceptability.

Two important factors determine the acceptability of a risk: (1) the probability of the occurrence and (2) the severity or benefit of the occurrence. For example, in gambling, the probability of winning depends on the number of times an event probably will occur compared to the number of possible occurrences. The severity of losing depends on how much money is bet. Thus, an individual may be willing to bet $100 at 10:1 odds of winning, but unwilling to bet the $100 if the odds are less.

The Significance of Risk

A commonly used benchmark for defining a trivial excess lifetime cancer risk is one in one million (1×10^{-6}). This is the circumstance in which one person in an exposed population of one million for seventy years is estimated to develop cancer as a result of the exposure. A risk of 1×10^{-5} represents a risk that is higher

Table 19-6   Risks of various occurrences[a]

Occurrences	Deaths per 100,000[b]	Individual risk[c]
Motor vehicle accident	1,680	2×10^{-2}
Falls	434	4×10^{-3}
Drowning	252	3×10^{-3}
Fires	196	2×10^{-3}
Firearms	70	7×10^{-4}
Electrocution	37	4×10^{-4}
Tornados	4.2	4×10^{-5}
Floods	4.2	4×10^{-5}
Lightning	3.5	4×10^{-5}
Animal bites or stings	1.4	1×10^{-5}

[a] Adapted from: Crouch and Wilson 1984.
[b] Approximate number of deaths per 100,000 over a seventy-year period.
[c] Individual excess lifetime risk.

Table 19-7   Risks of various occupations[a]

Activity	Deaths per 100,000[b]	Individual risk[c]
General		
Manufacturing	568	6×10^{-3}
Trade	365	4×10^{-3}
Service and government	730	7×10^{-3}
Transport and utilities	2,555	3×10^{-2}
Agriculture	4,088	4×10^{-2}
Construction	4,380	4×10^{-2}
Mining and quarrying	6,813	7×10^{-2}
Specific		
Police	4,130	2×10^{-2}
Fire fighting	5,600	6×10^{-2}
Railroad employment	1,680	2×10^{-2}

[a] Adapted from: Crouch and Wilson 1984.
[b] Approximate number of deaths per 100,000 over a seventy–year period
[c] Individual excess lifetime risk.

Table 19-8   Several one in a million cancer risks[a]

Source of risk	Exposure over a lifetime
Cosmic rays	One transcontinental round trip by air
	Living 1.5 months in Colorado versus New York
	Camping at 15,000 feet for 6 days versus sea level
Other radiation	20 days at sea level, natural background
	2.5 months living in masonry building vs. wood
	1/7 of a chest X-ray using modern equipment
Eating or drinking	40 diet sodas (saccharine)
	6 pounds of peanut butter (aflatoxin)
	180 pints of milk (aflatoxin)
	200 gallons of drinking water in Miami
	90 pounds of charbroiled steak
Smoking	2 cigarettes

[a] Adapted from Crouch and Wilson 1984.

For reference, Table 19-8 lists several actions or conditions that result in a one in one million cancer risk.

Determining the significance of a given level of risk is the first step in determining its acceptability. Lave (1982) described five approaches for defining a 'significant' risk. The first approach involves the concept of no acceptable risk. This approach was taken by Congress in the Delany Clause of the original Food, Drug and Cosmetic Act. That clause stated that no food additive should be considered safe if it is found to induce cancer when ingested by man or animals. Thus, zero risk was required. However, when the Food and Drug Administration attempted to ban saccharine, Congress overruled the ban based on the substantial benefits to diabetics and others associated with use of nonnutritive sweeteners. Still, the controversy has continued for many years between those who advocate total elimination of carcinogens from commerce and those who want to balance the risks with the benefits.

by a factor of ten. Similarly, a risk of 1×10^{-7} implies a tenfold margin of safety from the benchmark. Because the background cancer rate in the U.S. is about one in four (250,000 per million), adding an additional risk of one in one million will not be detectable (i.e., 250,001 per million compared to 250,000 per million). Thus, other means must be relied on to determine the significance of an environmental cancer risk.

The second approach is to define a level of risk that is acceptable. This approach sometimes is legally mandated by governmental agencies (Hallenbeck and Cunningham 1986). EPA has adopted different cancer risk criteria in different programs. In response to the vinyl chloride decision by the U.S. Court of Appeals, an acceptable risk was defined by EPA[8] as ranging from 1×10^{-4} to 1×10^{-6}. However, for the nationwide water quality criteria program the acceptable level of risk is defined as 1×10^{-4} to 1×10^{-7}. Under the Safe Drinking Water Act, residual risks after application of 'feasible' technology to municipal sewage treatment systems range downward from 1×10^{-4}. In the Superfund program, acceptable risks can be as high as 1×10^{-4}, depending on the analysis of several factors. Outside EPA, the Occupational Safety and Health Administration accepted residual risks of 1×10^{-3} after requiring the application of 'feasible' controls, and the Nuclear Regulatory Commission adopted a level of 5×10^{-3} as guidance for whole-body radiation. Although all of these values were derived based on different laws, different population exposure situations, and other variables, they do provide benchmarks for comparison.

The third approach is a comparison of the risks related to the activity of concern with those generally prevailing in society. Tables 19-7 and 19-8 show the risk of death resulting from several occurrences and occupations. The individual risk estimates shown in Table 19-7 are based on measured deaths occurring in the U.S. population. It is important to keep in mind that some of these risks are voluntary (e.g., smoking cigarettes) and some are involuntary (e.g., being hit by lightning). Exposure to a toxic substance in the air normally is considered to result in an involuntary risk. Generally, the level of risk associated with breathing chemicals in the air in the U.S. is lower than the risk of being involved in a motor vehicle accident or having an accident while working in hazardous occupations. Largely because people understand driving and working and do not under-

stand chemical carcinogens, most people view air pollution cancer risks as more dangerous and important to reduce than those associated with driving and working.

Approach four is more technical. It involves multiplying the numerical size of the population at risk by the probability of adverse health effects. For a risk assessment of a chemical carcinogen, this involves multiplying the risks by the number of people potentially at those risks to arrive at an expected cancer incidence. If the expected value is less than 0.5 (i.e., less than one-half cancer case anticipated), the risk might be assumed to be insignificant on the grounds that the most likely outcome is no effect. This approach implies that smaller populations could legitimately be exposed to higher individual risks than larger populations. Although this is one way of thinking about risk, public health policy generally has not been based on this approach.

The fifth approach involves prevention of a significant increase in the occurrence of the adverse effect. If the adverse effect is caused only by exposure to the chemical, then one cancer case is significant. On the other hand, if there is a high background rate of the effect, a large number of cancer cases would have to occur before a significant increase could be noted.

References

Albert, R.E., R.E. Train, and E.L. Anderson. 1977. Rationale developed by the U.S. Environmental Protection Agency for the assessment of carcinogenic risk. *Journal of the National Cancer Institute* 58:1537.

Anderson, E.L. and the Carcinogen Assessment Group (CAG) of the U.S. Environmental Protection Agency. 1983. Quantitative approaches in use to assess cancer risk. *Risk Analysis* 3(4):277.

Cline, P.V. and D.R. Viste. 1984. Migration and degradation patterns of volatile organic compounds. National Conference of Uncontrolled Hazardous Waste Sites Proceedings, p. 217.

Crump, K.S. 1981. An improved procedure for low-dose carcinogenic risk assessment from animal

[8] 54 FR 38044, September 14, 1989.

data. *Journal of Environmental Pathology and Toxicology* 52:675.

Crump, K.S. and W.W. Watson. 1979. A Fortran program to extrapolate dichotomous animal carcinogenicity data to low doses. National Institute of Environmental Health Sciences. Contract No. 1-ES-2123.

Crump, K.S., H.A. Guess, and I.L. Deal. 1977. Confidence intervals and test of hypotheses concerning dose-response relations inferred from animal carcinogenicity data. *Biometrics* 33:437.

Hallenbeck, W.H. and K.M. Cunningham. 1986. *Quantitative Risk Assessment for Environmental and Occupational Health*. Chelsea, MI: Lewis Publishers.

Lave, L.B. 1982. Approaches to defining significant risk. April 1982 (unpublished document).

Moolgavkar, S.H. and A.G. Knudson. 1981. Mutation and cancer: A model for human carcinogenesis. *Journal of the National Cancer Institute* 66:1037.

National Academy of Sciences (NAS). 1983. *Risk Assessment in the Federal Government: Managing the Process*. Prepared by the Committee on the Institutional Means for Assessment of Risk to Public Health, Commission on Life Sciences. Washington, D.C. National Academy Press.

Parsons, F., P. Wood, and J. Demarco. 1984. Transformation of tetrachloroethane and trichloroethane in microcosms and groundwater. *Research Technology Management* p. 56.

Patrick, D.R. 1992b. The impact of exposure assessment assumptions and procedures on estimates of risk associated with exposure to toxic air pollutants. Paper 92-95.02. Presented at the 85th Annual Meeting of the Air & Waste Management Association. Kansas City, MO.

Patrick, D.R. and E.L. Anderson. 1990. Recent advances in the state of the art in risk assessment. Presented at the 1990 Hazmat International Conference. Atlantic City, NJ. June 1990.

Patrick, D.R. and W.D. Peters. 1985. Exposure assessment in setting air pollution regulations: ASARCO Tacoma, a case study. Presented at the Society for Risk Analysis Annual Meeting. Washington, D.C.

Ruckelshaus, W.D. 1983b. Statement before the Subcommittee on Oversight and Investigations, Committee on Energy and Commerce, U.S. House of Representatives. November 7, 1983.

Sehmel, G.A. and W.H. Hodgson. 1979. A model for predicting dry deposition of particles and gases to environmental surfaces. PNL-SA-6271-REV 1.

Prepared for the U.S. Department of Energy by Pacific Northwest Laboratory. Richland, WA.

Thorslund, T.W. 1990. Preliminary Report. Task 115 – Obtaining unit risk estimates for inhaled B[a]P. Task 116 – Development of a systematic approach for estimating PAH relative potencies. EPA Contract 68-02-4403. U.S. Environmental Protection Agency, Office of Health Assessment, Office of Research and Development. Washington, D.C.

Thorslund, T.W., G. Charnley, and E.L. Anderson. 1986. Innovative use of toxicological data to improve cost-effectiveness of waste cleanup. Presented at the Superfund 86: Management of Uncontrolled Hazardous Waste Sites, Washington, D.C.

Thorslund, T.W., C.C. Brown, and G. Charnley. 1987. Biologically motivated cancer risk models. *Risk Analysis* 7:109.

U.S. Environmental Protection Agency (EPA). 1976a. Interim procedures and guidelines for health risks and economic impact assessments of suspected carcinogens. 41 FR 21402, May 25, 1976.

U.S. Environmental Protection Agency (EPA). 1985b. Development of statistical distributions or ranges of statistical factors used in exposure assessments. OHEA-E-161. Final Report. Office of Health and Environmental Assessment. Washington, D.C.

U.S. Environmental Protection Agency (EPA). 1986b. Report of the dioxin update committee. Office of Pesticides and Toxic Substances. Washington, D.C.

U.S. Environmental Protection Agency (EPA). 1986c. Guidelines for carcinogen risk assessment. 51 FR 33992, September 24, 1986.

U.S. Environmental Protection Agency (EPA). 1986h. Integrated risk information system (IRIS) database. Appendix A. Reference Dose (RfD): Description and use in health risk assessments. Office of Health and Environmental Assessment. Washington, D.C.

U.S. Environmental Protection Agency (EPA). 1986i. Guideline on air quality models (Revised) and Supplement A (1987). EPA 450/2-78-027R. (NTIS PB 86-245248). Office of Air Quality Planning and Standards. Research Triangle Park, NC.

U.S. Environmental Protection Agency (EPA). 1988b. Superfund exposure assessment manual. EPA 540/1-88-001. OSWER Directive 9285.5-1. Office of Emergency and Remedial Response. Washington, D.C.

U.S. Environmental Protection Agency (EPA). 1989a. Exposure factors handbook. EPA 600/8-89-043. Office of Health and Environmental Assessment. Washington, D.C.

U.S. Environmental Protection Agency (EPA). 1989c. Air Superfund national technical guidance series. Volume II: Estimation of baseline air emissions at Superfund sites. Interim Final Report. EPA 450/1-89-002. Office of Air Quality Planning and Standards. Research Triangle Park, NC.

U.S. Environmental Protection Agency (EPA). 1989d. Air Superfund national technical guidance series. Volume III: Estimation of air emissions from cleanup activities. Interim Final Report. EPA 450/1-89-003. Office of Air Quality Planning and Standards. Research Triangle Park, NC.

U.S. Environmental Protection Agency (EPA). 1989e. Air Superfund national technical guidance series. Volume IV: Procedures for dispersion modeling and air monitoring for Superfund air pathway analysis. Interim Final Report. EPA 450/1-89-004. Office of Air Quality Planning and Standards. Research Triangle Park, NC.

U.S. Environmental Protection Agency (EPA). 1989f. Risk assessment guidance for Superfund. Volume 1. Human health evaluation manual. Part A. EPA 540/1-89-002. Office of Emergency and Remedial Response. Washington, D.C.

U.S. Environmental Protection Agency (EPA). 1989g. Report to Congress on indoor air quality. EPA 400/1-89-001A. Office of Air and Radiation. Washington, D.C.

U.S. Environmental Protection Agency (EPA). 1991a. Risk assessment. *EPA Journal.* March/April 1991.

Van den Berg, M., M. Van Greevenbroek, K. Olie, and O. Hutzinger. 1986. Bioavailability of PCDDs and PCDFs on fly ash after semi-chronic ingestion by rat. *Chemosphere* 15:509.

Wallace, L., E. Pellizzari, T. Hartwell, C. Sparacino, and H. Zelon. 1983. Personal exposures to volatile organics and other compounds indoors and outdoors – the TEAM Study. NTIS PB83-121357. U.S. Environmental Protection Agency, Office of Research and Development. Washington, D.C.

Wallace, L., E. Pellizzari, T. Hartwell, R. Whitmore, C. Sparacino, and H. Zelon. 1986. Total exposure assessment methodology (TEAM) study: Personal exposures, indoor-outdoor relationships and breath levels of volatile organic compounds in New Jersey. *Environment International* 12:369–387.

Wallace, L., E. Pellizzari, T. Hartwell, C. Sparacino, R. Whitmore, L. Sheldon, H. Zelon, and R. Perritt. 1987. The TEAM (Total Exposure Assessment Methodology) study: Personal exposures to toxic substances in air, drinking water and breath of 400 residents of New Jersey, North Carolina and North Dakota. *Environmental Research* 43:290–307.

Wallace, L., E. Pellizzari, T. Hartwell, R. Whitmore, H. Zelon, R. Perritt, and L. Sheldon. 1988. California TEAM study: Breath concentrations and personal air exposures to 26 volatile compounds in air and drinking water of 188 residents of Los Angeles, Antioch and Pittsburgh, California. *Atmospheric Environment* 22:2141–2163.

20

Cost-Benefit Approaches

David R. Patrick

INTRODUCTION

This chapter discusses past uses of cost-benefit approaches in the regulation of environmental pollutants and the extent to which these approaches could be used in the future for toxic air pollutants. The consideration by EPA of the costs of control in developing regulations under various sections of the 1970 Clean Air Act Amendments has been subject of intense debate and litigation. A particularly controversial use of cost information is the balancing of the cost of a regulation against the adverse health and environmental effects reduced by the regulation. Although it may be economically appealing to balance these factors, the disadvantages make the process extremely difficult. This chapter does not support either view with respect to cost-benefit analysis; it only describes the potential uses and their advantages and disadvantages.

THE USE OF COST INFORMATION IN AIR POLLUTION REGULATION

Cost-benefit analysis has been defined (Sassone and Schaffer 1978) as 'an estimation and evaluation of net benefits associated with alternatives for achieving defined public goals.' Many analysts argue that the costs of an environmental regulation should be balanced with the benefits derived from that regulation to utilize more efficiently the nation's finite resources. Although this clearly makes sense economically, in practice it has been difficult to apply. Others argue equally convincingly that adverse health effects should be above the economic equation (i.e., that we cannot attach a price to adverse human health effects and suffering).

Legislative guidance, unfortunately, has been inconsistent. In the early years of environmental legislation, the Clean Water Act and the Solid Waste Act generally focused on technical standards to be met. The Toxic Substances Control Act and the Federal Insecticide, Fungicide and Rodenticide Act later provided for economic approaches, but less in terms of balancing and more in terms of deciding whether the social costs outweigh the benefits of a given use, and whether use should be allowed at all (Pedersen 1981). On the other hand, those in EPA implementing the Safe Drinking Water Act argued that although costs were not balanced

362

explicitly against benefits, in fact it was done implicitly by the decision-makers (Kimm, Kuzmack, and Schnare 1980).

The Clean Air Act is widely recognized as a legislative command to provide the benefits associated with cleaner air without explicitly balancing these benefits with the associated costs (Barnes 1983). Section 108 of the Act requires EPA to set national primary air quality standards at levels 'allowing an adequate margin of safety' and secondary standards at levels 'requisite to protect the public welfare from any known or anticipated adverse effects.' EPA also is required to provide state and local air pollution control agencies information relating to the cost of installation and operation of emission control technologies. However, there is no suggestion of balancing the costs of reaching these levels with the purported or anticipated benefits (Barnes 1983). On the other hand, section 317(a) requires the Administrator to prepare an economic impact assessment for actions taken under sections 111(b), 111(d), regulations for ozone and stratospheric protection, regulations for prevention of significant deterioration, and regulations for certain mobile source.[1] However, section 317 also states that where costs specifically are required to be taken into account, the adequacy or inadequacy of an economic impact assessment may be taken into consideration, but for the purposes of judicial review cannot be treated as conclusive with respect to a requirement that cost is taken into account (Barnes 1983). Thus, although EPA clearly must take into consideration the cost of achieving emissions reductions, it is apparently enough for EPA simply to evaluate the financial impact of the standard and consider relevant factors and make no errors in judgment.[2] A

quantitative cost-benefit analysis is not required and protection of the public health is the 'paramount consideration.'[3] This message was repeated in EPA's guidance for State Implementation Plans published in 1976.[4] EPA clearly stated that a new source could not be allowed to worsen existing air quality where a cost-benefit analysis indicated that the economic costs of controls are excessive in relation to the resulting air quality benefits. Nonetheless, EPA stated that although cost considerations could not override public health concerns, the agency was sensitive to the cost impacts of the Clean Air Act and economic impacts would be assessed to determine whether adjustments could be made consistent with the law.

The use of cost-benefit analysis in environmental regulations was promoted by Executive Order 12291,[5] Executive Order 12498,[6] and the growing authority of the Office of Management and Budget during the 1980s. Executive Order 12291 requires that the benefits of regulation outweigh the costs to society and that regulatory impact analyses should be conducted for every major rule. A major rule is defined as any action likely to result in an annual cost to the economy of at least $100 million, including a major increase in costs or prices, or significant adverse effects on competition, employment, and productivity. EPA later established guidelines (EPA 1983a) for preparing the regulatory impact analyses required by Executive Order 12291. These analyses were to consist of a consideration of alternate approaches, assessment of the benefits, analysis of the costs, and a final comparison of benefits and costs.

The use of cost-benefit analysis in regulating hazardous air pollutants has been embraced by EPA since its initial vinyl chloride regulations

---

[1] Importantly, section 317(a) did not require economic impact assessment for hazardous air pollutants.

[2] *Portland Cement Association* v. Ruckelshaus, 486 F.2d 375, 387, (D.C. Cir. 1973), *cert. denied*, 417 U.S., 921, followed in *National Lime Association* v. *EPA*, 627 F.2d 416, 429 (D.C. Cir. 1980).

[3] *Union Electric Co.* v. *EPA*, 427 U.S. 246, 272 (1976).

[4] 41 FR 55527, December 21, 1976.

[5] 46 FR 13193, February 19, 1981.

[6] 50 FR 1036, January 8, 1985.

(discussed in Chapter 16). However, it was not until 1983, in testimony by Administrator Ruckelshaus, that EPA's Administrator explicitly asked Congress for a more flexible test for hazardous air pollutant decision-making, a test that weighed such factors as the costs of control and the benefits of the risk reduced (Ruckelshaus 1983b). Congress did not act on the request.

Although costs were not explicitly considered in hazardous air pollutants promulgations, it is clear that EPA did consider costs implicitly in establishing emission standards under section 112. When section 112 was enacted in 1970, Congress established a two-step process for identifying and regulating hazardous air pollutants. EPA was required to list as hazardous those air pollutants that may cause or contribute to an increase in mortality or serious illness and then publish 'national emission standards' for sources that emit those pollutants. The resulting regulations were to achieve an 'ample margin of safety to protect the public health.' Although the decision to list a pollutant was viewed strictly as a health-based decision, the lack of specific language defining 'ample margin of safety' resulted in much less agreement on the extent to which costs and other factors should be considered in making decisions concerning the sources of hazardous air pollutants to be regulated and appropriate levels of control. Many argued that factors other than health effects could not be considered because they were not specifically provided in the Act. EPA, however, generally took the position that although health effects were the major determinant of regulatory need, other factors could be considered.

In the early 1970s, many argued that when no safe threshold of effect for exposure to a carcinogenic pollutant was known, standards should be set at a level of zero emissions. This view led to the Delany Amendment to the Food, Drug and Cosmetics Act that banned the use of food additives for which there was any evidence

of cancer. However, in the proposed standards for vinyl chloride,[7] a carcinogen assumed to have no health effect threshold, EPA concluded that a zero emissions requirement was not appropriate because it would result in closing an entire industry in order to eliminate a poorly defined health risk. In the proposed standards, EPA interpreted section 112 to 'authorize setting emission standards that require emission reduction to the lowest level achievable by use of the best available control technology in cases involving apparent nonthreshold pollutants, where complete emission prohibition would result in widespread industry closure and EPA has determined that the cost of such closure would be grossly disproportionate to the benefits of removing the risk that would remain after imposition of best available control technology.'

This interpretation was challenged by the Environmental Defense Fund but settled out of court when EPA agreed to propose new, more stringent standards and to set an eventual objective of zero emissions (Anderson and Beauchamp 1988). EPA did propose new vinyl chloride standards in 1977,[8] but withdrew them in 1985 on the basis that the costs were unreasonable and the technology to reduce significantly the emissions was not available, and that vinyl chloride emissions posed minor risks. The Natural Resources Defense Council subsequently petitioned the courts to review the decision.

In the early 1980s, EPA continued to articulate this view in its hazardous air pollutant actions. For example, in its promulgation of national emission standards for inorganic arsenic[9] EPA presented information on the cost, energy, and other effects of hazardous air pollutant regulations and appeared to utilize that information along with risk estimates in reaching decisions on whether, and the extent to which, to regulate specific source categories. However, no decision-making rule was specified by EPA.

[7] 40 FR 59532, December 24, 1975.
[8] 42 FR 28154, June 2, 1977.
[9] 51 FR 27956, August 4, 1986.

This came to an end in 1987 when the U.S. Court of Appeals announced a decision on vinyl chloride.[10] The court concluded that section 112 did not allow EPA to consider economics and technical feasibility in initially establishing hazardous air pollutant standards. The court found that EPA is required to make an initial determination of what is 'safe', and only then use costs and other information in establishing the margin of safety.

Before proceeding further, it may be useful to review the regulatory process that EPA generally used at that time for hazardous air pollutants. This process typically involved the following steps.

1. Air pollutants were identified that could be present in sufficient quantity in the ambient air to result potentially in an increase in mortality or serious health effects in humans.

2. Preliminary assessments of the possible health effects associated with, and potential sources of, the pollutants were conducted to estimate rough exposure and risk levels.

3. Based on the rough exposure and risk estimates, a decision was made whether to assess the candidate pollutant further. If the risks appeared sufficient to warrant regulatory concern and there were significant sources of the pollutant, further studies were initiated. However, no absolute risk level of concern was identified.

4. For the hazardous air pollutant candidates selected for further assessment, detailed health assessments and a thorough review of sources and estimated emissions then were conducted.

5. The detailed health assessments underwent scientific peer review. The source assessments identified significant sources and locations, expected emissions and availability of controls. However, cost and energy impact analyses normally were not conducted at this time.

6. A detailed exposure and risk assessment was conducted based on the peer-reviewed health assessment and the more thorough knowledge of the sources and estimated emissions.

7. A decision was made whether the population risks associated with any source categories were sufficient to warrant full-scale development of National Emission Standards for Hazardous Air Pollutants (NESHAPs) under section 112.

8. For source categories for which NESHAPs were determined to be warranted, in-depth engineering studies were conducted that included detailed analyses of the cost, energy and other impacts; available control technologies were identified considering all of these impacts.

9. On completion of these studies, the pollutant was listed as 'hazardous' under section 112 and regulations were proposed. Following public comment, final regulations were promulgated.

This standard setting process was complex and lengthy but viewed by EPA as essential to the effective use of both science and scarce resources in making decisions on the very large numbers of potentially hazardous air pollutants and source categories emitting those pollutants.

A TYPICAL COST-BENEFIT ANALYSIS

The goal of a typical cost-benefit analysis is to organize available information on costs, benefits, and economic impacts to clarify the trade-offs among various control options. These analyses can vary in the amount of detail and precision of the information. For air pollutants, the primary benefit is a reduction in adverse health effects such as lung cancer,

---

[10] *Natural Resources Defense Council, Inc.* v. *EPA*, 824 F.2d 1146[1987].

decreased lung function, and respiratory disease. Controlling some air pollutants also has the benefit of reducing soiling and nuisance, improving visibility, and reducing other adverse effects such as climate modification and acid deposition.

In performing a complete cost-benefit analysis, a substantial amount of information ideally is needed. Rarely is complete data available, however, and compromises must be made to develop the most accurate estimates possible given the available data. First, baseline costs are developed to provide an assessment of what happens without regulation. The costs of the regulation then are estimated. Three cost elements typically are estimated: (1) installed capital cost, (2) annual operating cost, and (3) cost-effectiveness. Cost-effectiveness is the annualized operating cost divided by the annual amount of pollution controlled and frequently is expressed in dollars per ton (English or metric) of pollutant. The control technology normally also is evaluated to determine its adequacy. In addition, the completeness of the cost estimate, the cost bases and retrofit costs are determined. Control system reliability should be evaluated to estimate the risk of system failure, that is one measure of the uncertainty of the cost estimate. In addition to the capital and operating costs to achieve a regulation, other costs also should be estimated to the extent that information is available. These can include, for example, the cost to the government to enforce compliance and the cost of unemployment programs where jobs are lost.

On the benefit side, attempts must be made to quantify health and environmental effects. Mortality often is valued on the basis of a statistical life saved. If mortality is valued directly, the compensation generally accepted for small risks typically is calculated. A common comparison is to the occupational setting where risks typically range from 1 in 100,000 to 1 in 1,000. Based on typical annual wages in that setting, the cost of a statistical life saved in this setting has been reported to range from $400,000 to $7,000,000 (1982 dollars). Where mortality is valued indirectly, the cost of the regulation is subtracted from the monetized benefits and the result is divided by the estimated number of statistical lives saved.

Morbidity (illness) typically is valued by estimating the direct cost, unless it is feasible to use willingness to pay. Direct cost considers medical costs, loss of work and earnings, and effects on future productivity. Because it usually does not consider pain or suffering, the benefits of reduced morbidity generally are underestimated. Environmental benefits typically also are based on direct costs (e.g., improvements in fishing, forests, and agriculture), increased travel stimulated by improvements in the environment, increased property values, or improvements in visibility or preservation of wildlife areas.

In attempting to quantify health effects, particular attention is paid to such factors as the following.

■ Level, frequency, and duration of exposure
■ Number of people exposed and their demographic composition
■ For noncarcinogens, expected numbers of adverse health effects and margins of safety
■ For carcinogens, risk to the most exposed individual(s) and the total population risks
■ Uncertainties

For broad-scale evaluations, intergenerational equity (i.e., the effects on the future) and larger-scale economic studies to assess the effects on product price often are considered. To the extent possible, marginal costs should be compared with marginal benefits. If a sufficient number of control options and data are available, a graph of costs and benefits should identify an optimum level (i.e., at the 'knee' of the curve). In addition, a cost-effectiveness curve should be developed, again if sufficient control option data are available. This helps to show if there is an optimum cost/control point. Cost-effectiveness analysis is particularly useful in identifying the levels of benefits and incremental trade-offs between successively more stringent levels of control.

PROBLEMS ASSOCIATED WITH COST-BENEFIT ANALYSES

There are a number of problems associated with the use of cost-benefit analysis. One of the most significant is that quantifying benefits (i.e., reductions in adverse health effects) is considerably less precise than quantifying costs. In addition, the valuation of death or illness is difficult both quantitatively and philosophically. Although no one would seriously consider the expenditure of the U.S. gross national product to reduce the possibility of one minor illness, few would argue against a regulation that saved many lives at a trivial cost. The dilemma lies in finding the appropriate point in between where both economic rationality and responsible social action are achieved. Studies in the late 1970s estimated costs per life saved for typical regulations in the range of $200,000 to $2,000,000 (Sinder and Worrell 1979; Fischoff et al. 1981). An unusual analysis (Fischoff et al. 1981) sought to establish an acceptable price that a person might pay to take a pill that had a one in one thousand (0.001) probability of causing instant painless death. The agreed on price was reported to be in the range of $1,000,000 to $4,000,000. Obviously this was a hypothetical analysis and in real life a choice such as this would depend greatly on the personality and the life situation of the individual making the choice.

It is particularly difficult to establish with certainty the likelihood of adverse health effects both in terms of level and severity. As discussed in other chapters, there are significant uncertainties in quantifying the extent and severity of diseases resulting from environmental exposures. EPA Administrator Ruckelshaus enumerated many of these uncertainties in his Congressional testimony in 1983 (Ruckelshaus 1983b). Although EPA and other regulatory agencies generally have made regulatory risk decisions in the face of these uncertainties, they have been reluctant to attempt to quantitatively balance these risks against costs that can be precisely quantified.

Administrator Ruckelshaus noted that safety is not the absolute removal of risk but, instead, a balance in directing social resources toward reasonable levels of protection and he acknowledged that many of the uncertainties might never be resolved (Ruckelshaus 1984).

As an example of the uncertainties and their impact on the decision-making process, a risk estimate for a specific situation frequently can vary by one to three factors of ten or more, depending on the available data and the assumptions made. Kimm, Kuzmack, and Schnare (1980) noted that it is hard to imagine a cost-benefit analysis whose conclusions would not change if one side of the balance were changed by a factor of one thousand or even one hundred. Still, there is no widespread agreement on how and the extent to which risks should be quantified, and there is little argument that there are far greater uncertainties in the risk estimates than the cost estimates.

On the other hand, regulatory decision-making without a common measure of comparison has led EPA to promulgate regulations that were inconsistent both within and across legislative mandates. This was confirmed when Administrator Thomas commissioned a study in 1987 to determine if the agency's priorities were appropriate (EPA 1987a). The issue that was addressed was whether the agency was applying resources to, and taking actions on, the most important environmental problems. The findings of this study were surprising to some, although much less so to those who advocate more careful attention to the cost of environmental regulation. The study found that as a result of differing legislative mandates, schedules and requirements, the agency was expending disproportionate resources on some environmental problems with little health benefit and insufficient resources on other environmental problems with potentially large adverse health effects.

POSSIBLE COST-BENEFIT APPROACHES FOR AIR TOXICS

Given that there are no widely accepted methods for directly comparing the costs of control with the improved health resulting from the application of those controls, the question is whether cost-benefit analyses can play any useful role in the regulation of toxic air pollutants. In general, the answer is yes. Basala (1988) noted that one advantage of controlling of toxic air pollutants is that it often results in other improvements in air quality. Chapter 24 describes other environmental programs where the pollutants of concern often are the same as those designated as toxic air pollutants. For example, control of air toxics that exist as fine particles can benefit the attainment of the particulate matter National Ambient Air Quality Standard (40 CFR Part 50). Many air toxics also are volatile organic compounds that are precursors to ozone and whose control can benefit attainment of the ozone ambient air quality standard. Better control of air toxics also possibly can reduce the potential for accidental releases of the extremely hazardous chemicals (to be regulated under section 112(r) of the 1990 Amendments) that also are on the list of toxic air pollutants. Finally, reduced emissions of air toxics means fewer of these pollutants available for intrusion into indoor environments.

Cost-benefit analyses can provide useful information for priority ranking of control options that otherwise appear equal. For example, the process can eliminate from consideration options that clearly are not cost-effective. In other instances, a cost-benefit analysis can identify potential bottlenecks in the supply system for control systems that might be widely used. On the other hand, situations exist where some emissions of an air toxic are necessary largely for economic reasons. For example, in some instances control of an air toxic can be achieved only through elimination of an offending chemical for which there are few acceptable substitutes. One specific case is coke oven emissions that have been implicated as carcinogenic to humans but that are difficult to control to a level that might represent an insignificant

risk. The 1990 Clean Air Act Amendments provide a complex and long-term program to deal with coke oven emissions; this is described in Chapter 25. In 1991, EPA indicated they were considering regulating coke oven emissions through a negotiation process that will attempt to achieve the maximum degree of control and reduction of risk within the industry's economic capacity to do so. The problem is that coke is as essential raw material, at least for the foreseeable future, in steel production.

Another case is perchloroethylene that is the most widely used dry cleaning solvent and that has been implicated as an animal carcinogen. There currently is no chemical that can be substituted for perchloroethylene and provide the same combination of cleaning power, safety and low cost. Given that society continues to want their clothes dry cleaned, the use of perchloroethylene will be controlled but some excess risk must be accepted.

References

Anderson, R. and M. Beauchamp, American Petroleum Institute (API). 1988. Economic analysis and the regulation of toxic air pollutants. Presented at the 81st Annual Meeting of the Air Pollution Control Association. Dallas, TX.

Barnes D.W. 1983. Back door cost-benefit analysis under a safety-first Clean Air Act. *Natural Resources Journal* 23:827.

Basala, A.C. 1988. Benefit-cost analysis for regulating air toxics: help or hindrance? Presented at the 81st Annual Meeting of the Air Pollution Control Association. Dallas, TX.

Fischoff, B., S. Lichtenstein, P. Slovic, S. Derby, and R. Keeney. 1981. *Acceptable Risk.* Cambridge, England: Cambridge University Press.

Kimm, V.J., A.M. Kuzmack, and D.W. Schnare. 1980. The questionable value of cost-benefit analysis: The case of organic chemicals in drinking water. Internal Memorandum. U.S. Environmental Protection Agency, Office of Drinking Water. March 1980.

Pedersen, W.F. 1981. Why the Clean Air Act works badly. *University of Pennsylvania Law Review,* pg. 1059.

Ruckelshaus, W.D. 1983b. Statement before the Subcommittee on Oversight and Investigations, Committee on Energy and Commerce, U.S. House of Representatives. November 7, 1983.

Ruckelshaus, W.D. 1984. Address before the Commonwealth Club of California, San Francisco. October 26, 1984.

Sassone, P.G. and W.A. Schaffer. 1978. *Cost-Benefit Analysis: A Handbook.* New York: Academic Press.

Sinder, J.A. and A.C. Worrell. 1979. *Unpriced Values: Decisions Without Market Prices.* New York: John Wiley and Sons.

U.S. Environmental Protection Agency (EPA). 1983a. Guidelines for performing regulatory impact analysis. EPA 230/01-84-003. Office of Policy Analysis. Office of Policy, Planning, and Evaluation. Washington, D.C.

U.S. Environmental Protection Agency (EPA). 1987a. Unfinished business: A comparative assessment of environmental problems. Office of Policy Analysis. Office of Policy, Planning, and Evaluation. Washington, D.C.

V

Control Methods

Gaseous Toxic Air Pollutant Control Technologies

Joseph Laznow and Avi Patkar

INTRODUCTION

As discussed in Chapter 1, more than two billion pounds of toxic air pollutants are emitted annually by industrial facilities throughout the United States. Concern that exposure to these emissions may result in serious illness or death led Congress to expand regulations on pollutants that are considered to be hazardous to human health and the environment. The result was section 112 of the 1990 Clean Air Act Amendments. Section 112 establishes a two-phased strategy to control emissions of 189 listed substances from major and area stationary sources. The first phase requires technology-based standards based on the maximum achievable control technology (MACT) applicable to specific source categories. The second phase is risk-based and involves further reductions if necessary to protect the public health with an ample margin of safety.

MACT standards can include process changes in addition to end-of-pipe control treatment (air pollution control). The purpose of this chapter is to describe available end-of-pipe technologies. Because particulate matter control technologies have been in use for many years, and are well understood (as an example, see A&WMA 1992), this chapter focuses on control technologies for gases. Emissions of gaseous toxic air pollutants can be reduced by numerous add-on control techniques, generally divided into two categories – destruction (e.g., combustion) and recovery (e.g., adsorption, absorption, and condensation). These are installed downstream of the source of the pollutant.

Because of the high cost of add-on controls, both for new sources and for retrofit of existing sources, consideration should be given to the elimination or reduction of toxic air pollutants in the process itself. These waste reduction practices include process modification, substitution of material, and other pollution prevention measures, and are discussed in Chapter 23.

EMISSION CONTROL TECHNOLOGIES

Combustion control devices are used to destroy the contaminant. The process of combustion also is referred to as oxidation or incineration. These devices include thermal incinerators, with recuperative or regenerative heat recovery, and catalytic incinerators. Recovery control devices also are used to collect toxic air pollutants before their final disposition, and can include recovery, destruction, or disposal. Recovery devices encompass carbon adsorbers, liquid absorbers, and condensers.

Incineration and carbon adsorption are the most widely used methods of industrial waste gas control and, when designed and operated properly, are expected to meet MACT emission standard requirements. They provide effective and economical control of air toxics as well as other volatile organic compounds (VOCs). Combustion devices are applied more commonly because they are capable of high destruction efficiencies for almost any type of gaseous organic pollutant. The emission reductions typically achieved with the application of various add-on control technologies, over a range of toxic air pollutant concentrations, are shown in Figure 21-1.

The specific applicability of each control device depends on many factors, including emission stream characteristics, the chemical and physical properties of the pollutant, required removal efficiency, and others. Ideally, regulatory compliance should be sought at the lowest overall cost. The emission stream and toxic air pollutant characteristics that most affect the applicability of each control

technique are identified and limiting values for each are presented in Table 21-1.

Selecting the most appropriate emission control technology is based on critical selection factors including:

- Gas stream composition and concentrations
- Process exhaust volume (i.e., maximum, average, and minimum flow rate), temperature, pressure, humidity, and reactivity
- Number of individual emission sources
- Existence of problem pollutants (e.g., particulate matter, chlorinated compounds, and heavy hydrocarbons)
- Hours of annual operation (i.e., percent operating time)
- Equipment location (e.g., indoors, outdoors, ground level, roof, and available space)
- Auxiliary fuel or energy costs
- Overall economics (i.e., capital and annual operating costs)

Thermal Incinerators

Thermal incineration is the direct application of combustion to destroy exhaust gases. It is a rapid, high-temperature, gas-phase reaction in

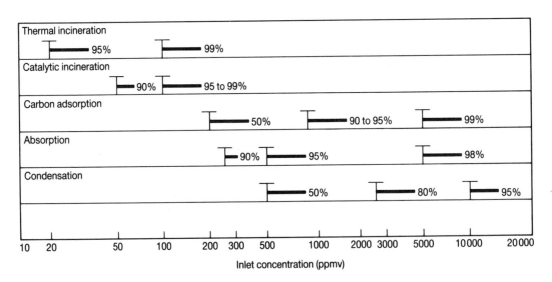

Figure 21-1. Approximate percent reduction ranges for add-on equipment. Source: EPA (1991f).

Table 21-1 Key emission stream and pollutant characteristics for selecting control techniques for organic vapors from point sources

Control device	Emission stream characteristics					Pollutant characteristics[a]			
	Organic concent.[b] (ppmv)	Heat content (BTU/scf)	Moisture content (%)	Flow rate (scfm)	Temp. (°F)	Molecular weight (lb/lb-mol)	Solubility	Vapor press (mmHg)	Adsorptive properties
Thermal incinerator	>20 (<25% LEL[c])			<50,000[d]					
Catalytic incinerator	50–10,000 (<25% LEL[c])			<50,000					
Flare		>300[e]		<2,000,000[f]					
Boiler/Proc. heater[g]		>150[h]		Steady					
Carbon adsorber	700–10,000 (<25% LEL[c])		<50%[i]	300–200,000	≤130	45–130			Use available sorbents
Absorber	250–10,000			1,000–100,000			Must be soluble		
Condenser	>5,000–10,000			<2,000				>10 at room temperature	

[a] Single organic or mixture.
[b] From organic content.
[c] For mixtures of organics and air; in some cases, LEL can be increased to 40–50 percent with proper monitoring and control.
[d] For packaged units; multiple package or custom units can handle larger flows.
[e] Based upon EPA guidelines for 98 percent destruction efficiency.
[f] lbs/hour.
[g] Applicable if unit is already on site.
[h] Total heat content.
[i] Relative humidity (organic concentration < about 1,000 ppmv).
Source: EPA 1991f.

which the pollutants are oxidized. Combustion is a chemical process resulting from the rapid combination of oxygen with various elements or chemical compounds resulting in the release of heat. Compared to other techniques, thermal incineration is broadly applicable. The constituents of the waste stream, excluding halogenated compounds, theoretically can be converted to carbon dioxide and water in the presence of sufficient heat, oxygen, and residence time. Early thermal incineration systems utilized direct combustion with no energy recovery. More recently, systems are designed with recuperative heat exchangers or regenerative systems that operate in a cyclic mode to achieve high energy recovery.

Thermal incineration is an effective destruction method for most organic pollutants in a gas stream. It is much less dependent on the pollutant and emission stream characteristics than other techniques. It also is applied to more serious emission problems that require higher destruction efficiencies, such as emissions from hazardous waste incinerators. However, exhaust streams containing large quantities of inorganic and metallic compounds should not be combusted and gas streams containing sulfur or chlorine compounds may be subject to additional control requirements. It is estimated that 80 to 90 percent of the 189 listed substances,

and approximately 75 percent of the total quantity of toxic air pollutant emissions, can be controlled effectively using thermal oxidation technologies.

In the thermal incineration process, the waste gas is brought to a sufficiently high temperature and held at that temperature for a sufficient residence time for the complete combustion (oxidation) of organic compounds in the waste stream. The temperature of a combustible material must be raised to 100°F or more above its ignition temperature (typically 1,000 to 1,400°F) and held at that temperature for 0.3 to 1.0 second (depending on the material). Incomplete combustion of many organic compounds can result in formation of aldehydes, organic acids, and other hazardous compounds that may require additional control, such as a scrubber, for capture before release to the atmosphere.

The primary components of a thermal incineration unit are a fan or blower that delivers the waste stream to the combustion chamber, burners that ignite the fuel and organics, the combustion chamber where mixing occurs and the air toxics are destroyed by oxidation, a heat recovery device, and the exhaust stack. Figure 21-2 is a schematic flow diagram of a typical thermal incineration system.

Because significant energy release accompanies the thermal oxidation process, heat

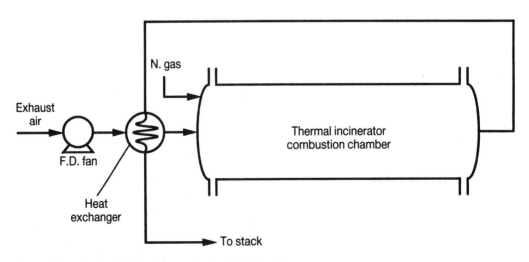

Figure 21-2. Schematic flow diagram of a thermal incineration system.

recovery systems generally are utilized to reduce the energy consumption of the incinerator. Thermal oxidation systems also incorporate heat recovery or catalytic operation (see below) to minimize auxiliary fuel requirements because fuel cost is the major operating expense for most incinerators.

Heat recovery is used to preheat the waste stream going to the combustion chamber by transferring heat from the incinerator exhaust before discharge to the stack. This is known as primary heat recovery and can be accomplished by using recuperative or regenerative devices that are differentiated by their ability to reuse the thermal energy released by the process itself. Secondary heat recovery involves the exchange of heat in the exhaust leaving the primary heat recovery device to another medium that is used in other processes. The level of thermal energy recovery can be as high as 95 percent, depending on the type of thermal oxidation system utilized. Because the heat content of the waste gas stream frequently is too low to support combustion, supplemental fuel usually is required to attain the temperatures necessary in the combustion chamber for the desired destruction efficiency. Natural gas is the preferred supplemental fuel, although distillate and residual oil is used, as well as liquified petroleum gas.

In order for the thermal incinerator to achieve the desired destruction efficiency, certain key parameters must be controlled. These include the combustion air flow rate, the waste stream flow rate, auxiliary fuel flow, residence (retention) time, combustion chamber operating temperature, oxygen availability, and the degree of turbulence between air and combustible materials. Residence time is the time required for the initiation and completion of the oxidation reactions. Operating temperature depends on the residence time, the oxygen concentration, the type and concentration of the contaminant involved, the type and amount of auxiliary fuel, and the degree of mixing. Good mixing within the chamber is necessary to enhance heat transfer and oxygen availability and to prevent stratification. Increasing the temperature offers the greatest potential for increased destruction

efficiency. The higher the temperature, the faster the oxidation reaction proceeds.

In general, the degree of oxidation depends on both temperature and residence time. Thus, at higher temperatures the same degree of oxidation can be achieved at shorter residence times, or a higher residence time allows the use of a lower temperature. This effect is demonstrated in graphical form in Figure 21-3. The choice between the higher temperature or longer residence time is based on economic considerations. Increasing residence time involves using a larger combustion chamber, resulting in higher capital costs. On the other hand, raising the operating temperature increases fuel usage, adding to the operating cost.

Combustion temperature, turbulent mixing, and residence time are the three most important variables affecting an incinerator's ability to destroy organic air pollutants. The temperature must be high enough to ignite the contaminant-fuel mixture, there must be sufficient turbulent mixing of the fuel, air and waste, and there must be adequate residence time for the reaction to occur. The effective destruction of organic compounds also is strongly dependent on the availability of oxygen in the thermal environment.

Although incinerators can accommodate minor fluctuations in flow, they are not well suited to waste streams with highly variable flow. This results because increased flow reduces residence time and results in poor mixing

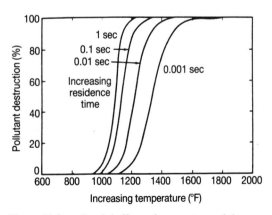

Figure 21-3.   Coupled effects of temperature and time on rate of pollutant. Source: A&WMA (1992).

and decreases the completeness of combustion. Additionally, increased flow can result in lower combustion chamber temperatures resulting in decreased destruction efficiency.

Thermal incinerators for general use operate at a temperature ranging from 1,200 to 1,600°F and are designed for a minimum residence time of 0.5 to 1 second in the combustion zone. Most hazardous waste incinerators, however, operate at 1,800 to 2,200°F in order to ensure complete destruction of the more hazardous materials in the waste stream.

Thermal oxidation at 1,800°F with retention times of 0.5 to 1 second provides destruction efficiencies of greater than 99 percent for such common toxic air pollutants as methanol, methyl ethyl ketone, methyl isobutyl ketone, and toluene. Other toxic air pollutants, including glycol ethers, xylene, and styrene, typically require temperatures of 1,800 to 2,000°F with retention times of 1 to 2 seconds to achieve destruction efficiencies of 99 percent or greater.

Halogenated toxic air pollutants, such as 1,1,1-trichloroethane, require a longer retention time in order to achieve the same level of destruction efficiency. In addition, downstream scrubbing often is required to control the hydrochloric acid that is formed in the combustion process. Sometimes caustic (sodium hydroxide solution) scrubbers are required and an afterburner may be needed for supplemental incineration, depending on the exhaust limitations imposed on the stack gases.

Thermal incinerators are commercially available as packaged units in many configurations and for flow rates ranging from a few hundred standard cubic feet per minute (scfm) to more than 100,000 scfm. They also can be custom designed for almost any application. The air toxics concentration of waste streams that can be effectively controlled by thermal incineration range from about 100 parts per million (ppm) up to 25 percent of the lower explosive level (LEL). Concentrations of flammable compounds typically cannot exceed 25 percent of the LEL for safety and insurance reasons. If the concentration of flammable compounds is greater than 25 percent, the gas stream should be diluted with air to reduce the potential for explosion.

In a recuperative incinerator the waste gas stream is preheated before entering the combustion chamber through use of heat exchangers that recover heat energy from the combustion chamber exhaust in a continuous, steady-state process. Plate-to-plate and shell-and-tube heat exchangers typically are used for recuperative heat recovery. The type chosen is based on waste gas flow rate, temperature of the incinerator exhaust stream, desired heat exchange efficiency, and economics. Depending on the configuration and the number of passes of heat exchange, that determines the thermal efficiency of the system, the thermal energy recovery can range from 60 to 75 percent. Fuel usage, therefore, can be reduced by up to 75 percent.

The primary advantages of the recuperative system are its simple design and lower initial capital cost, and the fact that these units typically are smaller and lighter in weight than other thermal devices. This allows space-saving configurations including the ability for installation on roofs. Recuperative systems are applicable especially to lower gas flows (under 15,000 scfm) and for intermittent operations.

On the negative side, recuperative systems can result in higher nitrogen oxides (NO_x) emissions, owing to direct flame contact at extremely high temperatures in the combustion chamber, and higher fuel cost as compared to other thermal oxidation units because of lower thermal recovery efficiencies. Autoignition or preignition of the gas stream, with its accompanying energy release, in the heat exchange plates or tubes also can cause an excessive temperature rise. This condition can result in thermal stress, tube deterioration, and exchanger failure, all significant maintenance problems. Because of the limitations and economics of the heat exchangers, the recuperative system typically is restricted to a maximum operating temperature of approximately 1,600°F; this can limit the ability to achieve maximum destruction of some critical pollutants.

In a regenerative incinerator the process exhaust is preheated through use of a ceramic heat exchange bed before entering the central combustion chamber. The energy from combustion is used to heat the ceramic bed to the combustion chamber outlet temperature. Large ceramic beds are used as massive heat sinks. When cool these beds first absorb heat energy from the exhaust gases exiting the central oxidation chamber; when hot they release this energy to the new incoming airstream. This continuous cycle of alternately storing and releasing heat permits an uninterrupted flow of contaminated gas through the system at all times. It is made possible by sequencing the gas flow through the beds via valve controls and by utilizing an odd number of chambers.

Aside from the ceramic media heat exchanger, regenerative systems operate in the same manner as conventional thermal incineration. Regenerative systems are recommended especially for larger gas flows (greater than 10,000 scfm), with low to medium concentrations (less than about 10 percent of the LEL), and in applications that operate a high percentage of the time. Regenerative systems can recover up to 95 percent of the energy in the exhaust gas, with a comparable reduction in fuel usage and cost savings. Streams with contaminant concentrations as low as 3 percent of the LEL can operate in a self-sustaining mode without auxiliary fuel. However, if the temperature of the exhaust stream after passing through the ceramic bed is not sufficient for combustion, auxiliary fuel must be added in the combustion chamber.

The primary advantages for the regenerative system arise from its basic design. The ceramic heat exchanger beds allow for operation at temperatures over 2,000°F, with retention times more than two seconds. The resulting greater destruction efficiency is beneficial for meeting regulatory requirements; it also is better at controlling odors. Induced autoignition in the ceramic heat exchange beds reduces the gas stream's exposure to high-temperature flames, minimizes or eliminates the generation of NO_x emissions, and reduces the amount of fuel required to reach combustion temperatures. Additionally, the ceramic construction of the heat exchange bed and internals of the thermal

oxidation system increases the ability to handle halogenated hydrocarbons with minimum concern for corrosion. However, the initial capital costs are higher and the units typically are larger in size and much heavier than catalytic or recuperative system, thus precluding installation on roofs.

Thermal incinerators can be used to control waste streams containing a wide variety of organic compounds and are technically feasible for controlling emissions with high destruction efficiencies for exhaust streams containing dilute concentrations of air toxics in a high volume of air. However, the costs associated with control of a dilute air stream can be very high owing to the supplemental fuel requirements.

Selection of the appropriate thermal oxidation system for a specific application involves evaluation of numerous factors. Analysis of operating conditions, such as air toxic concentrations and percent operating time, provides guidance for selecting the most economical system. Each candidate system for an application must be evaluated and reviewed on the basis of the advantages and disadvantages for both present and future requirements. Consideration must be given to thermal energy recovery efficiency because it limits the operating cost of a system. These efficiencies can range from 60 to 95 percent depending on the system design. Other considerations include maximum operating temperature capability, emissions of NO_x and carbon monoxide (both frequently regulated), system weight, installation flexibility, and overall capital, operating, and maintenance costs.

In summary, studies have indicated that well-designed and properly operated commercial thermal incinerators can achieve destruction efficiencies of most organic compounds greater than 99 percent. This destruction efficiency generally is achieved through minimum operation at 1,600°F with a nominal residence time of at least 0.75 seconds.

The major advantages of thermal incineration are that pollutants are permanently destroyed and virtually any gaseous organic stream can be incinerated safely and cleanly. The technology is mature, it generally provides dependable performance with low maintenance requirements, and it can handle fluctuations in loading. However, thermal incineration is not appropriate for batch operations. Supplemental fuel costs can be high and must be balanced against the additional capital cost of including heat recovery equipment. Also, gas streams containing halogenated compounds may require additional equipment downstream of the incinerator to control any resulting acidic gases.

The advantages and disadvantages of utilizing thermal incineration to control toxic air pollutant emissions are summarized below.

Advantages of Thermal Incineration

- Permanent destruction of gaseous pollutants
- Destruction efficiency greater than 99 percent
- Simple design concept
- Handles fluctuations in pollutant loadings and mix
- Low maintenance requirements

Disadvantages of Thermal Incineration

- High fuel (operational) costs unless heat recovery is utilized to the fullest extent
- Large quantities of halogenated organics cannot be handled without additional treatment (i.e., scrubber control system)
- High capital cost owing to large airflow rates

Catalytic Incinerators

Catalytic incineration is similar in design and operation to thermal incineration except that a catalyst is used to enhance the oxidation of air contaminants at much lower temperatures. Catalysts increase the rate of reaction and permit the reaction to occur at more favorable pressures. As preheated exhaust gases enter the catalytic bed, they are adsorbed onto the surface of the catalyst. The catalyst bed initiates and promotes the oxidation of the air toxics without being permanently altered itself. The combustion essentially is flameless, because the oxidation takes place on a catalyst bed surface rather than in a combustion chamber. A temperature of 600°F at the catalyst bed inlet typically results in an adequate initial reaction rate. However,

care must be taken to ensure complete combustion. The maximum outlet temperature also must be limited to prevent catalyst deactivation from overheating.

Typical catalysts include platinum, palladium, copper chromates, copper oxide, chromium, manganese, and nickel. Catalysts made of intermediate metals such as manganese dioxide, are less expensive and frequently can be substituted for the noble metals (i.e., platinum and palladium) when halogens or sulfur-containing compounds are not present. The catalyst beds are manufactured by depositing these materials in thin layers onto an inert substrate, usually a honeycomb-shaped ceramic. The beds can be monolithic or packed with spheres or pellets. The catalyst also can be configured in a plate and frame design, where blocks are held in place by a metal frame inside the incinerator.

The monolithic catalyst is a porous solid block containing parallel, nonintersecting channels aligned in the direction of the gas flow. Monoliths offer the advantage of minimal attrition resulting from thermal expansion and contraction during startup and shutdown, and they have a low overall pressure drop. Packed-bed systems contain catalyst particles supported either in a tube or a shallow tray through which the gases pass. Pelletized catalyst is superior where large amounts of contaminants such as phosphorus or silicon compounds are present.

For the catalyst to be effective, the active sites on which the organic gas molecules react must be accessible. The buildup of condensed polymerized material or reactions with certain metal particulate matter prevents effective contact from occurring. For example, in spray-coating operations catalyst effectiveness can be reduced from poisoning with coating aerosols. This occurs when the overspray enters the duct work and, subsequently, the incineration system.

Catalytic incinerators are preferred when large volumes of cool gas must be treated, or when the gas stream concentration is too low for combustion. They are highly effective with gas streams where emissions are well defined and are free of particulate matter. Catalytic incineration is not as broadly applied as thermal incineration because catalytic system performance is more sensitive to pollutant characteristics and process conditions. Coating formulations and process emissions must be carefully analyzed before selecting a system because some air toxics can coat or mask the catalyst. Additionally, fluctuations in flow rate must be kept to a minimum to prevent damage to the catalyst.

Combustion efficiencies of 95 to 98 percent can be expected in industrial applications such as printing, laminating, and coating operations. Higher efficiencies, up to 99 percent, require larger catalyst volumes and often higher temperatures. As with thermal incineration, hot exhaust gases can be passed through a heat exchanger to provide some of the preheat energy necessary to promote more efficient combustion. Because the catalyst allows the reaction to take place at lower temperatures, significant auxiliary fuel saving is achieved. Catalytic incineration requires dilution of the emission stream if the concentration of flammable gases is greater than 25 percent of the LEL. To prevent overheating of the catalyst bed, dilution also is required if the heat content of the emission stream is greater than 10 BTU per standard cubic foot (BTU/scf).

The primary components of the catalytic incineration unit are the fan, combustion mixing chamber, catalyst chamber, waste gas preheater (recuperative heating device), secondary heat recovery, and exhaust stack. The preheater is used to heat the incoming waste stream to the required oxidation temperature, usually about 600°F, but rarely exceeding 1,000°F because the catalysts cannot operate above this temperature. The mixing chamber is used to mix the hot combustion products from the preheat burner thoroughly with the process waste stream. This ensures that the stream sent to the catalyst bed has a uniform temperature. The combustion reaction takes place at the catalyst bed. A heat recovery device is used if supplemental fuel requirements are expected to be high. The maximum heat recovery for catalyst systems is about 70 percent, with an equivalent reduction in annual fuel (operating) costs.

Many parameters affect the performance of a catalytic incinerator. They include operating temperature, temperature rise across the catalyst bed, space velocity (i.e., the volume of gas entering the catalyst bed divided by the volume of the catalyst bed), pressure drop across the bed, contaminant concentration and species, and catalyst characteristics. The optimum operating temperature is dependent on the type of catalyst as well as the concentration and type of pollutant. Temperatures that are too high can reduce catalyst activity. Because space velocity is highly dependent on operating temperature, as it increases, the destruction efficiency will decrease. Additionally, the amount and type of pollutant determine the heating value of the waste stream and, thus, the amount of supplemental fuel required to maintain the desired operating temperature.

The type of catalyst used for a specific application is determined by the pollutants in the waste stream. Contaminants such as mercury, iron oxide, lead, phosphorus, bismuth, arsenic, antimony, tin, zinc, sulfur, halogens, silicon, organic solids, chlorinated compounds, heavy hydrocarbons, and particulate matter in the emission stream can poison or foul the catalyst and severely reduce its performance and useful lifetime. Some chlorine-resistant catalysts have been developed that can handle emission streams containing halogenated compounds.

Aerosols or solid particles that deposit on a catalyst also can mask, plug, or coat its surface. This reduces the catalyst's effectiveness by preventing contact between the pollutant and the catalyst surface and, thus, minimizing sites available for the oxidation reaction to occur. Therefore, it is important that the catalyst surfaces are clean and active to ensure optimum performance. The catalyst can be cleaned and regenerated by blowing superheated steam or air across it and elevating the unit temperature. Particulate matter in the waste stream generally is removed using filtration.

Thermal aging, resulting from erosion, attrition, and vaporization at higher temperatures, reduces catalyst life and activity. These processes deteriorate the catalyst's effectiveness at the required preheat temperature and reduce flow through the system. Consequently, additional fuel must be used to increase the gas stream temperature to achieve oxidation, resulting in a shortened catalyst life, lower destruction efficiencies, and higher operating costs. With proper operating temperature and adequate temperature control, thermal aging normally is slow and satisfactory system performance can be maintained for three to five years before replacement of the catalyst is necessary.

A well-operated and maintained catalytic incinerator can achieve air toxic destruction efficiencies of 95 to 98 percent, comparable to thermal incineration. Higher destruction efficiencies (98 to 99 percent) can be achieved but require larger catalyst volume or higher temperatures, and are usually designed for site-specific applications.

The contact time required between the contaminant and the catalyst for complete oxidation to occur normally is 0.3 second. The excess air requirements for catalytic incineration units usually are only 1 to 2 percent greater than the stoichiometric requirements. The pollutant content of the waste stream may be in the parts per million range to 25 percent of the LEL. Catalyst incinerators are commercially available as packaged units in sizes up to 100,000 scfm. The capital costs are comparable with those of thermal incinerators.

A potential concern associated with the use of catalytic incineration is variability of the waste gas flow rate and pollutant concentrations. A constant gas flow rate and concentration is recommended for optimum operation. Large fluctuations in flow rate or concentration will cause fluctuations in the control of the pollutants. If operating fluctuations are typical and cannot be avoided, other control methods may need to be considered.

Factors to consider in determining if catalytic incineration is suitable for control of air toxic emissions for a specific application include waste gas flow rate, the concentration and characteristics of the contaminants, and the presence of catalyst poisons and particulate matter. Catalytic incineration units can be designed to control high-volume, low-concentration waste streams. As with thermal incineration units, heat recovery and volume reduction techniques can be applied to constructing smaller units and decreasing capital and operating costs.

Catalytic incineration has several advantages over thermal incineration. The retention times for gases in catalytic units are significantly less, with the retention requirement for thermal incineration on the order of twenty to fifty times longer. As a consequence, catalytic incinerators are much smaller than thermal incineration units. The smaller size results in lower capital equipment and operating costs, and lower weight that can accommodate space savings configurations including roof mounting. The most attractive advantage is the lower supplemental fuel requirement resulting in fuel costs as much as half of that for thermal incineration. Catalytic incineration, therefore, can be less expensive than thermal incineration in treating emission streams with low pollutant concentrations.

Despite these advantages, catalytic incineration does not have the wide acceptance and use of thermal incineration. This is due primarily to the potential for catalyst poisoning. The poisoning or fouling of catalyst surfaces decreases performance and increases maintenance; spent catalyst replacement cost also is significant. The lower operating cost (i.e., lower fuel consumption) can be offset easily by this increased maintenance. Catalyst plugging and poisoning limits the applications and increases the risk of future replacement costs. The catalyst system also has limitations in inlet pollutant concentration because of the potential excessive temperature rise that can occur, and fluctuations in the air toxic content of the emission stream must be minimized to prevent catalyst damage. The major concern and limitation of the catalytic system, and probably the most critical, is that catalyst destruction efficiency deteriorates with time.

The advantages and disadvantages of utilizing catalytic incineration to control toxic air pollutant emissions are summarized below.

Advantages of Catalytic Incineration

■ Destruction efficiency greater than 95 percent
■ Permanent destruction of air toxics
■ Lower fuel requirements
■ Simple design concept
■ Handles fluctuations in loading and mix
■ Low maintenance requirements
■ Low combustion temperatures required across the catalyst bed
■ Small equipment size requirements

Disadvantages of Catalytic Incineration:

■ High fuel (operational) costs without heat recovery utilized to the fullest extent (however, much lower than thermal incineration)
■ Eventual catalyst replacement required; life is limited to three to five years
■ Poisoning of the catalyst
■ Incompatible with halogenated organics
■ Initial higher cost, although lower than thermal incineration

Carbon Adsorption

Vapor phase adsorption has been shown to be an effective means of recovering and reducing many chemicals from atmospheric discharge. It is a surface-related phenomenon that occurs when a gas is brought into contact with a solid substance, resulting in collection, capture, and retention of the contaminant (adsorbate) from the gas phase by the adsorbing granular solid (adsorbent). Diffusion mechanisms control the transfer of the adsorbate from the gas phase to the external surface of the adsorbent, from the external surface of the adsorbent to internal pores, and finally to an active site in the pores. The adherence of adsorbate to adsorbents result from the intermolecular electrostatic attractive (van der Waals') forces between them. The level of adsorption depends on the mass transfer gradient from the gas phase to the solid surface. Adsorbent particles are highly porous and have very large exposed surface area to volume ratios, thus allowing them to take up appreciable volumes of gases. As can be expected, an increase in the surface area of the adsorbent increases the total amount of gas that can be adsorbed (adsorption efficiency). Because adsorption is reversible, in that the adsorbate can be removed from the adsorbent by increas-

ing temperature or lowering pressure, it is widely used for solvent recovery.

Adsorption is particularly useful when the contaminant has sufficient value to warrant recovery. It can be used on waste gas streams of extremely low pollutant concentration and for contaminant gases that are noncombustible. Adsorption systems are capable of controlling a wide variety of organic compounds with high removal efficiencies (95 to greater than 99 percent).

The adsorption systems used to control organic emissions from waste gas streams involve two sequential processes. The first process involves the adsorption cycle, in which the waste gas stream is passed over an adsorbent bed for contaminant removal. The gas stream comes into contact with the adsorbent onto which the adsorbate is adsorbed. Gas molecules penetrate the pores of the adsorbent contacting the large surface area available for adsorption. The second process involves regeneration of the adsorbent bed, in which contaminants are removed using a small volume of steam or hot air, so that the adsorbent can be reused. The system includes multiple beds so that one bed is adsorbing the organics from the gas stream while at least one other bed is being regenerated. Multiple adsorber vessels are used for on-line regeneration of the bed material. Gas flows through one vessel, where the pollutants are removed, while another vessel is regenerated. A typical flow diagram for an adsorption/stripping system is shown in Figure 21-4.

Adsorbents are either natural or synthetic materials of microcrystalline structure, whose internal pore surfaces are readily accessible to gas molecules. Adsorbent materials most typically used are activated carbon, activated alumina (hydrated aluminum oxide), silica gels, and molecular sieves (synthetic zeolites). Activated carbon has greater affinity for nonpolar molecules. Silicon and aluminum oxides are utilized to adsorb polar molecules such as water, ammonia, hydrogen sulfide, and sulfur dioxide.

Commercial grades of activated carbon are produced from coconut shells, bituminous and lignite coals, petroleum coke, sawdust, and

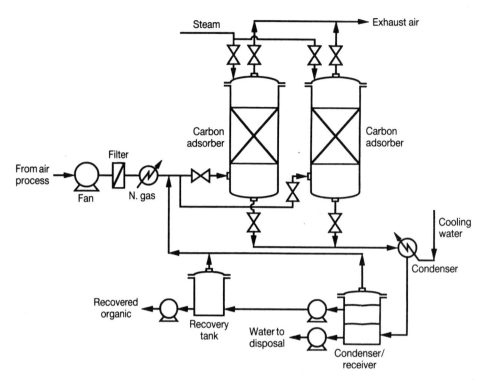

Figure 21-4.   Schematic flow diagram of an adsorption/stripping system.

other wood materials, that are subjected to heat treatment without air to remove volatile non-carbon materials. Activated alumina is produced by special heat treatment of precipitated or native aluminas or bauxite. It is used mainly for drying gases (dehydration) under pressure. The manufacture of silica gel consists of the neutralization of sodium silicate by mixing it with dilute mineral acid and washing the gel free of the salts produced during the neutralization reaction. It is used primarily for dehydration of gases in water. Molecular sieves are crystalline, dehydrated zeolites (aluminosilicates), in which atoms are arranged in a specific pattern. They have an internal complex structure of regularly spaced cavities with interconnecting pores of definite size. Adsorption takes place with molecules that are small enough in size and of suitable shape to enter the cavities through the pores. They are used most often for control of sulfur dioxide, nitrogen oxides, and mercury. Hydrated molecular sieves for adsorption of organic compounds also have been developed in recent years.

Activated carbon is the most versatile of the solid adsorbents and the most widely used for applications in air pollution control, solvent recovery, and odor elimination. Activated carbon systems have been developed that can handle large volumes of air and low concentrations of pollutants, and they can operate within reasonable pressure drop limitations. Both the internal and external surfaces of the carbon are used as adsorption sites.

Activated carbons vary in their affinity for different gases and their adsorption capacities. The adsorption capacity generally increases with molecular weight. Typical applications include gas streams containing nonpolar adsorbates with molecular weights over 45 but not greater than 130. These compounds predominantly are chlorinated compounds, alcohols, ketones, and aromatic compounds. High molecular weight compounds that are characterized by low volatility are more completely adsorbed on activated carbon. However, the affinity of carbon for these compounds makes it difficult to remove (desorb) them during regeneration of the carbon bed. Hence, carbon adsorption is not applied to compounds with boiling points above 300°F and molecular weights greater than 130. Additionally, because highly volatile materials with molecular weights less than about 45 do not adsorb readily on carbon, adsorption typically is not used for controlling emission streams containing such compounds. The properties of typical granular gas adsorbent carbons are given in Table 21-2.

Because activated carbons are sensitive to emission stream conditions, pretreatment of the emission stream may be necessary using such methods as filtration, cooling, or dehumidification. When liquid or solid particles, high boiling organics, or polymerizable substances are present, pretreatment procedures are required. Entrained solids such as dust, lint, and other particulate matter also will plug the carbon bed over a period of time. Particulate matter can be controlled easily by a cloth or fiberglass filter. Entrained liquids can cause operating problems and should be removed by means of mist eliminators or packed filters. Ideal adsorption conditions for activated carbons are a relative humidity less than 50 percent and gas stream temperatures greater than 130°F.

Regeneration is the process of desorbing the contaminants from the carbon. Regeneration of the carbon bed usually is initiated before breakthrough. Breakthrough is that point in the adsorption cycle when the carbon bed approaches saturation and the concentration of organics in the effluent stream begins to increase dramatically, indicating a drop in system efficiency. Economic considerations dictate whether the adsorbent is regenerated or replaced after the adsorption efficiency begins to decline. Regeneration is accomplished by reversing the conditions that are favorable to adsorption (i.e., by increasing the temperature or reducing the system pressure). The regeneration time, the time needed to strip the carbon bed of all adsorbed organic compounds, typically ranges from thirty minutes to two hours.

The ease of regeneration depends on the magnitude of the forces holding the contaminant to the surface of the carbon. The most common method of regeneration is through steam stripping. Low pressure, superheated steam is introduced into the carbon. The steam is needed to provide a sufficient concentration gradient to promote mass transfer of the adsorbate from the carbon bed. The steam releases heat as it cools. That heat then is available for adsorbate vaporization. Another regeneration method is the use of hot inert gas or hot air. Compounds with boiling points higher than 300°F do not desorb well when low-pressure steam is used. In such cases, hot inert gas can be used. After regeneration, the desorbing stream then would consist of a gas stream with a high contaminant concentration. This gas stream then could be sent to an incinerator for final destruction of the pollutants. With either steam or hot air regeneration, the desorbing agent flows through the bed in the direction opposite to the waste stream. This desorption scheme allows the exit end of the carbon to remain contaminant free.

Separation must be considered when large quantities of organics dissolve in the condensate from the steam regeneration process because most local authorities will not allow such water to be discharged into municipal sewage treatment systems. A complex distillation system may be required to fractionate the mixture into separate components, or a single-batch still can be designed to fractionate the organics from the water. Desorbed organics and steam typically are condensed in a shell-and-tube heat exchanger. Water-insoluble substances can be removed by simple decantation; those that are water soluble can be recovered using distillation. After regeneration, the carbon bed is cooled and

Table 21-2   Properties of activated carbon

Property	Typical value
Bulk density	22–34 lb/ft³
Heat capacity	0.27–0.36 BTU/lb/°F
Pore volume	0.56–1.20 cm³/g
Surface area	600–1,600 m²/g
Average pore diameter	15–25 Å
Regeneration temperature (steaming)	100–140°C
Maximum allowable temperature	150°C

Source: A&WMA 1992.

dried. On-line regenerative systems have been shown to be more dependable, effective, and economically viable means of air pollution control than nonregenerative systems that require the removal of the adsorbent for off-site regeneration.

The most commonly used adsorption units for air pollution control are fixed, or rotating, regeneration carbon beds and disposable or rechargeable carbon canisters. The canister types are used for controlling low flow rates (i.e., less than 100 cfm) and would not be used to control high-volume flow rates.

The components of a typical fixed-bed carbon adsorption system to remove toxic air pollutants include a fan or blower (to convey the waste gas into the carbon beds and to overcome the pressure drop across the carbon bed), at least two fixed-bed carbon adsorption vessels (one vessel being used for adsorption while the other is being regenerated), a stack for the treated waste gas outlet, a steam valve for introduction of the desorbing stream, a condenser to condense the steam and captured pollutants and cool the condensate mixture to 100°F, and a decanter for separation of the condensate and water.

The degree of adsorption efficiency is dependent on the operational conditions of the exhaust gas system (i.e., system temperature, system pressure, humidity of waste gas, and gas flow rate), as well as characteristics of both the adsorbate and adsorbent. The type and concentration of contaminants in the waste stream determines the adsorption capacity of the carbon (pounds of material adsorbed per pound of carbon). In general, adsorption capacity increases with the compound's molecular weight or boiling point. There is a relationship between pollutant concentration and the carbon adsorption capacity. As concentration decreases, so does the adsorption capacity. However, the capacity does not decrease proportionately with the concentration decrease. Therefore, adsorption capacity still exists at very low pollutant concentration levels. Important factors when choosing an appropriate adsorbent include the amount of adsorbent needed, temperature rise of the gas stream due to adsorption, ease of

regeneration, and the useful life of the adsorbent.

In selecting adsorption equipment for gaseous toxic air pollutant control there must be a provision for sufficient residence time. The contaminants require sufficient contact with the active sites of the carbon to allow enough time for mass transfer to occur. This is especially true if there are many molecules (high concentration streams) competing for the same sites. Residence or contact time of the contaminants with the active sites can be increased by using larger carbon beds. However, the pressure drop across the system will increase, resulting in increased operating costs. Lower gas velocities result in increased retention time. The gas velocity also must be sufficiently low to prevent the movement of the adsorbent in the fixed bed.

The adsorption capacity of the carbon, and thus the performance of the adsorber, are directly related to the temperature of the gas stream. Adsorption efficiency decreases with increasing temperature. At elevated temperatures, the vapor pressure of the contaminants increase, reversing the mass transfer gradient. Contaminants then would be more likely to be desorbed back into the gas phase than to be retained on the carbon. At lower temperatures the vapor pressures are lower, favoring retention of the contaminants by the carbon. The temperature of the gas stream should be in the range of 60 to 120°F. Because adsorption is an exothermic process, some method of heat removal from the carbon may be necessary, depending on the amount of contaminant being removed from the gas phase. To prevent excessive bed temperatures resulting from the adsorption process and oxidation reactions in the bed, high concentrations frequently must be reduced. This usually is done by condensation or dilution of the emission stream ahead of the adsorption process.

Design considerations for fixed-bed carbon adsorption of a VOC include the following.

VOC Content

The VOC content of the gas to be treated can be defined in units of pounds of VOC per 1,000 scf. The conversion from ppmv (volume) can be made by using the first part of Figure 21-5 or the following relationship.

$$\frac{\text{lb of VOC}}{1,000 \text{ scf}} = \frac{\text{ppmv} \times \text{VOC molecular weight (lb/lb-mole)}}{359 \times 1,000}$$

(21-1)

Operating Capacity

The determination of the actual operating capacity of carbon is affected by many factors. For a particular VOC feed stream, the operating capacity can be determined from operating industrial units, carbon manufacturers, or pilot-plant studies. Some reported operating capacities are shown in Table 21-3. Usually, a safety factor of two is used between the saturated adsorption capacity and the operating capacity.

Carbon Requirements

Carbon requirement can be defined as pounds of carbon per 1,000 scf and depends on the VOC content of the gas to be treated and the operating capacity of the carbon.

$$\frac{\text{lb of carbon}}{1,000 \text{ scf}} = \frac{\text{VOC content (lb of VOC/1,000 scf)} \times 100}{\text{operating capacity (lb of VOC/100 lb carbon)}}$$

(21-2)

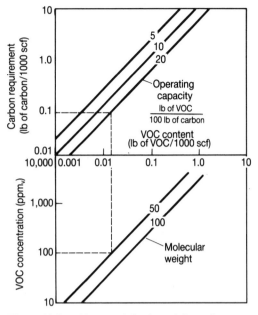

Figure 21-5. Nomograph for determining carbon requirement.

Figure 21-5 is a nomograph for determining the carbon requirement for systems with a variety of VOC concentrations, molecular weights, and operating capacities. Systems with high VOC operating capacities per pound of carbon and high concentrations of VOC in the process waste gas can have the same carbon requirement as systems

Table 21-3   Important properties for design of adsorption system for selected organics

Chemical	Boiling point (°F)	Liquid molar volume (cm³/mole)	Lower explosion limit (ppm)	Reported operating capacity (lb/100 lb C)
Benzene	176	95	14,000	6
Butyl acetate	260	152	14,000	8
n-Butyl alcohol	210	105	17,000	8
Cyclohexane	178	118	13,000	6
Ethyl acetate	171	106	25,000	8
Ethyl alcohol	173	61	33,000	8
Heptane	208	163	12,000	6
Hexane	154	140	13,000	6
Isopropyl acetate	199	129	18,000	8
Methyl acetate	135	83	31,000	7
Methyl alcohol	149	42	60,000	7
Toluene	231	118	14,000	7
Trichlorotrifluorethane	118	120	NA	8
Xylene	291	140	10,000	10

with low VOC operating capacities per pound of carbon and low concentrations of VOC in the waste gas. Typical carbon requirements will vary from 0.1 to 10 pounds per 1,000 scf.

Flow Rate
Typical flow rates of gas to be treated range from 300 to 100,000 scfm. Up to 10,000 scfm, single vertical adsorbers are used. Above 10,000 scfm, single horizontal adsorbers are preferred; above 30,000 scfm, multiple horizontal units are used. There is no practical limit to flow rate because multibed systems operate with beds in simultaneous adsorption cycles.

Gas Velocity (Superficial)
Linear velocities of 50 to 100 feet per minute (fpm) normally are used in bed designs. At higher velocities the bed pressure drop becomes too high for standard blowers. At lower velocities, the bed becomes too large and expensive to fabricate and transport. If inlet concentrations are low, the bed area required for the volume of carbon needed usually permits a velocity at the high end of this range.

Bed Depth
Fixed-bed carbon adsorption systems normally have bed depths of 1.5 to 6 feet, although under certain conditions the depth may be as small as a few inches. The minimum bed depth of 1.5 feet is assigned for the adsorption zone. The maximum bed depth of 6 feet is considered to be reasonable for the pressure developed by standard blowers.

Pressure Drop
The pressure of the feed gas affects the design of the system in terms of the power requirements of the blower. A pressure drop of 2 to 6 inches H_2O per foot of carbon bed normally is experienced with a flow velocity of 50 to 100 fpm through the carbon bed. Bed pressure drop versus superficial velocity for two different carbons is shown in Figure 21-6.

Loading Time Per Bed
Loading time (the time needed for carbon to become saturated with organics) is calculated

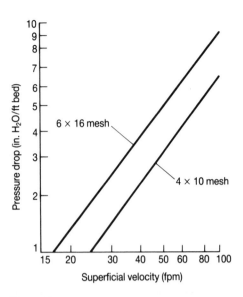

Figure 21-6.   Bed pressure drop for BPL carbon.

from bed depth, bulk density, gas velocity (superficial), and carbon requirement as

$$\text{Loading time} = \frac{\begin{array}{c}\text{bed depth (ft)}\\ \times \text{ density (lb/ft}^3)\\ \times 1{,}000\end{array}}{\begin{array}{c}\text{velocity (fpm)} \times 60 \text{ (min/h)}\\ \times \text{ carbon required}\\ (\text{lb/1,000 ft}^3)\end{array}}$$

(21-3)

Bed Area and Pounds of Carbon Per Bed
The bed area (per bed) is calculated by dividing the flow rate (scfm) by the superficial gas velocity through the bed. The pounds of carbon (per bed) are most conveniently determined by calculating the bed volume and multiplying by the bed density.

Bed area (ft^2) = flow rate (scfm)/gas velocity (fpm)

Carbon (lb) = bed area (ft^2) × bed depth (ft) × density (lb/ft^3)

Minimum Cycle Time

The minimum cycle time for the system normally is dependent on the time required to regenerate, dry, and cool the bed. Steam regeneration rates typically are 0.5 to 1 pound of steam per minute per square foot (lb/min/ft^2), but can be as high as 4 lb/min/ft^2. These rates correspond to regeneration times of thirty to ninety minutes at steam usage rates of 0.3 or 1 pound of steam per pound of carbon. Cooling and drying the beds normally can be done in fifteen minutes; however, the minimum possible time for regeneration, drying, and cooling is one hour. Because the bed is loaded for the same period of time, the total minimum cycle for one bed thus takes two hours. In addition, because two beds are usually employed, each carbon bed must have enough adsorption capacity to allow sufficient time for the other bed to be regenerated, dried, cooled, and placed back in adsorption service before breakthrough.

Spontaneous combustion of fixed beds of carbon can occur whenever the gas stream contains oxygen and compounds, such as ketones, aldehydes, or organic acids, that are easily oxidized in the presence of carbon. Hot spots exceeding the autoignition temperature of the carbon may develop in the bed. Hot spots also may form rapidly under abnormal bed conditions, such as uneven bed depth due to carbon movement and stagnant bed areas. To ensure safe operation with these pollutants, on-line monitoring of carbon monoxide and carbon dioxide concentrations in the effluent stream should be considered.

The system pressure also will influence adsorption efficiency. Increase in the gas phase pressure will promote more effective and rapid mass transfer of the contaminants from the gas phase to the carbon. Therefore, the probability that the contaminants will be captured is increased.

The relative humidity or moisture content of the gas phase also will affect adsorption efficiency. Although high relative humidity hinders adsorption, methods are available to reduce these effects. One method is to mix the gas stream with lower humidity ambient air.

This process will lower the cost of carbon regeneration, but will result in higher gas volume with an increased expenditure for capital equipment and higher power consumption. An alternate method is to cool and condense the water vapor and reheat the gas stream. Although water vapor is not adsorbed preferentially to the contaminants, the presence of water vapor in the gas phase has been demonstrated to have a detrimental effect on the adsorption capacity of the carbon. Humidity control may be necessary for gas streams with a relative humidity greater than 50 percent and lower emission stream concentrations (i.e., less than 1,000 ppm). The effect of humidity or moisture in the gas phase generally is insignificant for toxic air pollutant concentrations greater than 1,000 ppm.

The pollutant concentration of waste streams controlled by carbon adsorption units can range from the parts per billion level to as high as 25 percent of the LEL. If flammable gases are present, the pollutant concentrations again may be limited by insurance requirements to less than 25 percent of the LEL. There is no obvious practical limit to flow rate because multibed systems operate with multiple beds in simultaneous adsorption cycles. Carbon adsorption recovery efficiencies of 95 to greater than 99 percent have been demonstrated to be achievable in well-designed and well-operated units. To ensure that breakthrough does not occur, continuous monitoring of the outlet bed concentration is recommended.

The major advantage of carbon adsorption is that the product is recovered for reuse. Fuel and power costs are low, and a high air toxic removal efficiency can be attained with low inlet concentrations. The major technical disadvantages of carbon adsorption occurs when the gas stream contains water-soluble solvents, such as methyl ethyl ketone or methanol. Wastewater produced from the condensed regeneration steam containing these soluble organic compounds generally requires treatment before disposal.

The advantages and disadvantages of utilizing carbon adsorption to control emissions are summarized below.

Advantages of Carbon Adsorption

- Pollutant removal efficiency greater than 99 percent
- High overall control efficiency for exhaust gases containing low pollutant concentrations
- Products can be recovered for reuse in process
- Predictable pollutant composition in exhaust gases, thus simplifying design and condensate treatment
- Excellent control and response to process changes
- Fuel costs are minimal

Disadvantages of Carbon Adsorption

- Wastewater from carbon regeneration may require treatment, especially if water-soluble organics are present
- Adsorbent progressively deteriorates in capacity as the number of cycles increases
- Prefiltering of the gas stream may be required to remove particulate matter to prevent plugging of the adsorbent bed
- High operating and maintenance requirements

Absorption

Absorption is a gas-liquid contacting process that utilizes the preferential solubility of a contaminated gas in a liquid solvent where one or more of the constituents of the gas dissolve in a nonvolatile liquid or liquid-solid slurry. Control systems that employ liquid media to remove gaseous pollutants typically are referred to as wet scrubbers. The rate of absorption in these systems is determined by the rates of diffusion of the pollutant through both the nonabsorbed gas and the absorbing liquid.

Absorption of the gaseous contaminant by a liquid occurs when the liquid contains less than the equilibrium concentration of the gaseous component. The driving force in the absorption process is the gradient between the actual liquid concentration and the concentration in equilibrium with the gas phase concentration. For a given set of operating conditions, the rate of absorption is determined by the difference between the actual contaminant concentrations in the solvent and the corresponding equilibrium concentrations. When the liquid concentration is below the gas equilibrium concentration, the contaminants move from the gas into the liquid. The mass transfer rate (rate at which the pollutant mass is transferred from one phase to another) at which absorption occurs depends also on the amount of exposed gas-liquid surface area. The mass transfer rate is a function primarily of liquid recirculation rate, absorption system size and shape, and liquid distribution.

Wet scrubbing effectively removes gaseous pollutants by various mechanisms including: (1) condensation of gases from the airstream into a cold liquid contact column that utilizes spraying, packing, or other mechanisms of contact; (2) dissolution of the contaminant gas in the scrubbing liquid and then removal by absorption; and (3) reaction with chemicals such as strong oxidizing agents in the gas or liquid phase to form water-soluble, odorless, and potentially less toxic waste products.

Absorption can occur by physical and chemical means. Physical absorption occurs when the absorbed compound dissolves in the solvent. Chemical absorption occurs when a reaction occurs between the absorbed compound and the solvent. The contact between the gas and liquid phases occurs in units that are designed to enhance the transfer zone as much as possible. This zone usually is provided by utilizing a packed tower or a plate scrubber where the gas stream generally passes countercurrent to the liquid flow.

Absorption depends on a variety of specific physical and chemical properties as well as operating characteristics of the system. The absorption rate (i.e., efficiency of absorption) for removing pollutants from a gaseous stream depends on the physical properties of the gas-liquid system (e.g., gas stream concentrations, diffusivity, viscosity, and density) and the absorber operating conditions (e.g., temperature, flow rate of the gaseous and liquid streams, and liquid to gas ratio). The absorption rate can be enhanced by lower temperatures, a greater contacting surface, higher liquid to gas ratios, increased turbulence, larger mass diffusion coefficients, and higher concentrations in the gas stream. The most important factor in determining the applicability of this process is the availability of a suitable solvent. The pollutant to be removed should be readily soluble in the solvent for effective absorption rates.

The absorption process has significant potential for application to the control of volatile organic toxic air pollutants. The mechanism is used widely in controlling sulfur dioxide, hydrogen sulfide, and hydrogen chloride, in addition to gas streams containing aromatics, chlorinated compounds, fluorinated compounds, alcohols, acids, substituted aromatics, aldehydes, and esters.

Other important considerations in the application of absorption as an air pollution control technique are the disposal of the absorber effluent and auxiliary equipment requirements. The scrubbed gas generally is vented to the atmosphere through a stack. The scrubbing solvent can be used once or it can be recycled before final disposal. If the absorber effluent contains organic compounds and is discharged ultimately to a municipal sewer system, wastewater treatment may be required to remove chemical reaction products. This is to prevent converting an air pollution problem into a potential water pollution problem. Facilities where this is the case should consider regeneration of the spent solvent or disposal in an environmentally acceptable manner.

Absorption also is used widely as a raw material or product recovery technique by separating and purifying gaseous streams containing high concentrations of organics. As an air emission control technique, it is used less often for organic gases and much more commonly for inorganic gases, such as ammonia, hydrogen sulfide, carbonyl sulfide, carbon disulfide, metals with hydride and carbonyl complexes, chlorides, oxychlorides, and cyanides. With both types of contaminants, however, removal efficiencies greater than 98 percent are achievable with absorption systems.

The use of absorption as the primary control technique for organic pollutants is subject to a number of problems and limitations. For example, the applicability of absorption for organic emissions control is based on the availability of vapor/liquid equilibrium data for the specific organic/absorbing solvent system. Such data are necessary for proper design of absorber systems. Additionally, in organic control applications, attempting to meet emission reduction requirements with absorption alone could require impractically tall absorption towers, extremely long contact times, and high liquid to gas ratios; all of these options may not be cost-effective. Therefore, organic gas absorbers generally are more effective when they are used in combination with other control devices such as incinerators or carbon adsorbers.

As previously noted, the most important factor in the absorption process is the solubility of the pollutant gas in the absorbing solvent used. The ideal solvent should not be volatile, corrosive, flammable, or toxic, and it should be chemically stable, readily available, and inexpensive. Typical solvents used for inorganic gas control include water, low-volatility hydrocarbon oils, sodium hydroxide solutions, amyl alcohol, ethanolamine, weak acid solutions, and hypochlorite solutions.

Water is considered the ideal solvent for inorganic gas control. It offers distinct advantages over other solvents, the main one being cost. Additives commonly are used to increase chemical reactivity and absorption capacity. Water typically is used on a once-through basis and then discharged to a wastewater treatment system. Water-soluble organics also can be recovered by distillation or destroyed by incineration. For water-insoluble organics, high boiling point oils can be used as the absorbent. The organics then can be recovered by thermal stripping and condensing, and the oil can be recycled to the absorber. If fuel oil is a suitable solvent, it can be used to absorb the organics and then burned in a boiler. The effluent may require pH adjustment to precipitate metals and other pollutants as hydroxides or salts. These typically are less toxic and can be more easily disposed.

Absorption normally is conducted in a scrubbing device, where the liquid is designed to flow down the absorption chamber and the gas stream flows upward; this is called countercurrent flow. The types of equipment commonly used for gas–liquid contact operations include packed towers, plate or tray towers, spray chambers, and venturi scrubbers. A packed tower is filled with packing material that is designed to expose a large wetted surface area to the gas stream. Packed columns generally have lower pressure drops, are simpler and cheaper to construct than other types, and are preferable for liquids that foam excessively. Plate towers use plates or trays that are arranged so that the gas stream is dispersed through a layer of liquid on each plate. Plate columns are less susceptible to plugging, have a lower overall weight, have fewer problems with channeling, and are less susceptible to damage during temperature surges. In a spray tower, the gas mixture is contacted with a liquid spray. In a venturi scrubber, the gas and liquid streams come into contact at the throat of the venturi nozzle.

Scrubbers are designed to provide maximum contact between the gas and liquid streams in order to increase the mass transfer rate between the two phases. Significant gas-media contact can be achieved by utilizing various mixing mechanisms, including impingement, spraying, atomization, and agitation. Gas-media contact also can be enhanced by the use of special packing materials. This packing can be constructed of plastic or ceramic materials and must be replaced periodically. However, as surface area increases so does the pressure drop through the packed bed. This can increase operating costs.

Fixed-bed packed scrubbers are the most widely used absorption system. The efficiency of absorption depends on the height of the packed tower because the longer the gas path through the scrubbing medium, the greater the probability that contaminant gases will be transferred and absorbed. Determination of the appropriate absorber system variables (e.g., column diameter and height) is dependent on the individual gas–liquid equilibrium relationship for the specific organic/absorbing solvent system and the type of absorber to be used. The cross-sectional area of the tower determines the capacity of the gas flow rate, with design flow rates typically in the range of 1,000 to 10,000 cfm. Figure 21-7 provides a schematic flow diagram of a countercurrent packed-column absorption.

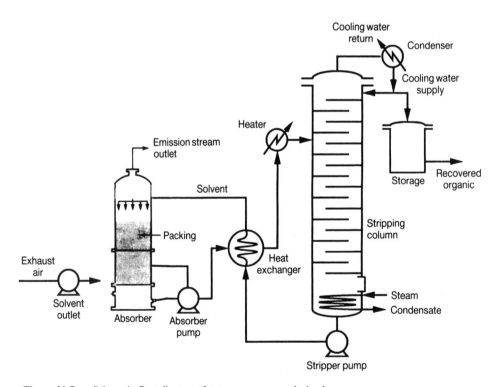

Figure 21-7. Schematic flow diagram of a countercurrent packed-column absorption/stripping system.

The spray-chamber control system is similar to the packed-bed tower but does not use packing to promote liquid to gas contact. Spray towers operate with pressure drops of one to two inches of mercury, whereas packed towers operate at pressure drops of one to eight inches of mercury.

Airflow through scrubbers can be counter-current, crosscurrent, or cocurrent. Counter-current flows operate with the gas and liquid streams flowing in opposite directions (i.e., the gas stream moves upward while the scrubbing liquid moves downward through the packing) and are the most commonly used. They provide the highest theoretical removal efficiencies. Crosscurrent systems have the gas flowing across the liquid stream. The airstream moves horizontally through a packed bed and is irrigated by the scrubbing liquid flowing verti-cally down the packing. These are less likely to clog, require lower liquid to gas ratios and are used most frequently when the gases are highly soluble in the liquid. Cocurrent systems operate with both streams in the same direction. These systems have lower pressure drops, do not flood, and are more efficient in removing small (i.e., submicron) particles. An even distribution of liquid across the cross-sectional area of the scrubber vessel is essential to achieve max-imum removal efficiency. The packing also must provide a high free-flow area to minimize the gas-side pressure drop. Both packing and liquid distribution systems must be resistant to plugging and fouling.

The packing is the critical component affect-ing the performance of this type of emission control system. Important factors in the packing selection process include durability and corro-sion resistance, free space per cubic foot of packed space, wetted surface area per unit volume of packed space, frictional resistance to the flow of gas, packing stability and structural strength, and weight per cubic foot of packed space.

The major advantages of absorption systems are removal efficiencies greater than 98 percent for gas streams with high organic concentra-tions and highly soluble contaminants as well as the possible recovery of contaminants for reuse

in the primary process. A 90 percent removal efficiency for lower-concentration streams (300 ppm) also is achievable. Wet scrubbing ideally is suited to the treatment of a wide range of airflows ranging up to more than 100,000 cfm. Particulate matter in the gas stream may cause fouling and plugging of the packing, and the blowdown stream from the scrubber may require treatment and proper disposal.

The advantages and disadvantages of utiliz-ing absorption to control emissions are summa-rized below.

Advantages of Absorption

- High removal efficiencies
- Low pressure drops
- Operation in highly corrosive atmospheres
- Capable of achieving high mass-transfer efficiencies
- Low capital cost
- Small space requirements
- Expandable and versatile
- Ability to collect particulate matter as well as gases
- Low energy consumption

Disadvantages of Absorption

- Blowdown stream disposal may cause problems
- Particulate matter deposition may cause plugging of the bed or plates
- Fiberglass-reinforced plastic (FRP) construction, generally used for absorbers, is sensitive to temperature
- High maintenance costs

Condensation

Condensation is an operation in which one or more volatile components of a vapor mixture are separated from the remaining vapor by being changed to the liquid phase through extraction of the heat of condensation. Con-densation as an emission control method often is used with auxiliary air pollution control equipment. For example, condensers can be located ahead (upstream) of absorbers, incin-erators, or carbon beds to reduce the con-taminant loading to the more expensive control device. Condensers also can remove vapor components that might adversely affect the operation of other equipment or cause corrosion problems, or they can be used simply to recover a valuable material that otherwise would be destroyed. When condensers are used alone, as in vapor control from bulk gasoline terminals, refrigeration often is employed to obtain the low temperatures necessary for acceptable removal efficiencies.

The suitability of condensation as an emission control technique is dependent on the following factors: pollutant concentration in the inlet (usually greater than 1 percent), removal efficiency required, and recovery value of the pollutant. In general, condensation should be considered for gas flows of 2,000 scfm or less, pollutant concentrations of 5,000 ppm or more, and required removal efficiencies of 90 percent or less. When a condenser is used to control emissions, it usually is operated at the constant pressure of the control source – normally atmospheric. The two most common types of condensers that operate at atmospheric pressure are surface and contact condensers.

Most surface condensers are of the shell-and-tube type shown in Figure 21-8. The coolant usually flows through the tubes, and the vapors condense on the outside tube surface (shell). The condenser vapor forms a film on the cool tube and drains away to a collection tank for storage or disposal. The coolant used depends on the temperature required for condensation. Chilled water, brine, and refrigerants normally are used in surface condenser operation. Air-cooled surface condensers also are available

and usually are constructed with extended surface fins. When the cool air passes over the finned tubes, the vapors condense inside the tubes.

In contrast to surface condensers, where the coolant does not contact the vapors or the condensate, contact condensers usually cool the vapor by spraying an ambient-temperature or slightly chilled liquid directly into the gas stream. The coolant usually is water, although in some situations an organic raw material can be used. These devices are uncomplicated, as shown by the typical design in Figure 21-9. Most contact condensers are simple spray chambers that usually are baffled to ensure good contact.

Both devices have advantages and disadvantages for a given operation. Final selection usually will be based on the following comparative information.

1. Contact condensers are simple in design, and are less expensive to install than surface condensers. They have advantages in corrosive situations and when particulate matter has to be removed.
2. Contact condensers can be more efficient than surface condensers in removing VOCs from a vent gas stream because they act as an absorber as well as a condenser when the VOCs are soluble in the coolant.

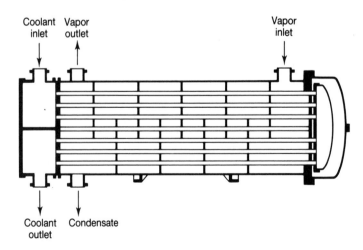

Figure 21-8.   Schematic diagram of a shell and tube surface condenser.

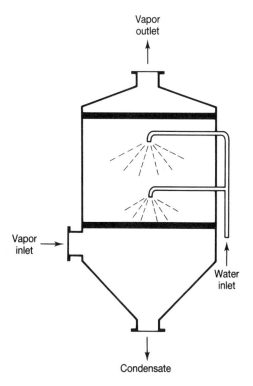

Figure 21-9.   Schematic diagram of a contact condenser.

3. Spent coolant from contact condensers usually cannot be reused directly and, therefore, can be a secondary emission source or a wastewater disposal problem.
4. The coolant from surface condensers is not contaminated and normally can be recycled.
5. Surface condensers may be equipped with more auxiliary equipment, such as a refrigeration unit, to supply the coolant; consequently, they generally require more maintenance than contact condensers.
6. Surface condensers can be used to recover valuable and marketable substances directly from the gas stream.

Although contact condensers can be highly efficient in removing organics from a vapor stream, they can create additional wastewater emission control problems downstream. The downstream treatment system may include steam stripping, biodegradation, or carbon adsorption. All of these treatment systems are expensive and increase capital costs significantly. Because of these liabilities, only surface condensers are evaluated further in this section. However, special cases may make contact condensers an attractive alternative.

Figure 21-10 illustrates one of the surface condenser configurations that can be utilized as an emission control device. The coolant is supplied to the condenser by a refrigeration unit. Temperatures as low as minus 80°F may be required in order to obtain the high removal efficiencies needed. The major equipment required for the condenser system includes a shell-and-tube heat exchanger, a coolant supply, a recovery tank for the condensate, and a pump to discharge the recovered material to storage or disposal.

The removal efficiency of a condenser system is determined by the amount of reduction in the organic partial pressure in the gas stream as it passes through the condenser. This is accomplished by reducing the temperature of the gas stream and condensing some of the organics. The temperature necessary to obtain a particular vapor concentration or removal efficiency is dependent on the vapor pressure of the

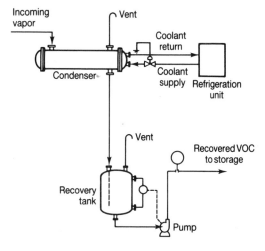

Figure 21-10.   Basic surface condenser of a volatile organic control system.

components. For a two-component vapor mixture in which one of the components is noncondensable (e.g., air), condensation will begin when the temperature reached is such that the vapor pressure of the volatile component is equal to its partial pressure. The point at which condensation first occurs is called the dew point. As the vapor is cooled further, condensation continues as long as the partial pressure stays equal to the vapor pressure. The less volatile a compound (i.e., the higher the normal boiling point), the smaller the amount that can remain as vapor at a given temperature. Figure 21-11 shows the vapor pressure dependence on temperatures for selected compounds.

The calculation methods for a gas stream containing multiple organics are complex, particularly when there are significant departures from the ideal behavior of gases and liquids. As a simplification, the temperature necessary for control by condensation can be approximated roughly by the weighted average of the temperatures necessary for condensation of each condensable organic in the gas stream at concentrations equal to the total organic concentration.

If water is present in the treated gas stream or if the organic has a high freezing point (e.g., benzene), normal design practice will require the use of an intermittent heating cycle for removal of ice or frozen hydrocarbons in a continuous system operated at low temperatures. Intermittently operated systems may be allowed simply to heat up, with ambient heat used for deicing. Alternatively, a separate chiller or condenser can be used to remove water, followed by a low-temperature chiller or condenser to control organic emissions.

Condenser design calculations are based on

$$Q = UA \, \Delta T_m \qquad (21\text{-}4)$$

Where Q = total heat transferred (BTU/hour)

U = overall heat transfer coefficient (BTU/hour/ft^2/°F)

A = heat transfer surface area (ft^2)

ΔT_m = log mean temperature difference between coolant and gas stream (°F)

In condenser design calculations, A is determined by reordering the equation.

$$A = \frac{Q}{U \, \Delta T_m} \qquad (21\text{-}5)$$

For the preliminary design of the condenser and refrigeration systems for VOC control, the following conditions can be assumed.

1. The temperature of the initial gas stream entering the condenser is 80°F, and the stream contains only VOCs and air (i.e., two components).
2. The gas stream outlet temperature is calculated assuming the partial pressure of the VOC in the outlet gas streams equals the vapor pressure of the VOC at that temperature.

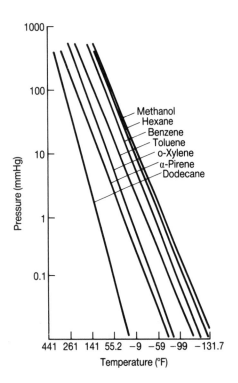

Figure 21-11.   Vapor pressures of selected compounds versus temperature. Source: Dreisbach (1952).

3. The air specific heat is 0.24 BTU/pound/°F; the VOC specific heat is 0.50 BTU/pound/°F; and the VOC latent heat is 200 BTU/pound.

4. The overall heat transfer coefficient is 5.0 BTU/hour/ft^2/°F.

5. The coolant inlet temperature is 10°F less than the gas stream final temperature.

6. The coolant temperature rise through the condenser is 25°F.

7. The system heat losses are negligible.

An optimum design requires the selection of a combination of equipment and operating conditions that will satisfy emission control requirements at minimum overall cost. Often a change that reduces the cost of one element (i.e., condenser size) must be balanced against the effect it has on other costs (i.e., refrigeration requirements). Although detailed design procedures are beyond the scope of this handbook, some of the more significant design considerations are discussed briefly in this section. For simplicity, these factors are discussed specifically as they apply to a condenser system using a surface condenser with a mechanical refrigeration unit designed to control VOC emissions at various concentrations in air.

The advantages and disadvantages of utilizing condensation to control emissions of toxic air pollutants are summarized below.

Advantages of Condensation

- Pure product recovery
- Water used as the coolant in an indirect contact condenser does not contact the contaminated gas stream and can be reused after cooling
- Small space requirements

Disadvantages of Condensation

- Low removal efficiency
- Coolant requirements may be extremely expensive
- Not applicable to low VOC concentrations

References[1]

Air & Waste Management Association (A&WMA). 1992. *Air Pollution Engineering Manual.* A.J. Buonicore and W.T. Davis, eds. New York: Van Nostrand Reinhold.

Dreisbach, R.R. 1952. *Pressure-Volume-Temperature Relationships of Organic Compounds, 3rd Edition.* Sandusky, OH: Handbook Publishers, Inc.

U.S. Environmental Protection Agency (EPA). 1991f. Handbook: Control technologies for hazardous air pollutants. EPA 625/6-91/014. Office of Research and Development. Research Triangle Park, NC.

---

[1] Additional reference material is provided in the bibliography.

22

Fugitive Emissions Controls[1]

Thomas A. Kittleman, Paul R. Jann, Jerry M. Schroy, and Seshasayi Dharmavaram

INTRODUCTION

Fugitive emissions are air releases not confined to a stack, duct, or vent. Fugitive emissions generally are low concentration but can be a substantial portion of the total air pollution discharge from a source. Generally, they come from small diffuse sources of large open surfaces and are not easily controlled. Examples of fugitive emissions are shown below. Only equipment leaks and secondary emissions from pits, ponds, and lagoons are considered in this chapter.

■ *Equipment leaks* – leakage of liquids or gases can occur at connections, joints, and interfaces. For instance, process fluids can escape from a piping system by permeation, diffusion, or laminar flow through valve stems, pump seals, and flanges. Equipment leakage generally is unintended.

■ *Secondary emissions* – the evaporation of a contaminant from a pit, pond, or lagoon is considered a fugitive loss because of the difficulty in controlling these emissions. The fugitive emissions are called secondary emissions because they involve losses due to handling the wastes and utilities associated with a production unit not emissions from the primary processing equipment. For instance, secondary emissions can occur by evaporation of materials discharged as a liquid spill on the ground or as wastewater to a ditch. Emissions from employees' cars and emissions from generating electricity for a process also are secondary emissions. In the extreme, all emissions except direct process emissions are secondary emissions.

■ *Windblown dust* – these emissions result when particulate matter is picked up and carried by the wind. Exposure of solid materials, allowing them to become windblown, can occur for many reasons. The most common source is agriculture, but activities such as mining and travel on dirt or gravel roads can generate dust that can become windblown.

■ *Other emissions* – many activities result in emissions that can be categorized as fugitive. Process operations such as tank filling, liquid

---

[1] Portions of this chapter have been excerpted from *Improving Air Quality: Guidance for Estimating Fugitive Emissions from Equipment* (CMA 1989) and are used with permission of the Chemical Manufacturers Association.

transferring, and packaging give rise to fugitive emissions. Many activities of everyday life also produce fugitive emissions. Examples are emissions from paints, deodorants, hair spray, and mothballs. Emissions from a single one of these are small; collectively, however, they account for substantial man-made fugitive emissions.

EQUIPMENT LEAKS

Background

Past Regulatory History
In the 1970s and early 1980s a number of equipment-leak field studies were conducted at industrial facilities. These included petroleum refineries (Boland et al. 1976; Eaton et al. 1978; Hanzevack 1978; Honerkamp et al. 1979; Hustvedt and Weber 1978; Rosebrook et al. 1978; Taback 1978; Wetherold and Provost 1979; Wetherold and Rosebrook 1980), synthetic organic chemical manufacturing industry (SOCMI) plants (Blacksmith, Harris, and Langley 1980; EPA 1980e; Langley et al. 1981; Langley and Provost 1981; Langley and Wetherold 1981), and natural gas processors (EPA 1983c). These studies led to several federal regulations, including New Source Performance Standards (NSPS) under section 111 of the Clean Air Act,[2] and National Emission Standards for Hazardous Air Pollutants (NESHAPs) under section 112 of the Act.[3] Except for the vinyl chloride NESHAP, all the federal requirements are similar and are based on leak detection and repair (LDAR) principles. Several states also have adopted requirements for existing sources in nonattainment areas[4] that are based on LDAR principles.

Leak Detection and Repair (LDAR)

All LDAR programs require that certain equipment (e.g., valves and pumps) be checked periodically for leakage. The check is made using a portable instrument such as a flame ionization detector (FID). Equipment found leaking must be repaired expeditiously. LDAR programs are considered work practices, in regulatory language, and violations occur when the prescribed procedures are not followed. Some regulations have quality control principles embodied in them allowing for less frequent inspection when monitoring shows consistent good performance; others do not.

EPA Method 21[5] is the test procedure typically used for LDAR inspections. A portable FID is the most common inspection device. The instrument probe is placed against the interface of a static seal (e.g., valve or flange), or one centimeter from a dynamic seal (e.g., pump), and moved all the way around the potentially leaking joint. A meter reading of 10,000 parts per million (ppm) generally has been considered as the definition of a leak. A reading below 500 ppm has been labeled as 'no detectable emission' for some regulatory purposes.[6]

When a leak is discovered, an initial repair attempt is required expeditiously (e.g., within one to five days). If the initial attempt fails, further repair attempts must be made shortly thereafter (e.g., within fifteen days). If all attempts fail, repairs can be delayed until the next shutdown if certain conditions are met (e.g., emissions for shutdown and repair would exceed emissions due to delayed repair).

---

[2] 40 CFR Part 60, Subpart VV (Synthetic Organic Chemical Manufacturing Industry), Subpart DDD (Polymers Manufacturing), and Subpart GGG (Petroleum Refining).

[3] 40 CFR Part 61, Subpart F (Vinyl Chloride Manufacturing) and Subpart V (Volatile Hazardous Air Pollutants).

[4] See the discussion on the National Ambient Air Quality Standards attainment program in Chapter 26.

[5] 40 CFR Part 60, Appendix A.

[6] A rigorous instrument calibration generally is not required and actual concentrations can be as much as a factor of 10 different from the meter reading.

Past Emission Estimates

Two types of field data have been collected and used for emission estimates: (1) screening and bagging studies, and (2) screening alone (CMA 1989; Radian 1986; Schaich 1991). Screening generally is based on EPA Method 21 and provides an indication of leakage but does not quantify it. Bagging involves enclosing and quantifying leakage from the component of interest. Several correlations between screening values and leak rates have been determined. However, the quality of these varies greatly.

Screening has been used in association with bagging in two ways. One way is to correlate the measured emission or bagging result with screening measurements made immediately before and after bagging. The relationship between leak rate and screening measurement can be used directly to predict emissions for a process where screening alone was done. Any given bagging and screening measurement may be substantially different from that predicted from the relationship derived from the measurements. However, when there are many individual components, the total or overall average becomes a good estimate. The second way is to

group components into screening measurement ranges and bag representative samples in each range (Tasher 1982).

Langley and Wetherold (1981) conducted one of the most thorough and useful early studies. Representative components were bagged in the range of screening values selected. The data obtained were used to quantify the relationship between screening values and leak rate (Figure 22-1 is an example). One of the reasons for the wide variability in Figure 22-1 is probably the fact that screening values were not response factor-corrected. Response factors are discussed below and in Appendix C. The data also have been used (Tasher 1982) to show how the portion leaking, at a leak definition of 10,000 ppm, relates to the average leak rate in a process (see Figure 22-2). Response factor correction is not as crucial in the accuracy of Figure 22-2. Screening values were used only to group components within a given plant and emissions were determined from bagging results for screening groups. Data were collected on such important factors as how often effective repairs can be expected, how often repaired components begin to leak again (shown in Table 22-1),

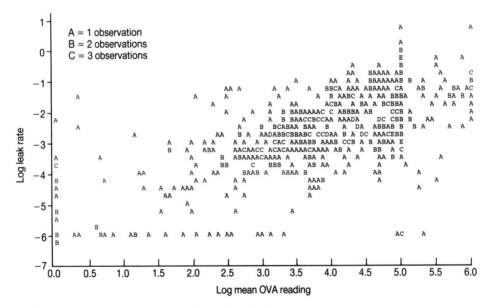

Figure 22-1. Log$_{10}$ leak rate versus log$_{10}$ OVA reading for all valves - zeroes set to one for OVA or 0.000 001 for leak rates. This figure is a rough interpretation of the original data. A = one observation, B = two observations, C = three observations.

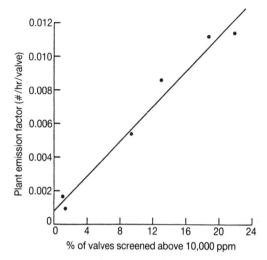

Figure 22-2. Relationship between emission and percent of valves leaking for six plants in EPA's maintenance study.

and how often components not leaking begin to leak in the future (shown in Table 22-2).

EPA used screening data from several studies (Blacksmith, Harris, and Langley 1980; EPA 1980e; Langley et al. 1981) to predict the 'average' leakage from different types of components (e.g., valves in gas/vapor service and valves in light liquid service). EPA concluded at

that time that, except for valves in gas/vapor service, there was no significant difference in the average leakage at chemical plants when compared to petroleum refineries. As a result, many of the early emission factors were derived from, and applied across, diverse industry sectors. Also, compound emission factors such as leak/no leak (i.e., one emission estimate for components that screen above 10,000 ppm and a different one for components that screen below 10,000 ppm) and stratified emission factors (e.g., three ranges) rely on the same data as the 'average' factors. More recent studies apply results more directly to processes where data are collected (Berglund, Romano, and Randall 1990; CMA 1989; Davis 1989; EPA 1988g; Enviroplan 1991; Hess and Kittleman 1983; Schaich 1991).

Leak Detection and Repair (LDAR) Programs

Most equipment leak regulations, including the planned maximum achievable control technology (MACT) regulations for hazardous air pollutants required under section 112(d)(3) of the 1990 Clean Air Act Amendments and developed as part of a negotiated regulation (EPA 1991s),

Table 22-1 Occurrence rate estimates for block and control valves by process[a]

	Number of sources followed[b]		30-day estimate	95% C.I.[d]	90-day estimate	95% C.I.[d]	180-day estimate	95% C.I.[d]
	<1,000[c]	1,000≤ X <10,000[c]						
Block valves								
Cumene units	110	4	2.0	(0.3,6.2)	6.0	(0.8,18)	11.7	(1.5,32)
Ethylene units	129	7	2.3	(0.8,4.8)	6.9	(2.3,14)	13.3	(4.6,26)
Vinyl acetate units	69	4	0	(0,0.7)	0	(0,2.1)	0	(0,4.1)
All units	308	15	1.2	(0.5,2.2)	3.6	(1.6,6.5)	7.0	(3.1,13)
Control valves								
Cumene Units	25	0	0	(0,2.6)	0	(0,7.5)	0	(0,15)
Ethylene Units	123	6	1.7	(0.4,3.9)	5.0	(1.2,11)	9.7	(2.5,21)
Vinyl acetate units	40	1	1.3	(0.2,3.2)	3.7	(0.7,9.3)	7.3	(1.3,18)
All units	188	7	1.6	(0.5,3.2)	4.6	(1.6,9.3)	9.0	(3.1,18)

[a] A leak is defined as having occurred if it initially screened <10,000 ppmv and later screened >10,000 ppmv.
[b] Unit 1 sources not identified by valve type and, therefore, not included in this count.
[c] ppmv = parts per million by volume.
[d] C.I. = Confidence Interval.

Source: Langley and Wetherhold 1981.

Table 22-2   Valve leak recurrence rate estimates

Time	Recurrence rate estimate (%)	95 percent confidence limits on the recurrence rate estimate
30-day	17.2	(5,37)
90-day	23.9	(7,48)
180-day	32.9	(10,61)

Source: Langley and Wetherhold 1981.

are based on LDAR principles. LDAR programs with fixed detection periods will reduce leakage at some plants. However, such programs become less effective as maintenance and operating practices improve. Newer regulations try to strike a balance between monitoring to promote and verify low leak rate performance and monitoring as an abatement technique. Such monitoring programs are based on quality control (QC) monitoring principles. Plants are encouraged to do better than the minimum required as a way to reduce the monitoring burden. If a plant's procedures fail to achieve good results, more monitoring is required.

Two things are needed to define any LDAR program: (1) a leak definition, and (2) the time allowed between leak checks. Where QC principles are used, a performance criterion also is needed to define the time between leak checks. There is more than one way, consistent with QC principles, to apply incentives and disincentives and to ensure that good performance levels are achieved.

The MACT for equipment leaks applies to a variety of process equipment. For hardware such as sample collectors and open-ended lines, compliance equipment is specified. For example, open-ended lines are not allowed and sample systems are not allowed to discharge line purges. In cases where small numbers of equipment are involved (e.g., agitators and compressors), fixed-period monitoring has been specified. With these types of equipment, the incentive for good performance may not be obvious (i.e., when small numbers are involved, incentives are not great). However, for the most numerous types of equipment, such as valves

and connectors, the incentives for good performance are clear.

Monitoring of all valves is required on a schedule ranging from once a month to once a year, depending on how large a portion exceeds the 500 ppm leak definition (i.e., better performance leads to less frequent monitoring). In contrast, connectors require monitoring on one- to four-year intervals. In this case, extended monitoring periods are allowed after successive demonstrations of good performance at the shorter periods. Less frequent monitoring of flanges than of valves is warranted because once flange problems are diagnosed and treated leaks are much less likely to recur, and monitoring costs for flanges are much greater than for valves.

State and local regulations vary widely but tend to emphasize frequent monitoring without benefits for good performance. Some even specify that any leakage is a violation. In the case of dynamic seals (e.g., pumps), some agencies specify hardware. Generally, however, a less intrusive and more cost-effective inspection program, similar to the planned MACT requirement, is allowed.

Emissions Estimating Techniques

Several techniques are available for estimating fugitive emissions from specific processes. The choice of a technique is determined by the cost and accuracy required for each combination of source (e.g., valves and pumps) and service categories (e.g., gas/vapor and heavy liquid). In general, the better the desired accuracy the higher the cost. Lower-cost techniques frequently give emission estimates that may be one to two factors of ten higher than actual emissions.

The available techniques, in increasing order of accuracy and cost, are: (1) SOCMI factors, (2) leak/no leak emission factors, (3) stratified emission factors, (4) established leak rate/ screening value curves, and (5) process-specific leak rate/screening value curves. These techniques and their underlying statistical bases are described below.

SOCMI Factors

EPA developed average emission factors for the SOCMI based on a study of twenty-four individual process units. This technique is the least costly but also the least accurate. The only requirement is a count of the number of components in the process within the source/service category of interest. The estimate of fugitive emissions using this option is obtained by multiplying the number of components within the source/service category of interest by the corresponding average SOCMI emission factor shown in Table 22-3.

The use of this technique to estimate fugitive emissions is predicated on three basic assumptions.

1. If the process were screened, the percentages of components that would screen above 10,000 ppm are shown in Table 22-4.
2. If the process were screened, the average screening values for the process that would screen below and above 10,000 ppm, respectively, are shown in Tables 22-5 and 22-6.
3. The relationships between screening value and leak rate are shown in Tables 22-8 and 22-9.

The only way to validate the first two assumptions is to screen the process components in the source/service category of interest. Note that this is the minimal requirement to use the leak/no leak, stratified, or established curves

Table 22-3   SOCMI emission factors

Source	Service	Average SOCMI emission factor[a]
Valves	Gas/vapor	0.0123
	Light liquid	0.0156
	Heavy liquid	0.0005
Pumps	Light liquid	0.1087
	Heavy liquid	0.0471
Compressor seals	Gas/vapor	0.5016
Pressure relief valves	Gas/vapor	0.2288
Connections (flanges)	All	0.0018
Open-ended lines	All	0.0037

[a]lbs per hour per source.

Table 22-4   Percentage of components expected to screen greater than 10,000 ppm when using the average SOCMI emissions factors

Source	Service	Percent of sources expected to screen above 10,000 ppm
Valves	Gas/vapor	11.4
	Light liquid	6.9
	Heavy liquid	[a]
Pumps	Light liquid	9.6
	Heavy liquid	2.2
Compressor seals	Gas/vapor	10.0
Pressure relief valves	Gas/vapor	3.6
Connections (flanges)	All	2.0
Open-ended lines	All	1.7

[a] See Table 22-6.

Table 22-5   Expected average screening values above and below 10,000 ppm when using the average SOCMI emissions factors or leak/no leak emission factors

Source	Service	Expected average screening values	
		<10,000 ppm	>10,000 ppm
Valves	Gas/vapor	717	53,676
	Light liquid	389	51,445
	Heavy liquid	a	a
Pumps	Light liquid	402	30,573
	Heavy liquid	1,178	29,792
Compressor seals	Gas/vapor	4,519	146,904
Pressure relief valves	Gas/vapor	1,960	156,087
Connections (flanges)	All	14	21,506
Open-ended lines	All	a	a

[a] See Table 22-6.

Table 22-6   Leak/no-leak emission factors

Source	Service	Leak/no leak emission factors[a]		
		Leak >10,000 ppm		No-Leak <10,000 ppm
Valves	Gas/vapor	0.0992		0.0011
	Light liquid	0.1874		0.0038
	Heavy liquid	–	0.0005	–
Pumps	Light liquid	0.9614		0.0264
	Heavy liquid	0.8547		0.0297
Compressor seals	Gas/vapor	3.5376		0.1967
Pressure relief valves	Gas/vapor	3.7202		0.0983
Connections (flanges)	All	0.0825		0.0001
Open-ended lines	All	0.0263		0.0033

[a] lbs per hour per source.

with default zeroes (described below). If default zeroes are used with established curves, all three screening options yield about the same estimate. The only way to validate the third assumption above is to bag (discussed below) some process components from the source/ service category of interest. If a bagging study is conducted, checking the default zeros should be given top priority (see below and Appendix B for procedures).

Use of the average factor approach requires the user to accept these assumptions. To the extent that these assumptions are valid, this option should generate reasonable estimates. However, these assumptions are restrictive and most modern processes deviate markedly from them.

Leak/No Leak Emission Factors

This technique is only slightly better than the use of SOCMI factors. It requires the use of a portable analyzer for screening all equipment. This approach is based on the assumption that sources can be grouped into two categories, those that leak (i.e., screening concentration greater than 10,000 ppm) and those that do not leak (i.e., screening concentration less than 10,000 ppm). Certain average mass emission rates are assumed for each category.

The assumptions underlying this technique are the same as for SOCMI factors discussed above except that the mass leak rate for components screened above 10,000 ppm can be apportioned on the basis of the number screened above 10,000 ppm. Therefore, the use of leak/

no leak factors to estimate fugitive emissions are based on two assumptions.

1. The average screening values for the process that screen above and below 10,000 ppm are shown in Table 22-5.
2. The functions used to transform screening values to leak rates (see Tables 22-8 and 22-9), on average accurately make this transformation.

This technique includes a complete screening of the process components for the source/ service category of interest. The one exception to this is for flanges. Because of their large number, a random sampling of flanges would be appropriate. However, the logistics often are so involved that a complete screening is easier. If a random sampling of the sources is done, the sample size[7] can be approximated by

$$n = N\left[1 - 0.05\left(\frac{50}{N}\right)\right] \quad (22\text{-}1)$$

Where n = sample size
N = the number of process flanges

Although it is necessary to determine the screening values as accurately as possible, all that is required is a count of the number screening above 10,000 ppm (n_L) and the number screening below this value (n_{NL}). The total emissions for the source/service category then are estimated using

$$\text{Total emissions} = N\left[\frac{n_L}{n_s}E_L + \frac{n_{NL}}{n_s}E_{NL}\right]$$
$$(22\text{-}2)$$

Where $n_L + n_{NL} = n_s$[8]
E_L = leaking emission factor
E_{NL} = nonleaking emission factor

Again, the term 'leaking' applies to components with screening values greater than 10,000 ppm and the term 'nonleaking' applies to components that screen below 10,000 ppm. The values of the leak/no leak emission factors are shown in Table 22-6.

The only assumption required for average SOCMI factors that is not required for leak/no leak factors is the fraction screened above 10,000 ppm. To the extent that the other assumptions are wrong for the average factors, they are wrong for the leak/no leak factors. The need for one less assumption, however, does not mean that this approach will give a more accurate emission estimate.

The first assumption above can be checked easily by a simple examination of the process screening values (i.e., see whether the average process screening values – above and below 10,000 ppm – are different from the assumed values given in Table 22-4). Validation of the second assumption requires bagging some components (see below). This is suggested when using the established leak rate/screening value curves and validating this assumption leads to a more accurate estimate than obtained when using the leak/no leak or stratified emission factors. To the extent that the assumptions are valid, this option should generate reasonable emissions estimates. The most critical assumption relates to the assumed emission rates for components screened at zero. Therefore, in terms of accuracy it is important to bag some zero-screening sources.

Stratified Emission Factors
The stratified emission factors approach is based on three concentration intervals, the leak/ no leak approach was limited to just two (i.e., >10,000 ppm or <10,000 ppm). The assumptions are the same. Table 22-7 lists the emission factors. As in the case of the leak/no leak

---

[7] See Appendix B for a more detailed discussion of statistical terms such as sample size and confidence intervals.
[8] n_s, the total number of components screened, often equals N, the total number of components.

Table 22-7   Stratified emission factors

| Source | Service | Emission factors[a] Screening value ranges[b] | | |
		0–1,000	1,000–10,000	>10,000
Compressor seals	Gas/vapor	0.0250	0.5821	3.5456
Pump seals	Light liquid	0.0044	0.0739	0.9636
	Heavy liquid	0.0084	0.2042	0.8566
Valves	Gas/vapor	0.0003	0.0036	0.0994
	Light liquid	0.0006	0.0212	0.1879
	Heavy liquid	0.0005	0.0005	0.0005
Flanges	All	0.00004	0.0193	0.0827
Pressure relief valves	Gas/vapor	0.0251	0.6152	3.7287
Open-ended line	All	0.0003	0.0193	0.0263

[a] lbs per hour per source.
[b] parts per million.

approach, screening values have to be determined and emission factors identified from the above table. The total emissions for a source/service category then are estimated using

$$\text{Total emissions} = N \left[\frac{n_1}{n_s} E_1 + \frac{n_2}{n_s} E_2 + \frac{n_3}{n_s} E_3 \right]$$

$$(22\text{-}3)$$

Where $n_1 + n_2 + n_3 = n_s$

n_s = the total number of components screened

N = the total number of components

E_1 = emission factor corresponding to screening value range 0 to 1,000 ppm

E_2 = emission factor corresponding to screening value range 1,000 to 10,000 ppm

E_3 = emission factor corresponding to screening value range >10,000 ppm

The corresponding number of components screened in each range are n_1, n_2, and n_3, respectively. The same procedure should be used in validating the underlying assumptions as was used for the leak/no leak factors.

Established Leak Rate/Screening Value Correlations

This technique requires screening process components as outlined for the leak/no leak and stratified emission factor options. It should include some bagging (see below) to improve the final fugitive emissions estimates. Tables 22-8 and 22-9 give the most commonly used 'established' correlations. For comparison, one other set of curves is given in Table 22-10. Tables 22-9 and 22-10 are based on the same amount of data but are statistically different for everything except pumps. The screening values used in Table 22-10 were corrected for response factor; those in Table 22-9 were not. However, the significant difference between the data sets exists whether screening values are corrected or not.

Using established correlations may require some bagging. Some, and preferably all, components that screen above the instrument saturation point must be bagged. This is because a valid concentration to associate with these components cannot be determined when the instrument 'pegs' out. We only know a minimum bound on the concentration for such sources; the actual concentration could be much higher than this value.

In addition to bagging components that screen at the instrument's saturation point, it is

Table 22-8   Functions transforming screening values to leak rates using SOCMI factors (refinery study except valves in gas/vapor service)[a]

Source	Service	Curve parameters		
		a	b	g
Valves	Gas/vapor	−5.3	0.79	3.39
	Light liquid	−4.9	0.80	2.53
	Heavy liquid	[b]	[b]	[b]
Pumps	Light liquid	−4.4	0.83	4.58
	Heavy liquid	−5.1	0.83	2.44
Compressor seals	Gas/vapor	−4.4	0.83	4.58
Pressure relief valves	Gas/vapor	−4.4	0.83	4.58
Connections (flanges)	All	−5.2	0.88	2.02
Open-ended lines	All	[c]	[c]	[c]

[a] All functions are curves of the form

$$\text{Average leak rate (\#/hour)} = g \times 10 \exp[a + b \log_{10}(\text{screening value})]$$

[b] There is no curve for valves in heavy liquid service; hence, the function transforming screening values to leak rates is independent of the screening value, and reduces to a constant of 0.0005 #/hour for all screening values.

[c] There is no curve for open-ended lines; however, the function transforming screening values is slightly more complex than a simple constant given by

$$\text{Average leak rate} = 0.0033, \text{ screening value} < 10{,}000 \text{ ppm}$$

$$\text{Average leak rate} = 0.0263, \text{ screening value} > 10{,}000 \text{ ppm}$$

Table 22-9   Functions transforming screening values to leak rates (maintenance study after removing 'pegged' source)[a]

Source	Service	Curve parameters		
		a	b	g
Valves	Gas/vapor	−5.350	0.693	3.766
	Light liquid	−4.342	0.470	8.218
Pumps	Light liquid	−5.340	0.898	2.932
Flanges	All	−4.733	0.818	2.020

[a] All functions are curves of the form

$$\text{Average leak rate (\#/hour)} = g \times 10 \exp[a + b \log_{10}(\text{screening value})]$$

Where the screening values were uncorrected results obtained from a Century Systems Organic Vapor Analyzer [OVA].

Note: EPA recommends using the curve for pumps in light liquid service for all other source/service categories not listed above.

worthwhile to bag some components that screen below the instrument's lower limit of detection (LLD). This is explained further below. In addition, a testing program also might bag a number of components that screen above the LLD but below the instrument's saturation point. These bags are used to assess the validity of the curves used to relate screening values to leak rates and to assess the number of bags required to fit a process specific curve. They also are used directly to estimate fugitive emissions (see Appendix B for details).

The estimation of fugitive emissions using this option is obtained by combining estimates for three groups of components: (1) those screening below the screening instrument's

Table 22-10   Functions transforming screening values to leak rates[a]

Source	Service	Curve parameters		
		a	b	g
Valves[a]	Gas/vapor	−6.3	0.69	6.3
	Light liquid	−6.0	0.76	3.0
Pumps[a]	Light liquid	−5.9	1.0	8.3
Flanges[a]	All	−6.3	0.95	3.5
Valves[b]	Gas/vapor	−6	0.873	4.11
	Light liquid	−6	0.797	14.10
Pumps[b]	Light liquid	−5	0.824	4.18
Flanges[b]	All	−6	0.885	6.71

[a] Berglund et al. 1990, Environplan 1991 (data were response corrected).
[b] Draft EPA document (based on a combination of EPA data used to get parameters in Table 22-9 and the data used to get parameters in the top of this table. Neither data set was response factor corrected).

LLD, (2) those at the saturation point, and (3) those in between. The total emissions for the source/service category of interest are estimated using

$$\text{Total emissions} = N\left[\frac{n_1}{n_s}E_1 + \frac{n_2}{n_s}E_2 + \frac{n_3}{n_s}E_3\right]$$

(22-4)

Where N = the total number of components for the source/service category

n_1 = the total number that screen below the LLD

n_2 = the total number that screen above the LLD, but below the instrument saturation point

n_3 = the total number that screen at the saturation point.

$n_1 + n_2 + n_3 = n_s$

n_s = the total number screened

E_1 = emission factor below the LLD

E_2 = emission factor above the LLD, but below the instrument saturation point

E_3 = emission factor above the saturation point

The following sections describe how to obtain the relevant emission factors for each of these groups.

Below the LLD

If no bagging information is available, the relevant emission factor E_1 can be estimated

using 'default zeros' as given in Table 22-11. However, the best approach involves bagging some components in the process for which emissions are being estimated. The study for which correlations are listed in Table 22-10 included considerable bagging for components screened below the LLD. Results for different types of components ranged from about a factor of ten to a factor of one thousand below those

Table 22-11   'Default zero' values for estimating the contribution of sources screening below the lower limit of detection (LLD)

Source	Service	'Default zero'[a]
Valves	Gas/vapor	0.00007
	Light liquid	0.00099
	Heavy liquid	0.00009
Pumps	Light liquid	0.00009
	Heavy liquid	0.00009
Compressor seals	Gas/vapor	0.00009
Pressure relief valves	Gas/vapor	0.00009
Connections (flanges)	All	0.00020
Open-ended lines	All	0.00009

[a] lbs per hour.

Note: These 'default zeroes' were established from limited data (i.e., eleven sources), exclusively from valves in gas/vapor service. Background is not accounted for in any of these bagging tests. The observed emission level is equivalent to 8 ppm on the curve defined by Table 22-9. The other values in Table 22-11 were obtained from the corresponding Table 22-9 curves at 8 ppm. In addition, there are only four curves in Table 22-9, which have been extrapolated across sources and service categories. These groupings appeared reasonable in early refinery studies, but have not been tested for chemical plants.

determined from the studies on which Tables 22-8 and 22-9 are based. The lower test results probably still are high because of the sensitivity of the analytical methods used. In tests where methods used were able to quantify emissions from all components bagged, the emissions from nondetectable screened components averaged about 10^{-9} lb/hour (Dharmavaram et al. 1991; Randall 1989). The very low measurements were made on inorganic chemical processes, however, and may not be appropriate for organic chemical processes.

Above the LLD, Below Saturation

The relevant emission factor for sources screening greater than LLD and less than saturation, E_2, is determined from the geometric mean screening value[9] of this group and the appropriate curve in one of Tables 22-8, 22-9, and 22-10, using

$$E_2 = g \times 10^a \times [GM(SV)]^b \quad (22\text{-}5)$$

Where $GM(SV)$ = the geometric mean (screening value)

The geometric mean screening value is given by

$$GM(SV) = \left[\prod_{i=1}^{n_2} SV_i \right]^{\frac{1}{n_2}} \quad (22\text{-}6)$$

Where SV_i = screening value for component i
a, b, g = appropriate values from Tables 22-8 and 22-9

Above Saturation

The emission factor for sources screened above saturation, E_3, is determined using

$$E_3 = 10^{\overline{Y}} SBCF \quad (22\text{-}7)$$

$$\overline{Y} = \frac{1}{n} \sum_{i=1}^{n} \log_{10}(LR_i) \quad (22\text{-}8)$$

$$SBCF = + \frac{(m-1)t}{m} + \frac{(m-1)^3 t^2}{m^2 2!(m+1)}$$
$$+ \frac{(m-1)^5 t^3}{m^3 3!(m+1)(m+3)} + \cdots \quad (22\text{-}9)$$

$$t = \frac{S^2 (\ln_{10})^2}{2} = 2.651 S^2 \quad (22\text{-}10)$$

Where LR_i = leak rate (bag) estimate from component i screening at saturation
$SBCF$ = scale bias correction factor
$m = n$ = number of sources bagged from components screening at saturation

$$S^2 = \frac{1}{n-1} \sum_{i=1}^{n} \left(\log_{10}(LR_i) - \overline{Y} \right)^2$$

The confidence interval for emissions for components screening at saturation then is calculated using

$$LCL = 10 \exp \left[\overline{Y} - t(0.975, n-1) \sqrt{\frac{S^2}{n}} \right]$$
$$\times SBCF \quad (22\text{-}11)$$

$$UCL = 10 \exp \left[\overline{Y} + t(0.975, n-1) \sqrt{\frac{S^2}{n}} \right]$$
$$\times SBCF \quad (22\text{-}12)$$

If only one such source is found and bagged, then simply setting E_3 equal to the bagging result for this component provides the best possible estimate of this factor.

Established Correlation Validation

The basic assumptions behind this correlation technique are that the curves accurately relate screening values to leak rates, and that the

---

[9]The geometric mean is used as a matter of convenience to simplify calculations. Emissions also can be predicted for individual screening values and then averaged.

factors applied to the sources screening below the LLD accurately account for this very often large group of components. This approach requires bagging when some components exceed instrument saturation. Therefore, no assumption about this class of source is needed. The above assumptions can be validated only with some bagging data. The discussion in Appendix B on estimating an appropriate emission factor for components screening below the LLD includes a procedure for validating the 'default zeros.'

If some sources covering the range of observed screening values between the LLD and saturation are bagged, then the accuracy of the curve can be validated. There is a simple decision rule for this validation. The following discussion of this decision rule is restricted to the situation where only four components (two high and two low) are available for testing the curve. (See Appendix B for an expanded discussion.)

The simplest way to validate the curves is to plot the bagging data with the respective line(s) on log-log graph paper. If all four points lie below the line, the established relationship should be conservative. If they all lie above the line, the established relationship will tend to underestimate true emissions. If even one result lies on the opposite side of the line from the other three, then the established relationship should give a reasonable estimate.

The validation procedure could result in one of three outcomes: an overestimate (all points below), a reasonable estimate (points on either side), or an underestimate (all points above). The two outcomes where the established curves do not estimate emissions well can be treated in a variety of ways.

Process-Specific Leak Rate/Screening Value Curves

The required effort for this technique includes all the requirements for the use of the established leak rate/screening value curves outlined above. In addition, the procedure often requires bagging of more process components. The required number of additional baggings to meet the desired precision may be determined from the following procedure:

- Fit a least-squares line in the log-log space, regressing the log (leak rate) on the log (screening values) for the components bagged, i.e., of those screening above the LLD, but below saturation (see Appendix B).
- Calculate the mean square error, $MSE_{process}$ (see Appendix B).
- Estimate $n_{2,*B}$, the required number of additional baggings, as

$$n_{2,*B} = \text{minimum of } [n_2, k \times MSE_{process} - n_{2,B}]$$

(22-13)

Where n_2 = the total number of components screening above the LLD but below saturation

$n_{2,B}$ = the number of components screening above the LLD but below saturation that were initially bagged (i.e., those used to fit the line in the first step above)

k = determined either from Table 22-12 for the appropriate source/service category, or set equal to 44.

The criterion for establishing the values of k given in Table 22-12 was to find n such that the maximum variance of log (leak rate) predicted by the process-specific curves is less than or equal to the minimum variance of log (leak rate) predicted by the established curves. This approach is based on achieving prediction accuracy for the process-specific curve at least as good as that obtained for the corresponding established curve (i.e., the same source/service category) used as a basis for establishing the emission factors for that source/service type. Using $k = 44$ is equivalent to requiring the upper 95 percent confidence limit (evaluated at the average screening value) to be less than twice the best estimate (at the average screening value). Table 22-12 shows that except for gas/vapor valves fewer baggings are necessary than the maximum of thirty suggested by EPA.

The additional components to be bagged again should be selected as close to the extremes of the observed screening value range

Table 22-12   Values of k for use in establishing requisite numbers of baggings for generating process-specific leak rate/screening value curves

Source	Service	k	Total no. of bags needed[a]
Valves	Gas/vapor	78.5	40
	Light liquid	53.4	27
	Heavy liquid	15.1	8
Pumps	Light liquid	32.3	17
	Heavy liquid	15.1	8
Compressor seals	Gas/vapor	32.3	17
Pressure relief valves	Gas/vapor	32.3	17
Connections (flanges)	All	17.6	9
Open-ended lines	All	15.1	8

[a] This calculation assumes $MSE_{process} = 0.5$.

as possible. Moreover, each end of the range should be approximately equally represented. This amounts to adding bagged components in pairs, from each end of the screening value distribution, with each as extreme as possible. In this way, those components near the middle of the screening value range will be the last candidates to be bagged.

The above assumes some preliminary bagging is done along with the initial screening of the process. If this is not done, no estimate of $MSE_{process}$ can be directly determined. If such data are not available, then using an estimate of $MSE_{process} = 0.5$ appears reasonable from the data obtained to date. Using this value (along with the above procedures) allows for concurrent bagging and screening; from a logistical viewpoint, this may make much more sense than doing only a limited initial bagging.

The estimate of fugitive emissions in this option is obtained in a manner consistent with the use of the established curves, except that the process-specific curve parameters a, b, and g are used instead of the values from Table 22-8 or 22-9. Hence, the estimation procedures given for the previous option apply with the only difference being the use of different values of a, b, and g. (Again, see Appendix B for a least-squares curve-fitting technique and more detail on assumptions.)

The use of the least-squares procedure to fit a line to two variables involves several assumptions.

- The relationship between the leak rate and screening value is well described by a line in the log-log space.
- The variability of the log (leak rate) values is the same across the log (screening value) range.
- The log (screening values) are fixed and known at the time that the associated component is bagged.

These assumptions are imbedded implicitly in all the previous options as well, for each of these options at some state in their development involved the fitting of a least-squares line attempting to predict log (leak rate) from log (screening value). As is shown below, there are ways to validate the above assumptions using the available data, and it is assumed that these validation procedures were undertaken in the fitting of the lines used in the other options. The assumptions above generally can be validated with simple plotting procedures. These plotting procedures, as well as possible adjustments to the analysis that can be made if the assumptions do not appear to be valid, are outlined in more detail in Appendix B.

The last assumption is the most difficult to validate and is probably never truly valid. However, it may in many cases be satisfied to the extent that the least-squares result still suffices. Appendix B gives some further detail on how to adjust the analysis so that this assumption will not be required.

Situations Where All Components Screened Above LLD are Bagged

There may be some processes where so few components screen above the LLD that all the components that screen above the LLD can be bagged. In this case, it is not necessary to fit a process-specific curve to the data. However, it can be done if more than two components are bagged. The total emissions estimate in this situation would be

$$\text{Total emissions} = N \left[\frac{N_1}{N_s} E_1 + \frac{N_*}{N_s} E_* \right]$$

(22-14)

Where $n_1 + n_* = n_s$

N, n_s, n_1 and E_1 are as defined previously

n_* = the total number of components that screened above the LLD (including any that screened at the saturation point)

The emission factor E_* can be determined using

$$E_* = 10^{\overline{Y}_*} \times g \left[S_*^2 \times \frac{(\ln_{10})^2}{2} \right]$$ (22-15)

$$\overline{Y}_* = \frac{1}{n_*} \sum_{i=1}^{n_*} \log_{10}(LR_i)$$ (22-16)

$$S_*^2 = \frac{1}{n_* - 1} \sum_{i=1}^{n_*} \left(\log_{10}(LR_i) - \overline{Y}_* \right)^2$$

(22-17)

$$g(t) = 1 + \frac{(m-1)t}{m} + \frac{(m-1)^3 t^2}{m^2 2!(m+1)}$$
$$+ \frac{(m-1)^5 t^3}{m^3 3!(m+1)(m+3)} + \cdots$$

(22-18)

Where LR_i = leak rate (bag) estimate, source i
$m = n_* -$ number of sources bagged
$t = 2.651\, S_*^2$

The confidence interval for nonzero screening sources (E_*) is calculated using

$$LCL = 10 \exp \left[\overline{Y} - t(0.975, n-1) \right]$$
$$\times g \left[\frac{S_*^2 \times (\ln_{10})^2}{2} \right]$$

(22-19)

$$UCL = 10 \exp \left[\overline{Y} + t(0.975, n-1) \right]$$
$$\times g \left[\frac{S_*^2 \times (\ln_{10})^2}{2} \right]$$

(22-20)

Confidence intervals for the estimates from all of the options (except the use of the average SOCMI emission factors) can be constructed.

Measurement Techniques

In most instances, measurement of fugitive emissions is necessary to quantify accurately the emission rate from individual process units. The terms 'screening' and 'bagging' have been mentioned in previous sections and are discussed in more detail here.

Screening

Screening is the process of identifying leaks from sources using a portable monitoring instrument. There are a number of monitoring devices capable of measuring leaks of organic compounds. To be used for regulatory compliance purposes, however, they have to meet the specifications and performance criteria in EPA Reference Method 21. In some cases, equipment may not meet all the requirements of Method 21; this often is the case when monitoring for inorganic compounds. Whether Method 21 instrumentation is used or not, it is important to calibrate for the compounds and concentrations to be observed.

Most of the screening instruments for organic compounds are designed to operate on one of three basic principles: ionization, infrared absorption, or combustion. In ionization detection, a sample is ionized and the number of ions (i.e., charges) is measured. Two types of ionization detectors are available: flame ionization detector (FID) and photoionization detector (PID). Instruments that are based on the principle of nondispersive infrared (NDIR) rely on the light absorption characteristics of gases. Many are subject to interferences from water vapor and carbon dioxide in the air. Combustion devices are based on measuring the thermal conductivity (or heat of combustion) of a gas. Hot wire detectors or catalytic oxidizers are combustion devices that can provide a non-specific measure of organic concentration. In choosing any instrument, one with the lowest practical detection limit generally is desired for emission estimates. Proper calibration procedure, as specified by the manufacturer, must be employed in using the instruments. The FID-based organic vapor analyzer (Foxboro OVA series) and the catalytic oxidizer (J.W. Bacharach TLV sniffer) are the most popular devices currently used for screening equipment leaks.

The sensitivity of a screening device depends on the composition and concentration of the compounds in an air sample. The meter reading is converted to concentration using what is called a response factor (RF). The RF is defined as a ratio of measured to actual concentration. Response factors should be used in selecting appropriate monitoring devices.

Many RFs are available from published literature (Brown et al. 1980); otherwise, they can be determined using calibration gases and making the necessary measurements.[10] Most equipment leak regulations require that screening instruments have RFs in a specified range (e.g., <3 at 500 ppm for MACT, <10 at 10,000 ppm for NSPS) if uncorrected screening values are to be acceptable (Bursey et al. 1991). Otherwise, screening values must be RF corrected. Response factors are discussed in more detail in Appendix C.

Mass Emissions Sampling

The screening process described above provides an estimate of the emissions from a leaking source by the use of factors or screening value/leak rate relationships. However, a rigorous measurement procedure is required for an accurate determination of the mass emission rate. Such a determination requires source enclosure (i.e., bagging) and sampling of the gas in the bag. Three bagging sampling methods – vacuum, blow-through, and recirculation – are discussed below.

EPA protocol requires that accuracy be demonstrated on a known leak rate source to be within ±20 percent near the lowest emission levels to be measured. Some studies indicate sensitivity down to 10^{-9} lbs/hour is needed. Various approaches to the known leak demonstration have been used and the desired sensitivity obtained (Dharmavaram et al. 1991).

The sensitivity of the procedure to be used is dependent largely on the method for analyzing the sample collected and, in this author's experience, on the method of collecting the sample. For example, vacuum techniques are more sensitive than the commonly used blow-through. The known leak rate test is the place where the effect of inadequacies in techniques or hardware will show up if the chemicals of interest, or similar ones, are used.

Bagging (Source Enclosure)

The process of enclosing a source in a tent is known as bagging. Bagging contains the leak to allow measurement of the mass emission rate. The enclosure material is not as important as the 'sample bag' but it may be appropriate to use the same material or verify the acceptability of a different material. Typically, an impermeable plastic such as Mylar® (polyethylene terephthalate) has been the material of choice. Mylar® has a high melting point (250°C) and does not adsorb most hydrocarbons. Other materials that have been used successfully include Tedlar®, Teflon®, and aluminum foil.

Figure 22-3 shows the bagging of a vertical pump. The size of the bag should be kept as small as possible to: (1) minimize the time required to reach equilibrium, (2) minimize

---

[10] See Appendix C for more information on response factors and their use.

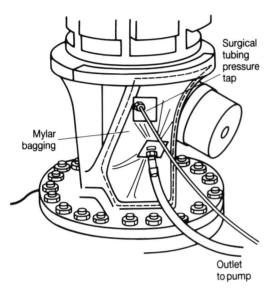

Figure 22-3.   Tent construction around the seal area of a vertical pump.

construction time, (3) obtain an effective seal, and (4) reduce/prevent condensation of heavy hydrocarbons.

Bagging (Sample Collection)
Samples collected for laboratory analysis can remain in the bag for an extended period before analysis. As a result, it is important to have sample stability for comparable periods. Bags made from Tedlar® or Teflon® have been shown to be acceptable in many situations. Conversely, polyethylene gas bags commonly used for span gases have been shown to be unacceptable for a variety of test samples. If there is any doubt about the acceptability of the sample bag material, it can be quickly checked. A sample of interest can be put in the bag and the concentration measured over time. No problems have been reported with glass or plastic containers for liquid samples.

Vacuum Method
Samples of gas are withdrawn by applying vacuum on the tent surrounding a source. A typical sampling train along with more detailed information is shown in Appendix D. Unless the leaks are very large, a pump that pulls 0.5 cfm is adequate. However, larger pumps (2.5 cfm)

also have been used. A slight vacuum (i.e., 0.1 inch or less of water) is maintained in the sampling tent. Samples are collected in sample bags or impingers and transported to the laboratory for analysis.

Blow-through Method
Blow-through refers to a technique where nitrogen is blown through an enclosure tent and the chemical concentration inside the tent is measured along with temperature and oxygen concentration. The pressure in the tent should never exceed 1 psig. This technique is best suited for use where leaks rates are very high and especially where hydrocarbons above their autoignition temperature will be collected. In this case, fires could occur if air is present (Davis 1989). Appendix E illustrates the method and provides more detailed information.

Recirculation Method
This method has not been used as frequently as the other two, but it should work and may have some practical advantages (Hosler, Grosshandler, and Marchman 1992). A schematic of one form of apparatus for this method is shown in Appendix F.

Data Analysis – Technical Studies

A comprehensive assessment and discussion of methods of data analysis is beyond the scope of this chapter. However, a brief overview is provided of what has been done and where procedures appear to be headed. Such an overview may help point those desiring greater detail in a productive direction. The reader should understand that, although the information in the remainder of this chapter is widely accepted, some of the points raised in this section are potentially controversial because they are new or narrowly focused.

Improved Leak Rate/Screening Value Relationships for Valves
Early relationships were based on limited data and on data weighted toward high screening values and leak rates. Based on information at that time, this appeared to represent the indus-

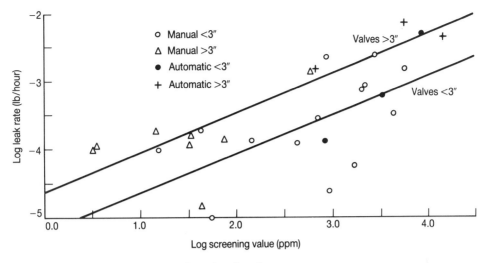

Figure 22-4.   Log leak rate versus screening values for valves.

try. More recent studies, however, with more data in the low screening value range have shown some substantial differences at low screening values (Berglund, Romano, and Randall 1990; Dharmavaram et al. 1991; Enviroplan 1991; Randall 1989).

The same early data were analyzed to see if the relationships depended on such factors as fluid phase, valve type, and line size. However, the only variable that showed a difference was fluid phase (i.e., gas or liquid). In contrast, more recent studies have not shown fluid phase to make a difference in the screening value/leak rate relationship.[11] Instead, the more recent studies show a strong dependence on factors such as line size. An example is shown in Figure 22-4 (Kittleman 1992).

Factors That Affect Valve Leak Rate

As pointed out in the previous paragraph, early studies indicated that only fluid phase affected valve leak rate. This was true not only for the relationship between screening value and leak rate, as discussed above, but for the actual average leakage as well. Factors such as valve type and line size appeared to make no difference. More recent screening data, however, have shown significant differences resulting from fluid phase as well as valve type, line size, and use (e.g., control or block).

What shows up as being significant likely depends on what is being tested. For instance, a significant difference was observed between gases and liquids for full-turn valves, but not for quarter-turn valves, and a significant difference was observed between full-turn and quarter-turn valves for liquids but not for gases (see Figure 22-5). This appears to indicate that other properties control the leak rate. One problem is that it has been difficult to obtain all of the supplemental data desired to make the comparison (e.g., information about packings and maintenance history). It also is known that differences in maintenance practices can mask many other features. Some better insight may be close at hand, because laboratory studies (Choi et al. 1992) and studies of large blocks of field data are under way.

---

[11] The early studies took place at several plants managed by different companies making different materials, and the results were not RF-corrected. The fluid-phase distinction could easily be an artifact of where or how the data were collected. The more recent revelations appeared in studies where data were RF-corrected and were collected at processes with less divergent characteristics (i.e., processes making the same product or under control of the same manufacturer).

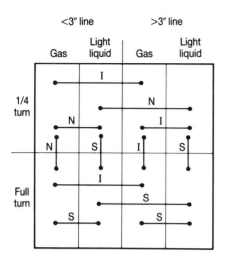

Figure 22-5.   Results of Z tests for significance between cell using nonzero screening data. S = significantly different, N = not significantly different, I = insufficient data for realistic test.

More Useful Ways to Present Results (Plant Specific)

As noted earlier, accuracy and cost increase as one moves up the hierarchy of emissions estimating techniques (i.e., from SOCMI factors

to process-specific screening valve/leak rate relationships). However, enough data may now exist to demonstrate the validity of another way to achieve the emission estimate quality at a lower cost. This method uses correlations of multiplant emission estimates based on studies of these plants.

The best screening and bagging study emission estimates, and the fraction of components above a specified leak definition, can be correlated for a number of plants. At high leak definition, a very good correlation exists (see Figure 22-6). As leak definition decreases, the quality of the correlation also decreases but still is good at 500 ppm (i.e., the MACT valve leak definition). An example of this relationship presented by EPA (Pahel-Short 1989) is provided below.

Data Base

Because of the large number of variables and interactions that may affect leakage, a sizable data base is necessary to identify and quantify effects. For instance, two processes on the same site may operate differently and two plants in the same company may operate even more differently. However, the difference between

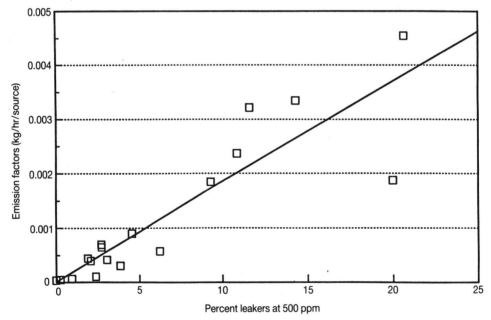

Figure 22-6.   Correlation of emission factors versus percent leakers (gas valves in ethylene oxide and butadiene service).

two company's procedures likely is to be greatest. Leakage, for instance, can be quantified and predicted where factors such as common practices and culture apply. However, even the most significant equipment variables cannot be quantified where more important differences (i.e., maintenance procedures) are not controlled. Uncontrolled studies appear to have led to improperly assigning the significance of certain independent variables. Once that occurs it becomes impossible to determine what, in fact, is important. This demonstrates the need for data analysis based on scientific principles, not merely on statistical association.

Considerable screening and bagging data have been added since the mid-1980s. At the time of this writing in 1992, the Chemical Manufacturers Association (CMA) had accumulated data through their POSSEE and CHEMSTAR programs and other organizations, such as American Petroleum Institute (API), also intended to compile data bases.

Control Techniques

Equipment and procedures are engineered for each process. As a result, there is no single leak prevention technique that is applicable everywhere. Furthermore, the same results can be achieved in a variety of ways. This was demonstrated by an EPA study of the phosgene production industry (Randall et al. 1989). LDAR programs are inspection requirements that by themselves do not necessarily provide good emission control. It is a QC axiom that you cannot inspect quality into a product. Quality, in this case a low-leak process, must be built into the hardware and worker practices. Monitoring is a way of keeping track. However, when monitoring becomes the control the result can be poor performance.

Different processes must be designed and engineered differently in order to make their desired products. Even for a specific product, there are many ways to achieve the same result (i.e., a low-leak process). As a result, it generally is counterproductive to try to specify any one aspect of emission control (e.g.,

hardware, operating practices, and maintenance practices). However, there are some general rules that can be applied. For instance, unnecessary equipment should be avoided as should 'open-ended lines.'

For any specific process, there will be cost-effective combinations of hardware and worker practices. Monitoring can help determine the effectiveness of the various combinations for specific plants.

LDAR Programs

As discussed earlier, these are monitoring programs that include a requirement to repair expediently all equipment found leaking. There are many variations. Some appear to be aimed at encouraging good performance; others are punitive and may replace existing practices, good or bad. LDAR programs consist of two functions – monitoring and repair.

Monitoring

An appropriate analytical instrument and technique must be selected for the application. Generally, EPA Method 21 is specified for hydrocarbons. When specifying the monitoring program, it is necessary to identify analyzers or analyzer characteristics (e.g., accuracy and response time). Equipment to be monitored (e.g., valves and pumps) and monitoring frequency must be specified. Monitoring frequency can be a fixed period or it can be the incentive type that varies with monitoring results. Equipment most likely to leak, such as pumps and compressors, generally are monitored much more often than equipment, such as flanges, that are unlikely to leak.

A leak definition is needed. The leak definition is the point where pertinent instrument characteristics are determined. For instance, the FID calibration is concentration dependent for many compounds.

Repair

Repair is required on all equipment when monitoring results exceed the leak definition. Attempts to repair generally must occur within a specified period. Replacement or major repair normally can be held until the next process shutdown only after at least two repair attempts fail when the process is operating. The timing and extent of repair attempts must be specified.

The leak definition is an important part of an LDAR program. The extremes commonly are 500 ppm and 10,000 ppm. The leak definition should be low enough to require repair where reasonable. Conversely, the leak definition should be high enough not to force repair where repair attempts likely will fail, could worsen leakage, or where normal background readings make it difficult to detect leaks. Screening values below a certain level come and go without explanation; in these cases, repair is unnecessary. Improper repair also can cause serious operating problems. For instance, over-tightening of packing may make a valve inoperable.

Maintenance

Equipment leaks generally occur because something has been installed improperly or has worn out. In contrast, permeation through gaskets and packings is undetectable with normal monitoring techniques. Consequently, one of the most important features of any leak prevention program is the maintenance practice. Even the best equipment will leak if improperly maintained. Many plants have a predictive or preventive maintenance program. These practices can be improved when accompanied by judicious monitoring. For instance, 'directed' maintenance has been demonstrated to stop leakage more effectively than maintenance alone. In 'directed' maintenance, a screening instrument is used while the equipment is being worked on (e.g., tightening a valve packing) to insure that leakage is minimized.

Operating Practices

Next to maintenance practices, operating practices probably are the most important factor in determining how much a process will leak. Operators are important when it comes to detecting leakage and stopping it on a day-to-day basis. For instance, small leaks of chemicals that are odorous or irritating can be sensed by operators on their normal rounds. In processes handling highly toxic chemicals or where corrosivity quickly changes small leaks into big ones, plant operators usually are the first to find a leak and initiate corrective action.

However, it is not enough to know that something is leaking. Operators must be able to find the leak. Many creative approaches have been devised to locate leaks. The toxicity and corrosivity of the compounds and the magnitude of a leak will determine how quickly leak repair is mandated. Some processes are shut down immediately at any sign of leakage and are not started again until repairs are made. The costs of these shutdowns can be great and they generally are not done unless a real danger is likely.

Design

Design practices, other than specifying high quality equipment, play a role in reducing potential leakage of highly toxic materials. Instead of producing large quantities of some feedstocks in large cost-effective processes, smaller plants sometimes are used and the products consumed on the same site. This not only reduces the possibility of leakage during transport and storage, but during production as well. These feedstocks can be produced and consumed so quickly that the opportunity for leakage is reduced.

Less radical design approaches also are practiced. For instance, minimizing the number of possible leak points can be expected to reduce leakage as well. However, there are practical limits to this. For instance, it may be possible to replace some flanges with welds; however, some flanges always are needed because cutting welded pipes can be dangerous.

Leakless Equipment

Process hardware (e.g., valves, flanges, and pipes) comes in a wide variety of types and designs. Some have demonstrated a tendency to leak less than others. Although potentially useful, these are not yet major factors in plant performance; for example, a low-investment but well-maintained and operated plant is likely to leak less than a high-investment but poorly maintained and operated plant. All types of equipment cannot be used in every process. In fact, much of the highly touted leakless equipment has limited practical applicability. For instance, there are several types of valves that do not have actuators extending through the body; pinch and diaphragm valves are examples. However, process pressures, fluid compatibility, precision in use, life expectancy, and cost are all factors that can limit their use.

Another example is the use of 'bellows' seals. These seals can be used on most valves, but leakage into the seal can render the valve inoperable and create a serious hazard. Nonetheless, some equipment designs, when appropriately used, are better than others. More about leakless equipment is included in the equipment discussions below.

Valves

There are many types of valves. The most commonly used can be divided into two generic types: full turn (see Figures 22-7 to 22–10) and quarter turn (see Figures 22-11 to 22–13). Full-turn valves generally are used for flow control. When they are adjusted frequently, as are control valves, they are prone to leakage unless routinely checked and maintained. There are claims that improved packings are making leakage less of a problem. However, field data to support the claims were not available as of this writing in 1992.

As noted above, valves can be fitted with a bellows seal intended to prevent stem leakage (see Figure 22-14). Although some successes have been reported, the use of bellows seals has

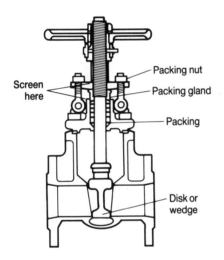

Figure 22-7.   Rising stem gate valve.

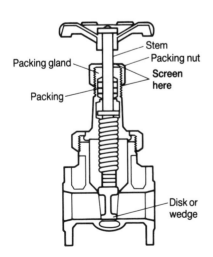

Figure 22-8.   Nonrising stem gate valve.

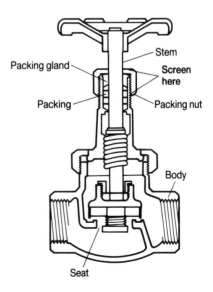

Figure 22-9.   Manual globe valve.

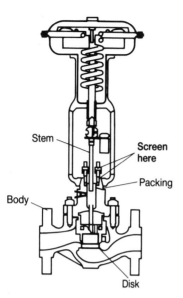

Figure 22-10.   Globe type control valve.

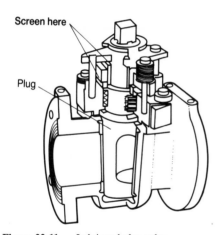

Figure 22-11.   Lubricated plug valve.

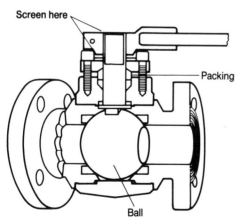

Figure 22-12.   Ball valve.

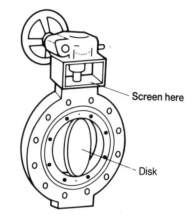

Figure 22-13.   Butterfly valve.

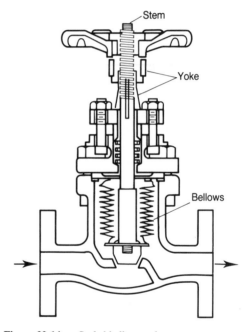

Figure 22-14.   Sealed bellows valve.

not yet reached wide acceptance in this country for a variety of reasons, including

■ Size limits, owing to the height of the bellows needed to get sufficient deflection
■ High cost
■ Bellows failure can lead to serious leaks
■ Bellows may not operate with polymeric fluids or slurries

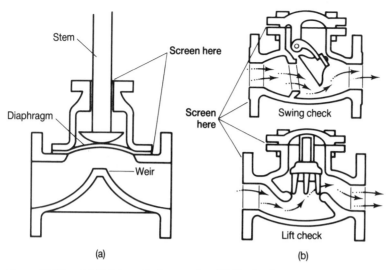

Figure 22-15.   (a) Weir type diaphragm valve. (b) Check valves.

Quarter-turn valves (e.g., ball and plug) leak less than full-turn valves in certain applications and are employed widely in plants requiring low leakage (e.g., plants processing extremely odorous or toxic chemicals). The most effective applications are at low temperatures where polymeric internal materials can be used. Quarter-turn designs also can be equipped with secondary seals (e.g., packing and bellows) to further limit leakage.

Other types of valves (e.g., diaphragm and pinch) exist that do not have actuators that contact the process fluid. These have limited applicability, but generally will not leak short of valve failure (see Figure 22-15).

Connectors

Flanges are the most common type of connector although screwed fittings and other types are used to a limited extent. In some processes, welded pipe is used to eliminate as many connections as possible. However, some flanged connections always are required to allow safe dismantling for maintenance, repair, and cleaning.

Flange designs are based on service (i.e., pressure). Heavier than required flanges supposedly lessen the probability of leakage. Gasket material may be important but permeation through a gasket is trivial compared to what can occur due to a scratched face or improper alignment. Vibration and improperly supported lines also can cause leakage.

Pumps

Packed pumps have a reputation for leakage (see Figure 22-16). These are no longer widely used in the chemical process industry. Some leakage in mechanical seals supposedly is necessary to prevent overheating (Figure 22-17 is a diagram of a single mechanical seal). Nonetheless, some recent cartridge designs reportedly are better than mechanical seals used in the past.

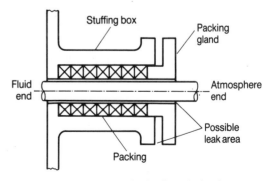

Figure 22-16.   Diagram of a simple packed seal.

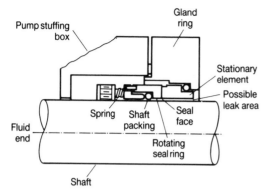

Figure 22-17.   Diagram of a basic single mechanical seal.

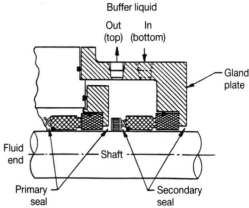

Figure 22-19.   Diagram of a double mechanical seal (tandem arrangement).

Dual mechanical seals can have a barrier fluid between the seals. The barrier fluid is maintained at a higher pressure than the process fluid (see Figure 22-18) so that leakage occurs into the process. When the process fluid is at a higher pressure than the barrier fluid, a tandem seal is used (see Figure 22-19). Dual seal applications require an appropriate barrier fluid and a system for handling the barrier fluid.

Canned or magnetically coupled pumps do not have a shaft through the housing; consequently, they have no shaft leakage. However, they are not practical in all applications. For example, they generally come in small sizes, they are more expensive, and heat buildup can be a problem. Other specialty pumps (e.g.,

diaphragms or squeezed tube designs) have much more limited applicability.

Agitators
Agitators are similar to pumps but have longer shafts. As a result, seals may see more vigorous action. Horizontal designs can use the same seal designs as pumps, but vertical designs generally do not.

Compressors
There are two principal types of compressors used: centrifugal (rotating) and reciprocating. Centrifugal compressors are more common, but

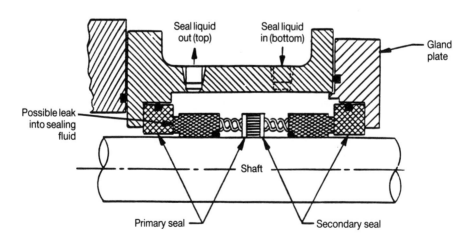

Figure 22-18.   Diagram of a double mechanical seal (back-to-back arrangement).

high-pressure requirements can mandate reciprocal designs. Most types of low-emission seals on reciprocating compressors are designed to contain leakage so that it can be treated. With centrifugal compressors there are several seal types to choose from.

As with pumps, seals by their nature allow some leaking. For example, some leakage generally is required for lubrication or cooling. Labyrinth seals are the simplest and most trouble free (see Figure 22-20). They also generally leak and typically are used only for operations such as compressing air where leakage is of no concern. Other types include liquid film (i.e., bushing) seals (see Figure 22-21), mechanical (contact) seals (see Figure

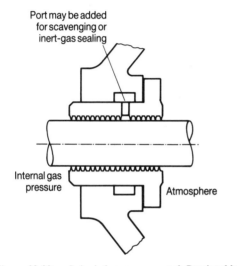

Figure 22-20.   Labyrinth compressor seal. Reprinted by courtesy of the American Petroleum Institute.

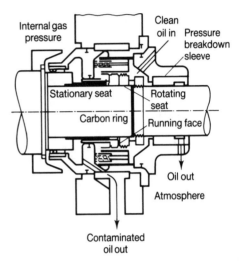

Figure 22-22.   Mechanical contact compressor seal. Reprinted by courtesy of the American Petroleum Institute.

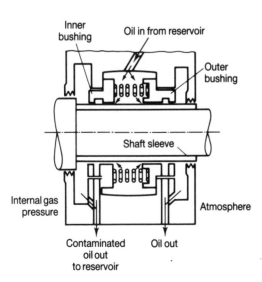

Figure 22-21.   Liquid film compressor seal. Reprinted by courtesy of the American Petroleum Institute.

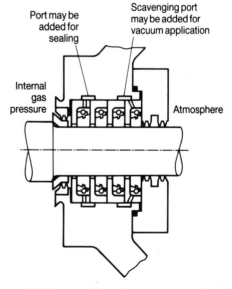

Figure 22-23.   Restrictive ring compressor seal. Reprinted by courtesy of the American Petrolem Institute.

22-22), restrictive-ring seals (see Figure 22-23), and mechanical seals. The liquid film (bushing) seal and the mechanical (contact) seal require a liquid lubricant. Pressure of the liquid can be maintained above that of the process preventing process gases from escaping. This is not a practical system, however, when seal oils cannot be tolerated in the gas stream. The restrictive-ring seal uses a series of rings made of a self-lubricating material. This seal can rely on its natural lubrication, in which case there will be some gas loss. It also can be operated with a seal liquid or buffer gas between the rings. Mechanical seals can use either liquids or gases for lubrication; the atmospheric side of the seal can be vented to a treating system. Double mechanical seals can be used with a buffer gas to prevent outward leakage.

Each type of seal has advantages and disadvantages. Selection normally depends on process gas conditions (i.e., pressure and temperature) and availability of compatible materials for use with the gas and the process. As with valves and pumps, there are diaphragm compressors that eliminate the need for seals and subsequently eliminate leakage; however, their applicability is limited.

Open-Ended Lines

When an open-ended pipe is sealed against the process fluid using only one valve, it is called an open-ended line. Because valves can leak, additional protection is required by most leak prevention regulations. This protection can be provided by installing a second valve or by capping the line with a cap, plug, or blind flange.

Pressure Relief Valves

Pressure relief valves are designed to allow process fluids to escape to prevent overpressure damage to equipment (see Figure 22-24). These valves are designed to 'reseat' when the pressure is relieved to prevent the continued loss of process fluids, as happens when a rupture disc relieves the pressure. Rupture discs also can be used alone or in series with relief valves. When the combination approach is used, no leakage occurs until the rupture disc fails. After the

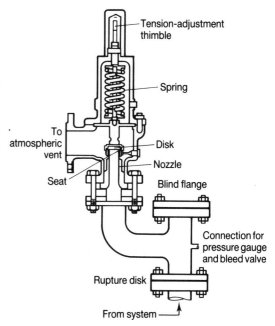

Figure 22-24.   Rupture disk installation upstream of a relief valve.

overpressure is vented, the relief valve reseats and prevents further loss of process fluids. The rupture disc then is replaced.

Emissions also can occur when the relief valve is set to open at a pressure too close to the process operating pressure. In such instances, repeated opening and closing of the valve (called simmering) can occur. Proper selection of valves and the margins between set pressure and process operating pressure are important to prevent these losses. Simmering is eliminated by proper selection of the set point.

In addition to simmering, leakage can occur when the valve does not seat well. These situations can be dealt with in a variety of ways. A seating problem often is eliminated through use of soft (i.e., polymeric) seats. In some cases, relief valve discharges are sent to a control device. This practice is not always considered to be prudent, however, and has been implicated in some explosions when valves failed to operate. Relief valves should be monitored after a release to verify that they have reseated properly. Most leak control regulations require such monitoring.

Samplers

Purging a sample line to obtain a pure process sample is prohibited by most leak control regulations. Instead, sample lines that recycle process fluids back to the process until the sample is to be taken are now required by several government regulations.

Add-On Controls

If fugitive emissions are contained, they can be treated much as stack emissions can. However, containment generally is not practiced for fugitive emissions control. When processes are enclosed it is generally done to protect the equipment and workers from the weather. In such cases the concentration of leaking compounds in the building exhaust generally is low. In other cases, the exhaust concentration may be high enough to analyze but not high enough to allow effective abatement. However, there are instances where containment is used. In those situations worker entry is not allowed. If a serious leak occurs in those situations, treatment of the exhaust may be practical.

Lines venting relief valves and barrier fluids from dual mechanical seal pumps can be routed to control devices. When liquids may be exhausted with gases, as when a disc rupture occurs, a catch tank frequently is used. For gases, flares are the most commonly used control device. EPA regulations place limits on flare design parameters based on hydrocarbon combustion properties. These limits are inappropriate when hydrogen is used as the fuel. Hydrogen is as good a flare fuel as hydrocarbons but requires a greater velocity because it burns faster.

EVAPORATIVE SECONDARY EMISSIONS

Secondary emissions are a type of fugitive emission that result from the evaporation of volatile materials from water or land surfaces. For water surfaces the magnitude of these emissions is proportional to the mass transfer rates of the system, the intensity and amount of turbulent flow, and the volatility (a function of vapor pressure and solubility) of the specific compounds in the mixture. Aeration processes such as mechanical surface agitators and cooling towers can strip volatiles from water media and release them to the atmosphere. From land surfaces (e.g., landfills and landfarms), materials also can volatilize below ground and migrate

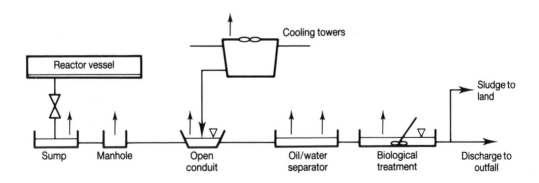

Sources indirectly related to the process stream include:

- Cooling towers
- Sewers
- Flumes/trenches
- Sumps
- Equalization basins
- Wastewater tankage
- Biological treatment systems
- Landfills
- Land treatment sites

Figure 22-25.   Secondary emission sources.

to the surface due to adsorptivity, changes in temperature, barometric pressure, and wind vortices. Cracks and fissures in the soil channel flow upward to the surface. Secondary emission rates from land surfaces generally are much lower than from water surfaces. Both are low enough, however, to strain the most sensitive analytical methods, with concentrations usually in the parts per billion by volume range. Figure 22-25 illustrates a variety of secondary emission sources.

Regulatory Framework

The regulatory framework surrounding secondary emissions began only recently. Through the 1980s, EPA and industry both studied secondary emissions to assess the magnitude and potential risk impact from the release of hazardous air pollutants. The first regulation was the NSPS published in 1989 for the petroleum refinery source category.[12] It required refiners to install airtight covers on oil–water separators, dissolved air flotation, and induced flotation units, and water seals on process drains. In 1990, EPA published a NESHAP for benzene.[13] The benzene NESHAP not only was directed at controlling vents and transfer operations, it also targeted benzene waste streams, including wastewater with greater than 10 ppm benzene. EPA also published a standard for treatment, storage, and disposal facilities (TSDF) under the Resource Conservation and Recovery Act (RCRA).[14] Only an initial phase of the RCRA TSDF standard was promulgated, however, targeting vents and fugitive emissions. In a later phase, control of secondary emissions has been proposed that targets surface impoundments, tanks, and containers containing organics at concentrations greater than or equal to 500 ppm.

An important NESHAP with an anticipated late 1992 proposal is the Hazardous Organic NESHAP (HON). In addition to vents, transfer operations, storage, and fugitive emissions, this NESHAP targets wastewater collection and treatment systems for any stream with >500 ppm of chemicals from a list of 112 provided in the regulations. In its preproposal form, the HON only targeted the SOCMI source category, but it likely will serve as a template for other regulations in the future.

Emission Estimation

Because measurement of secondary emissions is complicated and expensive, computer models have been used to predict the volatilization of organic chemicals from water and land surfaces. Two common software programs available from EPA are CHEMDAT 7 and SIMS 2.0. CHEMDAT 7 simulates both aerated and nonaerated wastewater units as well as landfills. SIMS 2.0 only simulates wastewater units but can link the entire chain of units into a series to represent a complete wastewater system. Input for both programs is similar, and both rely heavily on the idealistic Henry's Law constant of each specific compound for mass transfer estimation. Other vendors and universities have developed software that also models wastewater units and even collection systems (EPA 1989i; EPA 1990i; Berglund and Whipple 1987).

Measurement Techniques

Six measurement techniques generally can be used to assess secondary emissions of organic and inorganic compounds from open ditches, surface impoundments, wastewater treatment units, land treatment areas, and landfills. Some methods have more than one application, but no one method is universally appropriate for all situations. Table 22-13 lists the measurement methods and their possible applications. This list is by no means all inclusive. For example, several additional optical methods are in prototype stage.

---

[12] 40 CFR 60, Subpart QQQ.
[13] 40 CFR 61, Subpart FF.
[14] 40 CFR 260, 261, 264, 265, 270, and 271

Table 22-13   Air measurement methods and applications

Method	Type[a]	Class[b]	Application
Concentration-profile	I	OC, IC	Surface impoundments Land treatment Aeration basins Large open-top tanks
Transect-technique	I	OC, IC	Surface impoundments Land treatment Landfills Aeration basins Large open-top tanks
Isolation flux chamber	D	OC, IC	Surface impoundments Land treatment Landfills Large open-top tanks
Broad band infrared	D	OC	Open-top tanks Clarifiers Aeration units
Dual-beam laser (Dial)	D	OC, IC	Surface impoundments Land treatment Holding tanks Clarifiers Equilization basins Aeration units
Fourier transform infrared	D	OC, IC	Surface impoundments Land treatment Holding tanks Clarifiers Equilization basins Aeration units

[a] Measurement type: I = indirect, D = direct.

[b] Species: OC = organic compounds, IC = inorganic compounds.

The first two methods in Table 22-13, concentration-profile method and transect technique, are indirect measurement techniques; the last four methods, isolation flux chamber, broadband infrared, dual-beam laser, and Fourier transform infrared, directly measure secondary emissions from water and land surfaces. A brief description of each method and their limitations is presented below.

Concentration-Profile Method

The concentration-profile (C-P) method is an indirect sampling technique based on experimental measurements of wind velocity, volatile species concentration, and temperature profiles in the boundary layer above the emitting surface (Figure 22-26). A vertical series of air samples is collected in the boundary layer zone to determine the volatile species concentration as a function of height. The C-P method is used both for organic and inorganic species, by selecting the most appropriate collection/sampling system (e.g., cryogenic, solid absorbent, or liquid absorbent traps and trains). After collection and analysis, the slope determinations from concentration and velocity profiles can be extrapolated to the air/water interface to calculate the mass flux rate. Figure 22-27 illustrates the sampling system. It consists of a mast with a wind direction indicator, wind speed sensors, temperature sensors, and air collection probes spaced at six intervals.

The C-P method was developed by Thibodeaux, Parker, and Hick (1981) at the University of Arkansas. The literature, however, does not indicate frequent use. An EPA report

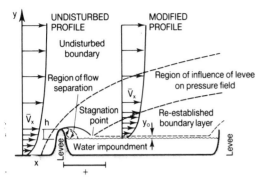

Figure 22-26. Boundary layers above a surface impoundment.

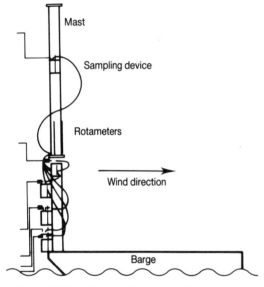

Figure 22-27. Concentration mast in place on barge.

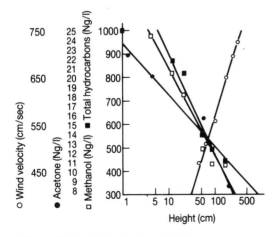

Figure 22-28. Field data from Mill 2.

by Thibodeaux, Parker, and Hick (1982) describes the C-P method, including the technical background of the micrometeorology, the limitations of the analytical model, results of a field study, and guidelines for proper sampling method. Figure 22-28 illustrates an example of field results from an aeration basin adjacent to a pulp and paper mill. Note that the concentrations are low and that they decrease with height as expected.

The C-P method can provide valid data for some compounds but not others from a surface impoundment or land treatment plot when compared to measurements in water. The difficulty of obtaining a valid logarithmic concentration profile in many cases must be attributed to the low concentrations of components and associated variability at such low detection levels. The method must be used in the turbulent air boundary layer very near the source (i.e., less than two meters above the surface) and far enough downwind from upwind flow disturbances so it is in the reestablished boundary layer region. The variability associated with the emission rates measured by this technique are typically in the range of factors of two to ten on the low side.

Transect Technique

The transect technique is an indirect emission measurement approach for measuring fugitive particulate and gaseous emissions from area and line sources. The method is analogous to traversing that is used in stack sampling procedures. Instead of traversing the diameter of a stack, however, the transect method uses an array of fixed instrumented towers to analyze the downwind plume of a source. Horizontal and vertical arrays of samplers measure concentrations of organic or inorganic compounds within the effective cross section of the fugitive emission plume. A normal concentration distribution or curve is fitted to the measured concentration. The volatile species emission rate is obtained by spatial integration of the concentration over the assumed plume area or by virtual-source Gaussian receptor modeling.

The sampling equipment typically consists of five sampling masts aligned perpendicular to

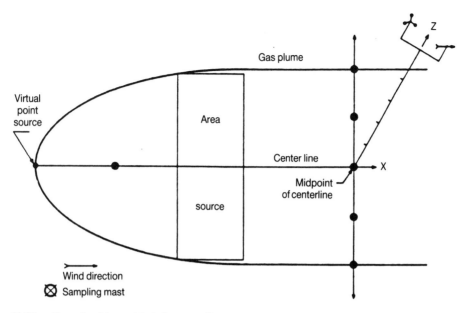

Figure 22-29.   Example of transect technique sampling.

the plume downwind and a sixth mast placed at an upwind location. The central mast downwind has multiple samplers in a vertical array (e.g., three to five collectors) and instrumentation that measures the direction, speed, and temperature of the wind. The central mast is aligned with the expected plume centerline. The remaining four downwind masts have single samplers and are placed at equal distances to the left and right of the central mast. The spacing of the downwind array is selected to cover the expected horizontal plume cross section, as defined by observation or profiling with a real-time total hydrocarbon (THC) analyzer (Balfour and Schmidt 1984). Figure 22-29 is an example of the transect technique.

In using the transect technique, valid definition of the plume cross section is difficult because of low component concentrations, upwind backgrounds, and variability in plume location/concentrations. Accuracy is estimated to be within one to two factors of ten. A recent article by Esplin (1988) describes a boundary layer emission monitoring method that is a variation of the transect method. In this case, an area source is sampled continuously at three or more different elevations while traversing the

plume with a balloon tethered to a sampling cart. Field studies of a kidney-shaped aeration lagoon for total reduced sulfur compounds showed a method precision of 15 percent.

Isolation Flux Chamber

The emission isolation flux chamber (EIFC) is a device used to directly measure emissions of organic and inorganic compounds from land and liquid surfaces (Balfour and Schmidt 1984). The approach uses an enclosure device (i.e., flux chamber) to sample gaseous emissions from a defined surface area. Clean, dry sweep or purge air is added to the chamber at a known controlled rate. The volumetric flow rate of sweep air through the chamber is recorded and the concentration of the species of interest is measured as it leaves the chamber after equilibration is achieved.

A diagram of the flux chamber apparatus used for measuring emission rates is shown in Figure 22-30. The sampling equipment consists of a stainless steel and acrylic chamber with mixing impeller, ultrahigh-purity sweep air, a rotameter for measuring flow into the chamber, and a sampling manifold for monitoring or collecting the species of interest. Concentra-

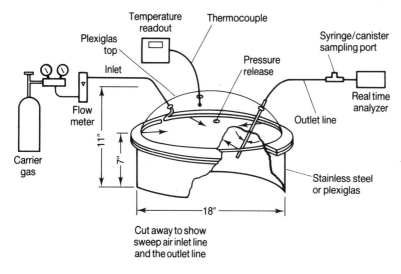

Figure 22-30.   Cutaway diagram of the emission isolation flux chamber and support equipment.

tions of total hydrocarbons are monitored continuously in the chamber outlet gas stream using a portable FID or PID. Samples are collected for subsequent gas chromatography (GC) analysis once a steady-state emission rate is obtained. Air, soil, and liquid temperatures are measured with a thermocouple. For large surface areas, a statistical sampling approach (e.g., gridding) is suitable for well-mixed impoundments or homogeneous land treatment. However, landfills with a variety of waste types can create highly variable data, making the representativeness of an 'average' emission rate questionable.

Several studies have been conducted to validate this method of measurement. Gholson et al. (1987) used a simulated surface impoundment (five foot square enclosed tank) to measure the emission rate of the flux chamber at several points on the surface. The results were compared to the emission rate from the total surface area of the enclosure.

The IR results showed that a consistent negative bias (from minus forty percent to minus 80 percent) exists for all flux chamber measurements. This bias became significantly more negative at a lower sweep rate (2 l/min). Precision was less than 5 percent under all conditions for the single component studies and

between 6 and 13 percent for the three-component study. Eklund et al. (1987) reported on a field study at an industrial wastewater lagoon, where average sampling variability was 25 percent and average analytical variability, using total nonmethane hydrocarbon (TNMHC) analysis, was 12 percent. There was no report of accuracy.

Broadband Infrared

This open-path technique uses infrared (IR) transmitters and receivers strategically positioned approximately one hundred meters apart. Increasing the spacing reduces the sensitivity. The first unit transmits a beam to the second receiver, which then transmits a beam to a third receiver, etc. The target gas must cross the IR beam to be detected. The transmitters and receivers must be positioned in sight of one another and at the same elevation. The IR beam passes through two filters in the receiver that compares the transmitted light to that absorbed by a known amount of the specific gas being monitored. The differential absorption of the beam can determine if the target gas is present at concentrations as low as 100 parts per million-meters (ppm-m) over one hundred meters. Specific gases must be detected by selecting a filter that matches the wavelength of maximum absorption of the gas. The reference wavelength is selected to eliminate interference from precipitation and dust. Heavy fog can scatter the beam, but the receiver is effective until approximately 90 percent of the beam is scattered.

Another open-path IR system uses a gas filtration correlation technique with two gas cells. One cell contains a sample of the target gas; the other is a reference cell. Incoming light is focused through the cells onto an IR detector. The difference in the spectral transmittance is an indication of the amount of target gas present in the path of the incoming radiation. It has a sensitivity of 20 ppm-m over the cell length and is intended to be used with an airborne platform for detecting hydrocarbon leaks from natural gas pipelines. It may be limited to daytime use because it requires a four-degree temperature contrast between ground and target gas.

The major problem with optical measurement systems is converting the analytical results into mass flux from the emitting surface. An IR analyzer reports concentrations as a function of path length. A 100 ppm-m concentration can be interpreted as 100 ppm over one meter or 2 ppm over fifty meters. It is necessary to know the optical path length. More importantly, one must know the overall plume dimensions and relate the measured line-of-sight concentration per meter to an emission rate. An open-top tank as depicted in Figure 22-31 has a defined air space above the surface. Measurements taken with no wind generally will be accurate with quasi-steady-state conditions. However, a moderate

wind partially dilutes air above the tank surface and causes large errors in measured versus actual emission rates. Under these conditions, it is better to measure just downwind from the tank and use a defined area and measured wind speed to quantify airflow through the measurement path. For aeration tanks, the rate of air sparging from the bottom of the tank can be used to convert the path length concentration to surface flux rate, analogous to flow in a stack.

Dual Beam Laser

This method is a direct measurement technique similar to the dual-beam IR, except that a laser beam has a much greater range (200 to 500 kilometers). Its principle of operation is differential absorption lidar (Dial) technology for remote sensing of fugitive emissions (McClenny et al. 1974; Measures 1984). A carbon dioxide laser with a wavelength coincident with the gas under investigation is used to transverse the monitoring area. The beam is returned to the system using optical mirrors and its signal level is measured. The difference between return signal and output signal provides a measure of the amount and concentration of gas present. Changes in the environment because of weather or other interfering material are taken into account by using a second laser whose wavelength is not absorbed by the gas (see Figure 22-32).

One method of laser deployment is a perimeter technique that is used to monitor a plant fenceline at a constant elevation. A more innovative method uses a scanning laser to detect chemicals at various elevations around a unit. The laser may be elevated to scan down at targets or located at ground level to scan upward at cube-corner retroreflectors (mirrors). Choice of wavelengths is crucial to accuracy. A full knowledge is needed of the gases that may be present at a specific site. This is important to remove possible effects due to interfering gases. Commercial units are gas specific and can detect a variety of organic compounds with minimum detection in 500 meters of 5 to 800 parts per billion (ppb). The reported accuracy is 5 percent (LASERSAFE). This again represents an accuracy over the measured path length.

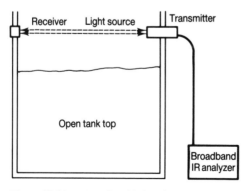

Figure 22-31.   Broadband infrared.

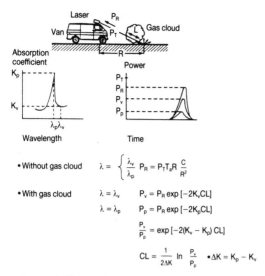

Figure 22-32.   Differential absorption technique.

Conversion of line-of-sight concentration per meter to emission flux rate can introduce considerable error when defining plume boundaries and integrating over spatial volume.

Another approach to converting line-of-sight concentrations to an emission flux rate uses a tracer gas. Winds disperse the calibrated release of a tracer at an upwind location. Near simultaneous analysis of both tracer and chemical species of interest allows calculation of path concentration and emission flux rate based on the direct proportionality of tracer dispersion measurement.

Fourier Transform Infrared (FTIR)

This optical method is a direct measurement technique that can measure both organic and inorganic emissions. FTIR evolved from the remote optical sensing of emissions (ROSE) system developed in the mid-1970s and extended its sensitivity by at least two factors of ten (spectral resolution = $0.06\,\mathrm{cm}^{-1}$). It can be used for long-path measurements and remote-stack sampling.

The FTIR differs from the conventional IR spectroscopic method in that the light source is remote from the analyzer. A light-source telescope is positioned so that the emission source is between the remote light-source and the analyzer, with associated tracking mirror, receiver telescope, and interferometer (see Figure 22-33). The nominal sensitivity of the FTIR system is 1 ppm-m (or 1 ppb-km) for absorption measurements. Emission measurements of warm sources such as stacks are less sensitive (by about a factor of ten) than lower temperature (100°C) sources. Accuracy is 10 percent for the absorption mode and 20 percent

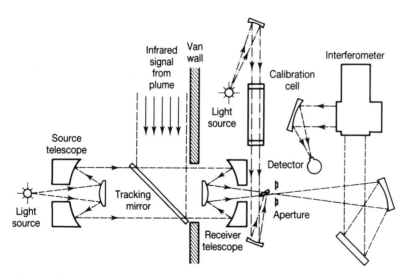

Figure 22-33.   Fourier transform infrared (FTIR) sampling.

for the emission mode (Herget 1978). As stated earlier, optical methods are accurate for the measurement of species in the path length. The accuracy of converting the line-of-sight concentration to a surface emission flux rate, however, depends on good experiment design to define the plume boundaries or use of a dispersed tracer.

Herget and Brasher (1979) and Herget and Brasher (1980) describe long-path absorption measurements made at sources such as a gypsum pond, oil refinery, pulp and paper mill, and a large brick kiln. These field studies included measurements of inorganics such as hydrogen fluoride, nitric oxide, sulfur dioxide, carbon monoxide, and carbon dioxide.

Conclusion

Some of the techniques discussed in this section are precise (i.e., repeatable) but lack accuracy. Unfortunately, there have not been enough rigorous comparative studies to provide a sufficient data base to validate these techniques. The isolation flux chamber (IFC) and laser remote sensing currently are the more popular methods because of their inherent lower variability and ease of use. However, the IFC is not well-suited to measuring emissions from extremely large bodies of varied composition or highly agitated liquid surfaces; laser and optical techniques also lose accuracy when converting line-of-sight path concentration to surface flux rate.

In selecting a method, the ultimate choice depends on the strength of emission source, the type and size of source, environmental factors, and cost. New methods and improved instruments are increasingly entering the market as interest in secondary emissions measurement techniques escalates. Baker and McCready (1988) provide a comprehensive overview of plant-wide remote sensing techniques, as of January 1988, which can be applied to fugitive and secondary emission measurements; the Research Triangle Institute (RTI) also published a review of methods for remote sensing from stationary sources (Saeger et al. 1988); and Minnich et al. (1989) presented a paper on the subject.

Controlling Secondary Emissions

Growing concerns over exposure to chemicals in the air and water have caused industry and regulatory agencies to focus on the reduction of volatile emissions from the collection and treatment of industrial wastewater. The chemical industry has initiated measures to assess and reduce risks associated with air emissions from its facilities. Measures such as process changes and installation of new and more efficient production units and specific emission control equipment have reduced air emissions significantly from chemical processes. Industry and others also are evaluating the significance of volatile emissions during the collection and treatment of industrial wastewater.

There are a variety of ways to collect and treat wastewater; this increases the difficulty in estimating air emissions of volatiles from wastewater collection and treatment facilities. The data needed to assess emissions from several types or variations of collection or treatment facilities also often are unknown, unavailable, incomplete, or unproven. Consequently, the effectiveness of various wastewater collection and treatment facilities in removing volatiles is being evaluated slowly. Because there is no single scheme used to collect and treat facility wastewater, it currently is impossible to develop model plants representing all possible scenarios. The section below presents generally accepted practices. Other useful references include Kincannon et al. (1983), Freeman et al. (1984), and EPA (1988k).

Application to Wastewater Treatment Facilities (WWTF)

As described earlier, emissions of volatiles from wastes can occur at a number of points during the transport, treatment, and disposal process. There are four main approaches to controlling these emissions.

- Operation and maintenance guidelines for the site
- In-situ modifications to process units in WWTF
- Design changes to new units in WWTF

- Pretreatment or elimination of some streams

Operation and maintenance guidelines include using subsurface injection at land treatment sites, minimizing turbulence in treatment units, and requiring submerged draining of tank trucks. In situ modifications include designing the process to provide only the oxygen necessary to support biodegradation by the microorganisms or requiring construction of tall, deep tanks or storage basins. Pretreatment includes any approach to remove the volatilizing material from the wastewater before the emission source. Removal could occur at the process unit or at the front of the WWTF.

Option Selection
Factors that should be considered when evaluating any of these options include:

- Application of the control option to the specific unit, but considering other downstream process units handling the same waste within the site.
- Impending changes to the chemical plant, the WWTF, or the wastewater load and composition.
- Source of emitting components that may need to be reduced (e.g., individual or multiple units, process wastewater, or tank truck).
- Tertiary generation of waste resulting from a control option.

Overall, it appears that if the waste components are to be biodegraded in a biological treatment system, options are favored that allow the chemicals to reach the biological system for ultimate disposal.

Minimization of Volatiles in the Wastewater
Foremost among control strategies employed to reduce volatile emissions to the air from industrial wastewater is to reduce volatiles in the wastewater stream before it contacts ambient air. Minimizing the volatiles before the wastewater contacts ambient air greatly reduces the potential for air emissions from the downstream collection and treatment units. Steam stripping is a proven technology that involves the fractional distillation of wastewater to remove volatile constituents. The operating principle of steam stripping is the direct contact of steam with wastewater. This contact provides heat for vaporization of the volatile constituents. The overhead vapor containing water and organics is condensed and separated, usually by decantation, to recover the constituents. The recovered materials usually are recycled or incinerated. In principle, a multistage steam stripper system can be designed to achieve almost any level of volatile removal. In practice, however, removal efficiencies are determined by practical limits in column design, operation, and cost.

Other technologies are available for removing volatile pollutants from wastewater. These include air stripping, carbon and ion exchange, adsorption, chemical oxidation, membrane separation, and liquid-liquid extraction. With the exception of air stripping, the removal efficiencies of these technologies are dependent on physical properties other than pollutant volatility in water. Although these technologies generally may not be as effective at reducing air emissions as steam strippers, they may be more effective at removing certain volatiles and also may be more practical than steam stripping for some situations.

Controlling Emissions
The second strategy that can be used to reduce volatile emissions to the air is to control the various sources located within wastewater collection and treatment systems. Typical wastewater collection and treatment facilities incorporate unit operations arranged in series or in parallel and culminate in a biological treatment system. These can include drains, junction boxes, lift stations, manholes, trenches, weirs, sumps and surface impoundments, collection facilities, oil/water separators, equalization basins, neutralization basins, clarifiers, aeration basins, pH adjustment tanks, and flocculation tanks. Many of these collection and treatment system units are open to the atmosphere and allow volatile-containing wastewater to contact ambient air. Whenever this happens, there is a

potential for emissions to the air. The pollutants volatilize in an attempt to reach equilibrium owing to differences in partial pressures above the wastewater. In doing so, they are emitted to the ambient air surrounding the collection and treatment units. The magnitude of emissions depends on factors such as the physical properties of the pollutants, the temperature of the wastewater, and the design of the individual collection and treatment units. All of these factors, as well as the general scheme used to collect and treat facility wastewater, have a major effect on emissions to the air.

Collection and treatment schemes are facility-specific. The flow rate and composition of wastewater streams at a particular facility are functions of the processes used. The wastewater flow rate and composition, in turn, influence the size and types of collection and treatment units that must be employed at a given facility. Figure 22-34 illustrates a typical scheme for collecting and treating process wastewater generated at a facility and the opportunity for volatilization of pollutants. Figure 22-34 illustrates wastewater being discharged from process equipment into a drain. Drains typically are open to the atmosphere and provide an opportunity for volatilization from the wastewater. The drain normally is connected to the process sewer line that carries the wastewater to the downstream collection and treatment units.

Figure 22-34 also illustrates the wastewater being carried past a manhole and onto a lift station where several process wastewater streams are joined. The manhole provides an escape route for volatiles in the sewer line, and the junction box, if open to the atmosphere, allows an additional opportunity for volatilization. Wastewater is discharged from the junction box to a lift station where it is pumped to a treatment system. Open lift stations provide additional opportunity for volatilization. The equalization basin, the first treatment unit shown in Figure 22-34, regulates the wastewater flow and pollutant compositions to the remaining treatment units. Equalization basins also typically provide a large area for wastewater contact with the ambient air. For this reason, emissions may be high from this unit. Suspended solids are removed in the clarifier and the wastewater then flows to the aeration basin where microorganisms act on the organic constituents.

Both the clarifier and the aeration basin typically are open to the atmosphere. Further, the aeration basin normally is aerated either mechanically or with diffused air. Wastewater leaving the aeration basin normally flows

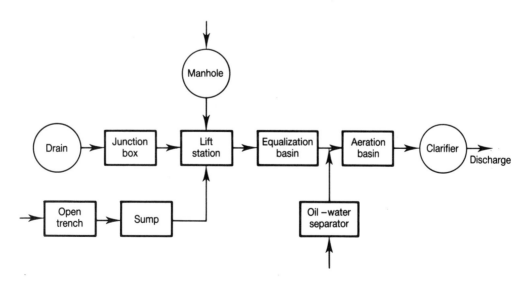

Figure 22-34.   Typical wastewater collection and treatment schemes.

through a secondary clarifier for solids removal before it is discharged from the facility. The secondary clarifier also likely is open to the atmosphere. Part of the solids that settle in the clarifier are discharged to a sludge digester; the remainder are recycled to the aeration basin. Finally, waste sludge from the sludge digester generally is landfilled or land farmed.

Performance of Emission Controls

The desired control of emissions from wastewater treatment facilities is possible only with a thorough understanding of the liquid and gas mass transfer processes, especially recognizing the concept of the rate-limiting step. Air emissions are limited by the transfer of the volatile compound through the bulk liquid, through the gas phase, or across the gas-liquid interface. For a specific chemical in a given system, the resistance to transfer through one of these phases usually dominates the volatilization process. Emissions occur predominantly by diffusive and convective mechanisms, influenced by wastewater stream composition, physical properties, unit design, and temperature. Pollutants in the wastewater volatilize into the ambient air just above the liquid surface in an attempt to reach equilibrium between the liquid and the upper phases of the contents and of the air above the liquid. Volatility of the pollutant in water is the most significant physical property affecting rate of emission.

Emissions from a unit or from a system may be controlled by eliminating or altering the volatile concentration in the waste stream, by changing the composition of the volatiles contained in the waste stream, sometimes by adding or combining streams and sometimes by separating streams, or by altering the amount or the mode of contact and turbulence between the waste stream and the ambient air. For emissions from a single source, it is important to recognize that effective controls cannot be developed for a particular waste treatment unit without considering other units within that facility, or anticipating the eventual fate of the waste stream. If a chemical is degraded eventually in a biological treatment system, recovered, or incinerated, then upstream controls to contain

the chemical within the waste will be highly effective. If a downstream unit contains a highly turbulent point from which significant emissions could occur, then the effectiveness of keeping the chemical in the waste stream may be low.

The following three steps are used to assess the need to control air emissions from a wastewater treatment facility.

- Characterize the waste
- Characterize the process units in the facility and the mass transfer mechanisms involved
- Determine the emissions potential of the units from measurements or from models

Failure to develop a sound basis for the emissions control methods selected could result in the wrong units being controlled or the development of inappropriate controls. The ultimate control strategy may be as simple as reducing the turbulence during waste addition to a tank or reducing the amount of mixing in a tank, or it may require additional controls on the discharge of a specific chemical at the source. Whatever controls are selected, they must be compatible with the specific conditions involved in the system under consideration.

In the end, the development and use of effective technology for controlling emissions to the air from a wastewater treatment facility is a facility-specific undertaking. Such assessments require knowledge of the chemical and physical properties plus the variability limits of the waste stream, the environmental conditions, the unit and facility design and operating characteristics, and the proper application of emissions measurements or predictive models.

References

Baker, G.L. and D.I. McCready. 1988. Plant-wide remote detection of chemical emissions – One company's perspective. Presented at the 81st Annual Meeting of the Air Pollution Control Association. Dallas, TX.

Balfour, W.D. and C.E. Schmidt. 1984. Sampling approaches for measuring emission rates from hazardous waste disposal facilities. Presented at the 77th Annual Meeting of the Air Pollution Control Association. San Francisco, CA.

Berglund, R.L., R.R. Romano, and J.L. Randall. 1990. Fugitive emissions from the ethylene oxide production industry. *Environmental Progress* 9(1):10–17.

Berglund, R.L. and G.M. Whipple. 1987. Predictive modeling of organic emissions. *Chemical Engineering Progress* 83(11):46–54.

Blacksmith, J.R., G.E. Harris, and G.J. Langley. 1980. Frequency of leak occurrence for fittings in synthetic organic chemical plant process units. Prepared under EPA Contract No. 68-02-3171–1 by the Radian Corporation for the U.S. Environmental Protection Agency, Office of Air Quality Planning and Standards. Research Triangle Park, NC.

Boland, R.F., T.E. Ctvrtnicek, J.L. Delaney, D.E. Earley, and Z.S. Khan. 1976. Screening study for miscellaneous sources of hydrocarbon emissions in petroleum refineries. EPA-450/3-76-041. U.S. Environmental Protection Agency, Office of Air Quality Planning and Standards. Research Triangle Park, NC.

Brown, G.E., D.A. DuBose, W.R. Phillips, and G.E. Harris. 1980. Response factors of VOC analyzers calibrated with methane for selected organic compounds. EPA-600/2-81-022. U.S. Environmental Protection Agency, Office of Research and Development. Research Triangle Park, NC.

Bursey, J.T., J.A. Sokash, R.A. McAllister, T. Pauling, and F.L. Sowers. 1991. Method 21 evaluation for the HON. EPA Publication DCN No. 91-275-026-30-02. U.S. Environmental Protection Agency, Office of Air Quality Planning and Standards. Research Triangle Park, NC.

Chemical Manufacturers Association (CMA). 1989. Improving air quality: Guidance for estimating fugitive emissions from equipment. Washington, D.C.

Choi, S.J., R.D. All, M.R. Overcash, and P.K. Lim. 1992. Capillary-flow mechanism for fugitive emissions of volatile organics from valves and flanges: Model development, experimental evidence, and implications. *Environmental Science and Technology* 26(3):478–484.

Davis, B.C. 1989. A comparison of the fugitive emissions from two butadiene plants. Paper No. 89-30.2. Presented at the 82nd Annual Meeting of the Air & Waste Management Assoc. Anaheim, CA.

Dharmavaram, S., T.A. Kittleman, G.N. Vander-Werff, and D.J. Lonczak. 1991. Techniques for measuring hydrogen fluoride and chlorine releases for leaking equipment. Paper No. 91-79.1.

Presented at 84th Annual Meeting of the Air and Waste Management Association. Vancouver, BC.

Eaton, W.S., K.W. Carlton, R.A. Meyer, and M.L. Duckworth. 1978. Experimental techniques for the determination of fugitive hydrocarbon emissions. Presented at the 70th Annual Meeting of the Air Pollution Control Association. Toronto, Ontario, Canada.

Eklund, B., M. Kienbusch, D. Ranum, and T. Harrison. 1987. Development of a sampling method for measuring VOC from surface impoundments. Proceedings of 1987 EPA/APCA Symposium on Measurement of Toxic and Related Air Pollutants. Research Triangle Park, NC.

Enviroplan. 1991. Fugitive emissions correlation for SOCMI based on bagging studies of BD/EO production units. Ref. No. 51-9043-952. Study for Chemical Manufacturers Association. Roseland, NJ.

Esplin, G.J. 1988. Boundary layer emission monitoring. *Journal of the Air Pollution Control Association* 38:1158-61.

Freeman, R.A., V.M. Schroy, J.R. Klieve, and S.R. Archer. 1984. Air stripping of acrylonitrile from waste treatment streams. *Environmental Progress* 3(1):26–33.

Gholson, A.R., J.R. Albritton, R.K.M. Jayanty, J.E. Knoll, and M.R. Midgett. 1987. Field evaluation of flux chamber method for measuring volatile organic emissions from hazardous waste surface impoundments. In: Proceedings of the 1989 EPA/A&WMA International Symposium Measurement of Toxic and Related Air Pollutants. Publication No. VIP-13. EPA 600/9-89-060, p. 358. Published by A&WMA, Pittsburgh, PA.

Hanzevack, K.M. 1978. Fugitive hydrocarbon emissions measurement and data analysis methods. Symposium/Workshop on Petroleum Refinery Emissions. Jekyll Island, GA.

Herget, W.F. 1978. *American Laboratory* 14:72–78.

Herget, W.F. and J.D. Brasher. 1979. *Applied Optics* 18:3404–20.

Herget, W.F. and J.D. Brasher. 1980. *Optical Engineering* 19:508–14.

Hess, J.L. and T.A. Kittleman. 1983. Predicting volatile emissions. *Chemical Engineering Progress* 79(11):40–43.

Honerkamp, R.L., L.P. Provost, J.W. Sawyer, and R.G. Wetherold. 1979. Valve screening study of six San Francisco Bay area petroleum refineries. Report No. DCN 79-219-370-03. Prepared by the Radian Corporation. Austin, TX.

Hosler, A. D., T.N. Grosshandler, R.L. Marchman. 1992. Fugitive emission: The need for more accurate measurement techniques. Paper No. 92-137.06. Presented at 85th Annual Meeting of the Air & Waste Management Association. Kansas City, MO.

Hustvedt, K.C. and R.C. Weber. 1978. Detection of volatile organic compound emissions from equipment leaks. Paper No. 78-36.1. Presented at the 71st Annual Meeting of the Air Pollution Control Association. Houston, TX.

Kincannon, D.F., E.L. Stover, V. Nichols, and D. Medley. 1983. Removal mechanisms for toxic priority pollutants. *Journal of the Water Pollution Control Federation* 55(2):157–163.

Kittleman, T.A. 1992. Evaluation of recent bagging test results for the chemical industry. Paper No. 61c. Presented at 1992 Summer National Meeting, American Institute of Chemical Engineers.

Langley, G.J., S.M. Dennis, L.P. Provost, and J.F. Ward. 1981. Analysis of SOCMI VOC fugitive emissions data. EPA-600/2-81-111. U.S. Environmental Protection Agency, Office of Research and Development. Research Triangle Park, NC.

Langley, G.J. and L.P. Provost. 1981. Technical note: Revision of emission factor for nonmethane hydrocarbons for valves and pump seals in SOCMI processes. Report No. DCN #222-018-04-41. Prepared under EPA Contact No. 68-02-3542. U.S. Environmental Protection Agency. Research Triangle Park, NC.

Langley, G.S. and R.G. Wetherold. 1981. Evaluation of maintenance for fugitive VOC emissions control. Report No. EPA-600/2-81-080. U.S. Environmental Protection Agency. Research Triangle Park, NC.

LASERSAFE Product Literature. Environmental Laser Systems, Inc. Atlanta, GA.

McClenny, W.A. et al. 1974. Methodology for comparison of open-path monitors with point monitors. *Journal of the Air Pollution Control Association* 24:1044–46.

Measures, R.M. 1984. *Laser Remote Sensing – Fundamentals and Applications.* New York: John Wiley & Sons.

Minnich, T.R., R.J. Bath, R.M. Naman, R.D. Spear, O.A. Simpson, J. Faust, W.F. Herget, D.H. Stedman, S.E.McLaren, and W.M. Vaughan. 1989. Remote sensing of air toxics for pre-remedial hazardous waste site investigation. Paper No. 89-108.1. Presented at the 82nd Annual Meeting of the Air & Waste Management Association. Anaheim, CA.

Pahel-Short, R. 1989. Presentation to equipment leak regulatory negotiation committee. Docket No. A-89-10. U.S. Environmental Protection Agency. Washington, D.C.

Radian Corporation. 1986. Emission factors for equipment leaks of VOC and HAP. EPA-450/3-86-002. U.S. Environmental Protection Agency, Office of Air Quality Planning and Standards. Research Triangle Park, NC.

Randall, J.L., G. Harris, S. Wevill, and M. Bishop. 1989. Field study: Estimating fugitive emission from the phosgene production industry. Report No. DCN 89-259-056-01. Report prepared for the phosgene panel of the Chemical Manufacturers Association. Washington, D.C.

Rosebrook, D.D., R.G. Wetherold, C.D. Smith, G.E. Harris, and I.A. Jefcoat. 1978. The measurement of fugitive hydrocarbon emissions from selected sources in petroleum refineries. Presented at the 71st Annual Meeting of the Air Pollution Control Association. Houston, TX.

Saeger, M.L., C. Sokol, S.J. Coffrey, R.S. Wright, W.E. Farthing, and K. Baughman. 1988. A review of methods for remote sensing of atmospheric emissions from stationary sources. EPA Contract No. 68-02-4442. U.S. Environmental Protection Agency. Research Triangle Park, NC.

Schaich, J.R. 1991. Estimate fugitive emissione from process equipment. *Chemical Engineering Progress.* August 1991, pp 31–35.

Taback, H.J. 1978. Petroleum refinery fugitive emissions measurement emission factors and profiles. Symposium/workshop on petroleum refinery emissions. Jekyll Island, GA.

Tasher, S.A. 1982. Letter to F. Porter. On: Draft guideline series for the control of volatile organic fugitive emissions from synthetic organic chemical, polymer and resin manufacturing equipment. February 15, 1982.

Thibodeaux, L.J., D.G. Parker, and H.H. Hick. 1981. Quantifying organic emission rates from surface impoundments with micrometeorological and concentration profile measurements. Presented at the annual meeting of the American Institute of Chemical Engineers. New Orleans, LA.

Thibodeaux, L.J., D.G. Parker, and H.H. Hick. 1982. Measurement of volatile chemical emissions from wastewater basins. EPA-600/2-82-095. U.S. Environmental Protection Agency, Office of Research and Development. Research Triangle Park, NC.

U.S. Environmental Protection Agency (EPA). 1980e. VOC fugitive emissions in synthetic organic chemicals manufacturing industry –

Background information for proposed standards. EPA 450/3-80-033a. Office of Air Quality Planning and Standards. Research Triangle Park, NC.

U.S. Environmental Protection Agency (EPA). 1983c. Guideline series: Control of volatile organic compound equipment leaks from natural gas/gasoline processing plants. EPA 450/3-83-007. Office of Air Quality Planning and Standards. Research Triangle Park, NC.

U.S. Environmental Protection Agency (EPA). 1988g. Protocols for generating unit-specific emission estimates for equipment leaks of VOC and VHAP. EPA 450/3-88-010. Office of Air Quality Planning and Standards. Research Triangle Park, NC.

U.S. Environmental Protection Agency (EPA). 1989i. Hazardous Waste Treatment, Storage and Disposal Facilities (TSDF) – Air Emission Models. EPA 450/3-87-026. Office of Air Quality Planning and Standards. Research Triangle Park, NC.

U.S. Environmental Protection Agency (EPA). 1990i. Background Document for the Surface Impoundment Modeling System (SIMS) Version 2.0. EPA 450/4-90-0196. Office of Air Quality Planning and Standards. Research Triangle Park, NC.

U.S. Environmental Protection Agency (EPA). 1991s. Negotiated regulation for leak detection and repair program for organic chemical industry. 56 FR 9315. March 6, 1991.

Wetherold, R.G. and L. Provost. 1979. Emission factors and frequency of leak occurrence for fittings in refinery process units. EPA 600/2-79-044. U.S. Environmental Protection Agency, Office of Research and Development. Research Triangle Park, NC.

Wetherold, R.G. and D.D. Rosebrook. 1980. Environmental assessment of atmospheric emissions from a petroleum refinery. EPA 600/2-80-075a, NTIS PB 80-225-253. U.S. Environmental Protection Agency, Office of Research and Development. Research Triangle Park, NC.

23

Pollution Prevention

James Cummings-Saxton

INTRODUCTION

As we entered the 1990s, pollution prevention became accepted as a central component of successful environmental protection. In October 1990, the Report to Congress (EPA 1991j) noted:

Environmental programs that focus on the end of the pipe or the top of the stack, on cleaning up after the damage is done, are no longer adequate. We need new policies, technologies, and processes that prevent or minimize pollution – that stop it from being created in the first place.

Shortly thereafter, Congress codified the importance of pollution prevention by enacting the Pollution Prevention Act of 1990 (U.S. Congress 1990). That Act stated that it is:

the national policy of the United States that pollution should be prevented or reduced at the source whenever feasible ...

Further, in accord with the expressed will of Congress and the Executive Branch, EPA asserted its commitment to pollution prevention by stating in the report to Congress:

... pollution prevention is the most effective way to reduce risks by reducing or eliminating the source of pollution entirely ... In short, pollution prevention offers the unique advantage of harmonizing environmental protection with economic efficiency.

HISTORICAL CONTEXT

Although acceptance has come recently, the concepts of pollution prevention and waste minimization are not new.[1] EPA actively sought to foster waste minimization in the early 1970s, at that time terming the approach 'alternatives to end-of-pipe treatment' and 'in-process change.' The Resource Conservation and Recovery Act (RCRA), passed by Congress in 1976 as the governing legislation for handling solid wastes, stated as its objective:

---

[1] In this chapter, the term pollution prevention is taken also to encompass the concepts associated with waste minimization. The relationships between the two are discussed.

to promote the protection of health and the environment and to conserve valuable material and energy resources ...

Although a principal motivation at that time was to conserve our natural resources in the wake of the 1973 oil embargo, the link with environmental protection was recognized clearly. The Hazardous and Solid Waste Amendments (HSWA) of 1984 further emphasized this link by stating:

The Congress hereby declares it to be the national policy of the United States that, wherever feasible, the generation of hazardous waste is to be reduced or eliminated as expeditiously as possible.

In spite of such initiatives, a 1986 report by the congressional Office of Technology Assessment (OTA 1986) found that:

Despite some claims to the contrary, industry has not taken advantage of all effective waste reduction opportunities that are available ... The attention and resources given to required pollution control activities limit the amount of thought, time, and money that industry can devote to waste reduction.

Two precursors motivated the OTA study and gave force to its findings – the approximately 3,000 fatalities and a larger number of serious injuries associated with the December 1984 toxic chemical release in Bhopal, India, and the continuing discovery in the U.S. of large numbers of uncontrolled hazardous waste disposal sites. The latter demonstrated that the conditions encountered in Love Canal, New York and Times Beach, Missouri, were not unique. Although the Bhopal and waste site situations were of a different nature, their combined effect on public consciousness was to highlight the potential dangers in handling toxic chemicals and to demonstrate that, once generated, many chemical wastes are difficult to treat safely and dispose.

An outcome of the importance attached to waste reduction in the aftermath of the OTA

report and associated Congressional discussions was inclusion of the Emergency Planning and Community Right-to-Know Act (EPCRA) as Title III of the Superfund Amendments and Reauthorization Act of 1986. A significant requirement of EPCRA was that a toxic release inventory (TRI) survey be undertaken each year. This survey, which began in 1987, requires most firms that manufacture, process, or otherwise use listed chemicals in quantities exceeding specified thresholds to report the quantities of these chemicals that they transfer off-site or release to the environment.

A second outcome of the concern regarding toxic chemical releases was the formation by EPA in 1987 of an Office of Pollution Prevention to coordinate federal activities in this area. In large part, EPA has promoted its pollution prevention agenda as an important step toward a more collaborative relationship with the regulated community. For example, many EPA pollution prevention initiatives set goals that industry is asked to pursue on a voluntary basis as good citizens, and they allow flexibility in the methods used to achieve the desired goals. The 33/50 Program, Green Lights Program, inclusion of pollution prevention alternatives in development of Clean Air Act maximum achievable control technology (MACT) requirements, and technology assistance efforts at the national and state levels illustrate EPA's efforts in this area. In these initiatives, EPA neither specifies how the goals are to be achieved nor attempts to estimate the costs of such actions. Thus, within the constraints of established environmental regulations, EPA has sought to invoke pollution prevention as an opportunity to achieve environmental objectives in a collaborative, rather than a command and control, relationship.

Even before the increased governmental emphasis on waste reduction, a number of industrial firms had begun to allocate significant levels of resources to waste minimization. The principles of total quality management (TQM)[2] provided an underlying impetus for

---

[2] TQM is representative of a number of management approaches derived primarily from the teachings of Dr. W. Edwards Deming (Deming 1982; Deming 1986) and others.

some of these early efforts in waste minimization. The goal in TQM and analogous approaches is to motivate workers to achieve their highest quality performance. Static annual goals are replaced by evolving goals based on 'continuous improvement.' Although TQM is at heart economically motivated, the explicit goals are framed in terms of achieving highest quality output utilizing all inputs effectively.[3] One aspect of utilizing inputs effectively is to minimize generation of environmental residuals. Many leading U.S. firms now have embraced TQM as an intrinsic part of their efforts to remain competitive on a domestic and global scale. Additional factors that have motivated waste minimization efforts include: concern regarding the uncertain magnitude of future waste-related costs, including potential liability, and the desire to counter or avoid adverse publicity.[4,5]

WHAT IS POLLUTION PREVENTION?

Congress specified a hierarchy of waste management alternatives in Section 6602(b) of the Pollution Prevention Act of 1990 (U.S. Congress 1990). The four members of the hierarchy are, in order of preference: (1) pollution prevention, (2) recycling, (3) waste treatment, and (4) safe disposal. These alternatives are defined as follows.

Pollution Prevention

EPA defines pollution prevention as preventing waste generation at the source of generation (i.e., source reduction). Methods for achieving source reduction include raw material substitution or pretreatment, product reformulation, process modification, improved housekeeping, and more effective management practices. EPA also includes in-process (closed-loop) recycling as a pollution prevention mode. The term in-process recycling applies to situations in which generated wastes are recovered within the immediate manufacturing or usage operation and recycled for use in that same operation.

Recycling

Recycling involves recovering a waste after it has been generated and recycling it into a useful application. Out-of-process recycling applies to the situation in which the waste is transferred, before or after recovery, and used in an unrelated manufacturing or usage operation. A further distinction in out-of-process recycling pertains to whether it occurs on site or off site. Examples of out-of-process recycling include use of hazardous waste as a raw material in a manufacturing operation or as an effective substitute for a commercial product.

Waste Treatment

Waste treatment seeks to render the waste nonhazardous or less hazardous, with or without

---

[3] Deming cautioned firms against basing their actions strictly on quantitative estimates of costs and benefits, stating that '... the most important figures are those that are unknown or unknowable. What about the multiplying effects of a happy customer, in either manufacturing or in service? ... What about the multiplying effect of getting better material to use in production? What about the multiplying effect that you get all along the production line? Do you know that figure?" (Peters and Austin 1985).

[4] Illustrative of the potential damages that eventually may be incurred as a result of mismanagement of chemical wastes is an estimate by Russell et al. (1991) that between 1990 and 2020 the U.S. may spend between $373 billion and $1.7 trillion to clean up waste sites.

[5] In particular, adverse publicity associated with the environmental releases reported in the toxics release inventory surveys has provided a significant incentive for many firms to undertake pollution prevention actions.

energy recovery. Treatment modifies the physical, chemical, or biological properties of the waste in such a way as to eliminate or reduce the waste's hazardous character. In many cases a sequence of waste treatment steps is involved, with the initial steps often serving mainly to shift the waste between media for subsequent treatment in a more confined context. For example, the first step may involve capturing toxic air pollutants via a technology such as a wetted-wall column, then subjecting the recovered material to further treatment. Two treatment steps often applied to deal with certain types of large volume waste streams are: (1) incineration of solid wastes, which substantially reduces the waste volume and may recover energy, but involves some air emissions and solid waste residuals; and (2) waste stream neutralization, in which the acidity or corrosiveness of the stream is chemically neutralized.

Disposal

Proper disposal requires that the waste be disposed in such a way that its toxic components are prevented from adversely affecting human health or the environment. The available options have been substantially reduced over the past few years by more rigorous regulatory constraints (e.g., land disposal restrictions). Hazardous wastes, such as those produced via the recovery of toxic air pollutants, must after completion of all appropriate treatment actions be handled, transported, and disposed as prescribed in RCRA.

Emphasis Within the Hierarchy

The third and fourth stages in the hierarchy – waste treatment and disposal – are the environmental protection steps emphasized previously. Their inclusion in the hierarchy indicates that, whereas the weight given to hierarchical order signals an adjustment in emphasis, it does not represent a total shift in priorities.

The first two waste management alternatives – pollution prevention and recycling – are

together termed waste minimization. Some industry representatives have argued strongly that pollution prevention should be defined to be waste minimization. Two arguments advanced in favor of this broader definition are that: (1) properly conducted, recycling accomplishes essentially the same objectives as source reduction; and (2) in manufacturing settings a significantly larger number of personnel may be in a position to identify recycling alternatives than to carry out source reduction. Nevertheless, EPA recently has affirmed its intention that pollution prevention include only source reduction and in-process recycling (Habicht 1992b). EPA bases the more restrictive definition on the fact that there have been historical incidences in which 'sham recycling' was used as a cover for improper disposal of toxic wastes, and legitimate concern exists regarding releases during waste transfer operations. EPA contends that difficulties in monitoring and enforcing effective recycling dictate that this waste management mode be given second priority. However, EPA does agree with industry that, when properly conducted, recycling achieves many of the results attained via source reduction.

It is important to recognize that pollution prevention is a conceptually different approach to environmental protection than the end-of-pipe waste treatment and disposal methods historically emphasized. For one thing, pollution prevention focuses on the waste generation step, not on preventing releases. Implicit in such a focus is the intent to modify the manufacturing process itself, which has previously been treated as sacrosanct. For this reason, EPA has approached pollution prevention from a collaborative perspective, recognizing that the necessary understanding of process operations lies within industry. A second important aspect of pollution prevention is its multimedia nature. For example, when you modify a chemical reaction in order to reduce the amount of waste by-products generated, you intervene in the manufacturing process at a point preceding ultimate determination of the nature of waste by-products and their media of release. This multimedia nature often may not be reflected in the originating impetus for pollution prevention actions. As discussed below, waste audit evaluations that target medium-specific waste releases are a central facet of pollution prevention planning. A third characteristic of comprehensive pollution prevention is that it encompasses the entire life cycle of the materials involved. The life cycle perspective of pollution prevention is discussed next.

Chemical Life Cycle Perspective

It is useful to step back and consider the nature of pollutant releases occurring during all phases of a chemical's life cycle. By doing so, we can gain an appreciation of the broader context within which pollution prevention is judged to perform a vital role. There is considerable evidence that we are approaching the limits of the earth's carrying capacity, as reflected by ozone depletion on a global scale, extensive environmental degradation in large areas of Eastern Europe, and increased evidence of widespread contamination of the marine environment. Looking ahead, the world's population is projected to double by 2030, and the pace of environmental degradation is expected to accelerate. As concern mounts regarding these developments, the concept of sustainability has gained increased attention. Applied to chemicals, sustainability requires that a chemical's 'cradle to grave' implications be taken into account. The cradle to grave concept includes the following:

- Raw materials resource consumption
- Energy consumption
- Packaging considerations
- Customer use
- Waste generation
- Reuse
- Recycling
- Disposal

In the future, a chemical's ability to satisfy these sustainability issues likely will determine its market acceptability.

Figure 23-1 provides a simplified schematic representation of a chemical's major life cycle phases. The chemical, or its precursor material(s), is obtained initially via a materials extraction process such as coal and mineral mining or oil and gas extraction wells. In the schematic, materials extraction (including materials transport) is referred to as premanufacturing. Energy is consumed during the extraction process and waste materials (e.g., mine tailings) are released to the environment. Preprocessing of ores at the mine mouth to reduce pollutant concentrations is one way to reduce pollutant generation during downstream processing and use. However, the quantities of toxic waste generated at the mine mouth may be comparable to the quantities that would be generated during manufacturing and usage operations. Whether preprocessing at the mine represents a net pollution prevention benefit depends on how effectively the waste can be handled at the

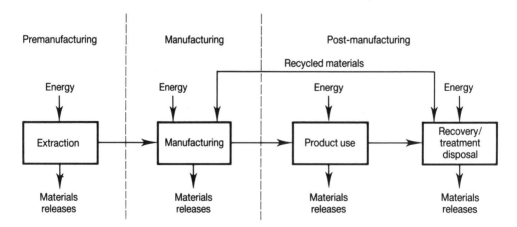

Figure 23-1.   Major life cycle phases for a chemical.

mine compared with the way those toxic components are dealt with in the manufacturing and usage arenas.

The chemicals then enter the manufacturing sector where they are refined and converted into intermediate and final products. Significant amounts of energy are consumed and large quantities of wastes are generated within the manufacturing sector. Once embodied within the manufactured product, the chemical contributes to the product's performance during its functional lifetime. The product's use may involve energy consumption and material releases to the environment. In fact, it has been estimated that 94 percent of commercial materials are released to the environment within a short time during extraction, manufacturing, and use in throwaway products (Ayres 1989). Only 6 percent is embodied in long-lived durable goods. At the conclusion of the useful lifetime, the material retained in such durable goods is recovered and recycled, treated in some way, or released to the environment via product disposal. In the end, unless converted into other materials, all materials eventually are released to the environment. Thus, sustainability requires that usage of nonrenewable materials be minimized and that their environmental properties not be harmful.

An initial step by industry toward addressing sustainability issues is the implementation of product stewardship, a manufacturer's accepting responsibility for the health, safety, and environmental consequences of its products during subsequent use and disposal. The Chemical Manufacturers Association (CMA) has introduced product stewardship as a central element of the Responsible Care® program. All CMA members are required to participate in the Responsible Care® program, and a code of management practices was drafted to delineate member responsibilities in this area (CMA 1991b). Among the responsibilities undertaken is to 'develop and produce chemicals that can be manufactured, transported, used and disposed of safely.' Pollution prevention represents an important element in the achievement of this goal.

Barriers to Pollution Prevention

In spite of the emerging governmental acceptance of pollution prevention as a preferred approach to environmental management, a number of regulatory barriers may be encountered as a facility seeks to carry out its pollution prevention plan. The genesis of these regulatory barriers is the historical context in which environmental protection was perceived initially to mean proper treatment and disposal of hazardous waste streams once generated, rather than prevention of their generation. Hazardous waste streams were judged to be the inevitable by-product of industrial and commercial processes. Further, initial legislative actions were aimed at the prevention of further harm to, and the regeneration of, air and water. The readily perceived degradation of these media led in the early 1970s to the Clean Air Act and the Clean Water Act. RCRA, when passed in 1976 amid the concerns fostered by the oil embargo, did invoke a broader perspective by introducing the concept of maximizing the effectiveness of materials use. However, RCRA retained the single medium perspective of the Clean Air and Clean Water Acts.

These three medium-specific environmental laws continue to provide the primary regulatory perspective within which industry facilities must function. Thus, although amendment of the Clean Air Act in 1990 did incorporate a number of pollution prevention oriented aspects, the regulatory context within which pollution prevention actions are carried out places primary emphasis on achieving single medium objectives. Such emphasis may not appear in concept to impede pollution prevention actions, and it may not in general. However, in specific instances the existing regulatory structure may impede sufficiently certain pollution prevention actions that their economics are rendered noncompetitive compared with established treatment and disposal alternatives.

Regulatory barriers to pollution prevention are recognized not only by industry but also by government, and a number of efforts are under way at the federal, regional, and state level to surmount these barriers. Adaptation of the regulatory system to emphasize pollution prevention promises to be gradual, occurring throughout the 1990s and perhaps beyond. The difficulty lies in knowing what aspects of the regulatory agenda, that has achieved many of its objectives, can be safely relinquished in shifting to a new, more flexible regulatory structure whose basic elements have yet to be fully defined. Environmental protection must be maintained and enhanced throughout this adjustment process. The adjustment is hampered by our lack of understanding of many of the mechanisms by which environmental harm occurs, as reflected by the continuing emergence of environmental problems of a broader or more intensive nature than anticipated just a few years ago.

As an industrial facility moves to carry out its pollution prevention plan within such an evolving regulatory context, it is essential that a continuing dialogue be maintained with state, and preferably also with federal, environmental personnel. By doing so, the facility can make sure not only that it is complying with emerging requirements, but also take advantage of regulatory approaches directed toward, and more compatible with, pollution prevention activities. In addition, the facility can communicate to the regulatory environmental personnel how regulations might be adapted to support achievement of pollution prevention objectives while maintaining the desired level of environmental protection.

One significant source of barriers to pollution prevention stems from the fact that an action that clearly satisfies environmental objectives from a multimedia perspective may not satisfy the letter of the law from one of the medium-specific regulatory perspectives. This type of difficulty is illustrated by the following example. Recovery and recycle of 95 percent of a volatile organic chemical (VOC) airborne stream may represent a desirable pollution prevention initiative from a life cycle perspective if the recovered material displaces a purchased raw material whose production generates a variety of releases to all three media. However, a state air emission limit for that VOC stream may require that 98 percent destruction and removal efficiency (DRE) be achieved through incineration or alternate technology. At this point in time, that state has no basis to waive the 98 percent DRE requirement, even though a life cycle analysis may indicate the recovery and recycle approach is more beneficial.

Not only are many regulatory requirements fixed, but life cycle evaluations are data intensive and complex. Sufficient knowledge rarely is available to evaluate the comparative significance of different pollutants in different media in different geographic contexts. Thus, such trade-offs are realistic only within fixed geographic settings, such as within a single facility or multiple facilities within a single geographic region. In order to consider such trade-offs even within this more restricted context, states and EPA need to establish trade-off protocols based on achieving desired multimedia region-specific environmental goals, such as those that might result if EPA focused a portion of its regulatory resources on areas promising greatest risk reduction.

RCRA-list hazardous wastes, in particular, pose significant barriers for establishing a broad sweep of recycling activities. One source of difficulty, as noted earlier, arises from the occasional practice of 'sham recycling' in which facilities attempt to circumvent RCRA-imposed requirements for proper handling, treatment, and disposal of hazardous wastes. Unfortunately, in order to prevent such sham recycling EPA has imposed an inclusive and rigid system for categorizing what is a hazardous waste. As a result, many facilities that have sought to accomplish waste minimization goals by recovering and reusing valued waste stream constituents have found that they must satisfy RCRA permitting requirements, even though use of an equivalent virgin material is not subject to such restrictions. The reporting requirements of RCRA are sufficiently demanding of resources and involve time lags such that recovery and reuse alternatives often become economically noncompetitive when compared with established treatment and disposal alternatives (Byers 1991).

In summary, regulatory constraints currently exist that bear on the selection of pollution prevention alternatives. Because these regulatory constraints are being reviewed by EPA and the states, longer-term planning by companies and facilities should consider pollution prevention approaches that are not currently viable. A dialogue should be maintained with the state and EPA to address the benefits of alternate courses of regulatory action.

Present Context

The impetus toward pollution prevention has been codified through a number of legislative actions and EPA initiatives undertaken over several years. The principal legislative actions are the requirements for toxic release reporting under EPCRA, the Pollution Prevention Act of 1990, the Clean Air Act Amendments of 1990, and a number of state legislative actions, such as the Toxic Use Reduction Acts of Massachusetts and Oregon. At the time of this writing in 1992, pending reauthorizations of the Clean Water Act and RCRA's promise to solidify further the legislative basis for emphasizing pollution prevention approaches. EPA initiatives have included establishing the Office of Pollution Prevention, implementing the Toxics Release Inventory (TRI) program, disseminating the results of the TRI survey, issuing a pollution prevention strategy, establishing a voluntary 33/50 Program to reduce the releases of 17 highly toxic substances, providing negotiated settlement provisions that take into account agreements to implement pollution prevention steps, conducting studies on how to utilize economic incentives as the basis for achieving required environmental objectives, and conducting pilot programs on more flexible regulatory approaches, such as integrated

permits.

The objective of the steps taken to date by EPA and state agencies has been to acquire data and to seek nonregulatory methods for achieving pollution prevention goals. The primary actions are described below.

Data Reporting Requirements

Toxic Release Inventory (TRI)[6]
As mandated in section 313 of EPCRA, larger chemical manufacturers, processors, or users of chemicals must report releases to each environmental medium or off-site transfer of over three hundred listed chemicals and chemical groups.[7] Since 1986, a few chemicals (e.g., alumina, sodium sulfate) have been delisted and other chemicals have been added to the list. Starting in 1991, TRI submissions had to include source reduction and recycling reports that identify what source reduction and recycling initiatives have been undertaken, compare releases in the given year with those in the prior year, and provide a basis for normalizing releases relative to production output. Other pending changes to the TRI reporting system include extending reporting requirements to selected nonmanufacturing sectors, and requiring that information on peak releases, such as those accompanying accidental releases, is provided.

Toxic Use Reduction Act (TURA)
TURA reporting, as represented by the program in Massachusetts, at the time of this writing in 1992 involves primarily an extension of TRI reporting. The same firms are required to report and many of the same chemicals are involved. However, there are two principal differences between TURA and TRI reporting. First, TURA reporting requires facility-wide data not specified by TRI, such as the quantity of listed

---

[6] See Chapter 26 for a more detailed dicussion.

[7] In 1989 and subsequent years, facilities must submit TRI reports if they: (1) manufacture or process 25,000 pounds or more, or otherwise use 10,000 pounds or more, of any listed chemical; (2) employ ten or more (full-time equivalent) employees; and (3) are classified within Standard Industrial Classification (SIC) sectors 20 to 39. In the 1987 and 1988 reporting years, the threshold quantities for reporting were slightly larger. The chemicals whose releases must be reported are listed in 40 CFR 372.

chemical that was manufactured, processed, otherwise used, generated as a by-product,[8] shipped as a product, or shipped in a product. Second, TURA requires that releases be reported on a production unit basis, rather than the facility-wide basis used in TRI. The product output of each production unit must be specified. A 'production unit' is defined as 'a process, line, method, activity, or technique, or a combination of series thereof, used to produce a product (or family of products).' (MADEP 1991) A given facility may encompass a number of production units (e.g., if it produces different chemical products or different electronic products). TURA requires that a by-product reduction index and an emissions reduction index be determined for each chemical in each production unit.

RCRA Biennial Census

The RCRA biennial census of hazardous waste generators provides data on the generation and handling of transportable hazardous wastes, but excludes air and water releases. These data potentially could shed light on the processes responsible for generating particular wastes, and changes in the quantities of secondary hazardous wastes generated by treatment of primary hazardous and nonhazardous wastes. However, use of the RCRA codes, which frequently lump together multiple chemical species in unspecified mixtures, makes the RCRA data difficult to correlate with the TRI data. EPA is seeking ways to work around this problem. States with TURA reporting requirements will provide data on the processes generating specific wastes, and can in this

regard help to link the RCRA and TRI data for those states.

Voluntary Programs

33/50 Program[9]

The 33/50 Program is a voluntary program in which participating companies agree to eliminate by 1995 half of their releases from all facilities of seventeen target chemicals identified by EPA. An intermediate goal of one-third reduction by 1992, together with the 1995 goal, provides the program with its 33/50 name. The chemicals were selected on the basis of having 'high production, high releases and off-site transfers relative to total production as indicated by TRI reports, potential for pollution prevention activities, and potential for a wide risk of health and environmental effects.' (EPA 1991k) The program offers participants maximum flexibility in how to achieve the desired targets – the reductions can be achieved at any subset of facilities, over any subset of chemicals, and within any medium or combination of media. Source reduction and recycling methods are strongly preferred by EPA, but the company is free to choose whatever methods it wishes to achieve the 33/50 goal. The benefits to the firm, in addition to the operating and financial benefits of the improved operation, are information and technical assistance and favorable publicity, including public recognition by EPA in the case of extraordinary efforts.

Supplemental Enforcement Projects

EPA has offered an option in some settlement proceedings for the defendant or respondent to

---

[8] By-product is defined in the TURA guidelines as 'a non-product output of a toxic substance generated by a production unit prior to handling, transfer, treatment, or release to the environment.' Thus, the quantity of by-product generated is specifically what source reduction seeks to minimize. In contrast, through 1990, TRI reporting identified only releases (i.e., the quantities released to the environment after treatment using pollution control equipment). For example, if 100 units of by-product were generated and sent to a 98.5 percent DRE incinerator, only 1.5 units were identified as released in the TRI report.

[9] See Chapter 26 for a more detailed discussion.

undertake a 'supplemental environmental project' in exchange for a reduction in the amount of the assessed civil penalty. Five categories of projects have been identified (Strock 1991)

- Pollution prevention projects
- Pollution reduction projects
- Projects remedying adverse public health or environmental consequences
- Environmental auditing projects
- Enforcement-related environmental public awareness projects

Among these five types of projects, EPA has stated that priority will be given to pollution prevention projects. Acceptable projects must provide environmental benefits that are related to the nature of the violation. The company can use the project to come into compliance, but must go substantially beyond compliance. Credit will not be granted to projects leading to substantial financial returns, because sound business practices dictate that such projects be undertaken anyway. Under no circumstances will an extension be granted in the time to achieve compliance.

Adjustments in Regulatory Approach

Regulatory Clusters

EPA has identified regulatory clusters in situations in which an industry is being subjected to significant multimedia regulatory burdens. For example, EPA currently is evaluating regulatory clusters for the petroleum refining and oil and gas extraction industries. In each case, pollution prevention approaches are being explored as a mechanism for meeting media-specific statutory requirements more cost effectively. EPA is working to provide industrial firms within each cluster with a clear perspective on the combined regulatory requirements to be imposed on that cluster over the next few years so that industry will have a sufficiently comprehensive planning framework to justify broader pollution prevention actions rather than reactive short-term, end-of-the-pipe-treatment responses.

Integrated Permits

EPA has been considering different forms of integrated permits for some time; a number of pilot programs have been established to test the effectiveness of such permitting approaches. Integrated permits may involve varying levels of integration, ranging from issuing a single multimedia permit for the facility at the upper end to coordinated review of the facility by a team of permitters leading to issuance of a harmonized set of single-medium permits. The advantages to the firm are the opportunity to get an overview of the permitting requirements and to provide that perspective to the regulating officials, the likelihood of identifying pollution prevention approaches that will satisfy requirements in more than one medium, and the probability of reducing the amount of paperwork and associated staff time.

Program Outreach

EPA has established a Pollution Prevention Information Clearinghouse (PPIC) to disseminate various types of pollution prevention information. The Clearinghouse consists of four components: (1) Pollution Prevention Information Exchange System (PIES); (2) Hotline; (3) Repository; and (4) Outreach. PIES, which operates as the central element, is a computerized network containing an array of pollution-prevention information: technologies, case studies, process-specific fact sheets, state and federal programs, training and educational materials, legislative profiles, equipment vendors, events and conferences, and directory of contacts. Using PIES, hard copy documents can be ordered or documents can be accessed and retrieved via modem. Users can add information to PIES and can direct queries to other PIES users via an electronic bulletin board. The Hotline amswers questions and directs queries on: how to use PIES, RCRA/Superfund, and small businesses. As its name implies, the Repository houses the hard copy data collected in support of PPIC activities. Outreach involves distributing PPIC-related information packets and bulletins.

METHODOLOGICAL FRAMEWORK

Program Overview

EPA has assembled a Waste Minimization Opportunity Assessment Manual (EPA 1988h) to assist companies in establishing an effective pollution prevention program. The EPA approach involves the following five activities.

- Management plan
- Opportunity assessment
- Feasibility analysis
- Implementation
- Monitoring and information sharing

The present discussion addresses planning of pollution prevention projects in existing facilities. It is equally important, and in many respects easier, to incorporate pollution prevention planning into the design of new facilities. Kraft identified a ten-step procedure and developed a set of checklists to follow in considering pollution prevention opportunities during plant design (Kraft 1992).

Management Plan

Support of company management provides an essential underpinning in most successful pollution prevention programs. This support should be formalized in a policy statement or management directive. Management also should make all company personnel aware of the goals of the pollution prevention activities and how a pollution prevention task force, as discussed below, can assist in attaining these goals.

The first step in the pollution prevention program is to establish a pollution prevention task force. Representatives from all technical and administrative departments in the company or facility that have a vested interest in establishment of such a program should be included on the task force. Process and safety engineers, environmental specialists, maintenance staff, public affairs officers, attorneys, information managers, administrators, and other specialists at the facility are needed to provide a sufficient breadth of expertise.

An important first task of the task force is to establish plant-level goals for the pollution prevention effort. These goals should be quantitative objectives that the task force believes the facility is capable of achieving within specified times. Representative goals include reducing fugitive emissions by half within eighteen months, reducing waterborne releases of chlorinated organics by two-thirds within four years, or eliminating the use of certain solvents within two years. Having identified the goals, a schedule of intermediate subgoals should be defined so that progress can be tracked over time. These goals can be refined or adjusted as information is gained from the opportunity assessment. It is essential that progress toward the goals be measurable.

Opportunity Assessment

The opportunity assessment involves establishing an inventory of waste streams, priority ranking waste streams, and identifying pollution prevention alternatives. The waste stream inventory should identify measured or estimated quantities of all wastes generated and of all releases to air, water, and land. Establishing the waste stream inventory requires that the following types of questions are answered.

- What waste streams are generated in the plant? What are the compositions and quantities?
- Which processes or operations generate these waste streams? What characteristic of the process or operation leads to generation of the waste stream? How efficient is the process or operation? What types of process controls are used to improve process or operation efficiency?
- What input materials contribute to the waste stream from each process or operation? How much of the input material enters the waste stream?
- What are the origins of fugitive losses? How much material of what composition is lost via fugitive emissions? What types of housekeeping practices are employed to limit fugitive losses?

Information to answer these questions can be obtained from a variety of sources, including materials purchasing receipts, product shipment billings, hazardous waste manifests, biennial RCRA reports, National Pollution Discharge Elimination System (NPDES) monitoring reports, and TRI reports. The information can be organized schematically in the form of flow diagrams that portray the interrelationships between processing steps, materials storage, and materials-handling operations.

Mass balances provide the framework for insuring that all materials are accounted for. On the basis of conservation of mass, at each process operation:

mass entering = mass exiting + mass accumulated

In other words, the mass of materials entering the operation must equal the mass of exiting materials, including wastes generated, adjusted by any mass that accumulates within the operation (e.g., additions to, or releases from, inventory). Mass balances should be established initially around parts of the facility that accomplish specific functional operations. More comprehensive mass balances then can be obtained by combining the function-specific mass balances. The time period being analyzed should encompass the range of expected facility operations.

Although it is desirable to evaluate all waste streams, resource limitations may dictate that initial attention be focused on the most important problems. Other problems can be addressed subsequently as time and resources permit. EPA suggests that the following factors be considered in setting pollution prevention priorities.

■ Quantity of waste
■ Compliance with current and future anticipated regulations
■ Potential environmental and safety liability
■ Hazardous properties of waste, including toxicity, flammability, corrosivity, and reactivity
■ Other safety hazards to employees
■ Costs of waste management

■ Ease of accomplishing pollution prevention
■ Potential recovery of valuable by-products

In some situations, the focus should remain on the entire facility (e.g., small facilities or facilities composed of similar operations) and activities, such as personnel training, that are facility-wide in nature.

Assessment teams should be formed to examine particular waste streams or particular parts of the plant. Each team should include members knowledgeable in the specific aspects of the plant being examined by that team. Outside experts also may be necessary. Such experts may include company personnel that have dealt with similar problems at other facilities, trade organization representatives that know how other organizations have dealt with such problems, and consultants who are knowledgeable in various aspects of pollution prevention. When outside experts are involved a structured site visit should be conducted during which the assessment team collectively characterizes the nature of the problems and the types of pollution prevention approaches that could be undertaken.

Pollution prevention alternatives of greatest promise then should be selected from those options identified by the assessment teams and from more broadly applicable alternatives identified by the task force. Ideas for pollution prevention can be obtained from trade associations, published literature, state and local environmental agencies, equipment vendors, and consultants. EPA operates a Pollution Prevention Information Exchange System (PIES) that provides information on pollution prevention approaches. Some state environmental agencies, in addition to providing information, provide technology assistance outreach by visiting plants and helping to identify effective pollution prevention alternatives. An overview of technological approaches to pollution prevention is presented below.

Although the focus of a pollution prevention task force is on source reduction, opportunities for recycling also should be identified and, in cases where neither source reduction nor recycling is effective, improved treatment and disposal approaches should be identified for separate consideration by plant management.

Feasibility Analysis

Having identified a number of pollution prevention alternatives, each must be evaluated with regard to its technical and economic viability. In addition to assuring that the alternative will work in the intended application, the technical evaluation should address the following questions (EPA 1988h).

■ Is the system safe for workers?
■ Will product quality be maintained?
■ Are the new items of equipment, materials, or procedures compatible with existing production procedures and desired production rates?
■ How long will production be interrupted during system installation?
■ Is special expertise required to operate or maintain the new system?
■ Does the system create other environmental problems?
■ Are the needed utilities available or must they be installed?
■ Is additional labor required?
■ Is space available?

Assessing cost-effectiveness is an important component of the selection process for pollution prevention alternatives. Factors that should be considered in the cost-effectiveness assessment of pollution prevention projects are discussed below.

Implementation

Project implementation involves selling the selected alternatives to the corporate hierarchy and overseeing implementation of the approved alternatives. If a pollution prevention task force has included members from all affected parts of the organization and has kept upper management informed during the developmental steps, management will be predisposed to approve at least a subset of the alternatives recommended by the task force. In most respects, project approval will be based on traditional resource allocation considerations. However, by forming a task force management signals its desire to undertake pollution prevention initiatives. Implementation aspects of pollution prevention projects are similar to those of other projects.

Monitoring and Information Sharing

Once the projects are implemented, they must be monitored to determine how effective they are in achieving their specified objectives. The monitoring process is especially important for pollution prevention projects because these are new types of projects for most companies. In many cases the economic payoff occurs over a longer time and intervention, if it is needed, should be undertaken quickly to ensure that the desired payoff will be achieved within the anticipated time frame. It should be confirmed that the projects are attaining their environmental objectives and that effects on production rate and on product quality are as predicted.

It is particularly important for pollution prevention projects that the results and experience gained be shared with other company facilities and with other members of the industry. The concepts of pollution prevention have yet to be widely accepted or understood. In order to move up the learning curve, it is important that experiences be shared by those who undertake leadership in this area. Both positive results achieved and difficulties encountered must be discussed in order to develop the understanding needed to move pollution prevention into the mainstream of project planning.

Technological Approaches

An important distinction when categorizing pollution prevention actions is that of intent. Is the objective to: (1) eliminate usage of certain chemicals, or (2) use the chemicals of concern more efficiently? Although the two objectives often overlap, attention conventionally is focused on the latter of these two alternatives. In recent years, however, a number of cases have arisen in which chemicals historically used because of their desirable functional properties and costs have been banned or restricted due to the environmental harm attending their use. Banning DDT use as a pesticide was an early instance of such a restriction in chemical use. The ongoing mandatory shift from stratospheric ozone-depleting chlorofluorocarbons and halons is a current example. Legislation such as the Toxic Use Reduction Act passed in Massachusetts, and being considered or passed in a number of other states, suggests that restrictions on toxic chemicals use will increase over the coming years. Thus, when embarking on pollution planning, not only should more efficient use of chemicals be targeted but opportunities for minimizing toxic chemical use also should be pursued actively.

Having decided on the overall objective in each application, elimination or improved efficiency, or both, the next question is method. All source reduction and recycling activities involve one, or a combination, of the following three actions: (1) material changes, (2) equipment changes, and (3) procedural changes. For example, establishing a management plan is an essential, overriding procedural change. As its mandates are carried out, it will involve in most cases a combination of material and equipment changes, as well as further procedural changes. Most process modifications involve both equipment and procedural changes, and often material changes as well. Computer management techniques and other procedural changes involve the acquisition of equipment that may be used to manage product purchasing, shipment, and internal handling more effectively. Computer techniques also may be used for closed-loop process control that with equipment changes can lead to increased product yields. Nelson (1990) provides a useful catalog of potential waste minimization actions, a number of which are discussed below.

Eliminating Chemical Use

A chemical's use may be 'eliminated' either by shifting to a chemical that is of lesser environ-mental concern, or by eliminating the chemical usage function entirely. An example of the former is the current effort to replace stratospheric ozone-depleting chemicals with non-ozone-depleting alternatives. Approaches to achieving the latter include reconfiguring the overall process so that the function performed is satisfied in another way or is transferred off site. Although some firms may reduce pollution by transferring certain usage functions off site, such actions do not represent an overall pollution prevention advance unless the usage function is accomplished more efficiently off site.[10]

By definition, all efforts to eliminate chemical use involve material changes. In many cases, equipment and procedural changes are involved as well. There are three principal ways to achieve source reduction objectives via eliminating chemical use: raw material substitution, product reformation, and process modification.

Raw Material Substitution

This involves reconfiguring the process to produce the same or a similar product using a less toxic raw material. In some cases, raw material specifications may be adjusted to permit displacement of the original raw material by a recycled by-product waste from another process. Illustrations of raw materials substitution include: using oxygen and peroxide bleaching processes instead of chlorine to bleach wood pulp during paper manufacture; and replacing chlorofluorocarbon solvent use in printed circuit board manufacture with aqueous-based cleaning agents (e.g., detergents or alkaline cleaners), and employing a follow-up chlorinated or hydrocarbon-based cleaning step for hard-to-remove deposits.

Product Reformulation

This involves adapting the nature of the product so that it performs the same or similar function,

---

[10]For example, an automobile manufacturer may reduce its volatile organic chemical (VOC) emissions by transferring some painting operations off site; if the off-site paint operation is less effective in confining VOC emissions, however, net pollution prevention is negative.

but uses less toxic material in its formulation. Illustrations of product reformulation include: shifting from organic solvent-based to aqueous-based coatings; adjusting product specifications where possible so that contaminants currently separated and discarded are retained with the product; switching from phosphate to non-phosphate fertilizers; and replacing chloro-fluorocarbon refrigerants with nonozone depleting substitute materials.

Process Modification

Process modifications that eliminate chemical use include reconfiguring a process so that the function performed by the chemical is no longer needed, is performed by another chemical, or is performed by a nonchemical alternative. Illustrations of these types of process modifications include: recycling a by-product stream from the same process (in-process) or another process (out-of-process) to satisfy a function currently performed by a purchased raw material; reducing algae inhibitor use in cooling towers by shielding water distribution from sunlight; and eliminating intermediate processing steps.

Product Substitution

A more general perspective on eliminating chemical usage includes the possibility of product substitution. Product substitution involves the accomplishment of the function performed by the current product by a different sort of product. Product substitutions generally occur over a longer time than considered in most pollution prevention assessments. Inevitable substitutions occur as technology advances and, as previously discussed from a chemical life cycle perspective, sustainability considerations will play a technology-forcing role in coming decades.

Improving Chemical Use Efficiency

Having ascertained that in a particular manufacturing context the functions accomplished by the toxic chemicals cannot at present be fulfilled via other means, the next focus of pollution prevention is to improve chemical use efficiency. This also occurs in three ways.

Materials Changes

The primary material change approach to improving chemical use efficiency is directed at raw material purity. Raw material purity may have a significant effect on the level of waste generation by the facility even though the contaminants appear to represent a small fraction of the entering feed stream. Approaches to improving raw material purity include: working with suppliers of purchased materials and with on-site facilities supplying feed streams; and installing preprocessing equipment.

Equipment Changes

Many types of equipment changes can be made to improve chemical use efficiency. They can be classed in two major categories

Chemical Reaction Yields These have a direct effect on the generation of by-product wastes. Having identified the optimal reaction conditions, the essential step in achieving high yields is to maintain these conditions uniformly over time and throughout the reactor. Approaches to achieving the desired reaction conditions include: maintaining effective physical mixing; achieving even distribution of entering raw materials; premixing reactants where possible; providing separate reactors for recycled streams; avoiding hot/cold spots through effective heating or cooling; using effective long-lasting catalysts; and using computer control to adjust immediately and compensate for process excursions.

Process equipment efficiency This affects the generation of wastes by minimizing material losses during raw material and product storage and materials handling, by effective recovery and recycling of postreaction wastes so that they can reenter the reaction phase, and by maximizing the efficiency of utilities in generation of electricity, steam, cooling water, and thermal energy. Examples of pollution prevention through improved process equipment efficiency include: increasing distillation reflux ratio to improve separation; using can-type or magnetically driven seal-less pumps; and minimizing waste product formation on heat exchanger walls by use of lower-pressure steam.

Procedural Changes

These can achieve improved chemical use efficiency in several ways.

Operating Procedure Changes   These include: an effective personnel training program that communicates the goals of pollution prevention, identifies how personnel should perform in accordance with planned pollution prevention protocols, and elicits personnel involvement in identifying further pollution prevention actions; producing or purchasing toxic materials on an as-needed basis so that on-site storage and handling is minimized; and reducing the number of samples taken and returning samples to processes after analysis.

Maintenance Actions   These are directed at insuring the uninterrupted operation of the equipment in their designed modes of operation. Actions of this type include: preventive maintenance to avoid unscheduled shutdowns or equipment operation excursions; systematic leakage detection and repair (LDAR) activities to minimize storage and fugitive losses; and maintaining painted equipment surfaces to prevent degradation and facilitate detection of emerging problems.

Housekeeping Actions   Examples include: an effective containment and cleanup protocol so that materials are not released to the sewer system; providing lighting adequate to maintain the alertness of staff and their ability to observe proper equipment functioning; properly labeling equipment and materials, and designing proper procedures; effectively sealing all containers; adding spill collection lips for bulk materials storage; designating proper materials-handling procedures; and providing recycling containers to motivate consideration of all materials as candidates for recycling.

Assessing Cost-Effectiveness

A central component of the screening process in selecting project alternatives is to assess the cost-effectiveness of pollution prevention projects compared with alternate project opportunities. This assessment often is challenging because some benefits of pollution prevention projects are subject to significantly greater uncertainty than cost factors typically encountered in project evaluations. Two examples of the uncertain aspects involved in evaluating pollution prevention projects are their potential beneficial effect on liability costs and possible increases in company revenues due to improved public image. In addition, the need to consider potentially long-lasting environmental effects often requires that the time for pollution prevention project evaluations be extended beyond the three to five years usually considered in project evaluations.

Of the three types of pollution prevention actions discussed above, procedural changes are the easiest to evaluate on the basis of cost effectiveness. These changes have the least effect on production processes, and the effect of accommodating such changes through modified scheduling can be assessed with relative ease. The implications of material or equipment changes are more difficult to evaluate to the extent that such changes affect the production process or product characteristics. Considering material changes, for example, a less volatile solvent may extend the time required for certain production activities, and a less toxic raw material may result in different product characteristics. Although the effect of slowing some production activities can be estimated with reasonable accuracy, the potential effect on product marketability of even seemingly minor changes in product characteristics may be uncertain.

Project evaluations can be conducted on an incremental or total cost basis. Incremental cost accounting addresses only those cost categories that are affected by the proposed project. Such an approach is appropriate in situations where the pollution prevention project is compared to existing operations. In this case, the objective is to identify the magnitude and direction of change in each cost element affected by the project. For example, the project may lead to savings in raw material and disposal costs, yet impose finance charges and utility costs, in addition to the costs currently being incurred. A project's net value over its lifetime can be evaluated using costing techniques such as net present value or internal rate of return.

Total Cost Accounting

Total cost accounting requires that all costs and benefits of the pollution prevention project, both direct and indirect, be estimated and compared to those estimated for competitive projects. In total cost accounting, all cost categories are evaluated, not only those that change. As a result, total cost accounting involves more effort. The difficult aspect in either total or incremental cost accounting, however, is estimating the costs and benefits that are highly uncertain. Once the most promising and easiest projects ('the low hanging fruit') have been implemented, it becomes essential to value properly the difficult-to-quantify economic benefits in order for pollution prevention projects to compete effectively in the allocation of corporate resources.

A recent report by the Tellus Institute (White, Becker, and Goldstein 1991) identifies four elements in total cost accounting for pollution prevention projects: (1) development of an expanded cost inventory, (2) consideration of an expanded time horizon, (3) use of a costing methodology that time discounts costs and revenues, and (4) activity-based cost allocation. The expanded cost inventory identifies four categories of costs: direct or usual costs, indirect or hidden costs, liability costs, and less tangible benefits. Usual costs, as their name implies, are those costs normally addressed in project evaluations (e.g., capital costs and operating and maintenance costs). Hidden costs are those that often are allocated to overhead rather than to specific production units (e.g., compliance costs, insurance, on-site waste management costs, and pollution control equipment operating costs). Liability costs are potential future waste-related costs that may be incurred in the event of personal injuries or property damages and penalties or fines that may be imposed by regulatory bodies. The most speculative cost category, less tangible benefits, includes elements such as increased revenue from enhanced product quality or company image, reduced health costs, and improved worker productivity.

The need for an expanded time horizon arises because some benefits of pollution prevention projects accrue over a longer period than the three to five years on which most project evaluations focus. As a result, pollution projects often take a longer time to achieve an acceptable return on investment. In this regard, they may be thought of as comparable to investments in long-term research. Examples of the potential longer-term benefits of pollution prevention projects include diminished probability of incurring high-liability costs and reduced vulnerability to significant increases in disposal costs. For similar reasons, it is important to use a costing methodology that properly accounts for costs and benefits accruing over time. Two effective alternatives are net present value and internal rate of return. The popular and easily visualized payback period is unacceptable because it does not discount costs and revenues over time.

Activity-based accounting is a prerequisite to properly assigning the costs and benefits of pollution prevention projects. If waste disposal charges are treated as an overhead item and allocated uniformly across all production activities, those production units responsible for the preponderance of wastes have limited incentive to allocate their share of corporate resources to reducing generation of these wastes. However, if waste disposal costs are allocated back to the activities generating the wastes, the production units capable of making the needed changes have the proper incentive. Similar reasoning holds for other waste-related activities, such as wastewater treatment and liability costs. Also, if some production units act to reduce their waste generation, under activity-based accounting the resulting benefits accrue to these units.

Support Materials

In response to the need for project evaluation procedures that assess pollution prevention projects on an equitable basis, EPA has sponsored a number of efforts to compile approaches for estimating costs that are unique to pollution prevention projects. EPA developed a four-tier approach for sequentially addressing cost elements of increasing uncertainty (EPA 1989h); General Electric developed a spreadsheet-supported approach termed Financial Analysis of

Table 23-1   Tier 1 – hidden cost worksheet

Item description	Cash flow information				Annualized cash flow[a]			
	Escal. rate (%)	First year C.F. (yrs)	Lifetime (yrs)	C.F. est.[a]	r_d = 5%	r_d = 10%	r_d = 15%	r_d = 20%
Depreciable Capital								
Expenditures								
A1 – Monitoring equipment								
A2 – Preparedness and protective equipment								
A3 – Additional technology								
A4 – Other								
Expenses								
B1 – Notification								
B2 – Reporting								
B3 – Monitoring/training								
B4 – Record keeping								
B5 – Planning/studies/modeling								
B6 – Training								
B7 – Inspections								
B8 – Manifesting								
B9 – Labeling								
B10 – Preparedness and protective equipment								
B11 – Closure/postclosure care								
B12 – Medical surveillance								
B13 – Insurance/special taxes								
B14 – Other								

[a] In thousands of year-0 dollars.

Waste Management Alternatives (GE 1987); and the George Beetle Company developed a computerized approach termed PRECOSIS (Peer 1990).

The EPA approach employs a series of worksheets as the framework for systematically evaluating four tiers of costs. In what EPA terms a Tier 0 evaluation, usual costs are used as the basis for project evaluation. If the pollution prevention project is judged cost-effective on this basis, further evaluation is unnecessary. If the pollution prevention project is not as cost-effective as competitive projects on the basis of Tier 0 costs, Tier 1 hidden costs are taken into account. To illustrate, the worksheet to be used in evaluating Tier 1 hidden costs is presented in Table 23-1. Project evaluation continues in this manner until the pollution prevention project is found to be cost-effective or all four cost categories have been considered and the project remains noncompetitive.

The strengths of the EPA approach are that it is well documented, easy to understand, available at no cost, and likely to be well maintained (i.e., analytic cost estimation formulas will be upgraded as understanding improves). In addition, the EPA approach provides guidance for accessing interrelated sources of information, such as EPA's Waste Minimization Assessment Opportunity Manual (EPA 1988h), and information on existing and prospective environmental regulatory requirements. For that reason, the EPA Manual is a good place to start in expanding the scope of project evaluations to incorporate pollution prevention-related costs and benefits. Subsequently, the General Electric or Beetle programs may be of value if a computerized approach employing more elaborate analytic formulations is desired. For example, a probabilistic approach may be employed to estimate liability costs. In many cases, however, a qualitative description of the nature of potential liabilities may suffice for management decisions.

References

Ayres, R.U. 1989. Industrial metabolism, In: *Technology and Environment*. National Academy of Engineering. Washington, D.C.

Byers, R.L. 1991. Regulatory barriers to pollution prevention: A position paper of the implementation council of the American Institute of Pollution Prevention. *Journal of the Air & Waste Management Association* 41(4):418.

Chemical Manufacturers Association (CMA). 1991b. Product stewardship code of management practices (Draft). Washington, D.C.

Deming, W.E. 1982. Quality, productivity and competitive position. MIT Center for Advanced Engineering Study. Cambridge, MA.

Deming, W.E. 1986. Out of the crisis. MIT Center for Advanced Engineering Study, Cambridge, MA.

General Electric Company (GE). 1987. Financial analysis of waste management activities. Corporate Environmental Programs.

Habicht, F.H. 1992b. EPA definition of 'Pollution Prevention.' Memorandum to EPA personnel. U.S. Environmental Protection Agency. Washington, D.C. May 28, 1992.

Kraft, R.L. 1992. Incorporate environmental reviews into facility design. *Chemical Engineering Progress*, p. 46.

Massachusetts Department of Environmental Protection (MADEP). 1991. Guidelines for classifying production units. May.

Nelson, K.E. 1990. Use these ideas to cut waste. *Hydrocarbon Processing*. March 1990.

Office of Technology Assessment (OTA). 1986. Serious reduction of hazardous waste for pollution prevention and industrial efficiency. September 1986.

Peer Consultants, P.C. and George Beetle Company. 1990. User's guide for pollution prevention economic assessment (PPEA), Version 1.1. U.S. Environmental Protection Agency, Center for Environmental Research Information. Cincinnati, OH.

Peters, T. and N. Austin. 1985. *A Passion for Excellence*. New York: Random House. p. 104.

Russell, M. et al. 1991. Hazardous waste remediation: The task ahead. University of Tennessee. December.

Strock, J.M. 1991. Interim policy on the inclusion of pollution prevention and recycling provisions in enforcement settlements. Memorandum to Regional Administrators. U.S. Environmental Protection Agency. Washington, D.C. February.

U.S. Congress. 1990. Pollution prevention act of 1990. Public Law 101-508. November 5, 1990.

U.S. Environmental Protection Agency (EPA). 1988h. Waste minimization opportunity assessment manual. EPA 625/7-88-003. National Environmental Research Center. Cincinnati, OH.

U.S. Environmental Protection Agency (EPA). 1989h. Pollution prevention benefits manual. Washington, D.C.

U.S. Environmental Protection Agency (EPA). 1991j. Pollution prevention strategy. Office of Pollution Prevention and Toxics. Washington, D.C.

U.S. Environmental Protection Agency (EPA). 1991k. The industrial toxics project, the 33/50 program: Forging an alliance for pollution prevention. EPA 560/1-91-003. Office of Pollution Prevention and Toxics. Washington, D.C.

White, A.L., M. Becker, and J. Goldstein. 1991. Total cost assessment: Accelerating industrial pollution prevention through innovative project financial analysis. Tellus Institute.

VI

Other Relevant Issues

24

Special Sources of Toxic Air Pollutants

David R. Patrick

INTRODUCTION

The hazardous air pollutant control program in section 112 of the 1990 Clean Air Act Amendments begins with the listing of 189 pollutants for regulation. It then requires EPA to identify major and area sources of these pollutants and to publish emission standards that reflect the maximum achievable control technology (generally available control technology for area sources). These standards are to be published on a schedule provided in the Amendments. At a later date, EPA is to evaluate the risks remaining after application of the specified technology and to require additional control if necessary to provide an 'ample margin of safety to protect the public health.' In mid-1992, EPA published its initial list of categories of major and area

sources for regulation under section 112(c)(1).[1] The lists of pollutants and sources, and their derivation, are presented in Chapter 3.

The 1990 Amendments also provided for studies and, in some cases, regulations for certain source categories of special interest to Congress. This interest was prompted in part by recognition of the complexity or potential cost in regulating some source categories. The source categories specially listed include electric utility steam generators, publicly owned treatment plants (i.e., sewage treatment plants), coke ovens, and incinerators. These source categories and the requirements provided in the 1990 Amendments are discussed in this chapter.

ELECTRIC UTILITY STEAM GENERATORS

Title IV of the 1990 Clean Air Act Amendments requires a significant reduction in the total nationwide emissions of sulfur dioxide (SO_2) and nitrogen oxides (NO_x) for control of acid deposition (also called acid rain). The primary impact of this Title is on the electric utility industry in the Eastern United States. By some

estimates, the annual cost to the utility industry of complying with these requirements will be in the range of $4 to 7 billion by the year 2000 (EPRI 1991b). The legislation does offer some flexibility in achieving compliance. For example, sulfur can be reduced through use of lower-sulfur fuels or stack gas scrubbers.

---

[1] 57 FR 31576, July 16, 1992.

461

Perhaps most importantly, SO_2 allowances[2] also may be purchased and traded across the country. EPA proposed regulations in late 1991[3] implementing the acid rain program.

Electricity is produced in the U.S. by a wide range of fossil fuel, nuclear, and cogeneration units. Of most concern with respect to air pollution are over one thousand coal-fired and over two hundred oil-fired units. For a variety of reasons, over eight hundred of the coal-fired units do not control emissions of SO_2. These rely instead largely on tall stacks to disperse the SO_2 away from the plant and the local area (Connor, Room, and Levin 1992).

Acid rain is caused when sulfur in coal and oil is oxidized to SO_2 and is emitted from power plant stacks. Also during the combustion process, nitrogen that makes up nearly four-fifths of the air we breathe is oxidized to various oxides of nitrogen (referred to as NO_x and emitted. The emissions of SO_2 and NO_x 'often undergo various chemical reactions that produce sulfates (e.g., ammonium sulfate), nitrates (e.g., ammonium nitrate), sulfuric acid (H_2SO_4), and nitric acid (HNO_3). Although some of these compounds may fall to the earth near the source, the tall stacks mean that more often they are transported many hundreds of miles before they are deposited. In many cases, this deposition occurs when the compounds are scrubbed from the atmosphere by rainfall, and the acidic nature of the compounds lowers the pH of the rain, hence the term acid rain.' (McCoy 1992). The term acid deposition is more correct because these acidic compounds can fall to the earth in such forms as snow or sleet. The full program provided in Title IV of the 1990 Amendments is extremely complex and beyond the scope of this handbook.

Of importance to this handbook, however, is the requirement in the 1990 Amendments that EPA conduct a study of electric utility steam generating units to investigate 'the hazards to public health reasonably anticipated to occur as a result of emissions by electric utility steam

Table 24-1   Compounds of concern in fossil fuel combustion emissions

Acetaldehyde	Hexachlorobenzene
Antimony compounds	Hydrochloric acid
Arsenic compounds	Hydrofluoric acid
Benzene	Lead compounds
Beryllium compounds	Manganese compounds
Biphenyl	Mercury compounds
Bis-(2-ethylhexyl)-	Naphthalene
phthalate	Nickel compounds
Cadmium compounds	Pentachlorophenol
Carbon disulfide	Phenol
Carbon tetrachloride	Phosphorus
Carbonyl sulfide	Polycyclic organic matter
Chlorine	Selenium compounds
Chlorobenzene	2,3,7,8-Tetrachlorodibenzo-
Chloroform	p-dioxin
Chromium compounds	Tetrachloroethylene
Cobalt compounds	Toluene
Dibenzofurans	Trichloroethylene
1,4-Dichlorobenzene	2,4,5-Trichlorophenol
Formaldehyde	

Source: EPRI 1991a.

generating units of pollutants listed under section 112(b) after imposition of the requirements of the Act.'[4] These sources, particularly coal-fired units, and to a lesser extent oil-fired units, may be significant emitters of toxic air pollutants, primarily various trace metals (e.g. arsenic, chromium, cadmium, mercury, beryllium, and others) in particulate matter, inorganic gases (e.g., hydrogen chloride), and some volatile organic products (e.g., uncombusted organics and organic combustion products). Table 24-1 lists the substances from the section 112(b) list of 189 hazardous air pollutants that have been detected in fossil-fuel power-plant exhaust gases (EPRI 1991a).

Even though emissions of particulate matter from a typical electric utility steam-generating plant are controlled, with the result that concentrations of particulate matter in the exhaust gases are relatively low, many of these units are large and exhaust large volumes of gases. This can result in substantial emissions of the toxic

---

[2] An allowance is the right to emit one ton of sulfur dioxide.
[3] 56 FR 63002, December 3, 1991.
[4] Section 112(n)(1).

air pollutants, although at low concentration, even after installation of what could be considered maximum achievable control technology. Thus, the ten tons per year emission limit on individual toxic air pollutants and the twenty-five tons per year emission limit on combined toxic air pollutants that defines a major source[5] likely are exceeded at almost any large coal-fired steam generating plant. What are uncertain, however, are the potential adverse effects of these toxic emissions. Utility industry studies only recently are beginning to assess the risks resulting in the populations or environment surrounding power plants (EPRI 1991a; Seigneur, Constantinou, and Levin 1991; Booth and McDaniel 1992; Levin et al. 1992). These early studies generally show that the individual risks from exposure to these emissions are relatively low.

In an attempt to reduce the uncertainties before imposing potentially costly new regulations on this industry, section 112(n)(1) of the 1990 Amendments requires EPA to conduct a detailed study of the hazards to public health anticipated to occur as a result of emissions of the listed toxic air pollutants from electric utility steam-generating units. EPA must report to Congress by November 15, 1993, with the results of the study. At the same time, EPA also is required to evaluate and report to Congress on alternative control strategies for emissions that warrant regulation.

The 1990 Amendments also require EPA and the National Institute of Environmental Health Sciences to study emissions of mercury from these and other source categories and to report on the effects, control technologies, and costs of control.[6] Mercury emissions are a particular concern for electric utility steam-generating plants because mercury is present in coal but is difficult to control, because of its volatility, using conventional technologies. Emissions of other trace metals also will be evaluated because they eventually deposit back to the earth's surface and many scientists are concerned with the potential for accumulation of these toxic materials on land, in water, and in the biota. EPA is required to study and report on the deposition of toxic air pollutants to the Great Lakes, the Chesapeake Bay, Lake Champlain, and the coastal waters.[7]

The requirement for these studies reflects Congress' concern for the large potential cost to the electric utility industry of further controls on emissions of toxic air pollutants in addition to the money that already will have been spent to control emissions to meet the requirements of Title IV. In particular, Title III could require a new round of expensive toxic air pollution regulation only a few years after the $4 to 7 billion expenditures estimated to be required to meet the acid deposition requirements. EPRI (1991a) estimated in one study that the cost to the utility industry to meet typical toxic air pollutants requirements could exceed another $7 billion. A major uncertainty is the extent to which the selected acid deposition control methods also reduce toxic air pollutant emissions. For example, some recent reports suggest that conversion to low-sulfur coal to meet SO_2 emission limits could result in significantly higher toxic metal releases (McIlvaine 1991). This conversion option appears likely to be broadly selected in early years to meet the requirements of Title IV (Kinsman, Evans, and Clendenin 1992).

An additional uncertainty is whether maximum achievable control technology for particulate matter and other pollutants, if it were installed to meet the requirements of section 112, would reduce emissions sufficiently to meet health-based standards that may be promulgated early in the next century.[8] Although initial studies indicate that maximum individual risks associated with exposures to emissions from coal-fired steam generating plants are relatively low (Levin et al. 1992), the large potential total emissions and broad geographic impact area could result in total population risks of concern to EPA.

---

[5] Section 112(a).
[6] Section 112(n)(1).
[7] Section 112(m).
[8] Section 112(f).

PUBLICLY OWNED TREATMENT WORKS

Section 112(e)(5) of the 1990 Amendments requires EPA to promulgate emission standards for publicly owned treatment works (POTWs) by November 15, 1995. Section 112(n)(3) also authorizes EPA to conduct, in cooperation with the owners and operators of POTWs, studies with respect to the release and control of toxic air pollutants. POTW is a term generally used to describe municipal wastewater (i.e., sewage treatment) plants.

POTWs clearly are emitters of toxic air pollutants. The Water Pollution Control Federation (WPCF) and EPA sponsored a workshop in July 1989 that focused on toxic air pollutant emissions from POTWs (WPCF 1990). As a result of this workshop, it was concluded that POTWs routinely emit over thirty of the listed air toxics. In addition, it was concluded, using cancer risk assessment guidelines and cancer potency values available at that time, that all POTWs handling more than fifty million gallons of wastewater per day potentially could exceed a one in one million (1×10^{-6}) cancer risk to the most exposed individual. Key air toxics identified by the attendees at the workshop included aromatic hydrocarbons (e.g., benzene, toluene, ethylbenzene and xylene) and chlorinated hydrocarbons (e.g., perchloroethylene, trichloroethylene, methylene chloride, chloroform, and carbon tetrachloride). More recent studies also have identified methyl chloroform, formaldehyde, acetaldehyde, styrene, p-dichlorobenzene, and polycyclic aromatic hydrocarbons (Buchan 1992). Many of the volatile chemicals result from both approved and unapproved releases by industrial, commercial, and residential sources. Major emission points include collector sewers, the treatment plant, and aeration basins. Control of volatile chemicals at the POTW generally is less effective than control at the source (i.e., reducing or stopping the release).

Other sources of toxic air pollutants at the POTW can include the potable water supply, where chlorinated organics can be formed in the chlorination process, and the wastewater treatment process, where chlorinated organics can be formed by reactions with bleaches and chlorine. Combustion of digester gas also can result in the formation of benzene through incomplete combustion of other aromatics, and vinyl chloride can be formed in anaerobic and biological processes occurring in slow-moving sewer lines and anaerobic digesters.

Many uncertainties exist in estimating and measuring emissions because the chemicals and their sources are so varied. For example, industrial stack emission measurement techniques are of little use where emissions arise from large surfaces, such as aeration basins, and multiple point sources, such as sewer lines. The variability in the releases also is compounded by variations in such factors as input rates, sampling locations, and weather conditions. Downwind ambient monitoring could resolve some of these uncertainties but adds substantially to costs.

Without effective, validated measurement techniques, emissions must be estimated. The estimation techniques for POTWs can involve use of mass balances, process models, and other methods. However, all of these ultimately are hampered by the same types of uncertainties facing the measurement techniques. Progress is being made in developing methods to overcome these problems. For example, researchers at the University of California at Davis have developed a group of semiempirical mass transport models for various of the sewage treatment processes (WPCF 1990) and the East Bay Municipal Utility District in California has successfully used VOC tracers in influent wastewater with downstream effluent analysis (Witherspoon, Bishop, and Card 1992).

POTW operators currently are working to develop more effective control strategies for these emissions. Some of the operational and design changes that hold promise were described at the 1989 Workshop. They include:

■ Reduced aeration rates
■ Covering and aerating some processes to facilitate removal
■ Discontinuing perchlorination

- Using pure oxygen rather than air in activated sludge treatment
- Blockage of incoming sewer gases
- Cover areas where chlorination/dechlorination occurs
- Reduce turbulence at discharge
- Promote biouptake
- Burn fugitive gases

In addition, owners and operators of POTWs are considering the use of advanced control technologies for the plants. These include advanced combustion processes, chemical oxidation, ultraviolet radiation, and cryogenic condensation. However, much more work is necessary before any could be recognized as 'maximum achievable control technology,' as likely to be defined by EPA.

The Clean Water Act requires that POTWs control wastewater discharges other than from domestic sources to protect the public health. This has been accomplished generally through pretreatment regulations established by sewer use ordinances. These, however, focus primarily on water pollutants. Substantial progress must be made in reducing the loadings of toxic chemicals to POTWs that also can become airborne. This could include source pretreatment control for more chemicals, material substitution, waste minimization, and even elimination of various industrial and consumer products.

COKE OVENS

Coke ovens have been sources of environmental concern for many years. In the coke production process, coal is destructively distilled in a reduced oxygen atmosphere and a large fraction of the organic constituents in the coal is driven off and typically recovered for sale or reuse. Literally hundreds of different chemicals are driven off and captured; these span the range from road tar to refined organic chemicals. The types and quantities of the recovered products are dependent largely on the coking temperature and time. After removal of the organics, a solid coke remains that is largely carbon, similar to charcoal.

Coke is a fuel that burns at high temperatures and is important in modern steelmaking. However, coke has been an article of commerce for over 2,000 years, with commercial production in coke ovens beginning in about 1620 (Shreve 1956). For many years the coal tars and other distillation by-products were largely discarded as wastes. However, in the 1840s the British began collecting naphtha for use as a solvent and the Germans began recovering tars to make roofing felt and wood preservatives. The use of coke by-products has grown significantly since these times. In the middle of this century, when the American steel industry was at its peak, over one hundred million tons per year of coal were converted to coke. Coke production has dropped substantially in recent years as the U.S. steel industry has decreased in size and as environmental concerns with coke oven and coke by-product emissions increased.

Emissions to the ambient air from coke ovens and coke by-product facilities comprise a wide range of particulate matter, aerosols, and gases. The most important components from a health effect standpoint generally are polycyclic aromatic hydrocarbons (PAHs), also known as polynuclear aromatics (PNAs). These substances are associated with cancer in occupationally exposed coke oven workers and some of the individual PAHs, most notably benzo[a]pyrene (B[a]P), cause cancer in laboratory test animals. At the same time, exposure to some PAHs appears to be associated only with noncarcinogenic effects or with no adverse health effects whatsoever (IARC 1983; IARC 1984a).

In response to a requirement in section 122 of the 1977 Amendments to the Clean Air Act, EPA listed coke oven emissions as a hazardous air pollutant in 1984.[9] EPA also later proposed,[10] but never promulgated, emissions

---

[9] 49 FR 36560, September 18, 1984.
[10] 52 FR 13586, April 23, 1987.

standards for coke oven batteries. However, EPA did propose and ultimately promulgate standards for benzene emissions from coke oven by-product plants.[11] The failure to promulgate emission standards for coke oven batteries resulted in part from the inability to identify coke oven control technologies that would reduce maximum individual cancer risks to individuals exposed to coke oven emissions to levels considered acceptable within any reasonable interpretation of the language of the 1970 Clean Air Act Amendments. As a result, emission standards could have forced closure of many, if not all, of the U.S. coke production facilities (Steiner 1991).

Because coke is a critical raw material in steelmaking, Congress concluded that coke ovens required special consideration in the 1990 Amendments. Section 112(d)(8) provides two options for reducing emissions of toxic air pollutants from coke ovens. The first is the process mandated for the bulk of toxic air pollutants and their sources. It involves application of maximum achievable control technology, in this case by 1995, and then meeting a risk-based standard[12] in the year 2003. Companies that expect to meet the risk standards or expect to shut down facilities by 2003 probably will choose this option. The substantial danger to the industry is that the level and application

of the risk-based standards will not be known until late in the 1990s or beyond.

The 1990 Amendments provide a second option that allows a coke oven facility to postpone the risk-based standards until the year 2020. In doing so, however, progressively tighter technology standards must be met in 1993, 1998, and 2010. Specific minimum requirements are provided for 1993 and 1998, although EPA may set tougher standards. EPA is required to promulgate standards for coke ovens by December 31, 1992, and compliance is required by December 31, 1995, regardless of EPA's promulgation date. At the time of this writing in 1992, EPA had announced its plan to develop national emissions standards for coke oven emissions through a regulatory negotiation process.

The steel industry has estimated that compliance with the requirements of the 1990 Amendments ultimately could cost in the range of $5 to 6 billion, an expenditure comparable to that spent by the steel industry for all environmental controls over the past 20 years (Steiner 1991). This cost and the continuing uncertainty with EPA's actions has led many companies in the steel industry to consider alterative process technologies for making steel without coke. One such, the Jewell-Thompson nonrecovery coke oven process, was specifically mentioned in the 1990 Amendments.

MUNICIPAL INCINERATORS

In the late 1980s, roughly one billion pounds of trash were being generated each day in the U.S., an average of four pounds per person per day. By the year 2010, EPA has estimated this daily average will increase about 25 percent. Simultaneously, the nation's landfill capacity is rapidly being eliminated. To deal with this growing concern, many communities have built or are planning to build incinerators. The construction and use of municipal incinerators have long been contentious. Although it is clearly recognized that they represent one solution to the growing trash problem, there is considerable

controversy about the appropriateness of the solution. Many individuals advocate incinerators, that reduce the volume of trash by 80 to 90 percent and can provide electricity or steam for other use. Properly designed and operated incinerators also have been found to be effective and safe (Anderson, Chrostowski, and Foster 1988; Clement 1990).

However, others are concerned with the safety and long-term potential effects on health and the environment from incinerator emissions. For example, those living near current or potential sites generally oppose the incinerators.

[11] 54 FR 38044, September 14, 1989.
[12] Section 112(f).

This not-in-my-backyard (NIMBY) thinking is understandable; even if the incinerator is safe, no one wants garbage trucks driving by all day and process upsets and truck accidents can and do occur. Others are dubious about the safety of the processes. Still others oppose incinerators for more subtle reasons. For example, most people embrace recycling and waste minimization although their success often depends on the convenience to the user. Because incinerators provide, in one sense, an easier disposal choice, proponents of recycling or waste minimization often oppose incinerators.

Air pollution standards for incinerators[13] have been in effect for many years; the primary pollutants controlled are particulate matter, nitrogen oxides, acid gases, and various organic compounds. In response to public pressures, many states and local air pollution agencies now are requiring incinerator owners to install more stringent controls in order to obtain operating permits. These added controls typically require more stringent design, operating, and add-on control technologies. For example, typical incinerators today are required to utilize specific combustion technologies and install acid gas scrubbers and baghouse filters. In addition, limitations are placed on many materials (e.g., lead) that can be introduced into the incinerator. Other requirements include automatic monitors for temperatures, gas flow rates and pressures, malfunction or process upset alarms, and twenty-four-hour telemetry monitoring by environmental agencies. In addition, disposal of the ash that remains after elimination of the combustibles generally is tightly regulated.

Still, concerns remain and the 1990 Amendments mandate a new and comprehensive regulatory program for solid waste incinerators. A new section 129 was added to the Clean Air Act that requires EPA to publish performance standards under section 111 for new solid waste incinerators and guidelines under section 111(d) for existing units. New source performance standards for units with a capacity greater than 250 tons per day (TPD) of municipal waste were to be promulgated by November 15, 1991. In fact, EPA promulgated these standards in early 1991[14] along with guidelines for the control of existing sources by states.[15] In the guidelines for existing sources, EPA established two categories of incinerators – large[16] and very large.[17] Different requirements apply to the two categories. States were required to submit to EPA by February 11, 1992, plans for implementing the existing source guidelines. Incinerators have three years from EPA approval of the state plans to comply, or until February 11, 1996, whichever occurs first.

Section 129(a) further required EPA to publish, by November 15, 1992, new source performance standards for municipal incinerators with a capacity of less than 250 TPD and for units combusting hospital waste, medical waste, and infectious waste. EPA also is required to propose, by November 15, 1993, and promulgate by November 15, 1994, new source performance standards for incineration units combusting commercial and industrial wastes. Finally, EPA is required to promulgate health-based standards for these source categories if necessary following the studies required under section 112(f) and the review by the National Academy of Sciences.

References

Anderson, E.L., P. Chrostowski, and S. Foster. 1988. Recent trends in health risk assessment: Impact on risk assessment of resource recovery projects. Presented at the 7th Annual Resource Recovery Conference. March 25, 1988.

Booth, R.B. and M.D. McDaniel. 1992. Summary of air toxic emission values from utility boilers firing

---

[13] See Subparts E and Ea, 40 CFR 60.

[14] 56 FR 5488, February 11, 1991.

[15] 56 FR 5514, February 11, 1991.

[16] Large incinerators have an aggregate capacity between 250 tons per day (TPD) and 1,100 TPD.

[17] Very large incinerators have an aggregate capacity greater than 1,100 TPD.

residual fuel oil and natural gas. Paper No. 92-132.01. Presented at the 85th Annual Meeting of the Air & Waste Management Association. Kansas City, MO.

Buchan, W.H. 1992. Comparative study of air toxic organic emissions from POTW processes. Paper No. 92-94.10. Presented at the 85th Annual Meeting of the Air & Waste Management Association. Kansas City, MO.

Clement International Corporation. 1990. Answers to questions commonly asked about trash-to-energy. Brochure prepared by Clement International Corporation. Fairfax, VA.

Connor, K.K., D.B. Room, and L. Levin. 1992. Methodology for assessing the utility contribution to air toxics risks. Paper No. 92-132.04. Presented at the 85th Annual Meeting of the Air & Waste Management Association Kansas City, MO.

Electric Power Research Institute (EPRI). 1991a. New focus on air toxics. *EPRI Journal*. March 1991.

Electric Power Research Institute (EPRI). 1991b. Responding to the clean air challenge. *EPRI Journal*. April/May 1991.

International Agency for Research on Cancer (IARC). 1983. Monographs on the Evaluation of Carcinogenic Risk of Chemicals to Humans. Polynuclear Aromatic Compounds. Volume 32. Part 1, Chemical, Environmental and Experimental Data. World Health Organization. Lyon, France.

International Agency for Research on Cancer (IARC). 1984a. Monographs on the Evaluation of Carcinogenic Risk of Chemicals to Humans. Polynuclear Aromatic Compounds. Volume 34. Part 3, Industrial Exposures in Aluminum Production, Coal Gasification, Coke Production and Iron and Steel Founding. World Health Organization. Lyon, France.

Kinsman, J.D., J.E. Evans, and J.H. Clendenin. 1992. Electric utility strategies for controlling SO_2

under Title IV (Acid Deposition Control) of the 1990 Clean Air Act Amendments. Paper No. 92-131.03. Presented at the 85th Annual Meeting of the Air & Waste Management Association. Kansas City, MO.

Levin, L., L. Gratt, K.D. Connor, D.B. Room, E. Constantinou, C. Seigneur, and T. Permutt. 1992. The comprehensive risk evaluation of air toxics: initial findings. Paper No. 92-132.06. Presented at the 85th Annual Meeting of the Air & Waste Management Association. Kansas City, MO.

McCoy and Associates, Inc. (McCoy). 1992. The proposed acid rain program – An in-depth analysis. *The Air Pollution Consultant* 2(2):4.1 (March/April).

McIlvaine Company. 1991. FAX Alert – Decisions should include air toxics factor. August 23, 1991.

Seigneur, C., E. Constantinou, and L. Levin. 1991. Multimedia health risk assessment of power plant emissions. Presented at the 84th Annual Meeting of the Air & Waste Management Association. Vancouver, BC.

Shreve, R.N. 1956. *The Chemical Process Industries*. New York: McGraw-Hill.

Steiner, B.A. 1991. The impact of the 1990 Clean Air Act Amendments on the U.S. iron and steel industry. Paper No. 91–121.5. Presented at the 83rd Annual Meeting of the Air & Waste Management Association. Pittsburgh, PA.

Water Pollution Control Federation (WPCF). 1990. *Air Pollution at Municipal Wastewater Treatment Facilities – Executive Summary and Workshop Report and Proceedings*. Alexandria, VA.

Witherspoon, J.R., W.J. Bishop and T.R. Card. 1992. Air emissions results, air toxic source control and control technology options for POTWs. Paper No. 92-94.01. Presented at the 85th Annual Meeting of the Air & Waste Management Association. Kansas City, MO.

25

Mobile Sources and Air Toxics

Michael P. Walsh

INTRODUCTION

Stationary industrial sources of air toxics historically have been the first subject of air toxics regulations and guidelines, in part because of permitting authority at the state and local level. However, as the vehicle population has grown in the U.S., mobile sources have become recognized as a major contributor to the overall health risks associated with air toxics. A variety of studies (EPA 1985e; EPA 1985f; EPA 1986t; Summerhays 1986; EPA-VII 1986) have documented that in many metropolitan areas mobile sources are one of the most important sources, and possibly the most important source, in terms of contributions to health risks associated with air toxics. Recently, EPA (1990b) concluded that motor vehicles cause approximately 54 to 58 percent of nationwide cancer cases associated with toxic air emissions. EPA noted that 'Area sources are found to contribute approximately 75 percent of the total number of annual cancer cases (including those from secondary formaldehyde) with point sources contributing approximately 25 percent of the total. Of the area sources, the major source is

mobile sources, contributing 78 percent of the total annual incidence attributed to area sources.'

Table 1-13 shows that diesel particulate emissions and various hydrocarbons stand out among motor vehicle sources of air toxics. EPA's projections also indicate that strategies designed specially to reduce diesel particulate and hydrocarbon emissions will lower many of the air toxics risks in the future.

In spite of the potential significance of mobile sources, before the 1990 Amendments there were no specific provisions in the Clean Air Act dealing solely with toxic emissions from mobile sources. In large part this was based on an assumption that traditional controls on the criteria pollutants, particularly carbon monoxide, ozone,[1] nitrogen oxides, and particulate matter, would as an inevitable by-product also lower the primary toxic emissions.

The one notable exception was a provision in section 202(a)(4) of the 1977 Clean Air Act Amendments that specifically prohibited the

---

[1] Ozone is produced in atmospheric reactions involving volatile organic compounds (VOC), which include hydrocarbons that are the major component of petroleum fuels, and nitrogen oxides.

use of any pollution control device that could result in any 'unreasonable risk to public health, welfare or safety.' This was in reaction to concerns associated with a potential increase in sulfates when catalytic converters were first used on cars.

This chapter focuses on mobile sources of air toxics. It starts with a review of the 1990 Clean Air Act Amendments, summarizing the major provisions that address toxic emissions from mobile sources either directly or indirectly. Then it reviews the nature of the concerns with individual pollutants from vehicles and their relative toxicities. Finally, it summarizes the factors that influence mobile-source air toxics, including fuels.

MOBILE SOURCE REQUIREMENTS IN THE 1990 AMENDMENTS

As the dominant national source of hydrocarbons (HC), carbon monoxide (CO), and nitrogen oxides (NO_x), motor vehicles again were singled out for special attention in the 1990 Clean Air Act Amendments. In addition to more stringent standards for emissions of these pollutants from cars, trucks and buses, the Amendments require substantial modification to conventional fuels, provide greater opportunity for the introduction of alternative fuels (but without mandating them), and extend the manufacturer's responsibility for compliance with auto standards in use to ten years or one hundred thousand miles. To the extent that they reduce HC and particulate emissions, all of these provisions also should lower toxic organic compounds. Importantly, the Amendments singled out toxic emissions from vehicles for special attention for the first time.

Conventional Vehicles and Fuels

Toxic Substances

EPA is required to complete a report by June 15, 1992, on the need for, and feasibility of, controlling unregulated motor vehicle toxic air pollutants including benzene, formaldehyde, and 1,3-butadiene. By June 15, 1995, EPA also is required to issue regulations to control hazardous air pollutants to the greatest degree achievable through technology that will be available considering cost, noise, safety, and necessary lead time. These regulations must apply, at a minimum, to benzene and formaldehyde. No effective date for the regulations is specified in the Amendments.

Light-Duty Vehicle Tailpipe Standards

Tier 1

The initial tailpipe standards (called Tier 1) for light-duty gasoline vehicles are 0.25 grams per mile (gpm) nonmethane hydrocarbons (NMHC), 3.4 gpm CO and 0.4 gpm NO_x for fifty thousand miles. These phase in over a three-year period (applicable to 40 percent of model year (MY) 1994 vehicles, 80 percent of MY 1995 vehicles, and 100 percent of MY 1996 vehicles) in certification. In use, starting in 1994, the standards will phase in 40 percent the first year, 80 percent the second, and 100 percent the third year (1996) for NO_x. For HC, in the first year, 40 percent will be required to meet an intermediate in-use standard (0.32 gpm NMHC) with the remainder achieving the current standard. This will rise to 80 percent in the second year. By the third year only 60 percent will meet the intermediate standard with the other 40 percent meeting the final (0.25 gpm NMHC) standard. In the fourth year, 80 percent will be required to meet the 0.25 gpm level, rising to 100 percent by MY 1998. Starting in 1996, new in-use standards are phased in for one hundred thousand miles (seventy-five thousand miles for recall testing) that allow for 25 percent higher emissions levels between fifty thousand and one hundred thousand miles for HC and CO, and 50 percent higher for NO_x. Diesel vehicles also are allowed to comply with a relaxed 1.0 gpm NO_x standard. Light trucks under 3,750 lbs loaded vehicle weight (LVW) will be required to achieve the same standards as cars.

Table 25-1   Certification emission standards for NMHC, CO, and NO$_x$ from light-duty vehicles (passenger cars) and light-duty trucks up to 6,000 lbs GVWR

Vehicle type	Standard[a]					
	NMHC	CO	NO$_x$	NMHC	CO	NO$_x$
	(5 year/50,000 mile)[b]			(10 year/100,000 mile)		
Non-Diesel						
LDTs (0–3,750 lbs LVW) and light-duty vehicles	0.25	3.4	0.4	0.31	4.2	0.6
LDTs (3,751–5,750 lbs LVW)	0.32	4.4	0.7	0.40	5.5	0.97
Diesel						
LDTs (0–3,750 lbs LVW) and light-duty vehicles	0.25	3.4	1.0	0.31	4.2	1.25
LDTs (3,751–5,750)	0.32	4.4	–	0.40	5.0	–

[a] Standards are in grams per mile (gpm).

[b] Applicable useful life is five years or fifty thousand miles, whichever first occurs.

Abbreviations: NMHC = nonmethane hydrocarbons; CO = carbon monoxide; NO$_x$ = nitrogen oxides; GVWR = gross vehicle weight rating; LDT = light-duty trucks; LVW = loaded vehicle weight.

Table 25-2   Particulate matter certification standards for light-duty vehicles and light-duty trucks up to 6,000 lbs GVWR[a]

Useful life period	Standard[b]
5 year/50,000 mile[c]	0.08
10 year/100,000 mile	0.10

[a] GVWR = Gross vehicle weight rating

[b] Standards are in grams per mile (gpm).

[c] Applicable useful life is five years or fifty-thousand miles, whichever first occurs.

For passenger cars (or 'light-duty vehicles' as they are called in the Act), certification useful life requirements were extended from five years/fifty thousand miles to ten years/one hundred thousand miles. Slightly less stringent certification emission standards apply after five years/fifty thousand miles for passenger cars and light-duty trucks (see Tables 25-1, 25-2 and 25-3). The extended useful life requirements will apply to other vehicle emissions unless specifically noted. The implementation schedule for the Table 25-1 and Table 25-2 standards calls for 40 percent of manufacturer's sales volume in MY 1994, 80 percent in MY 1995, and 100 percent after MY 1995. The implementation schedule for the Table 25-3 standards is 50 percent of the manufacturer's sales volume in MY 1996 and 100 percent in MY 1997.

Table 25-3   Emission certification for NMHC and CO from gasoline and diesel-fueled light-duty trucks of more than 6,000 lbs GVWR

Vehicle type	Standard[a]						
	NMHC	CO	NO$_x$	NMHC	CO	NO$_x$	PM
	(5 year/50,000 mile)[b]			(11 year/120,000 mile)[b]			
LDT test weight (3,751–5,750 lbs)	0.32	4.4	0.7[c]	0.46	6.4	0.98	0.10
LTD test weight (over 5,750 lbs)	0.39	5.0	1.1[c]	0.56	7.3	1.53	0.12

[a] Standards are in grams per mile (gpm).

[b] Applicable useful life is five years or fifty-thousand miles, whichever first occurs.

[c] Not applicable to diesel-fueled LDTs.

Abbreviations: NMHC = nonmethane hydrocarbons; CO = carbon monoxide; NO$_x$ = nitrogen oxides; GVWR = gross vehicle weight rating; LDT = light-duty trucks.

Tier 2

The Amendments include a second phase of standards (Tier 2) of 0.125 gpm NMHC, 1.7 gpm CO and 0.2 gpm NO$_x$ for passenger cars and light-duty trucks weighing 3,750 lbs LVW or less going into effect in MY 2004 if the EPA Administrator fails to reaffirm formally the existing standards or adopt alternative standards. It requires EPA to conduct a study on the need for and costs of the second-phase standards. The Administrator is empowered to retain the existing Phase I standards, implement the statutory Phase II standards, or adopt alternative standards (more stringent than Phase I) taking into consideration the need, technical feasibility, and cost-effectiveness of such standards.

Intermediate In-Use Standards

For the first two years that passenger cars and light-duty trucks are subject to the Tier I certification standards shown in Tables 25-1 and 25-2, less stringent, intermediate in-use emission standards apply for purposes of recall liability and the useful life period is only five years or fifty thousand miles, whichever occurs first.

The intermediate in-use standards for passenger cars and light-duty trucks up to 6,000 lbs gross vehicle weight rating (GVWR) are shown in Table 25-4 and the intermediate in-use standards for light-duty trucks above 6,000 lbs GVWR are shown in Table 25-5.

Table 25-5   Intermediate in-use standards for light-duty trucks more than 6,000 lbs GVWR

Vehicle type	Standard[a]		
	NMHC	CO	NO$_x$
LDTs (3,751–5,750 lbs LVW)	0.40	5.5	0.88[b]
LDTs (over 5,750 lbs LVW)	0.49	6.2	1.38[b]

[a] Standards are in grams per mile (gpm).
[b] Not applicable to diesel-fueled vehicles.

Abbreviations: NMHC = nonmethane hydrocarbons; CO = carbon monoxide; NO$_x$ = nitrogen oxides; GVWR = gross vehicle weight rating; LDT = light-duty trucks; LVW = loaded vehicle weight.

Final In-Use Standards

After two years, the final in-use standards take effect. These final in-use standards are the same as the certification standards shown in Table 25-1 and Table 25-2 and the in-use useful life periods are the same as the applicable certification useful life period (e.g., ten years/one hundred thousand miles for passenger cars).

If a vehicle is tested in-use prior to five years or fifty thousand miles, the fifty thousand mile certification standards apply. If the vehicle is tested after five years or fifty thousand miles, the one hundred thousand mile certification standards apply in the case of passenger cars and light-duty trucks up to 6,000 lbs. GVWR and 120,000 mile certification standard applies to light-duty trucks over 6,000 lbs GVWR.

Although passenger cars and light-duty trucks are expected to meet standards in-use for

Table 25-4   Intermediate in-use standards for passenger cars and light-duty trucks up to 6,000 lbs GVWR

Vehicle type	Standard[a]		
	NMHC	CO	NO$_x$
Passenger cars	0.32	3.4	0.4[b]
LDTs (0–3,750 LVW)	0.32	5.2	0.4[b]
LDTs (3,751–5,759 LVW)	0.41	6.7	0.7[b]

[a] Standards are in grams per mile (gpm).
[b] Not applicable to diesel-fueled vehicles.

Abbreviations: NMHC = nonmethane hydrocarbons; CO = carbon monoxide; NO$_x$ = nitrogen oxides; GVWR = gross vehicle weight rating; LDT = light-duty trucks; LVW = loaded vehicle weight.

Table 25-6   Schedule for implementation of final in-use standards for passenger cars and light-duty trucks

Vehicle type	Model year	Percent
Passenger cars and LDTs up to 6,000 GVWR	1996	40
	1997	80
	1998	100
LDTs over 6,000 GVWR	1998	50
	1999	100

Abbreviations: GVWR = gross vehicle weight rating; LDT = light-duty trucks.

their full useful lives, EPA may not conduct recall testing on passenger cars and light-duty trucks up to 6,000 lbs that are over seven years old or have been driven more than seventy-five thousand miles, or on light-duty trucks over 6,000 lbs GVWR that are over seven years old or have been driven more than ninety thousand miles. The schedules for implementing the final in-use standards are shown in Table 25-6.

Special in-use compliance rules apply to the NO$_x$ standard applicable to diesel-fuel passenger cars and light-duty trucks. For these vehicles the full life certification standards in Tables 25-1 and 25-2 also apply in use.

Trucks and Buses

The statutory standards created by the 1977 Amendments for HC, CO, and NO$_x$ for heavy-duty vehicles and engines are eliminated. In their place, there is a general requirement that standards applicable to emissions of HC, CO, NO$_x$, and particulate matter reflect 'the greatest degree of emission reduction achievable through the application of technology that the Administrator determines will be available for the model year to which such standards apply, giving appropriate consideration to cost, energy,

and safety factors associated with the application of such technology.' Standards adopted by EPA that are currently in effect will remain in effect unless modified by EPA.

EPA may revise (i.e., relax) such standards, including standards already adopted, on the basis of information concerning the effects of air pollutants from heavy-duty vehicles and other mobile sources, or the public health and welfare, taking costs into consideration. Unless EPA revises the standards, the Amendments establish statutory emission standards for heavy-duty trucks and buses as shown in Table 25-7.

A 4.0 gram per brake-horsepower-hour (g/BHP-h) NO$_x$ standard is established for MY 1998 and later model year gasoline and diesel-fueled heavy-duty trucks. The useful life provisions of California apply for light- and medium-duty trucks and EPA is authorized to delay the MY 1998 NO$_x$ standard for heavy trucks for two years based on lack of technological feasibility.

EPA is required to study the practice of rebuilding heavy-duty engines and the impact rebuilding has on engine emissions. EPA may establish requirements to control rebuilding practices, including emission standards. No deadlines are established.

Table 25-7   Emission standards for gasoline and diesel-fueled trucks and buses

Vehicle or engine type	Standard[a]				
	NMHC	HC	NO$_x$	CO	PM
Trucks (3,750 lbs or more but less than 5,750 lbs. LVW)	0.32	0.38	0.7	4.4	–
Trucks (5,750 lbs LVW or more but less than 8,500 lbs GVW)	0.39	0.46	1.1	5.0	0.12
Heavy-duty trucks[b]			4.0 (98)[c]		0.25 (91) 0.10 (94)
Heavy-duty buses[b]					0.10 (93)

[a] Standards are in grams per mile (gpm).

[b] Standards are in grams per brake-horsepower-hour (g/BHP-h).

[c] All standards apply in 1995 unless noted.

Abbreviations: NMHC = nonmethane hydrocarbons; HC = hydrocarbon; CO = carbon monoxide; NO$_x$ = nitrogen oxides; PM = particulate matter; GVW = gross vehicle weight; LVW = loaded vehicle weight.

Urban Buses

New Buses

The existing MY 1991 0.1 g/BHP-h particulate standard is relaxed to 0.25 g/BHP-h for MY 1991 and MY 1992. For MY 1993, the particulate standard is 0.1 g/BHP-h. Beginning with MY 1994, EPA is required to establish separate emission standards for urban buses including a particulate standard 50 percent more stringent than the 0.1 g/BHP-h (i.e., 0.05 g/BHP-h) standard. If EPA determines the 50 percent level is not technologically feasible, EPA is required to increase the allowable level of particulate to no greater than 70 percent of the 0.1 g/BHP-h level (i.e., 0.07 g/BHP-h).

EPA is required to conduct annual tests on a representative sample of operating urban buses to determine whether such buses meet the particulate standard over their full useful life. If EPA determines that buses are not meeting the particulate standard, EPA must require buses sold in areas with populations of 750,000 or more to operate on low-polluting fuels (e.g., methanol, ethanol, propane, natural gas, or any comparably low-polluting fuel). EPA may extend this requirement to buses sold in other areas if it determines there will be a significant benefit to the public health. The low-polluting fuel requirement is to be phased in over five model years commencing three years after EPA's determination.

Retrofit

Not later than twelve months after enactment, EPA was mandated to promulgate regulations requiring that urban buses, in major metropolitan areas, having their engines rebuilt after January 1, 1996, shall be retrofitted to meet emissions standards that 'reflect the best retrofit technology and maintenance practices.'

Diesel Fuel Sulfur Content

The 1990 Amendments also codify the recent EPA rule that sets the diesel fuel sulfur level at a maximum of 0.05 percent (weight) by October 1, 1993. It also requires the certification fuel for 1991 through 1993 model years to be 0.10 percent (weight) sulfur. Lowering the sulfur in the fuel will not only reduce sulfur emissions but also will facilitate the use of catalytic technology to reduce particulate emissions without significantly increasing sulfate emissions. EPA may exempt Alaska and Hawaii from the diesel fuel quality requirements.

Lead in Gasoline

As of January 1, 1996, lead is banned from all gasoline sold in the U.S. EPA is required, in cooperation with the U.S. Department of Agriculture, to develop procedures for testing the effectiveness of lead substitute additives for use in gasoline. Beginning in MY 1993, it will be unlawful to manufacture or sell new motor vehicle engines and nonroad engines that require the use of leaded gasoline.

Onboard Refueling Control Systems

By November 15, 1991, and after consultation with the U.S. Department of Transportation on safety issues, EPA was required to issue regulations requiring passenger cars to be equipped with vehicle-based (i.e., 'onboard') systems to control 95 percent of the evaporative HC emissions during refueling. The regulations take effect beginning in the fourth model year after the regulations are adopted and are to be phased in as follows: 40 percent of manufacturer's sales volume in the fourth year commencing after standards are promulgated and 80 percent in the fifth year commencing after standards are promulgated.

Reformulated Gasoline

Conventional gasoline will undergo significant modification during the 1990s as a result of the Amendments. There are two components – one for ozone problems and one for CO problems.

Gasoline in the Nine Most Serious Ozone Nonattainment Areas

Beginning January 1, 1995, cleaner, 'reformulated gasoline' must be sold in the nine worst ozone nonattainment areas with populations over 250,000. Other ozone nonattainment areas are permitted to 'opt in,' but EPA may delay on a limited basis requirements for reformulated gasoline in these areas if it determined the fuel will not be available in adequate quantities.

By November 15, 1991, EPA was required to establish requirements for reformulated gasoline requiring the greatest reduction of ozone-forming VOCs and toxic air pollutants achievable considering costs and technological feasibility.

Fuel Specification/Emission Performance Requirements

At a minimum, reformulated gasoline must: (1) not cause NO_x emissions to increase (EPA may modify other requirements discussed below if necessary to prevent an increase in NO_x emissions); (2) have an oxygen content of at least 2.0 percent by weight (EPA may waive this requirement if it would interfere with attaining an air quality standard); (3) have a benzene content no greater than 1.0 percent (by volume); and (4) contain no heavy metals, including lead or manganese (EPA may waive the prohibition against heavy metals other than lead if it determines that the metal will not increase on an aggregate mass or cancer risk basis, toxic air pollution emissions from motor vehicles).

In addition, toxic VOC emissions must be reduced by 15 percent over baseline levels beginning in 1995 and reduced to 25 percent beginning in the year 2000. EPA may adjust the 25 percent requirement up or down based on technological feasibility and cost considerations, but in no event may the percent reduction beginning in the year 2000 be less than 20 percent. Toxic air pollutants are defined by the 1990 Amendments in terms of the aggregate emissions of benzene, 1,3-butadiene, polycyclic organic matter (POM), acetaldehyde, and formaldehyde.

Oxygenated Fuel in CO Nonattainment Areas

Areas that have a CO design target of 9.5 parts per million (ppm) or above for 1988 and 1989 must have as part of their State Implementation

Plan a requirement that during that portion of the year in which the area is prone to high ambient concentrations of CO (primarily the winter months), gasoline sold may contain not less than 2.7 percent oxygen (by weight). Such requirements are to take effect no later than November 1, 1992 (or at such other date in 1992 as determined by the Administrator). For areas that exceed the 9.5 ppm CO design target for any two-year period after 1989, the 2.7 percent oxygen requirement must go into effect in that area no later than three years after the end of the two-year period.

Nonattainment areas classified as serious that have not achieved attainment by the date specified in the Act must require that the oxygen level in gasoline be 3.1 percent (by weight).

The EPA may waive the oxygenated fuel requirement in whole or in part for any area if it determines: (1) the use of oxygenated gasoline would present or interfere with the attainment by a given area with a federal or state ambient air quality standard; (2) mobile sources do not contribute significantly to the CO levels in the area; or (3) there is an inadequate domestic supply of, or distribution capacity for, oxygenated gasoline meeting the applicable requirements.

Gasoline Volatility

By June 15, 1991, the EPA was required to promulgate regulations limiting the volatility of gasoline to no greater than 9.0 pounds per square inch (psi) Reid vapor pressure (RVP). During the high ozone season (i.e., warmer months) EPA could establish a lower RVP in an individual nonattainment area if it determined that a lower level is necessary to achieve comparable evaporative emissions (on a per-vehicle basis) in nonattainment areas. The fuel volatility requirements were to take effect not later than the high ozone season for 1992.

For fuel blends containing 10 percent ethanol, the applicable RVP limitation may be one pound per square inch greater than for conventional gasoline.

Evaporative Controls

By June 15, 1991, EPA was required to establish evaporative hydrocarbon emission standards for all classes of gasoline-fueled motor vehicles. The standards, that were to take effect as expeditiously as possible, must require the greatest degree of reduction reasonably achievable of evaporative HC emissions during operation (i.e., 'running losses') and over two or more days of nonuse under ozone-prone summertime conditions.

Misfueling

The 1990 Amendments extend to individuals the prohibition against misfueling with leaded gasoline. Beginning October 1, 1993, it is prohibited for anyone to fuel a diesel-powered vehicle with fuel containing a sulfur content greater than 0.05 percent (weight) or that fails to meet a cetane index minimum of forty (or such equivalent aromatic level as prescribed by the Administrator).

Inspection and Maintenance (I/M)

The 1990 Amendments require enhanced I/M to be introduced in the most highly polluted areas. This will expand current coverage by forty new areas to a total of 110. The language requires annual, centralized programs or programs that achieve equivalent reductions in the serious nonattainment areas.

Alternative Fuels

The 1990 Amendments define 'clean alternative fuel' as any fuel including methanol, ethanol, or other alcohols (including any mixture thereof containing 85 percent or more by volume of such alcohol with gasoline of other fuels), reformulated gasoline, diesel, natural gas, liquefied petroleum gas, and hydrogen, or other power sources (including electricity) used in a clean-fuel vehicle that complies with the performance requirements in the Act.

Fleet Program

The 1990 Amendments require that serious, severe, and extreme ozone nonattainment areas with a 1980 population greater than 250,000 establish a clean-fuel program for fleets. A fleet is ten or more vehicles that are, or are capable of being, centrally refueled (excluding vehicles garaged at personal residences each night under normal circumstances). The program will mandate by 1998 California's low emission vehicle (LEV) standards of 0.075 gpm nonmethane organic material (NMOG), 3.4 gpm CO and 0.2 gpm NO_x for light-duty vehicles; and 0.1 gpm NMOG, 0.4 gpm NO_x, 4.4 gpm CO for light-duty vehicles and light-duty trucks below 6,000 lbs and for other trucks under 6,000 lbs GVWR, provided these vehicles are offered for sale in California. By 2001, these vehicles will be required without regard to availability in California.

EPA is mandated to establish an equivalent wraparound standard (exhaust, evaporative, and refueling emissions combined) for LEV's below 8,500 lbs GVWR. This wraparound standard is to be based on LEV vehicles using reformulated gasoline meeting the reformulated gasoline standards for the applicable time period. It will be left to the manufacturers to decide which standard to use – the LEV tailpipe standards or the wraparound standards.

Heavy duty fleet vehicles from 8,501 lbs GVWR to 26,000 lbs will be required to meet a combined NO_x and NMHC emission standard (3.15 g/BHP-h) that is a 50 percent reduction from a 1994 heavy-duty diesel engine baseline. If EPA determines that achievement of this standard is not possible for a clean diesel-fueled engine, it can be waived to only a 30 percent reduction.

Requirements Applicable to Clean Fuel Vehicles

By November 15, 1992, EPA was required to issue regulations implementing the Clean Fuel Vehicle Program. Congress followed California's lead in substituting a nonmethane organic gas (NMOG) standard in place of the total or nonmethane hydrocarbon standards currently used. A nonmethane organic gas is defined as the sum of nonoxygenated and oxygenated organic gases containing five or fewer carbon atoms and all known alkanes, alkenes, alkynes, and aromatics containing twelve or fewer carbon atoms. The test procedure for measuring NMOG is the recently adopted California NMOG Test Procedure. NMOG is a more appropriate substance to measure when evaluating the emissions performance of alternative fueled vehicles.

The emission standards applicable to dedicated clean-fuel vehicle passenger cars and light-duty trucks weighing up to 6,000 lbs GVWR, but not more than 3,750 lbs LVW are shown in Table 25-8. The emission standards for light-duty trucks weighing 3,751 to 5,750 LVW and up to 6,000 lbs GVWR are shown in Table 25-9. The emission standards for light-duty trucks greater than 6,000 lbs GVWR are shown in Table 25-10.

The emission standards for flexible-fueled passenger cars and light-duty trucks up to 6,000 lbs GVWR are shown in Table 25-11 and Table 25-12 (when operating on clean alternative fuel)

Table 25-8   Clean-fuel vehicle emission standards for light-duty trucks of up to 3,750 lbs LVW and up to 6,000 lbs GVWR and light-duty vehicles

Phase	Standard[a]				
	NMOG	CO_x	NO	PM[b]	HCOH
Phase I – 1996 MY					
50,000 mile standard[c]	0.125	3.4	0.4	–	0.015
100,000 mile standard[c]	0.156	4.2	0.6	0.08	0.018
Phase II – 2002 MY					
50,000 mile standard[c]	0.075	3.4	0.2	–	0.015
100,000 mile standard[c]	0.090	4.2	0.3	0.08	0.018

[a] Standards in grams per miles (gpm).
[b] Standards for PM apply only to diesel-fueled vehicles.
[c] Applicable useful life for purposes of certification.

Abbreviations: NMOG = nonmethane organic material; CO = carbon monoxide; NO_x = nitrogen oxides; PM = particulate matter; HCOH = formaldehyde; GVWR = gross vehicle weight rating; LVW = loaded vehicle weight; MY = model year.

Table 25-9   Clean-fuel vehicle emission standards for light-duty trucks of more than 3,750 lbs and up to 5,750 lbs LVW and up to 6,000 lbs GVWR

Phase	Standard[a]				
	NMOG	CO	NO_x	PM[b]	HCOH
Phase I – 1996 MY					
50,000 mile standard[c]	0.160	4.4	0.7	–	0.018
100,000 mile standard[c]	0.200	5.5	0.9	0.08	0.023
Phase II – 2001 MY					
50,000 mile standard[c]	0.100	4.4	0.4	–	0.018
100,000 mile standard[c]	0.130	5.5	0.5	0.08	0.023

[a] Standards in grams per miles (gpm).
[b] Standards for PM apply only to diesel-fueled vehicles.
[c] Applicable useful life for purposes of certification.

Abbreviations: NMOG = nonmethane organic material; CO = carbon monoxide; NO_x = nitrogen oxides; PM = particulate matter; HCOH = formaldehyde; GVWR = gross vehicle weight rating; LVW = loaded vehicle weight; MY = model year.

Table 25-10   Clean-fuel vehicle emissions for light-duty trucks greater than 6,000 lbs GVWR

| Test weight category | Standard[a] | | | | |
	NMOG	CO	NO$_x$	PM[b]	HCOH
Up to 3,750 lbs test weight					
50,000 mile std.[c]	0.125	3.4	0.4[d]	–	0.015
120,000 mile std.[c]	0.180	5.0	0.6	0.08	0.022
Above 3.750 lbs but not above					
5,750 lbs test weight					
50,000 mile std.[c]	0.160	4.4	0.7[d]	–	0.018
120,000 mile std.[c]	0.230	6.4	1.0	0.10	0.027
Above 5,750 lbs but not above					
8,500 lbs test weight					
50,000 mile std.[c]	0.195	5.0	1.1[d]	–	0.022
120,000 mile std.[c]	0.280	7.3	1.5	0.12	0.032

[a] Standards in grams per miles (gpm).
[b] Standards for PM apply only to diesel-fueled vehicles.
[c] Applicable useful life for purposes of certification.
[d] Standard not applicable to diesel-fueled vehicles.
Abvbreviations: NMOG = nonmethane organic material; CO = carbon monoxide; NO$_x$ = nitrogen oxides; PM = particulate matter; HCOH = formaldehyde; GVWR = gross vehicle weight rating; LVW = loaded vehicle weight; MY = model year.

Table 25-11   NMOG standards for flexible- and dual-fueled vehicles when operating on clean alternative fuel: light-duty trucks up to 6,000 lbs GVWR and light-duty vehicles

| Vehicle type | Standard[a] | |
	50,000 mile[b]	100,000 mile[b]
Beginning MY 1996		
LDTs (0–3,750 lbs LVW) and light-duty vehicles	0.125	0.156
LDTs (3,751–5,750 lbs LVW)	0.100	0.130
Beginning MY 2001		
LDTs (0–3,750 lbs LVW) and light-duty vehicles	0.075	0.090
LDTs (3,751–5,750 lbs LVW)	0.100	0.130

[a] Standards in grams per miles (gpm).
[b] Applicable useful life for purposes of certification.

Abbreviations: NMOG = nonmethane organic material; GVWR = gross vehicle weight rating; LDT = light-duty truck; LVW = loaded vehicle weight; MY = model year.

and Table 25-13 and Table 25-14 (when operating on conventional fuel).

The clean vehicle and flexible-fueled vehicle standards in the Amendments are based on standards recently adopted by California as part of its Low Emission Vehicles and Clean Fuel Program. The Amendments provide that if California adopts less stringent standards applicable to clean fuel and dual fueled passenger cars and light-duty trucks, the standards shown above are to be relaxed as well.

Heavy-Duty Clean Fuel Vehicles
MY 1998 and later heavy-duty vehicles and engines greater than 8,500 lbs GVWR and up to 26,000 lbs GVWR are required to meet a

Table 25-12 NMOG standards for flexible- and dual-fueled vehicles when operating on clean alternative fuel: light-duty trucks more than 6,000 lbs GVWR

Test weight category	Standard[a]	
	50,000 mile[b]	120,000 mile[b]
Beginning MY 1998		
LDTs (0–3,750 lbs test weight)	0.125	0.180
LDTs (3,751–5,750 lbs test weight)	0.160	0.230
LDTs (above 5,750 lbs test weight)	0.195	0.280

[a] Standards in grams per miles (gpm).
[b] Applicable useful life for purposes of certification.

Abbreviations: NMOG = nonmethane organic material; GVWR = gross vehicle weight rating; LDT = light-duty truck; MY = model year.

Table 25-13 NMOG standards for flexible- and dual-fueled vehicles when operating on conventional fuel: light-duty trucks up to 6,000 lbs GVWR and light-duty vehicles

Vehicle type	Standard[a]	
	50,000 mile[b]	100,000 mile[b]
Beginning MY 1996		
LDTs (0–3,750 lbs LVW) and light-duty vehicles	0.25	0.31
LDTs (3,751–5,750 lbs LVW)	0.32	0.40
Beginning MY 2001		
LDTs (0–3,750 lbs LVW) and light-duty vehicles	0.125	0.156
LDTs (3,751–5,750 lbs LVW)	0.160	0.200

[a] Standards in grams per miles (gpm).
[b] Applicable useful life for purposes of certification.

Abbreviations: NMOG = nonmethane organic material; GVWR = gross vehicle weight rating; LDT = light-duty truck; LVW = loaded vehicle weight; MY = model year.

Table 25-14 NMOG standards for flexible- and dual-fueled vehicles when operating on conventional fuel: light-duty trucks up to 6,000 lbs GVWR

Vehicle type	Standard[a]	
	50,000 mile[b]	100,000 mile[b]
Beginning MY 1998		
LDTs (0–3,750 lbs LVW)	0.25	0.31
LDTs (3,751–5,750 lbs LVW)	0.32	0.46
LDTs (over 5,750 lbs LVW)	0.39	0.56

[a] Standards in grams per miles (gpm).
[b] Applicable useful life for purposes of certification.

Abbreviations: NMOG = nonmethane organic material; GVWR = gross vehicle weight rating; LDT = light-duty truck; LVW = loaded vehicle weight; MY = model year.

combined NO_x and nonmethane hydrocarbon (NMHC) standard of 3.13 g/BHP-h (equivalent to 50 percent of the combined HC and NO_x emission standards applicable to conventional 1994 model year heavy-duty diesel-fueled vehicles or engines). EPA may relax this standard upon a finding that it is not technologically feasible for clean diesel-fueled vehicles. EPA must, however, require at least a 30 percent reduction from conventional-fueled vehicle and engine 1994 NO_x and NC standards.

Centrally fueled fleets with ten or more vehicles that are owned or operated by a single person and operate in a covered area (see below) are subject to the clean vehicle requirements. A number of vehicle fleets are exempted including rental fleets, emergency vehicles, enforcement vehicles and nonroad vehicles. Also, vehicles garaged at personal residences each night under normal circumstances are not covered.

Covered areas include any ozone nonattainment area with a population of 250,000 or more classified as serious, severe, or extreme, or any CO nonattainment area with 250,000 or more population and a CO design value at or above 16.0 ppm. States are required to implement clean-fuel vehicle phase-in as shown in Table 25-15.

In complying with this section, passenger cars and light-duty trucks up to 6,000 lbs GVWR must meet the Phase II emission standards shown in Table 25-9. Also, if passenger cars and light-duty trucks meeting the Phase II standards are sold in California in model years prior to 1998, then that model year becomes the applicable model year for phasing in clean-vehicle fleets. The fleet requirements also apply to federally owned fleets except vehicles exempted by the Secretary of the Department of Defense for national security reasons.

Conversions
The requirements for purchase of clean vehicles may be met through vehicle conversions of existing gasoline or diesel-powered vehicles to clean-fuel vehicles. EPA is required to establish regulations defining criteria for these conversions.

Urban Buses
The Amendments set a performance criteria, beginning in 1994, requiring clean buses operating more than 70 percent of the time in large urban areas, using any fuel, to reduce particulate emissions by 50 percent compared to conventional heavy duty vehicles (i.e., 0.05 g/BHP-h particulates). EPA is authorized to relax the control requirements to 30 percent. Beginning in 1994, EPA is to do yearly testing to determine whether buses subject to the standard are meeting the standard over their full useful life. If EPA determines that 40 percent or more of the buses are not, it must establish a low pollution fuel requirement. Essentially, this provision allows the use of exhaust after-treatment devices to reduce diesel particulate to a very low level provided they work in the field; if they fail, EPA will mandate alternative fuels.

Off-Highway Engines

The 1990 Amendments require EPA to study off-highway engine and off-highway vehicle emissions (other than locomotives), and if the study concludes that they are significant

Table 25-15   Clean-fuel vehicle phase-in requirements for fleets

Vehicle type	Percent phased in		
	MY 1998	MY 1999	MY 2000
LDT trucks up to 6,000 lbs GVWR and light-duty vehicles	30	50	70
Heavy-duty trucks above 8,500 lbs GVWR	50	50	50

Abbreviations: GVWR = gross vehicle weight rating; LDT = light-duty truck; MY = model year.

contributors to ozone or carbon monoxide problems in more than one area, EPA is required to regulate their emissions. In a very controversial provision, California would be precluded from adopting its own standards for engines smaller than 175 horsepower used in construction equipment or vehicles, or used in farm equipment or vehicles, or in locomotives. Within five years after enactment, EPA also is required to promulgate regulations containing standards applicable to emissions from new locomotives.

All Nonroad Engines/Vehicles Except Locomotives

By November 15, 1991, EPA was required to complete a study of the health and welfare effects of nonroad engines and vehicles (except locomotives). Within twelve months of completing the study, EPA must determine if HC, CO, or NO_x emissions from new or existing nonroad engines and vehicles significantly contribute to ozone or CO concentrations in more than one area in nonattainment for ozone or CO.

If EPA makes an affirmative finding, it must regulate those classes or categories of nonroad engines or vehicles by requiring the 'greatest degree of emission reduction achievable considering technological feasibility cost, noise, energy, safety and lead time factors.' EPA, in setting standards, must consider standards equivalent in stringency to on-road vehicle standards. No deadline for establishing standards was set.

EPA also may regulate other pollutants from nonroad engines and vehicles (e.g., diesel particulate) if it determines such standards are needed to protect the public health and welfare.

Locomotives

By November 15, 1995, EPA must establish standards for locomotive emissions that require the use of the best technology that will be available considering cost, energy and safety. The standards are to take effect at the earliest possible date considering the lead time needed to develop the control technology.

State Standards

States, including California, are prohibited from setting emission standards for new engines smaller than 175 horsepower used in construction equipment, or vehicles, or use in farm equipment, or in new locomotives or new engines used in locomotives.

TOXIC POLLUTANTS FROM MOBILE SOURCES

Diesel Particulate

Uncontrolled diesel-fueled engines emit approximately thirty to seventy times more particulate than gasoline-fueled engines equipped with catalytic converters and burning unleaded fuel. These particles are small and respirable (generally less than 2.5 microns) and consist of a solid carbonaceous core on which a myriad of compounds adsorb, including:

■ Unburned hydrocarbons
■ Oxygenated hydrocarbons
■ Polynuclear aromatic hydrocarbons
■ Inorganic species such as sulfur dioxide, nitrogen dioxide and sulfuric acid

Many of these emissions may cause cancer and exacerbate mortality and morbidity from respiratory disease (Walsh 1987, EPA 1990e). Diesel particulate has been identified as especially hazardous and toxic because of its composition. As noted by Ozkaynak and Spengler (1986), 'most of the toxic trace metals, organics, or acidic materials emitted from automobiles or fossil fuel combustion are highly concentrated in the fine particle fraction.' EPA has noted that up to ten thousand chemicals may be absorbed on the surface of diesel particles. Many of these chemical compounds are known to be mutagenic in short-term bioassays, and to be capable of causing cancer in laboratory animals. Epidemiological studies have tended to reinforce these concerns. For example, a study (CRC 1983) of heavy-construction workers found positive trends in lung cancer by length of union membership and a higher than expected rate among retirees. Further, a pilot study of U.S. railroad workers, conducted by researchers at Harvard, indicated that the risk ratio for respiratory cancer in diesel exposed subjects relative to unexposed subjects could be as great as 1.42 (i.e., the possibility of developing cancer may be 42 percent greater in individuals exposed to diesel exhaust than in individuals who are not exposed) (Schenker 1984). In addition, during late 1985 and 1986, the results of several animal studies were released that increased concerns regarding adverse health effects from diesel particulate emissions. In particular, a study carried out under the auspices of the European automobile manufacturers and conducted by Battelle-Geneva reported that unfiltered diesel exhaust produced an increase in lung tumor incidence from 1 to 40 percent; gasoline emissions reportedly showed no effect (Brightwell et al. 1986).

A comprehensive assessment of the available health information was carried out by a Working Group of the International Agency for Research on Cancer (IARC), at a meeting in Lyon, France, on June 14–21, 1988. The IARC Working Group concluded that diesel particulate is probably carcinogenic to humans. The term 'carcinogen' is used by the IARC to denote an agent that is capable of increasing the incidence of malignant tumors. In evaluating potential carcinogens, the IARC categorizes results into one of four categories. In deciding where to place an agent, IARC takes the total body of evidence into account. The categorization of an agent is the matter of scientific judgment, reflecting the strength of the evidence derived from studies in human and in experimental animals and from other relevant data.

Based on its evaluation of all available evidence regarding diesel exhaust, the IARC reached the following conclusions.

■ There is sufficient evidence for the carcinogenicity in experimental animals of whole diesel engine exhaust.
■ There is inadequate evidence for the carcinogenicity in experimental animals of gas-phase diesel engine exhaust (with particles removed).
■ There is sufficient evidence for the carcinogenicity in experimental animals of extracts of diesel engine exhaust particles.
■ There is limited evidence for the carcinogenicity in humans of diesel engine exhaust.

Based on these findings, IARC's overall evaluation was that diesel engine exhaust is probably carcinogenic to humans (Group 2A). Coincidentally, gasoline engine exhaust has been found to be possibly carcinogenic to humans (Group B2). In reaching these conclusions, several human epidemiological studies seemed to carry the most weight. As noted in the IARC report,

In the two most informative cohort studies (of railroad workers), one in the USA and one in Canada, the risk for lung cancer in those exposed to diesel engine exhaust increased significantly with duration of exposure in the first study and with increased likelihood of exposure in the second. Three further studies of cohorts with less certain exposure to diesel engine exhaust were also considered; two studies of London bus company employees showed elevated lung cancer rates that were not statistically significant, but a third, of Swedish dockers, showed a significantly increased risk of lung cancer.

In only two case-control studies of lung cancer (one of US railroad workers and one in Canada) could exposure to diesel engine exhaust be distinguished satisfactorily from exposures to other exhausts; modest increases in risk for lung cancer were seen in both, and in the first the increase was significant. In three further case-control studies, in which exposure to diesel engine exhaust in professional drivers and lung cancer risks were addressed, the Working Group considered that the possibility of mixed exposure to engine exhausts could not be excluded. None of these studies showed a significant increase in risk for lung cancer, although the risk was elevated in two.

In the three cohort studies (on railroad workers, bus company workers and dockers, respectively) in which bladder cancer rates were reported, the risk was elevated, although not significantly so. Four of the case-control studies of bladder cancer were designed to examine groups whose predominant engine exhaust exposure was assumed to be to that from diesel engines. Three showed a significantly increased risk for bladder cancer. In one of these, the large US study, a significant trend was also seen with duration of exposure; and in an analysis of one subset of self-reported diesel truck drivers, a substantial, significant relative risk was seen for bladder cancer.

Aldehydes

Formaldehyde and other aldehydes are emitted in the exhaust of both gasoline- and diesel-fueled vehicles. Formaldehyde is of particular interest both due to its photochemical reactivity in ozone formation and suspected carcinogenicity. EPA has classified formaldehyde as a probable human carcinogen (Group B1) (Adler and Carey 1989). Formaldehyde also can be a short-term respiratory and skin irritant, especially for sensitive individuals (Harvey et al. 1984).

Aldehyde exhaust emissions from motor vehicles correlate reasonably well with exhaust hydrocarbon (HC) emissions, and diesel vehicles generally produce aldehydes at a greater HC composition percentage rate than gasoline vehicles. EPA estimates that 32.6 percent of national formaldehyde emissions are from motor vehicles (EPA 1986s). By 1995, total formaldehyde emissions can be expected to drop, due to the increasing use of three-way and three-way plus oxidation catalyst-equipped, gasoline-fueled vehicles together with the phase-out of vehicles not equipped with catalysts. Formaldehyde emission inventories should be analyzed with caution, however, because formaldehyde can be generated by photochemical reactions involving other organic emissions.

Benzene

Benzene is present in exhaust, evaporative, and running loss emissions. Several epidemiology studies have associated benzene with an increased incidence of leukemia (as reported by the California Department of Health Services in CARB 1986). Mobile sources (including refueling emission) dominate the nationwide benzene

emission inventory, with roughly 70.2 percent of the total benzene emissions (EPA 1986s). Of the mobile-source contribution, 70 percent comes from exhaust and 14 percent comes from evaporative emissions (CARB 1986). EPA estimates that the fraction of benzene in the exhaust varies depending on control technology and fuel composition but is generally about 2 to 5 percent. The fraction of benzene in the evaporative emissions also depends on control technology and fuel composition but is generally about 1 to 2 percent (Adler and Carey 1989). Exposure to benzene during refueling includes self-service refueling, occupational exposure (i.e., service station attendants), and community exposure in an urban area.

Other Organics Associated with Nondiesel Particulate

Gasoline-fueled vehicles emit far less particulate than their diesel counterparts. It is thought that a number of nitro-polycyclic aromatic hydrocarbons (nitro-PAH) compounds account for much of the mutagenicity of diesel particulate emissions. Particulate emissions from gasoline-fueled vehicles contain significantly less of these nitro-PAHs; however, the mutagenicity of the gasoline soluble organic fraction (SOF), expressed as revertants/μg SOF, is greater than diesel SOF. Also, unlike diesel SOF, the mutagenic activity of gasoline SOF increases with the addition of S9 activation (indicating indirect-acting activity). This situation suggests that the nonnitro-PAHs may be responsible for the mutagenicity of gasoline SOF, rather than the nitro-PAHs. In any event, even if the overall mutagenicity and emissions of gasoline-fueled vehicle-exhaust particulate is less than that of diesel particulate, the overall impact from gasoline particulate might be significant, given the substantial travel (vehicle miles traveled or VMT) of gasoline-fueled vehicles. EPA estimates that approximately 43 percent of polycyclic organic matter (POM), a chemical grouping that includes many PAHs, may be attributed to mobile sources (EPA 1986s).

Dioxins

The major dioxin compound of interest is 2,3,7,8-tetrachlorodibenzo-*p*-dioxin. This dioxin compound exists in the particulate state or is adsorbed onto particulates. Dioxin is extremely toxic and has been the subject of health concern in a number of settings (U.S. Surgeon General 1980). Some qualitative analytical measurements have found dioxin to be present in the muffler scrapings of vehicles using either leaded or unleaded gasoline. As early as 1978, researchers from Dow Chemical Company reported both 2,3,7,8-TCDD and other isomers in particulates of both gasoline-fueled and diesel-fueled vehicles (reported in EPA 1980d). Recent evidence has indicated that the risk associated with dioxin may have been overestimated; as a result EPA has initiated a comprehensive review.

Asbestos

Asbestos is used in brake linings, clutch facings, and automatic transmissions. About 22 percent of the total asbestos used in the U.S. in 1984 was used in motor vehicles. Health effects of asbestos exposure have been known for many years, including cancer, asbestosis, and mesothelioma (Carton and Kauffer 1980, Enterline 1983).

Gasoline Vapors and Other Gas-Phase Volatile Organic Compounds

In a study by the Health Effects Institute, totally vaporized gasoline has been found to cause a statistically significant increase in kidney tumors in male rats and liver tumors in female mice (as reported in MVMA 1986). Although this study did not come to conclusions on the degree of risk posed to humans from gasoline vapors, a variety of compounds with potential toxic effects are known to be in gasoline vapors, including

benzene, xylene and toluene. EPA has classified gasoline vapors as a probable human carcinogen (Group B2) (Adler and Carey 1989). Other gas-phase organics, or VOCs, are present in both exhaust and evaporative emissions. In addition to the three compounds mentioned above, the majority of VOCs consist of unsaturated and saturated hydrocarbons along with alkyl benzenes, aliphatic aldehydes, and a variety of PAH (although most of the known mutagenicity of PAH motor vehicle emissions is associated with the particulate phase). Of all the VOCs emitted from motor vehicles, benzene, formaldehyde, benzo[a]pyrene, ethylene, and 1,3-butadiene are the pollutants most easily analyzed for their carcinogenic potential. EPA has classified 1,3-butadiene as a probable human carcinogen (Group B2) (Adler and Carey 1989). Because of the limited amount of published data, quantification of the potential carcinogenic effects of other VOCs is more difficult. EPA estimates that about 62 percent and 46 to 47 percent of toluene and xylene emissions nationwide, respectively, are emitted from motor vehicles (EPA 1986s).

Metals

The toxicological effects of metals, especially heavy metals, have been studied for many years (ACGIH 1971). In addition, many are now being analyzed for their carcinogenic potential. EPA has identified mobile sources as a significant contributor to nationwide metals inventories (EPA 1986s) including 1.4 percent of beryllium and 8.0 percent of nickel emissions. The California Air Resources Board also is analyzing these metals, as well as arsenic, manganese, and cadmium, as mobile source pollutants (Boyd 1986). Because of their relatively high unit risk value, emissions of chromium also may be a concern.

FACTORS THAT INFLUENCE MOBILE SOURCE AIR TOXIC EMISSIONS

Control Technology

To meet the relatively lenient HC and CO standards that applied in the early 1970s in the

U.S., auto manufacturers generally relied on enleanment of the air/fuel mixture and modification of spark timing. In addition, newer combustion chamber designs were introduced to reduce HC emissions and with faster flames to limit the increased NO$_x$. Even when HC and CO standards were tightened, the engine modification approach continued to predominate, with the addition of certain new wrinkles such as transmission-controlled spark timing and antidieseling throttle control. Attainment of initial HC and CO standards with limitations on NO$_x$ increases generally were possible without significant fuel consumption penalties. However, as emissions standards were tightened (especially in 1973 and 1974) it became more difficult to achieve low levels of CO, HC, and NO$_x$ without unacceptable compromises in performance or fuel economy. As a result, there was a fundamental shift in the technology to the catalytic converter.

Starting with 1975 model year cars, catalysts were installed on upwards of 80 percent of all new cars in the United States. Since 1981, they have been installed on all new gasoline-fueled cars. Whereas initial systems in 1975 primarily contained oxidation catalysts, over time the emphasis has gradually shifted to predominantly three-way systems. Model year 1991 is the seventeenth year that emissions control systems featuring catalytic converters have appeared in production quantities in light-duty vehicles. After an initial period of uncertainty, they have gained broad acceptance in the U.S. and Japan as the only practical way for automakers to comply with stringent exhaust emissions control standards. Although the primary motivation for the introduction of these controls has been reduction of the regulated pollutants, several advantages for controlling toxic emissions have also resulted.

Fuels

The relationship between fuels and air toxics emissions in many ways is not fully understood. Nevertheless, modifying fuels is one alternative to the implementation of technology-forcing emissions standards. Such an approach has become a major thrust of the 1990 Amendments as summarized above. Recent research has suggested that modifying the PAH content of diesel fuel would be an effective way of

reducing engine-out diesel PAH emissions (Ingham and Warden 1987). If metals such as chromium turn out to be important trace emissions as a result of impurities in diesel fuels, monitoring and even control of such trace substances could justifiably be considered. The remainder of this section is focused on a review of fuel issues regarding air toxics.

Gasoline

For legitimate health reasons, the use of lead additives is being severely curtailed or eliminated by government regulation (Ammest et al. 1980; House of Lords 1985; NRC 1980; Needleman et al. 1979). Partly in response to this and partly in response to energy needs, greater amounts of alcohols and ethers – either as high-octane blending components or as substitutes for gasoline – are being used. Other likely changes include the processing of heavier crudes and the likely decline in residual fuel demand. These could cause significant changes in emissions characteristics.

In the refining process, the required octane level for finished gasoline is achieved through the use of chemical additives, primarily lead (in the past) or by refinery processing. If the concentration of lead is reduced, fuel octane levels go down and the octane can be replaced only by increasing processing severity or using a substitute octane booster. To the extent that other metal-based additives are integrated into gasoline formulations, they like lead present the potential for mobile source emissions of metals with associated health effects.

Because alcohol fuels are high in octane, their use in gasoline has increased as the use of lead additives decreased. Some of the increased cost and energy penalty that is associated with the elimination of lead in gasoline also is reduced by the use of methyl tertiary butyl ether (MTBE), tertiary butyl alcohol (TBA), or other alcohols. Alcohols and ethers, however, also influence other fuel properties like volatility, material compatibility, and water sensitivity. The effect is dependent on the composition of the base gasoline. For example, a gasoline high in aromatics blended with alcohols will show increased volatility, more severe material problems, but less water sensitivity than a gasoline with less aromatics (e.g., a cracked gasoline).

Ethanol Blends

Blending ethanol with gasoline gives the fuel a higher octane value. The blend generally is no more corrosive to fuel-system metals than gasoline, although laboratory tests indicate potential reactions with elastomers (e.g., rubber and plastics). From an environmental standpoint, ethanol-gasoline blends have advantages and disadvantages. Although exhaust CO and HC levels tend to decrease, evaporative HC, NO_x, and photochemically reactive aldehydes tend to increase.

Methanol Blends

Methanol also can be blended easily with gasoline. If added alone to gasoline in a 10 percent blend, however, methanol may cause vapor lock, phase separation and corrosion of fuel-system parts. Blended properly, these problems can be eliminated. Methanol also is associated with well-known adverse health effects and it forms formaldehyde during combustion.

Impact of Blend Fuels on Emissions

The addition of oxygenates (e.g., low-level blends of approximately 6 percent oxygenates or less) to gasoline will alter the stoichiometric air/fuel ratio compared to pure gasoline. The leaner mixture may either reduce or increase the level of pollutants in the exhaust gas, depending on the carburetor setting. The leaning-out effect, provided the carburetor setting is unchanged, may reduce the emissions of CO and HC. If the mixture becomes too lean, the HC emissions could increase considerably due to misfiring. There also is a tendency toward increased evaporative emissions.

Pure Alcohol Vehicles

The use of pure alcohol engines will result in exhaust emissions with only a few components compared to gasoline. Aromatic hydrocarbons such as benzene are not formed and PAH emissions are very low. The characteristic emission components are CO, unburned alcohol, NO_x, and aldehydes. Engines designed for pure methanol tend to have low emissions of CO, NO_x, and unburned fuel and can be adapted easily to comply with existing emission standards around the world. Evaporative emissions also are low.

Aldehyde emissions, however, can be four to eight times higher than for gasoline vehicles. These compounds tend to be highly photochemically reactive and to contribute directly to eye irritation. As discussed above, there also is evidence that formaldehyde is a carcinogen. Fortunately, the data indicate that these emissions are reduced effectively by catalysts. Emission characteristics of the alcohol-fuel diesel engines also are good. They feature low emissions of NO_x, PAH, and particulates. Both concepts can be used together with an oxidation catalyst to effectively reduce the unburned fuel and aldehydes. The spark-ignited methanol engine shows virtually no particulate emissions.

Diesel Fuel

Probably the most important characteristic of diesel fuel is its ignition quality that is indicated by cetane number. The major challenge regarding diesel fuel is in maintaining current cetane number levels. This will be difficult in the future because of the expected reduction in the quality of crude oils and the increasing demand for diesel fuel.

Fuel issues for diesel engines revolve around fuel purity and its relationship to diesel particulate control. Impurities in diesel fuel are a concern in and of themselves, particularly metals like chromium, because of the potential for direct emission in the exhaust. With respect to diesel particulate control, fuel additives may be important. Two major approaches exist for meeting tight diesel particulate standards: (1) engine modifications to lower engine out emission levels, and (2) trap-oxidizers and their associated regeneration systems. Trap-oxidizer prototype systems have shown themselves capable of 70 to 90 percent reductions from engine-out particulate emissions rates and with proper regeneration the ability to achieve these rates for high mileage. One type of trap-oxidizer system that is being explored includes self-regeneration by means of metallic fuel additives. As discussed above, such additives could be a source of significant concern with regard to unregulated pollutants depending on the fuel additive used. Alternatively, an approach being actively pursued in Athens, Greece successfully uses cerium as the additive.

Inspection/Maintenance Programs

Motor vehicle emission inspection/maintenance (I/M) programs are oriented toward reducing CO and total HC emissions. However, as indicated above, emission control technology has been shown effective at reducing many specific HC emissions while reducing total HC. This situation strongly suggests that when I/M programs reduce total HC, they produce simultaneous benefits in reduction of other mobile source air toxics emissions. Work by the California Air Resources Board on benzene tends to support this conclusion (CARB 1986).

Transportation Controls

By reducing the use of vehicles or facilitating their smooth flow, transportation controls can reduce the total emissions of all the pollutants from vehicles. Additional reductions in vehicular emissions can be achieved by reducing dependence on individual cars and trucks and by making greater use of van and car pools, buses, trolleys, and trains. Improving urban traffic management by installing synchronized traffic lights, reducing on-street parking, switching to 'smart' roads, banning truck unloading during the day, and other measures also can improve transportation system fuel efficiency (U.S. Congress 1989).

Providing efficient, convenient, and affordable public transportation alternatives worldwide would produce multiple benefits. Greater use of public transportation would reduce congestion, cut fatalities and injuries from traffic accidents, and greatly improve air quality. Fortunately, such transportation improvements can be phased in over time. For example,

roadways initially dedicated to bus traffic can later be upgraded to light rail or heavy rail if circumstances warrant.

References

Adler, J.M. and P.M. Carey. 1989. Air toxic emissions and health risks from mobile sources. Paper No. 89-34A.6. Presented at the 82nd Annual Meeting of the Air & Waste Management Association. Anaheim, CA.

American Conference of Governmental and Industrial Hygienists (ACGIH). 1971. Documentation of the Threshold Limit Values for Substances in the Workroom Air. Third Edition. Cincinnati, OH.

Ammest, et al. 1983. Chronological trend in blood lead levels between 1976 and 1980. *New England Journal of Medicine* 308(23).

Boyd, J. 1986. State of California Air Resources Board-Toxic air contaminant program motor vehicle control strategies. Presented to the STAPPA/ALAPCO Air Toxics Conference. Washington, D.C.

Brightwell et al. 1986. Neoplastic and functional changes in rodents after chronic inhalation of engine exhaust emissions. Battelle-Geneva Institute. Geneva, Switzerland.

California Air Resources Board (CARB). 1986. Proposed benzene control plan and technical support document. Stationary Source Division. Sacramento, CA.

Carton, B. and E. Kauffer. 1980. The metrology of asbestos. *Atmospheric Environment* 14:1118–1980.

Coordinating Research Council (CRC). 1983. Cancer incidence among members of a heavy construction equipment operators union with potential exposure to diesel exhaust emissions. Submitted to CRC by Environmental Health Associates. April 18, 1983.

Enterline, P.E. 1983. Cancer produced by non-occupational exposure in the United States. *Journal of the Air Pollution Control Association* 33:318.

Harvey, Craig A. et al. 1984. Toxicologically acceptable levels of methanol and formaldehyde emissions from methanol-fueled vehicles. Paper No. 841357. Society of Automotive Engineers. Warrendale, Pennsylvania.

House of Lords. 1985. Report on lead in petrol and vehicle emissions. Select Committee on the European Communities. February 26, 1985.

Ingham, M.C. and R.B. Warden. 1987. Cost effectiveness of diesel fuel modifications for particulate control. Paper No. 870556. Society of Automotive Engineers. Warrendale, Pennsylvania.

Motor Vehicle Manufacturers Association (MVMA). 1986. Gasoline refueling vapors. *Journal of the Air Pollution Control Association* 36:230.

National Research Council (NRC). 1980. *Lead in the Human Environment.* Washington, D.C. National Academy Press.

Needleman et al. 1979. Deficits in psychologic and classroom performance of children with elevated dentine lead levels. *New England Journal of Medicine* 300(13).

Ozkaynak, H. and J.D. Spengler. 1986. Health effects of airborne particles. Cambridge, MA: Harvard University.

Schenker, J.D. 1984. Oral Statement at the American Lung Association Convention. May, 1984.

Summerhays, J. 1986. Presentation on the Southeast Chicago Air Toxics Study, Air Pollution Control Association, Air Toxics Workshop. Chicago, IL.

U.S. Congress. 1989. Advanced vehicle/highway systems and urban traffic problems. Staff Paper. Science, Education and Transportation Program. Office of Technology Assessment. September.

U.S. Environmental Protection Agency (EPA). 1980d. Volatile Organic Compound (VOC) Species Data Manual. Second Edition. EPA 450/4-80-015. Office of Air Quality Planning and Standards. Research Triangle Park, NC.

U.S. Environmental Protection Agency (EPA). 1985e. The air toxics problem in the United States: An analysis of cancer risks for selected pollutants. Office of Air Quality Planning and Standards and Office of Policy, Planning, and Evaluation. Research Triangle Park, NC, and Washington, D.C.

U.S. Environmental Protection Agency (EPA). 1985f.

Air toxics controllability study. Final Report submitted under Contract No. 68-01-7047 (Work Assignments No. 26). Office of Policy, Planning, and Evaluation and Office of Air Quality Planning and Standards. Washington, D.C. and Research Triangle Park, NC.

U.S. Environmental Protection Agency (EPA). 1986s. Compiling air toxics emission inventories. EPA 450/4-86-010b. Office of Air Quality Planning and Standards. Research Triangle Park, NC.

U.S. Environmental Protection Agency (EPA). 1986t. Santa Clara Valley integrated environmental management project: Revised stage one report. Office of Policy, Planning, and Evaluation. Washington, D.C.

U.S. Environmental Protection Agency (EPA). 1990b. Cancer risk from outdoor exposure to air toxics. Volume 1, Final Report. EPA 450/1-90-004a. Office of Air Quality Planning and Standards. Research Triangle Park, NC.

U.S. Environmental Protection Agency (EPA). 1990e. Health assessment document for diesel emissions. Draft Report. Office of Health and Environmental Assessment. Washington, D.C.

U.S. Environmental Protection Agency Region VII. (EPA-VII). 1986. Iowa air toxics emissions inventory: Phase 1. EPA 907/9-86-004. Kansas City, MO.

U.S. Surgeon General and the Congressional Research Service. 1980. Health effects of toxic pollution. Committee on Environment and Public Works. Serial No. 96-15. U.S. Government Printing Office, Washington, D.C.

Walsh, M.P. 1987. The benefits and costs of diesel particulate control: methanol fuel for the in-use urban bus. Paper No. 870013. Society of Automotive Engineers. Warrendale, Pennsylvania.

26

Other Programs That Control Toxic Air Pollutants

David R. Patrick

INTRODUCTION

Toxic air pollutants generally are defined as pollutants that have a potential to cause serious adverse human health effects or serious adverse environmental effects and are released or emitted in a routine (as opposed to unanticipated) manner to the ambient air. Section 112(b) of the 1990 Clean Air Act Amendments specifically lists 189 substances, compounds, and mixtures as toxic air pollutants for regulatory consideration. These pollutants are to be regulated initially through technology-based controls with additional control added later, if required to protect the public health with an 'ample margin of safety.' The pollutants, sources, and basic control programs have been described in previous chapters. The purpose of this chapter is to describe other environmental programs or activities that directly or indirectly result in the control of toxic air pollutants from stationary sources.

The most important of these programs are those currently in place in many states and local areas of the U.S. These programs evolved over the past decade both from frustration with EPA's slow progress in implementing the original hazardous air pollutant requirements of the 1970 Clean Air Act Amendments and in response to growing concerns in their respective jurisdictions. Although the scope and stringency of these programs vary widely, they clearly have resulted in significant reductions in emissions of air toxics that otherwise would not have occurred.

An important new federal toxic air pollutant control initiative, the Industrial Toxics Project, was announced by the Administrator of EPA in early 1991 and has become known as the 33/50 Program. This program seeks voluntary reductions in seventeen high-priority toxic air pollutants from major corporate emitters and is part of a nationwide pollution prevention strategy aimed at stopping pollution before it is generated.

A similar program, called the Early Reduction Program, is mandated in section 112(i)(5) of the 1990 Amendments. Although not structurally different from the mandated control program for toxic air pollutants, it likely will result in earlier reductions in quantities of toxic air pollutants than otherwise would have occurred from some facilities.

Many toxic air pollutants also are indirectly controlled as a result of the requirements of section 313 of the Emergency Planning and Community Right-to-Know Act of 1986 (also known as Title III of the Superfund Amendments and Reauthorization Act, or SARA). That program requires many manufacturing facilities to report annually the releases to the environment of over three hundred toxic chemicals. The publication of the results, the Toxic Release Inventory, is a substantial incentive to reduce reported releases. This is particularly strong for companies that are near the top of the list either nationally or locally.

In addition, there are other pollutants and public health concerns that have resulted in regulatory programs or activities that either directly or indirectly reduce the emissions of toxic air pollutants. An important example is the regulation of particulate matter and ozone precursors under sections 108–110 of the Clean Air Act to attain and maintain the national ambient air quality standards (NAAQS). Particulate matter can contain many of the trace metals listed in section 112(b) of the 1990 Amendments and many ozone precursors are volatile organic compounds (VOCs) also listed in section 112(b). Other important examples are the new accidental release prevention program required under section 112(r) of the 1990 Amendments and the efforts under way nationwide to reduce exposures to indoor air pollutants causing adverse human health effects. Many pollutants to be controlled in these programs also are listed specifically as toxic air pollutants. Each of these programs are described below and their interrelationships with toxic air pollutant control discussed.

STATE AND LOCAL AIR TOXICS PROGRAMS

A comprehensive survey of state and local agency toxic air pollution activities was carried out by the State and Territorial Air Pollution Program Administrators (STAPPA) and the Association of Local Air Pollution Control Officials (ALAPCO) published in 1989 (STAPPA/ALAPCO 1989). In that survey, fifty state and forty local agencies responded to a broad range of questions concerning their toxic air pollution programs and plans. The survey showed that air toxics control activities had increased substantially since a previous survey in 1984 and that air toxics control programs now exist in some form in every state in the U.S.

The approaches used by states vary widely but generally fall into one of three categories: formal regulatory programs, comprehensive policies, and informal programs. The 1989 survey found that only six states have legislation that specifically addresses air toxics. Most rely on general air pollution control legislation. States also use various methods to define the scopes of their programs, including specific pollutants, source categories, and size exemptions.

The level of effort devoted by states to air toxics programs also varies, with a median of 5.4 work years per state, up from one work year in 1984. Yearly expenditures also vary, with a median of $333,000 per state, up from $52,500 in 1984. These resources come both from state funds and federal grants under section 105 of the Clean Air Act.

The approaches used by local agencies generally mirror those of the states in terms of enabling legislation, scope, and implementation. The median level of effort devoted by local agencies was 1.8 work years for the agencies reporting, up from 0.5 work years in 1984. The median for yearly expenditures was $186,000, up from $20,000 in 1984.

Tables 16-3 and 16-4 summarize the programs in place in 1989. Clearly, these programs have resulted in significant reductions in emissions of air toxics in the past decade. The precise amount, however, is impossible to determine. A number of specific examples of successful state and local control activities are described in the STAPPA/ALAPCO survey results. The growth in state/local air toxics activity and resource utilization from 1984 to 1989 is expected to continue for a few years. However, the full actualization of the federal toxic air pollution program under Title III of the 1990 Amendments, and the new federal/state operating permit program established under Title V of the Amendments, should result in stabilized state and local air toxics efforts by the mid-1990s.

EPA'S 33/50 PROGRAM

In early 1991, the Administrator of EPA formally asked over six hundred U.S. companies to reduce voluntarily emissions of seventeen high-priority toxic air pollutants. The goal of the program is to reduce the total releases and transfers of these chemicals by 33 percent by 1992; and to reduce them by 50 percent by 1995 (hence, the 33/50 Program). This effort is a key component of the Agency's efforts to promote pollution prevention. Pollution prevention seeks to reduce the emissions of pollutants by reducing their generation, as opposed to end-of-pipe controls. Where successful, pollution prevention is much more cost-effective than destroying the pollutants or capturing and disposing of them.

The priority chemicals identified by EPA for the 33/50 Program are shown in Table 26-1, along with their annual reported releases in 1988. Although the list was derived from the Toxics Release Inventory (see below), all of the chemicals also are on the list of hazardous air pollutants in section 112(b) of the 1990 Amendments. These chemicals are widely used in many industries.

The 33/50 Program necessarily is voluntary because most of the companies involved were emitting the chemicals in compliance with applicable air pollution regulations. However, EPA believed that the added reduction in emissions to the air of these pollutants would greatly benefit public health. Importantly, many companies joined the program and began efforts to reduce emission of these chemicals. These included such giants as Bethlehem Steel, General Electric, Dupont, and Chevron.

Table 26-1   High-priority toxic chemicals for 33/50 Program

Chemical[a]	1988 Air releases (lbs)[b]
Benzene	28,117,955
Cadmium and compounds	119,412
Carbon tetrachloride	3,683,121
Chloroform	22,974,156
Chromium and compounds	1,181,482
Cyanide and compounds	1,881,210
Methylene chloride	126,796, 287
Lead and compounds	2,587,790
Mercury and compounds	25,629
Methyl ethyl ketone	127,676,717
Methyl isobutyl ketone	30,523,897
Nickel and compounds	539,864
Perchloroethylene	32,277,372
Toluene	273,752,712
Methyl chloroform	170,420,900
Trichloroethylene	49,071,464
Xylenes	155,888,584

[a] Source: EPA.
[b] Toxics Release Inventory, 1988.

EPA reported (EPA 1992b) that 734 companies had responded by February 1992 expressing commitments to voluntary reductions in releases and transfers of toxic chemicals. The actual reductions pledged by these companies were 304 million pounds. Important adjuncts to the 33/50 Program are more aggressive technical assistance and technology transfer efforts by EPA. EPA has conducted workshops, expanded its free computer bulletin board, called the Pollution Prevention Exchange System, published bibliographic reports and resource guides, and identified areas of new research.

EARLY REDUCTION PROGRAM

Section 112(i)(5) of the 1990 Amendments requires EPA or states to issue a permit for an existing source, for which the owner or operator demonstrates that the source has achieved a reduction of 90 percent or more of emissions of hazardous air pollutants (95 percent for particulate pollutants) by January 1, 1994. The reduction is to be based on 'verifiable and actual' emissions from a year generally no earlier than calendar year 1987. If a source can demonstrate this reduction, the permit applies for a period of six years 'from the compliance date for the otherwise applicable standard, provided that such reduction is achieved before the otherwise applicable standard is first proposed.' Given the possibility that the maximum achievable control standards (MACT) ultimately promulgated by EPA could require more stringent reductions, this program provides an incentive for sources to reduce emissions before the MACT standards are promulgated.

The status of the early reduction program was reviewed by Laznow and Daniel (1992). In June 1991, EPA proposed regulations[1] and released guidelines (EPA 1991d) for the implementation of the early reduction program. The guidance provides the procedures for applying and qualifying for the six-year extension and defines key aspects of the program. EPA outlined two options for participating. The first requires that emission reductions be achieved before proposal of the applicable MACT standard. The second allows sources potentially subject to MACT standards before 1994 to qualify, if the source makes an 'enforceable commitment, before proposal of the applicable MACT standard, to achieve the 90 (95) percent reduction prior to January 1, 1994.' The commitment is binding on the source and enforceable under section 113 of the Act. Importantly, emissions reductions can be demonstrated through control, reduction in production or use (i.e., pollution prevention), or 'permanent' source shutdown or curtailment in production. This latter is controversial since it raises the threat of job and economic losses resulting from clean air regulation.

An important feature of the early reduction program is the definition of source since a company's ability to achieve the required reduction can rest on how broadly or narrowly EPA defines the applicable source. EPA's proposed rule attempts to provide flexibility in the definition of source in order to encourage the broadest possible participation. This is done by allowing companies in some instances to define selected portions of a plant for inclusion into the program. EPA has defined source[2] in the following general ways:

■ Equipment leaks at organic chemical plants.

Table 26-2   List of high risk pollutants

Pollutant	Weighting factor[a]
Carcinogens	
2,3,7,8-Tetrachlorodibenzo-*p*-dioxin	100,000
Benzidine	1,000
Bis(chloromethyl)ether	1,000
Asbestos	100
Chromium compounds	100
Inorganic arsenic compounds and arsine	100
Chloromethyl methyl ether	10
Cadmium compounds	10
Heptachlor	10
Beryllium compounds	10
Acrylamide	10
Coke oven emissions	10
Hexachlorobenzene	10
Chlordane	10
Dichloroethyl ether	10
1,3-Butadiene	10
Benzotrichloride	10
Ethylene dibromide	10
Ethylene oxide	10
Vinyl chloride	10
Acrylonitrile	10
1,1,2,2-Tetrachloroethane	10
Vinylidene chloride	10
Benzene	10
1,2-Propylamine	10
Noncarcinogens	
2,4-Toluene diisocyanate	10
Acrolein	10
Acrylic acid	10
Chloroprene	10
Dibenzofurans	10
Mercury compounds	10
Methyl isocyanate	10
Methylene diphenyl diisocyanate	10
Phosgene	10

[a] The weighting factor is a measure of relative toxicity for offset purposes.

Source: 56 FR 27354, June 13, 1991.

■ The entire contiguous facility.
■ Any stationary source that is a facility, building, structure, or installation at a plant.
■ Any combination of buildings, structures, facilities, or installations within the

[1] 56 FR 27338, June 13, 1991.
[2] 56 FR 27341, June 13, 1991.

contiguous property where emission reductions from the aggregation of sources constitutes significant reduction in emissions from the entire contiguous property.[3]

■ Any individual emission point or combination of points within a contiguous property provided that emission reductions constitute a significant reduction in emissions from the entire contiguous property.[3]

These definitions generally allow facilities to aggregate sources along both functional and geographic lines. Although this approach is broadly supported by industry, some members of Congress, public interest groups, and states argue that it is inconsistent with the definition of source that will be used in the MACT standards program to come and it conflicts with many existing and proposed state programs. These issues were expected to be resolved when EPA promulgated the final program guidance, scheduled in mid-1992.

Another aspect of the early reduction program that is controversial arises from section 112(i)(5)(E) of the 1990 Amendments that specifies that EPA limit the use of offsetting reductions in emissions of other hazardous air pollutants from the source as counting toward the 90 and 95 percent reductions in pollutants that pose high risks of adverse public health effects. Congress specifically identified chlorinated dioxins and chlorinated furans as high-risk pollutants. In EPA's proposed regulations, EPA listed thirty-five such pollutants and assigned weighting factors to each that are a measure of their relative toxicity. The list is shown in Table 26-2.

By mid-1992, EPA reported that almost sixty companies had submitted 'enforceable commitments' to EPA. The majority of the submittals were from chemical plants. A base year total of 30,289,500 pounds of emissions of hazardous air pollutants were involved – a 90 percent reduction would be over 27 million pounds. However, EPA had approved only a small fraction owing to deficiencies in the submittals, although these were in process of being addressed. Companies that are participating include Exxon, Allied-Signal, Monsanto, Dow, and Union Carbide.

TOXICS RELEASE INVENTORY

As noted in Chapter 1, Congress enacted the Emergency Response and Community Right-to-Know Act in 1986 in part because of the tragic accident in Bhopal, India. In section 313, Congress specified that a toxic chemical release inventory program be initiated to make public the releases by industry of toxic chemicals to the nation's air, water, and land. As described by Arbuckle, Vanderver, and Wilson (1992), the intent was to provide information to EPA to assist in developing and priority ranking sources and chemicals for control. Many felt that the public pressures resulting from knowledge of 'large' toxic chemical releases also would promote better control and more responsible use of dangerous chemicals. Congress specified an initial list of 'toxic chemicals' and instructed EPA to establish a program whereby manufacturing facilities that release these and other toxic chemicals to the environment are required to report on a yearly basis their releases to air, water, and land. Annual reports were required to be submitted beginning in 1988. This formed the basis for the well-known Toxic Release Inventory (TRI) Program.

Importantly, releases of 170 chemicals from the list of 189[4] hazardous air pollutants in the 1990 Amendments are reported to TRI.

---

[3]Emission reductions are considered significant if they are made from base year emissions of at least 10 tons per year from an entire contiguous facility of at least 25 tons per year, or at least 5 tons per year from a contiguous facility of less than 25 tons per year.

[4]The TRI Report referenced here was published before the enactment of the Clean Air Act Amendments of 1990 at a time when there were 191 chemicals listed in draft Amendments. When finally enacted, the list contained 190 entries – ammonia was removed. Hydrogen sulfide (H_2S) also was to have been removed from the list, but inadvertently was left in the bill signed by the President. In late 1991, H_2S finally was removed from the list.

However, TRI reports are required only from facilities with ten or more full time employees in the manufacturing sector of U.S. business (SIC Codes 20–39) and there are many smaller manufacturing facilities and nonmanufacturing sources of toxic air pollutants. The TRI summary report is published annually by EPA. At the time of this writing in 1992, the 1988 Toxic Release Inventory National Report (EPA 1990a) was the most recently available published summary, although 1990 data were available through the National Library of Medicine's TOXNET system.

Key statistics from the TRI are summarized in Tables 1-8 and 1-9. The TRI provides the results of reports detailing total (i.e., air, water, and land) releases of over 300 chemicals from almost 20,000 U.S. manufacturing facilities. Total releases of these chemicals to the air were reported in 1990 to be about 2.2 billion pounds.

The chemical industry continued to lead all manufacturing categories but reported a 35 percent reduction in total releases from 1987 to 1990. Some of this reduction is real; however, some is a result of more precise estimation.

As noted above, 170 of the 189 hazardous air pollutants listed in the 1990 Amendments are reported in the TRI database. These 170 accounted for over 80 percent of the total TRI emissions to the air in 1988. Seven of these chemicals were released in quantities exceeding 100,000,000 pounds each in 1988 and accounted for over 70 percent of the total air releases of the 170 chemicals. Appendix D in the TRI Report matches the TRI chemicals with those on the Clean Air Act list, and provides their point source and fugitive release quantities. Table 1-10 summarizes the air emissions information for the top twenty-five chemicals on the Clean Air Act list.

NATIONAL AMBIENT AIR QUALITY STANDARDS PROGRAM

Relationship to Air Toxics

As noted earlier, two pollutants regulated with national ambient air quality standards (NAAQS) can include substances that normally also are considered to be toxic air pollutants. Particulate matter includes a broad spectrum of solid metallic, carbonaceous matter, or organic matter, often with higher molecular weight organics adsorbed onto the surface. Both trace metals and many higher molecular weight organics are listed in section 112(b) of the 1990 Amendments. Ozone also is produced in a complex photochemical reaction that involves volatile organic compounds (VOCs) and nitrogen oxides. Many VOCs are on the list of hazardous air pollutants. Examples of particulate matter and ozone precursors that also are toxic air pollutants are shown in Table 26-3.

The key to the production of ozone is the reactivity of the individual VOCs (i.e., the rapidity at which a compound breaks down under photochemical action to form ozone and other pollutants). In early years, it generally was thought that only the most highly reactive

VOCs were important. However, it soon became clear that almost all VOCs contribute to tropospheric ozone and that only a limited number do not break down in the troposphere under photochemical actions. Of course, the stable VOCs include chlorofluorocarbons that now are known to remain intact until they migrate to the stratosphere where they participate in the elimination of the 'good' stratospheric ozone. Table 26-4 lists the VOCs considered by EPA to be negligibly reactive for purposes of the ozone nonattainment program.

Clearly, many programs aimed at the reduction of emissions of particulate matter and VOCs will have a beneficial effect on toxic air pollutants. That already has occurred with the implementation of the particulate matter and ozone control programs of the past twenty years and it will improve further in the years to come. The specific degree of reduction in air toxics in any geographic area that results from control of particulate matter and VOCs, however, is impossible to estimate. The amount and composition are highly dependent on the sources, the emission rates, and the specific pollutants

Table 26-3   Selected criteria air pollutants that also can be toxic air pollutants

Particulate matter	Volatile organic compounds	Volatile organic compounds
Antimony compounds	Acetaldehyde[b]	Hexane
Arsenic compounds	Acetamide	Methanol
Beryllium compounds	Acetonitrile	Methyl chloride
Cadmium compounds	Acetophenone	Methyl ethyl ketone
Chromium compounds	Acrolein[b]	Methyl methacrylate
Cobalt compounds	Acrylamide	Methyl-*t*-butyl ether
Coke oven emissions[a]	Acrylic acid	Methylene chloride
Lead compounds	Acrylonitrile	Nitrobenzene
Manganese compounds	Allyl chloride	4-Nitrophenol
Mercury compounds	Aniline	2-Nitropropane
Nickel compounds	Benzene	Phenol
Polycyclic organic matter[a]	Benzotrichloride	Phenylene diamine
Selenium compounds	Benzyl chloride	Propionaldehyde
	1,3-Butadiene	Propylene dichloride
	Caprolactam	Propylene oxide
	Chlorobenzene	Styrene
	Chloroform	Styrene oxide
	Chloroprene	1,1,2,2-Tetrachloroethane
	Cresols	Tetrachloroethylene
	Cumene	Toluene
	Diazomethane	2,4-Toluene diamine
	1,4-Dichlorobenzene	1,2,4-Trichlorobenzene
	Diethanolamine	Trichloroethylene
	Diethyl sulfate	Triethylamine
	Diethyl sulfate	2,2,4-Trimethylpentane
	Epichlorohydrin	Vinyl acetate
	Ethyl acrylate	Vinyl bromide
	Ethyl benzene	Vinyl chloride
	Ethylene dichloride	Vinylidene chloride
	Ethylene oxide	Xylenes
	Formaldehyde[b]	

[a] Includes high molecular weight organic compounds.
[b] Also a product of photochemical reactions.

emitted. Although these can vary widely, the more important components of the control programs that result in air toxic emissions reductions are discussed below.

Criteria Air Pollutant Control

The Clean Air Act Amendments of 1970 defined two principal types of air pollutants to be regulated: criteria air pollutants and hazard-ous air pollutants. Although the definition of a hazardous air pollutant changed with the 1990 Amendments, it did not for criteria air pollutants. Criteria air pollutants are defined in section 108(a)(1) as air pollutants, the emissions of which can cause or contribute to '... air pollution which may reasonably be anticipated to endanger public health or welfare[5] ... and the presence of which in the ambient air results from *numerous or diverse mobile or stationary*

[5] Section 302(h) of the Act defines welfare effects as including, but not limited to, ' ... effects on soils, water, crops, vegetation, man-made materials, animals, wildlife, weather, visibility and climate, damage to and deterioration of property, and hazards to transportation, as well as effects on economic values and on personal comfort and well-being.'

Table 26-4   Negligibly reactive VOCs

Methane
Ethane
Methylene chloride (dichloromethane)
1,1,1-Trichloroethane (methyl chloroform)
Trichlorofluoromethane (CFC-11)
Dichlorodifluoromethane (CFC-12)
Chlorodifluoromethane (CFC-22)
Trifluoromethane (CFC-23)
1,1,1-Trichloro-2,2,2-trifluoroethane (CFC-113)
1,2-Dichloro-1,1,2,2-Tetrafluoroethane (CFC-114)
Chloropentafluoroethane (CFC-115)
1,1,1-Trifluoro-2,2-dichloroethane (HCFC-123)
2-Chloro-1,1,1,2-tetrafluoroethane (HCFC-124)
1,1-Dichloro-1-fluoroethane (HCFC-141b)
1-Chloro-1,1-Difluoroethane (HCFC-142b)
Pentafluoroethane (HFC-125)
1,1,2,2-Tetrafluoroethane (HFC-134)
1,1,1,2-Tetrafluoroethane (HFC-134a)
1,1,1-Trifluoroethane (HFC-143a)
1,1-Difluoroethane (HFC-152a)
Perfluorocarbon compounds of four types (see source for details)

Source: 40 CFR 51.100(s)(2).

sources [emphasis added] . . .' The term 'criteria' does not imply standards or rules but, rather, a compilation of information that supports standards. EPA was required in section 109(a)(1) to publish regulations prescribing a national primary ambient air quality standard and a national secondary ambient air quality standard for each air pollutant for which air quality criteria have been issued.

National primary ambient air quality standards are defined in section 109(b)(1) as standards '. . . the attainment and maintenance of which . . . allowing an adequate margin of safety, are requisite to protect the public health.' A national secondary ambient air quality standard is specified in section 108(b)(2) as a level of air quality '. . . the attainment and maintenance of which . . . is requisite to protect the public welfare[6] from any known or anticipated adverse effects associated with the presence of such air pollutant in the ambient air.'

National ambient air quality standards, or NAAQS for short, currently are established for six pollutants. Criteria pollutants can be emitted directly or formed in the atmosphere through secondary reactions. As noted earlier, two criteria pollutants – particulate matter and ozone – also can include toxic air pollutants listed in section 112(b) of the 1990 Amendments. Table 26-5 lists the current ambient air quality standards for particulate matter and ozone.

The NAAQS generally are set based on clinical (i.e., human) tests. These are possible for criteria pollutants since there is a presumed

Table 26-5   National ambient air quality standards (NAAQS) for particulate matter and ozone

| Pollutant | National ambient air quality standards (in µg m³) | |
	Primary	Secondary
Ozone (O₃)	1-hour = 235[a] (0.12 ppm)	1-hour = 235[a]
Particulate matter (measured as PM₁₀)[c]	Annual = 50[b] 24-hour = 150[a]	Annual = 50[b] 24-hour = 150[a]

[a] Not to be exceeded more than once per year.
[b] Annual arithmetic mean.
[c] Particulate matter equal to or less than 10 microns in diameter.

Source: 40 CFR Part 50.

[6] Welfare effects of most concern include injury to agricultural crops, visibility reduction at scenic vistas, damage to and the deterioration of property, and hazards to air and ground transportation.

Table 26-6   Summary of the health basis for NAAQS for particulate matter and ozone[a]

Pollutant	Health basis for NAAQS
Ozone	Affects individuals with preexisting respiratory disease, abnormal responders and heavy exercisers. Effects include pulmonary alteration, aggravation of preexisting respiratory disease and increased susceptibility to respiratory infection. Appears to function as an accelerator in aging.
Particulate matter	Focus now is on respirable particles, i.e., those less than a nominal 10 microns in diameter. Results in respiratory irritation and disease, reduced lung function in children and long-term deterioration of the respiratory system. Epidemiology studies indicate increases in human mortality with exposure to higher concentrations.

[a] This table is intended only as a summary and is not to be construed as complete or definitive.

threshold of effect (i.e., the ambient air standard).[7] A number of community epidemiological studies also have been conducted to evaluate human populations exposed to the various criteria pollutants. Particulate matter is an unusual criteria pollutant since it has no consistent chemical identity; instead, it can consist of one or more of many different solids, liquids, and even adsorbed gases. Importantly, however, exposure to particulate matter is associated with characteristic adverse health effects, unrelated to the effects associated with the toxic nature of specific chemicals that might make up the particulate matter. Table 26-6 briefly summarizes the key health information supporting the particulate matter and ozone NAAQS. Note that only the inhalable fraction of particulate matter (i.e., that is equal to or less than 10 microns in diameter – referred to as PM_{10}), is regulated.

Criteria Air Pollutant Regulatory Program

Sources of criteria air pollutants are both stationary and mobile and the 1970 and 1977 Amendments established regulatory programs for both. The Amendments also require federal and state governments[8] to play different roles in the process. The federal government (i.e., EPA) is required to establish the NAAQS, publish standards for new stationary sources of these air pollutants, regulate mobile sources, and oversee state activities. State governments are to regulate new and existing sources and enforce many federal regulations through authorities delegated by EPA. The basis for the state actions is measurement by the states and EPA of air quality in individual air quality control regions (AQCR)[9] to determine whether the AQCR is in attainment with each NAAQS. Areas in which the NAAQS are not being met, or nonattainment areas, receive special regulatory attention. However, even areas in attainment are regulated in many instances to ensure that growth in population and industry does not result in nonattainment in the future (i.e., that the NAAQS is maintained).

Although the general structure of the NAAQS attainment and maintenance program remains the same as provided in 1970 and 1977, the 1990 Amendments change the process in

[7] Clinical tests are not appropriate for pollutants, such as carcinogens, that have a potential to cause much more serious health effects.

[8] In many instances, local air pollution control agencies take responsibility for implementing the Clean Air Act. Use of the term 'state government' here is not meant to exclude local governments, only to simplify the discussion.

[9] AQCRs reflect the fact that air pollution does not adhere to governmental or geographic boundaries, and that the air quality in each specific region is a function of the sources, the geography, the climate, and the population in that region. Over 250 separate AQCRs have been designated.

several important ways. These changes resulted because although significant improvements occurred in nationwide air quality in the 1970s, air quality worsened in the 1980s. By the end of the 1980s over one hundred AQCRs with over half the country's population were not attaining the NAAQS. The ozone and carbon monoxide NAAQS were violated most often, influenced by the huge growth in the nation's automobile fleet in the last twenty years. The 1990 Amendments make substantial changes in several programs aimed at ultimate attainment and maintenance of the NAAQS. The Amendments also provide significant sanctions for failure to meet the standards by the required dates (Creekmore 1992).

These programs will significantly reduce emissions of toxic air pollutants in many instances. The three components of the criteria air pollutant control program most likely to result in reduction of toxic air pollutants emissions are described below.

State Implementation Plans

As described in section 110 of the Clean Air Act, state implementation plans (SIPs) are the primary tool used to attain and maintain the NAAQS in the U.S. The 1970 Clean Air Act mandated attainment of the NAAQS to ensure protection of the public health. However, it also recognized that population and economic growth meant greater emissions of air pollutants as time passed and special efforts were needed to counterbalance the growth and maintain the NAAQS. This meant increasingly stringent controls on old sources and greater restrictions on both siting and control of new sources. In fact, Congress enacted in the 1977 Amendments special requirements for nonattainment areas, including special programs in sections 171 through 178 for new or modified sources in nonattainment areas. The purpose was to move nonattainment areas toward compliance with the NAAQS and simultaneously make room for further economic growth.

Stensvaag (1991) describes how the resulting SIPs require application of reasonably available control technology (RACT) on existing sources and reasonable progress toward meeting the NAAQS by deadlines specified in the Act. Recognizing that application of RACT to existing sources could eventually result in air quality substantially better than the NAAQS and, thus, provide a 'windfall' for new sources, the 1977 Amendments specified additional requirements, called offsets, for new sources to prevent that windfall. The offset program is extremely complex and beyond the scope of this book.

The 1977 Amendments also introduced a new program in sections 160 through 169 aimed at prevention of significant deterioration (referred to as PSD) of air quality in attainment areas. The goal of this program was to oversee industrial growth in such a way as to prevent nonattainment from occurring at some future date. Under the PSD program, attainment areas are designated into three classes, depending largely on proximity to and size of national parks, wilderness areas and international parks. This program, too, is extremely complex and beyond the scope of this book.

The 1990 Amendments significantly strengthen the current programs dealing with NAAQS nonattainment. For example, instead of one category of nonattainment for ozone, the 1990 Amendments establish five increasingly stringent categories of nonattainment (in increasing severity called marginal, moderate, serious, severe, and extreme) and the degree of the required control at a facility depends on the nonattainment status where a facility is located. As nonattainment worsens, more stringent control is required on sources and increasingly smaller sources are covered. Boundaries of nonattainment areas also are expanded well beyond those in the past to include many areas that have not been affected in the past. For example, areas that produce ozone precursors, but are themselves in attainment, may have to add controls now to help reduce ozone levels downwind.

In the 1970 Amendments, the nonattainment control focus was on 'major sources,' defined as sources emitting more than one hundred tons per year of any criteria pollutant or precursor. The 1990 Amendments redefine major sources depending on the nonattainment status of the area. In the extreme category, for example, major sources are now defined as those emitting ten tons per year of a criteria air pollutant. A service station in an extreme ozone non-attainment area selling more than 200,000 gallons of gasoline per month now could qualify as a 'major source.'

All of these SIP changes aim at dramatically reducing emissions of criteria pollutants and their precursors. A beneficial side effect will be substantial reductions in emissions of toxic air pollutants as many more industrial and commercial facilities are controlled than in the past and required control levels generally tighten both in response to the requirements of the 1990 Amendments and because regulators often seek to achieve added reductions from industrial and commercial facilities that are viewed as more easily regulated and more able to pay the cost.

Mobile Source Pollution Control Program

Another important component of the criteria air pollutant control program is control of mobile sources. The growing contribution of motor vehicles to poor air quality has been recognized since the mid-1940s. Internal combustion engines, fueled with volatile petroleum liquids, contribute significant emissions of hydrocarbons and nitrogen oxides, both of which are precursors to ozone, as well as significant emissions of carbon monoxide and particulate matter. In addition, the fuels evaporate during transfer and storage and contribute additional hydrocarbons for ozone producing photochemical reactions. Clearly, reductions in the pollutants associated with motor vehicles are a critical component in the improvement in ambient air quality.

Because motor vehicles are nationally produced and distributed, Congress recognized very early that federal control was more appropriate than local control. Sections 202 through 216 of the 1970 Amendments enacted major new requirements on both vehicular emissions and fuels. These requirements forced drastic reductions in emissions of pollutants from motor vehicles and significant changes in the fuels used in the U.S. The initial focus was on passenger vehicles, owing to their large and rapidly growing number. Other controls were

added later for heavy-duty vehicles (i.e., trucks and buses), motorcycles, and aircraft. The goal of the mobile source provisions of the 1970 Amendments was to reduce emissions of hydrocarbons and carbon monoxide by 90 percent by 1975 and emissions of nitrogen oxides by 90 percent by 1976. These deadlines were later extended and for many years were the subject of modifications and delay.

Although the 1970 Act provided in section 116 that each state may establish and implement its own, more stringent standards, Congress recognized the difficulty that automobile manufacturers would have meeting varying state standards. Therefore, states were prohibited from enacting more stringent automotive standards unless they already existed at the time of enactment. This was the case only for California. The Act did provide that other states could adopt the California standards; however, none did under the 1970 Act.

Because of the continuing nationwide non-attainment of the ozone and CO NAAQS, the 1990 Amendments substantially tighten both motor vehicle and fuels requirements. The new emission standards are summarized here but discussed more completely in Chapter 25. First, emission standards for motor vehicles are being reduced in a program phased in over the 1990s, with a possibility of further reductions in the next decade. Trucks and buses also must meet tougher standards. Longer warranty periods also are required on motor vehicle emission control equipment (e.g., up to one hundred thousand miles and ten years for specified emission control equipment in automobiles). In addition, fuel producers must produce and sell reformulated fuels that meet specified physical and composition requirements, including reduction in volatility, aromatics and air toxics, elimination of lead, and addition of gasoline detergents. Finally, automobile manufacturers are required to begin producing vehicles capable of using alternative fuels (e.g., alcohol and propane) for sale to fleet operators in the serious, severe, and extreme ozone nonattainment areas. In this program, fleets of ten or more vehicles, capable of being fueled at a central location (excluding rental, law enforcement, and emergency fleets), must switch to alternative fuels late in this decade. This program, of course, requires the production of these vehicles, the production and distribution of the fuels, and purchase or conversion of vehicles by covered organizations.

All of these efforts will reduce emissions of toxic air pollutants that are either in gasoline (e.g., benzene, toluene, and xylene) or that are produced in the combustion process and at least partially escape catalytic conversion (e.g., benzene and formaldehyde). Emissions of particulate matter also will be reduced significantly from trucks and buses. Particulate matter from vehicles contain polycyclic organic matter and various trace metals.

New Source Performance Standards
The third important component of the criteria air pollutant control program that reduces toxic air pollutant emissions is the regulation by EPA under section 111 of new stationary sources. The intent of new source standards is to require facilities, as they are constructed, to be fitted with high levels of control technology. This will result in a gradual reduction in nationwide air pollution as more new sources are constructed. EPA has published over sixty new source performance standards (NSPS) since enactment of the 1970 Amendments. Unfortunately, industrial and automotive growth were more rapid than initially predicted and new air pollution in the 1970s and 1980s completely outstripped the gains obtained through the NSPS program. On the other hand, without the program air pollution levels in the U.S. likely would be significantly worse.

EPA establishes performance standards for new sources of criteria air pollutants and requires new plants to utilize the best system of

Table 26-7   Selected new source performance standards that reduce emissions of toxic air pollutants

Source categories	40 CFR Part 60[a]
Fossil fueled boilers	Subparts D, Da, Db, Dc
Incinerators	Subparts E, Ea
Asphalt	Subpart I, UU
Petroleum facilities	Subparts J, K, Ka, Kb, XX GGG, QQQ
Secondary smelters	Subparts L, M
Steel production	Subparts N, Na, Z, AA, AAa
Sewage treatment plants	Subpart O
Primary smelters	Subparts P, Q, R, S
Kraft pulp mills	Subpart BB
Surface coating processes	Subparts EE, MM, QQ, RR, SS, TT, WW, FFF, TTT
Organic chemical facilities	Subparts VV, III, NNN
Residential wood heaters	Subpart AAA
Rubber tire manufacturing	Subpart BBB
Polymers, resins, fibers	Subparts DDD, HHH
Petroleum drycleaners	Subpart JJJ

[a] Code of Federal Regulations, Title 40, Part 60.

emission reduction that EPA determines has been adequately demonstrated. This level of control is often referred to as best demonstrated control technology (BDT). EPA has issued new source performance standards covering most major industrial source categories emitting significant quantities of criteria air pollutants. These are published in the Code of Federal Regulations, Title 40, Part 60. The 1990 Amendments did not substantively affect the NSPS program, although section 112(c)(1) specifies that EPA should choose source categories of toxic air pollutants that '[to] the extent practicable ... shall be consistent with the list of source categories established pursuant to section 111.' Some source categories for which NSPS have been published that also can emit significant quantities of toxic air pollutants are listed in Table 26-7.

ACCIDENTAL RELEASE PROGRAM

The 1990 Amendments provide in section 112(r) a new program aimed at preventing accidental releases of 'extremely hazardous substances' (EHS) and minimizing the consequences of these releases. Accidental releases are defined as 'unanticipated emissions of regulated substances'. Within twenty-four months of enactment, EPA was to publish an initial list of one hundred EHS that, in case of accidental release, could cause death, injury, or serious adverse effects to human health and the environment. Sixteen substances that EPA is to

Table 26-8   Extremely hazardous substances

Listed in Title III	Other possible candidates
Ammonia	Benzotrichloride
Anhydrous ammonia	Benzal chloride
Anhydrous hydrogen chloride	Chloroethanol
Anhydrous sulfur dioxide	Dimethyl sulfide
Bromine	Fluorine
Chlorine[a]	Hydrocyanic acid
Ethylene oxide[a]	Methacrylonitrile
Hydrogen cyanide	Methyl disulfide
Hydrogen fluoride[a]	Methyl mercaptan
Hydrogen sulfide[a]	Methyl vinyl ketone
Methyl chloride[a]	Nickel carbonyl
Methyl isocyanate[a]	Phosphorous
Phosgene[a]	Phosphorous pentoxide
Sulfur trioxide	Potassium cyanide
Toluene diisocyanates[a]	Sodium cyanide
Vinyl chloride[a]	Titanium tetrachloride

[a] Also listed in Section 112(b)(1).

include are listed in section 112(r)(3). These are listed in Table 26-8 along with a number of other possible candidates for listing as EHS. Hazardous air pollutants from section 112(b) that also are listed as EHS are noted.

For each listed EHS, EPA is to establish a threshold quantity, taking into account the toxicity, reactivity, volatility, dispersability, combustibility, flammability, and amount of the substance. Within three years of enactment, EPA is to promulgate regulations and guidance to provide for the detection and prevention of accidental releases of EHS and for appropriate response to such releases by the sources. Section 112(r)(7)(B)(ii) also specifies that sources, at which a regulated substance is present in more than a threshold quantity to be defined by EPA, must prepare and implement a risk management plan to detect and prevent or minimize accidental releases, and provide prompt emergency response to any such releases in order to protect the public health and the environment. The plan is to include a 'hazard assessment' of the facility that assesses the potential effects of an accidental release, and provides an estimate of release quantities, a determination of possible down-

wind effects, potentially exposed populations, and historical data.

The 1990 Amendments also required in section 304 that the Occupational Safety and Health Administration (OSHA) promulgate chemical process safety standards, within twelve months of enactment, designed to protect employees from hazards associated with accidental releases in the workplace. Those regulations were published in early 1992.[10] OSHA published its own list of 131 'highly hazardous chemicals' that pose a particular threat of serious injury or fatality in the event of accidental release in the workplace.

The important elements of the OSHA standard are requirements for:

- Maintaining written safety information
- Performing workplace hazard assessments
- Consulting with employees and their representatives regarding the hazard assessments
- Establishing a system to respond to hazard assessment findings
- Reviewing hazard assessments
- Implementing written operating procedures
- Providing written safety and operating information to employees
- Providing contractors/contract employees with appropriate information and training
- Training employees and contractors
- Establishing a quality assurance program for process equipment and spare parts
- Establishing maintenance systems for critical process equipment
- Conducting pre-start-up safety reviews for new and modified equipment
- Implementing written procedures to manage change
- Investigating incidents that result or could have resulted in a major accident

At the time of writing this handbook in 1992, the nature and scope of EPA's program to implement the accidental release provisions of the 1990 Amendments were not known. Preliminary indications are that a large number of sources in the U.S. could be affected. Depending

[10] 57 FR 6356, February 24, 1992.

on the threshold planning quantities selected by EPA, sources such as sewage treatment plants, drinking water treatment facilities, and large swimming pools could be covered because of their use of chlorine for disinfection. In addition, large refrigeration facilities at food processing plants and warehouses also could be covered where they use ammonia as a refrigerant. Early indications also are that EPA is considering requirements for hazard assessments that are both detailed and resource intensive, particularly given that these assessments likely will be required for each process in a facility that is covered. Large chemical plants and refineries may have hundreds of processes. The hazard assessment also must be updated periodically.

Implementation of the accidental release program likely will result in reductions of toxic air pollutant emissions. For example, many chemical plants will duct more emission points to control devices such as flares to prevent sudden unconfined releases. Many plants also will tighten up their equipment and processes to reduce fugitive leaks and the chances of equipment failure; many also may purchase less leak-prone or more reliable equipment. The requirements for hazard assessments also may have substantial impacts. Just the act of preparing and releasing a hazard assessment that could result in adverse community reaction in all likelihood will induce many sources to improve control measures.

INDOOR AIR POLLUTION PROGRAM

Indoor air complaints have increased dramatically in the past twenty years in almost direct proportion to growing energy conservation measures (Neet, Warren, and Cowing 1992). Before that time, most buildings were designed and operated to provide 10 to 15 cubic feet per minute (cfm) of outside air for each building occupant; buildings also were not 'tight.' Still further back in time, windows often were opened, interior materials of construction were predominantly wood, tile, and other non-synthetics, and heating systems relied on steam or circulating hot water. After the 1973 oil embargo, building design and operation underwent substantial change. Less fresh air was used and buildings were sealed more tightly to reduce air flow in and out. Although these contributed to significant reductions in fuel use, they also trapped air pollutants and increased pre-existing levels of other air constituents such as carbon dioxide (EPA 1988e).

The two major problems associated with poor indoor air quality are building-related illness (BRI) and sick building syndrome (SBS). The first refers to specific illnesses (e.g., Legionnaires' disease) related to building design or operation. The second refers to less well-defined physical and psychological conditions associated with some buildings that generally disappear on leaving the building. Some studies indicate that as many as one-third to one-half of the buildings in the U.S. have had SBS incidences. Other studies have estimated that poor indoor air quality is responsible for tens of billions of dollars in lost productivity and direct medical costs.

In response to these growing concerns, section 403 of Title IV of the Superfund Amendments and Reauthorization Act of 1986 required EPA to establish a research program with respect to indoor air quality (and radon), to gather data and information, to coordinate efforts, and to assess appropriate actions to mitigate risks associated with indoor air quality problems. It also required EPA to report to Congress within two years of enactment respecting the activities undertaken and to make appropriate recommendations.

The required report was transmitted to Congress in mid-1989 (EPA 1989g). EPA concluded in the report that 'indoor air pollution represents a major portion of the public's exposure to air pollution and may pose serious acute and chronic health risks.' However, the report also stated that indoor air research and policy programs had not yet sufficiently characterized the problems and solutions to be able to define an appropriate long-term federal role. In part as a result of the conclusions and recommendations in the report to Congress, EPA initiated

Table 26-9   Principal indoor air pollutants of concern

Radon and radon daughters[a]
Environmental tobacco smoke
Biological contaminants
Volatile organic compounds[a]
Formaldehyde[b]
Polycyclic organic compounds[b]
Pesticides[a]
Asbestos[b]
Combustion gases[a]
Particulate matter[a]

[a] Typically contains substances listed in section 112(b)(1).
[b] Listed in section 112(b)(1).

Source: EPA 1989h.

more detailed and extensive studies concerning indoor air pollution and its ultimate control.

Indoor air pollutants span a wide range of materials and substances and can include chemicals, biological materials, radioactive substances, and others. The report to Congress identified several indoor air pollutants of concern; these are shown in Table 26-9. Those that also meet the definition of toxic air pollutant in section 112(b) of the 1990 Amendments, or can contain toxic air pollutants, are noted.

Significantly, as shown in Table 26-9 many indoor air pollutants are the same substances that are listed in section 112(b) as toxic air pollutants in the outdoor environment. In fact, many of the toxic air pollutants in the indoor environment are present as a result of outdoor air intrusion. Perhaps more significantly, recent studies (Wallace et al. 1983; Wallace et al. 1986; Wallace et al. 1987; Wallace et al. 1988) have shown that indoor air exposures to toxic air pollutants quite often are greater than outdoor air exposures. A major reason for this is the fact that most people spend up to 80 to 90 percent of their lives indoors (NRC 1981) and there are a number of sources of air pollutants indoors.

EPA's more recent views concerning indoor air pollution were provided by EPA's Deputy Administrator in testimony before Congress on April 10, 1991 (Habicht 1991). The Deputy Administrator emphasized EPA's conclusion that indoor air legislation was premature at that time but agreed that there were legitimate concerns that people are being exposed to levels of pollutants indoors that may be significant in terms of long-term risks. He also expressed concern with the more immediate effects of indoor exposures, as shown by the increasing incidence of health complaints and illness associated with the indoor environments in buildings, schools and residences. However, there are many unknowns and EPA, along with many other public and private organizations, is currently studying all aspects of indoor air pollution, from cause to cure. A number of the research and policy programs under way by EPA were described by the Deputy Administrator. Some of these include additional exposure and health effects research, more study to identify and characterize indoor air pollutants and to evaluate mitigation strategies, and more efforts to characterize the nature and pervasiveness of indoor air health effects. Research in each of these areas could provide useful information with respect to outdoor air toxics.

Although EPA argued against new indoor air legislation at that time, bills have been introduced in the Congress that would enhance the federal government's response to the issues. These bills contain a number of provisions concerning research and information dissemination, would focus resources on a number of specific pollutants, and would require health advisories and product labeling.

At the time of writing this handbook in 1992, the future regulatory status of indoor air pollution was unknown. If events follow EPA's lead, few regulations will be promulgated and mitigation of the problems likely will focus more on better building design and use. However, if indoor air problems such as the 'sick building syndrome' increase, we may see congressional intervention.

References

Arbuckle, J.G., T.A. Vanderver, and P.A.J. Wilson. 1992. Emergency Planning and Community Right-to-Know Act Handbook. Fourth Edition. Rockville, MD: Government Institutes, Inc.

Creekmore, T.A. 1992. Failure to meet requirements of the Clean Air Act for ozone and carbon monoxide. Paper 92-88.05. Presented at the 85th Annual Meeting of the Air & Waste Management Association. Kansas City, MO.

Habicht, F.H. 1991. Statement before the Subcommittee on Health and the Environment, Committee on Energy and Commerce, U.S. House of Representatives, April 10, 1991.

Laznow, J. and J. Daniel. 1992. Overview of the U.S. Environmental Protection Agency's hazardous air pollutant early reduction program. *Journal of the Air & Waste Management Associaiton* 42(1):31.

National Research Council (NRC). 1981. *Formaldehyde and Other Aldehydes*. Washington, D.C. National Academy Press.

Neet, J.O., M.S. Warren, and M.W. Cowing. 1992. Indoor air quality: The legal landscape. Paper No. 92-82.03. Presented at the 85th Annual Meeting of the Air & Waste Management Association. Kansas City, MO.

State and Territorial Air Pollution Program Administrators and the Association of Local Air Pollution Control Officials (STAPPA/ALAPCO). 1989. *Toxic Air Pollutants: State and Local Regulatory Strategies – 1989*. Washington, D.C.

Stensvaag, J. 1991. *Clean Air Act 1990 Amendments – Law and Practice*. New York: Wiley Law Publications, John Wiley & Sons.

U.S. Environmental Protection Agency (EPA). 1988e. Indoor air facts No. 4: Sick Buildings. Washington, D.C.

U.S. Environmental Protection Agency (EPA). 1989g. Report to Congress on indoor air quality. EPA 400/1-89-001A. Office of Air and Radiation. Washington, D.C.

U.S. Environmental Protection Agency (EPA). 1990a. Toxics in the community: national and local perspectives. EPA 560/4-90-017. Office of Pesticides and Toxic Substances. Washington, D.C.

U.S. Environmental Protection Agency (EPA). 1991d. Enabling document for regulations governing compliance extensions for early reductions of hazardous air pollutants. EPA 450/3-91-013. Office of Air Quality Planning and Standards. Research Triangle Park, NC.

U.S. Environmental Protection Agency (EPA). 1992b. EPA's 33-50 Program - Second Progress Report. Office of Pollution Prevention and Toxics. Washington, D.C.

Wallace, L., E. Pellizzari, T. Hartwell, C. Sparacino, and H. Zelon. 1983. Personal exposures to volatile organics and other compounds indoors and outdoors – the TEAM Study. NTIS PB83-121357. U.S. Environmental Protection Agency, Office of Research and Development. Washington, D.C.

Wallace, L., E. Pellizzari, T. Hartwell, R. Whitmore, C. Sparacino, and H. Zelon. 1986. Total exposure assessment methodology (TEAM) study: Personal exposures, indoor-outdoor relationships and breath levels of volatile organic compounds in New Jersey. *Environment International* 12:369–387.

Wallace, L., E. Pellizzari, T. Hartwell, C. Sparacino, R. Whitmore, L. Sheldon, H. Zelon, and R. Perritt. 1987. The TEAM (Total Exposure Assessment Methodology) study: Personal exposures to toxic substances in air, drinking water and breath of 400 residents of New Jersey, North Carolina and North Dakota. *Environmental Research* 43:290–307.

Wallace, L., E. Pellizzari, T. Hartwell, R. Whitmore, H. Zelon, R. Perritt, and L. Sheldon. 1988. California TEAM study: Breath concentrations and personal air exposures to 26 volatile compounds in air and drinking water of 188 residents of Los Angeles, Antioch and Pittsburgh, California. *Atmospheric Environment* 22:2141–2163.

27

Control of Air Toxics in Other Countries

Si Duk Lee and Toni Schneider

INTRODUCTION

Control of air toxics in countries other than the U.S. is not yet widespread but it is growing. Much like the U.S. air toxics program before the 1990 Clean Air Act Amendments, efforts in other countries have been more reactive than proactive. However, the experience of the U.S. in implementing the new air toxics provisions of the 1990 Clean Air Act Amendments could stimulate other countries to move ahead more aggressively. The purpose of the chapter is to summarize three international programs and show how they differ at this time from efforts in the U.S.

Discussion of the status of air toxics control in other countries requires use of terms such as 'criteria pollutant' and 'air toxic' that are not widely used worldwide but serve as a frame of reference. As described earlier, criteria pollutants are the air pollutants (sulfur dioxide, ozone, nitrogen oxides, carbon monoxide, particulate matter, and lead) regulated under sections 108 and 109 of 1970 Clean Air Act, amended in 1977 and 1990. This legislation required EPA to prepare criteria documents as scientific bases for regulatory decision-making; this was the origin of the term 'criteria pollutant.' All other chemicals and substances that may pollute the ambient air generally are called air toxics, although other terms are used, including noncriteria pollutants, hazardous air pollutants, and toxic air pollutants.

CONTROL OF AIR TOXICS IN JAPAN

A brief survey shows that Japan is leading the control of environmental pollution by air toxics in the Pacific rim countries. Table 27-1 provides a chronology of air pollution control in Japan since 1962. The primary concern of the Japanese air pollution program has been the criteria pollutants, as defined in the U.S., from both stationary and mobile sources. Lead, however, is not included in this group of pollutants, but is included in a group of 'special particulate' as noted below. Other air pollutants of concern fall in the category of toxic substances or unregulated pollutants.

Toxic Substances

Japan's Air Pollution Control Law designates the following four groups of chemical

Table 27-1   Chronology of environmental pollution problems and pollution prevention in Japan

1962	First National legislation for air pollution control introduced (concentration regulation for SO_x, soot, etc.)
1964	Established Environmental Pollution Division in the Ministry of Health and Welfare.
1964	Basic Law for Environmental Pollution Control (responsibilities of the Government and industries, and environmental standards prescribed).
1968	Air Pollution Control Law (K-value regulation, automobile exhaust gas regulation introduced).
1969	Ambient Air Quality Standards for SO_x.
1970	Emission standards of SO_x strengthened (second regulation).
1970	Ambient Air Quality Standards for CO.
1970	Law Concerning Special Measures for Relief of Pollution-Related Patients.
1970	Revision of Air Pollution Prevention Law (Fuel usage regulations and particulate dust regulations introduced, toxic substances such as Cd, Pb added to regulated substances).
1970	Regulation for Automotive Exhaust Emission.
1971	Regulations for SO_x (third regulation), soot and dust (second regulation) strengthened. Fuel usage standard, emission standard for Pb, Cd, Cl_2, HCl, F, structure, maintenance and management standards for abatement of particulate dust established.
1971	Environment Agency established.
1972	Emission standard for SO_x strengthened (fourth regulation).
1972	Ambient Air Quality Standards for Suspended Particulate Matter.
1972	Fuel Usage Regulations strengthened.
1973	Emission Standards for SO_x strengthened (fifth regulation).
1973	Ambient Air Quality Standards for NO_2 and Ozone established.
1973	Pollution Related Health Damage Compensation Law (amended in 1988).
1974	Fuel usage regulations strengthened.
1974	Emission standards for SO_x strengthened (sixth regulation).
1974	Air Pollution Control Law revised ('Total Emission Control' approach).
1974	Ambient Air Quality Standards for Photochemical oxidants (NO_x, ozone, etc.).
1975	Additional total emission control areas for SO_x designated (third regulation: five areas) Emission standards for NO_x strengthened (second regulation).
1975	Report on Emission Guideline for Hydrocarbons by Central Council for Environmental Pollution Control.
1976	Additional total emission standard areas for SO_x designated (third regulation: five areas). Emission standards for SO_x strengthened (eighth regulation). Fuel usage regulation strengthened.
1977	Emission standard for NO_x strengthened (third regulation). HCl regulation for waste incinerators.
1978	Ambient Air Quality Standards NO_x revised.
1978	Pollution Related Health Damage Compensation Law revised.
1979	Emission standards for NO_x strengthened (fourth regulation)
1981	Total emission control system for NO_x introduced: three areas designated.
1982	Emission standard for soot and dust strengthened (third regulation).
1983	Emission standards for NO_x strengthened (third regulation).
1987	Asbestos Handling Guidelines by Environment Agency and the Ministry of Health and Welfare.
1988	Environment Agency Report on Seasonal Air Pollution Control Measures.
1988	Environment Agency Report on 'New Middle-Term Outlook on Measures for Nitrogen Oxides in Large Cities.'
1988	The Environment Agency releases a 'Fundamental Plan to Make the Low Pollution Cars Prevailing.'
1989	Environment Agency Interim Report on Biotechnology and Environmental Conservation.
1989	Partial revision of the Cabinet Order for the Implementation of the Law Concerning the Examination and Regulation of Manufacture and etc. of Chemical Substances (trichloroethylene, tetrachloroethylene and carbon tetrachloride are added to the list of Class 2 Chemical Substances).

substances as special particulate (i.e., toxic substances), generated from soot- and smoke-emitting facilities and mandates control of their emissions: (1) cadmium and its compounds, (2) chlorine and hydrogen chloride, (3) fluorine, hydrogen fluoride and silicon fluoride, and (4) lead and its compounds.

The emission standards for these four categories of toxic substances are established on a limited number of soot and smoke emitting facilities, for each chemical or group of chemicals as shown in Table 27-2. The emissions of these toxic substances are considered to be associated with specific manufacturing

Table 27-2   Harmful substances (June 22, 1971)

Substance	Facility	Standard value (mg/Nm^3)
Cadmium and its compunds	Baking furnace and smelting furnace for manufacturing glass using cadmium sulfide or cadmium carbonate as raw materials.	1.0
	Calcination furnace, sintering furnace, smelting furnace, converter and drying furnace for refining copper, lead or cadmium.	
	Drying facility for manufacturing cadmium pigment, or cadmium carbonate.	
Chlorine and hydrogen chloride	Chlorine quick cooling facility for manufacturing chlorinated ethylene.	30 (chlorine)
	Dissolving tank for manufacturing ferric chloride.	80 (HCl)
	Reaction furnace for manufacturing activated carbon using zinc chloride.	
	Reaction facility and absorbing facility for manufacturing chemical products	
	Waste incinerator (HCl)	700
Fluorine, hydrogen fluoride and silicon fluoride	Electrolytic furnace for smelting aluminium (harmful substances are emitted from discharge outlet).	3.0
	Electrolytic furnace for smelting aluminum (harmful substances are emitted from top).	1.0
	Baking furnace and smelting furnace for manufacturing glass using fluorite or sodium silicofluoride as raw material.	10
	Reaction facility, concentrating facility and smelting furnace for manufacturing phosphoric acid.	
	Condensing facility, absorbing facility and distilling facility for manufacturing phosphoric acid.	
	Reaction facility, drying facility and baking furnace for manufacturing sodium triple-phosphate.	
	Reaction furnace for manufacturing superphosphate of lime.	15
	Baking furnace and open-hearth furnace for manufacturing phosphoric acid fertilizer.	20

operations as noted in the table. Other particulate matter emitted in small amounts is classified as 'soot and dust,' irrespective of the chemical composition.

Unregulated Substances

In order to prevent possible air pollution from unregulated substances, emission concentrations and environmental effects of potentially toxic substances are being surveyed. Based on the results obtained so far, some substances have been found to have a long-term impact on the environment, although the present levels in the air are not anticipated to cause imminent problems. Monitoring for asbestos, formalde-

hyde, and dioxin was conducted in 1985, 1986, and 1987, respectively.

Asbestos

Asbestos in occupation settings is regulated in Japan under the Industrial Safety and Health Law. The Environment Agency conducted investigations regarding health risks associated with exposure to asbestos in accordance with a three-year plan starting in 1981. From these, it was concluded that the risk to the public from asbestos in ambient air was very small, but asbestos is a substance with high potential for accumulation in the environment. Moreover, because of the pervasive use of asbestos, the

Environment Agency decided to conduct monitoring every two years after 1985 in order to keep abreast of the changes in the environment.

The agency conducted surveys in and around asbestos emission sources in 1987. These surveys showed high concentrations of asbestos on the grounds of some manufacturing plants of asbestos products. This indicated that asbestos discharges were not being controlled sufficiently. Based on this, the Environment Agency directed the public sectors and relevant ministries to promote further asbestos control measures. The Environment Agency and the Ministry of Health and Welfare also forwarded in October 1987 and February 1988 notices on how to handle asbestos used in school buildings.

In light of the above events, and to prevent further environmental pollution by asbestos, the Central Anti-Pollution Measure Council submitted in March 1989 a report on the steps to restrict asbestos discharge from the manufacturing plants. In order to regulate such unregulated air pollutants, a 'Draft of the Law Concerning the Amendment of Part of the Air Pollution Control Act' was submitted to the Diet in March 1989.

The Environment Agency also has initiated monitoring of a number of air toxics such as chloroform, formaldehyde, trichloroethylene, and tetrachloroethylene to examine the patterns of environmental pollution by these chemicals.

CONTROL OF AIR TOXICS IN SOME EUROPEAN COUNTRIES

Sweden

An action program has been developed in Sweden to reduce or ban the use of harmful chemicals. In December 1988 the Swedish government asked the National Chemicals Inspectorate (KEMI) and the Environmental Protection Agency (SNV) to draw up an action program to curb the use of substances harmful to human health or the environment. The result was a list of thirteen substances, the use of which should be reduced or banned altogether by the year 2000. The list includes three chlorinated solvents: methylene chloride, trichloroethylene, and tetrachloroethylene. All three are considered to be carcinogenic and are highly volatile.

Methylene chloride is used in the metal goods, mechanical engineering, and related sectors for metal degreasing and in paint removers. It also is used as a cleaning agent in the electronics industry and polyurethane manufacturing and is used as a solvent in the pharmaceuticals industry. The aim is that all uses of methylene chloride should be phased out in stages over a period of five to ten years.

Trichloroethylene also is used in metal degreasing and is stable and toxic in the aquatic environment. The same phased withdrawal plan is proposed as for methylene chloride, partly to avoid switching from one compound to the other. The use of trichloroethylene is to be banned altogether within ten years.

Tetrachloroethylene (or perchloroethylene) is used for dry cleaning and at present no substitutes exist. It is stable toxic in the aquatic environment. The first step will be to remove it from retailers' shelves and to impose a levy on it. Attempts are being made in the major industrial countries to develop new, less hazardous dry cleaning solvents.

Lead

Lead and lead compounds occur in many different chemicals and other products. Large amounts of lead are used in storage batteries, plastics and glass manufacture, for cable sheathing, in cartridges, and as a gasoline additive.

Lead is a toxic heavy metal that does not break down in the natural environment. Elevated levels of it, therefore, have been found in air, soil and water. It is proposed that in the long term the use of lead is to be banned. To accelerate the change, the European countries in general use a tax differential between leaded and unleaded gasoline.

Organotin Compounds

Organotin compounds are widely used and highly toxic. The largest quantities are used in Sweden in the production of polyvinyl chloride (PVC). Antifouling paint for ships hulls and cooling water chemicals also contain these compounds. The aim is to eliminate environmentally hazardous organotin compounds. It is already illegal to use hull paints containing tin on Swedish fishing and coastal vessels and on Swedish ships operating in the Baltic and North Seas. Tin-based paints for leisure craft also are banned.

Chloroparaffins

Chloroparaffins do not readily degrade and are resistant to oxidation, have been shown to bioaccumulate, and have been shown in animal tests to cause cancer. Chloroparaffins are highly toxic to crustaceans and to certain marine bacteria and fish species. The action program calls for all chloroparaffins to be withdrawn within ten years.

Phthalates

Worldwide, large quantities of phthalates are used, especially as plasticizers for polyvinyl chloride plastics. The main emissions occur to the atmosphere; the primary concerns are the scale of emissions and the slow degradation rate of phthalates. The aim is to reduce the quantities of phthalates entering the environment. In Sweden, manufacturers and importers of these chemicals have to submit reports on the health and environmental damage their products may cause.

Arsenic

Arsenic can have serious effects on human health and the environment. At the North Sea Conference in March 1990, the Environment Ministers of the states bordering the North Sea agreed to cut inputs of arsenic into the North Sea by at least 50 percent by 1995. Atmospheric emissions are to be halved by 1995, or 1999 at the latest. One key application of arsenic is in timber preservatives.

Creosote

Creosote can have serious effects on both human health and the environment. It has highly irritating effects on the skin and eyes. Skin contact combined with sunshine (UV radiation) can give rise to severe eczema. Creosote oil also may cause cancer. Creosote is used to treat timber for long-term protection against fungi, insects, and marine organisms. In Sweden alone there are ten million creosoted electricity and telephone poles. Creosote is not stabilized in the treated wood, and remains available on the surface. There is, therefore, considerable risk of human contact.

Cadmium

Cadmium has no known beneficial effect on living organisms. In man, it can cause kidney damage and it affects aquatic organisms even in low concentrations. The use of cadmium, for instance, in electroplating and in plastics and pigments has been illegal for many years. Despite the ban, cadmium remains a serous environmental problem in Sweden. The two most important products still containing cadmium are rechargeable nickel-cadmium batteries and fertilizers. To reduce cadmium levels in soil, a bounty scheme encourages the return of nickel-cadmium batteries and a levy on cadmium in fertilizers makes it financially advantageous to produce and import fertilizers with a very low cadmium content.

Mercury

Special action has been taken on the mercury problem. Mercury released into the environment is converted into methyl mercury that is extremely toxic to living organisms. The main route for human intake of mercury from the external environment is through food. The crucial factor is how much fish with a high mercury content we eat. Mercury damages the central nervous system, as was shown by a catastrophe in Japan. It also can harm fetuses. The Swedish National Food Administration, therefore, recommends pregnant women avoid eating fresh water fish. A large number of lakes where the mercury content of pike exceeds

1 milligram/kilogram (mg/kg) have been black-listed. Fish from blacklisted lakes may not be sold or given away, and people catching fish there are advised not to eat it. The present elevated levels of mercury in fish are due largely to the accumulation of many decades of atmospheric deposition. Mercury in soil is still increasing, owing to the current deposition, although at a slower rate. If mercury deposition remains as high as it is today, the situation in the aquatic environment will continue to deteriorate.

Table 27-3   Framework for risk management in the Dutch environmental policy

Framework	Individual cancer risk/substance (mortality, man)	Individual risk/substance (effects with threshold levels, man)	Collective risk/substance (function of ecosystem)
Unacceptable risks			
Maximum acceptable level	10^{-6}/year	'No effect concentration' taking into account sensitive groups	Concentration at which 95 percent of the species in an ecosystem are protected
Risk Reduction Desirable			
Negligible level (target value)	10^{-8}/year limit	1 percent of upper limit	1 percent of upper limit

Table 27-4   Priority substances

Metals and metalloids
Arsenic
Cadmium
Chromium(VI)
Copper
Lead
Mercury
Zinc

Nonhalogenated organics
Mineral oil and (gaseous) hydrocarbons
Acrolein
Acrylonitrile
Benzene
Ethylene
Ethylene oxide
Methanol (formaldehyde)
Phenol(s)
Phthalates
Polycyclic aromatic hydrocarbons
Propylene oxide
Styrene
Toluene

Halogenated aromatics
Chloroanilines
Chlorobenzenes
Chlorophenols
Dioxins
PCPs and PCTs

Other halogenated compounds
Chlorofluorocarbons
1,2-Dichloroethane
Dichloromethane
Hexachlorocyclohexane
Methylbromide
Tetrachloroethylene
Tetrachloroethane
1,1,1-Trichloroethane
Trichloroethylene
Trichloromethane
Vinyl chloride

Other substances
Asbestos
Fluorides
Carbon monoxide
Ozone
Particulates (coarse and fine)
Hydrogen sulfide

Acidifying and fertilizing substances
Ammonia
Phosphates
Nitrate
Nitrogen oxides
Sulfur oxides

The environmental goal of relevance to the effects of mercury on human health is as follows. In the longer term, the level of methyl mercury in fish should not exceed 0.5 mg/kg. If concentrations in fish from polluted lakes are to fall to this level, mercury deposition in Sweden must be cut by an estimated 80 percent. Efforts so far to deal with the mercury problem have aimed at reducing emissions from point sources. Besides this, however, a great deal must be done to achieve better control of certain waste products containing mercury.

The Swedish Environmental Protection Agency has proposed to reduce to the absolute minimum the use of mercury and mercury-containing products in Sweden. The plan to phase out mercury means the following.

1. Mercury-based methods in the chlor-alkali industry are to be replaced by methods not requiring mercury by 2010 at the latest.
2. The sale of mercuric oxide batteries is to be banned by the year 2000.
3. Sales of medical thermometers containing mercury will be banned from 1992 and for other mercury thermometers from 1993.
4. The National Board of Health and Welfare is to study the feasibility of phasing out amalgam (alloys of mercury) as a dental filling material.
5. Mercury is not to be used in switches in new cars. The sale of other mercury-containing switches, medical instruments, and industrial measuring instruments will be banned from 1993.
6. The use of mercury in fluorescent lighting tubes and mercury vapor lamps may continue for the time being. A labeling system, however, is being considered. In the long term, these applications also should be phased out.
7. Laboratory use of mercury will be reduced. However, some use of mercury will continue to be necessary for research purposes.

At the international level endeavors should be made to bring about reductions not only in emissions but also in the use of mercury.

Table 27-5 **Target and limit values for priority substances and percent reductions needed to achieve them** (quantities in $\mu g/m^3$ for air or $\mu g/l$ for water)

Substance	Target value[a]	Limit value[a]	Average	Concentration around sources	Percent reduction For target	For limit
Trichloroethane	50	50	0.65	80	35–40	35–40
Surface water	0.1		2.0		95	
Tetrachloroethane	25	2,000	1.0	30	20	
Surface water	0.1		3.5		98	
Benzene	1	10	2	40 (185)	97.5	75
Phenol	1	100	0.008	2	50	
Styrene	8	100	0.1	20	60	
Acrylonitrile	0.1	10	0.01	3	97	
Toluene		$3\,mg/m^3$	3	±100 20–40[b]		
1,2-Dichloroethane	1		0.2	1.5	33	
Ethylene oxide	0.03		0.02	0.5–1	90–95	
Methylbromide	1 (year) 100 (hour)			>100	95	
Vinyl chloride	1		0.2	5	80	
Propylene oxide	1		7×10^{-3}	1	0	
Dichloromethane	20		0.6–1.5	10–100	±80	
Trichloromethane	1		0.16	1.0	90	
Tetrachloroethane	1		0.8	3 (2–20)	65	
Epichlorohydrin	2		<0.05	0.25	0	0

[a] Quantities in $\mu g/m^3$ for air or $\mu g/l$ for water.

[b] Indoor environment.

The Netherlands

The Dutch strategy for the control of air toxics is contained in the National Environmental Policy Plan (NEPP) and the subsequent National Environmental Policy Plan Plus (NEPP+). One of the NEPP objectives is to obtain sustainable development through the reduction to acceptable or, if possible, negligible levels of

the risks posed to humans and the environment by an individual substance or groups of toxic substances.

A framework for risk management has been developed in which the various kinds of environmental risks have to be dealt with, as much as possible, in a similar way. An upper limit for the risks is indicated above which the risk of effects on humans and the environment is considered unacceptable, and a lower limit

Table 27-6   Example toxic air pollution regulations in other countries

Country	Air toxic[a]	Regulation[b]	Comment
Australia	Cd, Hg and compounds	3.0 mg/m^3	See Note 1
	Total As, Sb, Cd, Pb, Hg, and V	10.0 mg/m^3	
	Ni and compounds	20.0 mg/m^3	
	Nickel carbonyl	0.5 mg/m^3	Tentative
	Be and compounds	10.0 mg/m^3	24-hour
	Vinyl chloride		See Note 1
	Asbestos		See Note 1
France	Carcinogens		See Note 2
Germany	VHH, 20 metals, various inorganics, and organics		See Note 3
South Korea	CS$_2$	100 ppm/30 ppm	See Note 4
	CHOH	20 ppm	
	HCN	10 ppm	
	Br and compounds	5 ppm	
	Hg and compounds	5 mg/Sm^3	
	C$_6$H$_6$ and compounds	50 ppm	
	Phenol and compounds	10 ppm	
	As and compounds	3 ppm	
	Cr and compounds	1 mg/Sm^3	
	Cd and compounds	1 mg/Sm^3	
	Ni and compounds	20 mg/Sm^3	
Singapore	Sb and compounds	0.01 g/Nm^3	
	Cd and compounds	0.01 g/Nm^3	
	Hg and compounds	0.01 g/Nm^3	
	As and compounds	0.02 g/Nm^3	
United Kingdom	Any pollutant		See Note 5
	Metals, metalloids, asbestos, halogens, phosphorus and compounds		See Note 6

[a] Cd = cadmium, Hg = mercury, As = arsenic, Sb = antimony, Pb = lead, Ni = nickel, Be = beryllium, VHH = volatile halogenated hydrocarbons, CS$_2$ = carbon disulfide, CHOH = formaldehyde, HCN = hydrogen cyanide, C$_6$H$_6$ = benzene, Cr = chromium.
[b] mg/m^3 = milligrams per cubic meter, ppm = parts per million, mg/Sm^3 = milligrams per standard cubic meter, g/Nm^3 = grams per normal cubic meter.

Note 1: Standards derived from U.S. NESHAP (40 CFR 61).
Note 2: Based on the 1982 Seveso Directive, a 28 December 1983 circular defines use of risk assessment; over three hundred installations are subject to risk assessment studies.
Note 3: Specific industrial processes regulated by Technical Instructions on Air Quality Control (TI Air).
Note 4: Viscose rayon/other facilities.
Note 5: Integrated national control of processes with a potential for pollution to air, land, or water. Authorization (i.e., permit) is required; operator must install best practical means of control, meaning best available techniques not entailing excessive cost to minimize pollution.
Note 6: Local control required.

below which the risks are considered negligible (see Table 27-3).

The substances posing the highest risks to man and the environment are designated 'priority substances' in The Netherlands. About fifty substances have been designated. These are shown in Table 27-4.

Environmental standards usually are developed for the priority substances on the basis of all relevant information, collected in a so-called 'basic document' (emission sources, dispersion, monitoring methods, effects, abatement techniques, control costs). All environmental compartments (air, water, and soil) and their interrelationships are considered in such a document. In the basic document, a 'no-effect-level' will be indicated for human as well as ecosystem exposure, about which the National Health Council will provide an advisory report. This 'no-effect-level' is related to the upper limit of risk. The lower limit of risk is related to the so-called 'target value.' In principle, standards are set in the range between the target value and the maximum acceptable level; they are the result of the trade-offs among the detrimental effects to humans and the environment, social and economic aspects, and technical possibilities (see Table 27-5).

OTHER COUNTRIES

Control of toxic air pollution is not yet the focus of specific regulatory control programs in most countries other than the U.S. That results from such factors as differing views on the control of air pollutants, differing legislation, and lack of resources. In many countries, pollution control has not received significant attention because of the pressing need to solve other social problems. Air toxics also is an advanced form of air pollution, in the sense that most industrialized countries will attempt first to control broader national problems such as particulate matter (i.e., smoke stacks), sulfur oxides, and automobiles, before dealing with more subtle and localized issues such as air toxics.

The discussions above on Japan, Sweden, and the Netherlands are illustrative of the progress being made in Asia and Europe. Table 27-6 lists some specific air toxics related activities and regulations in other countries. For specific details, the reader is directed to IUAPPA (1991).

Reference
International Union of Air Pollution Prevention Associations (IUAPPA). 1991. *Clean Air Around the World, 2nd Edition.* Loveday Murley, ed. Brighton, England: International Union of Air Pollution Prevention Associations.

28

Risk Communication on Toxic Air Pollutants

Sharon M. Friedman

INTRODUCTION

Interest in the environment is at an all-time high. People are concerned about and involved in environmental issues on the local, regional, national, and global level. Children in schools are learning about global warming, acid rain, and recycling. Environmental groups are urging individuals to become good environmental citizens and protect the Earth.

Much of this environmental interest has been sparked by coverage in the mass media, where there has been a rebirth of environmental writing beats and even the development of entire new publications devoted to environmental issues. Within the past several years, three new national magazines about the environment have begun – *Buzzworm, E* and *Garbage: A Practical Journal for the Environment*. There is even a new national environmental magazine for youngsters called *P3, The Earth-based Magazine for Kids* (Friedman 1990).

The media's interest in the environment has made it the topic of this decade. In 1988, *Time* named the earth its 'Planet of the Year' and made a commitment to increase its coverage of environmental concerns. So have other leading media organizations. Cable News Network, for example, has given environmental

coverage high priority and its reporters take an activist stance on environmental protection issues. National Public Radio regularly provides strong in-depth environmental coverage. So do large prestigious newspapers such as the *New York Times* and the *Los Angeles Times*. Even smaller newspapers have indicated they are now giving more coverage to environmental issues than they did one or two years ago.

These are just a few examples of the ambitious environmental reporting now going on at newspapers, magazines, television stations, and other media in the United States. This trend does not appear to be abating. 'We feel that environmental issues are going to be the issues of the '90s,' says a vice president of a television station in Utah. A senior account executive at a marketing research firm adds: 'In half the markets we cover, the environment is of more interest to viewers than health and medicine ...' (Friedman 1990).

The public's resurgence of environmental interest also is being stimulated by numerous environmental groups that, particularly on the national scene, are producing ever more sophisticated information campaigns for public consumption. Greenpeace, for example, has produced a high-quality, professional videotape about environmental problems related to incinerators for community members to view.

One result of all of this public, media, and environmental group interest is increased regulation as legislators respond to what they perceive to be public demand for environmental protection. There is no doubt that the public is demanding protection from environmental hazards. They are worried about many issues, but toxic and radiation hazards are on the top of their list, as pointed out by a public opinion poll by the Roper Organization published in January 1990 in the *Washington Post*.

Leading the list of the public's most serious environmental risks was active hazardous waste sites, followed by worker exposure to toxics, nuclear accident radiation, and radioactive waste. Industrial accident pollution was halfway down the list, with 51 percent indicating it was a serious environmental risk, while 48 percent indicated a similar concern about industrial air pollution (Russell 1990).

This environmental concern is not new. The accidents at Three Mile Island, Bhopal, and Chernobyl all brought to the public's attention the potential dangers of chemical plants and nuclear power plants. Love Canal in New York and Times Beach in Missouri focused public concern on toxic wastes. With these concerns grew the phenomenon now known as the NIMBY Syndrome – not in my backyard – it's too dangerous.

While public concern about environmental risks was growing, industries and federal and state governments tried to deal with the risks by developing programs in risk assessment and risk management. While many of these programs theoretically should have worked well, they often were not accepted by the communities affected by the environmental risks. So, to deal with these community and public responses, a whole new field called risk communication was born.

DEFINING RISK COMMUNICATION

Since the mid-1980s, risk communication has moved into the forefront of concern among people and groups dealing with risk of all types. The U.S. Environmental Protection Agency (EPA), for example, considers risk communication strategically important in both its regulatory activities and its research agenda. Many industries regulated by the EPA also see risk communication as a key policy and management issue (Krimsky and Plough 1988).

What exactly is risk communication? Numerous definitions have been offered and, as the field has grown in depth and complexity, so too have the definitions. One of the first offered was by sociologist Vincent Covello of Columbia University and several colleagues, who defined it as 'any purposeful exchange of information and interaction between interested parties regarding health, safety or environmental threats.' After conducting five in-depth case studies of risk communication issues, Sheldon Krimsky and Alonzo Plough (1988), sociologists at Tufts University, broadened this

definition to include any type of 'communication that informs individuals about the existence, nature, form, severity, or acceptability of risk.'

In its volume on the subject, the National Research Council (NRC) in Washington, D.C., provides the most complete definition. It defines risk communication (NRC 1989) as 'an interactive process of exchange of information and opinion among individuals, groups and institutions. It involves multiple messages about the nature of risk and other messages, not strictly about risk, that express concerns, opinions, or reactions to risk messages or to legal and institutional arrangements for risk management.' The NRC report also notes that 'risk communication is successful only to the extent that it raises the level of understanding of relevant issues or actions and satisfies those involved that they are adequately informed within the limits of available knowledge.'

Clearly, risk communication deals with how people are informed about risks and what they have to say about them. Such risks can come from natural or technological hazards, such as earthquakes or air pollution, or from health practices such as unsafe sex that leads to AIDS or even eating foods with high levels of fat that can contribute to heart disease.

UNCERTAINTY FACTORS

Communicating about risk is not an easy task. Uncertainties abound and three types of uncertainties have a heavy impact on the communication process. The first concerns uncertainties about the scientific data being used in a risk assessment. Experts frequently disagree about various theories of how toxic substances and other risks affect people. A good example of this is the threshold theory, that says generally that there is a point below which exposure to a toxic chemical will not cause a long-term health effect. A number of scientists support this theory, but many also believe that there always will be some potential long-range health effect, no matter how low the level of exposure is. An expert who believes in the threshold theory will have a different interpretation of data in a risk assessment than an expert who does not.

In addition, in conducting risk assessments experts have to deal with assumptions and uncertainty. According to Covello (1991), 'the uncertainty stems from the series of estimates made in the assessment process: scientists must estimate the population's actual exposure to the hazards, assume animal tests will accurately predict the effect on humans, and assume high exposure rates typically used in animal studies can be modified to predict effects of low-level exposures on humans.'

Risk assessments involve models and these models are based on theory and current knowledge; sometimes they can be flawed. Peter Montague (1991) of Princeton University says that the process of risk assessment is 'fatally flawed because in the vast majority of cases, essential information is missing. People who do risk assessments work with vast gaps of knowledge and this will always be true. With 500 to 1,000 new chemicals entering commercial channels each year, we may never have an adequate base of information about the toxicity of the chemicals to which we routinely expose our neighbors. We also will continue to be ignorant about exposure levels.' The discussion on risk assessment in Chapter 19 adopts a more optimistic view, noting that while there are many uncertainties, the science is rapidly improving.

The second set of uncertainties that affect the risk communication process concern how to manage the various scientific, economic, social, and legislative trade-offs that need to be made. For example, should leaking drums of toxic waste be removed from a Superfund site or should they be contained as effectively as possible and buried there? Residents near that site will want the drums removed to protect themselves and their families; the EPA might decide that the drums are best managed by sealing them, capping the site, and then monitoring the situation periodically. Different value systems clash in such a decision – community values for health and a desire for the least risk possible versus governmental values for economic efficiency and minimal, controlled risk to the community.

The third area of uncertainty rests less with the risks and more with how the public views and reacts to various risks. Both sociologists and communications researchers are finding unexpected reactions to various risk communication situations. For example, a massive information campaign was conducted by the EPA and a number of state environmental organizations to alert citizens to dangers about radon. As a part of the campaign, Dr. C. Everett Koop, then Surgeon General, labeled radon a major health hazard, second only to cigarette smoking as a cause of lung cancer. The EPA, assisted by the New York-based Advertising Council, promoted radon testing in houses via television commercials, while numerous newspapers and magazines provided extensive coverage of the radon issue (Friedman et al. 1987; Krimsky and Plough 1988; Sandman, Weinstein, and Klotz 1987). Yet, with all of this activity, very little interest was stirred up in the

American public about this serious environmental risk.

However, mention the word 'dioxin' and link it to burning wastes in an incinerator and watch many members of the community rise up in outrage about a risk that many experts believe will have little effect on surrounding communities (Wirth 1991). These two current environmental issues – radon in people's homes versus dioxin in dilute amounts in their air – are perceived quite differently. One causes apathy and the other causes outrage. Why this occurs is what risk communication specialists are trying to find out.

PUBLIC VALUES AND RISK

One of the factors that risk communication researchers have identified in public response to environmental risk situations involves a fundamentally different interpretation from that of technical experts of what is important. In the past, many experts involved in risk assessment assumed that evaluating risk was a technical matter that required better, more accurate scientific information. Once they had this information, they felt they simply had to present it to people in a clear way and it would be accepted because it was the 'rational' way to proceed. According to New York University sociologist Dorothy Nelkin (1988), 'It was assumed that if public fear of risk is defined in terms of inadequate information, then increased technical evidence and better communication of this evidence by knowledgeable experts will allay public concern.'

However, in their efforts to promote a 'rational' view of risk, these experts ignored many factors that could not be measured, including differential perceptions about various risks, societal and social values, hidden agendas, and even certain characteristics of the risk itself that influence how people respond. Nelkin points out that researchers found that public concerns about risk could depend more on political, economic, social, and cultural factors than on actual danger. She also explains that

'while some people talk of risk in terms of cost-effective solution, of efficiency; others use the language of 'rights,' emphasizing moral issues and questions of social responsibility, justice and obligation. Some individuals evaluate risk in statistical terms; others talk of 'victims' or 'real people.' Some define risk as a problem that requires expert solutions; others seek more participatory controls.'

Studies in this field by pyschologist Paul Slovic (1991) and Fischhoff et al. (1981) have identified a number of factors that affect the way people accept risk. The size of risk is only one of a number of factors that count. Others include fairness, benefits, alternatives, control, and whether the risk is voluntary or not. People evaluate whether the risk is fairly distributed. They also want to know whether they and their neighbors are going to benefit from the risk or whether such benefits will accrue to others who are not at risk at all. They want to evaluate whether there are other ways to handle the situation, such as an alternative technology that eliminates the risk – even though that technology may be more expensive. Finally, they want some control over the risk and a choice in whether to accept it. Swanson et al. (1991) noted, 'A risk that the parties at risk assess and decide to accept is more acceptable than a risk that is imposed on them.'

PERCEPTIONS OF RISK

To be effective in risk communication, according to Slovic (1986), one 'must recognize and overcome a number of obstacles that have their roots in the limitations of scientific risk assessment and the idiosyncrasies of the human mind. Doing an adequate job of communicating means finding comprehensible ways of presenting complex technical material that is clouded by uncertainty, and is inherently difficult to understand.'

Slovic, Fischhoff and others have investigated psychological and sociological factors that affect perceptions of risk. Slovic notes that 'perceived risk can be characterized as a battleground marked by strong and conflicting views.' His research has shown that perceptions of risk can be described in terms of numerous characteristics or dimensions, analogous to personality dimensions that characterize people. Important in such perceptions are the degree to which a risk is perceived to be known or understood and the degree to which it evokes perceptions of dread, uncontrollability, and catastrophe. Other key factors invoked in such perceptions include whether the risk is fatal rather than injurious, whether it is offset by compensating benefits, and whether its effect is delayed in time so the risk will affect future generations (Swanson et al. 1991).

In contrast, Slovic says (1991), 'experts' perceptions of risk are not closely related to any of these various risk characteristics. Instead, experts appear to see riskiness as synonymous with expected annual mortality. As a result, many of the conflicts over 'risk' may result from experts and laypeople having different definitions of the concept. Expert recitations of 'risk statistics' often do little to change people's attitudes and perceptions.'

Risk perception research also has compared perceptions of risks and benefits for many activities and technologies and found that these are determined by the specific context in which one is exposed to substances. For instance, while people exhibit great fear of radiation from nuclear power plants, they do not fear it from diagnostic X-rays. Slovic explains that they also see certain types of risk as signals of great potential social impact. Swanson et al. notes (1991), 'For example, an accident that takes many lives but produces little social disturbance at large and is familiar and well understood – such as an airplane crash – does not have much signal value. But a small accident in an unfamiliar system or one that is poorly understood, such as in a nuclear reactor, has an important signal of future potentially catastrophic mishaps that could have major social consequences.'

Slovic (1991) argues that perceptions of risk influence the priorities and legislative agenda of agencies such as the EPA. When something goes wrong, such as 'an accident, discovery of pollution or product tampering, perceptions, interacting with social and institutional forces, can trigger massive social, economic and political consequences. An unfortunate event can be thought of as a stone dropped in a pond. The ripples spread outward, encompassing first the directly affected victims, then the responsible company or agency, and in the extreme, reaching other companies, agencies and industries.'

Many risk communication specialists advocate that those involved in handling various environmental risks such as air toxics accept that 'perception is reality.' Whether technical experts feel there is an actual danger or not from a risk situation, if the public perceives there is danger, that perception and concern is enough to have a serious impact on a business, industry or government agency. Organizations need to understand public risk perceptions and incorporate them into their decision-making processes.

THE PUBLIC'S RIGHT TO KNOW

The need to communicate about environmental risks such as air toxics is not something that will eventually go away. Rather, it will continue to increase, particularly since new legislation often requires procedures be devised to report information to the public. The Emergency Planning and Community Right-to-Know Act (EPCRA), also known as Title III of the Superfund Amendments and Reauthorization Act of 1986, requires industries and various other facilities to report to the EPA, as well as to states and local communities, about their use, storage, and emissions of more than three hundred toxic chemicals. This requirement is based on the belief that the more information citizens have about environmental conditions, the better equipped they will be to ensure their protection from unacceptable risks to their health and safety, and the more quickly facilities will reduce potentially hazardous situations.

The information that industries and other organizations file each year on their emissions under EPCRA is compiled into the Toxic Release Inventory (TRI) and is available to anyone. Citizens can call the EPA Right-to-Know Information Hotline[1] to get general information about the TRI. They also can get copies of the original reports companies submit from either the EPA or from state agencies. The national information is on computer and the TRI data base can be accessed by personal computer through the National Library of Medicine in Bethesda, MD. It will be available in local libraries in the future. A computer disk of state data also can be purchased.

Journalists not only have these resources but they also have an excellent guidebook, *Chemicals, the press, and the public*, that discusses how to use and interpret the TRI information. Published for the media by the Environmental Health Center of the National Safety Council with funding from EPA, the volume also has an accompanying videotape. In addition, staff from the Environmental Health Center have held workshops around the country for reporters to help them understand how to use the TRI. Citizens also can get information on the TRI, including a book from Greenpeace called *A Citizen's Toxic Waste Audit Manual* and another from the National Toxics Campaign called *From Poison to Prevention*.

The TRI information is used. There have been major newspaper and magazine series based on TRI data. Two widely discussed sets of articles were published in the *San Jose Mercury News* and in *The Courier Journal* of Louisville, KY (Environmental Health Center 1989). The California story was spurred by a report from the Silicon Valley Toxics Coalition that revealed that twenty-five major corporations with plants in Santa Clara County legally dumped more than twelve million pounds of toxic and potentially cancer-causing pollutants into the air, land and water annually. The data used by the environmental group came from the TRI.

In the Kentucky story, the reporter worked directly with the TRI data to compile the toxic release reports for the state. One story dealt with the perils of air pollution, reporting that 'Kentucky's major industries emit a wider variety of potentially hazardous air pollutants than previously believed.' It also pointed out that many were unregulated. A second story was on the usefulness of the TRI for helping local and state government officials check for compliance with permits and other regulatory information. It also recommended use of the data by citizens groups.

As shown by the California situation, environmental groups at all levels are making use of the data to issue reports that they turn over to the media and the public. At the national level, the National Wildlife Federation published a large report on the 1987 TRI data called *The Toxic 500*. It described in detail the top five hundred industrial polluters in the country, what they were emitting and where it was being emitted. This report was released at a news conference in Washington, D.C., attended by representatives of all major media outlets. Television coverage was heavy.

State and local environmental organizations also have assessed the TRI data for their regions and provided reports to the media and the public. For example, the Delaware Valley Toxics Coalition issued a report that broke down all of the emission data about various areas and industries in Eastern Pennsylvania.

The conclusion is that the private citizen, media reporter or environmental group – just about anyone with an interest – can find out with a minimum of trouble what pollutants industries say they are emitting. According to Winston and Strawn (1991), 'Congress' interest in air toxics was prompted, in part, by the public outcry raised when EPA released its Toxic Release Inventory (TRI) that reported that the vast majority of toxic pollutants released into the environment were released into air.'

---

[1] Phone number (800)535-0202.

IMPLICATIONS OF THE 1990 CLEAN AIR ACT

AMENDMENTS

Although Title III of the 1990 Amendments does not specify a regular public reporting mechanism for industries and others regulated under it, such as the TRI, it does include aspects that will be open to and debated by the public and the media. These include the requirement for public review for most regulatory actions and EPA's development of a national strategy to reduce public health risk in terms of cancer incidence attributed to area sources. The national strategy must be submitted to Congress and is certain to be a source of debate.

Of more direct interest to large and small companies is the requirement under the accidental release provisions of section 112(r) for owners and operators of stationary sources to prepare and implement a risk management plan that must be made available to the public. The plan is to include a hazard assessment, an accident prevention program, and an accident response program. Since this requirement overlaps with the general requirements of EPCRA, some coordination is expected that may involve even more public disclosure than currently is anticipated.

Another requirement of section 112(r) that potentially will involve the public and the media is that concerning the Chemical Safety and Hazard Investigation Board. This board is required to investigate accidental releases that cause fatalities or serious injury among the public or have the potential to cause substantial property damage or a number of deaths or injuries among the public. With one exception concerning protecting an organization's competitive position, any records, reports, or information obtained by this board will become available to the public.

Beyond these specific requirements, there will be public expectations that municipal waste combustors, chemical companies, and many other identified major sources of toxic air pollutants will be working hard to implement the 1990 Amendments. Disclosures of increases in emissions of toxic air pollutants through the TRI or evidence of foot-dragging in meeting legal requirements could bring about public displeasure and distrust of an organization. Evidence of lack of goodwill in carrying out the provisions of this law – however difficult that may be – could result in community concern and perhaps even public outrage.

Being in a community that is fearful, hostile, and distrustful works against an organization's well-being. Such an atmosphere generates controversies over environmental risks that often bring with them extensive media coverage, heated public meetings, lawsuits, and sometimes additional regulation. Frequently they dissolve the goodwill that an organization has worked hard to create. Disputes find their ways into courtrooms where individuals sue for compensation. Risk controversies subject organizations to public condemnation, put employees under great mental stress, and affect future policies and plans (Swanson et al. 1991).

All of these factors make it important that organizations that are regulated under Title III of the 1990 Amendments begin to plan early to communicate with the media and the community about the environmental risks. Effective communication of risk is critically important because it helps increase the likelihood of finding a solution that is acceptable to citizens

RISK COMMUNICATION MODELS

As noted above, in its early stages of development as a field, risk communication meant to many people that a message that described or characterized hazards or risk was developed by technical experts and addressed to nonexperts. This was a one-way message about the nature of risk from experts who either wanted to enlighten, but most likely wanted to persuade, an uninformed and passive public.

This led risk communication researchers to start with a simple model of how risk communication works. There were four main factors in the process.

1. The information source (i.e., an organization)
2. The channel the source uses (e.g., a news release to the media)
3. The message sent out by the information source
4. The receiver of the information (e.g., community members)

This simple model meant that if an organization had something to say or explain about risk, all it had to do was decide what the message should be, send it out in some way, and have people listen to and understand it. It quickly became apparent, however, that this model was oversimplistic and would not work. It was oversimplistic because it is a linear, one-way model and it really is not representative of the whole risk communication picture. Risk communication is not a one-way activity; if it is set up that way, it will fail. There must be feedback from the receivers to the source and, at a minimum, it must be a circular process.

In complicated risk controversies, the model gets even more complex than two-way communication. Studies have shown that risk communication takes many forms and that multiple generators of risk information – including nonofficial sources – play a key role in the overall risk communication scenario. Based on their analysis of five risk communication case studies, Krimsky and Plough (1988) suggest that risk communications resemble tangled webs rather than parallel series of interactions between senders and receivers of risk messages.

They chose the tangled web metaphor to highlight three important characteristics of a risk communication event. First, no one can anticipate which of several possible sources of communication will dominate a risk controversy. Official or scientific risk communicators are only one of many potentially influential voices. Second, risk communication messages

become entangled and may result in unpredictable social outcomes. Since there is a constant interplay among messages, communications from any source may change over time. Third, risk messages come in many forms (i.e., from literal to symbolic forms; from print to television) reflecting the diverse modes of communication.

All of this means that under particular circumstances it is quite difficult for any single communicator to establish the boundaries of risk communication for an issue. In their case studies, Krimsky and Plough (1988) found that technical information often did not play a dominant role in a risk communication controversy. They maintain, as much as technical experts would like science to drive the risk communication process, this is unlikely to be the case when there is uncertainty in the risk assessment and important social and economical contexts that share the risk debate.

Because of these factors, the NRC report (1989) explains that 'risk communication includes all messages and interactions that bear on risk decisions. Thus, risk communication includes announcements, warnings, and instructions moving from expert sources to nonexpert audiences. . . . But it also includes other kinds of messages – about risk information and information sources, about personal beliefs and feelings concerning risks and hazards, and about reactions to risk management actions and institutions. Not all these messages are strictly about risk, but all are material to risk management.'

The NRC report also notes (1989):

Increased efforts to 'get the message across' by describing the magnitude and balance of attendant costs and benefits or by telling people which option provided the greatest net benefit to society will have little effect for several reasons. First, costs and benefits are not equally distributed across a society. Those who bear more than a proportionate share of the costs of one of the options want to convince others that the selection of that alternative would be unfair to them. Other political participants want to make similar arguments on their own behalf or to consider the arguments of all the interested parties. Thus an important aspect of conflicts about technological issues is that these are often conflicts between different interest groups. These conflicts cannot be resolved simply by knowledge about likely effects of each alternative on the society as a whole or on various groups.

Second, people do not agree about which harms are most worth avoiding or which benefits are most worth seeking. They want to argue for the protection of what they value and to consider which values are most worth preserving or advancing in each decision context. Because conflicts about technological issues pit values against each other, it is impossible to calculate net benefit to society – or even to subgroups of the society – on any scale that will satisfy all the participants. Values need to be debated and weighed in a political process.

Third, citizens of a democracy expect to participate in debate about controversial political issues and about the institutional mechanisms to which they sometimes delegate decision-making power. A problem formulation that appears to substitute technical analysis for political debate, or to disenfranchise people who lack technical training, or to treat technical analysis as more important to decision making than the clash of values and interests is bound to elicit resentment from a democratic citizenry. Because of such reactions to them, problem formulations that attribute technological conflict to widespread public ignorance only exacerbate the conflict.

As part of this tangled web of risk communication, the mass media play a special role. In fact, Slovic identified the mass media and interpersonal communication as the two most important ways people learn about risk issues. Many people blame the media for stirring up risk controversies, yet the NRC (1989) report notes that 'It is mistaken ... to view journalists and the media always as significant, independent causes of problems in risk communication. Rather, the problem is often at the interface between science and journalism.'

The media exert a powerful influence on how people perceive risk. Study after study has shown that the media are the major conveyers of environmental, medical, and scientific information to the public. They provide a wealth of information about these subjects, but because of a number of journalistic constraints such as short deadlines and lack of newspaper space or television airtime, this information is limited. It is limited in depth, in scope, and in treatment of complexity (Friedman 1988).

Despite these limitations, the media decide what information will get to the public and the way they present this information affects how readers and viewers interpret it. Elements that come into play include verbal and visual images, tone, placement, and timing. Communication researchers theorize that the mass media help set the public's agenda. What this means is that by their coverage, they communicate the relative importance of a set of issues to the public, therefore forming what the public thinks it should be concerned about – the public agenda. This public agenda, in turn, influences the policy agenda of decision-makers.

According to the NRC (1989), 'scientists and risk managers should recognize the importance of the part journalists play in identifying disputes and maintaining the flow of information during resolution of conflicts.' Corporate officers must make efforts to understand how the media operate and learn how to work with them (Friedman 1988; Swanson et al. 1991). Risk communication to both the media and the public is an important part of corporate communication that requires an enlightened attitude on the part of top management and advanced policy and program planning.

To help organizations develop risk communication programs, EPA published a pamphlet entitled 'Seven Cardinal Rules of Risk Communication' (EPA 1988a), that can be applied usefully in both the public and private sector. However, the agency emphasizes that there are no easy prescriptions for successful risk communication. The rules are listed below but this listing only includes a summary of the guidelines that accompany each rule.

Rule 1: *Accept and involve the public as a legitimate partner.*
A basic tenet of risk communication in a democracy is that people and communities have a right to participate in decisions that affect their lives, their property, and the things they value.

Rule 2: *Plan carefully and evaluate your efforts.*
Risk communication will be successful only if carefully planned.

Rule 3: *Listen to the public's specific concerns.*
If you do not listen to people, you cannot expect them to listen to you. Communication is a two-way activity.

Rule 4: *Be honest, frank, and open.*
In communicating risk information, trust and credibility are your most precious assets.

Rule 5: *Coordinate and collaborate with other credible sources.*
Allies can be effective in helping you communicate risk information.

Rule 6: *Meet the needs of the media.*
The media are a prime transmitter of information on risks: they play a critical role in setting agendas and in determining outcomes.

Rule 7: *Speak clearly and with compassion.*
Technical language and jargon are useful as professional shorthand, but they are barriers to successful communication with the public.

IMPLEMENTING A RISK COMMUNICATION PROGRAM

To develop a risk communication program, organizations first need to make a commitment to risk communication as a management function and a tool for better community understanding. Putting a risk communication program in place also means that a risk assessment and management program should be there as well. Organizations should not send messages that are not backed up by actions. For example, a company cannot tell the public that it is going to reduce air toxic emissions if it will not be able to do just that. That is the quickest way to lose the community's trust and the company's credibility, both of which are key factors in successful risk communication.

Beyond matching actions to words, organizations will need to take a proactive stance about informing the media and the community about risk issues. This has some costs, but it has many more benefits. It requires a more open information stance, even to the point of releasing information early and going beyond what is required by the law. For example, a number of proactive companies release their TRI reports, with interpretation, to the media and the community at the time they submit the reports to the EPA.

Other ways to be proactive include preparing educational materials for the community and the media, training company employees in risk communication, conducting surveys and interviews, and holding open houses and plant tours

for community officials and citizens. Companies need to build bridges to public and private organizations such as working with opinion leaders. They also need to find uninvolved experts – third party sources – who can help provide information to the community. Accepting and practicing EPA's Seven Cardinal Rules is a critical factor in any risk communication program.

As the EPA rules state, risk communication requires planning. For a risk communication program to succeed, it must address many tasks: providing access to citizen participation; measuring citizen's perceptions of the risks involved; giving careful consideration to these perceptions; training personnel; working with the mass media; developing clear risk messages; evaluating whether these risk messages are successful; and finding ways to ensure community feedback. Therefore, risk communication programs cannot just be thrown into place when a public controversy occurs. A risk communication plan must be developed well in advance, with careful analysis of how to involve the public effectively.

Many companies are aware of this and have developed and tested such plans. Entire manuals have been developed by consultants for trade associations and their industries. *Risk Communication, Risk Statistics and Risk Comparisons* (Covello, Sandman, and Slovic 1988) was published by the Chemical Manufacturers

Association (CMA) for the chemical industry. It includes chapters on effective risk communication, guidelines for providing and explaining risk-related numbers and statistics, guidelines for using risk comparisons, and concrete examples of risk comparisons to use.

The Electric Power Research Institute (EPRI) also published a two-volume work, *Risk Communication Manual for Electric Utilities* (Swanson et al. 1991). Volume 1 is a practitioner's guide with chapters on the importance of risk communication, perceptions of risks, and risk communication program planning. It also includes an extensive chapter called 'Do's and Don'ts,' that points out helpful and harmful actions in risk communication situations. Volume 2 is a series of risk communication case studies that includes references back to actions described in the Do's and Don'ts chapter in Volume 1.

Both the utility and chemical industries take risk communication issues and programs very seriously. Besides publishing its manual, EPRI sponsored a number of risk communication workshops, with special emphasis on the problem of communicating about risks related to electromagnetic fields. CMA also developed a program involving risk communication called *Responsible Care*, which it requires all member companies to carry out. Adopted in 1988, the program commits member companies 'to improve performance in response to public concerns about the impact of chemicals on health, safety and environmental quality.' It has a Public Advisory Panel that directly involves the public in shaping the program.

Responsible Care built on an earlier volunteer program run by CMA called the *Community Awareness and Emergency Response* (CAER) program. CMA started CAER in 1985 to respond to public concerns about accidental chemical releases. Within two years, more than 1,100 facilities had CAER activities to improve joint community and industry emergency response efforts. Currently, members participate in CAER by taking part in *Responsible Care*.

The philosophy of *Responsible Care* is set down in ten guiding principles. These strive for better, safer performance in every part of the industry. They also promise better communication within the industry and with the public and government about the industry's operations and products. The program places high value in improving performance in health, safety, and environmental quality, listening and responding to public concerns, companies assisting each other to achieve optimum performance, and communicating with the public to report on progress. CMA also has numerous audiovisual and printed materials to help both members and nonmembers learn more about risk communication issues. It publishes a monthly CAER newsletter that features success stories, tips on how to work with various groups, resource information, and status reports about CAER and EPCRA compliance.

In the long run, these organizations and others find that developing risk communication plans and putting them into practice is just smart business. While it is a mistake to expect risk communication always to reduce conflict and smooth risk management, the NRC (1989) report recommends that 'risk managers need to consider risk communication as an important and integral aspect of risk management. In some instances, risk communication will, in fact, change the risk management process itself.'

Solving the nation's air toxics problems is a massive and complex task – scientifically, economically, and socially. There is no question that industries and businesses will find that a long-term view that supports openness about emissions, problems in controlling them, and future compliance plans and expenditures is the surest policy for community acceptance. Such openness requires developing a risk communication plan and committing to follow it. Risk communication is not easy. There are problems in establishing and keeping credibility and trust, in analyzing public risk perceptions, in developing risk messages and making them understandable to citizens, and in getting feedback about and involving the public in these activities. A trained corporate communication staff can aid in dealing with these problems, but top-level management must be involved. Risk communication is not something to be attempted lightly. Yet it clearly is something that organizations must encourage, practice, and plan for as part of their efforts to meet the air toxics requirements of the 1990 Clean Air Act Amendments.

References

Covello, V.T. 1991. How to understand a risk assessment. In: *Environmental Risk Reporting: The Science and the Coverage*. S.M. Friedman and C.L. Rogers, eds. Lehigh University. Bethlehem, PA.

Covello, V.T., P.M. Sandman, and P. Slovic. 1988. Risk communication, risk statistics and risk comparisons: A Manual for Plant Managers. Chemical Manufacturers Association. Washington, D.C.

Environmental Health Center. 1989. Chemicals, the press, and the public: A journalist's guide to reporting on chemicals in the community. National Safety Council. Washington, D.C.

Fischoff, B., S. Lichtenstein, P. Slovic, S. Derby, and R. Keeney. 1981. *Acceptable Risk*. Cambridge, England: Cambridge University Press.

Friedman, S.M. 1988. The journalist's world. In: *Scientists and Journalists: Reporting Science as News*. S.M. Friedman, S. Dunwoody and C.L. Rogers, eds. American Association for the Advancement of Science Books (2nd printing). Washington, D.C.

Friedman, S.M. 1990. Two decades of the environmental beat. *Gannett Center Journal* 4(3):13–23. Reprinted in *Media and the Environment*. C. LaMay and E. Dennis, eds. Washington, D.C. Island Press. 1991.

Friedman, S.M., J.F. Post, M.B. Vogel, and W.F. Evans. 1987. Reporting on radon: The role of local newspapers. *Environment* 29(2):4–5, 45.

Krimsky S. and A. Plough. 1988. *Environmental Hazards: Communicating Risk as a Social Process*. Dover, Mass: Auburn House.

Montague, P. 1991. What questions to ask experts about risk assessments. In *Environmental Risk Reporting: The Science and the Coverage*. S.M. Friedman and C.L. Rogers, eds. Lehigh University. Bethlehem, PA.

National Research Council (NRC). 1989. *Improving Risk Communication*. National Academy Press. Washington, D.C.

Nelkin, D. 1988. Communicating the risks and benefits of technology. Unpublished paper presented at Symposium on Science Communication: Environmental and Health Research, Los Angeles.

Russell. C. 1990. The riskiest of times? *Washington Post*, Health Section, January 30, p. 9.

Sandman, P.M., N.D. Weinstein, and M.L. Klotz. 1987. Public response to the risk from geological radon. *Journal of Communication* 37(3):93–108.

Slovic, P. 1986. Informing and educating the public about risk. *Risk Analysis* 6(4):403–415.

Slovic, P. 1991. Perceptions of risk: paradox and challenge. In: *Environmental Risk Reporting: The Science and the Coverage,* eds. S.M. Friedman and C.L. Rogers. Lehigh University. Bethlehem, PA.

Swanson, S., S. Friedman, P. Bridgen, V. Covello, P. Slovic, and J. Cohn. 1991. Risk communication manual for electric utilities, Volumes 1 and 2. Electric Power Research Institute. Palo Alto, CA.

U.S. Environmental Protection Agency (EPA). 1988a. Seven cardinal rules of risk communication. Office of Policy Analysis. Washington, D.C.

Winston and Strawn. 1991. Playing by the Rules: Surviving Clean Air in the 1990s. Unpublished paper presented at seminar, New York. February 5.

Wirth, P. 1991. Waste-burning controversy pits Keystone against residents. *The Morning Call*, Allentown, PA., pp. A1 and A20, May 19.

Appendices

A

Noncancer Toxicity of Inhaled Air Pollutants: Available Approaches for Risk Assessment and Risk Management

Annie M. Jarabek and Sharon A. Segal

INTRODUCTION

The notational computations for each of the dosimetric adjustments for the scenarios based on the type of agent (particle or gas) and on the observed toxic effect (respiratory tract or remote toxicity) described in the section in Chapter 7, entitled Dosimetric Adjustment to Human Equivalent Concentration (HEC), are provided in this appendix to impart an appreciation for the parameters involved. The derivation of these default models and explanations of their limitations, assumptions, and input parameters are provided more fully elsewhere (Jarabek et al. 1989; Jarabek et al. 1990; Overton and Jarabek 1989a; Overton and Jarabek 1989b). Release of revised methods is anticipated in late 1993.

Individual case studies of verified RfCs are provided as examples for each HEC scenario and to illustrate the evaluation of the data array and synthesis of studies to create a compelling dose-response assessment. Since the verified RfC values that appear in EPA's Integrated Risk Information System (IRIS) represent a consensus agreement of involved agency scientists, only verified values are provided in the following case examples. Note that the verification date for each particular assessment is provided for reference. The reader is encouraged to access IRIS for the most up-to-date versions of the RfC estimates and for examples of other dosimetry adjustments not available at the time of this writing.

EPA'S INTEGRATED RISK INFORMATION SYSTEM

RfCs that have been verified by EPA's Reference Dose/Reference Concentration Work Group are contained in IRIS, that is a publicly available, on-line database established by EPA. IRIS is organized on a chemical-specific basis and, in addition to the RfCs, contains other health risk (e.g. oral Reference Doses, RfDs, cancer unit risk estimates) and regulatory information, such as standards, guidelines, and ongoing EPA regulatory activities. The information contained in Section I, Chronic Health Hazard Assessment for Noncarcinogenic Effects (RfC and RfD), and Section II, Carcinogenicity Assessment for Lifetime Exposure (cancer unit risk estimates) for each chemical has been verified following rigorous EPA-wide work-group review and, therefore, represents EPA consensus. In addition to these health-based criteria, a brief summary of the data supporting the derivation of the criteria (i.e., selected primary literature) as well as a listing of critical EPA documents (e.g., criteria documents, health advisories, health assessment documents, health and environmental effects profiles, and reportable quantity, RQ, documents) relevant to each chemical is contained in each record.

Each IRIS file is arranged in the following general categories.

- Substance Identification and Use Information
- Chemical and Physical Properties
- Noncarcinogenic Assessment for Lifetime Exposure (Inhalation RfC and Oral RfD)
- Carcinogenicity Assessment for Lifetime Exposure (Unit Risk Estimates)
- Drinking Water Health Advisories and Acute Toxicity Information
- Aquatic Toxicity Assessment
- EPA Exposure Standards and Regulations under the Clean Air Act; Safe Drinking Water Act; Clean Water Act; Federal Insecticide, Fungicide, and Rodenticide Act; Toxic Substances Control Act; Resource Conservation and Recovery Act; and Superfund
- References

Contact names and telephone numbers for EPA personnel involved in the development of each record also are provided if the user wishes to obtain more detailed information on a chemical.

IRIS is a tool to provide quick reference for hazard identification and dose-response information, but should not be construed as guidance for specific exposure situations. However, together with specific exposure information, the data contained in IRIS are useful for characterizing public health risks for a particular chemical in specific exposure situations, and, therefore, can aid in the formulation of risk management decisions to protect public health. IRIS is accessible through the following channels.

1. EPA's electronic mail system (E-Mail) for employees of the EPA
2. The Public Health Network (PHN) of the Public Health Foundation for employees of state and local health departments
3. National Library of Medicine's TOXNET system (Bethesda, MD)
4. DIALCOM, Inc. (Rockville, MD)
5. Chemical Information Systems, Inc. (CIS) (Baltimore, MD)

INHALATION RfC DOSIMETRIC ADJUSTMENTS AND CASE EXAMPLES

HEC Scenario One. Particle:Respiratory Effect (Default)

As mentioned in Chapter 7, the optimal approach for any scenario is to use a dosimetry or PBPK model for the chemical of interest. An example of this optimal approach for the particle:respiratory effect scenario is the use of the Yu and Yoon (1990) model that incorporates both particle deposition and clearance which was recently used to estimate an HEC in humans for diesel emissions. When such models are not available, however, the dosimetric adjustments outlined in this appendix provide a reasonable default for the optimal modeling approach. The definitions of the parameter

symbols used in this appendix to illustrate the dosimetric adjustments are provided in Table A-1.

The Regional Deposited Dose (RDD), expressed as milligrams per minute per square centimeter of lung region, can be computed as

$$\text{RDD} = \frac{10^{-3} Y \dot{V}_E}{S} \sum_{i=1}^{n} P_i E_i \quad \text{(A-1)}$$

Where P_i = the particle mass fraction in the ith size range of an aerosol with a given particle mass median aerodynamic diameter (MMAD) and geometric standard deviation (σ_g)

E_i = the deposition efficiency for that species and respiratory tract region (i.e., ET, TB, PU, TH, or TOTAL) of interest

Y = exposure level (mg/m^3)

$\dot{V}_E$ = minute volume (l/min)

S = regional surface area (cm^2)

Human fractional deposition estimates can be calculated by an approach similar to that used for the animal data using empirically based logistic regression and theoretical models (Jarabek et al. 1990). These estimates then are applied to the same exposure (MMAD, σ_g) and concentration (mg/m^3) as that to which the experimental animals were exposed and adjusted for ventilatory parameters and respiratory tract surface areas to calculate the human RDD.

The RDD for the laboratory animal species in question then can be divided by the corresponding RDD for humans to calculate the ratio of deposition in that species about the deposition in humans. In other words, the regional deposited dose ratio (RDDR) is calculated as

$$RDDR = \frac{(RDD)_A}{(RDD)_H} \qquad (A\text{-}2)$$

Where $(RDD)_A$ = regional deposited dose in experimental animal species of interest, adjusted for surface area and ventilatory volumes

$(RDD)_H$ = regional deposited dose in humans, adjusted for surface area and ventilatory volumes

The appropriate RDDR to calculate is dictated by the observed toxicologic effect the risk assessor is evaluating. For example, the RDDR of the pulmonary (PU) region alone would be calculated for an effect involving only that region, whereas the RDDR$_{ET}$ would be used to determine the dose to the extrathoracic (ET) region in order to assess an effect in that region. The RDDR then is used to scale the exposure

Table A-1   Definitions of parameter symbols used in the dosimetric adjustments

ET	Extrathoracic region (ET) of the respiratory tract; extends from just posterior to the external nares to just anterior to the trachea
LOAEL	Lowest-Observed-Adverse-Effect Level
LOAEL(ADJ)	LOAEL adjusted to continuous exposure duration (e.g., adjusted for exposure regimen by h hours/24 hours and d days/7 days)
LOAEL(HEC)	Human equivalent concentration to LOAEL
MVa	Experimental animal minute volume (m^3/day)
MVh	Human ambient minute volume (m^3/day)
MVho	Human occupational minute volume (m^3/day)
MW	Molecular weight
NOAEL	No-Observed-Adverse-Effect Level
NOAEL(ADJ)	NOAEL adjusted to continuous exposure duration (e.g., adjusted for exposure regimen by h hours/24 hours and d days/7 days)
NOAEL(HEC)	Human equivalent concentration to NOAEL
PU	Pulmonary region of the respiratory tract; includes the terminal bronchioles and alveolar sacs
Sa	Surface area of respiratory tract region in experimental animal (cm^2)
Sh	Surface area of respiratory tract region in humans (cm^2)
TB	Tracheobronchial region of the respiratory tract; from trachea to the terminal bronchioles
TH	Thoracic region of the respiratory tract; the tracheobronchial region and the pulmonary region together

concentration associated with the observed effect to an equivalent concentration that reflects the dosimetric differences between humans and the experimental species in question. That is, the RDDR provides a factor for scaling the observed NOAEL of a study to a NOAEL(HEC), so

$$NOAEL(HEC) = NOAEL(ADJ) \times RDDR$$

(A-3)

Where NOAEL(HEC) = the human equivalent concentration NOAEL expressed in mg/m^3

NOAEL(ADJ) = the experimental NOAEL, expressed in mg/m^3, adjusted for continuous exposure duration (e.g., adjusted for exposure regimen by h hours/24 hours and d days/7 days)

The NOAEL(HEC) then is arrayed with others to determine the 'most sensitive' species and the critical effect, as discussed.

Case Study: Diphenylmethane Diisocyanate (MDI) (CAS 101-68-8)

Lesions of the nasal cavity observed in a chronic rat inhalation study serve as the basis of the RfC for MDI. In a study by Reuzel et al. (1990), SPF-bred Wistar rats were exposed to MDI aerosols 6 hours/day, 5 days/week for 24 months, with a satellite group sacrificed for histopathological evaluation after 12 months of exposure. There were no exposure-related effects on mortality, clinical chemistry, hematology, urinalysis, body weight gain, or clinical signs. Histopathological examination of tissues taken from the animals sacrificed at 6 months revealed exposure-related changes in the nasal cavity, lungs, and mediastinal lymph nodes including statistically significant slight disarrangement of the olfactory epithelium, basal

cell hyperplasia, accumulation of macrophages with yellow pigment, prevalence of Type II alveolar cells at the site of accumulation, localized fibrosis, alveolar duct epithelialization, and interstitial pneumonitis. Animals sacrificed after 24 months of exposure exhibited lesions of increased severity in the same tissues. Effects observed included basal cell hyperplasia of the olfactory epithelium that often was accompanied by Bowman's gland hyperplasia, increased severity in the accumulation of pigment-laden macrophages in alveolar duct lamina, and in localized fibrotic changes, and increased incidence of localized alveolar duct and bronchial epithelialization. The information obtained in this chronic study identifies a NOAEL of 0.2 mg/m^3 and a LOAEL of 1.0 mg/m^3. The critical effects identified in the Reuzel et al. (1990) study are corroborated by adverse effects noted in other animal inhalation studies. For example, degenerative lesions of the lungs and nasal cavity of rats occurring at exposure levels of 4 mg/m^3 and above were noted by Reuzel et al. (1985a) and Reuzel et al. (1985b).

MDI is used extensively in making flexible polyurethane foam. While it is well known that toluene diisocyanate (TDI) causes asthma and pulmonary function decline, these effects have not been as well characterized for MDI. There are a number of case reports describing occupational asthma and hypersensitivity pneumonitis, but exposure levels are unknown. Specific IgE and IgG antibodies have been demonstrated in occupationally exposed workers with diagnosis of occupational asthma. Few occupational studies have examined exposure-response relationships with respect to pulmonary function decline as a result of MDI (for a discussion of these studies, see IRIS summary sheet for MDI). However, it appears that adverse respiratory effects may be associated with exposure to MDI in humans; therefore, the choice of respiratory toxicity in animals as the critical effect is supported by available information in humans.

The RfC for MDI was calculated as follows.

Critical Effect:

Lesions of the nasal cavity (24-month Rat Inhalation Study, Reuzel et al. 1990)

NOAEL:

= 0.2 mg/m^3

NOAEL(ADJ):

= NOAEL × 6 hours/24 hours
× 5 days/7 days

= 0.2 mg/m^3 × 6/24 × 5/7

= 0.036 mg/m^3

NOAEL(HEC):

Calculated for a particle:respiratory effect in the extrathoracic region using

NOAEL(HEC) = NOAEL(ADJ) × RDDR(ET)

Where RDDR(ET) = 0.1460 for MMAD
= 0.068 μm
σ_g = 9.23, based on dosimetric modeling as described above

NOAEL(HEC) = 0.036 mg/m^3 × 0.1460
= 0.005 mg/m^3

LOAEL(HEC):

Using the same formula

LOAEL(HEC) = 0.024 mg/m^3.

RfC:

= NOAEL(HEC)/UF

Where UF = 100 (uncertainty factors of 10 for intraspecies variability, 3[1] for database deficiencies and 3 for interspecies extrapolation. The fact that MDI is not likely to enter the systemic circulation was used to reduce the magnitude of UF for data base)

RfC = 0.005 mg/m^3/100
= 0.0005 mg/m^3

CONFIDENCE:

Study: High
Data base: Medium
RfC: Medium

VERIFICATION:

December 1990

HEC Scenario Two. Particle:Remote Toxicity (Default)

In the case of toxic effects of particles observed outside the respiratory tract, the equivalent internal dose across species is assumed to be the mass of particles (mg) deposited per minute (min) per body weight unit (kg). The following equation is used to calculate the RDD expressed as mg/min per kg.

$$RDD = \frac{10^{-3} Y \dot{V}_E}{BW} \sum_{i=1}^{n} P_i E_i \qquad \text{(A-4)}$$

Where P_i = the particle mass fraction in the ith size range of an aerosol with a given particle diameter (MMAD) and geometric standard deviation (σ_g)

E_i = the deposition efficiency for that species and the total respiratory tract

Y = exposure level (mg/m^3)

$\dot{V}_E$ = minute volume (l/min)

BW = body weight (kg)

Until clearance and distribution parameters can be incorporated systematically, 100 percent of the deposited dose to the entire respiratory system (TOTAL) is assumed to be available for uptake to the systemic circulation. This assumption may result in less conservative HEC estimates than using retained dose and accounting for differential uptake from various respiratory tract regions, but is more accurate than using the exposure concentration. The ratio of the extrarespiratory RDD calculated for the experimental animal to that of the human (RDD$_{ER}$) then is used to calculate the HEC for extrarespiratory effects.

At the time of this writing in 1992, no RfC using the default for remote toxicity of particles had been derived.

---

[1] The logarithmic midpoint between 0 and 10; two of these multiplied together are assumed to equal 10.

HEC Scenario Three. Gas:Respiratory Effect (Default)

An analogous approach to that of HEC Scenario One can be used for gases that are reactive and have their effect in the respiratory system. The equivalent internal dose across species again is assumed to be the mass (mg) of toxic agent per minute (min) per surface area (cm^2) of the respiratory region of concern. Ventilatory parameters and regional respiratory surface areas are used to adjust dosimetrically for the species differences, as in the adjustment factor for HEC Scenario One, but the particle distribution and deposition efficiency integration term is dropped.

The regional gas dose (RGD) is calculated as

$$RGD = \frac{10^{-3} Y \dot{V}_E}{S} \quad \text{(A-5)}$$

Where Y = exposure level (mg/m^3)
  $\dot{V}_E$ = minute volume (l/min)
  S = regional surface area (cm^2)

It should be noted that this approach assumes that the entire inspired concentration goes to the region of concern, whereas not all inspired gas is necessarily deposited. For example, an alveolar ventilation rate would be appropriate to use with a strictly pulmonary effect. As in the case of the RDD for particles, the manifest toxic effect will dictate the RGD to calculate. In other words, the surface area used to normalize the internal dose should correspond with the region of observed toxicity. The regional gas dose ratio (RGDR) of the appropriate RGD values then is derived by

$$RGDR = \frac{(RGD)_A}{(RGD)_H} \quad \text{(A-6)}$$

Where $(RGD)_A$ = regional gas dose in experimental animal species of interest, adjusted for surface area and ventilatory volumes
  $(RGD)_H$ = regional gas dose in humans, adjusted for surface area and ventilatory volumes

The RGDR then is used to adjust dosimetrically the observed NOAEL to a human equivalent concentration gas exposure in the same fashion as the RDDR is used to adjust particle exposures (equation A-3).

Case Study: Bromomethane (CAS: 74-83-9)

The RfC developed for bromomethane serves as a good example of the integration of a large data base of information obtained from studies in experimental animals. Inspection of the health effects data base for bromomethane reveals that the major end points associated with exposure to this chemical in experimental animals are neurotoxicity, irritation of the upper respiratory tract, and cardiovascular effects. However, the degenerative and proliferative lesions of the olfactory epithelium of the nasal cavity observed in 29-month inhalation studies by Reuzel et al. (1987) and Reuzel et al. (1991) clearly yield the lowest LOAEL(HEC) and, thus, serve as the critical effect for the derivation of an RfC for bromomethane. The LOAEL identified in this study is supported by the effects seen in rats in a subchronic NTP (1990) study and mice in a chronic NTP (1990) study, as well as in subacute and subchronic studies in rats (Hastings 1990; Hurtt et al. 1987; Hurtt et al. 1988). The chronic NTP (1990) study in B6C3F1 mice supports the choice of the Reuzel et al. (1987 and 1991) study as the critical study with respect to effect and LOAEL, even though the LOAEL(HEC) for these effects in the NTP study was 13 mg/m^3, because a statistically significant increase in the incidence of nasal lesions was not observed in the critical study at 24 months at 3 ppm. The incidence of this lesion did not achieve statistical significance until the 29-month sacrifice in the critical study, and the NTP study was carried out only for 24 months. Therefore, it appears that bromomethane induces acute damage in the nasal cavity at low concentrations that is repaired and is not manifested again until the animals are exposed for longer than 24 months. This hypothesis is supported by the findings of Hurtt et al. (1988) and Hastings (1990). The data base is given a high confidence rating because there is a chronic inhalation study in two species supported by subchronic inhalation studies in several species, and because data are available on the developmental and reproductive effects of bromomethane as well as its pharmacokinetics following inhalation exposure.

The RfC for bromomethane was calculated as follows.

Critical Effect:
  Degenerative and proliferative lesions of the olfactory epithelium of the nasal cavity (29-month Rat Inhalation Study, Reuzel et al. (1987 and 1991))

NOAEL:
  None

LOAEL (mg/m^3):
  = (3 ppm $\times$ MW)/24.45
  Where: MW = 94.95

    LOAEL = (3 ppm $\times$ 94.95)/24.45
          = 12 mg/m^3

LOAEL(ADJ):
  = LOAEL $\times$ 6 hours/24 hours $\times$ 5 days/7 days
  = 12 mg/m^3 $\times$ 6/24 $\times$ 5/7
  = 2 mg/m^3

LOAEL(HEC):
  Calculated for a gas:respiratory effect in the extrathoracic region using

  LOAEL(HEC) = LOAEL(ADJ) $\times$ RGDR(ET)

  Where RGDR(ET) = (Mva/Sa)/(Mvh/Sh)
    Mva   = 0.30 m^3/day
    Mvh   = 20 m^3/day
    Sa(ET) = 11.6 cm^2
    Sh(ET) = 177 cm^2.
    RGDR(ET) = (0.42/11.6)/(20/177)
             = 0.23

    LOAEL(HEC) = 2 $\times$ 0.23 mg/m^3
               = 0.46 mg/m^3

RfC:
  = LOAEL(HEC)/UF

  Where: UF = 100 (the uncertainty factor of 100 reflects a factor of 10 for intraspecies variability, 3 for the use of a minimal LOAEL,

and 3 for interspecies extrapolation).

  RfC = 0.46 mg/m^3/100
      = 0.005 mg/m^3

CONFIDENCE:
  Study: Medium
  Data base: High
  RfC: Medium

VERIFICATION:
  December 1991

HEC Scenario Four. Gas:Remote Toxicity (Default)

As discussed in Chapter 7, physiologically based pharmacokinetic (PBPK) models have been applied successfully to estimating internal target tissue dose surrogates that have served as the basis of risk assessments. Although PBPK modeling is the procedure of choice for dose extrapolation and considered to be the optimal approach for risk assessment, such modeling is not possible without the values of physiologic and biochemical parameters used in the modeling process and an understanding of the agent's underlying biologic mechanisms of action. Unfortunately, these data generally are not available for most compounds. One of two methods is recommended, depending on the supporting pharmacokinetic data base.

Optimal Approach

The approach of choice for this scenario is to use published PBPK models to perform the interspecies and 'dose' extrapolations. For example, when extrapolating from a laboratory animal exposure that employed an intermittent regimen, the experimental animal exposure would be simulated with a PBPK model and the integrated average arterial (leaving the lung compartment in the model) concentration of the parent compound (or appropriate metric, such as a toxic metabolite, if known) calculated. Concentration in the target tissue of the appropriate dose surrogate also could be calculated if the dose-effect relationship is established. The human exposure, a constant 24-hour exposure as dictated by assumptions in the methodology, would then be simulated to find the exposure concentration that resulted in an equivalent arterial concentration (or other appropriate internal metric) to that of the laboratory animal. This exposure would be designated as the HEC. This type of extrapolation is illustrated in Figure A-1 with dichloromethane as the parent compound.

Default Approach

In order to be able to estimate HECs for gas exposures with remote toxicity effects in those cases for which the use of PBPK models is not possible owing to data gaps, a procedure has been developed to be used with compounds for which modeling would be applicable but for which some or all of the important parameters (e.g., partition coefficients and metabolic rate constants) are missing. The procedure was derived to estimate equivalent or more conservative values for HECs than would be obtained using a PBPK model based on Ramsey and Andersen (1984). A detailed description of the derivation is provided elsewhere (Overton and Jarabek 1989a; Overton and Jarabek 1989b).

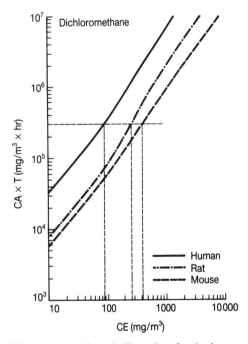

Figure A-1. Schematic illustration of optimal dosimetric extrapolation using PBPK model simulation. Source: Jarabek and Hasselblad (1991).

An important assumption in the approach, since the noncancer effects observed in chronic bioassays are the basis for the determination of HECs, is that 'lifetime' chronic animal exposure scenarios are equivalent to human 'lifetime' (i.e., 70 years) exposures. By definition of the RfC, the human exposure is continuous and constant. The animal exposure scenario is experiment-dependent, although usually intermittent and periodic (e.g., 6 hours/day, 5 days/week repeated over many weeks). NOAEL(HECs) are defined in terms of the requirement that the concentration of the inhaled compound in the pulmonary arterial blood (leaving the lung) of the human is no greater than the average of the arterial blood concentration leaving the lung of the laboratory species. In other words, it requires that the time integrated arterial blood concentration of the inhaled gas ($CA \times T$) of the human be equivalent to that of the laboratory animal for the same time length.

Because the specification of a biologically effective toxic dose to a given target tissue is not yet possible for a large number of compounds, the definition of a surrogate dose must be arbitrary. However, the toxic effects of some compounds are expected to be directly related to the inhaled parent compound; furthermore, $CA \times T$ is an internal dose and is closer to being the target dose than is the exposure concentration. Basing the effective dose extrapolation on another surrogate (e.g., metabolite) would require knowledge of the interspecies metabolism for the compound of interest and additional information about human and animal physiological parameters.

Another assumption is that the concentration of the inhaled compound within the animal achieved periodicity with respect to time. In other words, the internal concentration of the inhaled agent achieved a consistent pattern over the weeks of exposure. An illustration of periodicity is provided in Figure A-2. Periodicity of the arterial concentration of the agent was not achieved until the fifth week for the plotted theoretical exposure simulation.

Practically, periodicity conditions should be met during most of the time. For example, if

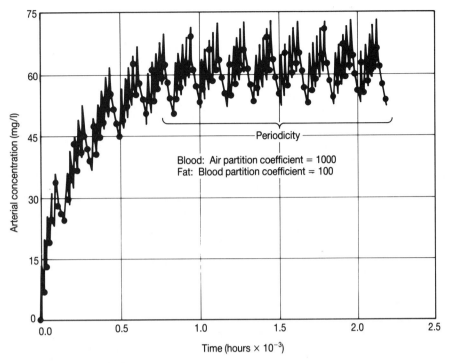

Figure A-2.   Time course of periodicity for 6 hours/day, 5 days/week exposure at 100 ppm in F-344 rats. Source: Jarabek et al. (1990).

this condition is met for nine-tenths of the exposure time (e.g., periodic during the last 90 weeks for a 100-week experiment), then estimates of average concentrations will be in error by less than 10 percent. Guidance on the relationship of the blood-to-air and fat-to-blood partition coefficients with respect to achieving periodicity of an inhaled agent in an animal species has been provided. The relationship is particularly significant when extrapolating from studies of different duration. In some short-term exposure cases, periodicity is not achieved and the criterion for the extrapolation procedure would be violated. This may have particular relevance to the development of risk assessment methods for short-term exposures.

Assuming the animal alveolar blood concentrations were periodic with respect to time for the majority of the experimental duration, the NOAEL(HEC) for extrarespiratory effects of gases is calculated as

$$NOAEL(HEC) = NOAEL(ADJ) \times \frac{\lambda_A}{\lambda_H}$$

(A-7)

Where NOAEL(HEC) = the NOAEL human equivalent concentration (mg/m^3)

NOAEL(ADJ) = the NOAEL adjusted for continuous exposure (e.g., adjusted for exposure regimen by h hours/24 hours and d days/7 days)

$\dfrac{\lambda_A}{\lambda_H}$ = the ratio of the blood to air partition coefficient of the chemical for the animal species to the human value, used only if $\lambda_A \leq \lambda_H$.

For the cases where $\lambda_A > \lambda_H$, model results have shown that the adjustment may not provide conservative estimates. The detailed derivation of boundary limits on λ is given elsewhere. For the situation in which $\lambda_A > \lambda_H$ and in the case where λ values are unknown, the default value of $\lambda_A/\lambda_H = 1$ is recommended.

Since blood-to-air partition coefficients are more readily available than are complete physiological parameter data (e.g., tissue partition coefficients and metabolic rate constants), the proposed method represents a reasonable alternate approach when PBPK modeling is not feasible.

Case Study: Ethylbenzene (CAS: 100-41-4)
The RfC for ethylbenzene is based on the occurrence of developmental toxicity in rats and rabbits following inhalation exposure to 1,000 ppm (Andrew et al. 1981; Hardin et al. 1981). In the cocritical studies, marginally adverse skeletal variants were seen at both 100 and 1,000 ppm in rats with the effects at 100 ppm being reduced compared with those occurring at 1,000 ppm. This RfC file is a good example of a situation where the critical effects identified in the co-critical studies are themselves marginally adverse, but are supported by the weight of evidence. A cluster of other mild effects (e.g., reduced litter size in rabbits) observed at 1,000 ppm, and an increase in the percentage of skeletal retarded fetuses observed at a higher concentration in rats in another study (Ungvary and Tatrai 1985) supports the choice of 1,000 ppm as the LOAEL. In addition, elevated maternal liver, kidney, and spleen weights were reported in the rats exposed to 1,000 ppm (Andrew et al. 1981).

The human data available on the health effects associated with exposure to this compound do not indicate any adverse effects; however, these studies are inadequate because of lack of quantitative exposure information and concurrent exposure to other substances. No chronic inhalation toxicity studies currently are available for ethylbenzene. No exposure-related toxic effects were reported in several subchronic inhalation studies conducted with ethylbenzene. Therefore, the constellation of mild

developmental effects observed in rats and rabbits is the best choice to serve as the basis for the ethylbenzene RfC.

The RfC for ethylbenzene was calculated as follows.

Critical Effect:
Developmental toxicity (Rat and Rabbit Inhalation Developmental Studies, Andrew et al. (1981) and Hardin et al. (1981))

NOAEL:
= 100 ppm

NOAEL (mg/m³):
= (100 ppm × MW)/24.45

Where: MW = 106.18

$$NOAEL = (100 \text{ ppm} \times 106.18)/24.45$$
$$= 434 \text{ mg/m}^3$$

NOAEL(ADJ):
= 434 mg/m³ (The experimental concentration is not duration adjusted for developmental toxicity studies).

NOAEL(HEC):
Calculated for a gas:extrarespiratory effect assuming periodicity was attained using

$$NOAEL(HEC) = NOAEL(ADJ) \times (\lambda_A/\lambda_H):$$

Where: λ_A/λ_H is the ratio of the blood-to-air partition coefficient for ethylbenzene in rats and the blood-to-air partition coefficient for ethylbenzene in humans. Since λ_A and λ_H are not known, a default value of 1.0 was used for this ratio.

$$NOAEL(HEC) = NOAEL(ADJ) \times 1$$
$$= 434 \text{ mg/m}^3$$

LOAEL (mg/m³):
= (1,000 ppm × 106.18)/24.45
= 4,340 mg/m³

LOAEL(ADJ):
= 4,340 mg/m³

LOAEL(HEC):
= 4,340 mg/m³

RfC:
  = NOAEL(HEC)/UF

Where UF = 300 (the uncertainty factor of 300 reflects a factor of 10 for intraspecies variability, 3 for interspecies extrapolation, and 10 to adjust data base deficiencies including the absence of multigeneration reproductive and chronic studies).

$$RfC = 434 \, mg/m^3/300$$
$$= 1 \, mg/m^3$$

CONFIDENCE:
  Study: Low
  Data base: Low
  RfC: Low

VERIFICATION:
  December 1990

HEC Scenario Five. Human Occupational

When human data are available and adequate to derive an RfC, adjustments are usually required to account for differences in exposure scenarios (e.g., extrapolation from an 8-hour per day occupational exposure to a continuous chronic exposure). The optimal approach again is to use a biologically motivated mathematical or PBPK model. An occupational exposure can be extrapolated in the same fashion as described to extrapolate intermittent exposure regimens from experimental laboratory animals, using particle deposition or PBPK models with human exertion (work) ventilation rates and exposure durations appropriate to the occupational setting.

In the event that a PBPK model or required physicochemical (e.g., σ_g, MMAD) and physiological parameters are not available, the default approach for human exposure scenarios is to adjust by the default occupational ventilation rate and for the intermittent workweek schedule, as

$$NOAEL(HEC) = NOAEL \times (MVho/MVh)$$
$$\times 5 \, days/7 \, days$$

(A-8)

Where: MVho = human occupational default minute volume ($10 \, m^3$/days)
  MVH = default ambient human minute volume ($20 \, m^3$/day)
  NOAEL is in mg/m^3

Case Study: Inorganic Mercury (CAS: 7439-97-6)

The data base for inorganic mercury is unusual in that it contains a substantial amount of human data. Analysis of the data base clearly shows that the mercury levels reported to be associated with preclinical and symptomatic neurological dysfunction generally are lower than those found to affect kidney and pulmonary function. Therefore, the RfC for inorganic mercury was based on four human occupational inhalation studies that demonstrate a constellation of neurological effects (hand tremor, increases in memory disturbances, slight subjective and objective evidence of autonomic dysfunction) occurring at comparable exposure levels. The LOAEL values derived from these studies can be corroborated by other human epidemiologic studies. Furthermore, the adverse effects reported in these studies are in accord with the well-documented effects of mercury poisoning.

This file provides a good example of evaluating and interpreting the data base as a whole to develop an RfC when the individual studies alone are not adequate to serve as the basis for an RfC because of methodological limitations. For example, little detail was presented concerning the measurement of the exposure levels in one of the cocritical studies (Fawer et al. 1983), and it is likely that there were variations in the mercury air levels during the period of exposure. Furthermore, the tremors may have resulted from intermittent exposure to concentrations higher than the time-weighted average (TWA). Air mercury levels in the other three cocritical studies (Piikivi and Hanninen 1989; Piikivi and Toulonen 1989; Piikivi 1989) were extrapolated from blood levels ($\mu g/l$) based on a conversion factor of 1:4.5 (air:blood) provided by Roels et al. (1987). Together, and with support from numerous other occupational studies, they clearly demonstrate the occurrence of neurological effects from exposure to inorganic mercury vapor at lower levels than those required to produce other adverse effects (i.e., renal and pulmonary).

The RfC for inorganic mercury was calculated as follows.

Critical Effect:

Hand tremor; increases in memory disturbances; slight subjective and objective evidence of autonomic dysfunction. Human occupational studies (Fawer et al. (1987); Piikivi and Hanninen (1989); Piikivi (1989); Piikivi and Toulonen (1989))[2]

NOAEL:
None

LOAEL (Occup.)
= 0.025 mg/m^3

LOAEL (Comm.)
= LOAEL (Occup.) $\times$ conversion factors[3]
= 0.025 $\times$ 10/20 $\times$ 5/7
= 0.009 mg/m^3

RfC:
= LOAEL/UF

Where UF = 30 (an uncertainty factor of 10 was used for intraspecies variability and for the use of a LOAEL together, and a factor of 3 applied to account for the inadequate data base on developmental and reproductive effects).

RfC = 0.009 mg/m^3/30
= 0.0003 mg/m^3

CONFIDENCE:
Study: Medium
Data base: Medium
RfC: Medium

VERIFICATION:
April 1990

---

[2] Air mercury levels in the Piikivi (1989), Piikivi and Hanninen (1989), Piikivi and Toulenen (1989), studies were extrapolated from blood levels (μg/l) based on a conversion factor of 1:4.5 (air to blood) provided by Roels et al. (1987).

[3] Conversion factors: occupational exposure (ventilatory volume of 10 m^3/8 hour for 5 days/week) converted to general population exposure (ventilatory volume of 20 m^3/24 hour for 7 days/week).

References

Andrew, F.D., R.L. Buschbom, W.C. Cannon, R.A. Miller, L.F. Montgomery, D.W. Phelps, et al. 1981. Teratologic assessment of ethylbenzene and 2-ethyoxyethanol. PB-83-208074. Battelle Pacific Northwest Laboratory. Richland, WA.

Fawer, R.F., U. DeRibaupierre, M.P. Guillemin, M. Berode, and M. Lobe. 1983. Measurement of hand tremor induced by industrial exposure to metallic mercury. *Journal of Industrial Medicine* 40:204–208.

Hardin, B.D., G.P.L. Bond, M.R. Sikov, F.D. Andrew, R.P. Beliles, and R.W. Niemeier. 1981. Testing of selected workplace chemicals for teratogenic potential. *Scandanavian Journal of Work, Environment and Health* 7(suppl 4):66–75.

Hastings, L. 1990. Sensory neurotoxicology: Use of the olfactory system in the assessment of toxicity. *Neurotoxicology and Teratology* 12:455–459.

Hurtt, M.E., K.T. Morgan, and P.T. Working. 1987. Histopathology of acute toxic responses in selected tissues from rats exposed by inhalation to methyl bromide. *Fundamental and Applied Toxicology* 9:352–365.

Hurtt, M.E., D.A. Thomas, P.K. Working, T.M. Monticello, and K.T. Morgan. 1988. Degeneration and regeneration of the olfactory epithelium following inhalation exposure to methyl bromide: Pathology, cell kinetics, and olfactory function. *Toxicology and Applied Pharmacology* 94:311–328.

Jarabek, A.M. and V. Hasselblad. 1991. Inhalation reference concentration methodology: Impact of dosimetric adjustments and future directions using the confidence profile method. Paper No. 91-173.3. Presented at the 84th annual meeting of the Air Waste Management Association. Vancouver, B.C.

Jarabek, A.M., M.G. Menache, M.L. Dourson, J.H. Overton, Jr., and F.J. Miller. 1989. Inhalation reference dose (RfD): An application of interspecies dosimetry modeling for risk assessment of insoluble particles. *Health Physics* 57(1): 177–183.

Jarabek, A.M., M.G. Menache, M.L. Dourson, J.H. Overton, Jr., and F.J. Miller. 1990. The U.S. Environmental Protection Agency's inhalation

RfD methodology: Risk assessment for air toxics. *Toxicology and Industrial Health* 6(5):279–301.

National Toxicology Program (NTP). 1990. Toxicology and carcinogenesis studies of methyl bromide (CAS No. 74-83-9) in B6C3F1 mice (inhalation studies). NTP TR 385, NIH Publication No. 91-2840. Peer Review Draft.

Overton, J.H. and A.M. Jarabek. 1989a. Estimating human equivalent No-Observed-Adverse-Effect-Levels: A comparison of several methods. *Experimental and Toxicologic Pathology* 37:89–94.

Overton, J.H. and A.M. Jarabek. 1989b. Estimating human equivalent No-Observed-Adverse-Effect-Levels for VOCs based on minimal knowledge of physiological parameters. Paper No. 89-91-8. Presented at the 82nd Annual Meeting of the Air & Waste Management Association. Anaheim, CA.

Piikivi, L. 1989. Cardiovascular reflexes and low long-term exposure to mercury vapor. *International Archives of Occupational and Environmental Health* 61:391–395.

Piikivi, L. and L. Hanninen. 1989. Subjective symptoms and psychological performance of chlorine-alkali workers. *Scandanavian Journal of Work, Environment and Health* 15:69–74.

Piikivi, L. and U. Toulonen. 1989. EEG findings in chlor-alkali workers subjected to low long term exposure to mercury vapor. *British Journal of Industrial Medicine* 46:370–375.

Ramsey, J.C. and M.E. Andersen. 1984. A physiologically based description of the inhalation pharmacokinetics of styrene in rats and humans. *Toxicology and Applied Pharmacology* 73:159–175.

Reuzel, P.G.J., C.F. Kuper, L.M. Appelman, A.W.J. de Jong, and J.P. Bruyntjes. 1985a. Sub-chronic (13-week) inhalation toxicity of polymeric MDI aerosol in rats (Part B1). Final Report. Prepared by Civo Institute for the International Isocyanate Institute.

Reuzel, P.G.J., C.F. Kuper, L.M. Appelman, and R.N. Hooftman. 1985b. Sub-chronic (13-week) inhala-

tion toxicity of polymeric MDI aerosol in rats (Part B2). Final Report. Prepared by Civo Institute for the International Isocyanate Institute.

Reuzel, P.G.J., C.F. Kuper, H.C. Dreef-van Der Meulen, and V.M.H. Hollanders. 1987. Chronic (29-month) inhalation toxicity and carcinogenicity study of methyl bromide in rats. Report No. V86.469/221044. Netherlands Organization for Applied Scientific Research, Division for Nutrition and Food Research, TNO. EPA/OTS Document #86-8700001202.

Reuzel, P.G.J., J.H.E. Arts, M.H.M. Kuypers, and C.F. Kuper. 1990. Chronic toxicity/carcinogenicity inhalation study of polymeric methylene diphenyl diisocyanate aerosol in rats. Final Report. Prepared by Civo Institute for International Isocyanate Institute.

Reuzel, P.G.J., C.F. Dreef-van Der Meulen, V.M.H. Hollanders, C.F. Kuper, V.J. Feron, and C.A. van Der Heijden. 1991. Chronic inhalation toxicity and carcinogenicity study of methyl bromide in Wistar rats. *Food and Chemical Toxicology* 29(1): 31–39.

Roels, H., S. Abdeladim, E. Ceulemans, and R. Lauwreys. 1987. Relationships between the concentrations of mercury in air and in blood or urine in workers exposed to mercury vapour. *Annals of Occupational Hygiene* 31:135–145.

Ungvary, G. and E. Tatrai. 1985. On the embryotoxic effects of benzene and its alkyl derivatives in mice, rats, and rabbits. *Archives of Toxicology* Suppl. 8:425–430.

U.S. Environmental Protection Agency (EPA). 1990k. Science Advisory Board's comments on the use of uncertainty and modifying factors in establishing reference dose levels. EPA/SAB/EHC/90-005. Letter from L.R. Loehr and A. Upton to W. Reilly. January 17, 1990.

Yu, C.P. and K.J. Yoon. 1990. Retention model of diesel exhaust particles in rats and humans. Amherst, NY: Health Effect Institute, Research Report No. 40.

B

Supplementary Procedures for Obtaining Fugitive Emission Estimates

Thomas A. Kittleman, Paul R. Jann, and
Seshasayi Dharmavaram

Appendix B1

Determining the Significance of an Observed Difference Between Actual and Assumed Process Average Screening Values

This procedure might be useful when evaluating the first assumption underlying the leak/no leak option. That assumption is that the average process screening values for components that screen above and below 10,000 ppm are as represented in Table 22-3.

A large sample (i.e., $n_i \geq 30$) approach to this problem is simply to construct a confidence interval for each group's average, then see if the interval includes the value assumed by the leak/no leak approach. The confidence intervals can be calculated using

$$\overline{X}_i \pm t(n_i - 1, 0.975) \times \frac{s_i}{\sqrt{n_i}} \quad \text{(B1-1)}$$

Where i = leak, no leak
$\overline{X}$ = the average screening value for group i

s_i = the standard deviation of the screening values for group i
n_i = the number of components in group i

The associated Student's t value should be obtained from an appropriate table. If no table is immediately available, then a value of $t = 2$ generally will suffice.

It is important to note that these values should include those components screening below the lower limit of detection (LLD) for the no leak group (i.e., setting them equal to the LLD). Those components screening at the saturation point for the leak group should be included in the leak group (i.e., setting them equal to whatever their corrected-for-response factor value gives).

Procedures for small sample sizes do exist. However, discussion of these is beyond the scope of this document. If there is sufficient need, the user may seek assistance for these situations, but generally sophisticated analyses are not required to assess the significance of the difference (Bazovsky 1961; Mood, Graybill, and Boes 1974; Patterson 1966; Searle 1971).

Appendix B2

Decision Rules for Validating Established Leak Rate/Screening Value Curves

The decision rule for validating the use of established curves relies on the Wilcoxon and slope tests. These tests require an underlying distribution that is basically symmetrical. Hence, all evaluation is done in the log-log space (Bazovsky 1961; Mood, Graybill, and Boes 1974; Searle 1971). If neither the Wilcoxon or slope test show the established curve to be different from the data, then the process-specific data should be combined with the data used to establish the existing curves (see Appendix B4 below).

WILCOXON TEST

The basic procedure is to calculate the difference between the log (leak rate) determined by bagging and that determined by the established curve. These values then are ranked in order of absolute magnitude (i.e., irrespective of sign) and ranks associated with the positive differences (i.e., bagging > curve) are summed.

If the number of baggings is 4, and all the bagging results lie above the line, then the value of the Wilcoxon statistic will be

$$W = (4 + 3 + 2 + 1) = 10 \qquad \text{(B2-1)}$$

The probability of observing such a value if the line truly represents the process is approximately 6.25 percent. Hence, the inference drawn is that the line does not adequately represent the process and its use may lead to underestimation of emissions. Any other patterns could occur with much higher probabilities (except all falling below the line, i.e., $W = 0$, where the inference would be that use of the curve could lead to overestimation of emissions). When data fall on both sides of the line, the line cannot be rejected without more data. Similar critical values for up to 12 baggings are shown in Table B2-1.

Table B2-1   Critical Wilcoxon values for up to 12 baggings

Number of bags	$W <$ value (curve will overestimate)	$W >$ value (curve will underestimate)
4	0	10
5	1	14
6	2	19
7	4	24
8	6	30
9	8	37
10	11	44
11	14	52
12	18	60

For sample sizes larger than 12, a normal approximation to the Wilcoxon has been shown to be more than adequate. This approximation leads to the following decision procedure.

$$Z = \frac{W - \dfrac{n(n + 1)}{2}}{\sqrt{\dfrac{n(n + 1)(2n + 1)}{24}}} \qquad \text{(B2-2)}$$

Where n is the number of baggings
    W is calculated as above

■ If $Z < -1.645$, then the curve will underestimate for the process under study
■ If $Z > 1.645$, then the curve will overestimate for the process under study
■ If $-1.645 \le z \le 1.645$, then the data do not provide enough evidence to indicate that the curve is inadequate to represent the process under study

SLOPE TEST

If the Wilcoxon test indicates that the existing curve is consistent with process specific data, then a test for slope also can be made to insure adequacy of the curve. To do this the user must first determine b, g, and MSE using the methods in Appendix B3. Continuing with the notation in this section, calculate.

$$\sum (X_i - \overline{X})^2 \qquad \text{(B2-3)}$$

The sample variance of b is

$$S^2(b) = \frac{MSE}{\sum (x_i - X)^2} \qquad \text{(B2-4)}$$

with a standard error

$$S(b) = \sqrt{S^2(b)} \qquad \text{(B2-5)}$$

The test computes

$$t^* = \frac{(b - b_{\text{est}})}{S(b)} \qquad \text{(B2-6)}$$

Where b_{est} = the slope of the established curve

If the absolute value of $t^* > t(0.975, n-2)$, then the slope of the curve is significantly different from the process under study. In this case, a process-specific curve should be fit to the data.

Appendix B3

Least-Squares Estimation of a Linear Relationship

The fitting of a line to describe the general relationship between two variables (X and Y) via the least-squares algorithm involves the determination of Y-intercept, a, and slope coefficient, b. These parameters are estimated so that the average residual, shown as

$$r_i = Y_i - a - bX_i, \; i + 1, \ldots, n \qquad \text{(B3-1)}$$

is zero and the sum of the residuals squared, MSE, is minimized.

Let $Y_i = \log_{10}$ (leak rate determined by bagging for component i), and let $X_i = \log_{10}$ (maximum screening value, at time of bagging, for component i). Then the least-squares estimators can be given by

$$b = \frac{(\overline{XY}) - (\overline{X})(\overline{Y})}{(\overline{X^2}) - (\overline{X})^2}, \; a = Y - bX \qquad \text{(B3-2)}$$

Where

$$\overline{A} = \frac{1}{n} \sum_{i=1}^{n} A_i, \text{ with } A_i = X_i, Y_i, X_iY_i, X_i^2 \qquad \text{(B3-3)}$$

as appropriate. Once these have been calculated, then the mean square error, MSE, can be calculated using

$$MSE = \frac{1}{n-2} \sum_{i=1}^{n} r_i^2, \text{ with } r_i = Y_i - a - bX_i \qquad \text{(B3-4)}$$

In order to transform the results from the log-log space back to arithmetic scales, a scale bias correction factor, $SBCF$, is required. This value can be obtained by summing a sufficient number (generally less than 15) of terms of an infinite series. More specifically, the $SBCF$ can be estimated by

$$g = g(\ln_{10})^2 \frac{MSE}{2} \qquad \text{(B3-5)}$$

Where

$$g(t) = 1 + \frac{(m-1)t}{m} + \frac{(m-1)^3 t^2}{m^2 2!(m+1)}$$
$$+ \frac{(m-1)^5 t^3}{m^3 3!(m+1)(m+3)} + \cdots \qquad \text{(B3-6)}$$

$$m = (\text{number of sources bagged} - 1) = n - 1$$

$$t = \frac{2.65}{MSE}$$

The a, b, and g are the same as used elsewhere in this section. The equation in arithmetic scale is

$$\text{Leak rate} = g \times 10^a \times (\text{screening value})^b$$

$$\text{(B3-7)}$$

Appendix B4

Combining Available Bagging Data with That Used to Develop the Established Leak Rate/Screening Value Curves

Retaining the notation of Appendix B3, the formulas for modifying the curve parameters to include the additional data collected for the process under study are

$$
b = \left[\frac{D + \sum_{i+1}^{n} X_i Y_i - \dfrac{1}{A+N} \times \left(B + \sum_{i=1}^{n} X_i \right)\left(C + \sum_{i=1}^{n} Y_i \right)}{E + \sum_{i=1}^{n} X_i^2 - \dfrac{1}{A+n} \times \left(B + \sum_{i=1}^{n} X_i \right)^2} \right]
$$

(B4-1)

$$
a = \frac{1}{A+n}\left[C + \sum_{i=1}^{n} Y_i - \left(B + \sum_{i=1}^{n} X_i \right) \times b \right]
$$

(B4-2)

$$
MSE = \frac{1}{A+n-2}\left\{ F + \sum_{i=1}^{n} Y_i^2 - \frac{1}{A+n} \right.
$$
$$
\times \left(C + \sum_{i=1}^{n} Y_i \right)^2 - b^2 \left[E + \sum_{i=1}^{n} X_i^2 \right.
$$
$$
\left. \left. - \frac{1}{A+n}\left(B + \sum_{i=1}^{n} X_i \right)^2 \right] \right\}
$$

(B4-3)

where the constants A to F are given in Table B4-1 (p. 546), and the $SBCF$ or g can be obtained from the MSE as in Appendix B3, using $m = A + n - 1$. Note the formulas for a and MSE above each contain the parameter b. The proper b to use in these formulas is the result from the first formula above.

Appendix B5

Validating the Assumptions of the Least-Squares Analysis

Most of the major assumptions underlying a least-squares analysis can be validated by simple plotting procedures. To assess the assumption that the relationship between two variables can be described adequately by a line, the user should plot the residuals $r_i = Y_i - a - bX_i$, r_i is residual i (see Appendix B3 for more explanation) versus their corresponding X values. If a line is not sufficient to describe the relationship, then this plot should show some curvature (Finney 1941; Patterson 1966). Otherwise it should appear to be simply random noise (i.e., no pattern whatsoever). If a simple line does not appear to be an adequate way to describe the relationship, then higher-order terms (e.g., quadratic) could be incorporated into the curve.

To assess the assumption of homogeneous variance, the user could plot the residuals versus their corresponding predicted values, r_i versus $a + bX_i$. If variances are heterogeneous then this plot should show some type of wedged shape. Otherwise, as before, it should appear to be simply random noise. If the variance did appear to be heterogeneous, then an appropriate weighting of the data could generate the desired homogeneity.

The assessment of the assumption of non-stochastic X values is more difficult. However, since a component that is bagged will generally have been screened both before and after bagging, the differences between these two screening values can be assessed to obtain some general idea about the magnitude of variability in the X values that is within component versus

Table B4-1   Constants for combining bagging data with existing data

Source/service type	Constants					
	A	B	C	D	E	F
Valves-gas vapor	99	399.86	−252.55	−963.26	1696.93	733.33
Valves-light liquid	129	501.87	−324.24	−1202.95	2076.92	945.80
Pumps-light liquid	52	184.71	−111.81	−359.66	697.85	295.24

the amount of variability that is between components. The user would most likely want to obtain professional assistance for this analysis. However, if it is found that the assumption of nonstochastic X values is seriously violated, then errors-in-variables regression techniques (as opposed to least-squares techniques) can be utilized.

The important point to be made is that there are a number of situations in which standard least-squares techniques may not generate adequate estimates. However, these situations generally can be detected. Any apparent dilemmas generally can be overcome with more sophisticated analysis procedures. We assume these tests were performed on data used for the established curves. Subsequent data have not given us reason to question the appropriateness of this approach. The user may find it preferable to seek professional assistance if the conditions for a least-squares curve are not met.

Appendix B6

Determining an Appropriate Sample Size

In estimating emissions for a given process unit, all equipment components must be surveyed for each class of components. For example, all valves in gas/vapor service must be screened to establish the number of components in each of the given ranges of screening values. The one exception to this 'total component screening' criterion is flanges and other connectors. In typical process units, flanges and other connectors represent the largest count of individual

equipment components, making it costly to screen all components. The purpose of this appendix is to present a methodology for determining how many flanges or other connectors must be screened to constitute a large enough sample size to identify the actual screening value distribution of flanges and other connectors in the entire process unit.

The basis for selecting the sample population to be screened is the probability that at least one 'leaking' flange will be in the screened population. The 'leaker' is used as a representation of the complete distribution of screening values for the entire class of sources. The following binomial distribution was developed to approximate the number of flanges and other connectors that must be screened to ensure that the entire distribution of screening values for these components is represented in the sample.

$$n \geq N \left[1 - \frac{1 - p}{D} \right] \quad \text{(B6-1)}$$

Where D may be taken as (fraction of leaking flanges) $\times$ N, and $p \geq 0.95$

Currently available data gathered for flanges in VOC service in chemical process units show an estimated 2.09 percent flanges leaking. EPA recommends starting with the factor 0.0209 for the fraction of leaking flanges, since the actual leak frequency for flanges in a chemical process unit probably will not be known before the selection of a sample size for screening. A larger sample size will be required for units exhibiting a lower fraction of leaking flanges and other connectors. If a lower leak frequency for a process unit is known or can be estimated, it may be substituted into the equation above in place of 0.0209.

After 'n' flanges have been screened, an actual leak frequency should be calculated as

$$\frac{\text{Leaking}}{\text{frequency}} = \frac{\text{number of leaking flanges}}{n}$$

(B6-2)

Then, the confidence level of the sample size can be calculated using the following equation, based on the hypergeometric distribution.

$$p = 1 - \frac{(N - D')!(N - n)!}{N!(N - D' - n)!}$$ (B6-3)

Where N = total population of flanges and other connectors i

$\quad n$ = sample size

$\quad D'$ = number of leaking flanges and connectors $\times$ N/n

If 'p' calculated in this manner is less than 0.95, then a less than 95 percent confidence exists that the screening value distribution has been properly identified. Therefore, additional flanges/connectors must be screened to achieve a 95 percent confidence level. The number of additional flanges required to satisfy the requirement for a 95 percent confidence level can be calculated by solving the first equation again, using the leak frequency calculated in the second equation and subtracting the original sample size. After this additional number of flanges/connectors have been screened, the revised fraction of leaking components and the confidence level of the new sample size (i.e., the original sample size plus the additional flanges/connectors screened) should be recalculated using the third equation. EPA requires sufficient screening to achieve a 95 percent confidence level, until a maximum of 50 percent of the total number of flanges and other connectors in the process unit have been screened. EPA believes that 50 percent of the total flange/connector population is a reasonable upper limit for a sample size. If half of the total number of flanges and other connectors are screened, no further flange screening is necessary, even if a 95 percent confidence level has not been achieved.

Appendix B7

Handling of Zero Screening Sources

In the case of no bagging data, the default zeros of Table 22-11 should be used. When bagging data exist, the initial step is to adjust for background by subtracting the background bag from the source bag, and calculating

$$X_i = LR_{S,i} - LR_{B,i}$$ (B7-1)

Where $LR_{S,i}$ = leak rate from source bag i

$\quad LR_{B,i}$ = leak rate from background bag i

It has been observed that a linear function in log-log space well describes the leak rate/screening value relationship. To compare with the curve established default zeroes, we use the necessary logarithmic conversions. Thus, we take the logarithm of the data and add a constant if necessary to avoid taking the logarithm of a nonpositive X_i. The following procedure is used.

CASE 1

If all of the data are positive,

$$Y_i = \log (X_i)$$ (B7-2)

CASE 2

If all of the data are nonnegative with at least one zero,

$$Y_i = \log (X_i - C)$$ (B7-3)

Where C = largest power of 10 less that the smallest nonzero value of x (e.g., if smallest nonzero is .0008, then C = .0001)

CASE 3

If some of the data are negative, the smallest value (K_1) is first subtracted from each observation in the data set (note K_1 is a negative number). This results in all nonnegatives, with at least one zero value. Then K_2 is added to each observation in the set, where K_2 = largest power of 10 less than the smallest nonzero value of the updated nonnegative data set. Hence, the logarithmic expression used is

$$Y_i = \log(X_i + C) \qquad \text{(B7-4)}$$

Where $C = K_2 - K_1$

The average, variance, and scale bias correction factor $SBCF$ are calculated for the Y_i as

$$\bar{Y} = \frac{1}{n} \sum_{i=1}^{n} Y_i \qquad \text{(B7-5)}$$

$$S^2 = \frac{1}{n-1} \sum_{i=1}^{n} (Y_i - \bar{Y})^2 \qquad \text{(B7-6)}$$

$$SBCF = 1 + \frac{n-1}{n} t + \frac{(n-1)^3 t^2}{n^2!(n+1)}$$

$$+ \frac{(n-1)^5 t^3}{n^3!(n+1)(n+3)} + \cdots \qquad \text{(B7-7)}$$

$$t = (\ln 10)^2 \frac{s^2}{2} = 2.651 S^2 \qquad \text{(B7-8)}$$

Approximately 10 to 15 terms are required to obtain accuracy of 10^{-1}. The SBCF is necessary to convert log-log scales back to arithmetic scales.

The estimate of emissions for components screening below the LLD (E_{1*}) is then

$$E_{1*} = \text{Max}[(10^{\bar{Y}} SBCF) - C, 0] \qquad \text{(B7-9)}$$

Where $C = 0$ for Case 1, and previously stated for Case 2 and Case 3.

A statistical test is used to compare the calculated E_{1*} with the established 'default zero.' The results of this test determine which value (E_{1*} or 'default zero') is used to represent emissions for components screening below the LLD (E_1). The test involves two cases.

Case A

If $E_{1*} <$ 'default zero,' then use E_{1*} if $E_{ul} <$ 'default zero.'
Where

$$E_{ul} = \left[10 \exp\left(\bar{Y} + t(0.95, n-1) \sqrt{S^2/n} \right) \right. $$
$$\left. \times SBCF \right] - C$$
$$\text{(B7-10)}$$

and represents the 95 percent upper-confidence limit.

Use 'default zero' if $E_{ul} \geq$ 'default zero.'

Case B

If $E_{1*} >$ 'default zero,' then use E_{1*} if $E_{ll} >$ 'default zero.'

Where

$$E_{ll} = \left[10 \exp\left(\bar{Y} - t(0.95, n-1) \sqrt{S^2/n} \right) \right. $$
$$\left. \times SBCF \right] - C$$
$$\text{(B7-11)}$$

and represents the 95 percent lower-confidence limit.

Use 'default zero' if $E_{ll} \leq$ 'default zero.'

A confidence interval for E_{1*} can be calculated as

$$\left[10 \exp\left(\bar{Y} - t(0.95, n-1) \sqrt{S^2/n} \right) \times SBCF \right] - C$$
$$\text{(B7-12)}$$

$$\left[10 \exp\left(\bar{Y} + t(0.95, n-1) \sqrt{S^2/n} \right) \times SBCF \right] - C$$
$$\text{(B7-13)}$$

This interval should only involve the nonnegative region. In the case where the above computation includes negative values, it should be truncated to zero.

References

Bazovsky, I. 1961. *Reliability Theory and Practice.* Englewood Cliffs. NJ: Prentice-Hall.

Finney, D.J. 1941. On the distribution of a variate whose logarithm is normally distributed. *Journal of the Royal Statistical Society.* Series B(7):155–161.

Mood, A.M., F.A. Graybill, and D.C. Boes. 1974. *Introduction to the Theory of Statistics*, Third Edition. New York: McGraw-Hill.

Patterson, R.L. 1966. Difficulties involved in the estimation of a population mean using transformed sample data. *Technometrics* 8(3):535–537.

Searle, S.R. 1971. *Linear Models.* New York: John Wiley & Sons.

C

Response Factors

Thomas A. Kittleman, Paul R. Jann, and
Seshasayi Dharmavaram

DESCRIPTION

Response factors are used to convert an instrument reading into a concentration of the chemical that the instrument is sensing. For a given method of detection (e.g., flame ionization detection (FID) or photoionization detection (PID)), response factors can vary greatly from one chemical to another. For instance, an FID calibrated on methane (CH_4) (i.e., response factor set to 1.0 for methane) will not detect formaldehyde (CH_2O). Also, response factors can vary greatly for different concentrations of the same chemical. For aliphatic hydrocarbon compounds, the concentration variations in FID response factors generally are not great. For many compounds, however, there is no reason to believe that response factors obtained at one concentration (i.e., 10,000 ppm) will approximate response factors at a widely different (i.e., 500 ppm) concentration.

Response factors for many compounds can be found in the literature (Brown et al. 1980; Bursey et al. 1991). When no appropriate response factor can be found, it may be necessary to determine one experimentally. This is done by obtaining a known concentration and using the instrument to get the equivalent reading. If a response factor must be determined experimentally it frequently is desirable to

evaluate three concentrations likely to span the region of interest. If only the response factor at a specific leak definition is of interest, only that concentration would have to be measured. However, if screening data are to be used for emission estimation, multiple concentrations should be checked.

In the past there has not been uniformity in the way manufacturers have expressed and used response factors. Response factor can be defined in either of two ways. In the first method

$$\text{Response factor (RF)} = \frac{\text{actual concentrations (AC)}}{\text{instrument reading (IR)}}$$

(C-1)

Where the correction is AC = RF $\times$ IR.

In the second method

$$\text{Response factor (RF)} = \frac{\text{instrument reading (IR)}}{\text{actual concentration (AC)}}$$

(C-2)

Where the correction is AC = IR/RF.

550

**Table C-1   Response factors for mixtures –
Bacharach sampler**

Mixture designation[a]	Bacharach sampler	
	Either calculation method	Actual measured
A	2.53	2.4
B	1.97	1.92
C	1.66	1.56
D	1.52	1.49

[a] A = 500 ppm 1,1,1-trichloroethane, 125 ppm hexane.
  B = 100 ppm hexane, 200 ppm 1,1,1-trichloroethane, 200 ppm 1,3-dichlorobutane.
  C = 200 ppm hexane, 200 ppm 1,1,1-trichloroethane, 100 ppm 1,3-dichlorobutane.
  D = 200 ppm hexane, 100 ppm 1,1,1-trichloroethane, 200 ppm 1,3-dichlorobutane.

Source: Unpublished data collected by Dow Chemical Company.

**Table C-2   Response factors for mixtures – TE 580
sampler**

Mixture designation[a]	TE 580 sampler	
	Either calculation method	Actual measured
A	2.619	2.60
B	2.619	2.58
C	1.86	1.81
D	1.86	1.96

[a] A = 500 ppm 1,1,1-trichloroethane, 125 ppm hexane.
  B = 100 ppm hexane, 200 ppm 1,1,1-trichloroethane, 200 ppm 1,3-dichlorobutane.
  C = 200 ppm hexane, 200 ppm 1,1,1-trichloroethane, 100 ppm 1,3-dichlorobutane.
  D = 200 ppm hexane, 100 ppm 1,1,1-trichloroethane, 200 ppm 1,3-dichlorobutane.

Source: Unpublished data collected by Dow Chemical Company.

The response factors for mixtures can be calculated from the response factors of the individual chemicals in the mixture. For this calculation, the response factors are assumed to be a molar property. The response factors for mixtures can be determined using the equation appropriate for that response factor definition. For example, if AC = RF × IR, then

$$\frac{1}{\text{mixture RF}} = \frac{\text{mole fraction}_1}{RF_1}$$

$$+ \frac{\text{mole fraction}_2}{RF_2}$$

$$+ \cdots + \frac{\text{mole fraction}_n}{RF_n}$$

$$(C\text{-}3)$$

Where 1, 2, . . . , n are individual chemicals in the mixture.

Also, if AC = IR/RF, then

Mixture RF = $(\text{mole fraction}_1)(RF_1)$

$+ (\text{mole fraction}_2)(RF_2)$

$+ \cdots + (\text{mole fraction}_n)(RF_n)$

$$(C\text{-}4)$$

This approach has been demonstrated experimentally to give good results as shown by the comparison in Tables C-1 and C-2 for two different types of instruments.

References
Brown, G.E. et al. (Radian Corp.) 1980. Response factors of VOC analyzers calibrated with methane for selected organic compounds. EPA-600/2-81-022. U.S. Environmental Protection Agency, Office of Research and Development. Research Triangle Park, NC.
Bursey, J.T. et al. (Radian Corp.) 1991. Method 21 evaluation for the HON. EPA Publication DCN No. 91-275-026-30-02. U.S. Environmental Protection Agency, Office of Air Quality Planning and Standards. Research Triangle Park, NC.

D

Vacuum Bagging

Thomas A. Kittleman, Paul R. Jann, and
Seshasayi Dharmavaram

DESCRIPTION

Figure D-1 is a diagram of the vacuum bagging apparatus and the basic steps in the field procedure follows. CMA (1989) includes a specific detailed hardware list. In areas where electrical equipment cannot be used, air driver equipment may be used. The vacuum sampling procedure consists of the following steps when using hardware shown in the figure (specific applications for inorganics are discussed in Dharmavaram et al. 1991).

Before sampling, the sampling train should be checked for leaks. The vacuum pump should be set to leak check at a few inches (a maximum of 10 inches) of Hg. During normal testing, the dry gas meter vacuum gage should not exceed 3 inches Hg. The vacuum test should be at a greater vacuum than occurs during testing. The resulting flow when sampling should be about 0.5 cfm. A stopwatch and dry gas meter should be used to determine flow when leak checking.

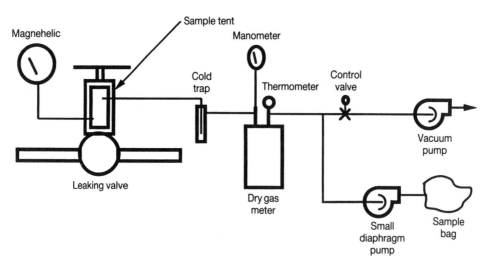

Figure D-1.   Vacuum method for bagging.

By timing the rate of flow, the leak rate can be quantified and should be less than 0.01 cfm. The leak check should include the flask because the flask cork is the most likely place for a leak. The following procedure should be used.

1. Obtain screening value of component to be bagged.
2. Enclose the component in a tight shroud. Take care to ensure that only the component of interest is enclosed (e.g., flanges should not be enclosed with a valve). When bagging a hot component, it may be necessary first to wrap silicone fabric before taping down tenting material, or to use alternate tenting material, such as aluminum foil and steel bands.
3. Connect the sampling train to the tent.
4. Immerse the cold trap in an ice bath.
5. Record the initial reading of the dry gas meter.
6. Start the vacuum pump and a stopwatch simultaneously.
7. Record the temperature and pressure at the dry gas meter.
8. Observe the VOC concentration at the vacuum pump exhaust with the hydrocarbon detector (indicates source variability).
9. Record the temperature, pressure, dry gas meter reading, outlet VOC concentration and elapsed time every 2 to 5 minutes. Barometric pressure is needed to reduce the field data. Unless other arrangements have been made, the lab should keep a log of barometric pressure (hourly) when field testing is in progress. Record appropriate barometric pressures on the field data sheets for each field test.
10. When the outlet VOC concentration stops increasing, the system is at equilibrium and a gas sample bag is filled from the discharge of the Teflon-lined diaphragm pump (fluctuations in VOC reading are common and do not mean equilibrium has not been reached).
11. Another bag is filled with ambient air near the tent area to correct for background VOC concentration. When bagging a 'zero' screening value component, it is extremely important to take a background bag. Data obtained without a background bag may greatly bias emission factors on the high side.
12. A final set of temperature and pressure readings is taken and the vacuum pump and stopwatch are simultaneously stopped.
13. Record stopwatch time and final dry gas meter reading.
14. The cold trap is removed, sealed, and transported to the laboratory along with the two bag samples and the data sheet. Be sure all samples are clearly labeled and identified on data sheets. Be sure all needed data are recorded.
15. Remove tent.
16. Rescreen the component bagged and record reading.

All of the above data and any pertinent comments should be recorded in a permanent laboratory notebook. Field and lab data are used to calculate the emission rate using the following equations.

CALCULATIONS FOR VACUUM BAGGING METHOD

Gas Portion of Sample

$$\frac{\text{std. ft}^3}{\text{hour}} = \frac{\Delta \text{meter reading} \times \dfrac{530}{(460 + T)} \times \dfrac{\text{barometer } P - P}{760} \times 60 \text{ min/hour}}{\text{minutes sampled}}$$

(D-1)

Where pounds/hour = $\sum$ [(std. ft^3/hour) $\times$ (concentration ppmv) $\times$ (10^{-6}) $\times$ MW/387]

Liquid Portion of Sample

pounds/hour = [($\sum$ pounds liquid in trap)/ (minutes sampled)] $\times$ 60 min/hour

(D-2)

Total Sample

pounds/hour = pounds/hour in gas

+ pounds/hour in liquid

(D-3)

Where T = °F in the gas meter
P = mm Hg vacuum on gas meter

meter reading is in ft^3 (difference of readings showing total gas pulled through sample apparatus)

concentration ppmv is concentration of each individual compound measured in the gas portion of the sample (i.e., the pound/hour calculation is done separately for each chemical and then summed). When a background bag is analyzed, its concentration should be subtracted.

MW is molecular weight for which the pound/hour is being calculated

pounds liquid in trap is the sum of the mass of chemicals of interest measured in the liquid sample (does not include condensed water)

References

Chemical Manufacturers Association (CMA). 1989. Improving air quality: Guidance for estimating fugitive emissions from equipment. Washington, D.C.

Dharmavaram, S., T.A. Kittleman, G.N. Vander-Werff, and D.J. Lonczak. 1991. Techniques for measuring hydrogen fluoride and chlorine releases for leaking equipment. Paper No. 91-79.1. Presented at 84th Annual Meeting of the Air and Waste Management Association. Vancouver, BC.

E

Blow-Through Bagging

Thomas A. Kittleman, Paul R. Jann, and
Seshasayi Dharmavaram

DESCRIPTION

Figure E-1 shows the basic hardware for field analysis. Samples can be withdrawn from the tent and taken to the laboratory for greater sensitivity, but the equipment and procedures are not discussed further below (Davis 1989; Enviroplan 1991). The procedure below is based on using the screening instrument for leak rate measurements. This requires additional calibration of the screening instrument for nitrogen-rich atmospheres. However, the most serious drawback is probably that the lower limit of sensitivity is high for several reasons, including: (1) high nitrogen flows can be required, (2) it is necessary to use the diluter for the analysis, thus raising the detection limit by an additional factor of 10, and (3) it is limited to

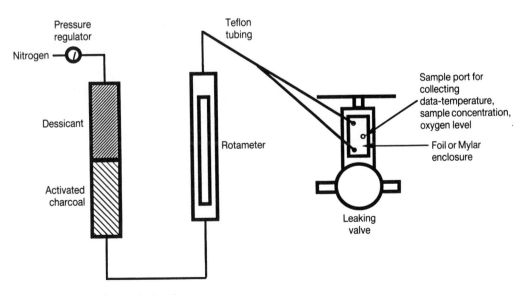

Figure E-1. Blow through for bagging.

the capability of portable equipment. A diagram of the equipment for this variant of the method is shown. Some of these problems can undoubtedly be overcome by using lab analysis.

Safety has been a key factor in the development of the blow-through procedure. Nitrogen is used as a dilution gas instead of air to prevent an explosive atmosphere within the tent. All of the instruments used are battery operated and approved for Class I, Division I use. In summary, the blow-through method consists of the following steps

1. Interview the unit operator to determine the composition (in weight or volume percent) of the material in the designated equipment component and the operating conditions of the pump.
2. Screen the component by placing a Tygon nozzle on the end of the Foxboro/Century OVA (Organic Vapor Analyzer), holding the end of the nozzle within 1 cm of the leak interface, and recording the highest concentration seen on the OVA readout. The procedure for valves is the same; however, the analyzer does not need to be 1 cm away from the valve.
3. Cut a tent from appropriate material (see source enclosure section in Chapter 22) that will fit easily over the equipment component.
4. Connect tubing from the nearest low-pressure nitrogen station to a rotameter stand, that includes a regulator, desiccant, activated charcoal, and a rotameter in series.
5. Run tubing from the rotameter outlet to a 'Y' that splits the nitrogen flow into two pieces of tubing. Insert the tubes into openings located on either side of the tent.
6. Turn on nitrogen at the utilities station and regulate it at the rotameter to approximately 40 liters/minute.
7. After the nitrogen is flowing, wrap aluminum foil around those parts of the equipment component where air could enter the tent-enclosed volume.
8. Use duct tape, wire, or rope to secure the tent to the component.
9. Put a third hole in the tent roughly equidistant from the two nitrogen-fed holes.
10. Measure the oxygen concentration in the tent by inserting the lead from an O_2 meter into the third hole. Adjust the tent (by adding additional tape, foil, rope, etc.) until the O_2 concentration is less then 5 percent.
11. Measure the temperature in the tent.
12. Calibrate the OVA to methane or hexane at a known concentration in nitrogen using the OVA dilution probe. Remember to correct for the dilution before inserting the OVA concentration reading into the calculation below.
13. Check the VOC concentration at several points in the tent with the OVA to ensure that the tent contents are at steady state and at least two separate times.
14. Measure the hydrocarbon VOC concentration in the tent with the OVA at three different nitrogen flow rates. Typically, the flow rates will be 40, 30 and 20 liters/minute if there is no OVA response at the higher rates (i.e., the mass leak rate from the valve is very low). After each adjustment of the nitrogen flow, check the O_2 concentration to ensure it stays below 5 percent. Alternatively, collect samples in Tedlar® or aluminized sample bags by drawing the sample out of the bag with a portable sampling pump.
15. Remove the tent and any plugs from the component and collect any condensate on the inside of the tent in a plastic graduated cylinder. Record the amount collected and the elapsed time the tent was on the component.
16. If there is liquid dripping from the component, collect the drips for a timed period that produces enough collected material for accurate volume measurement. Record the amount in the report, but do not add it to the vapor leak rate.

The equation to be used with this approach is

Tent rate (lb/hour) =

$$4.836 \times 10^{-5} \times \frac{Q \times MW \times OVA \times RF}{T + 460}$$

(E-1)

Where

Q = flow rate into tent in m^3/hour

$$= \frac{N_2 \text{ flow rate in litres per minute}}{1 - \dfrac{\text{tent oxygen concentration in \%}}{21}}$$

$$\times \left(0.06 \; \frac{m^3}{\text{liters}} \frac{\text{minutes}}{\text{hour}} \right)$$

MW = molecular weight of gas in lb/lb-mole[4]

T = temperature in tent in °F

OVA = instrument reading minus background reading, in ppmv

RF = response factor for leaking gas relative to calibration gas[5]

4.836×10^{-5} = a conversion factor taking into account the gas constant and assuming a pressure in the tent of 1 atmosphere

References

Davis, B.C. 1989. A comparison of the fugitive emissions from two butadiene plants. Paper No. 89-30.2. Presented at the 82nd Annual Meeting of the Air & Waste Management Association. Anaheim, CA.

Enviroplan. 1991. Fugitive emissions correlation for SOCMI based on bagging studies of BD/EO production units. Ref. No. 51-9043-952. Study for Chemical Manufacturers Association. Roseland, NJ.

---

[4] For mixtures: gas molecular weight = (mole fraction of gas 1)(molecular weight of gas 1) + (mole fraction of gas 2)(molecular weight of gas 2) + ...

[5] Response factor is calculated at the OVA concentration. See Appendix C for a more detailed discussion.

F

Recirculation Bagging

Thomas A. Kittleman, Paul R. Jann, and
Seshasayi Dharmavaram

DESCRIPTION

The equipment for a recirculation bagging test is shown in Figure F-1. The mini-impingers can be replaced with adsorbent tubes for certain application (Hosler et al. 1992). More detail on the method is not now available.

Reference

Hosler, A.D., T.N. Grosshandler, and R.L. Marchman. 1992. Fugitive emission: The need for more accurate measurement techniques. Paper No. 92-137.06. Presented at 85th Annual Meeting of the Air & Waste Management Association. Kansas City, MO.

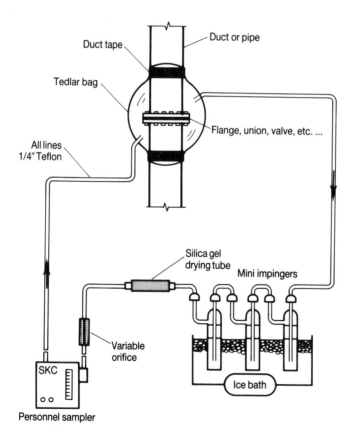

Figure F-1. Recirculation bagging equipment.

G

Exposure and Risk Models

Gerald E. Anderson

MODEL APPROACHES

A model of human health risk from inhalation of toxic air pollutants must, at a minimum, consist of a dispersion analysis for estimating source-receptor relationships and, thus, patterns of concentration of airborne pollutants, and a means of estimating population patterns in the polluted region. Exposure and risk models may include one or more of the following features, in addition to dispersion and population modules, as those features relate to application-specific problems or ancillary considerations.

The following discussion of modeling approaches addresses several specific models in turn. Although the models described here are only a sampling of those that have been developed by agencies or consulting firms, they are thought to represent the range of model 'styles' in use.

AERAM

Initial development of AERAM (Air Emission Risk Assessment Model; Eschenroeder, Magil, and Woodruff, 1985) for EPRI was by Arthur D. Little, with the intention to develop a capability directly applicable to the assessment of human risk from hazardous pollutants emitted by coal-fired power plants. Further model development

effort was carried out by IWG Corp. The approach to the essential components of a Human Exposure/Risk model – dispersion, population, and a convolution procedure for computing human exposure – and additional modules for developing input emissions and output risk are as follows.

Emissions

AERAM contains an emissions characterization module that performs three functions; it produces estimates of uncontrolled emissions of gaseous pollutants and fly ash; it calculates the volume flow rate, temperature, and exit speed of the flue gases; and it estimates the distribution of fly ash particle sizes.

Dispersion

Dispersion computations are carried out using the EPA's Industrial Source Code Long Term (ISCLT). ISCLT is a multipoint, steady-state, Gaussian plume model that is made publicly available by the EPA through their UNAMAP model set, and described in EPA (1986i). ISCLT has features to address momentum and thermal fluxes in the plume, gradual plume rise, stack tip downwash, building wake effects, dry deposition, first-order decay processes, and stack, area and line sources. It addresses moderate terrain in a crude fashion. Terrain, if addressed, must be lower than the elevation(s) of the source(s).

Use of ISCLT provides the advantage that it is approved by the EPA for purposes of showing industrial compliance with NAAQS and PSD increments. Although evaluations for applications to peak impact identification and quantification do not assure performance for long-term mean patterns of concentration, no evaluations have shown any other model to have superior performance for similar applications. EPA's approval assures that many, if not all, state and local air agencies will not question, and may require, the use of ISCLT. ISCLT is discussed in more detail in Chapter 10.

Population Exposure

The population data used is Census Tract data from the U.S. Census Bureau. Modeling is done on a user-defined grid of exposure cells in an application-specific modeling region. The cells need not be of uniform size, or in a regular (e.g., Cartesian) form. Population from Census centroids are aggregated into their respective exposure districts. ISCLT-computed concentration is averaged at all receptors (i.e., centroids) within each exposure district. User supplied data on population cohorts may be used to define the exposed subgroups.

Dose

The mean concentration (μg/m^3) of an exposure district is taken as the exposure level for all people residing in the exposure district. Dose to an individual is computed as the product of the exposure concentration and an exposure factor. Exposure factors comprise an estimated mean inhalation rate (m^3/day). Exposure factors may be defined for subgroups of the population having different inhalation rates because of different mean age, size, activity level, etc. Dose rate in μg/day is aggregated for each exposure district, each population subgroup, and the total population.

Risk

Risk can be computed for each population subgroup. Risk is computed from dose through a dose/response (D/R) function. EPA reviews data and literature on species carcinogenic potency, analyzes D/Rs, provides for public review, and publishes recommended values and forms of D/Rs to be used in federally reviewed analyses. EPA generally relies most heavily on the Office of Human Health Assessment (formally the Carcinogen Assessment Group) in determining carcinogenicity. EPA generally uses conservative D/R functions. These are discussed in Chapter 6.

The formulation of AERAM envisions the use of internally computed D/Rs directly from laboratory data. The AERAM risk module, RISK, addresses the scaling of carcinogenicity data obtained from laboratory tests on animals to equivalent potency for humans. The scaling addresses the differences in body weight, surface area, absorption, metabolic rate and excretion rate. Three alternate data-fitting methods are made available – one hit, log-probit, and multistage.

HEM

Initial development of HEM (Human Exposure Model) was carried out by Anderson and Lundberg (1983) for the EPA to provide them with their first Human Exposure/Dose/Risk model. The capability was needed by the EPA to support their NESHAP program under section 112 of the 1970 Clean Air Act. HEM is a set of two models developed separately with different concepts, structures, and features. The two models were called SHED (Systems Applications' Human Exposure and Dose) and SHEAR (Systems Applications' Human Exposure And Risk).

SHED was developed as a national screening model to allow the EPA to carry out prioritizations for NESHAP attention on the basis of relative national aggregate risk. It was necessary to omit some model features to provide efficient computations and manageable input data requirements when applied to inventories of many thousands of sources.

SHED addressed each industrial source independently, with no attention to overlapping impact patterns. SHED addressed each species independently, so impacts could be expressed in terms of dose rather than risk, with risk computed externally. Since each source was addressed independently, no grid or locational framework was needed except a standardized, polar grid oriented relative to the site of the source being addressed. The model included national meteorological and population data bases so that it could be applied to a national inventory of sources in a single batch run.

SHEAR was developed to complement SHED and provide a capability to address multisource, multispecies risk to estimate net individual and community risk from all community sources of risk. To provide this capability, SHEAR defined the same source and dispersion algorithms as SHED on a single grid, defined by the user for a selected modeling region.

Both models address individual sources (e.g., emissions from a single property – an industrial plant, a field of facilities with fugitive sources, or a dry cleaning shop) and area sources (e.g., community distributed sources such as motor vehicle traffic, home chimneys, etc.) In addition, both models address individual sources that are only defined generically, by prototype. SHED merely multiplies exposure and dose for such facilities. SHEAR includes algorithms for locating such facilities within the defined modeling region either randomly or by user-specified rules. Prototype treatment was used for source categories for which each source is a large emitter, with a significant exposure pattern, but for which there were so many examples that it would not be feasible to develop a complete inventory (e.g., gasoline service stations.)

HEM does not contain an emissions modeling module because it would not be feasible to develop such a capability for the wide range of source types to which it was to be applied. Even for a single source, it is not clear that there is any advantage to integrating an emission model into a risk model. The tasks seem to be independent and can be executed independently with no loss of utility.

Dispersion

With most versions of HEM, the model can be operated with internal dispersion algorithms or with concentration patterns input as the results of an external modeling run. Using the external option, HEM could be run with the ISCLT model, as is AERAM. In cases of special source or site considerations, any other model could serve to provide input concentrations. For example, in sites with high surrounding terrain, a complex terrain dispersion model such as RTDM could be used (see Chapter 10).

Population Exposure and Dose

The population data used is BG/ED (see Chapter 14) data from the Census Bureau. The entire national Census Bureau BG/ED data base is resident within HEM. BG/EDs have an average population of about one thousand, producing population patterns much more finely resolved than is possible with census tracts. Further, the HEM exposure algorithms generate exposure estimates on either the polar impact grid cells or on the BG/ED centroids, whichever is locally the limiting resolution factor. SHEAR exposures can be aggregated into the user-defined grid cells for display purposes. Exposure and dose are determined for individuals or for the exposed population, aggregated by the concentration at which exposure occurs and in sum.

Risk

EPA unit risk factors are filed in HEM. Tables of individual or population risk are aggregated and tabulated by the hazard level (individual risk) at which they occur.

LIFETIME1

LIFETIME1 (Paul 1984) is an unfinished model based on SHED (HEM). The intent was to enhance SHED capability by adding the following features.

- Address residential migration patterns of people over a seventy-year lifetime
- Address source emission variations over facility lifetime

■ Address indoor/outdoor concentration differentials
■ Incorporate ISCLT
■ Cohort-specific exposures by activity patterns

The model was developed to an experimental test stage without some planned features. Lifetime variations of exposure and risk accumulation addressed movement between risk areas and no-risk areas; movement between areas of different risk levels was planned. Population net migration between two areas was defined as a function of existing population, area growth rate, the distance between the areas and the economic base of each area. Migration data came from Census Bureau county and state files. Indoor concentrations were related to outdoor concentrations by a simple, linear, attenuation model.

NEM

NEM was developed by EPA for use in assessing health risks associated with new or modified NAAQS. Since the model was intended to address exposures to criteria pollutants, the focus was on short time periods (periods measured in hours) and threshold exceedances rather than lifetime accumulations of trace concentrations.

Because of this focus, there was a great sensitivity to identifying the particular short-term events in the lives of the exposed when their exposure would be maximized. To identify such events, NEM addresses the moment-by-moment activity patterns of people as they move from exposure district to district about an urban area and from microenvironment to environment in any given district. Microenvironments are defined as characteristic locations within a district in which the concentration bears a definable bias with respect to the mean district ambient concentration.

NEM addresses a large number of cohorts, or age-occupation groups making up the regional population. NEM results are separately aggregated over these cohorts as well as in total. Systems Applications International (SAI)

developed a proprietary version of NEM with a substantial number of unique features. This version was given to EPA. Both the EPA model and the SAI model, NEMSAI, are discussed here. Among other features, the NEMSAI model was recoded in Fortran 77 (NEM is coded in PL1.) NEMSAI has been run on micro- and minicomputers with greatly reduced (a factor of ten) computer running time requirements.

Dispersion

Dispersion patterns in the EPA model are not modeled, but are determined by hourly records of a monitoring network. Monitoring data is not easily relatable to emission rates from specific sources; also, as modeling networks are not practically as dense as modeling receptor points, the concentration data may not be sufficient to define significant variations in the concentration patterns.

Monitoring data can only represent past occurrences. In particular, the method cannot be easily applied to the analysis of the effects of control programs, if those programs would be expected to lead to patterns that are not scalars of prior patterns. Rollback methods are used to apply NEM to control strategies, but rollback is often incorrect in practical situations. Missing concentration data are filled in through the use of estimates obtained with Fourier transforms.

NEMSAI has been adapted to accept modeled concentration patterns as input. NEMSAI has been applied, in particular, to the analysis of ozone risk using ozone concentrations obtained with SAI's photochemical grid model, the Urban Airshed Model (UAM).

In NEM microenvironment concentrations are prescribed as linear functions of the regional monitoring data. In NEMSAI the treatment for the indoor microenvironment has been replaced with an indoor/outdoor model that addresses ventilation, filtration, and source/sink effects for user-specified stocks of building types and qualities.

Population

Population is defined on a case-specific basis by user-provided data at the SMSA, county, or census tract level. Within a population district, neighborhood types are defined to adjust activity patterns for the resident cohorts. As the required population data bases are not generally available for specific cities, data must either be acquired or an existing database must be adapted to the study city. Age-occupation groups definitions, and activity patterns are usually accepted as universally valid.

Exposure/Risk

In the basic NEM program, three different measures are used to characterize exposure to various levels of air pollution.

- Pollutant encounters: the number of times a given pollutant concentration level is experienced by one individual
- People exposed: the number of people who experience a given pollutant concentration level or higher during a one-year period
- Peak exposure: the number of people who experience their highest concentration level within a given concentration interval

SHAPE

SHAPE is an exposure model developed by Systems Architects, Inc. (Ott 1981). It is applicable to estimating the exposures of a distribution of cohorts to concentrations within a single microenvironment. Concentrations are neither modeled nor measured; instead, they are input by sampling from a log normally distributed random variable. Geometric mean and variance are user-specified.

Further assumptions include the following.

- Selection of each human activity is based on a probability distribution.
- The time duration of each activity is a random variable, conditioned on the activity selected.
- Once an activity is selected, it remains in effect until the end of its time duration.
- Activity selection can be conditioned on the outcomes of previous activities and on the time of day.

SuperSHEAR

SuperSHEAR is a proprietary model developed by SAI. It represents a substantially advanced version of SHEAR. The internal national data files of dispersion meteorology and BG/ED populations, as well as the dispersion modules, are the same as in SHEAR. The advanced feature of SuperSHEAR is that all concentrations, and therefore exposures, doses and risk are computed at each BG/ED centroid. No reference is made to a computational grid.

As a result of this modification, SuperSHEAR aggregations are made on a vector of BG/EDs that are extracted from the national file for a user-specified impact zone about each source location in an inventory of sources. If more than one source impacts the same BG/ED(s), then each source's exposure is added to the total aggregation for that (those) BG/ED(s).

Since there would be less than total overlap of the impact patterns of any two sources that are not colocated, some, but not all, BG/EDs would have aggregated impacts from both. With this approach, multisource, multispecies problems can be addressed with any general inventory of sources without defining a modeling region. The entire United States is, potentially, the modeling region.

SCREAM

The South Coast Risk and Exposure Analysis Model (SCREAM) is a version of the HEM/SHEAR human exposure, dose, and risk model that includes advancements and also includes special data sets applicable to the South Coast Air Basin (Los Angeles). SCREAM is undergoing additional development that will substantially change its structure and capability.

SCREAM was first installed on SCAQMD computers in 1986. At that time, SCREAM was structured similarly to SuperSHEAR, but had three separate levels of resolution of the population patterns. Population data were made available for each individual city block in the district. BG/ED and census tract level population data was available from the SuperSHEAR-formatted U.S. Census Bureau files.

By user's option, SCREAM aggregations of exposure, dose, or risk could be made at the city block level of resolution, at the BG/ED level, or at the census tract level. In the default mode of operation, aggregations are made at the block level for each block within 5 km of a particular source, at the BG/ED level between 5 km and 10 km, and at the census tract level at distances beyond 10 km. The model structure for any one level of resolution is identical with that for any other. That is, when operating on block-level data, each block is treated as a pseudo-BG/ED; similarly, census tract are so treated when they are addressed. In any exercise, the final vector of impacted districts includes blocks, BG/EDs, and census tracts.

Additional data sets made available for the SCAQMD application include STAR dispersion climatology for thirty District meteorological stations, population growth data for years since 1980, and growth projections for the future.

Currently, further SCREAM development is under way for the SCAQMD. In that program, algorithms and data are being incorporated to address the following issues.

■ Population mobility: A version of NEM (NEMSAI) is being incorporated into SCREAM, together with updated data bases on age-activity patterns for the Los Angeles area population.

■ Diurnal dispersion climatology: the District STAR data arrays have been reprocessed to provide a STAR array for each of the twenty-four hours of the diurnal pattern for each of the thirty stations. The twenty-four diurnal arrays are used to provide occurrence frequencies of dispersion conditions that are correlated with the diurnal pattern of human activity defined for each cohort in the NEMSAI data base.

■ Multimedia transport coefficients and data developed for the SCAQMD were being incorporated into it so as to add exposures

and risk from other routes to the inhalation exposures from airborne pollutants.

■ An Indoor Air Quality Model was being incorporated into SCREAM to address indoor microenvironment concentrations in much more detail than is provided by NEMSAI.

■ A Personal Air Quality Model (PAQM) developed by SAI was being incorporated into SCREAM to provide a user-selectable, simplified alternative to the use of NEMSAI. With PAQM, the effects of age-activity patterns on exposure and risk are determined at any selected receptor location. The actual community aggregated risks that depend on intermigrations between exposure districts are not addressed as they are in the SCREAM/NEMSAI model.

References

Anderson, G.E. and G.W. Lundberg. 1983. User's Manual for SHEAR. Publication No. SYSAPP-83/RY. Prepared for U.S. Environmental Protection Agency, Office of Air Quality Planning and Standards. Research Triangle Park, NC.

Eschenroeder, A,Q., G.C. Magil, and C.R. Woodruff. 1985. Assessing the health risks of airborne carcinogens. Report prepared by Author D. Little, Inc. for the Electric Power Research Institute. Palo Alto, CA.

Ott, W.R. 1981. Computer simulation of human air pollution exposures to carbon monoxide. Paper No. 81-57.6. Presented at the 74th Annual Meeting of the Air Pollution Control Association. Philadelphia, PA.

Paul, R. 1984. Memoranda on the feasibility of a Lifetime Exposure Model. To Michael Dusetzina, Pollutant Assessment Branch. U.S. Environmental Protection Agency, Office of Air Quality Planning and Standards. Research Triangle Park, NC.

U.S. Environmental Protection Agency (EPA). 1986i. Guideline on air quality models (Revised) and Supplement A (1987). EPA 450/2-78-027R. (NTIS PB 86-2452 48). Office of Air Quality Planning and Standards. Research Triangle Park, NC.

Acronyms and Abbreviations

Acronym or Abbreviation	Meaning
AAC	Acceptable Ambient Concentration
AAL	Acceptable Ambient Level
AAS	Atomic Absorption Spectroscopy
ACGIH	American Council of Governmental Industrial Hygienists
ADD	Average Daily Dose
ADI	Acceptable Daily Intake
AEERL	Air and Energy Engineering Research Laboratory, U.S. EPA
AIC	Acceptable Intake Chronic (Exposure)
AIHA	American Industrial Hygiene Association
AIHC	American Industrial Health Council
AirRISC	Air Risk Information Support Center
AIRS	Aerometric Information Retrieval System
AIS	Acceptable Intake for Subchronic exposure
ALAPCO	Association of Local Air Pollution Control Officials
API	American Petroleum Institute
APS	Activity Pattern Simulator
AQCR	Air Quality Control Region
AQMD	Air Quality Management District
ARAR	Applicable or Relevant and Appropriate Requirement
AREAL	Atmospheric Research and Exposure Assessment Laboratory, U.S. EPA
ASTM	American Society for Testing and Materials
AT	Averaging Time
ATSDR	Agency for Toxic Substances and Disease Registry
A&WMA	Air and Waste Management Association
BACT	Best Available Control Technology
BAT	Best Available Technology
BBS	Bulletin Board Services
BCF	Bioconcentration Faetor
BDT	Best Demonstrated Technology
BHP	Brake Horsepower
BID	Background Information Document
BIF	Boilers and Industrial Furnaces
BIO	Bioavailability Factor
BRI	Building Related Illness
BW	Body Weight
CAA	Clean Air Act
CAAA	Clean Air Act Amendments
CAER	Community Awareness and Emergency Response
CAG	Cancer Assessment Group, U.S. EPA (now Human Health Assessment Group – HHAG)
CARB	California Air Resources Board
CAS	Chemical Abstract Services
CDI	Chronic Daily Intake
CD-ROM	Compact Disc – Read Only Memory
CDU	Chronic Daily Uptake
CEAM	Center for Exposure Assessment Modeling, U.S. EPA
CEGL	Continuous Exposure Guidance Level
CEL	Continuous Exposure Limit
CEMS	Continuous Emission Monitoring Systems
CERCLA	Comprehensive Environmental Response, Compensation and Liability Act
CERI	Center for Environmental Research Information, U.S. EPA
CFR	Code of Federal Regulations
CHIEF	Clearinghouse for Inventories and Emission Factors
CLP	Contract Laboratory Program (Superfund)
CMA	Chemical Manufacturers Association
CO	Carbon Monoxide
COT	Committee on Toxicology (National Research Council)
CR	Contact Rate
CRAVE	Carcinogen Risk Assessment Verification Endeavor
CTC	Control Technology Center
CTG	Control Technology Guideline
CVAAS	Cold Vapor Atomic Absorption Spectroscopy

Acronym or Abbreviation	Meaning
CWA	Clean Water Act
DHS	Department of Health Services (California)
DIAL	Differential Absorption Lidar
DOD	Department of Defense
DOE	Department of Energy
DQO	Data Quality Objective
DRE	Destruction and Removal Efficiency
EA	Endangerment Assessment
EAG	Exposure Assessment Group, U.S. EPA
ECAO	Environmental Criteria and Assessment Office, U.S. EPA
ECD	Electron Capture Detector
ECETOC	European Chemical Industry Ecology and Toxicology Centre
ED	Exposure Duration
EDF	Environmental Defense Fund
EEGL	Emergency Exposure Guidance Level
EEI	Emergency Exposure Indices
EEL	Emergency Exposure Limit
EF	Exposure Frequency
EIFC	Emission Isolation Flux Chamber
EMMI	Environmental Monitoring Methods Index
EMTIC	Emission Measurement Technical Information Center
EPA	U.S. Environmental Protection Agency
EPCRA	Emergency Planning and Community Right-to-Know Act
EPRI	Electric Power Research Institute
ERPG	Emergency Response Planning Guidelines
ET	Exposure Time
FDER	Florida Department of Environmental Regulation
FEV	Forced Expiratory Volume
FID	Flame Ionization Detection
FIFRA	Federal Insecticide, Fungicide and Rodenticide Act
FPD	Flame Photometric Detection
FOIA	Freedom of Information Act
FR	Federal Register
FTIR	Fourier Transform Infrared (Spectroscopy)
GACT	Generally Available Control Technology
GAO	General Accounting Office (Congress)
GC	Gas Chromatography
GC/MS	Gas Chromatography/Mass Spectrometry
GEAP	Good Exposure Assessment Practices
GEMS	Graphical Exposure Model System
GEP	Good Engineering Practice (Stack Height)
GEPD	Georgia Environmental Protection Division
GFAAS	Graphite Furnace Atomic Absorption Spectroscopy
GFC	Gas Filter Correlation
GIS	Geographic Information Systems
GLP	Good Laboratory Practices
GVWR	Gross Vehicle Weight Rating
HAAS	Hydride Atomic Absorption Spectroscopy
HAD	Health Assessment Document
HAP	Hazardous Air Pollutant
HAPPS	Hazardous Air Pollutant Prioritization System
HC	Hydrocarbon
HEA	Health Effects Assessment
HEAST	Health Effects Assessment Summary Table
HEC	Human Equivalent Concentration
HECD	Hall Electrolytic Conductivity Detector
HEED	Health and Environmental Effects Document
HEEP	Health and Environmental Effects Profile
HEM	Human Exposure Model
HERL	Health Effects Research Laboratory, U.S. EPA
HHAG	Human Health Assessment Group, U.S. EPA

Acronym or Abbreviation	Meaning
HHEM	Human Health Evaluation Manual (Superfund)
HI	Hazard Index
HON	Hazardous Organic NESHAP
HPLC	High-Performance Liquid Chromatography
HQ	Hazard Quotient
HRGC	High-Resolution Gas Chromatography
HRMS	High-Resolution Mass Spectrometry
IARC	International Agency for Research on Cancer
IC	Ion Chromatography
IC/PCR	Ion Chromatograph with Postcolumn Reactor
ICP	Inductively-Coupled Argon Plasma (Emission Spectroscopy)
IDLH	Immediately Dangerous to Life and Health
IEC	Inhalation Exposure Concentration
IFC	Isolation Flux Chamber
I/M	Inspection/Maintenance
IR	Ingestion Rate
IR	Infrared
IRIS	Integrated Risk Information System
ISCLT	Industrial Source Complex Long-Term Model
ISCST	Industrial Source Complex Short-Term Model
JAPCA	Journal of the Air Pollution Control Association
JAWMA	Journal of the Air and Waste Management Association
LADD	Lifetime Average Daily Dose
LAER	Lowest Achievable Emission Rate
LARTS	Los Angeles Regional Transportation Simulation
LC	Liquid Chromatography
LC_{50}	Median Lethal Concentration (concentration at which 50 percent of animals die)
LD_{50}	Median Lethal Dose (dose at which 50 percent of animals die)
LDAR	Leak Detection and Repair
LEF	Leaking Emission Factor
LEV	Low-Emission Vehicle
LIF	Laser Induced Fluorescence
LLD	Lower Limit of Detection
LOAEL	Lowest Observed Adverse Effects Level
LOEL	Lowest Observed Effects Level
LRMS	Low-Resolution Mass Spectroscopy
LVW	Loaded Vehicle Weight
MACT	Maximum Achievable Control Technology
MDD	Maximum Daily Dose
MEI	Maximum Exposed Individual
MF	Modifying Factor
MGLC	Maximum Ground-Level Concentration
MHAPPS	Modified Hazardous Air Pollutant Prioritization System
MI/FTIR	Matrix Isolation/Fourier Transform Infrared
MIR	Maximum Individual Risk
MLE	Maximum Likelihood Estimate
MMAD	Mass Mean Aerodynamic Diameter
MPI	Multiphoton Ionization
MRL	Minimum Risk Level
MS	Mass Spectroscopy
MSE	Mean Square Error
MTBE	Methyl Tertiary Butyl Ether
MY	Model Year
NAAQS	National Ambient Air Quality Standard
NAMCS	National Ambulatory Medical Care Survey
NAPCA	National Air Pollution Control Administration (pre-EPA)
NAPCTAC	National Air Pollution Control Techniques Advisory Committee
NAS	National Academy of Sciences
NASN	National Air Sampling Network
NATICH	National Air Toxics Information Clearinghouse
NCDC	National Climatic Data Center

Acronym or Abbreviation	Meaning
NCDEM	North Carolina Division of Environmental Management
NCHS	National Center for Health Statistics
NDIR	Non-Dispersive Infrared
NEDS	National Emission Data System
NEM	National Ambient Air Quality Standards Exposure Model
NESHAPs	National Emission Standards for Hazardous Air Pollutants
NIMBY	Not In My Backyard
NIOSH	National Institute of Occupational Safety and Health
NLEF	Nonleaking Emission Factor
NMHC	Nonmethane Hydrocarbons
NMOG	Nonmethane Organic Material
NOAA	National Oceanographic and Atmospheric Administration
NOAEL	No Observed Adverse Effects Level
NOEL	No Observed Effects Level
NO_x	Nitrogen Oxides
NPD	Nitrogen Phosphorus Detector
NPDES	National Pollution Discharge Elimination System
NPL	National Priorities List (Superfund)
NRC	National Research Council
NRC	Nuclear Regulatory Commission
NRDC	Natural Resources Defense Council
NSPS	New Source Performance Standard
NTIS	National Technical Information Service
NTL	No Threat Level
NTP	National Toxicology Program
OAQPS	Office of Air Quality Planning and Standards, U.S. EPA
OAR	Office of Air and Radiation, U.S. EPA
OEL	Occupational Exposure Limit
OHEA	Office of Health and Environmental Assessment, U.S. EPA
OMMSQA	Office of Modeling, Monitoring Systems and Quality Assurance, U.S. EPA
OPPE	Office of Policy, Planning and Evaluation, U.S. EPA
OPTS	Office of Pesticides and Toxic Substances, U.S. EPA
ORD	Office of Research and Development, U.S. EPA
OSHA	Occupational Safety and Health Administration
OSWER	Office of Solid Waste and Emergency Response, U.S. EPA
OSTP	Office of Science and Technology Policy
OTA	Office of Technology Assessment (Congress)
OTTRS	Office of Technology Transfer and Regulatory Support, U.S. EPA
OVA	Organic Vapor Analyzer
PAH	Polycyclic Aromatic Hydrocarbons
PBPK	Physiologically Based Pharmacokinetic
PCB	Polychlorinated Biphenyls
PEL	Permissible Exposure Limit
PID	Photoionization Detection
PM	Particulate Matter
PM_{10}	Particulate Matter (10 micron or less, aerodynamic diameter)
PNA	Polynuclear Aromatic Hydrocarbons
POHC	Principal Organic Hazardous Compounds
POM	Polycyclic Organic Matter
POTW	Publicly Owned Treatment Works
PSD	Prevention of Significant Deterioration
QA/QC	Quality Assurance/Quality Control
QAPjP	Quality Assurance Project Plan
RACT	Reasonably Available Control Technology
RCRA	Resource Conservation and Recovery Act
RDD	Regional Deposited Dose
RDDR	Regional Deposited Dose Ratio
RF	Response Factor
RfC	Risk Reference Concentration
RfD	Risk Reference Dose
RfD_{dt}	Developmental Toxicant Reference Dose

Acronym or Abbreviation	Meaning
RfDs	Subchronic Reference Dose
RGD	Regional Gas Dose
RGDR	Regional Gas Dose Ratio
RME	Reasonable Maximum Exposure
ROSE	Remote Optical Sensing of Emissions
RQ	Reportable Quantity
RTECS	Registry of Toxic Effects of Chemical Substances
RVP	Reid Vapor Pressure
SAB	Science Advisory Board, U.S. EPA
SARA	Superfund Amendments and Reauthorization Act
SBS	Sick Building Syndrome
SCAG	South Coast Association of Governments
SCC	Source Classification Code
SCRAM	Support Center for Regulatory Air Models
SCRS	Source Category Ranking System
SDI	Subchronic Daily Intake
SEAM	Superfund Exposure Assessment Manual
SemiVOST	Semivolatile Organic Sampling Train
SF	Slope Factor
SHAPE	Simulation of Human Air Pollutant Exposure
SIC	Standard Industrial Classification
SIP	State Implementation Plan
SMSA	Standard Metropolitan Statistical Area
SOCMI	Synthetic Organic Chemical Manufacturing Industry
SOF	Soluble Organic Fraction
SO_x	Sulfur Oxides
SPEGL	Short-Term Public Emergency Guidance Level
SPEL	Short-Term Public Emergency Limit
SPHEM	Superfund Public Health Evaluation Manual
SQAQMD	South Coast Air Quality Management District (California)
STAPPA	State and Territorial Air Pollution Program Administrators
STAR	Stability Array
STEL	Short-Term Exposure Limit
STF	Summary Tape File
SVOC	Semivolatile Organic Compound
TAC	Toxic Air Contaminant
TACB	Texas Air Control Board
TBA	Tertiary Butyl Alcohol
T-BACT	Best Available Control Technology – Toxics
TCD	Thermal Conductivity Detector
TEAM	Total Exposure Assessment Methodology
TEF	Toxic Equivalency Factor
TEM	Transmission Electron Microscopy
THC	Total Hydrocarbon
THEES	Total Human Environmental Exposure Study
TLV	Threshold Limit Value
TNMHC	Total Nonmethane Hydrocarbon
TPQ	Threshold Planning Quantity
TQM	Total Quality Management
TRI	Toxic Release Inventory
TSCA	Toxic Substances Control Act
TSDF	Treatment Storage and Disposal Facilities
TTN	Technology Transfer Network
TURA	Toxic Use Reduction Act
TWA	Time-Weighted Average
UAM	Urban Airshed Model
UF	Uncertainty Factor
USDA	U.S. Department of Agriculture
USGS	U.S. Geologic Survey
UV	Ultraviolet (Spectroscopy)
VMT	Vehicle Miles Traveled

Acronym or Abbreviation	Meaning
VOC	Volatile Organic Compound
VOST	Volatile Organic Sampling Train
WDNR	Wisconsin Department of Natural Resources
WPCF	Water Pollution Control Federation
WWTF	Wastewater Treatment Facilities
XATEF	Crosswalk/Air Toxic Emission Factor (Data Base)

Bibliography

CHAPTER 8 – EXPOSURE ASSESSMENT PRINCIPLES

Sources for Analytical Methods

Aleckson, K.A., J.W. Fowler, and Y.T. Lee. 1986. Inorganic Analytical Methods Performance and Quality Control Considerations. In: Quality Control in Remedial Site Investigation: Hazardous Industrial Solid Waste Testing. Fifth Volume. ASTM DTP 925.

American Public Health Association, American Water Works Association, Water Pollution Control Federation. 1975. Standard Methods for Examination of Water and Wastewater, 14th Edition.

American Society for Testing Materials. Annual Book of Standards. Vol. 11.01 Waster (I); Vol. 11.02 Water (II); Vol. 11.03 Atmospheric Analysis, Occupational Health and Safety; Vol. 11.04 Pesticides, Waste disposal, etc.

Flotard, R.D., M.T. Homshen, J.S. Wolff, and J.M. Moore. 1986. Volatile Organic Analytical Methods Performance and Quality Control Considerations. In: Quality Control in Remedial Site Investigation: Hazardous and Industrial Solid Waste Testing. Fifth Volume. ASTM STP 925.

Garner, F.C., M.T. Homsher, and J.G. Pearson. 1986. Performance of EPA Method of Analysis of 2,3,7,8-tetrachlorodibenzo-p-dioxin in Soils and Sediments by Contractor Laboratories. In: Quality Control in Remedial Site Investigation: Hazardous and Industrial Solid Waste Testing. Fifth Volume. ASTM STP 925.

Martin, T.D. 1973. Inductively Coupled Plasma – Atomic Emission Spectrometric Method of Trace Elements Analysis of Water and Waste, Method 200.7, Modified by CLP Inorganic Data/Protocol Review Committee. U.S. Environmental Protection Agency, Environmental Monitoring Systems Laboratory. Cincinnati, OH.

U.S. Environmental Protection Agency (EPA). 1973. Handbook for Monitoring Industrial Wastewater. EPA Technology Transfer Office. Washington, D.C.

U.S. Environmental Protection Agency (EPA). 1974. Methods for Chemical Analysis of Water and Wastewater. EPA Technology Transfer Office. Washington, D.C.

U.S. Environmental Protection Agency (EPA). 1979. Handbook for Analytical Quality Control in Water and Wastewater Laboratories. EPA 600/4–79–019. Office of Research and Development. Washington, D.C.

U.S. Environmental Protection Agency (EPA). 1979. Methods for Chemical Analysis of Water and Wastes. EPA 600/4–79–20. Office of Research and Development. Washington, D.C.

U.S. Environmental Protection Agency (EPA). 1979. Organochlorine Pesticides and PCBs, Method 608; 2,3,7,8-TCDD, Method 613; Purgeables (Volatiles), Method 6224; Base/Neutrals, Acids and Pesticides, Method 625. 44 FR 69501, 44 FR 69526, 44 FR 69532, and 44 FR 69540, December 3, 1979.

U.S. Environmental Protection Agency (EPA). 1981. Users Guide for the Continuous Flow Analyzer Automation System. Environmental Monitoring Systems Laboratory. Cincinnati, OH.

U.S. Environmental Protection Agency (EPA). 1983. Technical Assistance Document for Sampling and Analysis of Toxic Organic Compounds in Ambient Air. EPA 600/4-83-G27. Office of Research and Development. Washington, D.C.

U.S. Environmental Protection Agency (EPA). 1984. Solid Waste Leaching Procedure. SW-924. Office of Solid Waste. Washington, D.C.

U.S. Environmental Protection Agency (EPA). 1984. Toxicity Characteristic Leaching Procedure (TCLP) – Draft Method 13. TK0703. Office of Solid Waste. Washington, D.C.

U.S. Environmental Protection Agency (EPA). 1984. User's Guide to the Contract Laboratory Program. Office of Emergency and Remedial Response. Washington, D.C.

U.S. Environmental Protection Agency (EPA). 1984. Test Methods for Evaluating Solid Waste, Physical/Chemical Methods. SW-846. Office of Solid Waste. Washington, D.C.

U.S. Environmental Protection Agency (EPA). 1984. Calculation of Precision, Bias and Method Detection Limit for Chemical and Physical Measurements. (QAMS Chapter 5). Office of Research and Development. Washington, D.C.

U.S. Environmental Protection Agency (EPA). 1984. Compendium of Methods for the Determination of Toxic Organic Components in Ambient Air. EPA 600/4-84-041. Office of Research and Development. Washington, D.C.

U.S. Environmental Protection Agency (EPA). 1986. Demonstration of a Technique for Estimating Detection Limits with Specified Assurance Probabilities. Contract No. 68-01-6939. Office of Research and Development. Washington, D.C.

Winefordner, J.D. Trace Analysis: Spectroscopic Methods for Elements. Chemical Analysis 46:41-42

Winge, R.K., Peterson, V.J., and Fassel, V.A. 1970. Inductively Coupled Plasma – Atomic Emission Spectroscopy Prominent Lines. EPA-600/4-79-017.

Wollf, J.S., Homsher, M.T., Flotard, R.D., and Peterson, J.G. 1986. Semivolatile Organic analytical Considerations. In: Quality Control in Remedial Site Investigation. Hazardous and Industrial Solid Waste Testing, Fifth Volume. ASTM 925.

CHAPTER 9 – SOURCE SAMPLING AND ANALYSIS

Available Source Sampling Methodologies

Bursey, J.T., M. Hartman, J. Homolya, R. McAllister, J. McGaughey, and D. Wagoner. 1985. Laboratory and Field Evaluation of the SemiVOST Method, Volumes I and II. Prepared by Radian Cororation under Contract No. 68-02-4119, NTIS Nos. PB 86 123551/AS and PB 86 123569/AS. September.

Clark, A.E., M. Lataille, and E.L. Taylor. 1983. The Use of a Portable PID Chromatograph for Rapid Screening of Samples for Purgeable Organic Compounds in the Field and in the Laboratory. EPA Region I Laboratory, Lexington, MA. June.

de Vera, E.R., B.P. Simmons, R.D. Stephens, and D.L. Storm. 1980. Samplers and Sampling Procedures for Hazardous Waste Streams. Publication No. EPA-600/2-80-018, NTIS No. PB 80-135-353. U.S. Environmental Protection Agency. Cincinnati, OH.

Driscoll, J.N. 1974. *Flue Gas Monitoring Techniques*. Ann Arbor, MI: Ann Arbor Science Publishers.

Environment Canada. 1989. Reference Method for Source Testing: Measurement of Selected Semi-Volatile Organic Compounds from Stationary Sources. Reports EPA 1/RM/2, Environment Canada, Conservation and Protection. Ottawa, Canada.

Hansen, E.M. 1984. Protocol for the Collection and Analysis of Volatile POHCs Using VOST. Publication No. EPA 600/8-84-007. Report prepared by Envirodyne Engineers for U.S. Environmental Protection Agency, Office of Research and Development. Research Triangle Park, NC.

Harris, J.C., D.J. Larsen, C.E. Rechsteiner, and K.E. Thrun. 1984. Sampling and Analysis Methods for Hazardous Waste Combustion. EPA 600/8-84-002, NTIS #PB 84-155-845. U.S. Environmental Protection Agency, Office of Research and Development. Research Triangle Park, NC.

Homolya, J., J. McGaughey, D. Wagoner, M. Hartman, J. Margeson, J. Knoll, and M. Midgett. 1985. Validation of the Semi-Volatile Organic Sampling Train Method for Measuring Emissions from Hazardous Waste Incinerators. Presented at

78th Annual Meeting of the Air Pollution Control Association. Detroit, MI.

Johnson, L.D. Trial Burns: Methods Perspective. *Journal of Hazardous Materials* 22:143.

Johnson, L.D. Detecting Waste Combustion Emissions. *Environmental Science and Technology* 20(3):223.

Keith, L.H., ed. 1991. *Compilation of EPA's Sampling and Analysis Methods.* Compiled by W. Mueller and D.L. Smith. Boca Raton, FL: Lewis Publishers, Inc.

Logan, T.J., R.G. Fuerst, M.R. Midgett, and J. Prohaska. 1986. Validation of the Volatile Organic Sampling Train (VOST) Protocol: Volume I, Laboratory Validation Phase and Volume II, Field Validation Phase. EPA 600/4–86–014a and EPA 600/4–86–014b. U.S. Environmental Protection Agency, Office of Research and Development. Research Triangle Park, NC.

National Council for Air and Stream Improvement (NCASI). 1987. Optimization and evaluation of an impinger capture method for measuring chlorine and chlorine dioxide in pulp bleach plant vents. NCASI Technical Bulletin No. 520, New York.

Owens, M.H., S.A. Mooney, and T. Lachajczyk. 1987. Development of VOST Sample Analysis Protocol for Water-Soluble Volatile POHCs and PICs. EPA 600/8–87–008. U.S. Environmental Protection Agency, Office of Research and Development. Research Triangle Park, NC.

Schuetzle, D., ed. 1979. *Monitoring Toxic Substances*, ACS Symposium Series 94, American Chemical Society, Washington, D.C.

Sampling/Analytical Concerns

Adams, R.E., R.H. James, L.B. Farr, M.M. Thomason, H.C. Miller, and L.D. Johnson. Analytical Methods for Determination of Selected Principal Organic Hazardous Constituents in Combustion Products. *Environmental Science and Technology* 20(7):711.

Ahearn, A., ed. 1972. *Trace Analysis by Mass Spectrometry, 1st Edition.* New York: Academic Press.

Benchley, D.L., C.D. Turley, and R.G. Yarmac. 1973. *Industrial Source Sampling, 1st Edition.* Ann Arbor, MI: Ann Arbor Science Publishers.

Budzikiewicz, H., C. Djerassi, and D. Williams. 1976. *Mass Spectrometry of Organic Compounds, 1st Edition.* San Francisco, CA: Holden Day, Inc.

Dorsey, J., C. Lochmuller, L. Johnson, and R.

Statnick. 1975. Guidelines for Environmental Assessment Sampling and Analysis Programs – Level I. U.S. Environmental Protection Agency, Office of Research and Development. Research Triangle Park, NC.

James, R.H., R.E. Adams, J.M. Finkel, H.C. Miller, and L.D. Johnson. Evaluation of Analytical Methods for the Determination of POHC in Combustion Products. *Journal of the Air Pollution Control Association* 35(9).

Lodge, J.P., Jr., ed. 1989. *Methods of Air Sampling and Analysis, 3rd Edition.* Chelsea, MI: Lewis Publishers, Inc.

Maddalone, R.F. and S.C. Quinlivan. 1976. Technical Manual for Inorganic Sampling and Analysis. Prepared under EPA Contract No. 68–02–1412, Task Order No. 16, U.S. Environmental Protection Agency.

Mulik, J.D. and E. Sawicki. 1979. *ION Chromatographic Analysis of Environmental Pollutants, Volume 2.* Ann Arbor, MI: Ann Arbor Science Publishers, Inc.

Reed, R.I. 1966. *Applications of Mass Spectrometry to Organic Chemistry, 1st Edition.* New York: Academic Press.

Sawicki, E., J.D. Mulik, and E. Wittgenstein. 1978. *ION Chromatographic Analysis of Environmental Pollutants.* Ann Arbor, MI: Ann Arbor Science Publishers, Inc.

Vandergrift, A., L. Shannon, E. Lawless, P. Borman, E. Sailee, and M. Reichel. 1971. Particulate Pollutant System Study, Volume III – Handbook of Emission Properties. Report prepared under EPA Contract No. 22–69–104, U.S. Environmental Protection Agency.

Varian. 1978. Basic Liquid Chromatography, Varian Associates, Inc. Palo Alto, CA.

White, C.M., ed. 1985. *Nitrated Polycyclic Aromatic Hydrocarbons.* New York: Huething Publishers.

Williams, D.H. and I. Fleming. 1966. *Spectroscopic Methods in Organic Chemistry.* London: McGraw-Hill Publishing Co. Limited.

Method Validation

Youden, W.J. and E. H. Steiner. 1975. Statistical Manual of the Association of Official Analytical Chemists. Association of Official Analytical Chemists. Arlington, Virginia.

Detection Limits

Currie, L.A. 1968. Limits for Qualitative Detection and Quantitative Determination. Analytical Chemistry 40:587.

Fugitive Emissions

Englehart, P.J. and C. Cowherd, Jr. 1986. Assessment of Fugitive Particulate Emissions from Hazardous Waste Treatment Storage and Disposal Facilities: An Overview of Survey/Sampling Efforts with Preliminary Analytical Results. Paper No. 86–20.2. Presented at the 79th Annual Meeting of the Air Pollution Control Association. Minneapolis, MN.

Ginnity, B. 1988. Toxic Gas Monitor for Accidental Releases and Process Emissions. Paper No. 88–46.1. Presented at the 81st Annual Meeting of the Air Pollution Control Association, Dallas, TX.

Hopke, P.K., S.N. Chang, G.E. Gordon, and S.W. Rheingrover. 1986. Identification of Fugitive Emissions Compositions by Target Transformation Factor Analysis of Wind-Trajectory-Selected Samples. Paper No. 86–21.4. Presented at the 79th Annual Meeting of the Air Pollution Control Association. Minneapolis, MN.

Kalika, P.W. 1975. Development of Procedures for Measurement of Fugitive Emissions. Prepared under EPA Contract No. 68–02–1815. U.S. Environmental Protection Agency.

Kenson, R.E. and P.T. Bartlett. 1976. Technical Manual for the Measurement of Fugitive Emissions: Roof Monitor Sampling Method for Industrial Fugitive Emissions. EPA 600/2–76–089b. U.S. Environmental Protection Agency, Office of Research and Development. Washington, D.C.

Schmidt, C.E. and M. McDonough. 1989. Guidance Document on Air Pathway Analysis and Estimating Emission Rates for Air Contaminants from Hazardous Waste Sites. In: Proceedings of the 1989 EPA/A&WMA International Symposium Measurement of Toxic and Related Air Pollutants, published by Air & Waste Management Association. Publication No. VIP-13, EPA Publication No. 600/9–89–060, p. 161. Pittsburgh, PA.

Alternatives to Emission Measurement

Anderson, D. 1973. Emission Factors for Trace Substances. EPA 450/2–73–001, NTIS Publication No. PB 230–894/AS. U.S. Environmental Protection Agency, Office of Air Quality Planning and Standards. Research Triangle Park, NC.

U.S. Environmental Protection Agency. 1988. Protocols for Generating Unit-Specific Emission Estimates for Equipment Leaks of VOC and VHAP. EPA 450/3–88–010. Office of Air Quality Planning and Standards. Research Triangle Park, NC.

U. S. Environmental Protection Agency. 1990. Crosswalk/Air Toxic Emission Factor Data Base Management System User's Manual. EPA 450/2–90–018. Office of Air Quality Planning and Standards. Research Triangle Park, NC.

U.S. Environmental Protection Agency. 1991. VOC/PM Speciation Data System Documention and User's Guide, Version 1.32a. EPA 450/2–91–002. Office of Air Quality Planning and Standards. Research Triangle Park, NC.

Quality Assurance/Quality Control

Taylor, K.J. 1987. *Quality Assurance of Chemical Measurements*. Chelsea, MI: Lewis Publishers, Inc.

CHAPTER 21 – GASEOUS TOXIC AIR POLLUTANT CONTROL TECHOLOGIES

Basdekis, H.S., C.S. Parmele, D.G. Erikson, and R.L. Standifer. 1980. Organic chemical manufacturing. Vol. 5: Adsorption, condensation, and absorption devices. EPA 450/3–80–027. U.S. Environmental Protection Agency, Office of Air Quality Planning and Standards. Research Triangle Park, NC.

Caine, J.C. and E.L. Biedell. 1991. VOC Control through thermal oxidation. Presented at the Environmental Technology Expo. Chicago, IL. April.

California Air Resources Board (CARB). 1989b. Fact sheet on air toxics 'hot spots' information and assessment act of 1987 (AB 2588) and emission inventory criteria and guidelines. Technical Support Division. November.

Casill, R.P. and J. Laznow. 1991. Technology based solutions for industrial air toxic emissions. Presented at the Air & Waste Management Association Specialty Conference on Air Toxic Issues in the 1990s. King of Prussia, PA. April.

Casill, R.P. and J. Laznow. 1991. Air toxic control technologies for industrial applications. Presented at the 84th Annual Meeting of the Air & Waste Management Assocation. Vancouver, BC.

Cheremisinoff, P.N. 1976. Control of gaseous air pollutants. *Pollution Engineering*, May.

Cheremisinoff, P.N. 1991. Air pollution control and management. *Pollution Engineering*, June.

Godish, T. 1991. *Air Quality*. Chelsea, MI: Lewis Publishers.

Hodel, A.E. 1992. Catalytic oxidation controls VOC emissions. *Chemical Processing*, June.

Newton, J. 1991. Controlling toxic chemicals in the air. *Pollution Engineering*, August.

Patkar, A.N. and J. Laznow. 1992. Hazardous air pollutant control technologies. *HAZMAT WORLD*, February.

Patkar, A.N., J. Laznow, and R.P. Casill. 1991. Clean air act title III: Review of air toxic control technologies and their application. Presented at the Summer National Meeting of AIChE. Pittsburgh, PA. August.

Patkar, A.N., J. Laznow, and G. McCarthy. 1992. Hazardous air pollution control technologies: An overview. Presented at the Air & Waste Management Association Specialty Conference on New Hazardous Air Pollutant Laws and Regulations. King of Prussia, PA. April.

Pennington, R.L. 1991. Thermal oxidation for air toxics control. Presented at the Air & Waste Management Association Specialty Conference on Air Toxics Issues in the 1990s. King of Prussia, PA. April.

Pennington, R.L. 1992. Clean air compliance through thermal oxidation. Presented at the Air & Waste Management Association Specialty Conference on New Hazardous Air Pollutant Laws and Regulations. King of Prussia, PA. April.

Renko, R.J. 1988. Fume Incineration Systems. *Plastics Technology*, July.

Renko, R.J. 1989. Controlling fume emissions by incineration. *Plant Engineering*, October.

Seiwert, J.J. 1991. Advances in air pollution control technology. Presented at the Environmental Technology Expo. Chicago, IL. April.

Shen, T.T. 1986. Hazardous waste incineration: Emissions and their control. *Pollution Engineering*, July.

Sholtens, M.J. 1991. Air pollution control: A comprehensive look. *Pollution Engineering*, May.

Tessitore, J.L., J.G. Pinion, and E. DeCresie. 1990. Thermal destruction of organic air toxics. *Pollution Engineering*, March.

U.S. Environmental Protection Agency (EPA). 1990. OAQPS Control cost manual. EPA 450/3–90–006. Office of Air Quality Planning and Standards. Research Triangle Park, NC.

U.S. Environmental Protection Agency (EPA). 1991. Handbook: Control technologies for hazardous air pollutants. EPA 625/6–91/014. Office of Research and Development. Research Triangle Park, NC.

Vatavuk, W.M. 1990. *Estimating Costs of Air Pollution Control*. Chelsea, MI: Lewis Publishers.

Index